Clay Mineral Catalysis of Organic Reactions

Clay Mineral Catalysis of Organic Reactions

Benny K.G. Theng
Manaaki Whenua—Landcare Research
Palmerston North
New Zealand

CRC Press
Taylor & Francis Group
Boca Raton London New York

CRC Press is an imprint of the
Taylor & Francis Group, an **informa** business

CRC Press
Taylor & Francis Group
6000 Broken Sound Parkway NW, Suite 300
Boca Raton, FL 33487-2742

© 2019 by Taylor & Francis Group, LLC
CRC Press is an imprint of Taylor & Francis Group, an Informa business

No claim to original U.S. Government works

Printed on acid-free paper

International Standard Book Number-13: 978-1-4987-4652-6 (Hardback)

This book contains information obtained from authentic and highly regarded sources. Reasonable efforts have been made to publish reliable data and information, but the author and publisher cannot assume responsibility for the validity of all materials or the consequences of their use. The authors and publishers have attempted to trace the copyright holders of all material reproduced in this publication and apologize to copyright holders if permission to publish in this form has not been obtained. If any copyright material has not been acknowledged, please write and let us know so we may rectify in any future reprint.

Except as permitted under U.S. Copyright Law, no part of this book may be reprinted, reproduced, transmitted, or utilized in any form by any electronic, mechanical, or other means, now known or hereafter invented, including photocopying, microfilming, and recording, or in any information storage or retrieval system, without written permission from the publishers.

For permission to photocopy or use material electronically from this work, please access www.copyright.com (http://www.copyright.com/) or contact the Copyright Clearance Center, Inc. (CCC), 222 Rosewood Drive, Danvers, MA 01923, 978-750-8400. CCC is a not-for-profit organization that provides licenses and registration for a variety of users. For organizations that have been granted a photocopy license by the CCC, a separate system of payment has been arranged.

Trademark Notice: Product or corporate names may be trademarks or registered trademarks, and are used only for identification and explanation without intent to infringe.

Library of Congress Cataloging-in-Publication Data

Names: Theng, B. K. G., author.
Title: Clay mineral catalysis of organic reactions / Benny K.G. Theng.
Description: Boca Raton : CRC Press, Taylor & Francis Group, 2018. | Includes bibliographical references.
Identifiers: LCCN 2018008642 | ISBN 9781498746526 (hardback : alk. paper)
Subjects: LCSH: Clay catalysts. | Catalysts. | Mineralogical chemistry. | Chemistry, Organic.
Classification: LCC QD501 .T575 2018 | DDC 541/.395--dc23
LC record available at https://lccn.loc.gov/2018008642

Visit the Taylor & Francis Web site at
http://www.taylorandfrancis.com

and the CRC Press Web site at
http://www.crcpress.com

Dedication

To my wife, Judith

Contents

Preface xi
Author xiii

Chapter 1 Clays and Clay Minerals: Structures, Compositions, and Properties 1

    1.1   Concepts and Definitions 1
    1.2   Structural Aspects 7
    1.3   Specific Structural and Surface Properties 16
            1.3.1   Kaolinite 26
            1.3.2   Halloysite 30
            1.3.3   Chrysotile 32
            1.3.4   Smectite 34
            1.3.5   Vermiculite 38
            1.3.6   Chlorite 41
            1.3.7   Mica, Illite, Synthetic Mica-Montmorillonite, and Fluorotetrasilicic Mica 42
            1.3.8   Sepiolite and Palygorskite 45
            1.3.9   Allophane and Imogolite 48
    References 53

Chapter 2 Surface Acidity and Catalytic Activity 85

    2.1   Introduction 85
    2.2   Acids and Bases 85
            2.2.1   Acid Strength 86
    2.3   Brønsted Acidity 95
    2.4   Lewis Acidity 101
    2.5   Brønsted–Lewis Acid Combination and Synergy 106
    2.6   Concentration and Distribution of Surface Acid Sites 109
            2.6.1   Amine Titration 109
            2.6.2   Adsorption and Desorption of Basic Probe Molecules 114
    References 118

Chapter 3 Surface Activation and Modification 131

    3.1   Introduction 131
    3.2   Acid Activation 133
    3.3   Thermal Activation and Related Treatments 155
            3.3.1   Heating and Calcination 155
            3.3.2   Microwave and Ultrasound Irradiation 157
    3.4   Pillared Interlayered Clays and Porous Clay Heterostructures 163
    3.5   Organically Modified Clay Minerals 176
            3.5.1   Organoclays and Related Materials 176
            3.5.2   Surface Grafting and Silylation 181
    References 183

Chapter 4 Organic Catalysis by Clay-Supported Reagents 221

    4.1   Introduction 221
    4.2   Clay-Supported Metal Salts 221
            4.2.1   Clay-Supported Zinc(II) Chloride and Other Metal Chlorides 222
            4.2.2   Clay-Supported Iron(III) Nitrate and Other Metal Nitrates 226
    4.3   Clay-Supported Metal Oxides and Metal Sulfides 229
    4.4   Clay-Supported Metal Nanoparticles 230
    4.5   Clay-Supported Organic and Metal-Organic Reagents 236
    4.6   Clay-Supported Heteropolyacids 241
    References 244

Chapter 5 Clay Mineral Catalysis of *Name* Reactions 261

    5.1   Introduction 261
    5.2   Alder–Ene Reaction 261
    5.3   Baeyer–Villiger Condensation and Oxidation 262
    5.4   Bamberger Rearrangement 263
    5.5   Baylis–Hillman Reaction 263
    5.6   Beckmann Rearrangement 264
    5.7   Biginelli Reaction 265
    5.8   Diels–Alder Reaction 266
    5.9   Ferrier Rearrangement 269
    5.10   Fischer Glycosidation/Glycosylation 270
    5.11   Fischer–Hepp Rearrangement 270
    5.12   Fischer Indole Synthesis 270
    5.13   Fischer–Tropsch Synthesis 271
    5.14   Friedel–Crafts Reaction 271
    5.15   Friedländer Synthesis 273
    5.16   Fries Rearrangement 274
    5.17   Hantzsch Dihydropyridine Synthesis 274
    5.18   Heck Reaction 275
    5.19   Knoevenagel Condensation 276
    5.20   Mannich Reaction 277
    5.21   Markovnikov Addition Rule 277
    5.22   Michael Addition 278
    5.23   Mukaiyama Aldol Reaction 279
    5.24   Nicholas Reaction 280
    5.25   Oppenauer Oxidation 280
    5.26   Paal–Knorr Synthesis 280
    5.27   Pechmann Condensation 281
    5.28   Prins Reaction 281
    5.29   Ritter Reaction 282
    5.30   Sakurai Allylation Reaction 282
    5.31   Sonogashira Reaction 283
    5.32   Strecker Reaction 283
    5.33   Suzuki Reaction 284
    5.34   Wacker Oxidation 284
    5.35   Wittig Reaction 285
    References 285

Chapter 6 Clay Mineral Catalysis of Isomerization, Dimerization, Oligomerization, and Polymerization Reactions 299

    6.1   Introduction 299
    6.2   Isomerization 299
            6.2.1   Hydrocarbons 299
            6.2.2   Non-Hydrocarbons 304
    6.3   Dimerization..306
            6.3.1   Hydrocarbons 306
            6.3.2   Non-Hydrocarbons 307
    6.4   Oligomerization 310
            6.4.1   Hydrocarbons 310
            6.4.2   Non-Hydrocarbons 311
    6.5   Polymerization..312
            6.5.1   Hydrocarbons 319
            6.5.2   Non-Hydrocarbons 322
            6.5.3   *In Situ* Polymerization of Monomers and Polymer-Clay Nanocomposite Formation 325
    References 328

Chapter 7 Clay Mineral Catalysis of Redox, Asymmetric, and Enantioselective Reactions 347

    7.1   Introduction 347
    7.2   Oxidation Reactions 355
            7.2.1   Epoxidation and Oxygenation 361
    7.3   Reduction, Hydrogenation, and Deoxygenation 362
    7.4   Asymmetric/Enantioselective Reactions and Syntheses.......................366
    References 370

Chapter 8 Clay Mineral Catalysis of Natural Processes and Prebiotic Organic Reactions.......389

    8.1   Introduction 389
    8.2   Clays and Clay Minerals as Geocatalysts 389
            8.2.1   Hydrocarbon Cracking 389
            8.2.2   Carboxylic Acid Transformation...391
            8.2.3   Kerogen Transformation and Pyrolysis...................................394
    8.3   Clay Mineral Catalysis of Prebiotic Organic Reactions.......................398
            8.3.1   Polypeptide Synthesis...399
            8.3.2   Selective Adsorption and Polymerization of Amino Acid Enantiomers..401
            8.3.3   Dimerization and Oligomerization of Nucleotides 402
    References 404

Index.. 417

Preface

Because of their fine particle size, extensive surface area, layer structure, and peculiar charge characteristics, clay minerals have a large propensity for taking up both small and polymeric organic molecules (Theng 1974, 2012). What is not generally appreciated, however, is that clays and related layer silicates can also catalyze a large range and variety of organic conversions and transformations although acid-treated clays have long been used as cracking catalysts in petrochemical processing.

Clays are abundant, inexpensive, noncorrosive, and nonpolluting. Furthermore, clay minerals can efficiently promote numerous organic conversions under solvent-free conditions with the assistance of microwave and ultrasound irradiation. Clay catalysts are also easy to separate from the reaction mixture, allowing them to be recycled without a significant loss of activity. For these reasons, clay-based and clay-supported catalysts are well suited for conducting and practising eco-friendly (*green*) chemistry.

The catalytic activity of clay minerals is linked to their capacity to act as solid acids in either the Brønsted or Lewis sense as influenced by sample pretreatment and experimental conditions. More importantly, layer silicates offer a reduced *dimensionality* of reaction space, making it easier for reactants to meet and collide than would be the case in the three-dimensional space of a flask or reactor under homogeneous conditions (Balogh and Laszlo 1993).

Clay minerals with a layer, a layer-ribbon, or a short-range order structure and having different particle shapes, have all featured as heterogeneous acid catalysts. In this regard, the smectite group of layer silicates has received special attention. Within this group, montmorillonite has long held the limelight because of its superior intercalating ability and the ease with which its surface properties can be modified for specific catalytic purposes.

We might add that natural clays are inhomogeneous in terms of layer charge distribution, layer-stacking order, and other surface properties, while their particles often contain *associated minerals* as well as transition metal impurities. Because of this variability between similar samples, the experimental results are not always self-consistent. The use of commercially available acid-activated montmorillonites and related surface-modified forms can overcome, or at least minimize, the negative effect of sample variability on data reproducibility.

The surface acidity of smectites can be enhanced by washing with mineral acids or impregnation with superacids and heteropolyacids, while their porosity can be adjusted by acid treatment or modified by intercalation of cationic surfactants and silylating agents. A well-known example here is the intercalation of oligomeric hydroxy-metal cations by ion exchange, and the subsequent conversions to their corresponding metal oxides, by calcination. The resultant *pillared interlayered clays* are acidic and have a two-dimensional pore system, opening the way to shape-selective catalysis. Smectites and their acid-treated derivatives can also support catalytically active reagents such as metal salts, metal-organic complexes, and metal nanoparticles.

This book describes the ability of cation-exchanged, acid-activated, and surface-modified clay minerals to catalyze organic conversions and syntheses, including a series of *name* reactions and various isomerization, dimerization, oligomerization, and polymerization reactions. These materials are similarly efficient in promoting a number of redox, asymmetric, and enantioselective reactions. Clay minerals can also serve as catalysts of certain natural processes and may play an important role in the prebiotic synthesis of bioorganic molecules.

The idea of writing a book on clay catalysis was suggested by E.H. (Ed) Edelson following my visit to Exxon Production Research Company in Houston, Texas (USA), in 1985. Since a reference book on clay-catalyzed organic reactions was not available at the time, I responded positively to his suggestion. It was in the summer of 1987 that I began to search the relevant literature, using *Chemical Abstracts* and *Current Contents* as the primary sources of information. That activity, however, had to be abandoned, or at least put on hold, when my employment status became insecure toward the end of 1988. For the same reason, Ed Edelson decided to withdraw from the coauthorship role.

It was not until the spring of 2015 that I was in a position to approach CRC Press/Taylor & Francis Group with the proposal of writing a reference book on clay-catalyzed organic reactions. In the intervening years, the book *Organic Chemistry Using Clays* by Balogh and Laszlo (1993) came out in print as did a number of review articles on a similar topic. The present volume is essentially a summary of the published literature on clay mineral catalysis that has accumulated over the past four or so decades. In order to keep its size within manageable limits, however, the patent literature has been omitted. At the same time, I have attempted to provide a critical account, rather than a neutral assessment, of the extensive but scattered information. Also by doing it alone, I am solely responsible for whatever merits and deficiencies the book may have.

As a reference source, this book is primarily directed to research scientists and industrial chemists who use, or contemplate using, clay minerals as catalysts for targeted organic conversions and syntheses. The present volume would also be of interest and value to environmental and soil chemists as well as teachers and postgraduate students of organic geochemistry and solid acid catalysis.

I am grateful to my colleagues, G.J. Churchman and R.W. McCabe, for their encouragement and moral support. I am indebted to Nicolette Faville and Cissy Pan for drawing up the numerous equations, figures, and reactions schemes. I also thank the reception staff at Palmerston North for assistance in formatting text and tables. The library staff at Lincoln, especially Edward Doonerwind, was very helpful in obtaining and providing hundreds of papers and articles on interlibrary loan. Thanks are also due to scientific societies and publishers for granting permission to reproduce some figures and diagrams. Finally, I wish to acknowledge the steadfast and unwavering support of my wife, Judith, to whom this book is dedicated.

Benny K.G. Theng
Palmerston North, New Zealand
March 2018

REFERENCES

Balogh, M. and P. Laszlo. 1993. *Organic Chemistry Using Clays*. Berlin: Springer-Verlag.
Theng, B.K.G. 1974. *The Chemistry of Clay-Organic Reactions*. London: Adam Hilger.
Theng, B.K.G. 2012. *Formation and Properties of Clay-Polymer Complexes*, 2nd edition. Amsterdam: Elsevier.

Author

Benny K.G. Theng is an honorary research associate with Manaaki Whenua—Landcare Research in Palmerston North, New Zealand (NZ). Before joining the research staff of MW-LC in 1992, he was a scientist with the New Zealand Soil Bureau, DSIR. His research has focused on the behavior and reactivity of small and polymeric organic compounds at clay mineral surfaces. He was born in Indonesia but did all his undergraduate and postgraduate studies at the University of Adelaide, Australia, gaining a PhD degree in soil science. He was a research fellow, lecturer, and visiting scientist at several universities and research institutes in Australia, Belgium, Germany, Japan, France, Chile, and China. There is an open invitation for him to serve as a visiting professor at Zhaoqing University, Guangdong, China. He also did some consulting work for clay-based companies in America and Europe. He is the recipient of the Adam Hilger Prize (UK) for his first book, the ICI Prize (NZ) for chemical research, and the Bailey Distinguished Member Award from The Clay Minerals Society (USA). He is a fellow of the New Zealand Institute of Chemistry, New Zealand Society of Soil Science, and the Royal Society of New Zealand. He likes reading historical books and thrillers, playing contract bridge, and listening to classical and chamber music.

1 Clays and Clay Minerals
Structures, Compositions, and Properties

1.1 CONCEPTS AND DEFINITIONS

Clay may signify a rock, a sedimentary deposit, or the alteration (weathering) products of primary silicate minerals (Moore 1996). It is in one or other of these senses that such terms as *ball clay*, *fire clay*, *bentonite*, *bleaching earth*, and *fuller's earth* have been used in the literature (Grim 1962; Hosterman 1973; Weaver and Pollard 1973), including some early references to clay catalysis (Rideal and Thomas 1922; Gurwitsch 1923; Broughton 1940; Endell et al. 1941; Brooks 1948; Grenall 1948; Robertson 1948; MacCarter et al. 1950; Thomas et al. 1950; Holmes and Mills 1951; Hansford 1952; Theng 1974, 1982; Ballantine 1986; Rupert et al. 1987). Ball clay, fire clay, and refractory clay are largely of sedimentary origin, dominantly comprising the clay mineral kaolinite of varying particle size and (usually) low crystallinity (Range et al. 1969; Jepson 1984; Patterson and Murray 1984). On the other hand, bentonite is a rock formed by diagenetic and hydrothermal alteration of compacted volcanic material in an aqueous environment (shallow seas, brackish lakes), having montmorillonite as its principal clay mineral constituent (Hofmann 1968; Weaver and Pollard 1973; Grim and Güven 1978; Borchardt 1989; Kaufhold et al. 2002; Galán 2006; Christidis and Huff 2009). Bleaching earth and fuller's earth are fine-grained, earthy materials capable of decolorizing vegetable and mineral oils (Siddiqui 1968; Fahn 1973; Grim and Güven 1978; Morgan et al. 1985; Robertson 1986; Beneke and Lagaly 2002; Hussin et al. 2011). The name *fuller's earth* derives from *fulling*, a process of removing grease and dirt from raw wool by means of an aqueous slurry of crude clays (Grim 1962; Robertson 1986; Beneke and Lagaly 2002). Being purely functional, such definitions carry no reference to composition. Although some samples of fuller's earth are largely composed of palygorskite (*attapulgite*), the majority of deposits are montmorillonite-rich (Wright 1968; Hosterman 1973; Grim and Güven 1978; Robertson 1986; Murray 1999, 2000). Because of their diversity and compositional variability, the catalytic activity of natural clay materials is difficult to correlate with structural properties, and the reproducibility of experimental results becomes contentious. For these reasons, much of the published literature on clay mineral catalysis relates to the use of commercially available catalysts, and their cation-exchanged and metal salt-impregnated forms (Cornélis et al. 1983, 1990, 2004; Laszlo 1986, 1998; Clark et al. 1994; Cseri et al. 1995a, 1995b; Vaccari 1999; Varma 2002; Fernandes et al. 2012; Kaur and Kishore 2012; McCabe and Adams 2013).

The joint nomenclature committees (JNCs) of the Association Internationale pour l'Etude des Argiles (AIPEA) and the Clay Minerals Society (CMS) have defined the term *clay* as "…a naturally occurring material composed primarily of fine-grained minerals, which is generally plastic at appropriate water contents and will harden with (*sic*) dried or fired" (Guggenheim and Martin 1995). By this definition synthetic clays and clay-like materials are not regarded as *clay* even though they may be fine-grained and display the attributes of plasticity and hardening on drying and firing. On the other hand, the *clay fraction* of a soil or sediment denotes a textural class of minerals whose particles are smaller than a given dimension. Among soil scientists there is international agreement to set the upper limit of size at 2 μm equivalent spherical diameter (e.s.d.). Although this value is clearly arbitrary, the *non-clay constituents*, notably feldspar and quartz, are practically

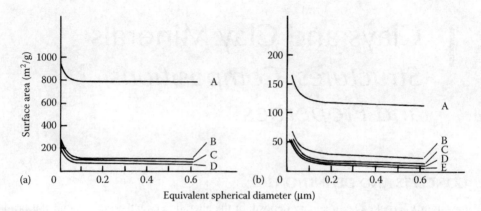

FIGURE 1.1 Relationship between external basal surface area and equivalent spherical diameter of clay mineral particles of varying thickness: (a), 2:1 type layer structures (e.g., montmorillonite) consisting of superposed 1 nm thick layers with a particle thickness of 1 nm (A), 8 nm (B), 10 nm (C), and 12 nm (D); and (b), 1:1 type layer structures (e.g., kaolinite) composed of superposed 0.7 nm thick layers with a particle thickness of 7 nm (A), 35 nm (B), 50 nm (C), 75 nm (D), and 100 nm (E). (From Caillère, S. et al., *Minéralogie des Argiles. 1. Structure et Propriétés Physico-Chimiques.* 2nd ed., Masson, Paris, France, 1982.)

concentrated in the >2 μm e.s.d. fraction (Grim 1968; Bonneau and Souchier 1982). We should add that soils contain many kinds of inorganic and organic *nanoparticles* with at least one dimension in the nanoscale range (<100 nm) (Brown et al. 1978; Theng and Yuan 2008; Calabi-Floody et al. 2009, 2011).

The lateral dimension and thickness of particles are related to their surface area, which in turn would affect their adsorptive and catalytic activity. Figure 1.1 shows that for a given thickness, there is a steep increase in surface area when the diameter of the particles falls below about 0.1 μm (100 nm). Equally important in this respect are the shape and mutual arrangement of the particles (cf. Figure 1.5). These factors influence the geometry and volume of pores, and hence the accessibility of the pore surface to organic molecules. For this reason, *texture* will be used here to connote both particle-size distribution and pore-size distribution. Apart from having a large and reactive surface, clays and clay minerals are attractive as catalysts *per se* and as catalyst supports because they are abundant, environmentally benign, inexpensive, noncorrosive, recoverable, and recyclable (Brown 1994; Clark et al. 1994, 1997; Varma 2002; Garrido-Ramírez et al. 2010; Nagendrappa 2011; Zhou 2011; Kaur and Kishore 2012; Kumar et al. 2014).

Many clay-catalyzed reactions take place at external particle, or intradomain, surfaces (Theng 1974; Ovcharenko 1982; Solomon and Hawthorne 1983; Barrer 1984), *domains* being regions of oriented clay particles arranged in turbostratic array (cf. Figure 1.5) (Aylmore and Quirk 1960; Quirk 1968). Accordingly, the rate-limiting step of the reaction is determined by the diffusion of the reactants into, and that of the products out of, the interparticle pore and interlayer space (e.g., Breen et al. 1985). Thus, any pretreatment that increases the accessible surface area and porosity of the system would promote catalytic activity. Such textural improvements may be achieved through acid treatment (*activation*), heating (*thermal treatment*), and microwave/ultrasound irradiation of the clay material as described in Chapter 3 (Bannerjee and Sen 1974; Ovcharenko 1982; Mahmoud et al. 2003; Noyan et al. 2006; Steudel et al. 2009a, 2009b; Korichi et al. 2012; Heller-Kallai 2013; Komadel and Madejová 2013). Indeed, acid activation (using mineral acids) lies behind the formation and commercial production of catalytically active montmorillonites such as the *K-catalysts* (Süd-Chemie), the Filtrol clay catalysts (Filtrol Corporation and Laporte Industries), and the F-series catalysts (Engelhard) (Robertson 1948; Balogh and Laszlo 1993; Cseri et al. 1995a, 1995b; Chitnis and Sharma 1997; Li et al. 1997; Vaccari 1999; Nikalje et al.

2000; Flessner at al. 2001; Jang et al. 2005; Pushpaletha et al. 2005; De Stefanis and Tomlinson 2006; Dasgupta and Török 2008; Fernandes et al. 2012; McCabe and Adams 2013; Kumar et al. 2014).

A large number and variety of catalyzed organic conversions take place in the interlayer space of expanding clay minerals, notably the cation-exchanged and surface-modified forms of smectites (Theng and Walker 1970; Thomas et al. 1977; Thomas 1982; Jones et al. 1983; Adams et al. 1983; Pinnavaia 1983; Van Damme et al. 1984; Adams 1987; Rupert et al. 1987; Rozengart et al. 1988; Purnell 1990; Eastman and Porter 2000). Besides being capable of intercalating organic molecules as well as metal salts and complexes, smectites can show a high selectivity in catalyzing organic reactions and syntheses (Pinnavaia et al. 1979; Adams et al. 1987). Further, their low or reduced *dimensionality* makes it easier for reactants to meet and collide than would be the case in three dimensions (Laszlo 1987a, 1987b; Balogh and Laszlo 1993; McCabe and Adams 2013). Related to this feature is the geometric constraint that the clay mineral interlayer can impose on the adsorption and stability of a steric isomer of certain organic compounds (Fripiat 1986). We should also add that organic compounds are packed in some particular orientation in the interlayer space (Theng 1974; Thomas et al. 1977; Thomas 1982; Pinnavaia 1983; He et al. 2014). Hydroxy methacrylates, for example, can polymerize spontaneously when intercalated into montmorillonite. On the other hand, intercalated methyl methacrylate and the amino methacrylates fail to polymerize because their orientation is presumably unfavorable to the formation of intermolecular bonds (Solomon and Loft 1968; Theng and Walker 1970).

It stands to reason that the organic molecules or monomers in question must be capable of penetrating the interlayer space of the clay mineral if they are to react or polymerize. Interlayer entry can be greatly facilitated by prior intercalation of oligomeric hydroxy-metal cations, notably the $[Al_{13}O_4(OH)_{24}(H_2O)_{12}]^{7+}$ *Keggin* ion, commonly abbreviated to $(Al_{13})^{7+}$. Subsequent calcination converts these bulky inorganic cations into their corresponding metal oxides, which then act as *pillars* to prop apart the individual silicate layers in a particle, giving rise to a series of microporous and mesoporous materials, and opening the way to shape-selective catalysis. The formation, properties, and catalytic activity of pillared interlayered or cross-linked smectites have been the topic of many reviews (Vaughan and Lussier 1980; Pinnavaia 1983; Barrer 1984; Pinnavaia et al. 1984; Occelli et al. 1985; Adams 1987; Figueras 1988; Vaughan 1988; Kloprogge 1998; Ding et al. 2001; Gil et al. 2000, 2008; 2010; De Stefanis and Tomlinson 2006; Dasgupta and Török 2008; Zhou et al. 2010; Vicente et al. 2013), and are described more fully in Chapter 3.

As might be expected, clays and clay bodies are largely composed of *clay minerals*, a term which is also difficult to define precisely. The definition that the JNCs have proposed is "…phyllosilicate minerals and minerals which impart plasticity to clay and which harden upon drying or firing" (Guggenheim and Martin 1995). Since the origin of the material is not part of the definition, clay mineral (unlike clay) may be synthetic. Since grain size does not feature as a criterion in the JNCs' definition of clay minerals, phyllosilicates of any size such as macroscopic mica, vermiculite, and chlorite may be included. Indeed, much of our basic and detailed understanding of clay mineral structures is derived from X-ray diffraction analysis of *macrocrystalline* forms (e.g., mica and vermiculite) (Radoslovich 1975; Walker 1975; Bailey 1980). A similar concept was advocated by Weaver (1989) who suggested the term *physils* for the whole family of phyllosilicates (including palygorskite and sepiolite) irrespective of grain size. The JNCs have further proposed that non-phyllosilicate minerals would qualify as clay minerals if they impart plasticity to clay, and harden on drying or firing.

Because of their ultrafine grain size, the characterization and identification of clay minerals call for a combination of various techniques. In addition to X-ray powder diffraction, chemical and thermal analyses, organic intercalation, together with electron microscopy and various spectroscopic methods, are used for this purpose (Grim 1968; Gard 1971; Mackenzie and Caillère 1975; Lagaly

1981; Sudo et al. 1981; Newman 1987; Dixon and Weed 1989; Środoń 2006; Bergaya and Lagaly 2013b). Even then, the occurrence of structural defects, layer-stacking disorders, and interstratified structures make for practical difficulties. Progress in this area has been considerably aided by the application of modern, sophisticated instrumental methods such as electron spin resonance (ESR), Mössbauer, nuclear magnetic resonance (NMR), neutron scattering, X-ray photoelectron, and X-ray absorption spectroscopies. Details of the underlying theories, and the quality of information that these techniques can provide, are given in the reviews and books, written or edited by Stucki and Banwart (1980), Fripiat (1981), Thomas (1984), Wilson (1994), Bergaya et al. (2006b), and Bergaya and Lagaly (2013b).

Perhaps a more intractable analytical problem than microcrystallinity *per se* is the presence in the clay sample of *associated minerals* such as carbonates, quartz, and iron/aluminum oxyhydroxides (Brown et al. 1978; Schwertmann 1979; Brown 1980; Dixon and Weed 1989; Guggenheim and Martin 1995). These non-phyllosilicate *impurities*, in the form of either discrete entities or surface-adsorbed species, are difficult to detect by X-ray diffractometry (XRD) because of their low concentration and short-range order nature. More importantly, such impurities could render clay materials unstable as catalysts and catalyst supports. The occurrence of associated minerals can be indicated only indirectly by means of selective dissolution, coupled with differential infrared spectroscopy and XRD (Wada and Greenland 1970; Campbell and Schwertmann 1985; Schulze 1981, 1994).

A classic example of a clay-size mineral with short-range order is allophane, a group of hydrated alumino-silicates with varying Al/Si ratios that occur in many soils derived from volcanic ash and weathered pumice (Wada and Harward 1974; Fieldes and Claridge 1975; Wada 1980, Parfitt 1990, 2009; Dahlgren et al. 2004; Harsh 2012). Although the X-ray diffraction pattern of allophane shows broad diffuse peaks, these minerals cannot be said to be amorphous (cf. Section 1.3.9).

Although some associated and short-range order minerals as such, or as supports of iron/copper oxides, are catalytically active (Garrido-Ramirez et al. 2010, 2012), the emphasis here is on catalysis over phyllosilicates having long-range order whose particles may adopt a variety of shapes. Kaolinite, for example, commonly occurs as polygonal platelets (Figure 1.9), halloysite as tubules (Figure 1.11), smectite as thin wavy flakes (Figure 1.15), while chrysotile (Figure 1.12), sepiolite, and palygorskite (Figure 1.19) appear as fibrils. Variations in particle shape may also occur within a species. For example, halloysite particles may adopt cylindrical, spheroidal, and platy (tabular) forms (Kunze and Bradley 1964; de Souza Santos et al. 1966; Kirkman 1977; Nagasawa 1978; Sudo et al. 1981; Tazake 1982; Churchman and Theng 1984; Dixon 1989; Joussein et al. 2005; Pasbakhsh et al. 2013). Similarly, the shape of smectite particles may vary from very thin plates to laths and fibers (Sudo et al. 1981; Güven 1988, 2009; Ras et al. 2007; Johnston 2010). For this reason, it is both difficult and impractical to classify clay minerals in terms of particle morphology.

A classification scheme for planar hydrous phyllosilicates, proposed by Martin et al. (1991) and Guggenheim et al. (2006), based on layer type, net layer charge, nature of the interlayer materials, and octahedral character, has found general acceptance. Table 1.1 shows a similar scheme but one that is more appropriate to clay catalysis. The meaning and significance of the various terms in this table, together with the basic features of clay mineral structures, are given in the following sections. Particular attention is directed to certain points that are important to understanding catalytic activity.

For more details of the chemistry, mineralogy, and structures of clay minerals than can reasonably be included here, the books edited or written by Grim (1968), Weaver and Pollard (1973), Gieseking (1975), Greenland and Hayes (1978), Brindley and Brown (1980), Newman (1987), Dixon and Weed (1989), Wilson (1994), Bergaya et al. (2006b), Carretero and Pozo (2007), Bergaya and Lagaly (2013a), and Fiore et al. (2013) should be consulted.

TABLE 1.1
Classification Scheme for Clay Minerals and Related Phyllosilicates Together with Some Structural-Chemical Properties

A. Minerals with Layer Structures

Layer Type	Group	Charge per Formula Unit (x)[1]	Octahedral Character	Layer Thickness (nm)	Interlayer Materials			Representative Species
					Cations	OH-Sheet	Water	
1:1	Kaolin	Zero	Dioctahedral	0.7	None	Absent	Absent	Kaolinite, dickite, nacrite
		Zero	Dioctahedral	0.7–1.0	None	Absent	Up to 2H$_2$O	Halloysite
	Serpentine	Zero	Trioctahedral	0.7	None	Absent	Absent	Chrysotile
2:1	Pyrophyllite	Zero	Dioctahedral	0.92	None	Absent	Absent	Pyrophyllite
	Talc	Zero	Trioctahedral	0.9	None	Absent	Absent	Talc
	Smectite	0.2–0.6	Dioctahedral	≥0.96	Various	Absent	Variable	Montmorillonite, beidellite, nontronite
		0.2–0.6	Trioctahedral	≥0.96	Various	Absent	Variable	Saponite, hectorite, sauconite
	Vermiculite	0.6–0.9	Dioctahedral	≥0.94	Various	Absent	Variable	Dioctahedral vermiculite
		0.6–0.9	Trioctahedral	≥0.94	Various	Absent	Variable	Trioctahedral vermiculite
	Interlayer-deficient mica	0.6–0.85	Dioctahedral	1.0	K	Absent	Variable	Illite[2]
	True mica	0.85–1.0	Dioctahedral	1.0	K	Absent	Absent	Muscovite, celadonite
		0.85–1.0	Trioctahedral	1.0	K	Absent	Absent	Phlogopite
	Brittle mica	1.8–2.0	Dioctahedral	1.0	Ca	Absent	Absent	Margarite
		1.8–2.0	Trioctahedral	1.0	Ca	Absent	Absent	Clintonite
	Chlorite	Variable	Dioctahedral	1.4	None	Present	Absent	Donbassite
		Variable	Di-trioctahedral	1.4	None	Present	Absent	Sudoite
		Variable	Trioctahedral	1.4	None	Present	Absent	Chamosite, clinochlore

(Continued)

TABLE 1.1 (Continued)
Classification Scheme for Clay Minerals and Related Phyllosilicates Together with Some Structural-Chemical Properties

B. Minerals with Layer-Ribbon Structures

Layer Type	Modulated Component	Charge per Formula Unit (x)[1]	Octahedral Character	Layer Thickness (nm)	Inter-Ribbon Materials			Representative Species
					Cations	OH-Sheet	Water	
2:1	Octahedral sheet	~0.1	Di-trioctahedral	1.27 ($c \sin \beta$)	Ca (common)	Absent	$2H_2O$	Palygorskite
	Octahedral sheet	~0.1	Trioctahedral	1.34	Ca (common)	Absent	$4H_2O$	Sepiolite

C. Minerals with Short-range Order Structures[3]

Group	Layer Thickness (nm)	Representative Species
Allophane (variable Al/Si ratios)	0.7–1.0 (spherule wall)	Allophane
Imogolite (Al/Si = 2.0)	~1.4 (tubule wall)	Imogolite

Sources: Adapted from Mackenzie, R.C. and Mitchell, B.D., *Earth-Sci. Rev.*, 2, 47–91, 1966; Brown, G. et al., The structures and chemistry of soil clay minerals, in *The Chemistry of Soil Constituents*, Greenland, D.J. and Hayes, M.H.B. (Eds.), John Wiley & Sons, Chichester, UK, pp. 29–178, 1978; Bailey, S.W., Structures of layer silicates, in *Crystal Structures of Clay Minerals and Their X-ray Identification*, Brindley, G.W. and Brown, G. (Eds.), Mineralogical Society, London, UK, pp. 1–123, 1980; Sudo, T. et al., *Electron Micrographs of Clay Minerals. Developments in Sedimentology 31*, Kodansha, Tokyo, Japan, 1981; Martin, R.T. et al., *Clays Clay Miner.*, 39, 333–335, 1991; Guggenheim, S. et al., *Clay Miner.*, 41, 863–877, 2006; *Clays and Clay Minerals* 54: 761–772, 2006.

[1] net negative layer charge, expressed as a positive number;
[2] now considered to be a *series* name;
[3] also denoted as *non-crystalline* or *poorly crystalline*.

1.2 STRUCTURAL ASPECTS

As already remarked on, clay minerals may adopt a variety of particle (*crystal*) shapes. Underlying these differences in particle morphology, however, is a layer structure comprising essentially two types of sheets. One of these is the tetrahedral ("T") sheet, which, as the name suggests, consists of linked silicon tetrahedra. As shown in Figure 1.2a, the silicon cation in each tetrahedron is in fourfold coordination with oxygen anions. Linkage between adjoining tetrahedra occurs through corner sharing such that the basal oxygens are approximately coplanar, forming an open hexagonal network, while the apical oxygens point in the same direction (Figure 1.2b). The other type of sheet, referred to as octahedral ("O"), consists of linked aluminum or magnesium octahedra in which the central cation (Al^{3+} or Mg^{2+}) is coordinated, and equidistant, to six hydroxyl anions (Figure 1.2c). In this case, linkage between octahedra occurs through edge sharing (Figure 1.2d). When aluminum is the central cation, only two out of every three octahedral sites are occupied for electrical neutrality to be maintained. The resultant sheet structure, as found in the mineral gibbsite, is referred to as dioctahedral. On the other hand, when magnesium is the central cation, all octahedral positions are filled. The brucite-like sheet structure that arises is accordingly designated as trioctahedral.

A silicate layer is formed when the two types of sheets combine in a given proportion by fitting the apical oxygens of the tetrahedral sheet into the vertices of the octahedral sheet. Combination in equal proportions gives rise to the 1:1 or "T-O" layer type (Figure 1.3a) of which kaolinite and serpentine represent the dioctahedral and trioctahedral prototypes, respectively. Similarly, a 2:1 or "T-O-T" layer type arises when this combination occurs in a 2:1 proportion, with the octahedral sheet being sandwiched between two inward-pointing tetrahedral sheets (Figure 1.3b). The corresponding dioctahedral and trioctahedral prototype structures are exemplified by pyrophyllite and talc, respectively.

Note that two types of octahedral sites can be distinguished (Figure 1.3b). Although each type is surrounded by two hydroxyls and four oxygens, the hydroxyls can occur in either a *cis* (M_2) or a *trans* (M_1) arrangement with respect to the cation site, the population of M_2 being twice as large as that of M_1 (e.g., Wolters and Emmerich 2007). Another noteworthy feature is that the *under-coordinated* metal ion, notably Al^{3+}, exposed at the layer edge, normally has a single water molecule and a terminal hydroxyl group attached to it (Johnston 2010). Such an $Al(H_2O)(OH)$ grouping (cf. Figure 1.7) may be considered as a Lewis acid site since the aluminum ion can act as an electron acceptor, especially under dehydrating conditions (Solomon 1968; Solomon et al. 1971; Almon and Johns 1976; Solomon and Hawthorne 1983; Sposito 1984; Rupert et al. 1987; Rozengart et al. 1988; Heller-Kallai 2002; Komadel and Madejová 2013; Liu et al. 2013d, 2013e), while the aluminol group

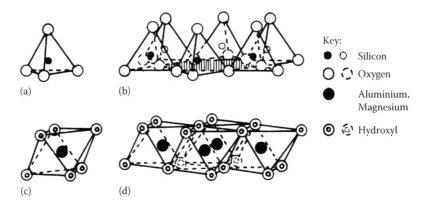

FIGURE 1.2 Diagram showing the basic building blocks of clay mineral structures: (a), a silica tetrahedron with the central silicon ion coordinated to four oxygens; (b), linkage between silica tetrahedra through corner-sharing, yielding a tetrahedral sheet with the basal oxygens forming a network of hexagonal cavities (hatched area); (c), an alumina octahedron with the central aluminum ion coordinated to six hydroxyls; and (d), an alumina octahedral sheet formed by linking individual octahedra through edge-sharing.

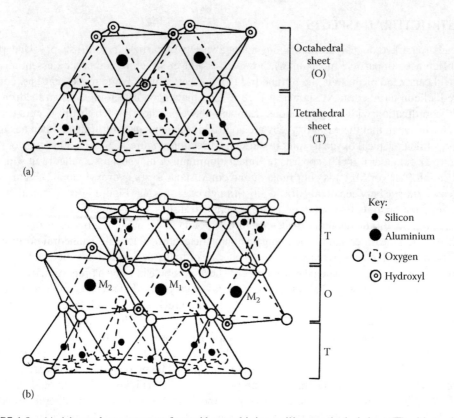

FIGURE 1.3 (a), 1:1 type layer structure formed by combining a silica tetrahedral sheet (T) with an alumina octahedral sheet (O) and (b), 2:1 type layer structure in which the alumina octahedral sheet is sandwiched between two silica tetrahedral sheets. Also shown are the two types of octahedral sites, denoted by M_1 and M_2, and their relative distribution.

can participate in protonation/deprotonation reactions depending on the medium pH (Tombácz and Szekeres 2004; Kaufhold et al. 2011).

Since the lateral dimension of the tetrahedral sheet is intrinsically greater than that of its octahedral counterpart, layer formation must be accompanied by mutual sheet adjustment. Adjustment in the (ideal) tetrahedral sheet largely occurs by the alternate rotation, in a clockwise and anticlockwise fashion, of adjoining tetrahedra through angles of α about an axis normal to the basal oxygen plane (*c*-axis). As a result, the *a*-dimensions and *b*-dimensions of the sheet are shortened by approximately cos α, and the (ideal) hexagonal geometry of the basal oxygen network becomes essentially ditrigonal. Tetrahedral rotation, however, cannot fully compensate for the intrinsic misfit between the tetrahedral and octahedral sheets. Additional adjustment can be effected by increasing the thickness of the tetrahedral sheet accompanied by a reduction in the basal area of each tetrahedron (as evaluated by the angle τ). Further compensation can occur by tetrahedral tilting (as evaluated by the Δ*z* parameter), as a result of which the basal oxygens of the tetrahedral sheet cease to be coplanar (Takéuchi 1965). The various tetrahedral sheet adjustments are illustrated in Figure 1.4. More details are to be found in Brown et al. (1978), Bailey (1980), Brindley and Brown (1980), Guggenheim (2011), and Brigatti et al. (2013).

In non-interstratified structures, the particles are made up of a finite number of either 1:1 or 2:1 type layers, which are continuous in the *a*-direction and *b*-direction and stacked in a more or less parallel manner along an axis perpendicular to the layer planes (*c*-axis). A typical particle of kaolinite, for example, consists of 70–200 individual layers, stacked on top of each other (Wan and Tokunaga 2002; Johnston 2010). Such an arrangement offers scope for layer displacement, by

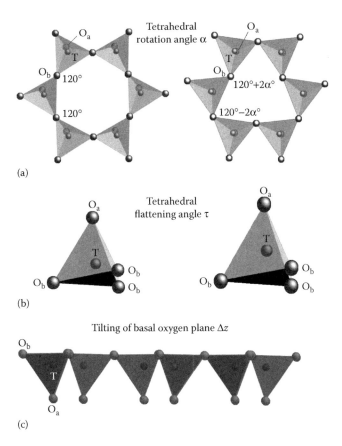

FIGURE 1.4 Diagram showing the various adjustments that the tetrahedral sheet may adopt in order to fit its larger lateral dimension to that of the smaller octahedral sheet. (a), through rotation of adjoining tetrahedra as evaluated by the angle α (deviation from 120° of each angle in the ring); (b), through tetrahedral flattening as evaluated by the angle τ (deviation from 109°28′ of Ox_a-T-Ox_b triads); and (c), through tetrahedral tilting as evaluated by the Δz parameter. Ox_a = apical oxygen; Ox_b = basal oxygen; T = tetrahedral cation. NB: the hexagonal cavity formed by the basal oxygens of the tetrahedral sheet (a) becomes ditrigonal. (From Brigatti, M.F. et al., Structure and mineralogy of clay minerals, in *Handbook of Clay Science,* 2nd ed. *Developments in Clay Science,* Vol. 5A, Bergaya, F., and Lagaly, G.(Eds.), Elsevier, Amsterdam, the Netherlands, pp. 21–81, 2013.)

rotation and/or translation (Figure 1.5), and hence for the occurrence of polytypes or structures with different layer-stacking sequences. Layer-stacking disorder in the kaolinite group of minerals (Giese 1988; Bookin et al. 1989; Plançon et al. 1989) has been elegantly demonstrated by the combined application of high-resolution transmission electron microscopy (HRTEM), selected area electron diffraction (ED), and low-temperature Fourier-transform infrared spectroscopy (FTIR) (Kogure and Inoue 2005a, 2005b; Johnston 2010; Balan et al. 2014).

The relative occupancy of the interlayer space by extraneous materials provides another source of structural variation. In the smectite and vermiculite groups of layer silicates, this space contains cations and water molecules (Table 1.1; cf. Figures 1.13 and 1.16). The nature of the interlayer cations and their complement of coordinated water molecules play an important part in Brønsted acid-catalyzed conversions by clay minerals (Fripiat and Cruz-Cumplido 1974; Adams et al. 1979; Theng 1982; Atkins et al. 1983; Ballantine et al. 1983; Jones et al. 1983). Since these factors together with the interlayer porosity may be modified within certain limits, clays serve as versatile catalysts and catalyst supports for a wide range and variety of organic reactions.

A silicate layer and its associated interlayer material (when present) constitute a *unit structure* while a *unit cell* denotes a three-dimensional repeating unit within this assembly. An important

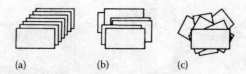

FIGURE 1.5 Diagram showing the different types of layer-stacking arrangement in an individual clay particle: (a), perfect order (as in an ideal smectite *tactoid* in aqueous suspension); (b), translational disorder; and (c), turbostratic disorder (as in a smectite *quasi-crystal*). NB: in a clay aggregate, each rectangle in (a) and (b) represents a single particle (instead of a silicate layer). An aggregate where the oriented particles are arranged in turbostratic array, as in (c), is referred to as a *domain*. (From Caillère, S. et al., *Minéralogie des Argiles. 1. Structure et Propriétés Physico-Chimiques.* 2nd ed., Masson, Paris, France, 1982.)

parameter in clay mineral identification by XRD is the repeat distance along the c-axis, which represents the first-order reflection from planes parallel to the layers. Referred to as the basal or $d(001)$ spacing, this parameter corresponds to the thickness of an individual silicate layer plus the interlayer distance (Δ-*value*) separating successive layers within a particle (Figure 1.6). Thus, for non-expanding phyllosilicates (e.g., kaolinite), the basal spacing is equal to the layer thickness (Figure 1.7). On the other hand, for structures containing interlayer materials (e.g., smectites), the basal spacing is invariably greater than the minimum van der Waals layer thickness of an individual silicate layer (ca. 0.96 nm), even after complete dehydration and layer collapse, because of the finite size of the interlayer counterions (Figure 1.13).

For the sake of simplicity, it is generally assumed that the unit structure composition of a given mineral is invariant, or at least homogeneous, throughout the particle. In the case of interstratified structures (MacEwan and Ruiz-Amil 1975; Reynolds 1980; Sawhney 1989; Brigatti et al. 2013) where two or more layer types are present within a single particle, this assumption is clearly invalid. Rectorite is an example of a regularly interstratified phyllosilicate that has featured as a catalyst (Liu et al. 2013b). Here a (dioctahedral) mica layer alternates with a (dioctahedral) smectite layer (Sawhney 1989; Guggenheim et al. 2006; Brigatti et al. 2013). Even in the absence of interstratification, however, strict unit structure homogeneity is probably more the exception than the rule. Indeed, both the extent and location of isomorphous substitution in smectite particles, as reflected by a net layer charge per formula unit, x (Table 1.1), commonly vary from layer to layer (Lagaly and Weiss 1970, 1976; Stul and Mortier 1974; Christidis and Dunham 1993; Janek et al. 1997; Christidis and Eberl 2003; Czímerová et al. 2006; Ertem et al. 2010). Another source of inhomogeneity arises from variations in layer-stacking arrangement to which we have already referred (Brown et al. 1978; Brindley 1980; Brigatti et al. 2013).

Isomorphous substitution in phyllosilicates is controlled more by the space and coordination requirement than by the valency of the cations involved. Accordingly, Al^{3+} and Mg^{2+} ions can readily interchange in these structures, giving rise to the dioctahedral and trioctahedral (character) series, respectively (Table 1.1). A good example already referred to is the replacement of every two octahedrally coordinated Al^{3+} ions in gibbsite by three Mg^{2+} ions to yield the brucite sheet structure. In clay mineral structures, however, this process is generally incomplete. In addition, the valency of the substituting ion is usually lower than that of its principal structural counterpart. A case in point is the partial replacement of Si^{4+} by Al^{3+} in the tetrahedral sheet of many 2:1 type layer silicates. ^{27}Al and ^{29}Si solid-state NMR spectroscopic studies have shown that this particular process generally obeys Loewenstein's (1954) rule in that adjacent tetrahedral sites may not be occupied by aluminum (Lipsicas et al. 1984; Serratosa et al. 1984; Herrero et al. 1985; Sanz 2006). Similarly, magnesium and iron may partially substitute for aluminum in octahedral positions.

Quantum-chemical calculations using a self-consistent charge extended Hückel model indicate that the ordering of successful occupancy of octahedral positions in (dioctahedral) smectites is $Al^{3+} > Fe^{3+} > Mg^{2+} > Fe^{2+} > Na^+ = Ca^{2+} > K^+$ (Aronowitz et al. 1982). That is to say, for this series of cations Fe^{3+} requires the smallest, and K^+ the greatest, energy in replacing octahedrally

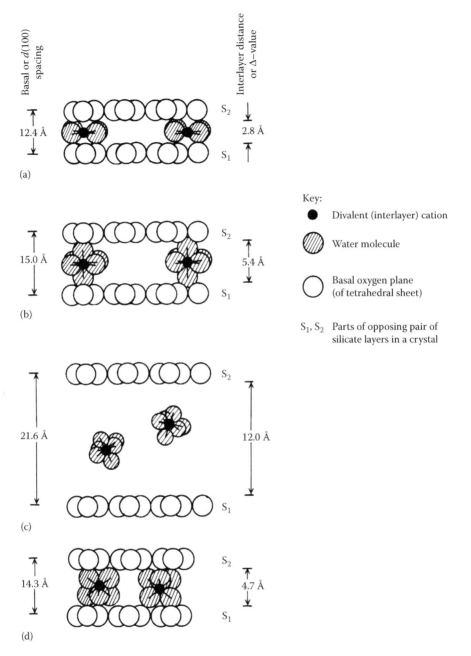

FIGURE 1.6 Diagram showing the arrangement of: (a), interlayer $Cu(H_2O)_4^{2+}$ counterions in a one-sheet hydrate of hectorite; (b), interlayer $Cu(H_2O)_6^{2+}$ counterions in a two-sheet hydrate of hectorite; (c), interlayer $Mn(H_2O)_6^{2+}$ counterions in a fully water-swollen (expanded) hectorite (water molecules not coordinated to counterions are omitted); and (d), interlayer $Cu(H_2O)_6^{2+}$ counterions in a two-sheet hydrate of vermiculite. Interlayer distance (Δ-value) is obtained by subtracting 9.6 Å from the corresponding basal spacing value. NB: 1 Å = 0.1 nm. (From Clementz, D.M. et al., *J. Phys. Chem.*, 77, 196–200, 1973; McBride, M.B. et al., *J. Phys. Chem.*, 79, 2430–2435, 1975; Pinnavaia, T.J., Applications of ESR spectroscopy to inorganic-clay systems, in *Advanced Chemical Methods for Soil and Clay Minerals Research*, Stucki, J.W. and Banwart, W.L. (Eds.), D. Reidel, Dordrecht, the Netherlands, pp. 391–421, 1980.)

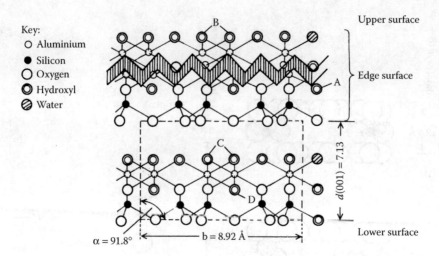

FIGURE 1.7 The structure of kaolinite, viewed along the *a*-axis, showing the superposition of adjacent layers within a particle. One external particle surface (upper) consists of a hydroxyl plane, while the other surface (lower) is made up of an oxygen plane. At the particle edge surface, each exposed aluminum ion normally carries a water molecule, while the edge silicon ion has an attached hydroxyl group. The outer hydroxyls are denoted by A and B, the inner-surface hydroxyls by C, and the inner hydroxyl by D. The unit cell dimension is indicated by the broken rectangle. NB: 1 Å = 0.1 nm. (From Dixon, J.B., Kaolinite and serpentine group minerals, in *Minerals in Soil Environments*, 2nd ed., Dixon, J.B. and Weed, S.B. (Eds.), Soil Science Society of America, Madison, WI, pp. 467–525, 1989.)

coordinated Al^{3+}. In accordance with this model, the vast majority of naturally occurring dioctahedral smectites contain some Fe^{3+} ions in octahedral positions, the extreme case being exemplified by the ideal nontronite layer structure (cf. Figure 1.13; Table 1.1). This point is particularly important to catalysis in that structural ferric ions can accept electrons from adsorbed organic compounds across the layer boundary, and hence act as Lewis acids (Solomon 1968; Theng 1971; Tennakoon et al. 1974; Almon and Johns 1976; Solomon and Hawthorne 1983). Strucural ferric (and ferrous) ions, exposed at clay mineral particle surfaces, can also catalyze the oxidation of organic molecules through a Lewis acid or Fenton-like reaction (Adams et al. 1987; Soma and Soma 1989; Cheng et al. 2008; Purceno et al. 2012). The activity of iron-exchanged and iron-pillared interlayered smectites as heterogeneous Fenton catalysts in degrading recalcitrant organic compounds, and in waste-water treatment, have been reviewed by Garrido-Ramírez et al. (2010), Herney-Ramirez et al. (2010), and Navalon et al. (2010). The catalytic activity of clays and pillared interlayered clays in Fenton-like reactions is summarized in Chapter 7.

Perhaps the most important consequence of isomorphous substitution, however, is that the layers become deficient in positive charges. In other words, the structure acquires a permanent negative charge that is independent of medium pH. The layer charge deficit is balanced by sorption of an equivalent amount of extraneous cations, commonly referred to as *charge-balancing* cations' or *counterions*. The latter term corresponds to *gegenions* and *compensatory cations* in the German and French literature, respectively. Most of these counterions are located in the interlayer space, and with the possible exception of those in mica and brittle mica, can be exchanged for other (inorganic or organic) cations in the external solution. Since the exchange process is stoichiometric, unless steric and chemical factors intervene, the term *exchangeable cations* has been used synonymously with counterions. The cation exchange capacity (CEC) of clay minerals is traditionally expressed in milliequivalents (meq) per 100 g of calcined clay (Bergaya et al. 2006a), and serves as an index of the extent of isomorphous substitution. The SI unit of CEC, recommended by the IUPAC, is $cmol_c/kg$, which is numerically equivalent to meq/100 g (Table 1.2).

TABLE 1.2
Cation Exchange Capacity (CEC), Layer Charge Density, and Swelling Characteristics of Some Common Clay Mineral Species as Affected by the Nature of the Interlayer Counterions

Species	CEC[1] (cmol$_c$/kg)	Mean Charge Density[2]			Interlayer Counterion	Basal Spacing in Water (nm)[3]	Swelling in Water[4]
		μeq/m²	C/m²	Area/Charge (nm²)			
Kaolinite	Nil	Nil	Nil	N.a.	None	0.71	Absent
Pyrophyllite	Nil	Nil	Nil	N.a.	None	0.92	Absent
Talc	Nil	Nil	Nil	N.a.	None	0.93	Absent
Montmorillonite (Wyoming)	98	1.22	0.118	1.36	Li	2.20; ind (>4 nm)	Restricted; extensive
					Na	1.90; ind (>4 nm)	Restricted; extensive
					K	1.50	Restricted
					Ca, Mg	1.90	Restricted
Vermiculite (Kenya)	124	1.58	0.152	1.04	Li	1.50; ind (>4 nm)	Restricted; extensive
					Na	1.48	Restricted
					K	1.10	Restricted
					Ca	1.54	Restricted
					Mg	1.48	Restricted
Illite (Fithian)	26	2.80	0.271	0.60	K	1.0	Absent
Mica (muscovite)	<5	3.54	0.342	0.46	K	1.0	Absent
Brittle mica (margarite)	<5	7.26	0.702	0.23	Ca	1.0	Absent

Sources: Norrish (1954); Quirk (1968), Suquet et al. (1975), Greenland and Mott (1978), MacEwan and Wilson (1980).

[1] In reality, kaolinite has a small CEC due to isomorphous substitution in its structure (cf. Table 1.7). The CEC of illite, mica, and brittle mica is low because their interlayer counterions are practically non-exchangeable.
[2] The values for montmorillonite, vermiculite, and illite are derived on the basis of external and interlayer surface areas, while those for muscovite and margarite are calculated on the basis of external particle areas; N.a. = not applicable.
[3] basal spacing can be indeterminate (ind); i.e., in excess of 4 nm.
[4] extensive ('*osmotic*') swelling is associated with the development of diffuse electric double layers on interlayer surfaces; restricted swelling is often referred to as '*crystalline*' swelling.

Alternatively, the (negative) layer charge may be expressed in terms of electron charge per unit cell, or more commonly as charge per formula unit (x). Expressed as a positive number, x is therefore an important parameter in the classification of phyllosilicates. In the absence or near-absence of isomorphous substitution ($x \approx 0$), the interlayer space is essentially empty. For the majority of 2:1 type phyllosilicates, however, $x > 0$, and the resultant negative layer charge is balanced by hydrated or non-hydrated cations, occupying interlayer positions. The interlayer space may also be filled with a hydroxide sheet to give the chlorite layer type. These points are summarized in Table 1.1.

In terms of surface reactivity and intercalating ability, however, the site (origin) of isomorphous substitution is as important as, if not more so than, its extent. Thus, substitution of Mg^{2+} for Al^{3+} in the octahedral sheet of smectites gives rise to a (positive) charge deficit that is delocalized over ≈ 8 oxygen atoms on each side of adjacent silicate layers. On the other hand, the charge deficit, arising from substitution of Al^{3+} for Si^{4+} in the tetrahedral sheet, is more or less confined to one side of the layer structure (Schoonheydt and Johnston 2013). Isomorphous substitution, especially in tetrahedral sites, also increases the (Lewis) base strength of the siloxane surface (Yariv 1992).

More interestingly, an acid-activated synthetic beidellite (with tetrahedral substitution) became *superacidic* after heating to 300°C due to the formation of Al-OH-Si linkages (Jacobs 1984; Schutz et al. 1987). In comparing the catalytic activity of illite-smectite clays in hydrocarbon-forming reactions, Johns and McKallip (1989) also found that the activity of a cation-exchange site resulting from tetrahedral substitution (of Al^{3+} for Si^{4+}) was 40 times greater than that arising from substitution in the octahedral sheet.

Depending on the extent of isomorphous substitution, the surface charge density of smectites ranges from 1.2 to 1.8 µeq/m^2 (0.117–0.175 C/m^2), corresponding to a negative charge separation of 0.9–1.35 nm (Środoń and McCarty 2008; Table 1.2). The effective charge density, however, is twice as large because the smectite layers are stacked one on top of another. As a result, the negative charge separation is reduced to 0.65–0.84 nm (Johnston 2010; Schoonheydt and Johnston 2013). This point is pertinent to the catalytic activity of pillared interlayered smectites since the distance between negatively charged sites has a determining effect on the interpillar separation and porosity in these materials.

In layer silicate structures with little, if any, isomorphous substitution (e.g., pyrophyllite and talc), the basal oxygens of the tetrahedral sheet behave as very weak electron donors or Lewis bases (Farmer 1978; Sposito and Prost 1982; Sposito 1984; Schoonheydt and Johnston 2013). Being nonpolar and hydrophobic, the siloxane surface of such minerals interacts only weakly with uncharged polar molecules, including water. Nevertheless, the *neutral siloxane surface*, exposed between negatively charged sites in the interlayer space of tetrahedrally substituted smectites, can show a high affinity for some aromatic hydrocarbons and neutral polar organic contaminants through ion-dipole and van der Waals interactions (Jaynes and Boyd 1991; Johnston 2010; Boyd et al. 2011; Schoonheydt and Johnston 2013).

In smectites of low to moderate charge, and where isomorphous substitution largely occurs in the octahedral sheet as in montmorillonite (Figure 1.14), the Lewis base character of the basal surface oxygens would be enhanced. In this instance, the negative layer charge is largely delocalized, allowing the counterions to form stable complexes with water molecules by ion-dipole interactions. Depending on the nature of the counterions and the ambient relative humidity, up to four sheets of water molecules may be intercalated (*nano-confined*), increasing the basal spacing from 0.96 to ca. 1.9 nm (Table 1.2). This restricted interlayer expansion is commonly referred to as *crystalline* swelling (Slade et al. 1991; Quirk 1994; Laird and Shang 1997; Laird 2006).

Because of the high enthalpies of ion hydration, ranging from −263 to −515 kJ/mol (for alkali metal ions), and from −1304 to −1922 kJ/mol (for alkaline earth metal ions) (Burgess 1978; Table 2.7), water molecules tend to coordinate to, and form hydration shells around, the counterions. These points are illustrated in Figure 1.6 showing the structures of one-sheet and two-sheet hydrates of Cu^{2+}-hectorite as deduced from XRD analysis and electron spin resonance (ESR) spectroscopy (Clementz et al. 1973; McBride et al. 1975). In relation to clay catalysis, we should add that the cation-water structures are a source of protons (Brønsted acidity), arising from the dissociation (hydrolysis) of coordinated water molecules (Mortland and Raman 1968; Theng 1982; Jones et al. 1983; Chapter 2).

In the one-sheet hydrate, each Cu^{2+} ion is coordinated to four water molecules in a square planar configuration, giving a basal spacing of 1.24 nm (Figure 1.6a). Here the $Cu(H_2O)_4^{2+}$ plane lies parallel to the silicate layers, while the copper ion binds to the surface without the intervention of water molecules. Similar *inner-sphere* coordination complexes are formed with monovalent (alkali metal) cation-smectite systems at a relative humidity (p/p_o) of up to 0.5 at 25°C. On the other hand, in a two-sheet hydrate (Figure 1.6b), each interlayer Cu^{2+} ion is octahedrally coordinated to six water molecules, the octahedron being oriented with its long symmetry axes perpendicular to the silicate layers, giving a basal spacing of 1.50 nm. The water molecules here serve both as ligands to the counterions and as dielectric bridges between these ions and the clay mineral surface. Such complexes in which the surface interacts with the counterions through a *water-bridge* are accordingly referred to as being of the *outer-sphere* type. A similar silicate layer-counterion arrangement applies to the two-sheet hydrate of Ca^{2+}-monmorillonite (Sposito 1984). Figure 1.6c shows the structure of Mn^{2+}-hectorite at maximum interlayer expansion (crystalline swelling), giving a basal spacing of 2.16 nm (Pinnavaia 1980, 1983). ESR spectroscopy indicated that under these conditions the $Mn(H_2O)_6^{2+}$ ions could tumble freely in the interlayer space. In other words, the cation-solvate does not adopt any particular orientation toward the silicate layers but behaves much like its counterpart in solution.

With rising isomorphous substitution involving the tetrahedral sheet, as is the case in vermiculite, the excess negative layer charge would be increasingly localized to those surface oxygens near the sites of substitution. As a result, the electron-donating character of the corresponding basal siloxane surface (ditrigonal cavities) would be further enhanced. In a two-sheet hydrate of Cu^{2+}-vermiculite, for example, this effect induces the $Cu(H_2O)_6^{2+}$ complex to adopt an orientation in which its ligand axis is inclined at a high angle (ca. 45°) to the silicate layers, giving a basal spacing of 1.43 nm (Clementz et al. 1974). In such a conformation, all six water ligands can be hydrogen-bonded to surface oxygens, allowing the central Cu^{2+} ion to get closer to the negatively charged site than would otherwise be possible (Figure 1.6d).

A similar picture has emerged from infrared (IR) spectroscopic studies on adsorbed water in smectite and vermiculite (Farmer and Russell 1971; Prost 1976; Sposito and Prost 1982). The IR spectra typically show a high-frequency (HF) band between 3650 and 3570 cm^{-1} besides a principal maximum in the region of 3430–3350 cm^{-1}, and a low-frequency shoulder near 3230 cm^{-1}. A comparable HF band is also found in the IR spectra of concentrated aqueous solutions and crystalline hydrates of perchlorate salts for which weak hydrogen bonding (ca. 2 kcal/mol) is known to exist between water and the perchlorate anions (Brink and Falk 1970). On the other hand, the HF band is absent from the spectra of pure water and of aqueous solutions of anions with stronger electron-donating properties. The position of the HF band is also influenced by the site of isomorphous substitution. In hectorite and montmorillonite, for example, this band appears at a higher frequency as compared with vermiculite (Sposito 1984). That the surface oxygens of layer silicates are only weak electron donors has also come to light by comparing the N–H stretching frequency of adsorbed ammonium and substituted ammonium compounds in smectite and vermiculite with that of the corresponding perchlorate, halide, and dichromate salts (Farmer and Mortland 1965; Laby and Walker 1970; Laby and Theng 1972). In this connection, we might also add that the vibrational spectra of ammonium in NH_4^+-exchanged smectites can provide information about interlayer charge heterogeneity (Chourabi and Fripiat 1981; Petit et al. 1999).

In common with water, uncharged polar organic compounds (as well as non-ionic polymers), adsorb to clay minerals either by direct coordination with the exchangeable cations (ion-dipole interaction), or indirectly through a *water bridge* (Mortland 1970; Theng 1974, 2008; Johnston 2010). In other words, organic molecules must successfully compete with water for the same (ligand) sites around the counterions if adsorption is to occur. The clay-organic interaction is accordingly influenced by the valency (and size) of the counterion as well as by the basicity and orientation (packing) of the organic species in the interlayer space (Chapter 3).

The extent of isomorphous substitution (in tetrahedral positions) in micas is even greater than that in vermiculite. As a result, the excess negative layer charge is now so highly localized to the ditrigonal cavities that the K^+ counterions become directly coordinated to the basal oxygens to form inner-sphere surface complexes. Indeed, each interlayer K^+ ion in muscovite and related structures is effectively twelvefold coordinated, six of the oxygens being from one silicate layer and the other six from the opposing layer (cf. Figure 1.18). Further, the K^+ ion can fit (*key*) very snugly into the space between two opposing ditrigonal cavities because its dimension is virtually identical with the cavity diameter (Sposito 1984). The strong electrostatic interactions between successive layers within a particle (crystal) account for the absence of interlayer swelling in water and very low CEC (Table 1.2). The resistance of micas to *weathering* by which the interlayer K^+

ions are replaced by other (hydrated) cations to yield vermiculite-like structures may be explained in similar terms (Norrish 1973; Fanning et al. 1989). If anything, the strength of interlayer bonding and cohesion is more pronounced in the brittle micas having twice the layer charge of true micas (Table 1.1).

1.3 SPECIFIC STRUCTURAL AND SURFACE PROPERTIES

The long-range order (*crystalline*) phyllosilicates fall into two broad (structural) groups, comprising the 1:1 (T-O) and 2:1 (T-O-T) layer types (Table 1.1). Among minerals in the former group, *kaolinite* (either as such or as a catalyst support, and often after acid/thermal treatment), leads the field in terms of its usage as a catalyst of organic reactions and syntheses, followed by *halloysite* and *chrysotile*. The various organic conversions catalyzed by these three clay mineral species are summarized in Table 1.3.

Among the 2:1 type phyllosilicates, the smectite group of minerals stand out as catalysts for a wide range and variety of organic conversions. Indeed, the vast majority (ca. 90%) of published work refers to catalysis over cation-exchanged and surface-modified *smectites* of both natural and synthetic origin. Smectites owe this prominence to their large propensity for intercalating organic compounds (Mortland 1970; Theng 1974, 2008, 2012; Yariv and Cross 2002; Lagaly et al. 2006; He et al. 2014) as well as to the ease with which their interlayer geometry and porosity may be modified by insertion of pillaring agents, as remarked on earlier (cf. Chapter 3). Within the smectite group of phyllosilicates (Table 1.1), *montmorillonite* (including *bentonite*) has held the limelight, followed distantly by *hectorite* (including *Laponite*), *saponite*, and *beidellite* (Thomas et al. 1977; Ovcharenko 1982; Theng 1982; Thomas 1982; Ballantine and Purnell 1984; Adams 1987; Ortego et al. 1991; Balogh and Laszlo 1993; Brown 1994; Eastman and Porter 2000; Kaur and Kishore 2012; McCabe and Adams 2013). The wide range and variety of organic reactions and syntheses catalyzed by smectites and their surface-modified forms will be described in the following chapters.

By comparison with smectites, organic catalysis by *vermiculite*, *chlorite*, *mica*, and *illite* is limited in both range and variety although a good number of publications are concerned with the use of *synthetic mica-montmorillonite* (SMM) and *fluorotetrasilicic mica* (TSM). The various organic conversions catalyzed by each of these six 2:1 type layer silicates are listed in Table 1.4.

Among the 2:1 type phyllosilicates with a *layer-ribbon* structure, only *sepiolite and palygorskite* (*attapulgite*) have featured as catalysts of organic reactions and syntheses (Table 1.5). Similarly, there are not many references to organic catalysis by layer silicates with short-range order, namely, *allophane* and *imogolite* (Table 1.6).

Accordingly, the following discussion is more concerned with the structural and surface properties of kaolinite, smectite, palygorskite, and sepiolite than with those of halloysite, chrysotile, mica, and vermiculite. The structure of chlorite, SMM, and TSM will only be briefly described because of their limited usage as catalysts. For the same reason, allophane and imogolite are included, while pyrophyllite, talc, and brittle mica are omitted.

TABLE 1.3
Examples of Organic Reactions Catalyzed by 1:1 Type Layer Silicates: Kaolinite, Halloysite, and Chrysotile

Reaction	Catalyzed by Kaolinite	Reference (in Chronological Order)
Synthesis of butadiene from ethanol	Thermally treated sample	Aleixandre and Fernandez (1960)
Decomposition of insecticides	Natural sample	Fowkes et al. (1960)
Decomposition of behenic acid	Natural sample	Eisma and Jurg (1969)
Thermal degradation of *n*-dodecanol to hexene isomers	Natural sample	Galwey (1969)
Polymerization of styrene	Natural sample Acid-activated sample	Solomon and Rosser (1965); Hawthorne and Solomon (1972) Njopwouo et al. (1987, 1988)
Decomposition of glycol, glycerol, digol, *n*-dodecane	Natural sample	Eltantawy and Arnold (1973)
Hydrolysis of organophosphorus pesticides	Homoionic samples	Saltzman et al. (1974); Mingelgrin et al. (1977)
Isomerization of α-pinene	Acid-activated sample	Battalova et al. (1975); De and Srivastava (1975); Nazir et al. (1976); Perissinotto et al. (1997); Volzone et al. (2005)
Decarboxylation/decomposition of fatty acids	Natural sample	Almon and Johns (1976)
Conversion of *n*-dodecanol and stearic acid	Natural sample	Wilson and Galwey (1976); Galwey (1977)
Polymerization of phenolic compounds	Natural sample	Wang et al. (1978)
Transformation of sterols	Fractionated heat-treated sample	Sieskind et al. (1979)
Decyclization and polymerization of octamethylcyclotetrasiloxane	Al-exchanged sample	Skobets et al. (1979)
Cracking of cumene	Acid-activated sample	Youssef (1979); Hanna and Khalil (1982)
Hydrogenation and isomerization of cyclohexene	Natural sample	Curtis et al. (1983)
Hydrolysis of 1-(4-methoxyphenyl)-2,3-epoxypropane and N-methyl p-tolyl carbamate	Homoionic sample	El-Amamy and Mill (1984)
Dehydration of isopropyl alcohol	Li-exchanged sample	Goncharuk et al. (1984)
Reaction of olefins with diethyloxomalonate to yield γ-lactones	Natural kaolin	Roudier and Foucaud (1984)
Synthesis of alkyl aryl ethers from phenols and alkanols	Natural sample	Matsuzaki et al. (1986)
Oxidative polymerization of hydroquinone	Size-fractionated sample	Shindo and Huang (1985); Wang and Huang (1989a)
Transformation of 1-pristene	Natural sample	Lao et al. (1989)
Ring cleavage of pyrogallol	Ca-kaolinite	Wang and Huang (1989b)
Diels–Alder cycloaddition	Natural sample	Collet and Laszlo (1991); Cornélis and Laszlo (1994)

(Continued)

TABLE 1.3 (*Continued*)
Examples of Organic Reactions Catalyzed by 1:1 Type Layer Silicates: Kaolinite, Halloysite, and Chrysotile

Reaction	Catalyzed by Kaolinite	Reference (in Chronological Order)
Degradation of chlorfenvinphos and methidathion	Homoionic sample	Cámara et al. (1992)
Isomerization of 2,2-dimethylbutane	Kaolinite-supported Pt	Balogh and Laszlo (1993)
Friedel-Crafts alkylation of benzene with benzyl chloride yielding diphenylmethane	Acid-activated metakaolinite	Sabu et al. (1993)
Tetrahydropyranylation and trimethylsilylation of alcohols	Natural sample	Upadhya et al. (1996)
Dehydration of isopropanol and isomerization of 1-butene	Acid-activated metakaolinite	Perissinotto et al. (1997); Lenarda et al. (2007)
Inhibition of human leucocyte elastase	Natural sample	Fubini et al. (1997)
Oligomerization of glycine	Natural sample	Zamaraev et al. (1997)
Regeneration of carboxylic acids from corresponding allyl or cinnamyl esters	Natural sample	Gajare et al. (1998)
Conversion of nitriles to 2-oxazolines	Natural sample	Jnaneshwara et al. (1998, 1999)
Transesterification of alcohols and phenols	Natural sample	Ponde et al. (1998)
Friedel–Crafts alkylation of benzene with benzyl chloride	Kaolinite-supported metal chlorides	Sukumar et al. (1998)
Formation of melanoidins from D-glucose and L-tyrosine	Natural sample (Zettlitz)	Arfaioli et al. (1999)
Debutylation of 2-tert-butylphenol	Acid-activated sample	Mahmoud and Saleh (1999)
Conversion of benzyl chloride to diphenylmethane	Acid-activated metakaolinite	Sabu et al. (1999)
Selective cleavage of thioacetals	Natural sample	Bandgar and Kasture (2000)
Protection of carbonyl group and its conversion to 1,3-dithiolanes	Natural sample	Banerjee and Laya (2000)
Conversion of isatoic anhydride to 2-(*o*-aminophenyl)oxazolines	Natural sample	Gajare et al. (2000)
Transesterification of β-keto esters	Natural sample	Bandgar et al. (2001a)
Conversion of aryl acetates to corresponding phenols	Natural sample	Bandgar et al. (2001b)
Addition of aliphatic amines to α,β-ethylenic compounds	Natural sample	Shaikh et al. (2001)
Benzylation of benzene by benzyl chloride	Kaolinite-supported $InCl_3$ and $GaCl_3$	Choudhary and Jana (2002)
Decomposition of hydrofluoroethers	Natural sample	Kutsuna et al. (2002)
Cracking of heavy oil	Natural sample	Rong and Xiao (2002)
Decomposition of methyl chloroform	Natural sample	Kutsuna et al. (2003)
Synthesis of dialkoxymethanes from alcohols and paraformaldehyde	Natural sample	Pathak and Gerald (2003)
Epoxidation of cyclooctene; hydroxylation of cyclohexane	Iron(III) organic anions immobilized on organically modified sample	Nakagaki et al. (2004); Bizaia et al. (2009); de Faria et al. (2012)
Fluid catalytic cracking (FCC)	Silicate/phosphate modified kaolin	Zheng et al. (2005)
Chlorination of arenes	Acid-activated sample	Jayachandran et al. (2006)

(*Continued*)

TABLE 1.3 (Continued)
Examples of Organic Reactions Catalyzed by 1:1 Type Layer Silicates: Kaolinite, Halloysite, and Chrysotile

Reaction	Catalyzed by Kaolinite	Reference (in Chronological Order)
Oxidation of cyclohexane and n-heptane	Iron porphyrins immobilized on hexylamine-intercalated kaolinite	Nakagaki et al. (2006a)
Formation of diamondoids from kerogen	Natural kaolinite	Wei et al. (2006)
Synthesis of triazenes from cyclic amines	Thermally treated sample	Dabbagh et al. (2007)
Esterification of carboxylic acids	Acid-activated sample	Konwar et al. (2008)
Photocatalytic degradation of toluene and D-limonene	Kaolinite-supported TiO_2 nanoparticles	Kibanova et al. (2009)
Synthesis of 5-substituted 1-H-tetrazoles from aromatic nitriles and sodium azide	Natural kaolinite	Chermahini et al. (2010)
Esterification of oleic acid with methanol	Acid-activated metakaolinite	do Nascimento et al. (2011a, 2011b); de Oliveira et al. (2013)
Synthesis of 14-aryl-14H-dibenzo[a,j] xanthenes	Kaolinite-supported 1,3-dibromo-5,5-dimethylhydantion	Shirini et al. (2012)
Wet oxidative degradation of chlorophenols	Cu(II)-kaolinite	Khanikar and Bhattacharyya (2013)
Thermal decomposition of 12-aminolauric acid	Natural sample	Liu et al. (2013b)
Synthesis of heterocyclic compounds	Natural sample	Sahu et al. (2013)
Oxidation of hydrocarbons and degradation of dyes	Metal-dipicolinate immobilized on organically modified kaolinite	Araújo et al. (2014)
Conversion of 2-methylbut-3-yn-2-ol	Natural samples	Novikova et al. (2014)
Esterification of waste from deodorization of palm oil with ethanol	12-Tungstophosphoric acid supported on metakaolinite	Pires et al. (2014)

Reaction	Catalyzed by Halloysite	Reference (in Chronological Order)
Hydrolysis of ethyl acetate	Acid-activated sample	McAuliffe and Coleman (1955)
Synthesis of butadiene from ethanol	Thermally treated sample	Aleixandre and Fernandez (1960)
Decomposition of glycol, glycerol, digol, n-dodecane	Natural sample	Eltantawy and Arnold (1973)
Asymmetric hydrogenation of acrylic acids	Halloysite-organic complex	Mazzei et al. (1980)
Polymerization of styrene	Acid-activated sample	Njopwouo et al. (1987, 1988)
Isomerization of α-pinene to camphene	Acid-activated sample	Findik and Gündüz (1997)
Decomposition of hydrofluoroethers	Natural sample	Kutsuna et al. (2002)
Cracking of heavy oil	Natural sample	Rong and Xiao (2002)
Decomposition of methyl chloroform	Natural sample	Kutsuna et al. (2003)
Degradation of polystyrene	Acid-activated sample	Tae et al. (2004); Cho et al. (2006)
Oxidation of cyclohexane, cyclooctene, and n-heptane	Halloysite-immobilized metalloporphyrins	Nakagaki and Wypych (2007); Machado et al. (2008)
Polymerization of methyl methacrylate	Halloysite-organic complex; CuBr complex with silanized halloysite	Barrientos-Ramírez et al. (2009, 2011)
Reduction of 4-nitrophenol	Halloysite-supported Ag	Liu and Zhao (2009)
Polymerization of aniline	Halloysite-supported hematin	Tierrablanca et al. (2010)
Degradation of methanol and acetic acid	Halloysite-supported TiO_2	Wang et al. (2011)
Esterification of lauric acid	Natural (raw) sample	Zatta et al. (2011)

(Continued)

TABLE 1.3 (*Continued*)
Examples of Organic Reactions Catalyzed by 1:1 Type Layer Silicates: Kaolinite, Halloysite, and Chrysotile

Reaction	Catalyzed by Halloysite	Reference (in Chronological Order)
Fischer–Tropsch synthesis	Halloysite-supported Co	Chen et al. (2012)
Synthesis of organosilicone resin	Halloysite-supported Pt	Zhang et al. (2012)
Decomposition of methylene blue	Halloysite-supported Ag	Zou et al. (2012)
Oxidation of toluene	Halloysite-supported Cu and Co oxides	Carrillo and Carriazo (2015)
Synthesis of α-(2,2-dimethylphenyl)-styrene by alkenylation of *p*-xylene with phenylacetylene	Calcined halloysite	Zhao et al. (2016)

Reaction	Catalyzed by Chrysotile	Reference (in Chronological Order)
Polymerization of eugenol to lignin-like polymers	Natural sample	Siegel (1957)
Decarboxylation/decomposition of fatty acids	Natural sample	Almon and Johns (1976)
Aromatization, hydroisomerization, and hydrocracking of hydrocarbons, naphtha reforming, methanation	Metal (Pt, Pd, Mo, Ni) impregnated samples	Swift (1977)
Oxidation of *N,N*-dimethylaniline	Acid-activated sample	De Waele et al. (1983)
Hydrogenation of alkenes	Natural and organically modified samples	Cozak and DeBlois (1984)
Conversion of 2-propanol	Synthetic sample	Suzuki and Ono (1984)
Oxidation of fluorene to fluorenone	Natural sample	Zalma et al. (1986)
Dimerization of ethylene	Ni-substituted sample (heated)	Ono et al. (1987)
Aldol condensation of acetone with formaldehyde to yield methylvinyl ketone	Synthetic sample	Suzuki et al. (1987)
Friedel–Crafts alkylation and acylation	Chrysotile-supported transition metal salts	Pinho et al. (1995)
Inhibition of human leucocyte elastase	Acid-leached sample	Fubini et al. (1997)
Esterification of aliphatic alcohols and fatty acids	Chrysotile-immobilized lipases	de Lima et al. (1996); Silva and Jesus (2003)
Enantioselective esterification of 2-methylpentanoic acid	Chrysotile-immobilized lipases	de Jesus et al. (1998)
Degradation (oxidation) of sodium dodecylbenzenesulfonate	Natural sample	Fachini and Joekes (2002); Fachini et al. (2007)
Oxidation of cyclohexane	Chrysotile-immobilized Iron(III) porphyrins	Nakagaki et al. (2006b)
Degradation of cationic and non-ionic surfactants	Natural sample	Fachini and Joekes (2007)
Steam reforming of benzene	Air-calcined sample	Sarvaramini and Larachi (2011)

Note: The clay mineral species may be natural (untreated), acid-activated, or heated; sometimes the clay mineral merely acts as a solid support as an immobilizing substrate of catalytically active molecules.

TABLE 1.4
Examples of Organic Reactions Catalyzed by Vermiculite, Chlorite, Mica, Illite, Synthetic Mica-Montmorillonite (SMM), and Fluoro-Tetrasilicic Mica (TSM)

Reaction	Catalyzed by Vermiculite	Reference (in Chronological Order)
Decomposition of glycerol	Natural (powdered) sample	Walker (1967)
Condensation of intercalated *l*-ornithine	Natural sample	Fornés et al. (1973)
Isomerization of α-pinene	Natural sample	Goryaev et al. (1975); Battalova et al. (1977)
Conversion of heavy oil fractions	Acid leached and calcined sample	Suquet et al. (1994)
Isopropylation of benzene	Acid-activated sample	Ravichandran and Sivasankar (1997)
Hydroisomerization of octane	Pt/Al-pillared sample	del Rey-Perez-Caballero et al. (2000)
Transesterification of β-keto esters, carbohydrate derivatives, and amines	Natural sample	da Silva et al. (2002); Silva et al. (2004)
Synthesis of 2-diazo-1,3-dicarbonyls from 1,3-dicarbonyls and mesylazide	Natural sample treated with NaOH	Rianelli et al. (2004)
Hydroisomerization of decane	Al/Ce-, Al/Hf-, Al/Zn-pillared samples	Campos et al. (2007, 2008b)
Hydroconversion of heptane	Hydrothermally treated Al-pillared sample	Campos et al. (2008a)
Degradation of triarylmethine dyes	Natural sample from Llano, Texas	Rytwo et al. (2009)
Photocatalytic degradation of azo dye (reactive brilliant orange X-GN)	Fe-pillared sample	Chen et al. (2010)
Oxidation of ethinylestradiol and indigo carmine in presence of H_2O_2	Ground, expanded samples	Purceno et al. (2012)
Oxidative polymerization of hydroquinone (to humic acids)	Size fractionated natural sample	Shindo and Huang (1985); Wang and Huang (1989a)
Reaction	**Catalyzed by Chlorite and Mica**	**Reference (in Chronological Order)**
Transformation of aromatic compounds	Natural chlorite	Burkow et al. (1990)
Acylation of alcohols and amines; cyclization of arylaldehydes; O-arylation of phenols	Natural ferrous chamosite (chlorite)	Sreedhar et al. (2009); Arundhathi et al. (2010, 2011)
Hydroxylation of benzene to phenol	Chlorite-supported vanadium oxide	Gao and Xu (2006)
Polymerization of eugenol to lignin-like material	Natural muscovite	Siegel (1957)
Polymerization of styrene	Natural muscovite	Solomon and Rosser (1965)

(Continued)

TABLE 1.4 (Continued)
Examples of Organic Reactions Catalyzed by Vermiculite, Chlorite, Mica, Illite, Synthetic Mica-Montmorillonite (SMM), and Fluoro-Tetrasilicic Mica (TSM)

Reaction	Catalyzed by Chlorite and Mica	Reference (in Chronological Order)
Isomerization and dehydrogenation of butene	Natural mica (muscovite?)	Löffler et al. (1979)
Isomerization of cyclopropane	Natural mica	Prada-Silva et al. (1979); Tsou et al. (1987)
Hydroisomerization of octane	Natural mica (phlogopite)	del Rey-Perez-Caballero et al. (2000)

Reaction	Catalyzed by Illite	Reference (in Chronological Order)
Polymerization of styrene	Natural sample	Solomon and Rosser (1965)
Decomposition of n-hexane, 1-butanol, 2-ethoxyethanol, glycerol, ethylene glycol	Natural sample	Eltantawy and Arnold (1973)
Conversion of n-dodecanol and stearic acid	Natural sample	Wilson and Galwey (1976); Galwey (1977)
Synthesis of humic substances (from phenolic compounds and urea)	Natural sample	Wang et al. (1980)
Formation of hydrocarbons during kerogen pyrolysis	Natural sample	Davis and Stanley (1982); Tannenbaum and Kaplan (1985); Huizinga et al. (1987)
Hydrogenation and isomerization of cyclohexene	Natural sample	Curtis et al. (1983)
Oxidative polymerization of hydroquinone (to humic acids)	Size fractionated natural sample	Shindo and Huang (1985); Wang and Huang (1989a)
Transformation of 1-pristene	Natural sample	Lao et al. (1988, 1989)
Transformation of aromatic compounds	Natural sample	Burkow et al. (1990)
Hydrolysis of carbamate pesticides	Natural sample	Wei et al. (2001)
Decomposition of hydrofluoroethers and methyl chloroform	Natural sample	Kutsuna et al. (2002, 2003)
Hydroxylation of benzene to phenol	Illite-supported vanadium oxide	Gao and Xu (2006)
Thermal decomposition of 12-aminolauric acid	Natural sample	Liu et al. (2013b)
Conversion of 2-methylbut-3-yn-2-ol	Natural sample (hydromica)	Novikova et al. (2014)
Isomerization of α-pinene	Natural sample	Sidorenko et al. (2014)

(Continued)

TABLE 1.4 (*Continued*)
Examples of Organic Reactions Catalyzed by Vermiculite, Chlorite, Mica, Illite, Synthetic Mica-Montmorillonite (SMM), and Fluoro-Tetrasilicic Mica (TSM)

Reaction	Catalyzed by Synthetic Mica-Montmorillonite (SMM)	Reference (in Chronological Order)
Isomerization of cyclopropane, methylcyclopropane, *n*-butenes	Cation-exchanged sample	Hattori et al. (1973)
Hydroisomerization and hydrocracking of hydrocarbons	Pd-exchanged and Ni/Co substituted samples	Swift and Black (1974); Giannetti and Fisher (1975); Heinerman et al. (1983); Röbschläger et al. (1984); van Santen et al. (1985)
Oligomerization of propene	NH_4^+- and metal cation (Ni, Co, Zn)-exchanged samples	Fletcher et al. (1986); O'Connor et al. (1988)
Polymerization of organic ammonium derivatives	Na^+-exchanged sample	Kunyima et al. (1990)
Oligomerization of iso-butene	Ni-substituted and metal cation (Ni, Co, Zn)-exchanged samples	Vogel et al. (1990)

Reaction	Catalyzed by Fluorotetrasilicic Mica (TSM)	Reference (in Chronological Order)
Conversion of methanol to methyl formate and dimethyl ether	Cu-and Ti-exchanged samples	Morikawa et al. (1982); Morikawa (1993); Matsuda et al. (1995); Ohtake et al. (2007)
Conversion of methanol to hydrocarbons	Ti-exchanged sample	Morikawa et al. (1983a)
Conversion of 1-butene to *cis*- and *trans*-2 butene	Cation (Mg, Al, Pt)-exchanged samples	Morikawa et al. (1983b)
Allylation of acetals with allylic silanes	Al-exchanged sample	Kawai et al. (1986)
Alkylation of toluene by methanol	Cation (La, Ce, Ca, Sr)-modified Al-pillared samples	Sakurai et al. (1988, 1989)
Cracking of cumene	Cation (La, Ce, Ca, Sr)-modified Al-pillared samples	Sakurai et al. (1990)
Alkylation of toluene by methanol	La-Al pillared sample	Balogh and Laszlo (1993)
Hydrogenation of dienes	Cu-exchanged sample	Nakayama et al. (1993)
Oxidation of alkanes with *tert*-butyl hydroperoxide	Cation (Mn, Cr, Co, Cu)-exchanged samples	Tateiwa et al. (1995)
Polymerization of propene (in presence of methylalumoxane)	Sample intercalated with organometallic complexes	Tudor et al. (1996)
Michael reaction of β-ketoesters with vinyl ketones	Fe^{3+}-exchanged sample	Shimizu et al. (2003, 2005)
Hydrogenation of unsaturated aldehydes	Surfactant modified sample intercalated with Pd nanoparticles	Divakar et al. (2008)

Note: The clay mineral species may be natural (untreated), acid-activated, or heated; sometimes the clay mineral merely acts as a solid support as an immobilizing substrate of catalytically active molecules.

TABLE 1.5
Examples of Organic Reactions Catalyzed by Sepiolite and Palygorskite

Reaction	Catalyzed by Sepiolite	Reference (in Chronological Order)
Conversion of ethanol to ethylene and buta-1,3-diene	Natural sample and sepiolite-supported Mn	Kitayama and Michishita (1981)
Hydrogenation of nitrobenzenes	Sepiolite-supported Pd	Aramendia et al. (1984)
Dehydration and dehydrogenation of ethanol	Natural sample and $KAlO_2$-treated sample	Dandy and Nadiye-Tabbiruka (1982); Gruver et al. (1995)
Oxidative degradation of hydrocortisone	Natural sample	Cornejo et al. (1983)
Skeletal isomerization of cyclohexene	Natural and acid-treated samples (calcined); Al-orthophosphate impregnated sample	Campelo et al. (1987a, 1987b; 1989, 1990)
Dehydration of ethanol	Al-exchanged sample	Corma and Perez-Pariente (1987)
Dehydrogenation and hydrogenolysis of methylcyclohexene	Ni-impregnated natural and acid-treated samples	Corma et al. (1988)
Condensation of benzaldehyde with ethyl cyanoacetate, ethyl acetoacetate, and ethyl malonate	Alkali metal-exchanged samples	Corma and Martin-Aranda (1991)
Cracking of cumene	Natural sample	Shuali et al. (1991)
Condensation of malenonitrile with ketones	Cs-exchanged sample	Corma and Martin-Aranda (1993)
Hydrogenation of styrene	Sepiolite-supported nickel with or without impregnation with lanthanum	Anderson et al. (1994); Damyanova et al. (1996)
Dehydration of n-butyl alcohol to dibutyl ether	Ni-exchanged sample	Urabe et al. (1994)
Dehydration of ethanol	Al-substituted sample	d'Espinose de la Caillerie et al. (1995)
Glycerolysis of triolein and rapeseed oil	Cs-exchanged sample	Corma et al. (1998)
Decomposition of high-density polyethylene	Acid-treated sample	Breen and Last (1999)
Conversion of diethylene glycol to 1,4-dioxane	Natural sample	Kitayama et al. (1999)
Heck and Suzuki cross-coupling reactions	Alkali metal-exchanged sepiolite containing $PdCl_2$	Corma et al. (2004)
Suzuki cross-coupling reaction (between aryl halides and aryl boronic acids)	Pd-exchanged sample	Shimizu et al. (2002, 2004)
Selective oxidation of toluene	TiO_2-sepiolite supported vanadium oxide	Bautista et al. (2007)
Hydrogenation of alkenes; Heck reaction of iodobenzene and methyl acrylate	Sepiolite-supported Pd nanoparticles	Tao et al. (2009)
Photocatalytic degradation of olive mill waste water, acid red G, p-nitrophenol	Sepiolite-supported TiO_2	Uğurlu and Karaoğlu (2011); Zhang et al. (2011a)
Polymerization of ethylene	Sepiolite-supported methylaminoxane	Núnez et al. (2014)
Reaction	**Catalyzed by Palygorskite** (*Attapulgite*)	Reference (in Chronological Order)
Polymerization of pentenes and hexenes	Natural sample	Gurwitsch (1912)
Polymerization of propylene	Heat-activated sample	Gayer (1933)
Isomerization of 1-hexene	Natural sample	Hay et al. (1945)

(*Continued*)

TABLE 1.5 (Continued)
Examples of Organic Reactions Catalyzed by Sepiolite and Palygorskite

Reaction	Catalyzed by Palygorskite (*Attapulgite*)	Reference (in Chronological Order)
Isomerization of α-pinene	Natural sample (calcined)	Wystrach et al (1957)
Decomposition of insecticides	Natural sample	Fowkes et al. (1960)
Polymerization of styrene	Co-exchanged sample	Solomon and Rosser (1965)
Thermal decomposition of terphenyls	Natural sample	Juppe and Rau (1969)
Isomerization of parathion	Commercial sample (calcined)	Gerstl and Yaron (1981)
Oxidative degradation of hydrocortisone	Commercial sample	Cornejo et al. (1980, 1983)
Liquid-phase hydrogenation of 1-hexene	Palygorskite-supported Rh^0	Herrero et al. (1989)
Cracking of cumene	Natural sample	Shuali et al. (1991)
Transesterification of β-keto esters, carbohydrate derivatives, and amines	Natural sample	da Silva et al. (2002); Silva et al. (2004)
Alkylation of aniline with ethyl alcohol	Sample impregnated with Fe_2O_3 in the presence of SnO_2	Satyavathi et al. (2003)
Synthesis of 2-diazo-1,3-dicarbonyls from 1,3-dicarbonyls and mesylazide	Natural sample treated with NaOH	Rianelli et al. (2004)
Oxidation of ketones with H_2O_2	Sn-substituted sample	Lei et al. (2005, 2006)
Photodegradation of methylene blue	Cu- and Ag-substituted samples coated with TiO_2	Zhao et al. (2006, 2007)
Hydrogenation of cyclohexene	Palygorskite-supported Rh	Miao et al. (2007)
Hydrogenation of chloronitrobenzenes to chloroanilines	Palygorskite-supported Pt	Ma et al. (2009)
Oxidation of cyclohexane with O_2	Calcined Co-substituted sample	Zhang et al. (2009)
Esterification of *n*-butanol with acetic acid	Palygorskite-supported phosphotungstic acid	Zhang et al. (2010)
Decomposition of *o*-dichlorobenzene	Sample impregnated with VO_x-WO_y/TiO_2	He et al. (2011)
Acetylation of alcohols with acetic acid	Acid-activated sample	Pushpaletha and Lalithambika (2011)
Photodegradation of phenol	Sample coated with SnO_2-TiO_2	Zhang et al. (2011b)
Degradation of methylene blue with ozone	Palygorskite-supported Ce-Zr oxides	Li et al. (2012)
Cracking of biomass tars and benzene	Palygorskite-supported Fe and Ni	Liu et al. (2012, 2013c)
Photodegradation of methyl orange	Palygorskite-supported CuO and Pd-CuO nanoparticles	Huo and Yang (2013)
Photodecomposition of toluene	Palygorskite-supported TiO_2 nanoparticles	Papoulis et al. (2013)
Polymerization of ε-caprolactone	Natural sample	Wang et al. (2013)
Esterification of acetic acid with *n*-butanol	Palygorskite-supported sulfonated carbon	Jiang et al. (2014)
Conversion of 2-methylbut-3-yn-2-ol	Natural sample	Novikova et al. (2014)
Synthesis of mixed alcohols from syngas	Acid-activated palygorskite impregnated with Cu, Fe, and Co	Guo et al. (2015)
Oxidation of styrene	Pd-substituted sample	Wang et al. (2015)

Note: The clay mineral species may be natural (untreated), acid-activated, or heated; sometimes the clay mineral merely acts as a solid support as an immobilizing substrate of catalytically active molecules.

TABLE 1.6
Examples of Organic Reactions Catalyzed by Allophane and Imogolite

Reaction	Catalyzed by Allophane	Reference (in Chronological Order)
Oxidative polymerization of polyphenols	Natural sample	Kyuma and Kawaguchi (1964); Kumada and Kato (1970)
Polycondensation of catechol with glycine	Soil allophane	Fukushima et al. (2009); Miura et al. (2009); Okabe et al. (2011)
Decomposition of hemicellulose to xylose	Sulfonated sample	Ogaki et al. (2011)
Oxidation of phenol with H_2O_2	Synthetic allophane-supported Fe/Cu oxide	Garrido-Ramírez et al. (2012)
Oxidation of atrazine	Synthetic allophane-supported iron oxide	Garrido-Ramírez et al. (2013)
Photodecomposition of acetaldehyde	Acid-treated allophane/TiO_2 mixed powder	Ono and Katsumata (2014)

Reaction	Catalyzed by Imogolite	Reference (in Chronological Order)
Isomerization of 1-butene	Natural sample	Imamura et al. (1993)
Decomposition of *tert*-butyl hydroperoxide and 1,1-bis(*tert*-butyldioxy) cyclododecane	Cu-impregnated natural sample	Imamura et al. (1996)
Oxidation of aromatic hydrocarbons	Synthetic Fe-substituted sample	Ookawa et al. (2008)
Dihydroxylation of olefins	Synthetic silylated sample	Qi et al. (2008)
Conversion of methanol	Synthetic sample	Bonelli et al. (2009)

Note: The clay mineral species may be natural (untreated), acid-activated, or heated; sometimes the clay mineral merely acts as a solid support as an immobilizing substrate of catalytically active molecules.

1.3.1 Kaolinite

The kaolinite layer structure, viewed along the *a*-axis, together with the sequence of layers making up a particle, is shown in Figure 1.7. In the absence of interlayer materials, the repeat distance along the *c*-axis of 0.713 nm represents the basal or $d(001)$ spacing. This value is also equal to the thickness of an individual layer. The angle α of 91.8° arises from a displacement by $-\delta b$ of successive layers in the triclinic crystal.

Also shown are the four types of hydroxyl groups, designated A through D. X-ray and neutron diffraction studies have placed the inner-surface hydroxyls (C) nearly perpendicular to the layers (Brindley and Nakahira 1958; Newnham 1961; Adams 1983; Suitch and Young 1983; Brigatti et al. 2013; Balan et al. 2014). As such, the hydroxyl hydrogens of one layer could interact with the oxygens of the adjacent layer by hydrogen bonding. The inner-surface hydroxyl groups are also responsible for the three absorption bands near 3697, 3670, and 3650 cm^{-1} in the IR spectrum of kaolinite (Ledoux and White 1966; Farmer 1974; Balan et al. 2014). The group responsible for the band at ca. 3650 cm^{-1} apparently lies close to the layer plane whereas the hydroxyls giving rise to the two higher frequency bands are nearly perpendicular to the silicate layer (Rouxhet et al. 1977). The diffraction and infrared data may be reconciled by postulating that the three IR bands arise from coupling between essentially identical hydroxyls (Farmer and Russell 1964; Rouxhet et al. 1977). It is generally accepted that the fourth IR band at 3620 cm^{-1} is due to the stretching vibration of the inner hydroxyl group (D). This hydroxyl lies centrally over a ditrigonal cavity, with its O–H bond pointing toward a vacant octahedral site and making a small angle to the (001) plane (Ledoux and White 1964; Giese and Datta 1973; Adams 1983).

The octahedral vacancy in each kaolinite layer occurs at the same site while successive layers in a particle are displaced by $-a/3$. Two possible enantiomorphic structures can therefore arise, depending on whether the left-hand or right-hand site is selected to have the vacancy (Brown et al. 1978).

By contrast, in a particle of dickite (an ordered polytype of kaolinite), there is a left-right alternation of vacant sites between successive layers while the stacking sequence of layers is similar to that in kaolinite (Newnham 1961; Bailey 1963). On the other hand, nacrite (another ordered polytype) differs from kaolinite in its layer-stacking sequence as well as in the pattern of octahedral vacancies relative to successive ditrigonal cavities (Blount et al. 1969).

Many kaolinites also depart from ideality as to the manner in which the individual layers within a particle are stacked. Layer-stacking disorder may be due to displacements parallel to the b-axis in integral multiples of $b/3$, giving rise to *b-axis disordered* structures. Alternatively, disorder may be due to displacements parallel to the layers. As a result, certain diffraction bands in the XRD powder patterns are weakened or eliminated while other bands are broadened. Disorders of the type described and other more subtle irregularities in layer stacking have been discussed by Plançon and Tchoubar (1977), Brindley (1980), and Plançon et al. (1989).

Summation of the anionic and cationic charges in the ideal unit (formula) composition of $Al_2Si_2O_5(OH)_4$ shows that the layer structure is electrically neutral. In reality, however, most kaolinites carry a net negative charge. Although this charge may partly arise from the presence of an alumino-silicate gel coating over the surface (Ferris and Jepson 1975), or from the inclusion of mica and smectite layers in the particle (Lee et al. 1975; Lim et al. 1980; Jepson 1984; Ma and Eggleton 1999), chemical and spectroscopic (FTIR, Mössbauer, NMR, ESR, XANES) studies have provided compelling evidence for isomorphous substitution within the structure (Malden and Meads 1967; Angel and Hall 1973; Rengasamy et al. 1975; Bolland et al. 1976; McBride 1976; Komusiński et al. 1981; Fysh et al. 1983; Dixon 1989; Delineau et al. 1994; Newman et al. 1994; Gualtieri et al. 2000).

More often than not, Fe^{3+} substitutes for Al^{3+} in octahedral positions, while Al^{3+} may replace Si^{4+} in the tetrahedral sheet. Indeed, Fe^{3+} substitution in natural kaolinites is related to the degree of structural disorder (Mestdagh et al. 1980; Brindley et al. 1986; Gaite et al. 1997). For a number of kaolinites from different sources, the resultant net negative charge, expressed in terms of cation exchange capacity (CEC) varies between 1.4 and 3.6 $cmol_c/kg$ (Bolland et al. 1976, 1980). This range corresponds to a charge density of 0.073–0.145 C/m^2 (Table 1.7), which is comparable to that of smectite (montmorillonite), and not much lower than that of vermiculite (Table 1.2). Subsequently, Lim et al. (1980) reported a range of 0–1 $cmol_c/kg$ for the CEC of *pure* kaolinite, while Tari et al. (1999) measured values of 2.7–3.2 $cmol_c/kg$. Further, the cation exchange sites in kaolinite are apparently confined to the tetrahedral basal surface rather than being uniformly distributed throughout the whole particle (Weiss and Russow 1963; McBride 1976; Tombácz and Szekeres 2006). On the other hand, Ma and Eggleton (1999) have suggested that the negative charge of kaolinite is associated with the edges and basal hydroxyl surfaces, and that isomorphous substitution of Al^{3+} for Si^{4+} is not significant.

TABLE 1.7
Some Surface Properties of Kaolinite Samples

Sample Name	Net Negative Charge (CEC) at pH 7 ($cmol_c/kg$)	BET-N_2 Surface Area (m^2/g)	Charge Density ($\mu eq/m^2$)	Charge Density (C/m^2)
Greenbushes	3.6	24	1.50	0.145
API-9	1.5	20	0.75	0.073
Goomalling J	3.5	28	1.25	0.121
Cornish China 436	3.0	25	1.20	0.116
Birdwood	1.5	11	1.40	0.135
U.S. Clay	1.4	13	1.10	0.106
Cornish Clay	1.7	13	1.30	0.126

Source: Bolland, M.D.A. et al., *Australian Journal of Soil Research* 14, 197–216, 1976.

Of greater importance than the probable presence of a permanent negative layer charge is the variable charge that develops at the edge surface of kaolinite particles as the pH of the ambient solution is varied (Swartzen-Allen and Matijević 1974; van Olphen 1977). This pH-dependent charge arises from the protonation (under acid conditions) and deprotonation (under alkaline conditions) of incompletely coordinated (*under-coordinated*) aluminol groups exposed at particle edges (Schoonheydt and Johnston 2013). The pH at which the net edge charge is zero is referred to as the *point of zero (net) charge* (PZC). The PZC for kaolinites ranges from pH 6 to 7.5 (Rand and Melton 1975; Herrington et al. 1992; Blockhaus et al. 1997), depending on sample history and origin, and mode of measurement. Acid-base titration of aqueous suspensions of Zettlitz kaolinite, for example, yielded a PZC of 6–6.5 for the edge surface (Tombácz and Szekeres 2006), while Liu et al. (2013a) derived a pK_a value of 5.7 from first principle molecular dynamics (FPMD) simulations.

Earlier, Rand and Melton (1975) used a rheological method for determining the PZC of the kaolinite edge surface. They suggested that the PZC was identifiable with the point at which the curves relating Bingham yield values to pH (measured at varying ionic strengths) intersected. For Cornish kaolinite, the point of intersection occurred at pH 7.3. The validity of Rand and Melton's (1975) model was further supported by the results of Theng and Wells (1995) on a series of New Zealand halloysites with different particle shapes. In the case of a spheroidal halloysite from Opotiki (Kirkman and Pullar 1978), the plots of Bingham yield against pH failed to intersect, presumably because the particles were essentially *edgeless*. For the same reason, its phosphate sorption capacity was much smaller than that of the tubular and rolled forms of halloysite (Theng et al. 1982).

Thus, the edge surface of kaolinite particles is positively charged when the solution pH is more acidic than the PZC, and negatively charged when the solution pH exceeds the PZC. In highly alkaline solutions (pH > 9), the edge silanol groups can also lose their protons and contribute to the overall negative surface charge (Brady et al. 1996; Ma and Eggleton 1999; Schoonheydt and Johnston 2013). These points are illustrated in Figure 1.8, based on the early work by Schofield and Samson (1953). A similar scheme has been proposed by Alkan et al. (2005) and Zhu et al. (2016b) on the basis of electrokinetic measurements, potentiometric titration, ^1H NMR and FTIR spectroscopies, and field emission scanning microscopy.

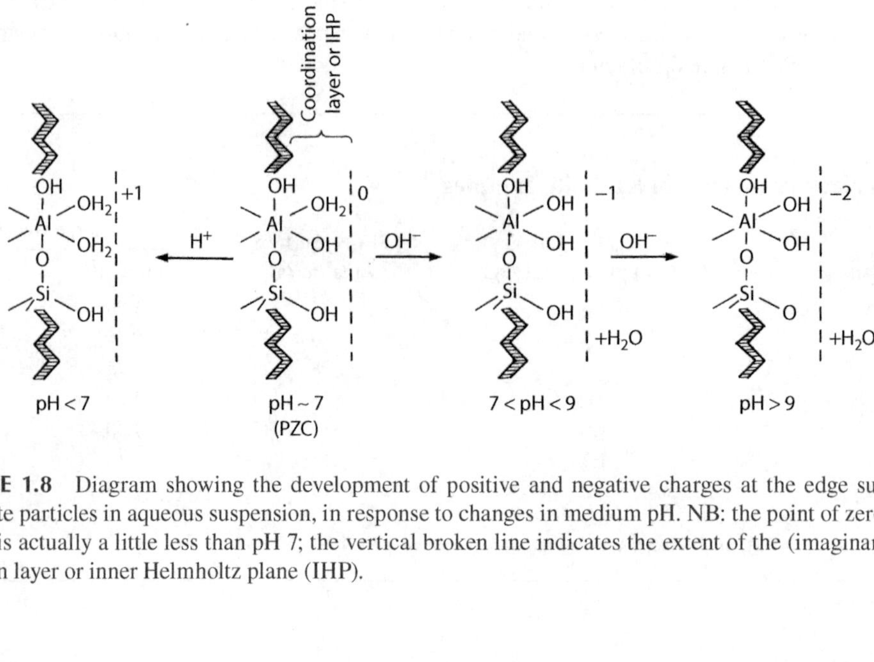

FIGURE 1.8 Diagram showing the development of positive and negative charges at the edge surface of kaolinite particles in aqueous suspension, in response to changes in medium pH. NB: the point of zero charge (PZC) is actually a little less than pH 7; the vertical broken line indicates the extent of the (imaginary) coordination layer or inner Helmholtz plane (IHP).

Since the edge surface of kaolinite is negatively charged when the suspension pH exceeds the PZC, anions in solution would be effectively repelled or *negatively adsorbed* (Schofield and Samson 1953; Quirk 1960; van Olphen 1977), while cations would be electrostatically attracted. For kaolinite (and also halloysite), the CEC derived from this source is comparable in magnitude (Sumner and Reeve 1966) to that associated with basal surfaces which, however, is independent of medium pH. Further, the pH-dependent negative charge of smectites of about 5 $cmol_c$/kg (Borchardt 1989) is not much larger than that of kaolinites. In the case of smectites, however, this amount makes up only a small fraction of the basal surface charge which, expressed in terms of CEC, is of the order of 98 $cmol_c$/kg (Table 1.2).

When the suspension pH falls below the PZC, the edge surface of kaolinite particles would acquire positive charges (by protonation), and hence can take up extraneous anions by electrostatic interactions or *anion exchange*. Such anions (e.g., chloride) remain essentially outside the coordination layer or inner Helmholtz plane (IHP) and can exchange stoichiometrically for other species (e.g., nitrate) in solution (Hingston et al. 1967, 1972). The anion exchange capacity of kaolinites, measured by chloride adsorption at pH 3, ranges from 2 to 20 $cmol_a$/kg (Quirk 1960; Sumner and Reeve 1966). On the other hand, such anions as phosphate and sulphate are *specifically adsorbed*; that is, taken up in amounts out of proportion to their concentration in solution (Hingston et al. 1972). In this process, the anion enters the IHP (cf. Figure 1.8) and replaces the hydroxyl group coordinated to edge aluminum through *ligand exchange* (Hingston et al. 1972; Parfitt 1978), as indicated in Equation 1.1:

$$[edge - Al(H_2O)(OH)]^0 + H_2PO_4^- \rightarrow [edge - Al(H_2O)(HPO_4)]^{-1} + H_2O \qquad (1.1)$$

When the system is then dehydrated, specifically adsorbed anions may form binuclear-bridging or bidentate complexes involving adjacent edge hydroxyl groups (Parfitt and Russell 1977; Parfitt 1980; Rao and Sridharan 1984). Unlike anion exchange, ligand exchange can occur even when the particle edge is negatively charged (Blockhaus et al. 1997). Whatever the initial charge of the edge surface, the overall effect of ligand exchange is one of reducing the positive charge of, or adding a negative charge to, this surface (Equation 1.1). A good example of this effect has been described by Yuan et al. (2000) for the interactions of allophane with humic acid.

Of greater relevance to catalysis is the ability of edge aluminum to serve as a Lewis acid or electron acceptor in organic reactions (Solomon 1968; Theng 1971). In an aqueous medium, however, this ability may be largely suppressed because water, being a Lewis base, would attach to such *under-coordinated* aluminum (Figure 1.7) and effectively mask its activity (Solomon and Hawthorne 1983). By the same token, the electron-accepting propensity of edge aluminum would be enhanced by excluding water from the system. In practice, this may simply be achieved by prior drying of the clay, or by conducting the reaction in a non-polar organic solvent.

Kaolinite commonly appears as pseudo-hexagonal platelets, ranging in thickness from 0.05 to 2 µm, forming *book-like* aggregates (Figure 1.9). The edge area of kaolinite particles typically makes up 12%–36% of the total (external particle) surface (Liétard et al. 1980; Lee et al. 1991; Yariv 1992; Brady et al. 1996; Wan and Tokunaga 2002). Accordingly, the influence of the particle edge on catalytic reactions, notably those involving electron transfer between edge surface and adsorbed organic compound, is much more in evidence with kaolinite than with smectite. On the other hand, smectites are more active in catalyzing organic reactions, which occur on, and involve, interlayer surfaces.

In this context, we should also mention that kaolinites can intercalate a range of polar organic compounds by: (1) contacting the mineral with the intercalant in the form of a liquid, a melt, or a concentrated aqueous solution; (2) adding an entraining agent to separate the kaolinite layers; and (3) displacement of a previously intercalated species (Theng 1974; MacEwan and Wilson 1980; Lagaly et al. 2006; Detellier and Schoonheydt 2014). By comparison with hydrated halloysites and smectites, organic intercalation into kaolinites is slow and often incomplete (Olejnik et al. 1970; Theng et al. 1984; Lagaly et al. 2006).

FIGURE 1.9 (a), Scanning electron micrograph of kaolinite showing parallel stacking of particles to form book-like aggregates (white arrows) and (b), transmission electron micrograph of kaolinite showing platy particles with a hexagonal outline. (From Carretero, M.I. and Pozo, M., *Mineralogía Aplicada. Salud y Medio Ambiente*, Thomson, Madrid, Spain, 2007.)

1.3.2 Halloysite

Hydrated halloysite or halloysite-(10 Å) has essentially the same layer structure as kaolinite (Table 1.1), but the individual layers in a particle are separated by a single sheet of water molecules, giving an ideal composition of $Al_2Si_2O_5(OH)_4 \cdot 2H_2O$, and a basal spacing of 1.01 nm. We might interpolate here that the formation of halloysites in nature is contingent on a constant presence and supply of water, and is favored by acid pH conditions. When the supply of water is intermittent and drying occurs, kaolinite is formed in preference (Churchman et al. 2016).

Most of the interlayer water in halloysite can be lost on standing in dry air, evacuation at room temperature, or mild heating (70°C) in air, reducing the basal spacing to 0.72 nm (Churchman and Carr 1972; Brindley 1980). The thickness of the interlayer water sheet is equal to the van der Waals diameter of a water molecule. Once dehydrated, halloysite does not rehydrate when exposed to water vapor, not even after prolonged contact with liquid water. Nevertheless, dehydrated halloysite or halloysite-(7 Å), like kaolinite, is capable of intercalating a number of polar organic compounds (MacEwan 1948; Carr and Hwa Chih 1971; Churchman and Theng 1984; Joussein et al. 2005). Intercalation occurs at a faster rate, and to a more complete extent as compared with kaolinite. On this basis, a simple and rapid intercalation method using formamide has been developed by Churchman et al. (1984) for differentiating halloysite from kaolinite in mixtures. In this context, we might mention that when halloysite intercalates certain polar organic molecules, the layers within a particle rearrange, giving rise to additional [*hkl*] reflections in the X-ray diffractogram (Churchman et al. 2016).

In addition to the 001 reflections, the XRD pattern of halloysite shows well-defined two-dimensional *hk* diffraction bands. This feature, which halloysite shares with smectite, is indicative of disordered (random) layer stacking. Dehydrated halloysite may therefore be regarded as a disordered polytype of kaolinite, the disorder being due to a mismatch of successive layers in a particle (Brindley 1980). On the other hand, Chukrov and Zvyagin (1966) have argued for the acceptance of halloysite as a distinct species.

Early on, Hendricks and Jefferson (1938) proposed that the interlayer water in halloysite-(10 Å) formed a more or less rigid hexagonal network in which the molecules were hydrogen bonded to each other and to opposing silicate layers (Figure 1.10). This concept of a rather uniform ice-like structure for interlayer water, however, does not appear to accord with reality. Bailey (1990) has postulated that the interlayer water molecules are coordinated to counterions, balancing the negative layer charge arising from isomorphous substitution of Al^{3+} for Si^{4+} in the tetrahedral sheet. Subsequent ^{27}Al NMR spectroscopic analysis by Newman et al. (1994), however, has shown that the Al(IV) content of many halloysites is similar to that of a standard kaolinite. X-ray photoelectron spectroscopy (XPS) of five halloysites from New Zealand has also provided good evidence for the replacement of Al^{3+} by Fe^{3+} in octahedral sites. This process, however, is apparently nonstoichiometric in that approximately two Al^{3+} ions are lost for every Fe^{3+} ion gained, giving rise to cation vacancies in the octahedral sheet (Soma et al. 1992). The positive correlation between the cation exchange capacity (5–15 $cmol_c/kg$) of the samples and their structural iron content is consistent with the formation of such vacancies (Soma and Theng 1998). Like that of kaolinite (cf. Figure 1.8), the edge surface of halloysite particles can carry a pH-dependent charge. For two halloysite samples from New Zealand, Theng and Wells (1995) have derived an edge point of zero charge (PZC) of 6.0 and 7.1 from rheological measurements.

We have already mentioned that halloysite particles can assume curved, fibrous, and rolled particle shapes. Layer curling has been ascribed to the intrinsic misfit between the laterally larger silicon tetrahedral sheet and the smaller aluminum octahedral sheet (Bates et al. 1950). Although this condition is partly offset by the alternate rotation of tetrahedra in opposite directions, the silicate layers are under a certain amount of strain. Since successive layers are separated by a water sheet, the resultant strain can be relieved by layer curling with the basal oxygen plane on the convex side (Radoslovich 1963). In kaolinite this process does not take place,

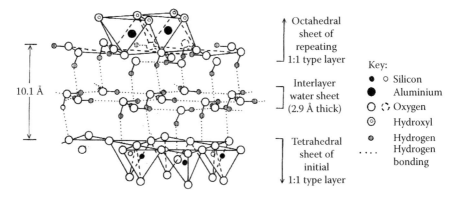

FIGURE 1.10 The structure of halloysite-(10 Å), as proposed by Hendricks and Jefferson (1938), showing the presence in the interlayer space of a single sheet of water molecules that are hydrogen-bonded (dotted line) to each other and to the hydroxyls and oxygens of opposing silicate layers. NB: 1 Å = 0.1 nm.

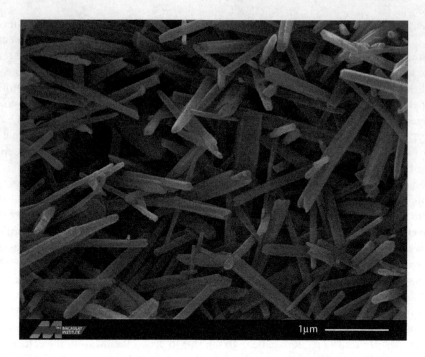

FIGURE 1.11 Scanning electron micrograph of a tubular halloysite (from China). (Reproduced from the "Images of Clay Archive" of the Mineralogical Society of Great Britain & Ireland and the Clay Minerals Society, www.minersoc.org/gallery.php?id=2.)

presumably because of the superposition of, and H-bonding between, adjacent layers within a particle (Hofmann 1968). Isomorphous substitution of iron for aluminum in the halloysite structure also affects particle shape, in that low structural iron contents tend to be associated with long tubular particles, intermediate contents with short and wide tubes (laths), and high contents with spheroidal and platy forms (Churchman and Theng 1984; Joussein et al. 2005).

Because of its potential in nanotechnological applications, including organic catalysis, the tubular forms of halloysite (Figure 1.11) have attracted a great deal of attention (Du et al. 2010; Rawtani and Agrawal 2012; Lvov and Abdullayev 2013; Tan et al. 2014; Pasbakhsh and Churchman 2015; Yuan et al. 2015; Hanif et al. 2016; Liu et al. 2016; Pasbakhsh et al. 2016). The length of halloysite tubules ranges from 50 to 5000 nm, the outer diameter from 20 to 200 nm, and the inner (*lumen*) diameter from 5 to 70 nm (Joussein et al. 2005; Pasbakhsh et al. 2013; Yuan et al. 2015; Zhang et al. 2016). Interestingly, a spherical halloysite from New Zealand shows a pronounced anti-inflammatory activity as indicated by its ability to inhibit edema in mice (Cervini-Silva et al. 2016). Table 1.3 lists a number of organic reactions catalyzed by halloysite.

1.3.3 Chrysotile

Chrysotile is a trioctahedral mineral species belonging to the serpentine group of 1:1 type layer silicates (Table 1.1). The ideal layer structure has the composition of $Mg_3Si_2O_5(OH)_4$, indicating that all octahedral sites are occupied by magnesium. As a result, the lateral dimension of the octahedral sheet is appreciably larger than that of the tetrahedral sheet. Since there is no tetrahedral rotation, nor substitution of Al^{3+} for Si^{4+} in the tetrahedral sheet (to increase its lateral dimension), layer curling occurs in order to compensate for the misfit between the constituent sheets (Roy and Roy 1954; Bates 1959). In this instance, however, the brucite-like octahedral sheet forms the outer side while

Clays and Clay Minerals 33

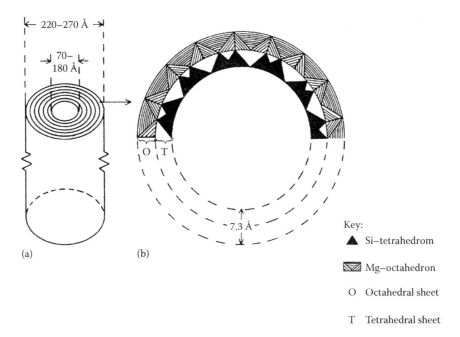

FIGURE 1.12 (a) Hollow chrysotile fibril (outer diameter: 220–270 Å; inner diameter: 70–80 Å; length: >1000 Å), consisting of T-O type layers, curled into concentric cylinders and (b) the layer structure of chrysotile viewed along the fibril axis. NB: 1 Å = 0.1 nm.

the tridymite-like tetrahedral sheet is on the inner side of the fibril (Figure 1.12), quite the opposite to the situation in rolled forms of halloysite. In the absence of isomorphous substitution in the layer structure, the interlayer space of chrysotile is devoid of counterions and associated water molecules (Table 1.1) that can give rise to Brønsted acid sites. Rather, the surface sites are basic in character and can act as electron donors or Lewis bases (Suzuki and Ono 1984; Bonneau et al. 1986).

Whereas halloysite particles can adopt a variety of shapes, naturally occurring chrysotiles almost invariably occur in the form of long (>100 nm), hollow fibrils or tubules. An individual fibril consists of a number of superposed 1:1 (T-O) type layers, each of which is ca. 0.73 nm thick, curled into concentric cylinders or spirals about the fibril axis (Figure 1.12). Because of layer curling, long O–H⋯O hydrogen bonding between contiguous layers in a particle (as in kaolinite) is no longer possible. Instead, a rather complicated steric packing arrangement is adopted (Wicks and Whittaker 1975).

Although the fibril thickness (cross section) varies between samples, the most frequently occurring dimension for the outer and inner diameter, estimated from electron micrographs, is 22–27 and 7–18 nm, respectively (Yada 1971), as indicated in Figure 1.12. On the other hand, Fripiat and Della Faille (1967) measured average values of 13.6–21 and 2.4–4.9 nm for the outer and inner diameter, respectively, for seven naturally occurring chrysotiles. These values are comparable to those reported by Sprynskyy et al. (2011). Except for the sample from Coalinga (California), the intratubular pores contained (up to 50%) amorphous materials. For the Coalinga chrysotile, the surface area accessible to nitrogen and water was 53.4 and 94.4 m^2/g, respectively, indicating that the intratubular pore space (*lumen*) was inaccessible to nitrogen. On this basis, approximately 19% and 22% of the void volume may be assigned to intertubular and intratubular pores, respectively. The sample used by Sprynskyy et al. (2011) has a nitrogen area of 15.3 m^2/g, which increases to 63.6 m^2/g following acid treatment, giving an average pore diameter of 9.8 and 3.9 nm, respectively.

The presence of mineral and trace metal impurities, isomorphous substitution, and surface defects in natural samples has led many workers to synthesize *stoichiometric* and metal-substituted chrysotiles under controlled hydrothermal conditions. (Roy and Roy 1954; Noll et al. 1958, 1960; Swift 1977; Suzuki and Ono 1984; Nitta et al. 1989a, 1989b; Falini et al. 2004, 2006; Foresti et al. 2005; Roveri et al. 2006; Piperno et al. 2007; Olson et al. 2008). Being relatively noncytotoxic, synthetic chrysotile can potentially serve as an attractive alternative to inorganic and carbon nanotubes for industrial and technological applications (Gazzano et al. 2005; Bloise et al. 2009; Foresti et al. 2009).

Ono et al. (1987) found that Ni-substituted chrysotile, $Ni_xMg_{3-x}Si_2O_5(OH)_4$, could catalyze ethylene dimerization after heating at high temperatures under vacuum. Earlier, Kibby et al. (1976) reported that the Ni^{2+} ions in partially and fully substituted chrysotile could be reduced in a flow of hydrogen. Using a similar procedure, Nitta et al. (1989a, 1989b) were able to obtain uniformly dispersed Ni/SiO_2 catalysts for the enantioselective hydrogenation of methyl acetoacetate, while reduction of cobalt-substituted chrysotile yielded cobalt-silica materials capable of catalyzing the hydrogenation of α,β-unsaturated aldehydes. More recently, Teixeira et al. (2013) prepared highly dispersed K-doped MgO catalysts for biodiesel synthesis by impregnating chrysotile fibrils with KOH and heating at 700°C. Table 1.3 lists the variety of organic reactions that have been reported to be catalyzed by chrysotile.

1.3.4 SMECTITE

The term *smectite* denotes a group of expanding 2:1 type layer silicates with isomorphous substitution in both octahedral and tetrahedral positions to give a total (net) negative layer charge (x) of 0.2–0.6 per formula unit or half unit cell (Table 1.1). The important dioctahedral members of the group are montmorillonite, beidellite, and nontronite together with hectorite and saponite, which are trioctahedral. Since montmorillonite is the most common species, these minerals have also been included in the *montmorillonite group*. To add to the confusion, *montmorillonite* may also refer to one of the end-members in the dioctahedral series of smectites (Brown et al. 1978). In this sense, its ideal composition may be written as $M^+_x \cdot nH_2O(Al_{2-x}Mg_x)_{oct}(Si_4)_{tet}O_{10}(OH)_2$, showing that the layer charge arises entirely from (partial) substitution of Mg^{2+} for Al^{3+} in the octahedral sheet. Beidellite, with the ideal composition of $M^+_x \cdot nH_2O(Al_2)_{oct}(Si_{4-x}Al_x)_{tet}O_{10}(OH)_2$ then represents the other end-member for which the layer charge is due to substitution of Al^{3+} for Si^{4+} in the tetrahedral sheet. In both instances, $M^+ \cdot nH_2O$ represents a monovalent counterion with its complement of coordinated water molecules (hydration shells). The corresponding ideal (extreme) composition for nontronite may be written as $M^+_x \cdot nH_2O(Fe_2^{3+})_{oct}(Si_{4-x}Al_x)_{tet}O_{10}(OH)_2$. The structures of beidellite, montmorillonite, and nontronite are schematically shown in Figure 1.13.

In reality, the composition of dioctahedral smectites is intermediate between that of the three end-members, and the layer charge distribution is heterogeneous (Stul and Mortier 1974; Lagaly and Weiss 1976; Janek et al. 1997; Czimerová et al. 2006). Thus montmorillonite, the most widely investigated smectite species for organic catalysis, nearly always contains some Al^{3+} and Fe^{3+} ions in tetrahedral and octahedral positions, respectively (Weaver and Pollard 1973). The Fe^{3+} ions in montmorillonite may occupy both M_1 and M_2 octahedral sites whereas those in nontronite are apparently confined to M_2 positions with a *cis* arrangement of octahedral hydroxyls (cf. Figure 1.3) (Goodman 1978, 1980; Johnston and Cardile 1985).

In this context, we might mention that the layer charge of montmorillonites may be varied, within rather broad limits, by placing Li^+ ions in interlayer cation-exchange sites, and heating the Li-exchanged minerals to 200°C–300°C (in air). This treatment induces the small Li^+ ions to migrate to the ditrigonal cavities in the tetrahedral sheet and/or the vacant octahedral sites in the structure (Hofmann and Klemen 1950; Greene-Kelly 1953; Quirk and Theng 1960; Glaeser and Méring 1971; Jaynes and Bigham 1987; Alvero et al. 1994; Theng et al. 1997; Komadel 2003; Skoubris et al. 2013). Since the negative layer charge is then partially neutralized, the resultant materials are referred to as *reduced-charge montmorillonites* (Brindley and Ertem 1971; Clementz et al. 1974; Maes and Cremers 1977; Sposito et al. 1983; Gates et al. 2000; Komadel et al. 2005). Some heavy metal ions

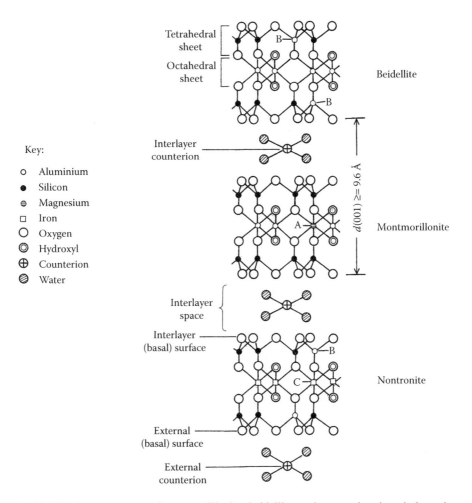

FIGURE 1.13 The layer structure of montmorillonite, beidellite, and nontronite viewed along the a-axis. The preponderant site of isomorphous substitution and the common substituting cation are indicated by the letters A, B, and C, respectively. The basal or $d(001)$ spacing varies according to the nature of the interlayer counterion and the ambient relative humidity. Under conditions of complete dehydration and layer collapse, this spacing is close to 9.6 Å. NB: 1 Å = 0.1 nm.

of small radius (e.g., Cu^{2+}, Ce^{3+}) can also migrate to the layer structure when the corresponding cation-exchanged montmorillonites are heated at ca. 200°C (Zhu et al. 2015, 2016a).

Hectorite with an ideal composition of $(M^+_x \cdot nH_2O)(Mg_{3-x}Li_x)_{oct}(Si_4)_{tet}O_{10}(OH)_2$ may be regarded as the trioctahedral analogue of montmorillonite in that its layer charge largely resides in the octahedral sheet due to partial substitution of Li^+ for Mg^{2+} ions. Many naturally occurring hectorites also have fluoride replacing hydroxyl in the structure. The synthesis of fluorhectorites, showing interesting rheological and shape-selective sorption properties, has been reported by Barrer and Jones (1970, 1971). A synthetic hectorite-like clay mineral, having the composition of $Na^+_{0.33}(Mg_{2.67}Li_{0.33})_{oct}(Si_4)_{tet}O_{10}(OH)_2$ (Neumann and Sansom 1970a, 1970b), is marketed under the trade name *Laponite* by Laporte Absorbents, UK and BYK-Chemie GmbH, Germany. Saponite is another trioctahedral species in the smectite group of phyllosilicates (Table 1.1) but in this case, the layer charge largely arises from isomorphous substitution of Al^{3+} for Si^{4+} in tetrahedral sites.

As already remarked on, the layer charge distribution in smectites generally varies among individual layers within a particle. Figure 1.14 gives the layer charge (deficit) of some representative

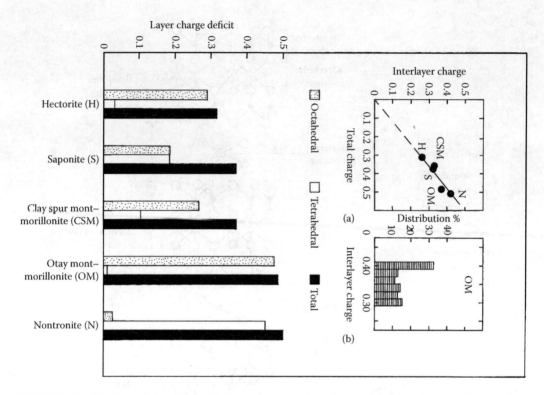

FIGURE 1.14 The layer charge characteristics of representative trioctahedral (hectorite, saponite) and dioctahedral (montmorillonite, nontronite) smectites, showing the relative contribution of octahedral and tetrahedral charges to the total charge. Inset (a) shows the relationship between interlayer charge and total charge for the five smectite species, while inset (b) shows the non-homogeneous interlayer charge distribution for Otay montmorillonite. (From Senkayi, A.L. et al., *Soil Sci. Soc. Am. J.,* 49, 1054–1060, 1985; Lagaly, G. and Weiss, A., The layer charge of smectitic layer silicates, in *Proceedings of the International Clay Conference 1975,* Bailey, S.W., (Ed.), Applied Publishing, Wilmette, IL, pp. 157–172, 1976.)

dioctahedral and trioctahedral smectites together with its partition into octahedral and tetrahedral positions (Senkayi et al. 1985). Inset (a) shows that out of the total charge deficit (0.30–0.50 per formula unit), 80%–90% can be assigned to interlayer sites. Accordingly, the interlayers of smectites account for the bulk of exchange sites and, by extension, of the total CEC and particle surface area. Inset (b) shows the heterogeneity in layer charge density for Otay montmorillonite, deduced from basal spacing measurements of the corresponding *n*-alkylammonium complexes (Lagaly and Weiss 1976). In this instance, about 35% of the layers have a deficit of 0.40 charges per formula unit, or following the notation of Lagaly and Weiss (1976), a cation density of 0.40 equivalents per $(Si,Al)_4O_{10}$ area unit; for the remaining 65% of the layers, this deficit ranges from 0.30 to 0.38. In addition to having an inhomogeneous charge density distribution, successive layers within a smectite *quasi-crystal* (Quirk and Aylmore 1971; Greene et al. 1973; Mystkowski et al. 2000) are randomly displaced (rotated) relative to each other, giving rise to a turbostratic layer arrangement (cf. Figure 1.5). Differences in microporosity and interlayer accessibility, arising from quasi-crystalline overlap, can cause variations in the BET-N_2 surface area of bentonites (Kaufhold et al. 2010).

By analogy with kaolinites, the aluminum ions exposed at the particle edge of (dioctahedral) smectites would be capable of acting as electron acceptors (Solomon 1968; Solomon and Hawthorne 1983). The Lewis acid character from this source, however, is not as pronounced as for kaolinites because the edge surface, at least of monovalent cation-exchanged smectites, takes up only a small fraction (<6%) of the total particle surface area (Wan and Tokunaga 2002; Macht et al. 2011). Also, as in kaolinite, the *under-coordinated* aluminol groups at the edge surface of smectite particles can acquire or lose protons in response to fluctuations in medium pH. The associated point of zero (net) charge varies from 6.5 to 8.1, depending on sample origin and pretreatment, and suspension ionic strength (Tombácz and Szekeres 2004; Tournassat et al. 2004; Rozalén et al. 2009). On the other hand, Kaufhold et al. (2011) measured an average pK of 4.73 for 36 samples of bentonite by titrating the clay suspensions (adjusted to pH 3) with alkali up to pH 12, while Liu et al. (2013a) obtained a value of 8.3 for montmorillonite, using first principle molecular dynamics (FPMD) simulations.

The small edge/basal surface ratio is related to layer aggregation in that smectites occur as *tactoids* in aqueous suspensions. A tactoid comprises a number of silicate layers in parallel (*face-to-face*) arrangement, adjacent layers being separated by 0.9–1.0 nm thick water sheets. Thus, scanning electron micrographs of smectites commonly show thin, flaky particles with warped edges (Figure 1.15). Under optimum conditions of dispersion, the layers of Li^+- and Na^+-montmorillonite tactoids may dissociate completely, due to extensive interlayer or *osmotic* swelling (Table 1.2), exposing a surface area close to 760 m^2/g. Recent analysis by Arndt et al. (2017), using atomic force microscopy, would indicate that the extent of osmotic swelling may vary significantly between different locations within a single tactoid. On the other hand, montmorillonites, containing counterions other than Li^+ or Na^+, tend to form tactoids of varying thickness (size) (Schramm and Kwak 1982; Whalley and Mullins 1991; Yariv and Michaelian 2002). The external or inter-tactoid surface areas, measured by the negative adsorption of anions, are generally in good agreement with the corresponding calculated values (Edwards et al. 1966; Banin and Lahav 1968; Cebula et al. 1979; Schramm and Kwak 1982). These points are summarized in Table 1.8.

In this context, we should point out that many catalytic reactions are carried out over smectites, which have previously been dehydrated. Under these conditions, the silicate layers within a particle collapse to a basal spacing of ca. 1 nm so that the exposed area would correspond to the interparticle (pore) surface. The extent of this area, commonly determined by nitrogen adsorption at 77 K and applying the Brunauer, Emmett, and Teller (BET) (1938) equation, varies from 31 to 130 m^2/g, depending on the particle size distribution, sample pretreatment, nature of the counterion, and the microporosity resulting from the turbostratic stacking of layers in a quasi-crystal. Unlike nitrogen, polar organic molecules (glycerol, ethylene glycol monomethyl ether, *p*-nitrophenol) can adsorb to both external particle and interlayer surfaces of dry and heat-collapsed smectites, yielding surface areas of 486–834 m^2/g (van Olphen and Fripiat 1979; Stul and Van Leemput 1982; Tiller and Smith 1990; Theng 1995; Kaufhold et al. 2010).

Besides being capable of taking up a wide range and variety of organic compounds, the interlayer space provides a peculiar micro-environment in terms of surface geometry, charge density, and the presence of hydrated counterions (Ballantine et al. 1983). For all these reasons, smectites are well suited

FIGURE 1.15 Scanning electron micrograph of a smectite (montmorillonite) from Miocene arkose, Madrid Basin, Spain, showing thin flakes with warped edges. (Reproduced from the "Images of Clay Archive" of the Mineralogical Society of Great Britain & Ireland and the Clay Minerals Society, www.minersoc.org/gallery.php?id=2.)

TABLE 1.8
Average Thickness (Size) of Montmorillonite Tactoids in Aqueous Suspension as Affected by the Nature of the Exchangeable Counterions Together with the Corresponding External and Interlayer Basal Areas

Counterion	Tactoid Thickness (n_i/n_{Li})			External Basal Area (m^2/g)			Interlayer Basal Area (m^2/g)			Tactoid Edge Area (% of External Basal)
				Calculated	Measured		Calculated	Measured		
(I)	(II)	(III)	(IV)	(V)	(VI)	(VII)	(VIII)	(IX)	(X)	(XI)
Li^+	1.0	1.0	1.0	763	650	625	0	113	138	1.3
Na^+	1.5	1.7	N.a.	509	570	560	254	193	203	2.0
K^+	2.0	2.7	2.0	382	310	436	381	453	327	2.6
Rb^+	2.2	N.a.	N.a.	347	N.a.	N.a.	416	N.a.	N.a.	2.9
NH_4^+	2.6	N.a.	N.a.	293	310	265	470	453	498	3.4
Cs^+	4.6	3.0	3.0	166	40	156	597	723	607	6.0
Mg^{2+}	9.6	N.a.	N.a.	79	N.a.	N.a.	684	N.a.	N.a.	12.7
Ca^{2+}	10.9	7.0	N.a.	70	100	114	693	663	649	14.3
Ba^{2+}	11.2	N.a.	N.a.	68	N.a.	N.a.	695	N.a.	N.a.	14.7

Columns (II) and (III): Number of silicate layers (n_i) relative to that for Li^+-montmorillonite (n_{Li}), assuming $n_{Li} = 1.0$; estimated from optical measurements by Banin and Lahav (1968) and Schramm and Kwak (1982), respectively.
Column (IV): Estimated from small-angle neutron scattering (Cebula et al. 1979).
Column (V): Taking 763 m^2/g as the total basal area of Li^+-montmorillonite, derived from unit cell dimension and weight (van Olphen 1977).
Columns (VI) and (VII): Derived from negative anion adsorption (Edwards et al. 1966; Schramm and Kwak 1982).
Columns (VIII), (IX), and (X): Difference between 763 and values in columns (V), (VI), and (VII), respectively.
Column (XI): Taking 10 m^2/g as the tactoid edge area, based on a tactoid dimension of 1 μm equivalent spherical diameter and a density of 2.5 g/cm^3.
N.a. = Not available.

to catalyze the conversion of intercalated organic molecules. In addition, the interlayer volume and pore geometry may be varied by inserting pillaring or cross-linking agents, providing scope for shape-selective sorption and catalysis (Barrer 1978, 1984; Shabtai et al. 1981; Occelli and Tindwa 1983; Pinnavaia 1983; Pinnavaia et al. 1984; Occelli et al. 1985; Figueras 1988; Vaughan 1988; Kloprogge 1998; Gil et al. 2000, 2008, 2010; De Stefanis and Tomlinson 2006; Vicente et al. 2013).

1.3.5 VERMICULITE

Vermiculites may occur as both clay-size and large particles. The microcrystalline forms (*clay vermiculites*) may either be dioctahedral or trioctahedral whereas macroscopic vermiculites, from which most of our information on structural and surface properties has been derived, are invariably trioctahedral (Walker 1975). The 2:1 or T-O-T type layer is similar in structure to that of biotites, with Mg^{2+} being the dominant cation occupying octahedral positions. The misfit between the octahedral and tetrahedral sheets is compensated by rotation of alternate tetrahedra through angles of 6°–10°. As a result, the surface oxygen (siloxane) network of the silicate layer assumes a ditrigonal form (Bailey 1980; Douglas 1989). However, unlike biotite where the octahedral iron usually occurs in the ferrous form, the iron substituting for magnesium in vermiculites is largely trivalent. In addition to Fe^{3+}, Al^{3+} may substitute for Mg^{2+} giving rise to an excess of positive charges in the octahedral sheet. However, the extensive replacement of Si^{4+} by Al^{3+} (usually to the extent of more than one Al^{3+} per $(O_{10}(OH)_2)$ in the tetrahedral sheet results in a net negative charge (x) of 0.6–0.9 per formula unit (Table 1.1), and a cation exchange capacity (CEC) of 120–250 cmol$_c$/kg (Brown et al. 1978). In naturally occurring vermiculites the counterions balancing this charge are commonly magnesium. These points are illustrated by the composition of two well-known and much investigated vermiculites:

$$Mg^{2+}_{0.32}.4.32\,H_2O(Mg_{2.36}Fe^{3+}_{0.48}Al_{0.16})_{oct}(Si_{2.72}Al_{1.28})_{tet}O_{10}(OH)_2 \text{ and}$$

$$Mg^{2+}_{0.48}.K_{0.01}.4.72\,H_2O(Mg_{2.83}Fe^{3+}_{0.01}Al_{0.15})_{oct}(Si_{2.86}Al_{1.14})_{tet}O_{10}(OH)_2$$

These represent the Kenya and Llano (Texas) specimens, respectively (Mathieson and Walker 1954; Shirozu and Bailey 1966; Douglas 1989).

The interlayer Mg^{2+} ions are positioned midway between every two opposing silicate layers and occupy one-third of the possible cation (exchange) sites. Under ambient conditions of humidity and temperature, the interlayer space also contains two superposed sheets of water molecules. These are arranged in a regular hexagonal network around each Mg^{2+} ion, and they interact mutually and with the surface oxygens of opposing silicate layers by means of hydrogen bonding (Sposito 1984; cf. Figure 1.6d). The basal spacing of this stable two-sheet hydrate of Mg^{2+}-vermiculite is 1.43–1.44 nm. A similar cation-water configuration has been proposed by Beyer and Graf von Reichenbach (2002) for the 1.485 nm hydrate of Na-vermiculite.

For the Kenya specimen, the position of interlayer magnesium together with its dehydration characteristics has been determined using single-crystal XRD (Mathieson and Walker 1954). As shown in Figure 1.16, up to five hydration states (*phases*) of varying stabilities, corresponding to basal spacings of 1.481, 1.436, 1.382, 1.150, and 0.902 nm, may be recognized. As the fully expanded 1.481 nm phase (obtained by prolonged immersion in water) gradually dehydrates, the interlayer space contracts. This creates vacant water sites around the Mg^{2+} ion and distorts the regular hexagonal network. At the same time, successive silicate layers are laterally displaced with respect to each other.

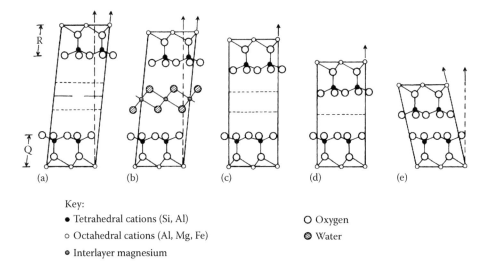

FIGURE 1.16 Projection normal to the *ac* plane for Mg-vermiculite at various stages of hydration-dehydration as indicated by the corresponding basal spacings: (a), 1.481 nm phase; (b), 1.436 nm phase; (c), 1.382 nm phase; (d), 1.159 nm phase; and (e), 0.902 nm phase. The attendant changes in stacking arrangement of successive silicate layers are also indicated. NB: Q and R denote silicate half layers, while dashed lines in diagrams (a), (c), and (d) represent water sheets. (From Walker, G.F., Vermiculites, in *Soil Components, Vol. 2. Inorganic Components*, Gieseking, J.E. (Ed.), Springer-Verlag, Berlin, Germany, pp. 155–189, 1975.)

Translational shifts by $\pm b/3$ of successive layers have also been observed for Llano vermiculite in the magnesium form. Besides having a relatively large net negative layer charge, this specimen shows partial tetrahedral cation ordering (Shirozu and Bailey 1966). Accordingly, the $Mg(H_2O)_6^{2+}$ octahedra tend to associate with, and occupy sites immediately adjacent to, the Al-rich (T_1) tetrahedra of opposing silicate layers (Figure 1.16). In the sodium and calcium forms, on the other hand, the ditrigonal cavities of opposing silicate layers face each other (De la Calle et al. 1976) as in muscovite (cf. Figure 1.18). The interlayer Na^+ ions, which are also sixfold coordinated to water molecules (Hougardy et al. 1976), can thus be located between (either of) the oxygen triads of the tetrahedral bases in adjacent layers. In calcium Llano vermiculite, the interlayer counterions may exist in both sixfold and eightfold coordination with water but only the Ca^{2+} ions in the (distorted) cubic arrangement with water are apparently positioned between the two ditrigonal cavities of opposing silicate layers (Slade et al. 1985).

As in smectites, the basal spacing of vermiculites is dependent on the nature of the inorganic counterion and the relative humidity of the atmosphere (Walker 1975; Huo et al. 2012). For practical purposes, however, vermiculites may occur in one of three stable hydration states: fully dehydrated, partially hydrated, and fully hydrated. The respective basal spacings are 0.9–1.0, ~1.2, and 1.4–1.5 nm corresponding to the presence in the interlayer space of zero, one, and two sheets of water molecules as illustrated in Figure 1.16 by the phases (e), (d), and (b), respectively. As in smectites, the layer charge density in vermiculite may vary among individual layers within a particle (Lagaly 1982).

Among the inorganic forms of vermiculite, only the lithium-saturated variety can show extensive interlayer (*macroscopic*) swelling in water (Table 1.2). Vermiculites saturated with *n*-propylammonium and *n*-butylammonium ions behave likewise when the complexes are immersed in water or dilute solutions of the corresponding *n*-alkylammonium chlorides (Walker 1960; Garrett and Walker 1962). The large interlayer separations, obtained by this means, would allow and facilitate the intercalation of relatively bulky cationic pillaring agents (e.g., hydroxy-Al polymers) through a simple exchange process. Surprisingly, however, the synthesis and catalytic activity of pillared vermiculites have received much less attention than their smectite counterparts (Kermarec et al. 1983). Although *hydroxy interlayered* vermiculites (HIV) appear to be less stable to heating than their smectite counterparts (Barnhisel and Bertsch 1989), the opposite situation applies to the respective pillared materials (del Rey-Perez-Caballero and Poncelet 2000). Vermiculites, however, have the disadvantage of having an appreciably higher charge density than smectites. Under conditions of complete exchange, the pillaring agents may therefore be so closely packed as to fill most of the interlayer space, precluding intercalation of the organic reactants. Such a situation is found in chlorites (described in the following) where the metal hydroxide sheet, balancing the negative layer charge, fills up the interlayer space (cf. Figure 1.17). Indeed, like chlorite, vermiculite pillared with polyoxoaluminum cations showed a basal spacing of ca. 1.4 nm as compared with the value of ca. 1.8 nm measured for $(Al_{13})^{7+}$-smectites under similar experimental conditions. However, when the layer charge of the vermiculite sample was first reduced (by acid treatment followed by calcination), the Al-pillared material gave a $d(001)$ spacing of 1.8 nm together with the high surface area and micropososity characteristic of its montmorillonite counterpart (del Rey-Perez-Caballero and Poncelet 2000). The negative layer charge of vermiculite may also be reduced by hydrothermal treatment, which is more benign to the mineral structure than acid activation (Cristiano et al. 2005). The effect of hydrothermal treatment on the catalytic activity of pillared vermiculites has been described by Campos et al. (2008a, 2008b). On the other hand, vermiculite has a distinct advantage over smectite in that macrocrystalline forms can be used for experimental purposes. As such, the interlayer orientation and bonding mode of the organic reactants and their corresponding transition state, if any, may be readily assessed by X-ray diffractometry (Garrett and Walker 1962, 1967; Slade et al. 1987; Vahedi-Faridi and Guggenheim 1997, 1999a, 1999b). The interactions of vermiculites with organic compounds have been summarized by Pérez-Rodríguez and Maqueda (2002).

1.3.6 Chlorite

The majority of chlorites have a trioctahedral layer structure, the particles being made up of successive 2:1 mica-type layers, which are interleaved with a continuous magnesium hydroxide sheet. The thickness of this interlayer sheet is ca. 0.4 nm, while that of the silicate layer is ca. 1.0 nm, giving a basal spacing close to 1.4 nm, which is a characteristic feature of the chlorite group of minerals (Figure 1.17). This value is comparable to the $d(001)$ spacing of a two-sheet hydrate of Mg^{2+}-vermiculite depicted in Figure 1.16b. Unlike vermiculite, however, the basal spacing of chlorite remains unchanged on treatment with ethylene glycol or heating up to 500°C.

As in vermiculites, there is much substitution of Al^{3+} for Si^{4+} in the tetrahedral positions. Although the resultant negative charge is partly offset by the substitution of Al^{3+} for octahedral Mg^{2+}, most of it is balanced by the interlayer hydroxide sheet. The positive charge of the brucite-like sheet arises from partial replacement of Mg^{2+} by Al^{3+} (Barnhisel and Bertsch 1989). With a composition of $[(Mg_{3-x}Al_x)(OH)_6]^{x+}$, the hydroxide sheet is strongly bound to opposing silicate layers by electrostatic attraction and hydrogen bonding, accounting for its stability to heating, and the failure of chlorites to expand in water and organic liquids. The requirement for (long) hydrogen bonding between the hydroxyl groups of the interlayer sheet and the oxygens of the tetrahedral bases of opposing silicate layers also restricts the number of possible layer-stacking arrangements (Bailey and Brown 1962).

In addition to the main trioctahedral subgroup, there are chlorites in which both the 2:1 type layers and the interlayer hydroxide sheets are dioctahedral. The mineral donbassite is a representative species of the dioctahedral chlorite subgroup. The third subgroup is referred to as di, trioctahedral since the silicate layers, in this case, are dioctahedral whereas the hydroxide sheets are trioctahedral, as exemplified by the mineral sudoite (Table 1.1). In all these structures, Mg^{2+} may be replaced by Fe^{2+} (and other suitable divalent cations) and Al^{3+} by Fe^{3+}. Chamosite, a trioctahedral ferrous chlorite with the composition of $(Fe_5^{2+}Al)(Si_3Al)O_{10}(OH)_8$, has featured as a heterogeneous catalyst for some organic syntheses (Table 1.4).

The hydroxy interlayers in chlorites can be partially extracted by a variety of reagents without materially affecting the silicate layer structure (Barnhisel and Bertsch 1989). Because of their intrinsic heat stability, such surface-modified chlorites are potentially useful for catalyzing the conversion of intercalated organic compounds. Structurally similar hydroxy interlayered smectites and

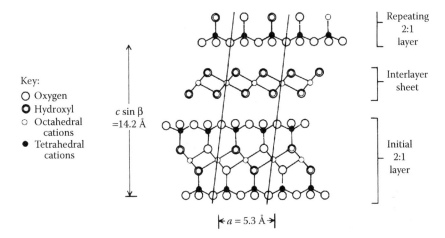

FIGURE 1.17 The structure of chlorite projected on the ac plane showing the position of the interlayer hydroxide sheet between two repeating (successive) 2:1 type layers. NB: 1 Å = 0.1 nm. (From Bailey, S.W., Chlorites, in *Soil Components, Vol. 2. Inorganic Components*, Gieseking, J.E. (Ed.), Springer-Verlag, Berlin, Germany, pp. 191–263, 1975.)

vermiculites occur widely in soil (Barnhisel and Bertsch 1989). Because of their compositional variability and limited availability, however, naturally occurring chlorite-like minerals are not suitable for large-scale catalytic usage. On the other hand, 2:1 layer silicates with different degrees of interlayer space filling (by hydroxy-metal cations), are readily obtainable (Sawhney 1968; Gupta and Malik 1969a, 1969b; Carstea et al. 1970; Brindley and Sempels 1977; Lahav et al. 1978; Yamanaka and Brindley 1978, 1979; Brindley and Kao 1980; Pinnavaia et al. 1984). As already mentioned, calcination of such *chloritized* materials gives rise to catalytically active pillared interlayered clay minerals as detailed in Chapter 3.

1.3.7 Mica, Illite, Synthetic Mica-Montmorillonite, and Fluorotetrasilicic Mica

Since the pioneering studies by such workers as Jackson and West (1930, 1933) and Pauling (1930), the mica group of minerals has received a great deal of attention. Research has been stimulated by the abundance of micas in rocks, sediments, and soils; by the wide variety of compositions; and the availability of well-crystallized macroscopic forms. The accumulated wealth of information on the micas has been the subject of many reviews (Radoslovich 1975; Bailey 1980, 1984; Fanning et al. 1989), which should be consulted for more details and key references to the extensive literature.

Figure 1.18 illustrates the structure of muscovite, showing the substitution of aluminum for silicon in the tetrahedral sheet (to the extent of one Al^{3+} out of every four Si^{4+}), giving an ideal composition of $K(Al_2)_{oct}(Si_3Al)_{tet}O_{10}(OH)_2$. The real structure departs from ideality in that the basal hexagonal rings are distorted through the rotation of alternate silicon tetrahedra in opposite directions, making the size of occupied octahedra smaller than that of their vacant counterparts. The interlayer potassium ions that balance the negative layer charge fit snugly in the center of the ditrigonal cavity between two opposing $(Si,Al)O_6$ rings, giving a basal spacing of 1.0 nm, which remains unchanged on treatment with ethylene glycol (or glycerol) or heating at 500°C. The high energy of layer separation (Giese 1978), the failure of mica crystals to expand (swell) in water, and the low exchange capacity (as measured by conventional methods) are in keeping with strong interlayer bonding. Nevertheless, there is scope, in these structures, for layer rotation and the occurrence of a

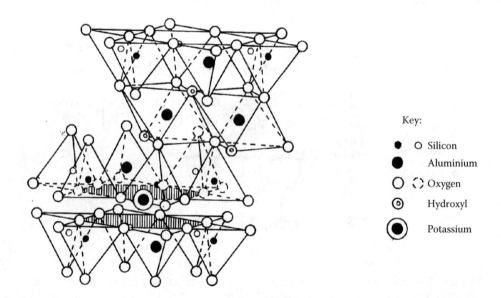

FIGURE 1.18 Idealized structure of muscovite mica. The interlayer K^+ ion fits snugly in the center between two opposing ditrigonal cavities (hatched), becoming effectively coordinated to twelve oxygens.

large variety of layer-stacking sequences. In practice, however, only a limited number of polytypes can be recognized.

Besides being coordinated to twelve oxygen atoms of two superposed ditrigonal rings, each interlayer K^+ ion is associated with two hydroxyl groups (belonging to the octahedral sheet), one lying directly above and one below the counterion. The orientation of these hydroxyls (with respect to the layer plane) affects the strength with which the interlayer K^+ ions are held, and hence the rate of mica weathering (Norrish 1973; Fanning et al. 1989). In dioctahedral structures, such as muscovite ($2M_1$ structure), the O−H bond is deflected away from the K^+ ion toward the vacant octahedral site, being inclined at an angle of 12° to the basal plane (Rothbauer 1971). As such, the distance separating the interlayer K^+ from the hydroxyl hydrogen is appreciably greater than that between it and each of the 12-ring oxygens. On the other hand, in trioctahedral micas (e.g., phlogopite) where there are no octahedral vacancies, the O−H bond is perpendicular to the silicate layer, making the K⋯H distance only slightly longer than the K−O bond (Rayner 1974). Since the hydrogen atoms of the OH groups in question carry an excess of positive charge, they exert a repulsive (*antibonding*) effect on neighboring potassium ions (Brown et al. 1978). This effect is less pronounced in muscovite than in phlogopite; accordingly, the energy of layer separation for muscovite is greater by ca. 9.8 kcal/mole than that for its trioctahedral counterpart (Giese 1977). Likewise, the substitution of the highly electronegative fluoride for hydroxyl has the effect of enhancing interlayer attraction by increasing the layer separation energy (Giese 1975). The inverse relationship between fluorine content and weathering rate, which applies to micas in general, may be explained in similar terms (Rausell-Colom et al. 1965; Newman 1969).

As introduced by Grim et al. (1937), the term *illite* denotes a group of clay-size or fine-grained micaceous minerals with a basal spacing of ca. 1 nm that does not increase when treated with ethylene glycol. A more precise description, proposed by Środoń and Eberl (1984), is that of "a non-expanding, dioctahedral, aluminous, potassium mica-like mineral" that occurs in the <4 μm fraction of soils and sediments. The term *illite layer* is then used for a 1.0 nm thick, non-expanding, potassium-rich layer, which makes up a component of dioctahedral, mixed-layer structures while *illitic material* becomes synonymous with the original *illite* of Grim et al. (1937).

As the description suggests, illite is structurally similar to muscovite. Apart from its particle size range, however, illites differ from muscovites in having a smaller negative layer charge, less potassium, and more water. An *average* composition for illite, derived from chemical analyses of a large number of samples (Weaver and Pollard 1973), may be written as $(Ca_{0.05}Na_{0.03}K_{0.63})_{int}(Al_{1.55}Fe^{3+}_{0.21}Fe^{2+}_{0.04}Mg_{0.28})_{oct}(Si_{3.41}Al_{0.59})_{tet}O_{10}(OH)_2$, indicating appreciable isomorphous substitution (of Al^{3+} for Si^{4+}) in the tetrahedral sheet, and some substitution (of Mg^{2+}, Fe^{3+}, and Fe^{2+} for Al^{3+}) in octahedral sites, with Ca^{2+}, Na^+, and K^+ occupying interlayer (int) positions. Any chemical formula for illite, however, may be misleading since many samples are essentially mixed-layer structures. More often than not, smectite makes up the other component although the presence of interstratified smectite layers is difficult to detect unless their proportion exceeds 10%.

The negative layer charge (0.6–0.85 per formula unit) of illite falls within the range given by vermiculite (Table 1.1). Unlike vermiculite, however, the interlayer counterions (largely K^+) in illite are practically non-exchangeable. Accordingly, the CEC of illite is much smaller than its layer charge deficit would suggest (Table 1.2). In this respect, and in showing a basal spacing of ca. 1.0 nm that does not change on immersing the particles in water or heating, illite is more akin to mica than to vermiculite. Since illite contains more water than mica, the terms *hydromica* and *hydrous mica* have sometimes been used as alternatives. The *extra* water in illite apparently occurs in the form of hydronium ions, occupying K^+-deficient interlayer positions (Loucks 1991; Nieto et al. 2010; Kühnel et al. 2017).

With respect to clay catalysis, the probable reduction of structural Fe^{3+} to Fe^{2+} is an important feature since octahedrally coordinated ferrous ions can act as electron acceptors in many organic conversions. Indeed, the increase in Fe^{2+}/Fe^{3+} ratios with increasing illitization in some bentonite-rich sediments has been ascribed to a redox reaction involving the oxidation of organic matter (Eslinger et al. 1979). Of greater significance, in this context, is the implication of the smectite →

illite transformation for the generation and migration of oil in sediments (Johns and Shimoyama 1972; Foscolos et al. 1976; Foscolos and Powell 1979; Seewald 2003; Wu et al. 2012). Many studies have also shown that clay minerals are involved in the catalytic conversion of sedimentary organic matter (*kerogen*) and specific bioorganic compounds (notably fatty acids) to petroleum-like hydrocarbons (Jurg and Eisma 1964; Andreev et al. 1968; Theng 1974; Almon and Johns 1976; Espitalié et al. 1980; Horsfield and Douglas 1980; Johns 1982; Goldstein 1983; Heller-Kallai et al. 1984; Heller-Kallai, 2002). The catalytic activity of clay minerals in cracking and related petroleum-forming reactions will be described in Chapter 8.

To conclude this subsection, we wish to outline the properties of synthetic mica-montmorillonite (SMM) and fluorotetrasilicic mica (TSM).

As its name suggests, SMM is essentially composed of mica (muscovite) layers, which are randomly interstratified with those of montmorillonite—in reality—beidellite (Granquist and Kennedy 1967; Granquist and Pollack 1967; Kellendonk et al. 1987). The synthesis is normally carried out by reacting silica, alumina, an alkali metal oxide or hydroxide, and a fluoride salt in specified proportions with excess water under hydrothermal conditions. The mineral crystallizes out in the form of ~5 nm thick platelets with an average diameter of 100 nm and a surface area of 134–165 m^2/g (Swift 1977; van Olphen and Fripiat 1979).

The general composition of SMM may be written as $M^+ \cdot nH_2O(Al_2)_{oct}(Si_{4-x}Al_x)_{tet}O_{10}(OH,F)_2$ where M^+ stands for a monovalent counterion, and the charge per formula unit (x) is ca. 0.65. In earlier preparations, Na$^+$ serves as the charge-balancing cation but subsequently NH$_4^+$ is the common counterion, giving a cation exchange capacity (CEC) of 140–170 cmol$_c$/kg (Wright et al. 1972). The ammonium-exchanged form (*Syn-1*) is manufactured in bulk by the Baroid Division of NL Industries and marketed under the trade name of Barasym SMM-100 with a CEC of 70 cmol$_c$/kg (barium method) and 140 cmol$_c$/kg (ammonium method) and a surface area of 134 m^2/g (van Olphen and Fripiat 1979). High-resolution magic-angle spinning (MAS) NMR spectroscopy of Syn-1 indicates that the interlayer space of the beidellite component is occupied by ammonium ions, while both ammonium and aluminum make up the counterions of the mica part in the interstratified layer structure (Alba et al. 2005). When the material is heated at 300°C–400°C (in air), the NH$_4^+$ ions decompose into ammonia and protons. As in zeolites, the high Brønsted acidity that develops when NH$_4^+$-SMM is subjected to mild calcination, can be directly ascribed to this process of deammoniation (*decationation*). On further heating to 500°C, Lewis acid sites are formed at the expense of Brønsted sites (Wright et al. 1972; Swift 1977; Kojima et al. 1986).

Besides being a potential source of protons, the NH$_4^+$ counterions can be exchanged by other cations (e.g., Pd^{2+}) to yield the corresponding metal ion-exchanged forms of SMM (Giannetti and Fischer 1975; van Santen et al. 1985). A more important modification for catalytic purposes is the partial or complete substitution of suitable metal ions (notably Ni^{2+}) for Al^{3+} in octahedral positions, with or without interlayer pillaring agents. Such metal-substituted (*impregnated*) forms of SMM, obtainable from the Harshaw-Filtrol Corporation, USA, show enhanced acidity and catalytic activity for the isomerization, cracking, and oligomerization of hydrocarbons. The variety of organic reactions, catalyzed by SMM, are listed in Table 1.4.

Like SMM, fluorotetrasilicic mica (TSM) is a synthetic, swelling layer silicate with the composition of $Na^+(Mg_{2.5}\Upsilon_{0.5})(Si_4)O_{10}F_2$ showing the replacement of hydroxyl by fluoride in the octahedral sheet, with sodium counterions (and water) occupying interlayer positions (Kitajima and Daimon 1975; Morikawa 1993). The negative layer charge, however, is not due to isomorphous substitution but derives from vacancies (defects) in the octahedral sheet as indicated by the open square symbol (Okada et al. 2007). The occurrence of similar vacancies has also been deduced by Soma and Theng (1998) from X-ray photoelectron spectroscopic analysis of some New Zealand halloysites. The measured CEC of Na-TSM ranges from 92 to 170 cmol$_c$/kg, which is appreciably lower than the value of 254 cmol$_c$/kg, deduced from its chemical composition. This discrepancy may be ascribed to the non-exchangeability of a proportion of the interlayer sodium ions (Soma et al. 1990; Kłapita et al. 2003). We should also add that Na-TSM is quite inert as a catalyst because it has few, if any, acid sites (Morikawa et al. 1983a, 1983b; Sivakumar et al. 2004; Ohtake et al. 2007). Catalytically active forms of TSM, however, can be prepared by replacing the interlayer Na^+ ions with multivalent cations or pillaring agents (Morikawa et al. 1982; Sakurai et al. 1988, 1989, 1990; Brody et al. 1989; Nakayama et al. 1993; Divakar et al. 2008). Na-TSM and cation-exchanged forms of TSM are obtainable from Topy Industries Co, Japan. Table 1.4 shows examples of organic reactions catalyzed by cation-exchanged and pillared TSM.

1.3.8 SEPIOLITE AND PALYGORSKITE

Sepiolite and palygorskite and are hydrous magnesian silicates with a fibrous particle morphology (Figure 1.19) and a characteristic dehydration pattern. Many early publications used the term *attapulgite* for palygorskite (Grim 1968), but this name has been discredited (Guggenheim et al. 2006). Individual fibrils of sepiolite vary in length from 10 to 5,000 nm, in width from 10 to 30 nm, and in thickness from 5 to 10 nm. The length of palygorskite fibrils ranges from 1,000 to 2,000 nm. The fibrils are commonly arranged in bundles or sheaths (Jones and Galan 1988; Singer 1989). Each fibril consists of 2:1 type layers, arranged in ribbons (also referred to as *bands*, *chains*, or *strips*) along the fibril (*a*-) axis, with the apical oxygens of the silicon tetrahedra in adjacent ribbons pointing in opposite directions. For this reason, sepiolite and palygorskite have been referred to as layer-ribbon or chain-layer silicates (Mackenzie and Mitchell 1966; Sudo et al. 1981; Table 1.1). Being joined at the corners through oxygen ions, the rectangular ribbons alternate with channels of similar size and shape, running the full length of the fibril. Under ambient conditions of humidity and temperature, these channels contain free (*zeolitic*) water as well as exchangeable counterions (Brigatti et al. 2013).

Figure 1.20 shows the Brauner and Preisinger (1956) structure for sepiolite, which is generally preferred to the earlier model of Nagy and Bradley (1955). The *b*- and *c*-dimension of the unit cell is ca. 2.7 and 1.34 nm, respectively, and the ideal composition is $Mg_8Si_{12}O_{30}(OH)_4(OH_2)_4 \cdot 8H_2O$.

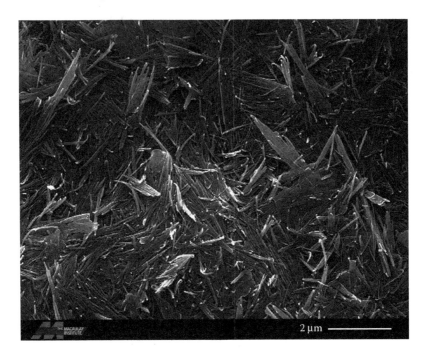

FIGURE 1.19 Scanning electron micrograph of palygorskite from Mexico. (Reproduced from the "Images of Clay Archive" of the Mineralogical Society of Great Britain & Ireland and the Clay Minerals Society, www.minersoc.org/gallery.php?id=2.)

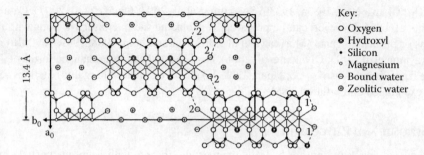

FIGURE 1.20 The structure of sepiolite, viewed along the fibril *a*-axis, according to Brauner and Preisinger (1956). The unit cell dimension is indicated by the solid rectangle. At the edge of the fibril the bound water molecules can interact with either neighboring silanol groups (a), or oxygen ions (b), as suggested by Serna et al. (1975). NB: 1 Å = 0.1 nm.

The structure contains four *bound* and eight zeolitic water molecules, the former being coordinated to octahedral magnesium ions, exposed at the ribbon edge. In practice, however, not all eight octahedral sites need be occupied. In addition, there is always some substitution of Fe^{3+} and Fe^{2+} for Mg^{2+} in the octahedral sheet, and of Al^{3+} and Fe^{3+} for Si^{4+} in tetrahedral positions. As a result, the structure carries a net negative charge, which is usually balanced by sorption of extraneous Ca^{2+} ions to the extent of 20–45 cmol$_c$/kg (Hénin and Caillère 1975; Singer 1989). Induced isomorphous substitution of Al^{3+} for Si^{4+} in the tetrahedral sheet increased both the thermal stability and CEC of sepiolite (Sun et al. 1995).

Figure 1.21 shows the structure of palygorskite as proposed by Bradley (1940). It has the ideal composition of $Mg_5Si_8O_{20}(OH)_2(OH_2)_4 \cdot 4H_2O$, indicating the presence of four zeolitic water molecules within the channel, and another four coordinated to Mg ions at the ribbon edge. A more important structural difference is that each ribbon in palygorskite consists of five octahedral sites, giving rise to a *b*-dimension of 1.8 nm as compared with ca. 2.7 nm in sepiolite. Accordingly,

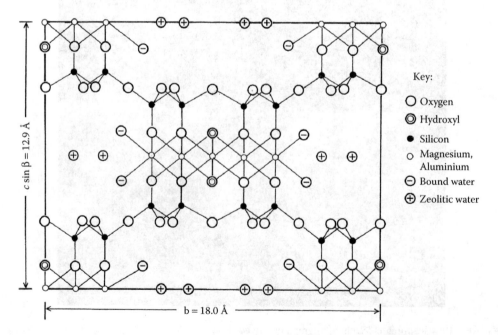

FIGURE 1.21 The structure of palygorskite, projected on the *bc* plane. The unit cell dimension is indicated by the solid rectangle. NB: 1 Å = 0.1 nm. (From Bradley, W.F., *Am. Mineral.*, 25, 405–410, 1940.)

the channel (tunnel) dimension in palygorskite (0.37 × 0.64 nm) is appreciably smaller than that in sepiolite (0.37 × 1.06 nm). In addition, there is relatively less isomorphous substitution within the palygorskite structure, giving rise to a CEC of 5–30 $cmol_c$/kg (Serna and Van Scoyoc 1979; Singer 1989).

Steric factors would be expected to intervene in the entry of some organic compounds into the channel structure. This is borne out by the observation that straight-chain hydrocarbons are more readily adsorbed than their branched counterparts (Nederbragt 1949). Kuang and Detellier (2005) have also found that methanol and ethanol can gain access to the channels of palygorskite whereas *n*-pentanol apparently fails to do so although partial channel penetration by methylene blue is possible (Zhang et al. 2015). For organic molecules, which are not sterically constrained, those with strongly polar groups are generally preferred to weakly polar species. Polar compounds (e.g., acetone) are taken up by displacing the zeolitic and bound water inside the channels, allowing the molecules to form H-bonds with structural water or coordinate directly to terminal Mg(II) cations (Barrer 1978; Serna and Van Scoyoc 1979; Fernandez Hernandez and Fernandez Alvarez 1983; Mikhail et al. 1983; Singer 1989; Kuang and Detellier 2004; Kuang et al. 2006). On the other hand, many non-polar compounds are excluded from, or can only gain limited entry into, the channel structure of these minerals (Barrer 1978; Serratosa 1979; Singer 1989). Since such compounds would interact only weakly through (non-specific) van der Waals forces, the extent of their sorption is measurably less than that shown by polar molecules of comparable size (Haden and Schwint 1967).

The characteristic dehydration behavior, mentioned earlier, refers to the stepwise loss of weight when sepiolite and palygorskite are heated at increasing temperatures (Caillère and Hénin 1957; Imai et al. 1969; Nagata et al. 1974; Singer 1989; Brigatti et al. 2013). As might be expected, the zeolitic water in sepiolite can be removed by mild heating (<200°C in air). At the completion of this step, the bound water molecules interact with neighboring hydroxyls or oxygen ions through hydrogen bonding as indicated by the broken lines, marked (1) and (2), in Figure 1.20. The second step corresponds to the loss of half of the bound water, that is, one molecule from each ribbon edge site. When this occurs (ca. 300°C in air or ca. 175°C under vacuum), the network of hydrogen bonds is disrupted, allowing the ribbons to rotate and the structure to fold. At this stage of dehydration, the original structure can be restored by allowing the mineral to take up water at ambient temperature and humidity. The third dehydration step occurs between 380°C and 680°C in air when the other half of the bound water is lost to yield the *anhydride*. This form can only be rehydrated by exposing the mineral to water vapor at elevated temperatures and pressures, and then often incompletely (Hayashi et al. 1969; Tarasevich and Ovcharenko 1971; Nagata et al. 1974; Serna et al. 1975; Van Scoyoc et al. 1979). The fourth step, which occurs between 680°C and 900°C, can be identified with dehydroxylation when the original structure is destroyed and a new crystalline or amorphous phase is formed (Preisinger 1959; Hayashi et al. 1969; Lokanatha et al. 1985). Palygorskite behaves similarly on heating, but the loss of bound water causing structural folding occurs at a lower temperature than in sepiolite. Further, step 3 may not always be clearly discernible in the differential thermal and thermogravimetric analysis patterns, probably because the high-temperature dehydration step in palygorskite overlaps with structural dehydroxylation. Fourier transform infrared spectroscopic analysis by Yan et al. (2012), for example, indicates that most of the hydroxyl groups in palygorskite are removed by heating at about 450°C, and complete structural dehydroxylation occurs at ~700°C accompanied by the formation of SiO_4 tetrahedral sheets. We might also add that although the channels in the anhydride form are no longer accessible to organic compounds, heating at ca. 600°C may actually increase the volume of pores between individual fibrils (Hénin and Caillère 1975).

The total surface area of sepiolite and palygorskite (including the channel area), calculated from structural models, is of the order of 850 m^2/g (Serna and Van Scoyoc 1979). Channel penetration by extraneous molecules, however, is often incomplete, even in the case of small, polar species (Barrer and Mackenzie 1954). For this reason, the experimentally derived surface area

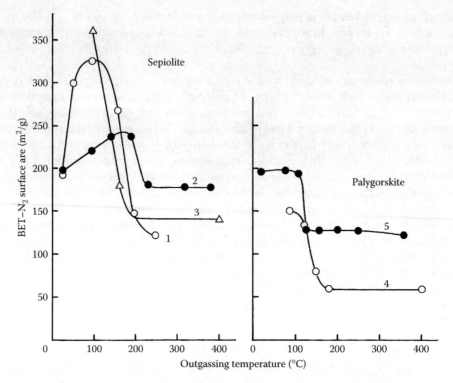

FIGURE 1.22 Relationship between the surface area accessible to nitrogen and the outgassing temperature for sepiolite and palygorskite samples. Closed circles: data of Barrer and Mackenzie (1954); open triangles: data of Dandy and Nadiye-Tabbiruka (1975); open circles: data of Fernandez Alvarez (1978). (From Serratosa, J.M., Surface properties of fibrous clay minerals (palygorskite and sepiolite), in *International Clay Conference 1978. Developments in Sedimentology 27*, Mortland, M.M. and Farmer, V.C. (Eds.), Elsevier, Amsterdam, the Netherlands, pp. 99–109, 1979.)

is appreciably lower than the theoretical value. Thus, the BET-nitrogen area of sepiolite varies between 230 and 380 m^2/g, while that of palygorskite ranges from 140 to 190 m^2/g (Serratosa 1979). These values refer to samples that have been outgassed at ca. 100°C and before the structure folds. Below this temperature, less surface is measured, presumably because nitrogen cannot displace the remaining surface-adsorbed water. Outgassing above this temperature leads to a steep decline in surface area until the point is reached when structural folding occurs (ca. 200°C). Further heating beyond this point has little effect on surface area, which remains practically constant at 140–190 m^2/g for sepiolites and 60–140 m^2/g for palygorskites. These points are illustrated in Figure 1.22. Variations in nitrogen areas among specimens may be ascribed to differences in particle size and crystallinity. Perhaps more relevant to catalysis is that as much as 60%–70% of the interparticle surface is contained in pores with a radius of 0.8–1.0 nm before folding occurs (Fernandez Alvarez 1978). These micropores collapse and become inaccessible to nitrogen when the structure folds.

Examples of organic reactions and transformations, catalyzed by sepiolite and palygorskite, are listed in Table 1.5.

1.3.9 Allophane and Imogolite

Allophane and imogolite are nanosize hydrous alumino-silicates of short-range order that abound, and frequently coexist, in the clay fraction of many soils, in particular those that derive from volcanic ash and weathered pumice. The formation, occurrence, and properties of allophane and imogolite have been reviewed by Fieldes and Claridge (1975), Wada (1989), Parfitt (1990, 2009), and Harsh (2012). Because of the diffuse nature of its XRD pattern, allophane has often been considered to be amorphous. High-resolution transmission electron microscopy (HRTEM), however, has consistently shown (Henmi and Wada 1976, Parfitt 1990, 2009; Calabi-Floody et al. 2009; Brigatti et al. 2013) that the unit particles of allophane consist of hollow spherules (*nanoballs*) as depicted in Figure 1.23. For this reason, the term *short-range order* (van Olphen 1971) is more appropriate than, and preferable to, *amorphous* (Wada and Harward 1974).

Allophane has a variable chemical composition. Two extreme forms of allophane have been recognized having the composition of $SiO_2Al_2O_3 \cdot 2.5H_2O$ (Al/Si = 2) and $2SiO_2Al_2O_3 \cdot 3H_2O$ (Al/Si = 1), respectively. On this basis, it seems likely that natural specimens with an Al/Si ratio between 1 and 2 represent physical mixtures of the two end-members, rather than distinct species within a series (Parfitt et al. 1980). By contrast, naturally occurring imogolites have compositions close to the ideal of $SiO_2Al_2O_3 \cdot 2H_2O$. Because of the similarity to imogolite, not only as to composition but also in terms of infrared spectral features, notably the presence of a well-defined band near 346 cm^{-1} (Farmer et al. 1977), allophane with an Al/Si ratio of ≈2 has been referred to as *imogolite-like* allophane (Parfitt and Henmi 1980).

Irrespective of composition, the structure of an allophane *nanoball* is a hollow spherule with an outer diameter of ca. 5 nm and a wall thickness of 0.7–1.0 nm (Kitagawa 1971; Henmi and Wada 1976; Wada and Wada 1977; Hall et al. 1985; Abidin et al. 2007). As in imogolite, the spherule wall is essentially made up of a 1:1 type layer, but in allophane the layer structure contains numerous vacancies (*defects*), particularly in the octahedral sheet. Clusters of vacant sites give rise to perforations of ca. 0.3 nm in diameter through which water can freely diffuse into and out of the intraspherule void space. These points are diagrammatically illustrated in Figure 1.24.

It is generally accepted that the gibbsitic sheet in imogolite-like allophane (Al/Si = 2) is on the outer or convex side of the spherule wall as depicted in Figure 1.24. Whether the same arrangement applies to the other end-member of allophane (Al/Si = 1), sometimes referred to as *halloysite-like* allophane, is still open to question (Wada and Wada 1977; Parfitt et al. 1980; Barron et al. 1982; van der Gaast et al. 1985). In this instance, about two-thirds of the possible octahredral sites are

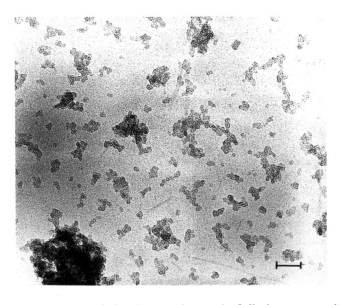

FIGURE 1.23 High-resolution transmission electron micrograph of allophane separated from the Kitakami pumice, Japan. NB: bar = 50 nm. (Courtesy of S.-I. Wada.) (From Brigatti, M.F. et al., Structure and mineralogy of clay minerals, in *Handbook of Clay Science*, 2nd ed. *Developments in Clay Science, Vol. 5A*, Bergaya, F. and Lagaly, G. (Eds.), Elsevier, Amsterdam, pp. 21–81, 2013.)

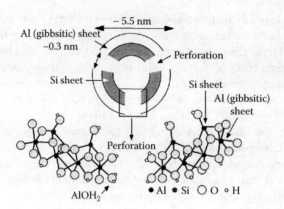

FIGURE 1.24 Diagram of an allophane *nanoball* showing the hollow spherule shape with a diameter of ca. 5 nm. The spherule wall (of ca. 0.7 nm in thickness) consists of an outer gibbsitic sheet and an inner silica sheet, and has a number of perforations with a diameter of ca. 0.3 nm. (From Hashizume, H. and Theng, B.K.G., *Clays Clay Miner.*, 55, 599–605, 2007.)

apparently vacant. As such, the tetrahedral sheet may conceivably form the outer spherule wall, and serve as the structural framework, as is the case for the *stream-deposit allophane* from Silica Springs, New Zealand (Wells et al. 1977; Childs et al. 1990).

In both imogolite-like and halloysite-like allophanes, there is appreciable substitution of Al^{3+} for Si^{4+} in tetrahedral positions. The proportion of fourfold coordinated aluminum (Al^{IV}) tends to rise as the material becomes more siliceous, making up as high as 50% of the total aluminum for allophane with an Al/Si ratio of ≈1. The resultant positive charge deficiency, however, is not reflected in the corresponding CEC values (at pH 7), which are always much smaller than would be expected from the Al^{IV} contents. In explanation, Henmi and Wada (1976) have suggested that the (permanent) negative charge from isomorphous substitution is internally compensated by proton adsorption. In addition, coatings of positively charged, amorphous hydrous oxides of aluminum and iron together with *allophane-like constituents* (Wada, 1980) may contribute toward balancing the negative layer charge. A recent Fe K-edge X-ray absorption fine structure (XAFS) study by Baker et al. (2014) indicates that iron in natural allophane and imogolite can both substitute for aluminum in octahedral positions and occur as a surface-adsorbed species. The presence of octahedrally coordinated Fe^{3+} in allophane has also been indicated by electron paramagnetic resonance spectroscopy (Cervini-Silva et al. 2015).

As might be expected, the charge characteristics of allophane are largely pH dependent, arising from the dissociation and protonation of $(OH)Al(OH_2)$ groups, exposed at defect sites in the spherule wall (Figure 1.24). These groups also exert a controlling influence on the reactivity of allophane toward charged organic species, such as amino acids, humic acid, and nucleotides (Hashizume and Theng 1999, 2007; Yuan et al. 2000; Hashizume et al. 2002). The PZC of allophanes, derived from ion-retention measurements, ranges from pH 4.5 to 6.5 (Wada 1978; Theng et al. 1982; Clark and McBride 1984). Electrokinetic (Horikawa 1975; Escudey and Galindo 1983) and rheological (Wells and Theng 1985) methods also yield sensible values. Interestingly, prior dehydration or heating often leads to a substantial (30%–50%) increase in CEC, suggesting a change in the coordination of some surface aluminum ions (Wada 1978). If so, such a process would also affect the surface acidity of allophane (Henmi and Wada 1974).

The external area of allophane, calculated on the basis of 4.0–5.5 nm for the (outer) diameter of a spherule, and 2.6 cm^3/g for the density, is 419–577 m^2/g (Kitagawa 1971). Retention studies, using ethylene glycol (EG) or ethylene glycol monoethyl ether (EGME), and assuming monolayer coverage and molecular areas applicable to planar surfaces, usually yield larger specific surface areas. On the other hand, the corresponding BET nitrogen areas are generally 30%–50% of the values derived from the retention of small, polar organic molecules (Egashira and Aomine 1974; Paterson 1977; Vandickelen et al. 1980; Hall et al. 1985; Hashizume and Theng 2007). This observation suggests that such molecules can penetrate the intra-spherule pore space, at least partially, whereas nitrogen is apparently incapable of doing so. The variation in surface area values, for any given method of measurement, reflects differences among specimens as to composition, origin, and pretreatment.

Pore size distribution studies, using nitrogen as adsorbate, indicate that the porosity of (dry) allophanes is largely contained in pores with a mean radius (r) of <1.0 nm (Rousseaux and Warkentin 1976; Paterson 1977; Vandickelen et al. 1980). Occupying a volume of 0.12–0.15 cm³/g, these micropores often show a bimodal distribution, one group of pores (class 1) having r = 0.3–0.6 nm and another (class 2) having r = 0.6–0.9 nm. Both classes are presumably associated with interspherule (intra-aggregate) pore spaces although class 1 pores may include wall perforations and defect structures. A third class of pores with r = 2.0–10 nm, making up ~30% of the total porosity, may be identified with inter-aggregate pores. This interpretation of the pore size distribution data is consistent with transmission electron micrographs of allophane (Figure 1.23) and with the results of adsorption studies using *n*-alkylammmonium chlorides (Theng 1972).

When examined by HRTEM, imogolite appears as slender, hollow tubules forming 10–30 nm thick bundles of several micrometres in length (Yoshinaga and Aomine 1962) as shown in Figure 1.25.

Figure 1.26 gives a diagram of the hollow tubular structure of imogolite, based on the proposal by Cradwick et al. (1972) as elaborated on by Brown et al. (1978). The imogolite *nanotube* has an outer diameter of ca. 2.1 nm and an inner diameter of ca. 0.64 nm. The tubule wall consists of a curved outer gibbsitic sheet to which (O_3SiOH) groups are attached on the inside. The unit formula composition is usually given as $(OH)_3Al_2O_3SiOH$, indicating the sequence of ions from the periphery to the center of the tubule where the orthosilicate group shares three oxygens with aluminum.

There seems to be little substitution of Al^{3+} for Si^{4+} in the imogolite structure and most, if not all, of the aluminum is in sixfold coordination (Henmi and Wada 1976; Wilson et al. 1984) although Fe^{3+} can substitute for Al^{3+} in octahedral sites of both imogolite and allophane (Baker et al. 2014). Indeed, Fe-substituted synthetic imogolite is an effective catalyst in the oxidation of aromatic

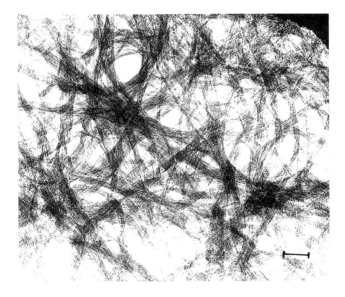

FIGURE 1.25 High-resolution transmission electron micrograph of imogolite separated from a gel film of the Kitakami pumice bed, Japan. NB: bar = 50 nm. (Courtesy of N. Yoshinaga.) (From Brigatti, M.F. et al., Structure and mineralogy of clay minerals, in *Handbook of Clay Science*, 2nd ed. *Developments in Clay Science, Vol. 5A*, Bergaya, F. and Lagaly, G. (Eds.), Elsevier, Amsterdam, pp. 21–81, 2013.)

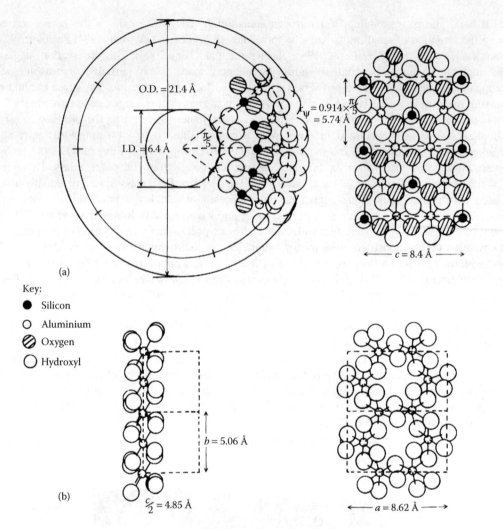

FIGURE 1.26 Diagram comparing the structure and unit cell dimensions of imogolite with those of gibbsite: (a), the structure of imogolite, viewed down the tubule axis, showing the atomic arrangement for two of the ten unit cells (left), and projected on a cylinder surface through the centers of the outer hydroxyl groups (right) and (b), the structure of a gibbsite sheet consisting of linked alumina octahedra (left), and projected on the ab plane (right). NB: 1 Å = 0.1 nm. (From Brown, G. et al., The structures and chemistry of soil clay minerals, in *The Chemistry of Soil Constituents*, Greenland, D.J. and Hayes, M.H.B (Eds.), John Wiley & Sons, Chichester, UK, pp. 29–178, 1978.)

hydrocarbons (Ookawa et al. 2006, 2008; Table 1.6). Imogolite can also develop pH-dependent negative and positive charges due to the dissociation and protonation, respectively, of hydroxyl groups attached to aluminum at the tubule ends. The point of zero charge (PZC), derived from ion-retention measurements at different pH values and acid-base potentiometric titration, occurs at pH 6–6.5 (Theng et al. 1982; Clark and McBride 1984; Tsuchida et al. 2005).

The total surface of imogolite, calculated on the basis of the tubule dimension, and a density of 2.65 g/cm³, amounts to 1400–1500 m²/g. By comparison, the specific surface area, derived from the retention of water, ethylene glycol (EG), or ethylene glycol monoethyl ether (EGME) ranges from 700 to 1100 m²/g (Egashira and Aomine 1974; Wada and Harward 1974). Like water, such small polar organic molecules can presumably enter the intratubular pores, but access to intertubular surfaces is apparently restricted, probably because the tubules tend to be aligned in parallel fashion to form bundles (Figure 1.25). In this context, we should add that a major uncertainty attaches to deriving surface areas from retention measurements. The problem lies in assigning the area occupied by a given adsorbate molecule at *monolayer* coverage when the adsorbing surface is nonplanar (Paterson 1977; Wada 1978; Tiller and Smith 1990). Further, strong hydrogen bonding and van der Waals interactions can be established at points where adjacent tubules come into close contact, inhibiting adsorbate entry into these regions. Similar interactions would occur between individual bundles of tubules where they overlap and intertwine. Indeed, such a network structure lies behind the pronounced gel-like character, shear-thinning flow behavior, and high-yield value of dilute aqueous suspensions of imogolite (Wells et al. 1980).

In terms of catalytic potential, the pore size distribution of imogolite (after drying) is perhaps more important than its particulate arrangement in suspension. Retention studies, using quaternary ammonium chlorides and water, indicate that slightly over half of the total volume of imogolite is contained in pores. Approximately 50% of these pores are intertubular, 25% intratubular with a volume of 0.13 cm^3/g, and the remainder of the porosity is associated with inter-bundle pores (Wada and Henmi 1972). Both the inter- and intra-tubular pores appear to be accessible to benzene (Wilson et al. 2002).

The catalytic activity of imogolite and related phases, such as proto-imogolite and heat-collapsed imogolite, has been summarized by Garrone and Bonelli (2016). The variety of organic reactions, catalyzed by allophane and imogolite, are listed in Table 1.6.

REFERENCES

Abidin, Z., N. Matsui, and T. Henmi. 2007. Differential formation of allophane and imogolite: Experimental and molecular orbital study. *Journal of Computer-Aided Materials Design* 14: 5–18.

Adams, J.M. 1983. Hydrogen atom positions in kaolinite by neutron profile refinement. *Clays and Clay Minerals* 31: 352–356.

Adams, J.M. 1987. Synthetic organic chemistry using pillared, cation-exchanged and acid-treated montmorillonite catalysts—A review. *Applied Clay Science* 2: 309–342.

Adams, J.M., T.V. Clapp, and D.E. Clement. 1983. Catalysis by montmorillonites. *Clay Minerals* 18: 411–421.

Adams, J.M., J.A. Ballantine, S.H. Graham, R.J. Laub, and J.H. Purnell. 1979. Selective chemical conversions using sheet silicate intercalates: Low-temperature addition of water to 1-alkenes. *Journal of Catalysis* 58: 238–252.

Adams, J.M., K. Martin, and R.W. McCabe. 1987. Clays as selective catalysts in organic synthesis. *Journal of Inclusion Phenomena* 5: 663–674.

Alba, M.D., M.A. Castro, P. Chain, M. Naranjo, and A.C. Perdigón. 2005. Structural study of synthetic mica-montmorillonite by means of 2D MAS NMR experiments. *Physics and Chemistry of Minerals* 32: 248–254.

Aleixandre, V. and T. Fernandez. 1960. Influence catalytique des différents minéraux argileux sur la synthèse du butadiène á partir de l'éthanol. *Silicates Industriels* 25: 243–248.

Alkan, M., Ö. Demirbaş, and M. Doğan. 2005. Electrokinetic properties of kaolinite in mono- and multivalent electrolyte solutions. *Microporous and Mesoporous Materials* 83: 51–59.

Almon, W.R. and W.D. Johns. 1976. Petroleum-forming reactions: Clay-catalyzed fatty acid decarboxylation. In *Proceedings of the International Clay Conference 1975*, (Ed.) S.W. Bailey, pp. 399–409. Wilmette, IL: Applied Publishing.

Alvero, R., M.D. Alba, M.A. Castro, and J.M. Trillo. 1994. Reversible migration of lithium in montmorillonite. *Journal of Physical Chemistry* 98: 7848–7853.

Anderson, J.A., L. Daza, S. Damyanova, J.L.G. Fierro, and M.T. Rodrigo. 1994. Hydrogenation of styrene over nickel/sepiolite catalysts. *Applied Catalysis A: General* 113: 75–88.

Andreev, P.F., A.I. Bogomolov, A.F. Dobryanskii, and A.A. Kartsev. 1968. *Transformation of Petroleum in Nature*. Oxford, UK: Pergamon Press.

Angel, B.R. and P.L. Hall. 1973. Electron spin resonance studies of kaolins. In *Proceedings of the International Clay Conference 1972*, (Ed.) J.M. Serratosa, pp. 47–60. Madrid, Spain: Division de Ciencias, C.S.I.C.

Aramendia, M.A., V. Borau, J. Gomez, C. Jiminez, and J.M. Marinas. 1984. Influence of substituents on the reduction of aromatic nitrocompounds over supported palladium catalysts. *Applied Catalysis* 10: 347–359.

Araújo, F.R., J.G. Baptista, L. Marçal et al. 2014. Versatile heterogeneous dipicolinate complexes grafted into kaolinite: Catalytic oxidation of hydrocarbons and degradation of dyes. *Catalysis Today* 227: 105–115.

Arfaioli, P., O.L. Pantani, M. Bosetto, and G.G. Ristori. 1999. Influence of clay minerals and exchangeable cations on the formation of humic-like substances (melanoidins) from D-glucose and L-tyrosine. *Clay Minerals* 34: 487–497.

Arndt, D.S., M. Mattei, C.A. Heist, and M.M. McGuire. 2017. Measurement of swelling of individual smectite tactoids *in situ* using atomic force microscopy. *Clays and Clay Minerals* 65: 92–193.

Aronowitz, S., L. Coyne, J. Lawless, and J. Rishpon. 1982. Quantum-chemical modeling of smectite clays. *Inorganic Chemistry* 21: 3589–3593.

Arundhathi, R., B. Sreedhar, and G. Parthasarathy. 2010. Chamosite, a naturally occurring clay as a versatile catalyst for various organic transformations. *Clay Minerals* 45: 281–299.

Arundhathi, R., B. Sreedhar, and G. Parthasarathy. 2011. Highly efficient heterogeneous catalyst for O-arylation of phenols with aryl halides using natural ferrous chamosite. *Applied Clay Science* 51: 131–137.

Atkins, M.P., D.J.H. Smith, and D.J. Westlake. 1983. Montmorillonite catalysts for ethylene hydration. *Clay Minerals* 18: 423–429.

Aylmore, L.A.G. and J.P. Quirk 1960. Domain or turbostratic structure of clays. *Nature* 187: 1046–1048.

Bailey, S.W. 1963. Polymorphism of the kaolin minerals. *American Mineralogist* 48: 1196–1209.

Bailey, S.W. 1975. Chlorites. In *Soil Components, Vol. 2. Inorganic Components,* (Ed.) J.E. Gieseking, pp. 191–263. Berlin, Germany: Springer-Verlag.

Bailey, S.W. 1980. Structures of layer silicates. In *Crystal Structures of Clay Minerals and Their X-ray Identification,* (Eds.) G.W. Brindley and G. Brown, pp. 1–123. London, UK: Mineralogical Society.

Bailey, S.W. (Ed.). 1984. *Micas. Reviews in Mineralogy,* Vol. 13. Washington, DC: Mineralogical Society of America.

Bailey, S.W. 1990. Halloysite—A critical assessment. In *Proceedings of the 9th International Clay Conference, 1989,* (Eds.) V.C. Farmer and Y. Tardy, pp. 89–98. Strasbourg, France: Sciences Géologiques, Mémoire 86.

Bailey, S.W. and B.E. Brown. 1962. Chlorite polytypism: I. Regular and semi-random one-layer structures. *American Mineralogist* 47: 819–850.

Baker, L.L., R.D. Nickerson, and D.G. Strawn. 2014. XAFS study of Fe-substituted allophane and imogolite. *Clays and Clay Minerals* 62: 20–34.

Balan, E., G. Calas, and D.L. Bish. 2014. Kaolin-group minerals: From hydrogen-bonded layers to environmental recorders. *Elements* 10: 183–188.

Ballantine, J.A. 1986. The reactions in clays and pillared clays. In *Chemical Reactions in Organic and Inorganic Constrained Systems,* (Ed.) R. Setton, pp. 197–212. Dordrecht, the Netherlands: D. Reidel.

Ballantine, J.A. and J.H. Purnell. 1984. Sheet silicates: Broad spectrum catalysts for organic synthesis. *Journal of Molecular Catalysis* 27: 157–167.

Ballantine, J.A., J.H. Purnell, and J.M. Thomas. 1983. Organic reactions in a clay micro-environment. *Clay Minerals* 18: 347–356.

Balogh, M. and P. Laszlo. 1993. *Organic Chemistry Using Clays.* Berlin, Germany: Springer-Verlag.

Bandgar, B.P. and S.P. Kasture. 2000. Natural kaolinite clay: A remarkable reusable solid catalyst for the selective cleavage of thioacetals without solvent. *Green Chemistry* 2: 154–156.

Bandgar, B.P., L.S. Uppalla, and V.S. Sadavarte. 2001a. Envirocat EPZG and natural clay as efficient catalysts for transesterification of β-keto esters. *Green Chemistry* 3: 39–41.

Bandgar, B.P., L.S. Uppalla, A.D. Sagar, and V.S. Sadavarte. 2001b. A mild procedure for rapid and selective deprotection of aryl acetates using natural kaolinite clay as a reusable catalyst. *Tetrahedron Letters* 42: 1163–1165.

Banerjee, A.K. and M.S. Laya. 2000. Reagents for the preparation and cleavage of 1,3-dithiolanes. *Russian Chemical Reviews* 69: 947–955.

Banin, A. and N. Lahav. 1968. Particle size and optical properties of montmorillonite in suspension. *Israel Journal of Chemistry* 6: 235–250.

Bannerjee, B.K. and M.K. Sen. 1974. Application of clay minerals in catalyst industry. *Bulletin of the Indian Society of Soil Science* 9: 247–253.

Barnhisel, R.I. and P.M. Bertsch. 1989. Chlorites and hydroxy interlayered vermiculite and smectite. In *Minerals in Soil Environments,* 2nd ed., (Eds.) J.B. Dixon and S.B. Weed, pp. 729–788. Madison, WI: Soil Science Society of America.

Barrer, R.M. 1978. *Zeolites and Clay Minerals as Sorbents and Molecular Sieves.* London, UK: Academic Press.

Barrer, R.M. 1984. Sorption and molecular sieve properties of clays and their importance as catalysts. *Philosophical Transactions of the Royal Society of London A* 311: 333–352.

Barrer, R.M. and D.L. Jones. 1970. Chemistry of soil minerals. Part VIII. Synthesis and properties of fluorhectorites. *Journal of the Chemical Society A* 1531–1537.

Barrer, R.M. and D.L. Jones. 1971. Chemistry of soil minerals. Part X. Shape-selective sorbents derived from fluorhectorites. *Journal of the Chemical Society A* 2594–2603.

Barrer, R.M. and N. Mackenzie. 1954. Sorption by attapulgite. Part 1. Availability of intracrystalline channels. *Journal of Physical Chemistry* 58: 560–568.

Barrientos-Ramírez, S., G.M.de Oca-Ramírez, E.V. Ramos-Fernández, A. Sepúlveda-Escribano, M.M. Pastor-Blas, and A. González-Montiel. 2011. Surface modification of natural halloysite clay nanotubes with aminosilanes. Application as catalyst supports in the atom transfer radical polymerization of methyl methacrylate. *Applied Catalysis A: General* 406: 22–33.

Barrientos-Ramírez, S., E.V. Ramos-Fernández, J. Silvestre-Albero, A. Sepúlveda-Escribano, M.M. Pastor-Blas, and A. González-Montiel. 2009. Use of nanotubes of natural halloysite as catalyst support in the atom transfer radical polymerization of methyl methacrylate. *Microporous and Mesoporous Materials* 120: 132–140.

Barron, P.F., M.A. Wilson, A.S. Campbell, and R.L. Frost. 1982. Detection of imogolite in soils using solid state ^{29}Si NMR. *Nature* 299: 616–618.

Bates, T.F. 1959. Morphology and crystal chemistry of 1:1 layer lattice silicates. *American Mineralogist* 44: 78–114.

Bates, T.F., F.A. Hildebrand, and A. Swineford. 1950. Morphology and structure of endellite and halloysite. *American Mineralogist* 35: 463–484.

Battalova, S. B., A.A. Likerova, and T.R. Mukitanova. 1975. Different catalytic systems in the isomerization of α-pinene. *Izvestiia Akademia Nauk Kazakh SSR Seriia Khimicheskaia* 25: 70–72.

Battalova, S. B., T.R. Mukitanova, and R.D. Dzhakisheva. 1977. Studies of different catalytic systems in α-pinene isomerization reactions. 3. *Izvestiia Akademia Nauk Kazakh SSR Seriia Khimicheskaia* 27: 71–73.

Bautista, F.M., J.M. Campelo, D. Luna, J. Luque, and J.M. Marinas. 2007. Vanadium oxides supported on TiO_2-sepiolite and sepiolite: Preparation, structural and acid characterization and catalytic behaviour in selective oxidation of toluene. *Applied Catalysis A: General* 325: 336–344.

Beneke, K. and G. Lagaly. 2002. From fuller's earth to bleaching earth: A historical note. *European Clay Group Association Newsletter* 5: 57–78.

Bergaya, F. and G. Lagaly (Eds.). 2013a. *Handbook of Clay Science,* 2nd ed. Developments in Clay Science, Vol. 5A. Amsterdam, the Netherlands: Elsevier.

Bergaya, F. and G. Lagaly (Eds.). 2013b. *Handbook of Clay Science,* 2nd ed. Developments in Clay Science, Vol. 5B. Amsterdam, the Netherlands: Elsevier.

Bergaya, F., G. Lagaly, and M. Vayer. 2006a. Cation and anion exchange. In *Handbook of Clay Science,* Developments in Clay Science, Vol. 1, (Eds.) F. Bergaya, B.K.G. Theng, and G. Lagaly, pp. 979–1001. Amsterdam, the Netherlands: Elsevier.

Bergaya, F., B.K.G. Theng, and G. Lagaly (Eds.). 2006b. *Handbook of Clay Science.* Developments in Clay Science, Vol. 1. Amsterdam, the Netherlands: Elsevier.

Beyer, J. and H. Graf von Reichenbach. 2002. An extended revision of the interlayer structures of one- and two-layer hydrates of Na-vermiculite. *Clay Minerals* 37: 157–168.

Bizaia, N., E.H. de Faria, G.P. Ricci et al. 2009. Porphyrin-kaolinite as efficient catalyst for oxidation reactions. *Applied Materials & Interfaces* 1: 2667–2678.

Blockhaus, F., J-M. Sequaris, H.D. Narres, and M.J. Schwuger. 1997. Adsorption-desorption behavior of acrylic-maleic acid copolymer at clay minerals. *Journal of Colloid and Interface Science* 186: 234–247.

Bloise, A., E. Belluso, E. Barrese, D. Miriello, and C. Apollaro. 2009. Synthesis of Fe-doped chrysotile and characterization of the resulting chrysotile fibres. *Crystal Research and Technology* 44: 590–596.

Blount, A.M., I.M. Threadgold, and S.W. Bailey. 1969. Refinement of the crystal structure of nacrite. *Clays and Clay Minerals* 17: 185–194.

Bolland, M.D.A., A.M. Posner, and J.P. Quirk. 1976. Surface charge on kaolinites in aqueous suspension. *Australian Journal of Soil Research* 14: 197–216.

Bolland, M.D.A., A.M. Posner, and J.P. Quirk. 1980. pH-independent and pH-dependent surface charges on kaolinites. *Clays and Clay Minerals* 28: 412–418.

Bonelli, B., I. Bottero, N. Ballarini, S. Passeri, F. Cavani, and E. Garrone. 2009. IR spectroscopic and catalytic characterization of the acidity of imogolite-based systems. *Journal of Catalysis* 264: 15–30.

Bonneau, L., H. Suquet, C. Malard, and H. Pezerat. 1986. Studies on surface properties of asbestos. I. Active sites on surface of chrysotile and amphiboles. *Environmental Research* 41: 251–267.

Bonneau, M. and B. Souchier. 1982. *Constituents and Properties of Soils*. London, UK: Academic Press.

Bookin, A.S., V.A. Drits, A. Plançon, and C. Tchoubar. 1989. Stacking faults in kaolin-group minerals in the light of real structural features. *Clays and Clay Minerals* 37: 297–307.

Borchardt, G. 1989. Smectites. In *Minerals in Soil Environments*, 2nd ed., (Eds.) J.B. Dixon and S.B. Weed, pp. 675–727. Madison, WI: Soil Science Society of America.

Boyd, S.A., C.T. Johnston, D.A. Laird, B.J. Teppen, and H. Li. 2011. Comprehensive study of organic contaminant adsorption by clays: Methodologies, mechanisms, and environmental implications. In *Biophysico-chemical Processes of Anthropogenic Organic Compounds in Environmental Systems*, (Eds.) B. Xing, N. Senesi, and P.M. Huang, pp. 51–71. New York: John Wiley & Sons.

Bradley, W.F. 1940. The structural scheme of attapulgite. *American Mineralogist* 25: 405–410.

Brady, P.V., R.T. Cygan, and K.L. Nagy. 1996. Molecular controls on kaolinite surface charge. *Journal of Colloid and Interface Science* 183: 356–364.

Brauner, K. and A. Preisinger. 1956. Struktur und Entstehung des Sepioliths. *Tschermaks Mineralogische und Petrographische Mitteilungen* 6: 120–140.

Breen, C., J.M. Adams, and C. Riekel. 1985. Review of the diffusion of water and pyridine in the interlayer space of montmorillonite: Relevance to kinetics of catalytic reactions in clays. *Clays and Clay Minerals* 33: 275–284.

Breen, C. and P.M. Last. 1999. Catalytic transformation of the gases evolved during the thermal decomposition of HDPE using acid-activated and pillared clays. *Journal of Materials Chemistry* 9: 813–818.

Brigatti, M.F., E. Galan, and B.K.G. Theng. 2013. Structure and mineralogy of clay minerals. In *Handbook of Clay Science*, 2nd ed. Developments in Clay Science, Vol. 5A, (Eds.) F. Bergaya, and G. Lagaly, pp. 21–81. Amsterdam, the Netherlands: Elsevier.

Brindley, G.W. 1980. Order-disorder in clay mineral structures. In *Crystal Structures of Clay Minerals and Their X-ray Identification*, (Eds.) G.W. Brindley and G. Brown, pp. 125–195. London, UK: Mineralogical Society.

Brindley, G.W. and G. Brown (Eds.). 1980. *Crystal Structures of Clay Minerals and Their X-ray Identification*. London, UK: Mineralogical Society.

Brindley, G.W. and G. Ertem. 1971. Preparation and solvation properties of some variable charge montmorillonites. *Clays and Clay Minerals* 19: 399–404.

Brindley, G.W. and C-C. Kao. 1980. Formation, compositions, and properties of hydroxyl-Al and hydroxyl-Mg montmorillonite. *Clays and Clay Minerals* 28: 435–443.

Brindley, G.W., C-C. Kao, J.L. Harrison, M. Lipsicas, and R. Raythatha. 1986. Relation between structural disorder and other characteristics of kaolinites and dickites. *Clays and Clay Minerals* 34: 239–249.

Brindley, G.W. and M. Nakahira. 1958. Further considerations of the crystal structure of kaolinite. *Mineralogical Magazine* 31: 781–786.

Brindley, G.W. and R.E. Sempels. 1977. Preparation and properties of some hydroxyl-aluminium beidellites. *Clay Minerals* 12: 229–237.

Brink, G. and M. Falk. 1970. Infrared studies of water in crystalline hydrates: $NaClO_4 \cdot H_2O$, $LiClO_4 \cdot 3H_2O$, and $Ba(ClO_4)_2 \cdot 3H_2O$. *Canadian Journal of Chemistry* 48: 2096–2103.

Brody, J.F., J.W. Johnson, G.B. Vicker, and J.J. Ziemiak. 1989. Olefin isomerization over an alumina-pillared fluoromica catalyst. *Solid State Ionics* 32/33: 350–353.

Brooks, B.T. 1948. Active-surface catalysis in formation of petroleum. *American Association of Petroleum Geologists Bulletin* 32: 2269–2286.

Broughton, G. 1940. Catalysis by metallized bentonites. *Journal of Physical Chemistry* 44: 181–184.

Brown, D.R. 1994. Clays as catalyst and reagent supports. *Geologica Carpathica, Series Clays* 45: 45–56.

Brown, G. 1980. Associated minerals. In *Crystal Structures of Clay Minerals and Their X-ray Identification*, (Eds.) G.W. Brindley and G. Brown, pp. 361–410. London, UK: Mineralogical Society.

Brown, G., A.C.D. Newman, J.H. Rayner, and A.H. Weir. 1978. The structures and chemistry of soil clay minerals. In *The Chemistry of Soil Constituents*, (Eds.) D.J. Greenland and M.H.B. Hayes, pp. 29–178. Chichester, UK: John Wiley & Sons.

Brunauer, S., P.H. Emmett, and E. Teller. 1938. Adsorption of gases in multimolecular layers. *Journal of the American Chemical Society* 60: 309–319.

Burgess, J. 1978. *Metal Ions in Solution*. Chichester, UK: Ellis Horwood.

Burkow, I.C., E. Jørgensen, T. Meyer, Ø. Rekdal, and L.K. Sydnes. 1990. Experimental simulation of chemical transformations of aromatic compounds in sediments. *Organic Geochemistry* 15: 101–108.

Caillère, S. and S. Hénin. 1957. The sepiolite and palygorskite minerals. In *The Differential Thermal Investigation of Clays*, (Ed.) R.C. Mackenzie, pp. 231–247. London, UK: Mineralogical Society.

Caillère, S., S. Hénin, and M. Rautureau. 1982. *Minéralogie des Argiles. 1. Structure et Propriétés Physico-Chimiques*. 2nd ed. Paris, France: Masson.

Calabi-Floody, M., J.S. Bendall, A.A. Jara et al. 2011. Nanoclays from an Andisol: Extraction, properties and carbon stabilization. *Geoderma* 161: 159–167.

Calabi-Floody, M., B.K.G. Theng, P. Reyes, and M.L. Mora. 2009. Natural nanoclays: Application and future trends—A Chilean perspective. *Clay Minerals* 44: 161–176.

Cámara, M.A., M. Navarro, S. Navarro Garcia, A. Barba, and C-M. Coste. 1992. Degradation catalytique du chlorfenvinphos et du methidathion deposés sur kaolinite et bentonite saturées par différents cations. *Journal of Environmental Science and Health Part B* 27: 293–306.

Campbell, A.S. and U. Schwertmann. 1985. Evaluation of selective dissolution extractants in soil chemistry and mineralogy by differential X-ray diffraction. *Clay Minerals* 20: 515–519.

Campelo, J.M., A. Garcia, D. Luna, and J.M. Marinas. 1987a. Catalytic activity of natural sepiolites in cyclohexene skeletal isomerization. *Clay Minerals* 22: 233–236.

Campelo, J.M., A. Garcia, D. Luna, and J.M. Marinas. 1987b. Surface properties of sepiolites from Vallecas-Madrid, Spain, and their catalytic activity in cyclohexene skeletal isomerization. *Reactivity of Solids* 3: 263–272.

Campelo, J.M., A. Garcia, D. Luna, and J.M. Marinas. 1989. Textural properties, surface chemistry and catalytic activity in cyclohexene skeletal isomerization of acid treated natural sepiolites. *Materials Chemistry and Physics* 24: 51–70.

Campelo, J.M., A. Garcia, D. Luna, and J.M. Marinas. 1990. New $AlPO_4$-sepiolite systems as acid catalysts, I. Preparation, texture, surface-chemical properties and cyclohexene skeletal isomerization conversion. *Journal of Materials Science* 25: 2513–2519.

Campos, A., B. Gagea, S. Moreno, P. Jacobs, and R. Molina. 2007. Hydroisomerization of decane on Pt/Al, Ce-pillared vermiculites. In *From Zeolites to Porous MOF Materials*, (Eds.) R. Xu, Z. Gao, J. Che, and W. Yan, pp. 1405–1410. Amsterdam, the Netherlands: Elsevier.

Campos, A., B. Gagea, S. Moreno, P. Jacobs, and R. Molina. 2008b. Decane hydroconversion with Al-Zr, Al-Hf, Al-Ce-pillared vermiculites. *Applied Catalysis A: General* 345: 112–118.

Campos, A., S. Moreno, S., and R. Molina. 2008a. Relationship between hydrothermal treatment parameters as a strategy to reduce layer charge in vermiculite, and its catalytic behaviour. *Catalysis Today* 133–135: 351–356.

Carr, R.M. and H. Chih. 1971. Complexes of halloysite with organic compounds. *Clay Minerals* 9: 153–166.

Carretero, M.I. and M. Pozo. 2007. *Mineralogía Aplicada. Salud y Medio Ambiente*. Madrid, Spain: Thomson.

Carrillo, A.M. and J.G. Carriazo. 2015. Cu and Co oxides supported on halloysite for the total oxidation of toluene. *Applied Catalysis B: Environmental* 164: 443–452.

Carstea, D.D., M.E. Harward, and E.G. Knox. 1970. Comparison of iron and aluminum hydroxy interlayers in montmorillonite and vermiculite. I. Formation. *Soil Science Society of America Journal* 34: 517–521.

Cebula, D.J., R.K. Thomas, and J.W. White. 1979. The structure and dynamics of clay-water systems studied by neutron scattering. In *International Clay Conference 1978. Developments in Sedimentology 27*, (Eds.) M.M. Mortland and V.C. Farmer, pp. 111–120. Amsterdam, the Netherlands: Elsevier.

Cervini-Silva, J., A. Nieto-Camacho, V. Gómez-Vidales, S. Kaufhold, and B.K.G. Theng. 2015. The anti-inflammatory activity of natural allophane. *Applied Clay Science* 105–106: 48–51.

Cervini-Silva, J., A. Nieto-Camacho, E. Palacios et al. 2016. Anti-inflammatory, antibacterial, and cytotoxic activity by natural matrices of nano-iron(hydr)oxide/halloysite. *Applied Clay Science* 120: 101–110.

Chen, Q., P. Wu, Z. Dang et al. 2010. Iron pillared vermiculite as a heterogeneous photo-Fenton catalyst for photocatalytic degradation of azo dye reactive brilliant orange X-GN. *Separation and Purification Technology* 71: 315–323.

Chen, S., J. Li, Y. Zhang, D. Zhang, and J. Zhu. 2012. Effect of preparation method on halloysite supported cobalt catalysts for Fischer–Tropsch synthesis. *Journal of Natural Gas Chemistry* 21: 426–430.

Cheng, M., W. Song, W. Ma et al. 2008. Catalytic activity of iron species in layered clays for photodegradation of organic dyes under visible irradiation. *Applied Catalysis B: Environmental* 77: 355–363.

Chermahini, A.N., A. Teimouri, F. Momenbeik et al. 2010. Clay-catalyzed synthesis of 5-substituted 1-*H*-tetrazoles. *Journal of Heterocyclic Chemistry* 47: 913–922.

Childs, C.W., R.L. Parfitt, and R.H. Newman. 1990. Structural studies of Silica Springs allophane. *Clay Minerals* 25: 329–341.

Chitnis, S.R. and M.M. Sharma. 1997. Industrial applications of acid-treated clays as catalysts. *Reactive and Functional Polymers* 32: 93–115.

Cho, K.-H., B.-S. Jang, K.-H. Kim, and D.-W. Park. 2006. Performance of pyrophyllite and halloysite clays in the catalytic degradation of polystyrene. *Reaction Kinetics and Catalysis Letters* 88: 43–50.

Choudhary, V.R. and S.K. Jana. 2002. Benzylation of benzene and substituted benzenes by benzyl chloride over $InCl_3$, $GaCl_3$, $FeCl_3$, and $ZnCl_2$ supported on clays and Si-MCM-41. *Journal of Molecular Catalysis A: Chemical* 180: 267–276.

Chourabi, B. and J.J. Fripiat. 1981. Determination of tetrahedral substitution and interlayer surface heterogeneity from vibrational spectra of ammonium in smectites. *Clays and Clay Minerals* 29: 260–268.

Christidis, G. and A.C. Dunham. 1993. Compositional variations in smectites: Part I. Alteration of intermediate volcanic rocks. A case study from Milos Island, Greece. *Clay Minerals* 28: 255–273.

Christidis, G.E. and D.D. Eberl. 2003. Determination of layer charge characteristics of smectites. *Clays and Clay Minerals* 51: 644–655.

Christidis, G.E. and W.D. Huff. 2009. Geological aspects and genesis of bentonites. *Elements* 5: 93–98.

Chukrov, F.W. and B.B. Zvyagin. 1966. Halloysite, a crystallochemically and mineralogically distinct species. In *Proceedings of the International Clay Conference 1966*, Vol. I, (Eds.) L. Heller and A. Weiss, pp. 11–25. Jerusalem, Israel: Israel Program for Scientific Translations.

Churchman, G.J. and R.M. Carr. 1972. Stability fields of hydration states of an halloysite. *American Mineralogist* 57: 914–933.

Churchman, G.J., P. Pasbakhsh, D.J. Lowe, and B.K.G. Theng. 2016. Unique but diverse: Some observations on the formation, structure and morphology of halloysite. *Clay Minerals* 51: 395–416.

Churchman, G.J. and B.K.G. Theng. 1984. Interactions of halloysites with amides: Mineralogical factors affecting complex formation. *Clay Minerals* 19: 161–175.

Churchman, G.J., J.S. Whitton, G.G.C. Claridge, and B.K.G. Theng. 1984. Intercalation method using formamide for differentiating halloysite from kaolinite. *Clays and Clay Minerals* 32: 241–248.

Clark, C.J. and M.B. McBride. 1984. Cation and anion retention by natural and synthetic allophane and imogolite. *Clays and Clay Minerals* 32: 291–299.

Clark, J.H., A.J. Butterworth, S.J. Tavener, and A.J. Teasdale. 1997. Environmentally friendly chemistry using supported reagent catalysts: Chemically-modified mesoporous soild catalysts. *Journal of Chemical Technology and Biotechnology* 68: 367–376.

Clark, J.H., S.R. Cullen, S.J. Barlow, and T.W. Bastock. 1994. Environmentally friendly chemistry using supported reagent catalysts: Structure–property relationships for clayzic. *Journal of the Chemical Society Perkin Transactions* 2: 1117–1130.

Clementz, D.M., M.M. Mortland, and T.J. Pinnavaia. 1974. Properties of reduced-charge montmorillonites: Hydrated Cu(II) ions as a spectroscopic probe. *Clays and Clay Minerals* 22: 49–57.

Clementz, D.M., T.J. Pinnavaia, and M.M. Mortland. 1973. Stereochemistry of hydrated copper (II) ions on the interlamellar surfaces of layer silicates. An electron spin resonance study. *Journal of Physical Chemistry* 77: 196–200.

Collet, C. and P. Laszlo. 1991. Clay catalysis of the non-aqueous Diels–Alder reaction and the importance of humidity control. *Tetrahedron Letters* 32: 2905–2908.

Corma, A., H. García, A. Leyva, and A. Primo. 2004. Alkali-exchanged sepiolites containing palladium as bifunctional (basic sites and noble metal) catalysts for the Heck and Suzuki reactions. *Applied Catalysis A: General* 257: 77–83.

Corma, A., S. Iborra, S. Miquel, and J. Primo. 1998. Catalysts for the production of fine chemicals. *Journal of Catalysis* 173: 315–321.

Corma, A. and R.M. Martín-Aranda. 1991. Alkaline-substituted sepiolite as a new type of strong base catalyst. *Journal of Catalysis* 130: 130–137.

Corma, A. and R.M. Martín-Aranda. 1993. Application of solid base catalysts in the preparation of prepolymers by condensation of ketones and malenonitrile. *Applied Catalysis* 105: 271–279.

Corma, A., A. Mifsud, and J. Perez-Pariente. 1988. Influence of the procedure of nickel deposition on the textural and catalytic properties of nickel/sepiolite catalysts. *Industrial & Engineering Chemistry Research* 27: 2044–2050.

Corma, A. and J. Perez-Pariente. 1987. Catalytic activity of modified silicates: I. Dehydration of ethanol catalysed by acidic sepiolite. *Clay Minerals* 22: 423–433.

Cornejo, J., M.C. Hermosin, J.L. White, J.R. Barnes, and S.L. Hem. 1983. Role of ferric iron in the oxidation of hydrocortisone by sepiolite and palygorskite. *Clays and Clay Minerals* 31: 109–112.

Cornejo, J., M.C. Hermosin, J.L. White, G.E. Peck, and S.L. Hem. 1980. Oxidative degradation of hydrocortisone in presence of attapulgite. *Journal of Pharmaceutical Sciences* 69: 945–948.

Cornélis, A., A. Gerstmans, P. Laszlo, A. Mathy, and I. Zieba. 1990. Friedel–Crafts acylations with modified clays as catalysts. *Catalysis Letters* 6: 103–110.

Cornélis, A. and P. Laszlo. 1994. Molding clays into efficient catalysts. *Synlett* 155–161.

Cornélis, A., P. Laszlo, and P. Pennetreau. 1983. Some organic syntheses with clay-supported reagents. *Clay Minerals* 18: 437–445.

Cornélis, A., P. Laszlo, M.W. Zettler, B. Das, and K.V.N.S. Srinivas. 2004. Montmorillonite K10. *e-EROS Encyclopedia of Reagents for Organic Synthesis*, pp. 3667–3671. Wiley Online Library.

Cozak, D. and C. DeBlois. 1984. Olefin hydrogenation catalysed by chrysotile asbestos fibres and supported titanocene. *Canadian Journal of Chemistry* 62: 392–394.

Cradwick, P.D.G., V.C. Farmer, J.D. Russell, C.R. Masson, K. Wada, and N. Yoshinaga. 1972. Imogolite, a hydrated aluminium silicate of tubular structure. *Nature Physical Science* 240: 187–189.

Cristiano, D.V., A.M. Campos, and R. Molina. 2005. Charge reduction in a vermiculite by acid and hydrothermal methods: A comparative study. *Journal of Physical Chemistry B* 109: 19026–19033.

Cseri, T., S. Békássy, F. Figueras, E. Cseke, L-C. de Menorval, and R. Dutartre. 1995a. Characterization of clay-based K catalysts and their application in Friedel–Crafts alkylation of aromatics. *Applied Catalysis A: General* 132: 141–155.

Cseri, T., S. Békássy, F. Figueras, and S. Rizner. 1995b. Benzylation of aromatics on ion-exchanged clays. *Journal of Molecular Catalysis A: Chemical* 98: 101–107.

Curtis, C.W., J.A. Guin, K.C. Kwon et al. 1983. Selectivity of coal minerals using cyclohexene as a probe reactant. *Fuel* 62: 1341–1346.

Czímerová, A., J. Bujdák, and R. Dohrmann. 2006. Traditional and novel methods for estimating the layer charge of smectites. *Applied Clay Science* 34: 2–13.

d'Espinose de la Caillerie, J-B., V. Gruver, and J.J. Fripiat. 1995. Modification of the surface properties of natural phyllosilicate sepiolite by secondary isomorphic substitution. *Journal of Catalysis* 151: 420–430.

da Silva, F.C., V.F. Ferreira, R.S. Rianelli, and W.C. Perreira. 2002. Natural clays as efficient catalyst for transesterification of β-keto esters with carbohydrate derivatives. *Tetrahedron Letters* 43: 1165–1168.

Dabbagh, H.A., A. Teimouri, and A.N. Chermahini. 2007. Environmentally friendly efficient synthesis and mechanism of triazenes derived from cyclic amines on clays, HZSM-5 and sulfated zirconia. *Applied Catalysis B: Environmental* 76: 24–33.

Dahlgren, R.A., M. Saigusa, and F.C. Ugolini. 2004. The nature, properties and management of volcanic soils. *Advances in Agronomy* 82: 113–181.

Damyanova, S., L. Daza, and K.L.G. Fierro. 1996. Surface and catalytic properties of lanthanum-promoted Ni/sepiolite catalysts for styrene hydrogenation. *Journal of Catalysis* 159: 150–161.

Dandy, A.J. and M.S. Nadiye-Tabbiruka. 1975. The effect of heating in vacuo on the microporosity of sepiolite. *Clays and Clay Minerals* 23: 428–430.

Dandy, A.J. and M.S. Nadiye-Tabbiruka. 1982. Surface properties of sepiolite from Amboseli, Tanzania, and its catalytic activity for ethanol decomposition. *Clays and Clay Minerals* 30: 347–352.

Dasgupta, S. and B. Török. 2008. Application of clay catalysts in organic synthesis. A review. *Organic Preparations and Procedures International* 40: 1–65.

Davis, J.B. and J.P. Stanley. 1982. Catalytic effect of smectite clays in hydrocarbon generation revealed by pyrolysis-gas chromatography. *Journal of Analytical and Applied Pyrolysis* 4: 227–240.

De, A.U. and S.P. Srivastava. 1975. Isomerization of α-pinene to camphene. *Journal of the Indian Chemical Society* 52: 164–165.

de Faria, E.H., G.P. Ricci, L. Marçal et al. 2012. Green and selective oxidation reactions catalyzed by kaolinite covalently grafted with Fe(III)pyridine-carboxylate complexes. *Catalysis Today* 187: 135–149.

de Jesus, P.C., P.L.F. da Silva, J.J. João, and M. G. Nascimento. 1998. Enantioselective esterification of 2-methylpentanoic acid catalysed via immobilized lipases in chrysotile and microemulsion-based gels. *Synthetic Communications* 28: 2893–2901.

De la Calle, C., J. Dubernat, H. Suquet, H. Pezerat, J. Gaultier, and J. Mamy. 1976. Crystal structure of two-layer Mg-vermiculites and Na, Ca-vermiculites. In *Proceedings of the International Clay Conference 1975*, (Ed.) S.W. Bailey, pp. 201–209. Wilmette IL: Applied Publishing.

de Lima, C., P.L.F. da Silva, M. G. Nascimento, and M.C. Rezende. 1996. The use of immobilized lipases on chrysotile for esterification reactions. *Journal of the Brazilian Chemical Society* 7: 173–175.

de Oliveira, A. N., L.R. da Silva Costa, L.H. de Oliveira Pires et al. 2013. Microwave-assisted preparation of a new esterification catalyst from wasted *flint* kaolin. *Fuel* 103: 626 – 631.

de Souza Santos, P., H. de Souza Santos, and G.W. Brindley. 1966. Mineralogical studies of kaolinite-halloysite clays: IV. A platy mineral with structural swelling and shrinking characteristics. *American Mineralogist* 51: 1640–1648.

De Stefanis, A. and A.A.G. Tomlinson. 2006. Towards designing pillared clays for catalysis. *Catalysis Today* 114: 126–141.

De Waele, J.K., E.F. Vansant, and F.C. Adams. 1983. Laser microprobe mass analysis of N,N-dimethylaniline and catalytic oxidation products adsorbed on asbestos fiber surfaces. *Mikrochimica Acta* III: 367–384.

del Rey-Perez-Caballero, F.J. and G. Poncelet. 2000. Microporous 18 Å Al-pillared vermiculites: Preparation and characterization. *Microporous and Mesoporous Materials* 37: 313–327.

del Rey-Perez-Caballero, F.J., M.L. Sanchez-Henao, and G. Poncelet. 2000. Hydroisomerization of octane on Pt/Al-pillared vermiculite and phlogopite, and comparison with zeolites. *Studies in Surface Science and Catalysis* 130: 2417–2422.

Delineau, T., T. Allard, J.-P. Muller, O. Barres, J. Yvon, and J-M. Cases. 1994. FTIR reflectance *vs.* EPR studies of structural iron in kaolinites. *Clays and Clay Minerals* 42: 308–320.

Detellier, C. and R.A. Schoonheydt. 2014. From platy kaolinite to nanorolls. *Elements* 10: 201–206.

Ding, Z., J.T. Kloprogge, R.L. Frost, G.Q. Lu, and H.Y. Zhu. 2001. Porous clays and pillared clays-based catalysts. Part 2. A review of the catalytic and molecular sieve applications. *Journal of Porous Materials* 8: 273–293.

Divakar, D., D. Manikandan, and T. Sivakumar. 2008. Tetra silicic mica—A synthetic support for nanoparticle generation and catalytic applications. *Catalysis Communications* 9: 2433–2438.

Dixon, J.B. 1989. Kaolinite and serpentine group minerals. In *Minerals in Soil Environments*, 2nd ed., (Eds.) J.B. Dixon and S.B. Weed, pp. 467–525. Madison, WI: Soil Science Society of America.

Dixon, J.B. and S.B. Weed (Eds.). 1989. *Minerals in Soil Environments*, 2nd ed. Madison, WI: Soil Science Society of America.

do Nascimento, L.A.S., R.S. Angélica, C.E.F. da Costa, J.R. Zamian, and G.N. da Rocha Filho. 2011a. Comparative study between catalysts for esterification prepared from kaolins. *Applied Clay Science* 51: 267–273.

do Nascimento, L.A.S., L.M.Z. Tito, R.S. Angélica, C.E.F. da Costa, J.R. Zamian, and G.N. da Rocha Filho. 2011b. Esterification of oleic acid over solid acid catalysts prepared from Amazon *flint* kaolin. *Applied Catalysis B: Environmental* 101: 495–503.

Douglas, L.A. 1989. Vermiculites. In *Minerals in Soil Environments*, 2nd ed., (Eds.) J.B. Dixon and S.B. Weed, pp. 635–674. Madison, WI: Soil Science Society of America.

Du, M., B. Guo, and D. Jia. 2010. New emerging applications of halloysite nanotubes: A review. *Polymer International* 59: 574–582.

Eastman, M.P. and T.L. Porter. 2000. Polymerization of organic monomers and biomolecules on hectorite. In *Polymer-Clay Nanocomposites*, (Eds.) T.J. Pinnavaia and G.W. Beall, pp. 65–93. Chichester, UK: John Wiley & Sons.

Edwards, D.G., A.M. Posner, and J.P. Quirk. 1966. Repulsion of chloride ions by negatively charged clay surfaces. *Transactions of the Faraday Society* 61: 2808–2823.

Egashira, K. and S. Aomine. 1974. Effects of drying and heating on the surface area of allophane and imogolite. *Clay Science* 4: 231–242.

Eisma, E. and J.W. Jurg. 1969. Fundamental aspects of the generation of petroleum. In *Organic Geochemistry*, (Eds.) G. Eglinton and M.J.T. Murphy, pp. 676–698. Berlin, Germany: Springer-Verlag.

El-Amamy, M.M. and T. Mill. 1984. Hydrolysis kinetics of organic chemicals on montmorillonite and kaolinite surfaces as related to moisture content. *Clays and Clay Minerals* 32: 67–73.

Eltantawy, I.M. and P.W. Arnold. 1973. Catalytic decomposition of organic molecules by clays. *Nature Physical Science* 244: 144.

Endell, J., R. Zorn, and U. Hofmann. 1941. Über die Prüfung auf Montmorillonit mit Benzidin. *Angewandte Chemie* 54: 376–377.

Ertem, G., A. Steudel, K. Emmerich, G. Lagaly, and R. Schuhmann. 2010. Correlation between the extent of catalytic activity and charge density of montmorillonites. *Astrobiology* 10: 743–749.

Escudey, M. and G. Galindo. 1983. Effect of iron oxide coatings on electrophoretic mobility and dispersion of allophane. *Journal of Colloid and Interface Science* 93: 78–83.

Eslinger, E.V., P. Highsmith, D. Albers, and B. De Mayo. 1979. Role of iron reduction in the conversion of smectite to illite in bentonites in the Disturbed Belt, Montana. *Clays and Clay Minerals* 27: 327–338.

Espitalié, J., M. Madec, and B. Tissot. 1980. Role of mineral matrix in kerogen pyrolysis: Influence on petroleum generation and migration. *American Association of Petroleum Geologists Bulletin* 64: 59–66.

Fachini, A. and I. Joekes. 2002. Interaction of sodium dodecylbenzenesulfonate with chrysotile fibers. Adsorption or catalysis? *Colloids and Surfaces A: Physicochemical and Engineering Aspects* 201: 151–160.

Fachini, A. and I. Joekes. 2007. Use of chrysotile fibres in the degradation of cationic and non-ionic surfactants in aqueous solutions. *Clean* 35: 100–103.

Fachini, A., M.A. Mendes, I. Joekes, and M.N. Eberlin. 2007. Oxidation of sodium dodecylbenzenesulfonate with chrysotile: On-line monitoring by membrane introduction mass spectrometry. *Journal of Surfactants and Detergents* 10: 207–210.

Fahn, R. 1973. Einfluβ der Struktur und der Morphologie von Bleicherden auf die Bleichwirkung bei Ölen und Fetten. *Fette, Seifen, Anstrichmittel* 75: 77–82.

Falini, G., E. Foresti, M. Gazzano et al. 2004. Tubular-shaped stoichiometric chrysotile nanocrytals. *Chemistry—A European Journal* 10: 3043–3049.

Falini, G., E. Foresti, I.G. Lesci, B. Lunelli, P. Sabatino, and N. Roveri. 2006. Interaction of bovine serum albumin with chrysotile: Spectroscopic and morphological studies. *Chemistry—A European Journal* 12: 1968–1974.

Fanning, D.S., V.Z. Keramidas, and M.A. El-Desoky. 1989. Micas. In *Minerals in Soil Environments*, 2nd ed., (Eds.) J.B. Dixon and S.B. Weed, pp. 551–634. Madison, WI: Soil Science Society of America.

Farmer, V.C. 1974. *The Infrared Spectra of Minerals*. London, UK: Mineralogical Society.

Farmer, V.C. 1978. Water on particle surfaces. In *The Chemistry of Soil Constituents*, (Eds.) D.J. Greenland and M.H.B. Hayes, pp. 405–448. Chichester, UK: John Wiley & Sons.

Farmer, V.C., A.R. Fraser, J.D. Russell, and N. Yoshinaga. 1977. Recognition of imogolite structures in allophanic clays by infrared spectroscopy. *Clay Minerals* 12: 55–57.

Farmer, V.C. and M.M. Mortland. 1965. Infrared study of complexes of ethylamine with ethylammonium and copper ions in montmorillonite. *Journal of Physical Chemistry* 69: 683–686.

Farmer, V.C. and J.D. Russell. 1964. The infrared spectra of layer silicates. *Spectrochimica Acta* 20: 1149–1173.

Farmer, V.C. and J.D. Russell. 1971. Interlayer complexes in layer silicates. The structure of water in lamellar ionic solutions. *Transactions of the Faraday Society* 67: 2737–2749.

Fernandes, C.I., C.D. Nunes, and P.D. Vaz. 2012. Clays in organic synthesis—Preparation and catalytic applications. *Current Organic Synthesis* 9: 670–694.

Fernandez Alvarez, T. 1978. Efecto de la deshidratacion sobre las propiedades adsorbentes de la palygorskita y sepiolita. I. Adsorcion de nitrogeno. *Clay Minerals* 13: 325–335.

Fernandez Hernandez, M.N. and T. Fernandez Alvarez. 1983. Effect of dehydration on the adsorption properties of sepiolite and palygorskite. II. Adsorption of organic substances. *Anales de química, Serie B* 79: 342–347.

Ferris, A.P. and W.B. Jepson. 1975. The exchange capacity of kaolinite and the preparation of homoionic clays. *Journal of Colloid and Interface Science* 51: 245–259.

Fieldes, M. and G.G.C. Claridge. 1975. Allophane. In *Soil Components*, Vol. 2. Inorganic Components, (Ed.) J.E. Gieseking, pp. 351–393. Berlin, Germany: Springer-Verlag.

Figueras, F. 1988. Pillared clays as catalysts. *Catalysis Reviews—Science and Engineering* 30: 457–499.

Findik, S. and G. Gündüz. 1997. Isomerization of α-pinene to camphene. *Journal of the American Oil Chemists' Society* 74: 1145–1151.

Fiore, S., J. Cuadros, and F.J. Huertas (Eds.). 2013. *Interstratified Clay Minerals. Origin, Characterization & Geochemical Significance*. Bari, Italy: Digilabs.

Flessner, U., D.L. Jones, J. Rozière et al. 2001. A study of the surface acidity of acid-treated montmorillonite clay catalysts. *Journal of Molecular Catalysis A: Chemical* 168: 247–256.

Fletcher, J.C.Q., M. Kojima, and C.T. O'Connor. 1986. Acidity and catalytic activity of synthetic mica-montmorillonite Part II: Propene oligomerization. *Applied Catalysis* 28: 181–191.

Foresti, E., E. Fornero, I.G. Lesci, C. Rinaudo, T. Zuccheri, and N. Roveri. 2009. Asbestos health hazard: A spectroscopic study of synthetic geoinspired Fe-doped chrysotile. *Journal of Hazardous Materials* 167: 1070–1079.

Foresti, E., M.F. Hochella, Jr., H. Kornishi et al. 2005. Morphological and chemical/physical characterization of Fe-doped synthetic chrysotile nanotubes. *Advanced Functional Materials* 15: 1009–1016.

Fornés, V., J.A. Rausell-Colom, A. Hidalgo, and J.M. Serratosa. 1973. Étude par spectroscopie infrarouge d'une reaction de condensation dans l'espace interlamellaire de la vermiculite. *Comptes rendus de l'Académie des Sciences Série B* 277: 635–637.

Foscolos, A.E. and T.G. Powell. 1979. Mineralogical and geochemical transformation of clays during burial-diagenesis (catagenesis): Relation to oil generation. In *International Clay Conference 1978. Developments in Sedimentology 27*, (Eds.) M.M. Mortland and V.C. Farmer, pp. 261–270. Amsterdam, the Netherlands: Elsevier.

Foscolos, A.E., T.G. Powell, and P.R. Gunther. 1976. The use of clay minerals and inorganic and organic indicators for evaluating the degree of diagenesis and oil generating potential of shales. *Geochimica et Cosmochimica Acta* 40: 953–966.

Fowkes, F.M., H.A. Benesi, L.B. Ryland et al. 1960. Clay-catalyzed decomposition of insectides. *Journal of Agricultural and Food Chemistry* 8: 203–210.

Fripiat, J.J. (Ed.). 1981. *Advanced Techniques for Clay Mineral Analysis*. Amsterdam, the Netherlands: Elsevier.

Fripiat, J.J. 1986. Internal surface of clays and constrained chemical reactions. *Clays and Clay Minerals* 34: 501–506.

Fripiat, J.J. and M.I. Cruz-Cumplido. 1974. Clays as catalysts for natural processes. *Annual Review of Earth and Planetary Sciences* 2: 239–256.

Fripiat, J.J. and M. Della Faille. 1967. Surface properties and texture of chrysotiles. *Clays and Clay Minerals* 15: 305–320.

Fubini, B., L. Mollo, S. Bodoardo, B. Onida, D. Oberson, and C. Lafuma. 1997. Evaluation of the surface acidity of some phyllosilicates in relation to their inactivating activity toward the enzyme human leucocyte elastase. *Langmuir* 13: 919–927.

Fukushima, M., A. Miura, M. Sasaki, and K. Izumo. 2009. Effect of an allophanic soil on humification reactions between catechol and glycine: Spectroscopic investigations of reaction products. *Journal of Molecular Structure* 917: 142–147.

Fysh, S.A., J.D. Cashion, and P.E. Clark. 1983. Mössbauer effect studies of iron in kaolin. I. Structural iron. *Clays and Clay Minerals* 31: 285–292.

Gaite, J.-M, P. Ermakoff, T. Allard, and J.-P. Muller. 1997. Paramagnetic Fe^{3+}: A sensitive probe for disorder in kaolinite. *Clays and Clay Minerals* 45: 496–505.

Gajare, A.S., N.S. Shaikh, G.K. Jnaneshwara, V.H. Deshpande, T. Ravindranathan, and A.V. Bedekar. 2000. Clay catalyzed conversion of isatoic anhydride to 2-(o-aminophenyl)oxazolines. *Journal of the Chemical Society, Perkin Transactions I*, 1(6): 999–1001.

Gajare, A.S., M.S. Shingare, V.R. Kulkarni, N.B. Barhate, and R.D. Wakharkar. 1998. Natural kaolinitic clay and EPZG[R]: Efficient solid catalysts for selective regeneration of carboxylic acids from their corresponding allyl or cinnamyl esters. *Synthetic Communications* 28: 25–33.

Galán, E. 2006. Genesis of clay minerals. In *Handbook of Clay Science*. Developments in Clay Science, Vol 1, (Eds.) F. Bergaya, B.K.G. Theng, and G. Lagaly, pp. 1129–1162. Amsterdam, the Netherlands: Elsevier.

Galwey, A.K. 1969. Heterogeneous reactions in petroleum genesis and maturation. *Nature* 223: 1257–1260.

Galwey, A.K. 1977. Compensation effect in heterogeneous catalysis. In *Advances in Catalysis*, Vol. 26, (Eds.) D.D. Eley, H. Pines, and P.B. Weisz, pp. 247–322. New York: Academic Press.

Gao, X. and J. Xu. 2006. A new application of clay-supported vanadium oxide catalyst to selective hydroxylation of benzene to phenol. *Applied Clay Science* 33: 1–6.

Gard, J.A. (Ed.). 1971. *The Electron-Optical Investigation of Clays*. London, UK: Mineralogical Society.

Garrett, W.G. and G.F. Walker. 1962. Swelling of some vermiculite-organic complexes in water. *Clays and Clay Minerals* 9: 557–567.

Garrett, W.G. and G.F. Walker. 1967. Chemical exfoliation of vermiculite and the production of colloidal dispersions. *Science* 156: 385–387.

Garrido-Ramírez, E.G., M.L. Mora, J.F. Marco, and M.S. Ureta-Zañartu. 2013. Charaterization of nanostructured allophane clays and their use as support of iron species in a heterogeneous electro-Fenton system. *Applied Clay Science* 86: 153–161.

Garrido-Ramírez, E.G., M.V. Sivaiah, J. Barrault et al. 2012. Catalytic wet peroxide oxidation of phenol over iron or copper oxide-supported allophane clay materials: Influence of catalyst SiO_2/Al_2O_3 ratio. *Microporous and Mesoporous Materials* 162: 189–198.

Garrido-Ramírez, E.G., B.K.G. Theng, and M.L. Mora. 2010. Clays and oxide minerals as catalysts in Fenton-like reactions. *Applied Clay Science* 47: 182–192.

Garrone, E. and B. Bonelli. 2016. Imogolite for catalysis and adsorption. In *Nanosize Tubular Clay Minerals: Halloysite and Imogolite*. Developments in Clay Science, Vol. 7, (Eds.) P. Yuan, A. Thill, and F. Bergaya, pp. 672–707. Amsterdam, the Netherlands: Elsevier.

Gates, W.P., P. Komadel, J. Madejová, J. Bujdák, J.W. Stucki, and R.J. Kirkpatrick. 2000. Electronic and structural properties of reduced-charge montmorillonites. *Applied Clay Science* 16: 257–271.

Gayer, F.H. 1933. The catalytic polymerization of propylene. *Industrial and Engineering Chemistry* 25: 1122–1127.

Gazzano, E., E. Foresti, I.G. Lesci, M. Tomatis, and C. Riganti. 2005. Different cellular responses evoked by natural and stoichiometric synthetic chrysotile asbestos. *Toxicology and Applied Pharmacology* 206: 356–364.

Gerstl, Z. and B. Yaron. 1981. Stability of parathion on attapulgite as affected by structural and hydration changes. *Clays and Clay Minerals* 29: 53–59.

Giannetti, J.P. and D.C. Fisher. 1975. Raffinate hydrocracking with palladium-nickel-containing synthetic mica-montmorillonite catalysts. In *Hydrocracking and Hydrotreating*. ACS Symposium Series 20, (Eds.) J.W. Ward and S.A. Qader, pp. 52–64. Washington, DC: American Chemical Society.

Giese, R.F. 1975. The effect of F/OH substitution on some layer silicate minerals. *Zeitschrift für Kristallographie* 141: 138–144.

Giese, R.F. 1977. The influence of hydroxyl orientation, stacking sequence, and ionic substitutions on the interlayer bonding of micas. *Clays and Clay Minerals* 25: 102–104.

Giese, R.F. 1978. The electrostatic interlayer forces of layer structure minerals. *Clays and Clay Minerals* 26: 51–57.

Giese, R.F. 1988. Kaolin minerals: Structures and stabilities. In *Hydrous Phyllosilicates (Exclusive of Micas)*. Reviews in Mineralogy, Vol. 19, (Ed.) S.W. Bailey, pp. 29–66. Washington, DC: Mineralogical Society of America.

Giese, R.F. and P. Datta. 1973. Hydroxyl orientations in kaolinite, dickite, and nacrite. *American Mineralogist* 58: 471–479.

Gieseking, J.E. (Ed.). 1975. *Soil Components*, Vol. 2. Inorganic Components. Berlin, Germany: Springer-Verlag.

Gil, A., L.M. Gandía, and M.A. Vicente. 2000. Recent advances in the synthesis and catalytic applications of pillared clays. *Catalysis Reviews—Science and Engineering* 42: 145–212.

Gil, A., S.A. Korili, R. Trujillano, and M.A. Vicente (Eds.). 2010. *Pillared Clays and Related Catalysts*. New York: Springer.

Gil, A., S.A. Korili, and M.A. Vicente. 2008. Recent advances in the control and characterization of the porous structure of pillared clay catalysts. *Catalysis Reviews—Science and Engineering* 50: 153–221.

Glaeser, R. and J. Méring. 1971. Migration des cations Li dans les smectites di-octahedriques (effet Hofmann-Klemen). *C R Hebdomadaires des Séances de l'Académie des Sciences Serie. D* 273: 2399–2402.

Goldstein, T.P. 1983. Geocatalytic reactions in formation and maturation of petroleum. *American Association of Petroleum Geologists Bulletin* 67: 152–159.

Goncharuk, V.V., N.G. Vasil'ev, M.V. Sychev, A.N. Chernaga, and M.A. Lukashevich. 1984. Study of physico-chemical, catalytic, and acidic properties of the lithium form of kaolinite in the dehydration of isopropyl alcohol. *Kinetika i Kataliz (Kinetics and Catalysis)* 25: 962–968.

Goodman, B.A. 1978. The Mössbauer spectra of nontronites: Consideration of an alternative assignment. *Clays and Clay Minerals* 26: 176–177.

Goodman, B.A. 1980. Mössbauer spectroscopy. In *Advanced Chemical Methods for Soil and Clay Minerals Research*, (Eds.) J.W. Stucki and W.L. Banwart, pp. 1–92. Dordrecht, the Netherlands: D. Reidel.

Goryaev, M.I., A.F. Artamonov, L.P. Petelina, R. Suleeva, and V.A. Yugai. 1975. Vermiculite-catalyzed isomerization of α-pinene. *Vestnik Akademii Nauk Kazakhskoy SSR*, No.3: pp. 59–62.

Granquist, W.T. and J.V. Kennedy. 1967. Sorption of water at high temperatures on certain clay mineral surfaces. Correlation with lattice fluoride. *Clays and Clay Minerals* 15: 103–117.

Granquist, W.T. and S.S. Pollack. 1967. Clay mineral synthesis. II. A randomly interstratified aluminian montmorillonoid. *American Mineralogist* 52: 212–226.

Greene, R.S.B., A.M. Posner, and J.P. Quirk. 1973. Factors affecting the formation of quasi-crystals of montmorillonite. *Soil Science Society of America Proceedings* 37: 457–460.

Greene-Kelly, R. 1953. Irreversible dehydration in montmorillonite. Part II. *Clay Minerals Bulletin* 2: 52–56.

Greenland, D.J. and M.H.B. Hayes (Eds.). 1978. *The Chemistry of Soil Constituents.* Chichester, UK: John Wiley & Sons.

Greenland, D.J. and C.J.B. Mott. 1978. Surfaces of soil particles. In *The Chemistry of Soil Constituents,* (Eds.) D.J. Greenland and M.H.B. Hayes, pp. 321–353. Chichester, UK: John Wiley & Sons.

Grenall, A. 1948. Montmorillonite cracking catalyst. *Industrial & Engineering Chemistry* 40: 2148–2151.

Grim, R.E. 1962. *Applied Clay Mineralogy.* New York: McGraw-Hill.

Grim, R.E. 1968. *Clay Mineralogy,* 2nd ed. New York: McGraw-Hill.

Grim, R.E., R.H. Bray, and W.F. Bradley. 1937. The mica in argillaceous sediments. *American Mineralogist* 22: 813–829.

Grim, R.E. and N. Güven. 1978. *Bentonites—Geology, Mineralogy, Properties and Uses.* Developments in Sedimentology 24. Amsterdam, the Netherlands: Elsevier.

Gruver, V., A. Sun, and J.J. Fripiat. 1995. Catalytic properties of aluminated sepiolite in ethanol conversion. *Catalysis Letters* 34: 359–364.

Gualtieri, A.F., A. Moen, and D.G. Nicholson. 2000. XANES study of the local environment of iron in natural kaolinites. *European Journal of Mineralogy* 12: 17–23.

Guggenheim, S. 2011. An overview of order/disorder in hydrous phyllosilicates. In *Layered Mineral Structures and their Application in Advanced Technologies.* EMU Notes in Mineralogy, Vol. 11, (Eds.) M.F. Brigatti and A. Mottana, pp. 72–121. London, UK: Mineralogical Society.

Guggenheim, S., J.M. Adams, D.C. Bain et al. 2006. Summary of recommendations of nomenclature committees relevant to clay mineralogy: Report of the Association Internationale pour l'Étude des Argiles (AIPEA) Nomenclature Committee for 2006. *Clay Minerals* 41: 863–877; *Clays and Clay Minerals* 54: 761–772.

Guggenheim, S. and R.T. Martin. 1995. Definition of clay and clay mineral: Joint report of the AIPEA nomenclature and CMS nomenclature committees. *Clay Minerals* 30: 257–259; *Clays and Clay Minerals* 43: 255–256.

Guo, H., H. Zhang, F. Peng et al. 2015. Mixed alcohols synthesis from syngas over activated palygorskite supported Cu-Fe-Co based catalysts. *Applied Clay Science* 111: 83–89.

Gupta, G.C. and W.U. Malik. 1969a. Transformation of montmorillonite to nickel-chlorite. *Clays and Clay Minerals* 17: 233–239.

Gupta, G.C. and W.U. Malik. 1969b. Chloritization of montmorillonite by its coprecipitation with magnesium hydroxide. *Clays and Clay Minerals* 17: 331–338.

Gurwitsch, L. 1912. Adsorption. *Zeitschrift für Chemie und Industrie der Kolloide* 11: 17–19.

Gurwitsch, L. 1923. Zur Kenntnis der heterogenen Katalyse. *Zeitschrift für physikalische Chemie Frankfurt* 107: 235–248.

Güven, N. 1988. Smectites. In *Hydrous Phyllosilicates (Exclusive of Micas).* Reviews in Mineralogy, Vol. 19, (Ed.) S.W. Bailey, pp. 497–559. Washington, DC: Mineralogical Society of America.

Güven, N. 2009. Bentonites—Clays for molecular engineering. *Elements* 5: 89–92.

Haden, W.L. and I.A. Schwint. 1967. Attapulgite—Its properties and applications. *Industrial & Engineering Chemistry* 59: 59–69.

Hall, P.L., G.J. Churchman, and B.K.G. Theng. 1985. Size distribution of allophane unit particles in aqueous suspensions. *Clays and Clay Minerals* 33: 345–349.

Hanif, M., F. Jabbar, S. Sharif, G. Abbas, A. Forooq, and M. Aziz. 2016. Halloysite nanotubes as a new drug-delivery system: A review. *Clay Minerals* 51: 469–477.

Hanna, B. and F.H. Khalil. 1982. Relation between the electric properties and the catalytic activity of some Egyptian clays. *Surface Technology* 17: 61–68.

Hansford, R.C. 1952. Chemical concepts of catalytic cracking. In *Advances in Catalysis and Related Subjects,* Vol. IV, (Eds.) W.G. Frankenburg, E.K. Rideal, and V.I. Komarewsky, pp. 1–30. New York: Academic Press.

Harsh, J. 2012. Poorly crystalline aluminosilicate minerals. In *Handbook of Soil Sciences. Properties and Processes,* 2nd ed., (Eds.) P.M. Huang, Y. Li, and M.E. Sumner, pp. 23-1–23-13. Boca Raton, FL: CRC Press.

Hashizume, H. and B.K.G. Theng. 1999. Adsorption of DL-alanine by allophane: Effect of pH and unit particle aggregation. *Clay Minerals* 34: 233–238.

Hashizume, H. and B.K.G. Theng. 2007. Adenine, adenosine, ribose, and 5'-AMP adsorption to allophane. *Clays and Clay Minerals* 55: 599–605.

Hashizume, H., B.K.G. Theng, and A. Yamagishi. 2002. Adsorption and discrimination of alanine and alanyl-alanine enantiomers by allophane. *Clay Minerals* 37: 551–557.

Hattori, H., D.L. Milliron, and J.W. Hightower. 1973. Mechanistic studies of hydrocarbon reactions over synthetic mica montmorillonite catalysts. *American Chemical Society, Division of Petroleum Chemisry Preprints* 28: 33.

Hawthorne, D.G. and D.H. Solomon. 1972. Catalytic activity of sodium kaolinites. *Clays and Clay Minerals* 20: 75–78.

Hay, R.G., C.W. Montgomery, and J. Coull. 1945. Catalytic isomerization of 1-hexene. *Industrial and Engineering Chemistry* 37: 335–339.

Hayashi, H., R. Otsuka, and N. Imai. 1969. Infrared study of sepiolite and palygorskite on heating. *American Mineralogist* 53: 1613–1624.

He, H., L. Ma, J. Zhu, R.L. Frost, B.K.G. Theng, and F. Bergaya. 2014. Synthesis of organoclays: A critical review and some unresolved isssues. *Applied Clay Science* 100: 22–28.

He, X., A. Tang, H. Yang, and J. Ouyang. 2011. Synthesis and catalytic activity of doped TiO_2-palygorskite composites. *Applied Clay Science* 53: 80–84.

Heinerman, J.J.L., I.L.C. Freriks, J. Gaaf, G.T. Pott, and J.G.F. Coolegem. 1983. The catalytic activity of nickel-substituted mica montmorillonite. *Journal of Catalysis* 80: 145–153.

Heller-Kallai, L. 2002. Clay catalysis in reactions of organic matter. In *Organo-Clay Complexes and Interactions*, (Eds.) S. Yariv and H. Cross, pp. 567–613. New York: Marcel Dekker.

Heller-Kallai, L. 2013. Thermally modified clay minerals. In *Handbook of Clay Science*, 2nd ed. Developments in Clay Science, Vol. 5A, (Eds.) F. Bergaya and G. Lagaly, pp. 411–433. Amsterdam, the Netherlands: Elsevier.

Heller-Kallai, L., Z. Aizenshtat, and I. Miloslavski. 1984. The effect of various clay minerals on the thermal decomposition of stearic acid under "bulk flow" conditions. *Clay Minerals* 19: 779–788.

Hendricks, S.B. and M.E. Jefferson. 1938. Structures of kaolin and talc-pyrophyllite hydrates and their bearing on water sorption of the clays. *American Mineralogist* 23: 863–875.

Hénin, S. and S. Caillère.1975. Fibrous minerals. In *Soil Components*, Vol. 2. Inorganic Components, (Ed.) J.E. Gieseking, pp. 335–349. Berlin, Germany: Springer-Verlag.

Henmi, T. and K. Wada. 1974. Surface acidity of imogolite and allophane. *Clay Minerals* 10: 231–245.

Henmi, T. and K. Wada. 1976. Morphology and composition of allophane. *American Mineralogist* 61: 379–390.

Herney-Ramirez, J., M.A. Vicente, and L.M. Madeira. 2010. Heterogeneous photo-Fenton oxidation with pillared clay-based catalysts for wastewater treatment: A review. *Applied Catalysis B: Environmental* 98: 10–26.

Herrero, C.P., J. Sanz, and J.M. Serratosa. 1985. Si, Al distribution in micas: Analysis by high-resolution ^{29}Si NMR spectroscopy. *Journal of Physics C: Solid State Physics* 18: 13–22.

Herrero, J., C. Blanco, and L.A. Oro. 1989. Preparation of rhodium-phyllosilicate catalysts without leaching in liquid-phase 1-hexene hydrogenation. *Applied Organometallic Chemistry* 3: 553–555.

Herrington, T.M., A.Q. Clarke, and J.C. Watts. 1992. The surface charge of kaolin. *Colloids and Surfaces* 68: 161–169.

Hingston, F.J., R.J. Atkinson, A.M. Posner, and J.P. Quirk. 1967. Specific adsorption of anions. *Nature* 215: 1459–1461.

Hingston, F.J., A.M. Posner, and J.P. Quirk. 1972. Anion adsorption by goethite and gibbsite. I. The role of the proton in determining adsorption envelopes. *Journal of Soil Science* 23: 177–192.

Hofmann, U. 1968. On the chemistry of clays. *Angewandte Chemie International Edition* 7: 681–692.

Hofmann, U. and R. Klemen. 1950. Verlust der Austauschfähigkeit von Lithiumionen an Bentonit durch Erhitzung. *Zeitschrift für anorganische und allgemeine Chemie* 262: 95–99.

Holmes, J. and G.A. Mills. 1951. Aging of a bentonitic cracking catalyst in air or steam. *Journal of Physical and Colloid Chemistry* 55: 1302–1320.

Horikawa, Y. 1975. Electrokinetic phenomena of aqueous suspensions of allophane and imogolite. *Clay Science* 43: 255–263.

Horsfield, B. and A.G. Douglas. 1980. The influence of minerals on the pyrolysis of kerogens. *Geochimica et Cosmochimica Acta* 44: 1119–1131.

Hosterman, J.W. 1973. Clays. *U.S. Geological Survey Professional Paper* 820: 123–131.

Hougardy, J., W.E.E. Stone, and J.J. Fripiat. 1976. NMR study of adsorbed water. 1. Molecular orientation and protonic motions in the two-layer hydrate of a Na-vermiculite. *Journal of Chemical Physics* 64: 3840–3851.

Huizinga, B.J., E. Tannenbaum, and I.R. Kaplan. 1987. The role of minerals in the thermal alteration of organic matter—III. Generation of bitumen in laboratory experiments. *Organic Geochemistry* 11: 591–604.

Huo, C. and H. Yang. 2013. Preparation and enhanced photocatalytic activity of Pd-CuO/palygorskite nanocomposites. *Applied Clay Science* 74: 87–94.

Huo, X., L. Wu, L. Liao, Z. Xia, and L. Wang. 2012. The effect of interlayer cations on the expansion of vermiculite. *Powder Technology* 224: 241–246.

Hussin, F., M.K. Aroua, and W.M.A.W. Daud. 2011. Textural characteristics, surface chemistry and activation of bleaching earth: A review. *Chemical Engineering Journal* 170: 90–116.

Imai, N., R. Otsuka, H. Kashide, and H. Hayashi. 1969. Dehydration of palygorskite and sepiolite from the Kuzuu district, Tochigi Pref., Central Japan. In *Proceedings of the International Clay Conference 1969*, Vol.1, (Ed.) L. Heller, pp. 99–108. Jerusalem, Israel: Israel Universities Press.

Imamura, S., Y. Hayashi, K. Kajiwara, H. Hoshino, and C. Kaito. 1993. Imogolite: A possible new type of shape-selective catalyst. *Industrial & Engineering Chemistry Research* 32: 600–603.

Imamura, S., T. Kokubu, T. Yamashita, Y. Okamoto, K. Kajiwara, and H. Kanai. 1996. Shape-selective copper-loaded imogolite catalyst. *Journal of Catalysis* 160: 137–139.

Jackson, W.W. and J. West. 1930. The crystal structure of muscovite–$KAl_2(AlSi_3)O_{10}(OH)_2$. *Zeitschrift für Kristallographie* 76: 211–227.

Jackson, W.W. and J. West. 1933. The crystal structure of muscovite–$KAl_2(AlSi_3)O_{10}(OH)_2$. *Zeitschrift für Kristallographie* 85: 160–164.

Jacobs, P.A. 1984. The measurement of surface acidity. In *Characterization of Heterogeneous Catalysts*, (Ed.) F. Delannay, pp. 367–404. New York: Marcel Dekker.

Janek, M., P. Komadel, and G. Lagaly. 1997. Effect of autotransformation on the layer charge of smectites determined by the alkylammonium method. *Clay Minerals* 32: 623–632.

Jang, B.-S., K.-H. Cho, K.-H. Kim, and D. -W. Park. 2005. Degradation of polystyrene using montmorillonite clay catalysts. *Reaction Kinetics and Catalysis Letters* 86: 75–82.

Jayachandran, B., P. Phukan, T. Daniel, and A. Sudalai. 2006. Natural kaolinitic clay: A remarkable catalyst for highly regioselective chlorination of arenes with Cl_2 or SO_2Cl_2. *Indian Journal of Chemistry* 45B: 972–975.

Jaynes, W.F. and J.M. Bigham. 1987. Charge reduction, octahedral charge, and lithium retention in heated, Li-saturated smectites. *Clays and Clay Minerals* 35: 440–448.

Jaynes, W.F. and S.A. Boyd. 1991. Hydrophobicity of siloxane surfaces in smectites as revealed by aromatic hydrocarbon adsorption from water. *Clays and Clay Minerals* 39: 428–436.

Jepson, W.B. 1984. Kaolins: Their properties and uses. *Philosophical Transactions of the Royal Society of London A* 311: 411–432.

Jiang, J., Y. Xu, C. Duanmu, X. Gu, and J. Chen. 2014. Preparation and catalytic properties of sulfonated carbon-palygorskite solid acid catalyst. *Applied Clay Science* 95: 260–264.

Jnaneshwara, G.K., V.H. Despande, and A.V. Bedekar. 1999. Clay-catalyzed conversion of 2,2-disubstituted malononitriles to 2-oxazolines: Towards unnatural amino acids. *Journal of Chemical Research (S)* 4: 252–253.

Jnaneshwara, G.K., V.H. Despande, M. Lalithambika, T. Ravindranathan, and A.V. Bedekar. 1998. Natural kaolinitic clay catalyzed conversion of nitriles to 2-oxazolines. *Tetrehedron Letters* 39: 459–462.

Johns, W.D. 1982. The role of the clay mineral matrix in petroleum generation during burial diagenesis. In *International Clay Conference 1981*. Developments in Sedimentology 35, (Eds.) H. van Olphen and F. Veniale, pp. 655–664. Amstrdam, the Netherlands: Elsevier.

Johns, W.D. and T.E. McKallip. 1989. Burial diagenesis and specific catalytic activity of illite-smectite clays from Vienna Basin, Austria. *American Association of Petroleum Geologists Bulletin* 73: 472–482.

Johns, W.D. and A. Shimoyama. 1972. Clay minerals and petroleum-forming reactions during burial and diagenesis. *American Association of Petroleum Geologists Bulletin* 56: 2160–2167.

Johnston, C.T. 2010. Probing the nanoscale architecture of clay minerals. *Clay Minerals* 45: 245–279.

Johnston, J.H. and C.M. Cardile. 1985. Iron sites in nontronite and the effect of interlayer cations from Mössbauer spectra. *Clays and Clay Minerals* 33: 21–30.

Jones, B.F. and E. Galán. 1988. Sepiolite and palygorskite. In *Hydrous Phyllosilicates (Exclusive of Micas)*. Reviews in Mineralogy, Vol. 19, (Ed.) S.W. Bailey, pp. 631–674. Washington, DC: Mineralogical Society of America.

Jones, W., D.T.B. Tennakoon, J.M. Thomas, L.J. Williamson, J.A. Ballantyne, and J.H. Purnell. 1983. The principles of chemical conversion of organic molecules using sheet silicate intercalates. *Proceedings of the Indian Academy of Sciences* 92: 27–41.

Joussein, E., S. Petit, J. Churchman, B. Theng, D. Righi, and B. Delvaux. 2005. Halloysite clay minerals—A review. *Clay Minerals* 40: 383–426.

Juppe, G. and H. Rau. 1969. Attapulgus clay-catalysed thermal decomposition of terphenyls. *Journal of Applied Chemistry* 19: 120–124.

Jurg, J.W. and E. Eisma. 1964. Petroleum hydrocarbons: Generation from a fatty acid. *Science* 144: 1451–1452.

Kaufhold, S., R. Dohrmann, M. Klinkenberg, S. Siegesmund, and K. Ufer. 2010. N_2-BET specific surface area of bentonites. *Journal of Colloid and Interface Science* 349: 275–282.

Kaufhold, S., R. Dohrmann, K. Ufer, and F.M. Meyer. 2002. Comparison of methods for the quantification of montmorillonite in bentonites. *Applied Clay Science* 22: 145–151.

Kaufhold, S., H. Stanjek, D. Penner, and R. Dohrmann. 2011. The acidity of surface groups of dioctahedral smectites. *Clay Minerals* 46: 583–592.

Kaur, N. and D. Kishore. 2012. Montmorillonite: An efficient, heterogeneous and green catalyst for organic synthesis. *Journal of Chemical and Pharmaceutical Research* 4: 991–1015.

Kawai, M., M. Onaka, and Y. Izumi. 1986. Solid acid-catalyzed allylation of acetals and carbonyl compounds with allylic silanes. *Chemistry Letters* 15(3): 381–384.

Kellendonk, F.J.A., J.J.L. Heinerman, and R.A. van Santen. 1987. Clay-activated isomerization reactions. In *Preparative Chemistry Using Supported Reagents*, (Ed.) P. Laszlo, pp. 455–468. New York: Academic Press.

Kermarec, M., M. Patel, P. Rabette, H. Pezerat, and D. Delafosse. 1983. Reactivity and structure of nickel-exchanged Prayssac vermiculite. *Journal of the Chemical Society Faraday Transactions 1* 79: 599–606.

Khanikar, N. and K.G. Bhattacharyya. 2013. Cu(II)-kaolinite and Cu(II)-montmorillonite as catalysts for wet oxidative degradation of 2-chlorophenol, 4-chlorophenol and 2,4-dichlorophenol. *Chemical Engineering Journal* 233: 88–97.

Kibanova, D., M. Trejo, H. Destaillats, and J. Cervini-Silva. 2009. Synthesis of hectorite-TiO_2 and kaolinite-TiO_2 nanocomposites with photocatalytic activity for the degradation of model air pollutants. *Applied Clay Science* 42: 563–568.

Kibby, C.L., F.E. Massoth, and H.E. Swift. 1976. Surface properties of hydrogen-reduced nickel chrysotiles. *Journal of Catalysis* 42: 350–359.

Kirkman, J.H. 1977. Possible structure of halloysite disks and cylinders observed in some New Zealand rhyolitic tephras. *Clay Minerals* 12: 199–216.

Kirkman, J.H. and W.A. Pullar. 1978. Halloysite in late pleistocene rhyolitic tephra beds near Opotiki, coastal Bay of Plenty, North Island, New Zealand. *Australian Journal of Soil Research* 16: 1–8.

Kitagawa, Y. 1971. The "unit particle" of allophane. *American Mineralogist* 56: 465–475.

Kitajima, K. and N. Daimon. 1975. Na-fluor-tetrasilicic mica [$Na_2Mg_{2.5}Si_2O_{10}F_2$] and its swelling characteristics. *Nippon Kagaku Kaishi* 6: 991–995.

Kitayama, Y., M. Kamimura, K-i Wakui, M. Kanamori, T. Kodama, and J. Abe. 1999. Cyclodehydration of diethylene glycol (DEG) catalyzed by clay mineral sepiolite. *Journal of Molecular Catalysis A: Chemical* 142: 237–245.

Kitayama, Y. and A. Michishita. 1981. Catalytic activity of fibrous clay mineral sepiolite for butadiene formation from ethanol. *Journal of the Chemical Society Chemical Communications* 88: 401–402.

Kłapita, Z., A. Gawel, T. Fujita, and N. Iyi. 2003. Structural heterogeneity of alkylammonium-exchanged, synthetic fluorotetrasilicic mica. *Clay Minerals* 38: 151–160.

Kloprogge, J.T., 1998. Synthesis of smectites and porous pillared clay catalysts: A review. *Journal of Porous Materials* 5: 5–41.

Kogure, T. and A. Inoue. 2005a. Determination of defect structures in kaolin minerals by high-resolution transmission electron microscopy (HRTEM). *American Mineralogist* 90: 85–89.

Kogure, T. and A. Inoue. 2005b. Stacking defects and long-period polytypes in kaolin minerals from a hydrothermal deposit. *European Journal of Mineralogy* 17: 465–473.

Kojima, M., J.C.Q. Fletcher, and C.T. O'Connor. 1986. Acidity and catalytic activity of synthetic mica-montmorillonite Part I: An infrared and temperature programmed desorption study. *Applied Catalysis* 28: 169–179.

Komadel, P. 2003. Chemically modified smectites. *Clay Minerals* 38: 127–138.

Komadel, P. and J. Madejová. 2013. Acid activation of clay minerals. In *Handbook of Clay Science*, 2nd ed. Developments in Clay Science, Vol. 5A, (Eds.) F. Bergaya and G. Lagaly, pp. 385–409. Amsterdam, the Netherlands: Elsevier.

Komadel, P., J. Madejová, and J. Bujdák. 2005. Preparation and properties of reduced-charge smectites—A review. *Clays and Clay Minerals* 53: 313–334.

Komusiński, J., L. Stoch, and S.M. Dubiel. 1981. Application of electron paramagnetic resonance and Mössbauer spectroscopy in the investigation of kaolinite-group minerals. *Clays and Clay Minerals* 29: 23–30.

Konwar, D., P.K. Gogoi, P. Gogoi et al. 2008. Esterification of carboxylic acids by acid activated kaolinite clay. *Indian Journal of Chemical Technology* 15: 75–78.

Korichi, S., A. Elias, A, Mefti, and A. Bensmaili. 2012. The effect of microwave irradiation and conventional acid activation on the textural properties of smectite: Comparative study. *Applied Clay Science* 59–60: 76–83.

Kuang, W. and C. Detellier. 2004. Insertion of acetone molecules in the nanostructured tunnels of palygorskite. *Canadian Journal of Chemistry* 82: 1527–1535.

Kuang, W. and C. Detellier. 2005. Structuration of organo-minerals: Nanohybrid materials resulting from the incorporation of alcohols in the tunnels of palygorskite. *Studies in Surface Science and Catalysis* 156: 451–456.

Kuang, W., G.A. Facey, and C. Detellier. 2006. Organo-mineral nanohybrids. Incorporation, coordination and structuration role of acetone molecules in the tunnels of sepiolite. *Journal of Materials Chemistry* 16: 179–185.

Kühnel, R.A., S.J. Van der Gaast, M.A.T.M. Broekmans, and B.K.G. Theng. 2017. Wetting-induced layer contraction in illite and mica-family relatives. *Applied Clay Science* 135: 226–233.

Kumada, K. and H. Kato. 1970. Browning of pyrogallol as affected by clay minerals. *Soil Science and Plant Nutrition* 16: 195–200.

Kumar, B.S., A. Dhakshinamoorthy, and K. Pitchumani. 2014. K10 montmorillonite clays as environmentally benign catalysts for organic reactions. *Catalysis Science & Technology* 4: 2378–2396.

Kunyima, B., K. Viaene, M.M. Hassan Khalil, R.A. Schoonheydt, M. Crutzen, and F.C. De Schryver. 1990. Study of the adsorption and polymerization of functionalized organic ammonium derivatives on a clay surface. *Langmuir* 6: 482–486.

Kunze, G.W. and W.F. Bradley. 1964. Occurrence of a tabular halloysite in a Texas soil. *Clays and Clay Minerals* 12: 523–527.

Kutsuna, S., L. Chen, K. Nohara, K. Takeuchi, and T. Ibusuki. 2002. Heterogeneous decomposition of $CHF_2OCH_2CF_3$ and $CHF_2OCH_2C_2F_5$ over various standard aluminosilica clay minerals in air at 313 K. *Environmental Science & Technology* 36: 3118–3123.

Kutsuna, S., L. Chen, K. Ohno et al. 2003. Laboratory study on heterogeneous decomposition of methyl chloroform on various standard aluminosilica clay minerals as a potential tropospheric sink. *Atmospheric Chemistry and Physics* 3: 1063–1082.

Kyuma, K. and K. Kawaguchi. 1964. Oxidative changes of polyphenols as influenced by allophane. *Soil Science Society of America Proceedings* 28: 371–374.

Laby, R.H. and B.K.G. Theng. 1972. Note on the infrared spectra of ammonium in montmorillonite and vermiculite. *New Zealand Journal of Science* 18: 535–539.

Laby, R.H. and G.F. Walker. 1970. Hydrogen bonding in primary alkylammonium-vermiculite complexes. *Journal of Physical Chemistry* 74: 2369–2373.

Lagaly, G. 1981. Characterization of clays by organic compounds. *Clay Minerals* 16: 1–21.

Lagaly, G. 1982. Layer charge heterogeneity in vermiculites. *Clays and Clay Minerals* 30: 215–222.

Lagaly, G., M. Ogawa, and I. Dékány. 2006. Clay mineral organic interactions. In *Handbook of Clay Science, Developments in Clay Science*, Vol. 1, (Eds.) F. Bergaya, B.K.G. Theng, and G. Lagaly, pp. 309–377. Amsterdam, the Netherlands: Elsevier.

Lagaly, G. and A. Weiss. 1970. Inhomogeneous charge distributions in mica-type layer silicates. In *Reunion Hispano-Belga de Minerales de la Arcilla*, (Ed.) J.M. Serratosa, pp. 179–187. Madrid, Spain: C.S.I.C.

Lagaly, G. and A. Weiss. 1976. The layer charge of smectitic layer silicates. In *Proceedings of the International Clay Conference 1975*, (Ed.) S.W. Bailey, pp. 157–172. Wilmette, IL: Applied Publishing.

Lahav, N., U. Shani, and J. Shabtai. 1978. Cross-linked smectites. I. Synthesis and properties of hydroxy-aluminum-montmorillonite. *Clays and Clay Minerals* 26: 107–115.

Laird, D.A. 2006. Influence of layer charge on swelling of smectites. *Applied Clay Science* 34: 74–87.

Laird, D.A. and C. Shang. 1997. Relationship between cation exchange selectivity and crystalline swelling in expanding 2:1 phyllosilicates. *Clays and Clay Minerals* 45: 681–689.

Lao, Y., J. Korth, J. Ellis, and P.T. Crisp. 1988. Mineral-catalysed transformations of terminal alkenes during pyrolysis. *Journal of Analytical and Applied Pyrolyis* 14: 191–201.

Lao, Y., J. Korth, J. Ellis, and P.T. Crisp. 1989. Heterogeneous reactions of 1-pristene catalysed by clays under simulated geological conditions. *Organic Geochemistry* 14: 375–379.

Laszlo, P. 1986. Catalysis of organic reactions by inorganic solids. *Accounts of Chemical Research* 19: 121–127.

Laszlo, P. (Ed.). 1987a. *Preparative Chemistry Using Supported Reagents*. New York: Academic Press.

Laszlo, P. 1987b. Chemical reactions on clays. *Science* 235: 1473–1477.

Laszlo, P. 1998. Heterogeneous catalysis of organic reactions. *Journal of Physical Organic Chemistry* 11: 356–361.

Ledoux, R.L. and J.L. White. 1964. Infrared studies of selective deuteration of kaolinite and halloysite at room temperature. *Science* 145: 47–49.

Ledoux, R.L. and J.L. White. 1966. Infrared studies of hydrogen bonding of organic compounds on oxygen and hydroxyl surfaces of layer lattice silicates. In *Proceedings of the International Clay Conference 1966*, Vol. 1, (Eds.) L. Heller and A. Weiss, pp. 361–374. Jerusalem, Israel: Israel Program for Scientific Translations.

Lee, L.T., R. Rahbari, J. Lecourtier, and G. Chauveteau. 1991. Adsorption of polyacrylamide on the different faces of kaolinites. *Journal of Colloid and Interface Science* 147: 351–357.

Lee, S.Y., M.L. Jackson, and J.L. Brown. 1975. Micaceous occlusions in kaolinite observed by ultramicrotomy and high resolution electron microscopy. *Clays and Clay Minerals* 23: 125–129.

Lei, Z., Q. Zhang, J. Luo, and X. He. 2005. Baeyer-Villiger oxidation of ketones with hydrogen peroxide catalyzed by Sn-palygorskite. *Tetrahedron Letters* 46: 3505–3508.

Lei, Z., Q. Zhang, R. Wang, G. Ma, and C. Jia. 2006. Clean and selective Baeyer-Villiger oxidation of ketones with hydrogen peroxide catalyzed by Sn-palygorskite. *Journal of Organometallic Chemistry* 691: 5767–5773.

Lenarda, M., L. Storaro, A. Talon, E. Moretti, and P. Riello. 2007. Solid acid catalysts from clays: Preparation of mesoporous catalysts by chemical activation of metakaolin under acid conditions. *Journal of Colloid and Interface Science* 311: 537–543.

Li, A.-X., T.-S. Li, and T.H. Ding. 1997. Montmorillonite K-10 and KSF as remarkable acetylation catalysts. *Chemical Communications* 15: 1389–1390.

Li, X., C. Ni, C. Yao, and Z. Chen. 2012. Development of attapulgite/$Ce_{1-x}Zr_xO_2$ nanocomposite as catalyst for the degradation of methylene blue. *Applied Catalysis B: Environmental* 117–118: 118–124.

Liétard, O., J. Yvon, J.F. Delon, R. Mercier, and J.-M. Cases. 1980. Determination of the basal and lateral surfaces of kaolins: Variation with types of crystalline defects. In *Fine Particles Processing*, Vol. 1, (Ed.) P. Somasundaran, pp. 558–582. New York: American Institute of Mining, Metallurgical and Petroleum Engineers.

Lim, C.H., M.L. Jackson, R.D. Koons, and P.A. Helmke. 1980. Kaolins: Sources of differences in cation-exchange capacities and cesium retention. *Clays and Clay Minerals* 28: 223–229.

Lipsicas, M., R.H. Raythatha, T.J. Pinnavaia et al. 1984. Silicon and aluminium site distributions in 2:1 layered silicate clays. *Nature* 309: 604–607.

Liu, D., P. Yuan, H. Liu et al. 2013e. Quantitative characterization of the solid acidity of montmorillonite using combined FTIR and TPD based on the NH_3 adsorption system. *Applied Clay Science* 80–81: 407–412.

Liu, H., T. Chen, D. Chang et al. 2013c. Characterization and catalytic performance of Fe_3Ni_8/palygorskite for catalytic cracking of benzene. *Applied Clay Science* 74: 135–140.

Liu, H., T. Chen, D. Chang, D. Chen, and R.L. Frost. 2012. Catalytic cracking of tars derived from rice hull gasification over goethite and palygorskite. *Applied Clay Science* 70: 51–57.

Liu, H., D. Liu, P. Yuan et al. 2013d. Studies on the solid acidity of heated and cation-exchanged montmorillonite using *n*-butylamine titration in non-aqueous system and diffuse reflectance Fourier transform infrared (DRIFT) spectroscopy. *Physics and Chemistry of Minerals* 40: 479–489.

Liu, H., P. Yuan, D. Liu, D. Tan, H. He, and J. Zhu. 2013b. Effects of solid acidity of clay minerals on the thermal decomposition of 12-aminolauric acid. *Journal of Thermal Analysis and Calorimetry* 114: 125–130.

Liu, M., R. He, J. Yang et al. 2016. Polysaccharide-halloysite nanotube composites for biomedical applications: A review. *Clay Minerals* 51: 457–467.

Liu, P. and M. Zhao. 2009. Silver nanoparticle supported on halloysite nanotubes catalyzed reduction of 4-nitrophenol (4-NP). *Applied Surface Science* 255: 3989–3993.

Liu, X., X. Lu, M. Sprik, J. Cheng, E.J. Meijer, and R. Wang. 2013a. Acidity of edge surface sites of montmorillonite and kaolinite. *Geochimica et Cosmochimica Acta* 117: 180–190.

Loewenstein, W. 1954. The distribution of aluminum in the tetrahedra of silicates and aluminates. *American Mineralogist* 39: 92–96.

Löffler, D., G.L. Haller, and J.B. Fenn. 1979. A molecular beam study of isomerization and dehydrogenation of butene on a mica surface. *Journal of Catalysis* 57: 95–104.

Lokanatha, S., B.K. Mathur, B.K. Samantaray, and S. Bhattacherjee. 1985. Dehydration and phase transformation in sepiolite. A radial distribution analysis study. *Zeitschrift für Kristallographie* 171: 69–79.

Loucks, R.R. 1991. The bound interlayer H_2O content of potassic white mica: Muscovite-hydromuscovite-hydropyrophyllite solutions. *American Mineralogist* 76: 1563–1579.

Lvov, Y.M. and E. Abdullayev. 2013. Functional polymer-clay nanotube composites with sustained release of chemical agents. *Progress in Polymer Science* 38: 1690–1719.

Ma, C. and R.A. Eggleton. 1999. Cation exchange capacity of kaolinite. *Clays and Clay Minerals* 47: 174–180.

Ma, H., K. Sun, Y. Li, and X. Xu. 2009. Ultra-chemoselective hydrogenation of chloronitrobenzenes to chloroanilines over HCl-acidified attapulgite-supported platinum catalyst with high activity. *Catalysis Communications* 10: 1363–1366.

MacCarter, W.S.W., K.A. Krieger, and H. Heinemann. 1950. Thermal activation of attapulgus clay—Effect on physical and adsorptive properties. *Industrial & Engineering Chemistry* 42: 529–533.

MacEwan, D.M.C. 1948. Complexes of clays with organic compounds. I. Complex formation between montmorillonite and halloysite and certain organic liquids. *Transactions of the Faraday Society* 44: 349–387.

MacEwan, D.M.C. and A. Ruiz-Amil. 1975. Interstratified clay minerals. In *Soil Components*, Vol. 2. Inorganic Components, (Ed.) J.E. Gieseking, pp. 265–334. Berlin, Germany: Springer-Verlag.

MacEwan, D.M.C. and M.J. Wilson. 1980. Interlayer and intercalation complexes of clay minerals. In *Crystal Structures of Clay Minerals and Their X-ray Identification,* (Eds.) G.W. Brindley and G. Brown, pp. 197–248. London, UK: Mineralogical Society.

Machado, G.S., K.A.D. de Freitas Castro, F. Wypych, and S. Nakagaki. 2008. Immobilization of metalloporphyrins into nanotubes of natural halloysite toward selective catalysts for oxidation reactions. *Journal of Molecular Catalysis A: Chemical* 283: 99–107.

Macht, F., K. Eusterhues, G.J. Pronk, and K.U. Totsche. 2011. Specific surface area of clay minerals: Comparison between atomic force microscopy measurements and bulk-gas (N_2) and liquid (EGME) adsorption methods. *Applied Clay Science* 53: 20–26.

Mackenzie, R.C. and S. Caillère. 1975. The thermal characteristics of soils minerals and the use of these characteristics in the qualitative and quantitative determination of clay minerals in soils. In *Soil Components*, Vol. 2. Inorganic Components, (Ed.) J.E. Gieseking, pp. 529–571. Berlin, Germany: Springer-Verlag.

Mackenzie, R.C. and B.D. Mitchell. 1966. Clay mineralogy. *Earth-Science Reviews* 2: 47–91.

Maes, A. and A. Cremers. 1977. Charge density effects in ion exchange. Part 1. Heterovalent exchange equilibria. *Journal of the Chemical Society Faraday Transactions 1* 73: 1807–1814.

Mahmoud, S., A. Hammoudeh, and M. Al-Noaimi. 2003. Pretreatment effects on the catalytic activity of Jordanian bentonite. *Clays and Clay Minerals* 51: 52–57.

Mahmoud, S. and S. Saleh. 1999. Effect of acid activation on the de-tert-butylation activity of some Jordanian clays. *Clays and Clay Minerals* 47: 481–486.

Malden, P.J. and R.E. Meads. 1967. Substitution by iron in kaolinite. *Nature* 215: 844–846.

Martin, R.T., S.W. Bailey, D.D. Eberl et al. 1991. Report on the Clay Minerals Society nomenclature committee: Revised classification of clay minerals. *Clays and Clay Minerals* 39: 333–335.

Mathieson, A.M. and G.F. Walker. 1954. Crystal structure of magnesium-vermiculite. *American Mineralogist* 39: 231–255.

Matsuda, T., K. Yogo, C. Pantawong, and E. Kikuchi. 1995. Catalytic properties of copper-exchanged clays for the dehydrogenation of methanol to methyl formate. *Applied Catalysis A: General* 126: 177–186.

Matsuzaki, T., Y. Sugi, J. Imamura, and K. Kizawa. 1986. Vapour phase synthesis over a kaolin catalyst of alkyl aryl ethers from phenols and alkanols. *Chemistry and Industry* 35–36.

Mazzei, M., W. Marconi, and M. Riocci. 1980. Asymmetric hydrogenation of substituted acrylic acids by Rh'-aminophosphine chiral complex supported on mineral clays. *Journal of Molecular Catalysis* 9: 381–387.

McAuliffe, C. and N.T. Coleman. 1955. H-ion catalysis by acid clays and exchange resins. *Soil Science Society of America Journal* 19: 156–160.

McBride, M.B. 1976. Origin and position of exchange sites in kaolinite: An ESR study. *Clays and Clay Minerals* 24: 88–92.

McBride, M.B., T.J. Pinnavaia, and M.M. Mortland. 1975. Electron spin resonance studies of cation orientation in restricted water layers on phyllosilicate (smectite) surfaces. *Journal of Physical Chemistry* 79: 2430–2435.
McCabe, R.W. and J.M. Adams. 2013. Clay minerals as catalysts. In *Handbook of Clay Science*, 2nd ed. Developments in Clay Science, Vol. 5B, (Eds.) F. Bergaya and G. Lagaly, pp. 491–538. Amsterdam, the Netherlands: Elsevier.
Mestdagh, M.M., L. Vielvoye, and A.J. Herbillon. 1980. Iron in kaolinite: II. The relationship between kaolinite crystallinity and iron content. *Clay Minerals* 15: 1–13.
Miao, S., Z. Liu, Z. Zhang et al. 2007. Ionic liquid-assisted immobilization of Rh on attapulgite and its application in cyclohexene hydrogenation. *Journal of Physical Chemistry C* 111: 2185–2190.
Mikhail, R.S., N.M. Guindy, and S. Hanafi. 1983. Surface properties of attapulgite. *Journal of Vacuum Science & Technology A* 1: 267–270.
Mingelgrin, U., S. Saltzman, and B. Yaron. 1977. A possible model for the surface-induced hydrolysis of organophosphorus pesticides on kaolinite clays. *Soil Science Society of America Journal* 41: 519–523.
Miura, A., R. Okabe, K. Izumo, and M. Fukushima. 2009. Influence of the physicochemical properties of clay minerals on the degree of darkening via polycondensation reactions between catechol and glycine. *Applied Clay Science* 46: 277–282.
Moore, D.M. 1996. Comment on: Definition of clay and clay mineral: Joint report of the AIPEA nomenclature and CMS nomenclature committees. *Clays and Clay Minerals* 44: 710–712.
Morgan, D.A., D.B. Shaw, M.J. Sidebottom, T.C. Soon, and R.S. Taylor. 1985. The function of bleaching earths in the processing of palm, palm kernel and coconut oils. *Journal of the American Oil Chemists' Society* 62: 292–299.
Morikawa, Y. 1993. Catalysis by metal ions intercalated in layer lattice silicates. In *Advances in Catalysis*, Vol. 39, (Eds.) D.D. Eley, H. Pines, and P.B. Weisz, pp. 302–327. New York: Academic Press.
Morikawa, Y., K. Takagi, Y. Moro-oka, and T. Ikawa. 1982. Cu-fluor tetra silicic mica. A novel effective catalyst for the dehydrogenation of methanol to form methyl formate. *Chemistry Letters* 11(11): 1805–1808.
Morikawa, Y., F.-L. Wang, Y. Moro-oka, and T. Ikawa. 1983a. Conversion of methanol to low molecular weight hydrocarbons over Ti ion exchanged form of layer silicate minerals. *Chemistry Letters* 12(7): 965–968.
Morikawa, Y., A. Yasuda, Y. Moro-oka, and T. Ikawa. 1983b. Catalytic activities of metal ion exchanged forms of fluoro tetrasilicic mica for the reaction of butenes. *Chemistry Letters* 12(12): 1911–1912.
Mortland, M.M. 1970. Clay-organic complexes and interactions. *Advances in Agronomy* 22: 75–117.
Mortland, M.M. and K.V. Raman. 1968. Surface acidity of smectites in relation to hydration, exchangeable cation, and structure. *Clays and Clay Minerals* 16: 193–198.
Murray, H.H. 1999. Applied clay mineralogy today and tomorrow. *Clay Minerals* 34: 39–49.
Murray, H.H. 2000. Traditional and new applications for kaolin, smectite, and palygorskite: A general overview. *Applied Clay Science* 17: 207–221.
Mystkowski, K., J. Środoń, and F. Elsass. 2000. Mean thickness and thickness distribution of smectite crystallites. *Clay Minerals* 35: 545–557.
Nagasawa, K. 1978. Kaolin minerals. In *Clays and Clay Minerals of Japan*. Developments in Sedimentology 26, (Eds.) T. Sudo and S.Shimoda, pp. 189–219. Tokyo, Japan: Kodansha and Amsterdsm the Netherlands: Elsevier.
Nagata, M., S. Shimoda, and T. Sudo. 1974. On dehydration of bound water of sepiolite. *Clays and Clay Minerals* 22: 285–293.
Nagendrappa, G. 2011. Organic synthesis using clay and clay-supported catalysts. *Applied Clay Science* 53: 106–138.
Nagy, B. and W.F. Bradley. 1955. The structural scheme of sepiolite. *American Mineralogist* 40: 885–892.
Nakagaki, S., F.L. Benedito, and F. Wypych. 2004. Anionic iron(III)porphyrin immobilized on silanized kaolinite as catalyst for oxidation reactions. *Journal of Molecular Catalysis A: Chemical* 217: 121–131.
Nakagaki, S., K.A.D. Castro, G.S. Machado, M. Halma, S.M. Drechsel, and F. Wypych. 2006b. Catalytic activity in oxidation reactions of anionic iron(III) porphyrins immobilized on raw and grafted chrysotile. *Journal of the Brazilian Chemical Society* 17: 1672–1678.
Nakagaki, S., G.S. Machado, M. Halma et al. 2006a. Immobilization of iron porphyrins in tubular kaolinite obtained by an intercalation/delamination procedure. *Journal of Catalysis* 242: 119–117.
Nakagaki, S. and F. Wypych. 2007. Nanofibrous and nanotubular supports for the immobilization of metalloporphyrins as oxidation catalysts. *Journal of Colloid and Interface Science* 315: 142–157.
Nakayama, T., S. Yoshida, and H. Kanai. 1993. Synthetic fluorotetrasilicic mica as a support for stereoselective catalysts. *Journal of Materials Science* 28: 4163–4166.

Navalon, S., M. Alvaro, and H. Garcia. 2010. Heterogeneous Fenton catalysts based on clays, silicas and zeolites. *Applied Catalysis B: Environmental* 99: 1–26.

Nazir, M., M. Ahmad, and F.M. Chaudhary. 1976. Isomerization of α-pinene to camphene using indigenous clays as catalysts. *Pakistan Journal of Scientific and Industrial Research* 19: 175–178.

Nederbragt, G.W. 1949. Separation of long-chain and compact molecules by adsorption to attapulgite-containing clays. *Clay Minerals Bulletin* 3: 72–75.

Neumann, B.S. and K.G. Sansom. 1970a. The formation of stable sols from laponite, a synthetic hectorite-like clay. *Clay Minerals* 8: 389–404.

Neumann, B.S. and K.G. Sansom. 1970b. *Laponite* clay—A synthetic inorganic gelling agent for aqueous solutions of polar organic compounds. *Journal of the Society of Cosmetic Chemists* 21: 237–258.

Newman, A.C.D. 1969. Cation exchange properties of micas. I. The relation between mica composition and potassium exchange in solutions of different pH. *Journal of Soil Science* 20: 357–373.

Newman, A.C.D. (Ed.). 1987. *Chemistry of Clays and Clay Minerals*. Monograph 6. London, UK: Mineralogical Society.

Newman, R.H., C.W. Childs, and G.J. Churchman. 1994. Aluminium coordination and structural disorder in halloysite and kaolinite by ^{27}Al NMR spectroscopy. *Clay Minerals* 29: 305–312.

Newnham, R.E. 1961. A refinement of the dickite structure and some remarks on polymorphism in kaolin minerals. *Mineralogical Magazine* 32: 683–704.

Nieto, F., M. Mellini, and I. Abad. 2010. The role of H_3O^+ in the crystal structure of illite. *Clays and Clay Minerals* 58: 238–246.

Nikalje, M.D., P. Phukan, and A. Sudalai. 2000. Recent advances in clay-catalyzed organic transformations. *Organic Preparations and Procedures International* 32: 1–40.

Nitta, Y., M. Kawabe, Y. Ohmachi, and T. Imanaka. 1989a. Preparation of uniformly dispersed nickel/silica catalysts from synthetic nickel-chrysotile. *Applied Catalysis* 53: 15–28.

Nitta, Y., K. Ueno, and T. Imanaka. 1989b. Selective hydrogenation of α,β-unsaturated aldehydes on cobalt-silica catalysts obtained from cobalt chrysotile. *Applied Catalysis* 56: 9–22.

Njopwouo, D., G. Roques, and R. Wandji. 1987. A contribution to the study of the catalytic action of clays on the polymerization of styrene: I. Characterization of polystyrenes. *Clay Minerals* 22: 145–156.

Njopwouo, D., G. Roques, and R. Wandji. 1988. A contribution to the study of the catalytic action of clays on the polymerization of styrene: II. Reaction mechanism. *Clay Minerals* 23: 35–43.

Noll, W., H. Kircher, and W. Sybertz. 1958. Adsorptionsvermögen und spezifische Oberfläche von Silikaten mit röhrenförmig gebauten Primärkristallen. *Kolloid Zeitschrift* 157: 1–11.

Noll, W., H. Kircher, and W. Sybertz. 1960. Über synthetischen Kobaltchrysotil und seine Beziehungen zu anderen Solenosilikaten. *Beiträge zur Mineralogie und Petrographie* 7: 232–241.

Norrish, K. 1954. Swelling of montmorillonite. *Discussions of the Faraday Society* 18: 120–134.

Norrish, K. 1973. Factors in the weathering of mica to vermiculite. In *Proceedings of the International Clay Conference 1972*, (Ed.) J.M. Serratosa, pp. 417–432. Madrid, Spain: Division de Ciencias, C.S.I.C.

Novikova, L., F. Roessner, L. Belchinskaya, M. AlSawalha, and V. Krupskaya. 2014. Study of surface acid-base properties of natural clays and zeolites by the conversion of 2-methylbut-3-yn-2-ol. *Applied Clay Science* 101: 229–236.

Noyan, H., M. Önal, and Y. Sarikaya. 2006. The effect of heating on the surface area, porosity and surface acidity of a bentonite. *Clays and Clay Minerals* 54: 375–381.

Núñez, K., R. Gallego, J.M. Pastor, and J.C. Merino. 2014. The structure of sepiolite as support of metallocene co-catalyst during in situ polymerization of polyolefin (nano)composites. *Applied Clay Science* 101: 73–81.

O'Connor, C.T., L.L. Jacobs, and M. Kojima. 1988. Propene oligomerization over synthetic mica-montmorillonite (SMM) and SMM incorporating nickel, zinc and cobalt. *Applied Catalysis* 40: 277–290.

Occelli, M.L., R.A. Innes, F.S.S. Hwu, and J.W. Hightower. 1985. Sorption and catalysis on sodium-montmorillonite interlayered with aluminum oxide clusters. *Applied Catalysis* 14: 69–82.

Occelli, M.L. and R.M. Tindwa. 1983. Physicochemical properties of montmorillonite interlayered with cationic oxyaluminum pillars. *Clays and Clay Minerals* 31: 22–28.

Ogaki, Y., Y. Shinozuka, T. Hara, N. Ichikuni, and S. Shimazu. 2011. Hemicellulose decomposition and saccharide production from various plant biomass by sulphonated allophane catalyst. *Catalysis Today* 164: 415–418.

Ohtake, T., T. Mori, and Y. Morikawa. 2007. Catalytic conversion of methanol by oxidative dehydrogenation. *Journal of Natural Gas Chemistry* 16: 1–5.

Okabe, R., A. Miura, M. Fukushima et al. 2011. Characterization of an adsorbed humin-like substance on an allophanic soil formed via catalytic polycondensation between catechol and glycine, and its adsorption capability to pentachlorophenol. *Chemosphere* 83: 1502–1506.

Okada, T., Y. Ehara, and M. Ogawa. 2007. Adsorption of Eu^{3+} to smectites and fluoro-tetrasilicic mica. *Clays and Clay Minerals* 55: 348–353.

Olejnik, S., A.M. Posner, and J.P. Quirk. 1970. The intercalation of polar organic compounds into kaolinite. *Clay Minerals* 8: 421–434.

Olson, B.G., J.J. Decker, S. Nazarenko et al. 2008. Aggregation of synthetic chrysotile nanotubes in the bulk and in solution probed by nitrogen adsorption and viscosity measurements. *Journal of Physical Chemistry C* 112: 12943–12950.

Ono, Y. and K.-I. Katsumata. 2014. Enhanced photocatalytic activity of titanium dioxide/allophane mixed powder by acid treatment. *Applied Clay Science* 90: 61–66.

Ono, Y., N. Kikuchi, and H. Watanabe. 1987. Preparation of nickel catalyst from nickel containing chrysotile. *Studies in Surface Science and Catalysis* 31: 519–529.

Ookawa, M., Y. Inoue, M. Watanabe, M. Suzuki, and T. Yamaguchi. 2006. Synthesis and characterization of Fe-containing imogolite. *Clay Science* 12(2): 280–284.

Ookawa, M., Y. Takata, M. Suzuki, K. Inukai, T. Maekawa, and T. Yamaguchi. 2008. Oxidation of aromatic hydrocarbons with H_2O_2 catalyzed by a nanoscale tubular aluminosilicate, Fe-containing imogolite. *Research on Chemical Intermediates* 34: 679–685.

Ortego, J.D., M. Kowalska, and D.L. Cocke. 1991. Interactions of montmorillonite with organic compounds—Adsorptive and catalytic properties. *Chemosphere* 22: 769–798.

Ovcharenko, F.D. 1982. Clay minerals as catalysts. In *International Clay Conference 1981*. Developments in Sedimentology 35, (Eds.) H. van Olphen and F. Veniale, pp. 239–251. Amsterdam, the Netherlands: Elsevier.

Papoulis D., S. Komarneni, D. Panagiotaras et al. 2013. Palygorskite-TiO_2 nanocomposites: Part 2. Photocatalytic activities in decomposing air and organic pollutants. *Applied Clay Science* 83–84: 198–202.

Parfitt, R.L. 1978. Anion adsorption by soils and soil materials. *Advances in Agronomy* 30: 1–50.

Parfitt, R.L. 1980. Chemical properties of variable charge soils. In *Soils with Variable Charge*, (Ed.) B.K.G. Theng, pp. 167–194. Lower Hutt, New Zealand: New Zealand Society of Soil Science.

Parfitt, R.L. 1990. Allophane in New Zealand—A review. *Australian Journal of Soil Research* 28: 343–360.

Parfitt, R.L. 2009. Allophane and imogolite; role in soil biogeochemical processes. *Clay Minerals* 44: 125–145.

Parfitt, R.L., R.J. Furkert, and T. Henmi. 1980. Identification and structure of two types of allophane from volcanic ash soils and tephra. *Clays and Clay Minerals* 28: 328–334.

Parfitt, R.L. and T. Henmi. 1980. Structure of some allophanes from New Zealand. *Clays and Clay Minerals* 28: 285–294.

Parfitt, R.L. and J.D. Russell. 1977. Adsorption on hydrous oxides. IV. Mechanism of adsorption of various ions on goethite. *Journal of Soil Science* 28: 297–305.

Pasbakhsh, P. and G.J. Churchman (Eds.). 2015. *Natural Mineral Nanotubes*. Oakville, ON: Apple Academic Press.

Pasbakhsh, P., G.J. Churchman, and J.L. Keeling. 2013. Characterisation of properties of various halloysites relevant to their use as nanotubes and microfibre fillers. *Applied Clay Science* 74: 47–57.

Pasbakhsh, P., R. de Silva, V. Vahedi, and G.J. Churchman. 2016. Halloysite nanotubes: Prospects and challenges of their use as additives and carriers—A focused review. *Clay Minerals* 51: 479–487.

Paterson, E. 1977. Specific surface area and pore structure of allophanic soil clays. *Clay Minerals* 12: 1–9.

Pathak, D.D. and J.J. Gerald. 2003. An efficient and convenient method for the synthesis of dialkoxymethanes using kaolinite as a catalyst. *Synthetic Communications* 33: 1557–1561.

Patterson, S.H. and H.H. Murray. 1984. Kaolin, refractory clay, ball clay, and halloysite in North America, Hawaii, and the Caribbean region. *U.S. Geological Survey Professional Paper 1306*, 56 p.

Pauling, L. 1930. The structure of micas and related minerals. *Proceedings of the National Academy of Sciences* 16: 123–129.

Pérez-Rodríguez, J.L. and C. Maqueda. 2002. Interactions of vermiculites with organic compounds. In *Organo-Clay Complexes and Interactions*, (Eds.) S. Yariv, and H. Cross, pp. 113–173. New York: Marcel Dekker.

Perissinotto, M., M. Lenarda, L. Storaro, and R. Ganzerla. 1997. Solid acid catalysts from clays: Acid leached metakaolin as isopropanol dehydration and 1-butene isomerization catalyst. *Journal of Molecular Catalysis A: Chemical* 121: 103–109.

Petit, S., D. Righi, J. Madejová, and A. Decarreau. 1999. Interpretation of the infrared NH_4^+ spectrum of the NH_4^+-clays: Application to the evaluation of the layer charge. *Clay Minerals* 34: 543–549.

Pinho, R.O., J.A.R. Rodrigues, P.J. Samenho Moran, and I. Joekes. 1995. Chrysotile-supported transition metal salts as Friedel–Crafts catalysts. *Journal of the Brazilian Chemical Society* 6: 373–376.

Pinnavaia, T.J. 1980. Applications of ESR spectroscopy to inorganic-clay systems. In *Advanced Chemical Methods for Soil and Clay Minerals Research*, (Eds.) J.W. Stucki and W.L. Banwart, pp. 391–421. Dordrecht, the Netherlands: D. Reidel.

Pinnavaia, T.J. 1983. Intercalated clay catalysts. *Science* 220: 365–371.

Pinnavaia, T.J., R. Raythatha, J.G.-S. Lee, L.J. Halloran, and J.F. Hoffman. 1979. Intercalation of catalytically active metal complexes in mica-type silicates. Rhodium hydrogenation catalysts. *Journal of the American Chemical Society* 101: 6891–6897.

Pinnavaia, T.J., M.-S. Tzou, S.D. Landau, and R.H. Raythatha. 1984. On the pillaring and delamination of smectite clay catalysts by polyoxo cations of aluminum. *Journal of Molecular Catalysis* 27: 195–212.

Piperno, S., I. Kaplan-Ashiri, S.R. Cohen et al. 2007. Characterization of geoinspired and synthetic chrysotile nanotubes by atomic force microscopy and transmission electron microscopy. *Advanced Functional Materials* 17: 3332–3338.

Pires, L.H.O., A.N. de Oliveira, O.V. Monteiro et al. 2014. Esterification of a waste produced from the palm oil industry over 12-tungstophosforic acid supported on kaolin waste and mesoporous materials. *Applied Catalysis B: Environmental* 160–161: 122–128.

Plançon, A., R.F. Giese, R. Snyder, V.A. Drits, and A.S. Bookin. 1989. Stacking faults in the kaolin group minerals: Defect structures of kaolinite. *Clays and Clay Minerals* 37: 203–210.

Plançon, A. and C. Tchoubar. 1977. Determination of structural defects in phyllosilicates by X-ray diffraction. *Clays and Clay Minerals* 25: 430–450.

Ponde, D.E., V.H. Deshpande, V.J. Bulbule, A. Sudalai, and A.S. Gajare. 1998. Selective catalytic transesterification, and protection of carbonyl compounds over natural kaolinitic clay. *Journal of Organic Chemistry* 63: 1058–1063.

Prada-Silva, G., D. Löffler, B.L. Halpern, G.L. Haller, and J.B. Fenn. 1979. The role of vibrational energy in surface isomerization of cyclopropane. *Surface Science* 83: 453–470.

Preisinger, A. 1959. X-ray study of the structure of sepiolite. *Clays and Clay Minerals* 6: 61–67.

Prost, R. 1976. Interactions between adsorbed water molecules and the structure of clay minerals: Hydration mechanism of smectites. In *Proceedings of the International Clay Conference 1975*, (Ed.) S.W. Bailey, pp. 351–359. Wilmette, IL: Applied Publishing.

Purceno, A.D., A.P. Teixeira, A.B. Souza, J.D. Ardisson, J.P. de Mesquita, and R.M. Lago. 2012. Ground vermiculite as catalyst for the Fenton reaction. *Applied Clay Science* 69: 87–92.

Purnell, J.H. 1990. Catalysis by ion-exchanged montmorillonites. *Catalysis Letters* 5: 203–210.

Pushpaletha, P. and M. Lalithambika. 2011. Modified attapulgite: An efficient solid acid catalyst for acetylation of alcohols using acetic acid. *Applied Clay Science* 51: 424–430.

Pushpaletha, P., S. Rugmini, and M. Lalithambika. 2005. Correlation between surface properties and catalytic activity of clay catalysts. *Applied Clay Science* 30: 141–153.

Qi, X., H. Yoon, S.-H. Lee, J. Yoon, and S.-J. Kim. 2008. Surface-modified imogolite by 3-APS-OsO_4 complex: Synthesis, characterization and its application in the dihydroxylation of olefins. *Journal of Industrial and Engineering Chemistry* 14: 136–141.

Quirk, J.P. 1960. Negative and positive adsorption of chloride by kaolinite. *Nature* 188: 253–254.

Quirk, J.P. 1968. Particle interaction and soil swelling. *Israel Journal of Chemistry* 6: 213–234.

Quirk, J.P. 1994. Interparticle forces: A basis for the interpretation of soil physical behavior. *Advances in Agronomy* 53: 121–183.

Quirk, J.P. and L.A.G. Aylmore. 1971. Domains and quasi-crystalline regions in clay systems. *Soil Science Society of America Journal* 35: 652–654.

Quirk, J.P. and B.K.G. Theng. 1960. Effect of surface density of charge on the physical swelling of lithium montmorillonite. *Nature* 187: 967–968.

Radoslovich, E.W. 1963. The cell dimension and symmetry of layer-lattice silicates. VI. Serpentine and kaolin morphology. *American Mineralogist* 48: 368–378.

Radoslovich, E.W. 1975. Micas in macroscopic forms. In *Soil Components*, Vol. 2. Inorganic Components, (Ed.) J.E. Gieseking, pp. 27–57. Berlin, Germany: Springer-Verlag.

Rand, B. and I.E. Melton. 1975. Isoelectric point of the edge surface of kaolinite. *Nature* 257: 214–216.

Range, K.-J., A. Range, and A. Weiss. 1969. Fire-clay kaolinite or fire-clay mineral? Experimental classification of kaolinite-halloysite minerals. In *Proceedings of the International Clay Conference 1969*, Vol. 1, (Ed.) L. Heller, pp. 3–13. Jerusalem, Israel: Israel Universities Press.

Rao, S.M. and A. Sridharan. 1984. Mechanism of sulfate adsorption by kaolinite. *Clays and Clay Minerals* 32: 414–418.
Ras, R.H.A., Y. Umemura, C.T. Johnston, A. Yamagishi, and R.A. Schoonheydt. 2007. Ultrathin hybrid films of clay minerals. *Physical Chemistry Chemical Physics* 9: 918–932.
Rausell-Colom, J.A., T.R. Sweatman, C.B. Wells, and K. Norrish. 1965. Studies in the artificial weathering of mica. In *Experimental Pedology. Proceedings of the 11th Easter School of Agricultural Science, University of Nottingham*, pp. 40–72. London, UK: Butterworth.
Ravichandran, J. and B. Sivasankar. 1997. Properties and catalytic activity of acid-modified montmorillonite and vermiculite. *Clays and Clay Minerals* 45: 854–858.
Rawtani, D. and Y.K. Agrawal. 2012. Multifarious applications of halloysite nanotubes: A review. *Reviews on Advanced Materials Science* 30: 282–295.
Rayner, J.H. 1974. The crystal structure of phlogopite by neutron diffraction. *Mineralogical Magazine* 13: 73–84.
Rengasamy, P., G.S.R. Krishna Murti, and V.A.K. Sarma. 1975. Isomorphous substitution of iron for aluminium in some soil kaolinites. *Clays and Clay Minerals* 23: 211–214.
Reynolds, R.C. 1980. Interstratified clay minerals. In *Crystal Structures of Clay Minerals and Their X-ray Identification,* (Eds.) G.W. Brindley and G. Brown, pp. 249–303. London, UK: Mineralogical Society.
Rianelli, R. S., M.C.B.V. de Souza, and V.F. Ferreira. 2004. Mild diazo transfer reaction catalyzed by modified clays. *Synthetic Communications* 34: 951–959.
Rideal, E.K. and W. Thomas. 1922. Adsorption and catalysis in Fuller's earth. *Journal of the Chemical Society Transactions* 121: 2119–2123.
Robertson, R.H.S. 1948. Clay minerals as catalysts. *Clay Minerals Bulletin* 1: 47–54.
Robertson, R.H.S. 1986. *Fuller's Earth: A History of Calcium Montmorillonite*. Hythe, UK: Volturna Press.
Röbschläger, K.-H.W., C.A. Emeis, and R.A. van Santen. 1984. On the hydroisomerization activity of nickel-substituted mica montmorillonite. *Journal of Catalysis* 86: 1–8.
Rong, T.-J. and J.-K. Xiao. 2002. The catalytic cracking activity of the kaolin-group minerals. *Materials Letters* 57: 297–301.
Rothbauer, R. 1971. Untersuchung eines $2M_1$-Muskovits mit Neutronenstrahlen. *Neues Jahrbuch für Mineralogie-Monatshefte* 143–154.
Roudier, J.-F. and A. Foucaud. 1984. Clay-catalyzed ene-reactions: Synthesis of γ-lactones. *Tetrahedron Letters* 25: 4375–4378.
Rousseaux, J.M. and B.P. Warkentin. 1976. Surface properties and forces holding water in allophane soils. *Soil Science Society of America Journal* 40: 446–451.
Rouxhet, P.G., N. Samudacheata, H. Jacobs, and O. Anton. 1977. Attribution of the OH stretching bands of kaolinite. *Clay Minerals* 12: 171–179.
Roveri, N., G. Falini, E. Foresti, G. Fracasso, I.G. Lesci, and P. Sabatino. 2006. Geoinspired synthetic chrysotile nanotubes. *Journal of Materials Research* 21: 2711–2725.
Roy, D.M. and R. Roy. 1954. An experimental study of the formation and properties of synthetic serpentines and related layer silicate minerals. *American Mineralogist* 39: 957–975.
Rozalén, M., P.V. Brady, and F.J. Huertas. 2009. Surface chemistry of K-montmorillonite: Ionic strength, temperature dependence and dissolution kinetics. *Journal of Colloid and Interface Science* 333: 474–484.
Rozengart, M.I., G.M. V'yunova, and G.V. Isagulyants. 1988. Layered silicates as catalysts. *Russian Chemical Reviews* 57: 115–128.
Rupert, J.P., W.T. Granquist, and T.J. Pinnavaia. 1987. Catalytic properties of clay minerals. In *Chemistry of Clays and Clay Minerals. Monograph 6,* (Ed.) A.C.D. Newman, pp. 275–318. London, UK: Mineralogical Society.
Rytwo, G., Y. Gonen, and R. Huterer-Shveky. 2009. Evidence of degradation of triarylmethine dyes on Texas vermiculite. *Clays and Clay Minerals* 57: 555–565.
Sabu, K.R., R. Sukumar, and M. Lalithambika. 1993. Acidic properties and catalytic activity of natural kaolinitic clays for Friedel–Crafts alkylation. *Bulletin of the Chemical Society of Japan* 66: 3535–3541.
Sabu, K.R., R. Sukumar, R. Rekha, and M. Lalithambika. 1999. A comparative study on H_2SO_4, HNO_3 and $HClO_4$ treated metakaolinite of a natural kaolinite as Friedel–Crafts alkylation catalyst. *Catalysis Today* 49: 321–326.
Sahu, P.K., P.K. Sahu, and D.D. Agarwal. 2013. Efficient and facile synthesis of heterocycles and their mechanistic consideration using kaolin. *RSC Advances* 3: 9854–9864.
Sakurai, H., K. Urabe, and Y. Izumi. 1988. New acidic pillared clay catalysts prepared from fluor-tetrasilicic mica. *Journal of the Chemical Society Chemical Communications* 1519–1520.

Sakurai, H., K. Urabe, and Y. Izumi. 1989. Acidity enhanced pillared clay catalysts. Modification of exchangeable sites on fluor-tetrasilicic mica by the fixed interlayer cations. *Bulletin of the Chemical Society of Japan* 62: 3221–3228.

Sakurai, H., K. Urabe, and Y. Izumi. 1990. Pillared tetrasilicic mica catalysts modified by fixed interlayer cations. Classification of fixation mode by cations. *Bulletin of the Chemical Society of Japan* 63: 1389–1395.

Saltzman, S., B. Yaron, and U. Mingelgrin. 1974. The surface catalyzed hydrolysis of parathion on kaolinite. *Soil Science Society of America Journal* 38: 231–234.

Sanz, J., 2006. Nuclear magnetic resonance spectroscopy. In *Handbook of Clay Science, Developments in Clay Science, Vol. 1*, (Eds.) F. Bergaya, B.K.G. Theng, and G. Lagaly, pp. 919–938. Amsterdam, the Netherlands: Elsevier.

Sarvaramini, A. and F. Larachi. 2011. Mössbauer spectroscopy and catalytic reaction studies of chrysotile-catalyzed steam reforming of benzene. *Journal of Physical Chemistry C* 115: 6841–6848.

Satyavathi, B., A.N. Patwari, and M.B. Rao. 2003. Regio-selective catalytic vapor phase alkylation of aniline: Preparation of 2,6-diethylaniline. *Applied Catalysis A: General* 246: 151–160.

Sawhney, B.L. 1968. Aluminum interlayers in layer silicates. Effect of OH/Al ratio of Al solution, time of reaction, and type of structure. *Clays and Clay Minerals* 16: 157–163.

Sawhney, B.L. 1989. Interstratification in layer silicates. In *Minerals in Soil Environments*, 2nd ed., (Eds.) J.B. Dixon and S.B. Weed, pp. 789–828. Madison, WI: Soil Science Society of America.

Schofield, R.K. and H.R. Samson. 1953. The deflocculation of kaolinite suspensions and the accompanying change-over from positive to negative chloride adsorption. *Clay Minerals Bulletin* 2: 45–51.

Schoonheydt, R.A. and C.T. Johnston. 2013. Surface and interface chemistry of clay minerals. In *Handbook of Clay Science*, 2nd ed. Developments in Clay Science, Vol. 5A, (Eds.) F. Bergaya and G. Lagaly, pp. 139–172. Amsterdam, the Netherlands: Elsevier.

Schramm, L.L. and J.C.T. Kwak. 1982. Influence of exchangeable cation composition on the size and shape of montmorillonite particles in dilute suspension. *Clays and Clay Minerals* 30: 40–48.

Schulze, D.G. 1981. Identification of soil iron oxide minerals by differential X-ray diffraction. *Soil Science Society of America Journal* 45: 437–440.

Schulze, D.G. 1994. Differential X-ray diffraction analysis of soil minerals. In *Quantitative Methods in Soil Mineralogy*, (Eds.) J.E. Amonette and L.W. Zelazny, pp. 412–429. Madison, WI: Soil Science Society of America.

Schutz, A., D. Plee, F. Borg, P. Jacobs, G. Poncelet, and J.J. Fripiat. 1987. Acidity and catalytic properties of pillared montmorillonite and beidellite. In *Proceedings of the International Clay Conference, Denver, 1985*, (Eds.) L.G. Schultz, H. van Olphen, and F.A. Mumpton, pp. 305–310. Bloomington, IN: The Clay Minerals Society.

Schwertmann, U. 1979. Non-crystalline and accessory minerals. In *International Clay Conference 1978. Developments in Sedimentology 27*, (Eds.) M.M. Mortland and V.C. Farmer, pp. 491–499. Amsterdam, the Netherlands: Elsevier.

Seewald, J.S. 2003. Organic-inorganic interactions in petroleum-producing sedimentary basins. *Nature* 426: 327–333.

Senkayi, A.L., J.B. Dixon, L.R. Hossner, and L.A. Kippenberger. 1985. Layer charge evaluation of expandable soil clays by an alkylammonium method. *Soil Science Society of America Journal* 49: 1054–1060.

Serna, C.J., J.L. Ahlrichs, and J.M. Serratosa. 1975. Folding in sepiolite crystals. *Clays and Clay Minerals* 23: 452–457.

Serna, C.J., and G.E. Van Scoyoc. 1979. Infrared study of sepiolite and palygorskite surfaces. In *International Clay Conference 1978. Developments in Sedimentology 27*, (Eds.) M.M. Mortland and V.C. Farmer, pp. 197–206. Amsterdam, the Netherlands: Elsevier.

Serratosa, J.M. 1979. Surface properties of fibrous clay minerals (palygorskite and sepiolite). In *International Clay Conference 1978. Developments in Sedimentology 27*, (Eds.) M.M. Mortland and V.C. Farmer, pp. 99–109. Amsterdam, the Netherlands: Elsevier.

Serratosa, J.M., J.A. Rausell-Colom, and J. Sanz. 1984. Charge density and its distribution in phyllosilicates: Effect on the arrangement and reactivity of adsorbed species. *Journal of Molecular Catalysis* 27: 225–234.

Shabtai, J., R. Lazar, and A.G. Oblad. 1981. Acidic forms of crosslinked smectites—A novel type of cracking catalysts. *Studies in Surface Science and Catalysis* 7: 828–840.

Shaikh, N.S., V.H. Deshpande, and A.V. Bedekar. 2001. Clay catalyzed chemoselective Michael type addition of aliphatic amines to α,β-ethylenic compounds. *Tetrahedron* 57: 9045–9048.

Shimizu, K-i., T. Kan-no, T. Kodama, H. Hagiwara, and Y. Kitayama. 2002. Suzuki cross-coupling reaction catalyzed by palladium-supported sepiolite. *Tetrahedron Letters* 43: 5653–5655.

Shimizu, K-i., R. Maruyama, S-I Komai, T. Kodama, and Y. Kitayama. 2004. Pd-sepiolite catalyst for Suzuki coupling reaction in water: Structural and catalytic investigations. *Journal of Catalysis* 227: 202–209.

Shimizu, K-i., M. Miyagi, T. Kan-no, T. Hatamachi, T. Kodama, and Y. Kitayama. 2005. Michael reaction of β-ketoesters with vinyl ketones by iron(III)-exchanged fluorotetrasilicic mica: Catalytic and spectroscopic studies. *Journal of Catalysis* 229: 470–479.

Shimizu, K-i., M. Miyagi, T. Kan-no, T. Kodama, and Y. Kitayama. 2003. Fe^{3+}-exchanged fluorotetrasilicic mica as an active and reusable catalyst for Michael reaction. *Tetrahedron Letters* 44: 7421–7424.

Shindo, H. and P.M. Huang. 1985. The catalytic power of inorganic components in the abiotic synthesis of hydroquinone-derived humic polymers. *Applied Clay Science* 1: 71–81.

Shirini, F., N.G. Khaligh, G.H. Imanzadeh, and P.G. Ghasem-Abadi. 2012. 1,3-Dibromo-5,5-dimethylhydantoin (DHB)/kaolin: An efficient reagent system for the synthesis of 14-aryl-14*H*-dibenzo[*a,j*]xanthenes. *Chinese Chemical Letters* 23: 1145–1148.

Shirozu, H. and S.W. Bailey. 1966. Crystal structure of a two-layer Mg-vermiculite. *American Mineralogist* 51: 1124–1143.

Shuali, U., L. Bram, M. Steinberg, and S. Yariv. 1991. Catalytic thermal reactions of cumene over sepiolite and palygorskite. *Journal of Thermal Analysis* 37: 1569–1578.

Siddiqui, M.K.H. 1968. *Bleaching Earths*. Oxford, UK: Pergamon Press.

Sidorenko, A.Y., G.M. Sen'kov, and V.E. Agabekov. 2014. Effect of acid treatment on the composition and structure of a natural aluminosilicate and its catalytic properties in α-pinene isomerization. *Catalysis in Industry* 6: 94–104.

Siegel, S.M. 1957. Catalytic and polymerization-directing properties of mineral surfaces. *Proceedings of the National Academy of Sciences* 43: 811–816.

Sieskind, O., G. Joly, and P. Albrecht. 1979. Simulation of the geochemical transformation of sterols: Superacid effect of clay minerals. *Geochimica et Cosmochimica Acta* 43: 1675–1679.

Silva, F.C., M.C.B.V. de Souza, V.F. Ferreira, S.J. Sabino, and O.A.C. Antunes. 2004. Natural clays as efficient catalysts for obtaining chiral β-enamino esters. *Catalysis Communications* 5: 151–155.

Silva, J.E.S. and P.C. Jesus. 2003. Evaluation of the catalytic activity of lipases immobilized on chrysotile for esterification. *Anais da Academia Brasileira Ciências* 75: 157–162.

Singer, A. 1989. Palygorskite and sepiolite group minerals. In *Minerals in Soil Environments*, 2nd ed., (Eds.) J.B. Dixon and S.B. Weed, pp. 829–872. Madison, WI: Soil Science Society of America.

Sivakumar, T., T. Krithiga, K. Shanthi, T. Mori, J. Kubo, and Y. Morikawa. 2004. Noble metals intercalated/supported mica catalyst—Synthesis and characterization. *Journal of Molecular Catalysis A: Chemical* 223: 185–194.

Skobets, I.E., F.D. Ovcharenko, M.T. Bryk, and N.G. Vasil'ev. 1979. Catalytic conversion of octamethylcyclotetrasiloxane on surface of Al form of kaolinite. *Kolloidnyi Zhurnal* 41: 501–506.

Skoubris, E.N., G.D. Chryssikos, G.E. Christidis, and V. Gionis. 2013. Structural characterization of reduced-charge montmorillonites. Evidence based on FTIR spectroscopy, thermal behavior, and layer-charge systematics. *Clays and Clay Minerals* 61: 83–97.

Slade, P.G., C. Dean, P.K. Schultz, and P.G. Self. 1987. Crystal structure of a vermiculite-anilinium intercalate. *Clays and Clay Minerals* 35: 177–188.

Slade, P.G., J.P. Quirk, and K. Norrish. 1991. Crystalline swelling of smectite samples in concentrated NaCl solutions in relation to layer charge. *Clays and Clay Minerals* 39: 234–238.

Slade, P.G., P.A. Stone, and E.W. Radoslovich. 1985. Interlayer structures of the two-layer hydrates of Na- and Ca-vermiculites. *Clays and Clay Minerals* 33: 51–61.

Solomon, D.H. 1968. Clay minerals as electron acceptors and/or donors in organic reactions. *Clays and Clay Minerals* 16: 31–39.

Solomon, D.H. and D.G. Hawthorne. 1983. *Chemistry of Pigments and Fillers*. New York: John Wiley & Sons.

Solomon, D.H. and B.C. Loft. 1968. Reactions catalyzed by minerals. Part III. The mechanism of spontaneous interlamellar polymerizations in aluminosilicates. *Journal of Applied Polymer Science* 12: 1253–1262.

Solomon, D.H. and M.J. Rosser. 1965. Reactions catalyzed by minerals. Part I. Polymerization of styrene. *Journal of Applied Polymer Science* 9: 1261–1271.

Solomon, D.H., J.D. Swift, and A.J. Murphy. 1971. The acidity of clay minerals in polymerization and related reactions. *Journal of Macromolecular Science: Part A—Chemistry* 5: 587–601.

Soma, M., G.J. Churchman, and B.K.G. Theng. 1992. X-ray photoelectron spectroscopic analysis of halloysites with different composition and particle morphology. *Clay Minerals* 27: 413–421.

Soma, Y. and M. Soma. 1989. Chemical reactions of organic compounds on clay surfaces. *Environmental Health Perspectives* 83: 205–214.

Soma, M., A. Tanaka, H. Seyama, S. Hayashi, and K. Hayamizu. 1990. Bonding states of sodium in tetrasilicic sodium fluor mica. *Clay Science* 8: 1–8.

Soma, M. and B.K.G. Theng. 1998. Surface and interlayer chemical compositions, and structure of clay minerals. In *The Latest Frontiers of the Clay Chemistry*, (Eds.) A. Yamagishi, A. Aramata, and M. Taniguchi, pp. 134–143. Sendai, Japan: The Smectite Forum of Japan.

Sposito, G. 1984. *The Surface Chemistry of Soils*. New York: Oxford University Press.

Sposito, G. and R. Prost. 1982. Structure of water adsorbed on smectites. *Chemical Reviews* 82: 554–573.

Sposito, G., R. Prost, and J.-P. Gaultier. 1983. Infrared spectroscopic study of adsorbed water on reduced-charge Na/Li-montmorillonites. *Clays and Clay Minerals* 31: 9–16.

Sprynskyy, M., J. Niedojadło, and B. Buszewski. 2011. Structural features of natural and acids modified chrysotile nanotubes. *Journal of Physics and Chemistry of Solids* 72: 1015–1026.

Sreedhar, B., R. Arundhathi, M. Amarnath Reddy, and G. Parthasarathy. 2009. Highly efficient heterogeneous catalyst for acylation of alcohols and amines using natural ferrous chamosite. *Applied Clay Science* 43: 425–434.

Środoń, J. 2006. Identification and quantitative analysis of clay minerals. In *Handbook of Clay Science, Developments in Clay Science, Vol. 1,* (Eds.) F. Bergaya, B.K.G. Theng, and G. Lagaly, pp. 765–787. Amsterdam, the Netherlasnds: Elsevier.

Środoń, J and Eberl, D.D. 1984. Illite. In *Micas. Reviews in Mineralogy, Vol. 13*, (Ed.) S.W. Bailey, pp. 495–544. Washington, DC: Mineralogical Society of America.

Środoń, J. and D.K. McCarty. 2008. Surface area and layer charge of smectite from CEC and EGME/H_2O-retention measurements. *Clays and Clay Minerals* 56: 155–174.

Steudel, A., L.F. Batenburg, H.R. Fischer, P.G. Weidler, and K. Emmerich. 2009a. Alteration of non-swelling clay minerals and magadiite by acid activation. *Applied Clay Science* 44: 95–104.

Steudel, A., L.F. Batenburg, H.R. Fischer, P.G. Weidler, and K. Emmerich. 2009b. Alteration of swelling clay minerals by acid activation. *Applied Clay Science* 44: 105–115.

Stucki, J.W. and W.L. Banwart (Eds.). 1980. *Advanced Chemical Methods for Soil and Clay Minerals Research*. Dordrecht, the Netherlands: D. Reidel.

Stul, M.S. and W.J. Mortier. 1974. The heterogeneity of the charge density in montmorillonites. *Clays and Clay Minerals* 22: 391–396.

Stul, M.S. and L. Van Leemput. 1982. Particle-size distribution, cation exchange capacity and charge density of deferrated montmorillonites. *Clay Minerals* 17: 209–215.

Sudo, T., S. Shimoda, H. Yotsumoto, and S. Aita. 1981. *Electron Micrographs of Clay Minerals. Developments in Sedimentology 31*. Tokyo, Japan: Kodansha and Amsterdam the Netherlands: Elsevier.

Suitch, P.R. and R.A. Young. 1983. Atom positions in highly ordered kaolinite. *Clays and Clay Minerals* 31: 357–366.

Sukumar, R., K.R. Sabu, L.V. Bindu, and M. Lalithambika. 1998. Kaolinite supported metal chlorides as Friedel–Crafts alkylation catalysts. *Studies in Surface Science and Catalysis* 113: 557–562.

Sumner, M.E. and N.G. Reeve. 1966. The effect of iron oxide impurities on the positive and negative adsorption of chloride by kaolinites. *Journal of Soil Science* 17: 274–279.

Sun, A., J.-B. d'Espinose de la Caillerie, and J.J. Fripiat. 1995. A new microporous material: Aluminated sepiolite. *Microporous Materials* 5: 135–142.

Suquet, H., C. De la Calle, and H. Pezerat. 1975. Swelling and structural organization of saponite. *Clays and Clay Minerals* 23: 1–9.

Suquet, H., R. Franck, J.-F. Lambert, F. Elsass, C. Marcilly, and S. Chevalier. 1994. Catalytic properties of two pre-cracking matrices: A leached vermiculite and a Al-pillared saponite. *Applied Clay Science* 8: 349–364.

Suzuki, T., S. Idemura, and Y. Ono. 1987. Methylvinyl ketone formation over synthetic chrysotile. *Catalysis Letters* 16(9): 1843–1846.

Suzuki, T. and Y. Ono. 1984. The conversion of 2-propanol over chrysotile. *Applied Catalysis* 10: 361–368.

Swartzen-Allen, S.L. and E. Matijević. 1974. Surface and colloid chemistry of clays. *Chemical Reviews* 74: 385–400.

Swift, H.E. 1977. Catalytic properties of synthetic layered silicates and aluminosilicates. In *Advanced Materials in Catalysis*, (Eds.) J.J. Burton and R.L. Garten, pp. 209–233. New York: Academic Press.

Swift, H.E. and E.R. Black. 1974. Superactive nickel-aluminosilicate catalysts for hydroisomerization and hydrocracking of light hydrocarbons. *Industrial & Engineering Chemistry Product Research and Development* 13: 106–109.

Tae, J.-W., B.-S. Jang, J.-R. Kim, Il. Kim, and D.-W. Park. 2004. Catalytic degradation of polystyrene using acid-treated halloysite clays. *Solid State Ionics* 173: 129–133.

Takéuchi, Y. 1965. Structures of brittle micas. *Clays and Clay Minerals* 13: 1–25.

Tan, D., P. Yuan, F. Annabi-Bergaya et al. 2014. Loading and *in vitro* release of ibuprofen in tubular halloysite. *Applied Clay Science* 96: 50–55.

Tannenbaum, E. and I.R. Kaplan. 1985. Role of minerals in the thermal alteration of organic matter—I: Generation of gases and condensates under dry condition. *Geochimica et Cosmochimica Acta* 49: 2589–2604.

Tao, R., S. Miao, Z. Liu et al. 2009. Pd nanoparticles immobilized on sepiolite by ionic liquids: Efficient catalysts for hydrogenation of alkenes and Heck reactions. *Green Chemistry* 11: 96–101.

Tarasevich, Yu. I. and F.D. Ovcharenko. 1971. Spectral study of the interaction of water with a palygorskite surface. *Doklady Akademii Nauk SSSR* 200: 897–900.

Tari, G., I. Bobos, C.S.F. Gomes, and J.M.F. Ferreira. 1999. Modification of surface charge properties during kaolinite to halloysie-7Å transformation. *Journal of Colloid and Interface Science* 210: 360–366.

Tateiwa, J.-i., H. Horiuchi, and S. Uemura. 1995. Cation-exchanged fluorotetrasilicic mica (M^{n+} = Mn^{2+}, Cr^{3+}, Co^{2+} and Cu^{2+})-catalysed oxidation of alkanes with *tert*-butyl hydroperoxide. *Journal of the Chemical Society, Perkin Transactions* 2: 2013–2017.

Tazake, K. 1982. Analytical electron microscopic studies of halloysite formation processes—Morphology and composition of halloysite. In *International Clay Conference 1981. Developments in Sedimentology 35*, (Eds.) H. van Olphen and F. Veniale, pp. 573–584. Amsterdam, the Netherlands: Elsevier.

Teixeira, A.P.C., E.M. Santos, A.F.P. Vieira, and R.M. Lago. 2013. Use of chrysotile to produce highly dispersed K-doped MgO catalyst for biodiesel synthesis. *Chemical Engineering Journal* 232: 104–110.

Tennakoon, D.T.B., J.M. Thomas, and M.J. Tricker. 1974. Surface and intercalate chemistry of layered silicates. Part II. An iron-57 Mössbauer study of the role of lattice-substituted iron in the benzidine blue reaction of montmorillonite. *Journal of the Chemical Society, Dalton Transactions* 20: 2211–2215.

Theng, B.K.G. 1971. Mechanisms of formation of colored clay-organic complexes. A review. *Clays and Clay Minerals* 19: 383–390.

Theng, B.K.G. 1972. Adsorption of ammonium and some primary *n*-alkylammonium cations by soil allophane. *Nature* 238: 150–151.

Theng, B.K.G. 1974. *The Chemistry of Clay-Organic Reactions*. London, UK: Adam Hilger.

Theng, B.K.G. 1982. Clay-activated organic reactions. In *International Clay Conference 1981. Developments in Sedimentology 35*, (Eds.) H. van Olphen, and F. Veniale, pp. 197–238. Amsterdam, the Netherlands: Elsevier.

Theng, B.K.G. 1995. On measuring the specific surface area of clays and soils by adsorption of *para*-nitrophenol: Use and limitations. In *Clays: Controlling the Environment. Proceedings 10th International Clay Conference, Adelaide, Australia, 1993*, (Eds.) G.J. Churchman, R.W. Fitzpatrick, and R.A. Eggleton, pp. 304–310. Melbourne, Australia: CSIRO Publishing.

Theng, B.K.G. 2008. Clay-organic interactions. In *Encyclopedia of Soil Science*, (Ed.) W. Chesworth, pp. 144–150. Dordrecht, the Netherlands: Springer.

Theng, B.K.G. 2012. *Formation and Properties of Clay-Polymer Complexes*, 2nd ed. Amsterdam, the Netherlands: Elsevier.

Theng, B.K.G., G.J. Churchman, J.S. Whitton, and G.G.C. Claridge. 1984. Comparison of intercalation methods for differentiating halloysite from kaolinite. *Clays and Clay Minerals* 32: 249–258.

Theng, B.K.G., S. Hayashi, M. Soma, and H. Seyama. 1997. Nuclear magnetic resonance and X-ray photoelectron spectroscopic investigation of lithium migration in montmorillonite. *Clays and Clay Minerals* 45: 718–723.

Theng, B.K.G., M. Russell, G.J. Churchman, and R.L. Parfitt. 1982. Surface properties of allophane, halloysite, and imogolite. *Clays and Clay Minerals* 30: 143–149.

Theng, B.K.G. and G.F. Walker. 1970. Interactions of clay minerals with organic monomers. *Israel Journal of Chemistry* 8: 417–424.

Theng, B.K.G. and N. Wells. 1995. The flow characteristics of halloysite suspensions. *Clay Minerals* 30: 99–106.

Theng, B.K.G. and G. Yuan. 2008. Nanoparticles in the soil environment. *Elements* 4: 395–399.

Thomas, C.L., J. Hickey, and G. Stecker. 1950. Chemistry of clay cracking catalysts. *Industrial & Engineering Chemistry* 42: 866–871.

Thomas, J.M. 1982. Sheet silicate intercalates: New agents for unusual chemical conversions. In *Intercalation Chemistry*, (Eds.) M.S. Whittingham and A.J. Jacobson, pp. 55–99. New York: Academic Press.

Thomas, J.M. 1984. New ways of characterizing layered silicates and their intercalates. *Philosophical Transactions of the Royal Society of London A* 311: 271–285.

Thomas, J.M., J.M. Adams, S.M. Graham, and D.T.B. Tennakoon. 1977. Chemical conversions using sheet-silicate intercalates. In *Solid State Chemistry of Energy Conversion and Storage. Advances in Chemistry Series 163*, (Eds.) J.B. Goodenough and M.S. Whittingham, pp. 298–315. Washington, DC: American Chemical Society.

Tierrablanca, E., J. Romero-García, P. Roman, and R. Cruz-Silva. 2010. Biomimetic polymerization of aniline using hematin supported on halloysite nanotubes. *Applied Catalysis A: General* 381: 267–273.

Tiller, K.G. and L.H. Smith. 1990. Limitations of EGME retention to estimate the surface area of soils. *Australian Journal of Soil Research* 28: 1–26.

Tombácz, E. and M. Szekeres. 2004. Colloidal behavior of aqueous montmorillonite suspensions: The specific role of pH in the presence of indifferent electrolytes. *Applied Clay Science* 27: 75–94.

Tombácz, E. and M. Szekeres. 2006. Surface charge heterogeneity of kaolinite in aqueous suspension in comparison with montmorillonite. *Applied Clay Science* 34: 105–124.

Tournassat, C., E. Ferrage, C. Poinsignon, and L. Charlet. 2004. The titration of clay minerals. II. Structure-based model and implications for clay reactivity. *Journal of Colloid and Interface Science* 273: 234–246.

Tsou, L., G.L. Haller, and J.B. Fenn. 1987. Effect of incident translational energy on the surface-induced isomerization of cyclopropane. *Journal of Physical Chemistry* 91: 2654–2658.

Tsuchida, H., S. Ooi, K. Nakaishi, and Y. Adachi. 2005. Effects of pH and ionic strength on electrokinetic properties of imogolite. *Colloids and Surfaces A: Physicochemical and Engineering Aspects* 265: 131–134.

Tudor, J., L. Willington, D. O'Hare, and B. Royan. 1996. Intercalation of catalytically active metal complexes in phyllosilicates and their application as propene polymerisation catalysts. *Chemical Communications* 17: 2031–2032.

Uğurlu, M. and M.H. Karaoğlu. 2011. TiO_2 supported on sepiolite: Preparation, structural and thermal characterization and catalytic behaviour in photocatalytic treatment of phenol and lignin from olive mill wastewater. *Chemical Engineering Journal* 166: 859–867.

Upadhya, T.T., T. Daniel, A. Sudalai, T. Ravindranathan, and K.R. Sabu. 1996. Natural kaolinitic clay: A mild and efficient catalyst for the tetrahydropyranylation and trimethylsilylation of alcohols. *Synthetic Communications* 26: 4539–4544.

Urabe, K., S.-I. Iida, and Y. Izumi. 1994. Ni-exchanged sepiolite as a fibrous clay catalyst for selective dehydration of *n*-butyl alcohol to dibutyl ether. *Studies in Surface Science and Catalysis* 83: 453–460.

Vaccari, A. 1999. Clays and catalysis: A promising future. *Applied Clay Science* 14: 161–198.

Vahedi-Faridi, A. and S. Guggenheim. 1997. Crystal structure of tetramethylammonium-exchanged vermiculite. *Clays and Clay Minerals* 45: 859–866.

Vahedi-Faridi, A. and S. Guggenheim. 1999a. Structural study of TMP-exchanged vermiculite. *Clays and Clay Minerals* 47: 219–225.

Vahedi-Faridi, A. and S. Guggenheim. 1999b. Structural study of monomethylammonium and dimethylammonium-exchanged vermiculite. *Clays and Clay Minerals* 47: 338–347.

Van Damme, H., H. Nijs, and J.J. Fripiat. 1984. Photocatalytic reactions on clay surfaces. *Journal of Molecular Catalysis* 27: 123–142.

van der Gaast, S.J., K. Wada, S.-I. Wada, and Y. Kakuto. 1985. Small-angle X-ray powder diffraction, morphology, and structure of allophane and imogolite. *Clays and Clay Minerals* 33: 237–243.

van Olphen, H. 1971. Amorphous clay materials. *Science* 171: 90–91.

van Olphen, H. 1977. *An Introduction to Clay Colloid Chemistry*, 2nd ed. New York: John Wiley & Sons.

van Olphen, H. and J.J. Fripiat (Eds.). 1979. *Data Handbook for Clay Materials and Other Non-Metallic Minerals.* Oxford, UK: Pergamon Press.

van Santen, R.A., K.-H.W. Röbschläger, and C.A. Emeis. 1985. The hydroisomerization activity of nickel-substituted mica montmorillonite clay. In *Solid State Chemistry in Catalysis. ACS Symposium Series 279*, (Eds.) R.K. Grasselli and J.F. Brazdil, pp. 275–291. Washington, DC: American Chemical Society.

Van Scoyoc, G.E., C.J. Serna, and J.L. Ahlrichs. 1979. Structural changes in palygorskite during dehydration and dehydroxylation. *American Mineralogist* 64: 215–223.

Vandickelen, R., G. de Roy, and E.F. Vansant. 1980. New Zealand allophanes: A structural study. *Journal of the Chemical Society, Faraday Transactions 1*, 76: 2542–2551.

Varma, R.S. 2002. Clay and clay-supported reagents in organic synthesis. *Tetrahedron* 58: 1235–1255.

Vaughan, D.E.W. 1988. Pillared clays—A historical perspective. *Catalysis Today* 2: 187–198.

Vaughan, D.E.W. and R.J. Lussier. 1980. Preparation of molecular sieves based on pillared interlayered clays (PILC). In *Proceedings 5th International Conference on Zeolites, Naples, 1980*, (Ed.) L.V.C. Rees, pp. 94–101. London, UK: Heyden.

Vicente, M.A., A. Gil, and F. Bergaya. 2013. Pillared clays and clay minerals. In *Handbook of Clay Science*, 2nd ed. Developments in Clay Science, Vol. 5A, (Eds.) F. Bergaya and G. Lagaly, pp. 523–557. Amsterdam, the Netherlands: Elsevier.

Vogel, A.P., C.T. O'Connor, and M. Kojima. 1990. Thermogravimetric analysis of the iso-butene oligomerization activity of various forms of synthetic mica-montmorillonite. *Clay Minerals* 25: 355–362.

Volzone, C., O. Masini, N.A. Comelli, L.M. Grzona, E.N. Ponzi, and M.I. Ponzi. 2005. α–Pinene conversion by modified-kaolinitic clay. *Materials Chemistry and Physics* 93: 296–300.

Wada, K. 1978. Allophane and imogolite. In *Clays and Clay Minerals of Japan. Developments in Sedimentology 26*, (Eds.) T. Sudo and S. Shimoda, pp. 147–187. Tokyo, Japan: Kodansha and Amsterdam the Netherlands: Eleveier.

Wada, K. 1980. Mineralogical characteristics of Andisols. In *Soils with Variable Charge*, (Ed.) B.K.G. Theng, pp. 87–107. Lower Hutt, New Zealand: New Zealand Society of Soil Science.

Wada, K. 1989. Allophane and imogolite. In *Minerals in Soil Environments*, 2nd ed., (Eds.) J.B. Dixon and S.B. Weed, pp. 1051–1087. Madison, WI: Soil Science Society of America.

Wada, K. and D.J. Greenland. 1970. Selective dissolution and differential infrared spectroscopy for characterization of "amorphous" constituents in soil clays. *Clay Minerals* 8: 241–254.

Wada, K. and M.E. Harward. 1974. Amorphous clay constituents. *Advances in Agronomy* 26: 211–260.

Wada, K. and T. Henmi. 1972. Characterization of micropores of imogolite by measuring retention of quaternary ammonium chlorides and water. *Clay Science* 4: 127–136.

Wada, K. and S.-I. Wada. 1977. Density and structure of allophane. *Clay Minerals* 12: 289–298.

Walker, G.F. 1960. Macroscopic swelling of vermiculite crystals in water. *Nature* 187: 312–313.

Walker, G.F. 1967. Catalytic decomposition of glycerol by layer silicates. *Clay Minerals* 7: 111.

Walker, G.F. 1975. Vermiculites. In *Soil Components, Vol. 2. Inorganic Components*, (Ed.) J.E. Gieseking, pp. 155–189. Berlin, Germany: Springer-Verlag.

Wan, J. and T.K. Tokunaga. 2002. Partitioning of clay colloids at air-water interfaces. *Journal of Colloid and Interface Science* 247: 54–61.

Wang, F., J. Zhang, C. Liu, and J. Liu. 2015. Pd-palygorskite catalysts: Preparation, characterization and catalytic performance for the oxidation of styrene. *Applied Clay Science* 105–106: 150–155.

Wang, G., R. Ma, T. Chen, C. Yan, J. Gao, and F. Bao. 2013. Palygorskite as efficient catalyst for ring-opening polymerization of ε-caprolactone. *Polymer-Plastics Technology and Engineering* 52: 1193–1199.

Wang, M.C. and P.M. Huang. 1989b. Pyrogallol transformations as catalyzed by nontronite, bentonite, and kaolinite. *Clays and Clay Minerals* 37: 525–531.

Wang, M.C. and P.M. Huang. 1989a. Catalytic power of nontronite, kaolinite and quartz and their reaction sites in the formation of hydroquinone-derived polymers. *Applied Clay Science* 4: 43–57.

Wang, R., G. Jiang, Y. Ding et al. 2011. Photocatalytic activity of heterostructures based on TiO_2 and halloysite nanotubes. *ACS Applied Materials & Interfaces* 3: 4154–4158.

Wang, T.S.C., M.-M. Kao, and P.M. Huang. 1980. The effect of pH on the catalytic synthesis of humic substances by illite. *Soil Science* 129: 333–338.

Wang, T.S.C., S.W. Li, and Y.L. Ferng. 1978. Catalytic polymerization of phenolic compounds by clay minerals. *Soil Science* 135: 350–360.

Weaver, C.E. 1989. *Clays, Muds, and Shales*. Amsterdam, the Netherlands: Elsevier.

Weaver, C.E. and L.D. Pollard. 1973. *The Chemistry of Clay Minerals. Developments in Sedimentology 15*. Amsterdam, the Netherlands: Elsevier.

Wei, J., G. Furrer, S. Kaufmann, and R. Schulin. 2001. Influence of clay minerals on the hydrolysis of carbamate pesticides. *Environmental Science & Technology* 35: 2226–2232.

Wei, Z., J.M. Moldowan, J. Dahl, T.P. Goldstein, and D.M. Jarvie. 2006. The catalytic effects of minerals on the formation of diamondoids from kerogen macromolecules. *Organic Geochemistry* 37: 1421–1436.

Weiss, A. and J. Russow. 1963. Über die Lage der austauschbaren Kationen bei Kaolinit. In *International Clay Conference 1963, Vol. 1*, (Eds.) I.Th. Rosenqvist and P. Graff-Petersen, pp. 203–213. Oxford, UK: Pergamon Press.

Wells, N., C.W. Childs, and C.J. Downes. 1977. Silica Springs, Tongariro National Park, New Zealand—Analysis of the spring water and characterization of the alumino-silicate deposit. *Geochimica et Cosmochimica Acta* 41: 1497–1506.

Wells, N. and B.K.G. Theng. 1985. Factors affecting the flow behavior of soil allophane suspensions under low shear rates. *Journal of Colloid and Interface Science* 104: 398–408.

Wells, N., B.K.G. Theng, and G.D. Walker. 1980. Behaviour of imogolite gels under shear. *Clay Science* 5: 257–265.

Whalley, W.R. and C.E. Mullins. 1991. Effect of saturating cation on tactoid size distribution in bentonite suspensions. *Clay Minerals* 26: 11–17.

Wicks, F.J. and E.J.W. Whittaker. 1975. A reappraisal of the structures of the serpentine minerals. *Canadian Mineralogist* 13: 227–243.

Wilson, M.A., P.F. Barron, and A.S. Campbell. 1984. Detection of aluminium coordination in soils and clay fractions using ^{27}Al magic angle spinning NMR. *Journal of Soil Science* 35: 201–207.

Wilson, M.A., G.S.H. Lee, and R.C. Taylor. 2002. Benzene displacement on imogolite. *Clays and Clay Minerals* 50: 348–351.

Wilson, M.C. and A.K. Galwey. 1976. Reactions of stearic acid, of *n*-dodecanol, and of cyclohexanol on the clay minerals illite, kaolinite and montmorillonite. *Journal de Chimie Physique* 73: 442–446.

Wilson, M.J. (Ed.). 1994. *Clay Mineralogy: Spectroscopic and Chemical Determinative Methods*. London, UK: Chapman & Hall.

Wolters, F. and K. Emmerich. 2007. Thermal reactions of smectites—Relation of dehydroxylation temperature to octahedral structure. *Thermochimica Acta* 462: 80–88.

Wright, A.C., W.T. Granquist, and J.V. Kennedy. 1972. Catalysis by layer lattice silicates. 1. The structure and thermal modification of a synthetic ammonium dioctahedral clay. *Journal of Catalysis* 25: 65–80.

Wright, P.C. 1968. The Meandu Creek bentonite—A reply. *Journal of the Geological Society of Australia* 15: 347–350.

Wu, L.M., C.H. Zhou, J. Keeling, D.S. Tong, and W.H. Yu. 2012. Towards an understanding of the role of clay minerals in crude oil formation, migration and accumulation. *Earth-Science Reviews* 115: 373–386.

Wystrach, V.P., L.H. Barnum, and M. Garber. 1957. Liquid phase catalytic isomerization of α-pinene. *Journal of the American Chemical Society* 79: 5786–5790.

Yada, K. 1971. Study of microstructure of chrysotile asbestos by high resolution electron microscopy. *Acta Crystallographica A* 27: 659–664.

Yamanaka, S. and G.W. Brindley. 1978. Hydroxy-nickel interlayering in montmorillonite by titration method. *Clays and Clay Minerals* 26: 21–24.

Yamanaka, S. and G.W. Brindley. 1979. High surface area solids obtained by reaction of montmorillonite with zirconyl chloride. *Clays and Clay Minerals* 27: 119–124.

Yan, W., D. Liu, D. Tan, P. Yuan, and M. Chen. 2012. FTIR spectroscopy study of the structure changes of palygorskite under heating. *Spectrochimica Acta Part A: Molecular and Biomolecular Spectroscopy* 97: 1052–1057.

Yariv, S. 1992. The effect of tetrahedral substitution of Si by Al on the surface acidity of the oxygen plane of clay minerals. *International Reviews in Physical Chemistry* 11: 345–375.

Yariv, S. and H. Cross (Eds.). 2002. *Organo-Clay Complexes and Interactions*. New York: Marcel Dekker.

Yariv, S. and K.H. Michaelian. 2002. Structure and surface acidity of clay minerals. In *Organo-Clay Complexes and Interactions*, (Eds.) S. Yariv and H. Cross, pp. 1–38. New York: Marcel Dekker.

Yoshinaga, N. and S. Aomine. 1962. Imogolite in some Ando soils. *Soil Science and Plant Nutrition* 8: 22–29.

Youssef, A.M. 1979. The adsorption and catalytic properties of some Egyptian clays. *Surface Technology* 9: 187–193.

Yuan, G., B.K.G. Theng, R.L. Parfitt, and H.J. Percival. 2000. Interactions of allophane with humic acid and cations. *European Journal of Soil Science* 51: 35–41.

Yuan, P., D. Tan, and F. Annabi-Bergaya. 2015. Properties and applications of halloysite nanotubes: Recent research advances and future prospects. *Applied Clay Science* 112–113: 75–93.

Zalma, R., J. Guignard, E. Copin, and H. Pezerat. 1986. Studies on surface properties of asbestos. IV. Catalytic role of asbestos in fluorene oxidation. *Environmental Research* 41: 296–301.

Zamaraev, K.I., V.N. Romannikov, R.I. Salganik, W.A. Wlassoff, and V.V. Khramtsov. 1997. Modelling of the prebiotic synthesis of oligopeptides: Silicate catalysts help to overcome the critical stage. *Origins of Life and Evolution of the Biosphere* 27: 325–337.

Zatta, L., J.E.F. da Costa Gardolinski, and F. Wypych. 2011. Raw halloysite as reusable heterogeneous catalyst for esterifaction of lauric acid. *Applied Clay Science* 51: 165–169.

Zhang, D., W. Huo, J. Wang et al. 2012. Synthesis of allyl-ended hyperbranched organic silicone resin by halloysite-supported platinum catalyst. *Journal of Applied Polymer Science* 126: 1580–1584.

Zhang, L., Q. Jin, L. Shan, Y. Liu, X. Wang, and J. Huang. 2010. $H_3PW_{12}O_{40}$ immobilized on silylated palygorskite and catalytic activity in esterification reactions. *Applied Clay Science* 47: 229–234.

Zhang, L., J. Liu, C. Tang et al. 2011b. Palygorskite and SnO_2-TiO_2 for the photodegradation of phenol. *Applied Clay Science* 51: 68–73.

Zhang, Y., Z. Li, W. Sun, and C. Xia. 2009. Co-doped attapulgite catalyzed solvent-free oxidation of cyclohexane using molecular oxygen. *Catalysis Letters* 129: 222–227.

Zhang, Y., A. Tang, H. Yang, and J. Ouyang. 2016. Applications and interfaces of halloysite nanocomposites. *Applied Clay Science* 119: 8–17.

Zhang, Y., D. Wang, and G. Zhang. 2011a. Photocatalytic degradation of organic contaminants by TiO_2/sepiolite composites prepared at low temperature. *Chemical Engineering Journal* 173: 1–10.

Zhang, Y., W. Wang, J. Zhang, P. Liu, and A. Wang. 2015. A comparative study about adsorption of natural palygorskite for methylene blue. *Chemical Engineering Journal* 262: 390–398.

Zhao Z., J. Ran, Y. Jiao, W. Li, and B. Miao. 2016. Modified natural halloysite nanotube solely employed as an efficient and low-cost solid acid catalyst for alpha-arylstyrenes production via direct alkenylation. *Applied Catalysis A: General* 513: 1–8.

Zhao, D., J. Zhou, and N. Liu. 2006. Characterization of the structure and catalytic activity of copper modified palygorskite/TiO_2 (Cu^{2+}-PG/TiO_2) catalysts. *Materials Science and Engineering: A* 431: 256–262.

Zhao, D., J. Zhou, and N. Liu. 2007. Surface characteristics and photoactivity of silver-modified palygorskite clays coated with nanosized titanium dioxide particles. *Materials Characterization* 58: 249–255.

Zheng, S.-Q., S.-H. Sun, Z.-F. Wang, X.-H. Gao, and X.-L. Xu. 2005. Suzhou kaolin as a FCC catalyst. *Clay Minerals* 40: 303–310.

Zhou, C.H. 2011. An overview on strategies towards clay-based designer catalysts for green and sustainable catalysis. *Applied Clay Science* 53: 87–96.

Zhou, C.-H., D. Tong, and X. Li. 2010. Synthetic hectorite: Preparation, pillaring and applications in catalysis. In *Pillared Clays and Related Catalysts*, (Eds.) A. Gil, S.A. Korili, R. Trujillano, and M.A. Vicente, pp. 67–97. New York: Springer.

Zhu, R.H., Q. Chen, R.L. Zhu et al. 2015. Sequestration of heavy metal cations on montmorillonite by thermal treatment. *Applied Clay Science* 107: 90–97.

Zhu, R.H., R.L. Zhu, F. Ge et al. 2016a. Effect of heating temperature on the sequestration of Ce^{3+} cations on montmorillonite. *Applied Clay Science* 121–123: 111–118.

Zhu, X., Z. Zhu, X. Lei, and C. Yan. 2016b. Defects in structure as the source of the surface charges of kaolinite. *Applied Clay Science* 124–125: 127–136.

Zou, M.L., M.L. Du, H. Zhu, C.S. Xu, and Y.Q. Fu. 2012. Green synthesis of halloysite nanotubes supported Ag nanoparticles for photocatalytic decomposition of methylene blue. *Journal of Physics D: Applied Physics* 45: 325302–325308.

2 Surface Acidity and Catalytic Activity

2.1 INTRODUCTION

The catalytic activity of clays and clay minerals is intimately related to their ability to act as solid acids (Benesi and Winquist 1978; Tanabe 1981; Theng 1982; Solomon and Hawthorne 1983; Laszlo 1987; Hattori and Ono 2015). Indeed, the inherent or acquired surface acidity of the smectite group of phyllosilicates (cf. Table 1.1) together with their propensity for adsorbing and intercalating organic compounds lies behind the exceptional capacity and versatility of these minerals to catalyze organic reactions and syntheses (Mortland 1970; Theng 1974; Thomas et al. 1977; Thomas 1982; Pinnavaia 1983; Adams 1987; Yariv and Cross 2002; McCabe and Adams 2013). Besides being abundant, inexpensive, noncorrosive, recoverable, and recyclable, clays and clay minerals are easy to separate from the reaction mixture, while the catalytic reaction can often be carried out at a moderately high temperature with minimum formation of by-products (Balogh and Laszlo 1993; Motokura et al. 2007; Dasgupta and Török 2008). The conversion and transformation of various organic compounds at smectite surfaces will be described in later chapters. Here we examine the relationship between surface acidity and catalytic activity. Before doing so, however, it seems appropriate to outline the concept of acidity in homogeneous systems.

2.2 ACIDS AND BASES

According to the Brønsted (1923) definition, an acid is a substance that is capable of donating a proton to a base acting as a proton acceptor. The dissociation of an acid (HA) may be represented by the following equilibrium:

$$HA \rightleftharpoons H^+ + A^- \tag{2.1}$$

The protonation of a base (B) is given by:

$$B + H^+ \rightleftharpoons BH^+ \tag{2.2}$$

Since species A^- in Equation 2.1 acts as a proton acceptor (i.e., a base), it is referred to as the conjugate base of HA, while in Equation 2.2 BH^+ acts as a proton donor (i.e., an acid), and hence represents the conjugate acid of B.

Thus, the acid-base interaction is essentially one of proton transfer from HA to B, represented by Equation 2.3:

$$HA + B \rightleftharpoons BH^+ + A^- \tag{2.3}$$

In the Lewis (1923) concept, on the other hand, an acid (A) is an ion or molecule capable of accepting a pair of electrons from a base (B), acting as an electron-pair donor. The formation of a coordinate bond is, therefore, implicit in the Lewis acid-base interaction and can be represented by Equation 2.4 (Busca 2007):

$$A + :B \rightleftharpoons A:B \tag{2.4}$$

where A : B is the acid-base complex (adduct)

We might add that some acid-base interactions cannot adequately be described in terms of either proton transfer or electron-pair acceptance. In order to account for such processes, the concept of hydrogen-bond donor (HBD) acidity has been introduced (e.g., Abraham et al. 1986) in which an acid is defined as a hydrogen-bond donor and a base as a hydrogen-bond acceptor. Although this type of acid-base interaction does occur in clay systems, such as between water molecules and the interlayer surfaces in hydrated halloysite (cf. Figure 1.10), the role of HBD acidity in clay catalysis awaits development and will not be considered here.

2.2.1 Acid Strength

With reference to Equation 2.1, the equilibrium constant (K_a) may be written as

$$K_a = \frac{a_{H^+} a_{A^-}}{a_{HA}} = \frac{a_{H^+}[A^-]}{[HA]} \cdot \frac{\gamma_{A^-}}{\gamma_{HA}} \tag{2.5}$$

where:
 a is the activity
 γ is the activity coefficient
 the quantities in square brackets denote the concentrations of the respective species

In dilute solutions, the ratio γ_{A^-}/γ_{HA} would approach unity, and Equation 2.5 simplifies to the following:

$$K_a = \frac{a_{H^+}[A^-]}{[HA]} \tag{2.6}$$

On taking the negative logarithm, Equation 2.6 converts to the familiar form:

$$pK_a = pH + \log\frac{[HA]}{[A^-]} \tag{2.7}$$

Accordingly, the strength or proton-donating power of an acid is directly related to its pK_a value; the smaller the numerical value, the stronger the acid strength of the medium in which the acid is dissolved. Similarly, the strength of a base is given by its pK_b value, although it is conventionally expressed in terms of the pK_a of the corresponding conjugate acid, according to Equation 2.8:

$$-pK_b = pK_a = pH + \log\frac{[BH^+]}{[B]} \tag{2.8}$$

In concentrated acid solutions, however, the proton-donating power of the medium cannot be as simply expressed as by Equation 2.7 because the activity coefficients must now be taken into account (Bender et al. 1984). In such a situation, the protonation of a given indicator base (B_1) is described by Equation 2.9:

$$(pK_a)_1 = pH + \log\frac{[B_1H^+]}{[B_1]} + \log\left(\frac{\gamma_{B_1H^+}}{\gamma_{B_1}}\right) \tag{2.9}$$

If a second indicator base (B_2) is now introduced into the system, and its $(pK_a)_2$ value is subtracted from that of B_1 we get:

$$\Delta pK_a = \log\frac{[B_1H^+]}{[B_1]} - \log\frac{[B_2H^+]}{[B_2]} + \log\left(\frac{\gamma_{B_1H^+}}{\gamma_{B_1}}\right)\left(\frac{\gamma_{B_2}}{\gamma_{B_2H^+}}\right) \tag{2.10}$$

For one and the same medium, however, $\gamma_{B1}/\gamma_{B1H^+}$ would vary in a similar way to $\gamma_{B2}/\gamma_{B2H^+}$ so that $\gamma_{B1}/\gamma_{B1H^+} \approx \gamma_{B2}/\gamma_{B2H^+}$, and the product $(\gamma_{B1H^+}/\gamma_{B1}) \times (\gamma_{B2}/\gamma_{B2H^+})$ would approach unity, making the last term in Equation 2.10 vanishingly small (Paul and Long 1957). By using a series of indicators, covering a range of pK_a values, we can thus construct a scale of acidity for concentrated acid solutions.

The underlying principle may be illustrated by writing the general (non-logarithmic) form of the equilibrium constant for Equation 2.9 as:

$$K_a = \frac{a_{H^+}[B]}{[BH^+]} \cdot \frac{\gamma_B}{\gamma_{BH^+}} \quad (2.11)$$

If we then substitute h_o for the product $a_{H^+} \cdot \gamma_B/\gamma_{BH^+}$ in Equation 2.11, and convert to the logarithmic form, we get Equation 2.12:

$$-\log K_a = -\log h_o - \log \frac{[B]}{[BH^+]} \quad (2.12)$$

Writing H_o for $-\log h_o$ yields:

$$pK_a = H_o + \log \frac{[BH^+]}{[B]} \quad (2.13)$$

H_o is known as the Hammett acidity function (Hammett and Deyrup 1932). Comparing Equation 2.13 with the equivalent expression for dilute aqueous solutions, given by Equation 2.7, shows the close analogy that exists between H_o and pH. Thus, H_o may be regarded as an extension of pH to media in which the activity of the proton ceases to be identical with its stoichiometric concentration (Benesi 1957).

Expressing the Hammett function in the form of Equation 2.14,

$$H_o = -\log \frac{a_{H^+}\gamma_B}{\gamma_{BH^+}} \quad (2.14)$$

as is commonly given in the literature (Walling 1950; Forni 1974; Benesi and Winquist 1978; Atkinson and Curthoys 1979; Tanabe 1981; Jacobs 1984; Anderson and Klinowski 1986) shows that solutions of strong acids with the same molarity can have markedly different H_o values (Figure 2.1, inset). The shape of the curve for sulfuric acid makes it convenient to use aqueous solutions of this acid as a basis for comparing the pK_a (H_o) values of *Hammett indicators* (Figure 2.1), some of which are listed in Table 2.1. Note that the less positive (or more negative) the pK_a value of the indicator base, the greater is the acid strength required for its protonation.

Thus, the simplest way of assessing the acid strength of solid catalysts, including clay minerals, is to note the change in color of Hammett indicators when they adsorb/intercalate (e.g., from a benzene solution) to the solid acid in question. If the adsorbed indicator base (B) assumes the color of its acid form, the H_o value of the surface is equal to, or lower than, the pK_a of its conjugate acid (BH^+). In other words, a given indicator would only convert into its corresponding conjugate acid or protonated form when the medium (here the clay surface) has an acid strength equal to, or greater than, that of the indicator base. For example, a medium (surface) with a pK_a between −5.6 and −8.0 (in the H_o scale) would yield a yellow color with benzalacetophenone but not with anthraquinone, while in a medium with a pK_a more negative than −8.2, all the indicators in Table 2.1 would give their respective acid colors. Likewise, if a solid acid turns purple in the presence of benzeneazodiphenylamine, and red with dicinnamalacetone, its acid strength falls in the range of $+1.5 < H_o < -3.0$.

Solid acid catalysts, including clay minerals, are therefore characterized by having a distribution, rather than a single value, of acid strength (Barthomeuf 1985) as shown in Figure 2.4 and Table 2.4. The spread (range) of acid strength could, in principle, be narrowed down by the judicial use of indicators. In practice, however, the choice of indicators is limited by the requirement that the acid

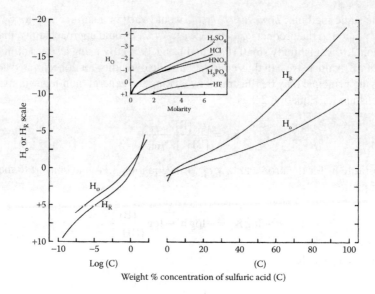

FIGURE 2.1 Relationship between H_o or H_R values and weight percent concentration [C] of aqueous sulfuric acid for weakly acid (a) and strongly acid (b) conditions. Inset: relationship between H_o values and molarity of aqueous solutions of different inorganic acids. (Adapted from Paul, M.A. and Long, F.A., *Chem. Rev.*, 57, 1–45, 1957. Jacobs, P.A., The measurement of surface acidity, In *Characterization of Heterogeneous Catalysts*, F. Delannay (Ed.), pp. 367–404, Marcel Dekker, New York, 1984.)

TABLE 2.1
Some Hammett Indicators for the Visual Estimation of Acid Strength of Solid Acid Catalysts, Listed in the Order of Increasing Acid Strength

Indicator (Compound)	pK_a (H_o)	$[H_2SO_4]^1$	Basic Color	Acid Color
Neutral red	+6.8	8×10^{-8}	Yellow	Red
Methyl red	+4.8	8×10^{-6}	Yellow	Red
4-Phenylazo-1-naphthylamine[2]	+4.0	5×10^{-5}	Yellow	Red
p-Dimethylaminoazobenzene (butter yellow)[2]	+3.3	3×10^{-4}	Yellow	Red
2-Amino-5-azotoluene	+2.0	5×10^{-3}	Yellow	Red
Benzeneazodiphenylamine[2]	+1.5	2×10^{-2}	Yellow	Purple
4-Dimethylaminoazobenzene[2]	+1.2	3×10^{-1}	Yellow	Red
Crystal violet	+0.8	1×10^{-1}	Blue	Yellow
p-Nitrobenzene-(p'-nitro)-diphenylamine	+0.4	~6	Orange	Purple
Dicinnamalacetone	−3.0	48	Orange	Red
Benzalacetophenone	−5.6	71	Colorless	Yellow
Anthraquinone	−8.2	90	Colorless	Yellow

Sources: Forni, L., *Catal. Rev.: Sci. Eng.*, 8, 65–115, 1974; Benesi, H.A. and B.H.C. Winquist., Surface acidity of solid catalysts, in *Advances in Catalysis*, Vol. 27 (Eds.) D.D. Eley, H. Pines, and P.B. Weisz, Academic Press, New York, pp. 97–182, 1978; Atkinson, D. and G. Curthoys, *Chem. Soc. Rev.*, 8, 475–497, 1979; Tanabe, K., Solid acid and base catalysts, in *Catalysis–Science and Technology*, Vol. 2 (Eds.) J.R. Anderson and M. Boudart, Germany, Springer-Verlag, Berlin, pp. 231–273, 1981; Unger, K.K. et al., *J. Chem. Technol. Biotechnol.*, 31, 453–469, 1981; Jacobs, P.A., The measurement of surface acidity, in *Characterization of Heterogeneous Catalysts* (Ed.) F. Delannay, Marcel Dekker, New York, pp. 367–404, 1984; Anderson, J.R. and K.C. Pratt, *Introduction to Characterization and Testing of Catalysts*, Academic Press, Sydney, Australia, 1985.

[1] Corresponding concentration of aqueous sulfuric acid solutions (% w/w).
[2] Reproducibility of ±0.005 mmol/g or better may be obtained with these indicators.

TABLE 2.2
Some Hammett Indicators for Determining the Acid Strength of Solid Acid Catalysts Using Ultraviolet Spectroscopy, Listed in the Order of Increasing Acid Strength

Indicator (Compound)	pK_a (H_o)	$[H_2SO_4]$[1]	Wavelength (nm) Basic Form	Acid Form
p-Dimethylaminoazobenzene	+3.3	3×10^{-4}	400	520
p-Nitroaniline	+1.1	$\sim 5 \times 10^{-2}$	380	323
o-Nitroaniline	−0.2	~17	378	270
o-Nitrodiphenylamine	−2.4	~41	350	440
2,4-Dinitroaniline	−4.4	~60	310, 365	235
Benzalacetophenone	−5.6	71	300	400
2,4,6-Trinitroaniline	−9.3	95	320, 390	340, 420
p-Nitrotoluene	−10.3	~98	260	340–380
Nitrobenzene	−11.3	~100	252	280

Sources: Same as for Table 2.1
[1] Corresponding concentration of aqueous sulfuric acid solution (% w/w).

color should not mask that of the basic form (Walling 1950). This constraint may largely be overcome by using fluorescent indicators and/or by determining the color change spectrophotometrically (Drushel and Sommers 1966). An extension of the latter approach is the use of indicators that absorb in the ultraviolet range of the spectrum, enabling the acid and basic forms of the compounds to be determined at the same time (Table 2.2).

A more serious drawback relates to the possibility that the indicator compound is physically adsorbed and/or coordinated to surface Lewis sites, causing a (bathochromic) shift to longer wavelengths (e.g., Jacobs 1984). As a result, the color produced might resemble that given by the conjugate acid or protonated form of the indicator (Lewis and Bigeleisen 1943; Drushel and Sommers 1966; Frenkel 1975; Anderson and Klinowski 1986). Indeed, the general consensus is that Hammett indicators are incapable of distinguishing between Brønsted and Lewis acidity. In other words, the acidity measured using these indicators represents the sum of both types of acidity (Atkinson and Curthoys 1979; Tanabe 1981; Solomon and Hawthorne 1983; Jacobs 1984; Anderson and Pratt 1985; Ghosh and Kydd 1985).

By analogy with Equation 2.14, the interaction between a Lewis acid (A) and an indicator base (B) may be expressed by the function:

$$H_o = -\log \frac{a_A \gamma_B}{\gamma_{AB}} \tag{2.15}$$

In principle, however, the relative strength of Lewis acid sites cannot be described by Equation 2.15 even if the color of the acid-base complex does mimic that of the protonated base (Benesi 1956). This problem arises because a coordinative complex formation between A and B, described by Equation 2.4, is strongly influenced by conformational and steric factors, and hence is specific to the Lewis site in question. Likewise, catalytic activity cannot be correlated with Lewis acid strength since no unique scale for this parameter can be constructed (Barthomeuf 1985).

Because of the inherent ambiguity of the Hammett indicator method, Hirschler (1963) has advocated the use of arylmethanol (arylcarbinol) indicators. On protonation, these compounds convert into the corresponding stable, *colored* carbocations (R⁺) according to the following equilibrium (Gold and Hawes 1951),

$$ROH + H^+ \rightleftharpoons R^+ + H_2O \tag{2.16}$$

TABLE 2.3
Some Arylmethanol Indicators for Estimating Brønsted Acidity of Solid Catalysts, Listed in the Order of Increasing Acid Strength

Indicator (Compound)	pK_{R^+}	$[H_2SO_4]$[1]
4-Dimethylamino-triphenylmethanol	+4.75	1.71×10^{-5}
4,4′,4″-Trimethoxy-triphenylmethanol	+0.82	1.2
4,4′-Dimethoxy-triphenylmethanol	−1.24	16
4-Methoxy-triphenylmethanol	−3.40	32
4,4′,4″-Trimethyl-triphenylmethanol	−4.02	36
4,4′-Dimethoxy-diphenylmethanol	−5.71	44
Triphenylmethanol	−6.63	49
4,4′,4″-Trichloro-triphenylmethanol	−10.40	64
3,3′,3″-Trichloro-triphenylmethanol	−11.03	68
Diphenylmethanol	−13.30	77
4,4′-Dichloro-diphenylmethanol	−13.96	80
4,4′,4″-Trinitro-triphenylmethanol	−16.27	88

Sources: As for Table 2.1; see also Deno, N.C. et al., *J. Am. Chem. Soc.*, 77, 3044–3051, 1955; Hirschler, A.E., *J. Catal.*, 2, 428–439, 1963.

[1] Corresponding concentration of aqueous sulfuric acid solution (% w/w).

where R represents a substituted diphenylmethyl or triphenylmethyl radical (Table 2.3). Equation 2.16 is described by the so-called H_R acidity function (Deno et al. 1955, 1959), which like H_o, is operationally defined as:

$$H_R = pK_R + \log \frac{[ROH]}{[R^+]} \tag{2.17}$$

The equivalent logarithmic form of Equation 2.17 is given by:

$$H_R = -\log a_{H^+} + \log a_{H_2O} + \log \frac{\gamma_{R^+}}{\gamma_{ROH}} \tag{2.18}$$

Equation 2.18 indicates that H_R approaches pH in dilute aqueous solutions. Since the H_R function includes a term for the activity of water, the measurement of acid strength using arylmethanol indicators is, if anything, more sensitive to the presence of physisorbed water than is the case with Hammett indicators.

The relationship between H_R and H_o, obtained by combining Equations 2.14 and 2.18, is given by:

$$H_R = H_o + \log a_{H_2O} - \log \frac{\gamma_{ROH}\gamma_{BH^+}}{\gamma_B\gamma_{R^+}} \tag{2.19}$$

According to Equation 2.19, different numerical values of H_R and H_o apply to the same weight percent concentration of sulfuric acid (Figure 2.1).

Despite the prototropic nature of Equation 2.16, Tarasevich and Fedorova (1979) have suggested that the ionization of arylmethanols, adsorbed on montmorillonite and palygorskite, can also be activated by Lewis sites. This suggestion is based on the preservation of absorption bands in the 400–450 nm region of the spectra, characteristic of the respective carbocations, for complexes that have been outgassed at 500°C. It seems more likely, however, that the presence of residual Brønsted sites is responsible for the protonation of the adsorbed indicators since the acidity of such sites, as we will see later, can be enhanced by neighboring Lewis (*defect*) sites through an inductive effect (Lunsford 1968; Theng 1982; Guisnet 1990; Corma and García 1997; Wilson and Clark 2000; Chizallet and Raybaud 2010; Camejo-Abreu et al. 2014).

Following Hirschler and Schneider (1961) and Hirschler (1963), Atkinson and Curthoys (1980) and Unger et al. (1981) have applied the Hammett and arylmethanol indicator methods in parallel in an attempt at distinguishing between Brønsted and Lewis acidity in silicas, aluminas, and synthetic zeolites. The concentration of (Brønsted and Lewis) acid sites is estimated by titration with n-butylamine—that is, from the amount of amine required to restore the indicator to its original (acid) color as described in Section 2.6.1. They argued that such a distinction should be possible if Hammett indicators measured the sum of Brønsted and Lewis acidities, while arylmethanols would only react with Brønsted sites.

Desjardins et al. (1999) have adopted the same approach to assess the acid strength and concentration of Brønsted and Lewis acid sites in Na^+-, Ca^{2+}-, and Fe^{3+}-exchanged montmorillonites (Table 2.4) with respect to their activity in catalyzing the oxidative oligomerization of chloro- and methoxy-phenol. The superior performance of Fe^{3+}-montmorillonite may be ascribed to the ability of interlayer ferric ions to act as electron acceptors (Lewis acid sites) as elaborated on in Section 2.4. The high activity of (anhydrous) Fe^{3+}-exchanged K10 montmorillonite in catalyzing toluene benzylation (Hart and Brown 2004), and cyclohexanol acetylation (Shimizu et al. 2008), has been explained in similar terms.

Some time ago, Weil-Malherbe and Weiss (1948) observed that some weakly basic indicators developed identical colors when they adsorbed to clays or were added to concentrated solutions of sulfuric acid. Since then, the Hammett indicator method has often been used to assess the surface acidity of clay minerals under various conditions of sample pretreatment (Benesi 1956; Fowkes et al. 1960; Hirschler and Schneider 1961; Solomon and Hawthorne 1971; Solomon et al. 1971; Frenkel 1974; Henmi and Wada 1974; Telichkun et al. 1977; Sivalov et al. 1980; Liu et al. 2013b). The commonly used procedure is to suspend the clay powder in a nonpolar solvent (e.g., benzene, cyclohexane, isooctane), add a few drops of the indicator (dissolved in the same solvent), and note the color of the adsorbed compound. The data for different clay minerals, obtained in this manner, are summarized in Table 2.5.

In common with other solid acids, individual species of layer silicates show a range of acid strengths. As might be expected, the acid strength of H^+-exchanged—in reality, mixed H^+/Al^{3+}-exchanged—samples is appreciably higher than that of the sodium-saturated counterparts. Despite differences in sample origin and pretreatment, however, individual sets of data are reasonably self-consistent. The rather wide distribution of acid strengths, shown by kaolinite and montmorillonite, may be partly due to the presence of residual water or water that is taken up from the atmosphere during handling. In the case of palygorskite, the apparent discrepancy may be ascribed to the

TABLE 2.4
Total, Brønsted, and Lewis Acidity for Na^+-, Ca^{2+}-, and Fe^{3+}-Exchanged Montmorillonites (M) Measured by the Hammett (H_o) and Arylmethanol (H_R) Indicator Methods Combined with n-Butylamine Titration

Strength	Strongly Acid Sites pK_a	Acid Site Number (mmol/g)			Moderately Acid Sites pK_a	Acid Site Number (mmol/g)			Weakly Acid Sites pK_a	Acid Site Number (mmol/g)		
		Na-M	Ca-M	Fe-M		Na-M	Ca-M	Fe-M		Na-M	Ca-M	Fe-M
Total acidity (H_o)	−2.0	0.11	0.16	1.0	1.5	0.31	0.39	1.43	3	1.32	1.40	1.82
Brønsted acidity (H_R)	−4.0	0.01	0.01	0.7	3.0	0.20	0.27	1.34	7	1.25	1.28	1.61
Lewis acid site number[1]		0.10	0.15	0.3		0.11	0.12	0.09		0.07	0.12	0.21

Source: Data from Desjardins, S. et al., *J. Soil Contamin.*, 8, 175–195, 1999.

Note: Measurements were carried out in iso-octane. Before use, the clay mineral suspensions were adjusted to pH 2, and dried at 89°C.

[1] Obtained by subtracting Brønsted from total acidity.

TABLE 2.5
Acid Strength of Representative Clay Minerals Measured by the Hammett Indicator Method and Expressed in Terms of the Hammett Acidity Function (H_o)

Clay Mineral	Exchangeable Cation (Counterion)	H_o Range[1]
Kaolinite	Natural	−3.0 to −5.6
	Na$^+$	−3.0 to −5.6
	H$^+$/Al^{3+}	−5.6 to −8.2
Halloysite	Na$^+$	−5.6 to −8.2
Pyrophyllite	None	−5.6 to −8.2
Talc	None	+1.5 to −3.0
Montmorillonite	Li$^+$	+3.3 to +1.5
	Na$^+$	+2.0 to +1.5
	K$^+$	+4.8 to +1.5
	Ca^{2+}	+1.5 to −3.0
	Mg^{2+}	−3.0 to −5.6
	H$^+$/Al^{3+}	−5.6 to −8.2
	K10[2]	−5.6 to −5.9
		−6 to −8
Palygorskite ('attapulgite')	Natural	+1.5 to −3.0
Sepiolite	Natural	+6.8 to −1.5
Allophane	Na$^+$	+1.5 to −5.6
	H$^+$/Al^{3+}	+1.5 to −5.6
Imogolite	Na$^+$	+1.5 to −3.0
	H$^+$/Al^{3+}	+1.5 to −5.6

Sources: Benesi, H.A., *J. Am. Chem. Soc.*, 78, 5490–5494, 1956; Benesi, H.A., *J. Phys. Chem.*, 61, 970–973, 1957; Fowkes, F.M. et al., *J. Agricul. Food Chem.*, 8, 203–210, 1960; Bailey, G.W. et al., *Soil Sci. Soc. Am. Proc.*, 32, 333–234, 1968; Henmi, T. and K. Wada., *Clay Min.*, 10, 231–245, 1974; Solomon, D.H. and D.G. Hawthorne., *Chemistry of Pigments and Fillers*, John Wiley & Sons, New York, 1983; Wu, D.Y. et al., *Clay Sci.*, 8, 367–379, 1992; Imamura, S. et al., *Indust. Eng. Chem. Res.*, 32, 600–603, 1993; Kitayama, Y. et al., *J. Mol. Catal. A: Chem.*, 142, 237–245, 1999; Liu, H. et al., *Phys. Chem. Min.*, 40, 479–489, 2013a.

[1] Values refer to samples that have been dried to near-zero water content, kept at zero relative humidity, or heated at 100°C–120°C.

[2] K10 is a commercially available acid-activated montmorillonite. Data from Laszlo, P., *Science*, 235, 1473–1477, 1987 and Cornélis, A. et al., Montmorillonite K10. *e-EROS Encyclopedia of Reagents for Organic Synthesis*, Wiley Online Library, pp. 3667–3671, 1999.

intervention of steric and structural factors. For example, anthraquinone (pK_a of −8.2) with a van der Waals diameter of ca. 0.8 nm would have difficulty in gaining entry into the channel structure of palygorskite (cf. Chapter 1). Similarly, none of the indicators in Table 2.1 would have unrestricted access to the interlayer space of montmorillonite under the commonly used experimental conditions (Kladnig 1979; Sivalov et al. 1981; Anderson and Klinowski 1986). The role of steric factors in the interactions of clay minerals with basic organic compounds will be described in the following.

Interestingly, both Frenkel (1974) and Liu et al. (2013a) failed to measure acid strengths more negative than $H_o = -4$ for a range of montmorillonites, irrespective of sample pretreatment (Figures 2.4 and 2.16). The higher acidity values previously reported (Table 2.5) may be partly due to incomplete desorption of the protonated indicators. Similarly, no acid sites stronger than the equivalent of 77% w/w sulfuric acid ($H_R = -13.30$) were detected by Tarasevich and Fedorova (1979), using arylmethanol indicators (Table 2.3).

Nevertheless, H_o values in the range of −5.6 to −8.2 have been measured for some clay minerals (Figure 2.2; Table 2.5) under dehydrating conditions when the water content of the system falls below 5% w/w (Hirschler and Schneider 1961; Solomon and Hawthorne 1971; Solomon et al. 1971). The work by Collet and Laszlo (1991) illustrates the importance of humidity control in the kaolinite-catalyzed Diels–Alder addition of cyclopentadiene to methylvinylketone. In the case of kaolinite containing 1.9% residual water, the reaction gave >75% yield with a stereoselectivity of 11. By comparison, the clay sample with 0.9% water gave >80% yields, but the endo preference decreased to 7.

Another example worth mentioning in this context is the protonation and demetallation of metalloporphyrins when their complexes with smectites are dehydrated to near-zero water content (Figure 2.3). An H_o value < −8.2 would be required for these processes to occur in an interlayer

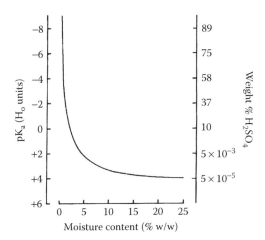

FIGURE 2.2 Relationship between surface acidity, expressed in terms of pK_a (H_o) or equivalent weight percent aqueous solution of sulfuric acid, and moisture content for kaolinite. Note that at 1% moisture content the pK_a is close to −3, equivalent to 48 weight % sulfuric acid, while at near-zero moisture content (after drying at 110°C) the pK_a = −8.1, equivalent to 90 weight % sulfuric acid. (Adapted from Solomon, D.H. et al., *J. Macromol. Sci. A*, 5, 587–601, 1971.)

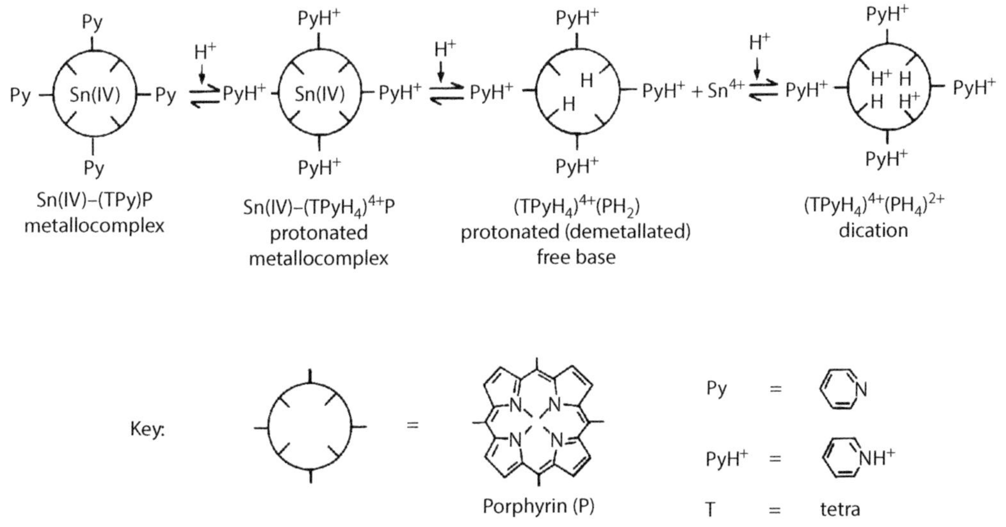

FIGURE 2.3 Sequence of protonation and demetallation of a Sn(IV)-tetrapyridyl-porphyrin complex, adsorbed on hectorite, as the system is dehydrated to near-zero water content. Protonation of the pyridyl substituents precedes both demetallation and protonation of the pyrrole groups inside the porphyrin (porphin-pyrrole) ring. (Modified from Abdo, S. et al., *Clays Clay Miner.*, 28, 125–129, 1980.)

TABLE 2.6
Protonation of Some Organic Species (Bases) Adsorbed on Montmorillonite Surfaces

Organic Species (in Alphabetical Order)	Exchangeable Cation (Counterion)	Remark	Reference
Alkenes	Al^{3+}, Cr^{3+}, Cu^{2+}, Fe^{3+}	Formation of carbocations, yielding either ethers/alcohols or dimers/oligomers	Adams et al. (1979, 1981); Adams and Clapp (1985)
Amides	H^+ (H^+/Al^{3+})[1]	Protonation commonly occurs on the oxygen atom, and influenced by substituent group	Tahoun and Mortland (1966); Cloos (1972)
Amines	Various	With some amines, protonation may yield colored surface complexes	Swoboda and Kunze (1968); Yariv et al. (1968); Harter and Ahlrichs (1969); Santos et al. (1970); Yariv and Heller (1970); Laura and Cloos (1975a, 1975b); Breen et al. (1995); Šnircová et al. (2009)
Aminopyridine	Ca^{2+}, K^+		Ikhsan et al. (2005)
3-Aminotriazole	Various	Extent of protonation increases with polarizing power of cation	Russell et al. (1968a)
Asulam (4-amino benzensulfonyl-methyl-carbamite)	H^+ (H^+/Al^{3+})[1]		Fusi et al. (1980)
Histidine	Various	Concomitant formation of cation hydroxide	Heller-Kallai et al. (1973)
Porphyrins and metalloporphyrins	Various	Metalloporphyrins may also be demetallated	Cady and Pinnavaia (1978); Van Damme et al. (1978); Abdo et al. (1980)
Pyridine	Mg^{2+}	Concomitant formation of $Mg(OH)_2$	Farmer and Mortland (1966); Swoboda and Kunze (1968)
s-Triazines	Various	Hydrolysis often follows protonation with loss of herbicidal activity Extent of protonation and hydrolysis influenced by substituent group	Cruz et al. (1968); Russell et al. (1968b); Brown and White (1969); White (1976)
Urea	H^+ (H^+/Al^{3+})[1]		Mortland (1966)

Note: The active protons either derive from dissociation of water molecules, coordinated to exchangeable cations, or introduced into exchange sites by acid washing/activation.

[1] The exchange complex of acid-washed/activated montmorillonite is occupied by both protons and aluminum ions.

Sn(IV)-porphyrin-hectorite complex (Van Damme et al. 1978; Canesson et al. 1979; Abdo et al. 1980) since this compound can only be demetallated in 100% w/w sulfuric acid, and then only incompletely (Buchler 1975). Furthermore, the protonation-demetallation process could be reversed by wetting the complex. Cady and Pinnavaia (1978) have also reported that Fe^{3+}- and VO^{2+}-montmorillonites can intercalate and protonate *meso*-tetraphenylporphyrin (Table 2.6). By treating montmorillonite with trifluoromethanesulfonic acid, Salmón et al. (1997) could also obtain a super-acidic sulfonic material ($H_o < -12.75$) capable of catalyzing the formation of alkyl cellosolves (glycol ethers) from propylene oxide and alcohols. More recently, Ruiz-Guerrero et al. (2006) were able to synthesize 1,3,5-triphenylbenzenes, β-methylchalcones, and 2,4,6-triphenyl pyrilium salts from acetophenones using a similar super-acidic montmorillonite.

2.3 BRØNSTED ACIDITY

A large number of clay-catalyzed organic reactions involve the transfer of a proton from the silicate surface (acting as a solid acid) to the adsorbed organic species (acting as a base). For this reason, a description of Brønsted acidity is important to understanding the catalytic activity of clay minerals. Brønsted acid-catalyzed organic conversions involving clay minerals have been the topic of several reviews (Theng 1982; Thomas 1982; Ballantine 1986; Adams 1987; Balogh and Laszlo 1993; Dasgupta and Török 2008). Some early examples of the capacity of montmorillonite for protonating adsorbed organic species are listed in Table 2.6. The large range and variety of clay-catalyzed organic reactions involving Brønsted acids have been summarized by McCabe and Adams (2013).

In the case of smectites, there is ample and persuasive evidence to show that Brønsted acid sites essentially arise from the dissociation of water molecules associated with exchangeable cations. The extent of dissociation is dependent on the polarizing power of the counterion in that the higher the valency (z), and the smaller the radius (r), the greater is the polarizing effect. The Brønsted acidity of cation-exchanged smectites is thus positively correlated with the ionic potential (z/r ratio) of the counterion (Graham 1964; Fripiat et al. 1965; Fripiat 1968; Mortland 1968; Mortland and Raman 1968; McBride 1994).

The proton-donating capacity of clay minerals is also greatly influenced by the water content of the system. At high water contents (>20% w/w), the polarizing effect of the counterion is presumably dissipated among water molecules making up its hydration sphere and those occupying interlayer vacancies (Calvet 1975). But as the water content of the clay mineral sample falls below ca. 5% w/w, the forces of polarization act on the few remaining molecules in the inner coordination sphere (hydration shell) of the cation. As a result, there is a marked increase in the dissociation of this *residual* water $(H_2O)_n$ and in the overall Brønsted acidity of the system (Poinsignon et al. 1978).

Such a situation may be described by Equation 2.20,

$$[M(H_2O)_n]^{z+} \rightleftharpoons [MM(H_2O)_{n-1}(OH)]^{(z-1)+} + H^+ \tag{2.20}$$

where M represents a metal counterion of valency z, and n is an integer. Using NMR spectroscopy, Touillaux et al. (1968) and Fripiat (1971) were able to determine the lifetime of a proton on an interlayer water molecule. For a monolayer hydrate of Na^+-montmorillonite and a bilayer hydrate of the Ca^{2+}-exchanged sample, they obtained a value of ca. 0.23×10^{-9} s at 25°C, which was 5–6 orders of magnitude shorter than for a proton in bulk water. In other words, the degree of dissociation of cation-coordinated water is 10^5–10^6 greater than that of bulk water. A similar picture has emerged from surface conductivity and dielectric measurements (Fripiat et al. 1965; Calvet 1975), and from first principles molecular dynamics simulation (Liu et al. 2011a).

In conformity with the earlier analysis, Frenkel (1974) observed that both the total titratable acidity and the concentration of relatively strong ($H_o = +1.5$) acid sites were closely related to the ionic potential of the counterions (Figure 2.4, inset [a]). By the same token, the surface acidity and concentration of acid sites of *dry* Al^{3+}-montmorillonite are appreciably larger than of its *wet* counterpart (Figure 2.4, inset [b]). More recently, Liu et al. (2013a) found that the Brønsted acidity of Li^+-, Na^+-, K^+-, Mg^{2+}-, and Al^{3+}-montmorillonites was positively correlated with the ionic potential of the cations. Jankovič and Komadel (2003) similarly observed that the activity of cation-exchanged montmorillonites in catalyzing the protection of aromatic aldehydes with acetic anhydride was related to the ionic potential of the cations. Earlier, Goldstein (1983) reported that the catalytic activity of cation-exchanged bentonites in decomposing *tert*-butyl acetate (to isobutylene and acetic acid) increased in the order $Na^+ < Mg^{2+} < Cd^{2+} < Th^{3+} < Al^{3+}$. Reddy et al. (2005, 2007) made a similar observation for the Brønsted acid-catalyzed esterification of succinic anhydride to di-(*p*-cresyl) succinate, and that of succinic acid with *iso*-butanol, as did Shimizu et al. (2008) with respect to the acetylation of cyclohexanol with acetic anhydride.

Kaolinite can also show a very high surface acidity ($H_o < -8.2$) when dehydrated to near-zero water content (Figure 2.2) although the Brønsted and Lewis acid sites here are associated more with particle edges than interlayer basal surfaces. Lao et al. (1989), for example, have reported that

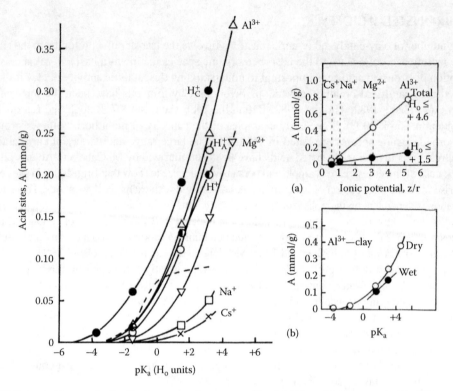

FIGURE 2.4 Acid strength distribution curves for montmorillonite (Wyoming) saturated with different counterions as determined by titration with *n*-butylamine. H$^+$ denotes proton-exchanged clay, H$_A^+$ refers to acid-activated clay, dried at 130°C, and H$_C^+$ denotes acid-activated clay, dried under ambient conditions. Inset (a) shows total acid site concentration (H$_o \leq$ +4.6) and concentration of relatively strong acid sites (H$_o \leq$ +1.5) as a function of ionic potential of the counterion. Inset (b) shows surface acidity distribution for Al^{3+}-montmorillonite after storage over water vapor (wet) and after drying at 130°C (dry). (All data are taken from Frenkel 1974, while the broken curve is from Benesi 1957 and refers to the natural form of Wyoming montmorillonite.)

kaolinite, kept at 25°C–50°C, can mediate the double-bond isomerization of 1-pristene to 2-pristene, while the sample, maintained at 100°C–150°C, can catalyze hydrogen transfer to yield pristane (cf. Figure 2.7),

In view of the specific affinity of arylmethanols for protons (Equation 2.18), it seems surprising that the H$_R$ indicator method has not been more widely applied to assessing the proton-donating capacity of clay surfaces. What little information is available refers almost exclusively to montmorillonite and then only with respect to its reactivity toward triphenylcarbinol (triphenylmethanol). Wilson and Galwey's (1976) investigation is a notable exception in that they titrated kaolinite, illite, and montmorillonite with *n*-butylamine in the presence of adsorbed Ph$_3$COH. The equivalence point for these minerals (in their natural form) was found to be 0.04, 0.1, and 0.4 mmol/g, respectively. To the best of our knowledge, no acid site distribution data comparable to what have been obtained for silica-aluminas and aluminas, using a series of arylmethanol indicators (Damon et al. 1977; Unger et al. 1981), are available for clay minerals.

Fripiat et al. (1964) and Tarasevich and Doroshenko (1983) have shown that *dry* montmorillonite is capable of protonating adsorbed and intercalated triphenylcarbinol (Ph$_3$COH) to yield the stable, yellow triphenylmethylcarbonium (Ph$_3$C$^+$) ion (Figure 2.5). The absorption spectrum of Ph$_3$C$^+$ consists of a doublet band centering around 420 nm with a molar extinction coefficient of 38 000 (Gold and Hawes 1951; Helsen 1970). The yield of Ph$_3$C$^+$ ions under a given set of conditions, derived by spectrophotometric means, may then be related to the concentration of accessible Brønsted sites with H$_R \leq$ −6.63 (Table 2.3).

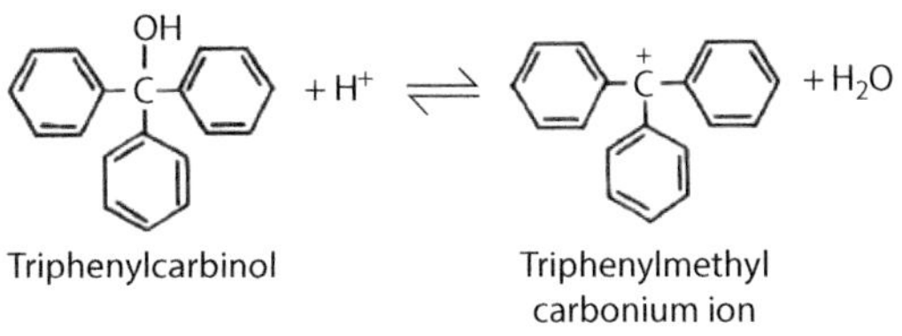

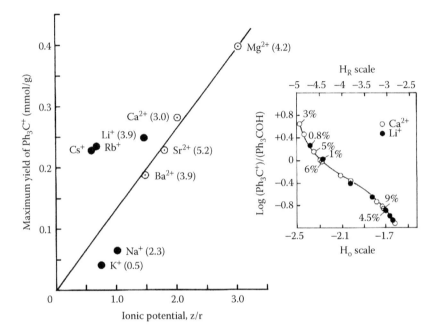

FIGURE 2.5 Protonation of triphenylcarbinol to give triphenylmethyl carbonium (carbocation) in the interlayers of some cation-exchanged montmorillonites under dehydrating conditions. The reaction can be reversed by adding water to the system.

FIGURE 2.6 Yield of triphenylmethylcarbonium (Ph_3C^+) ions by protonation of triphenylcarbinol (Ph_3COH) adsorbed to montmorillonite as a function of the ionic potential of the exchangeable cations (counterions). The numbers in brackets correspond to the number of water molecules associated with each counterion. Inset shows relationship between log Ph_3C^+/Ph_3COH and surface acidity (given by the H_o or H_R scale) for Li^+- and Ca^{2+}-exchanged montmorillonites containing different weight percent water. (Adapted from Helsen, J., *Bulletin Groupe Français des Argiles,* 22, 139–155, 1970. Chaussidon, J. and Calvet, R., Catalytic reactions on clay surfaces, *IUPAC Third International Congress of Pesticide Chemistry, Helsinki, 1974,* pp. 230–236, Georg ThiemeVerlag, Stuttgart, Germany.)

Figure 2.6 shows some data, obtained in this way, for montmorillonite saturated with different cations to which a specified number of water molecules are coordinated (Helsen 1982). For the divalent cation-exchanged samples, the yield of Ph_3C^+ is linearly related to the ionic potential of the respective counterion. This observation is further evidence for the proposal that the reactive protons essentially arise from the dissociation of water molecules coordinated to the counterions although no such regularity is observed for montmorillonite saturated with monovalent cations (Li^+, Na^+, K^+, Rb^+, Cs^+).

In explanation, Helsen et al. (1975) suggested that Na^+, and more so K^+ (Table 2.7), tended to *key* into the surface depressions of the silicate layer because either cation could fit rather snugly in a ditrigonal cavity with an equivalent spherical diameter of ca. 0.26 nm. On the other hand, Li^+ is too small, while Rb^+ and Cs^+ are too bulky, to obtain a good fit (cf. Figure 1.4). On keying, Na^+ and K^+ ions would have to shed most, if not all, of their coordinated water molecules, and hence effectively cease to act as a source of protons. This process, rather than the lack of intercalation on the part

TABLE 2.7
Ionic Radius, Ionic Potential (z/r), Hydration Enthalpy ($-\Delta H_{hyd}$), and Probable Number of Coordinated Water Molecules (N) for Some Cations

Cation	Radius[1] (pm)	z/r (×100)[2]	$-\Delta H_{hyd}$ (kJ/mol)[3]	N	$-\Delta H_{hyd}$/N(kJ)
Li^+	74	1.35	515	4	129
Na^+	102	0.98	405	4	101
K^+	138	0.72	321	4	80
Rb^+	149	0.67	296	4	74
Cs^+	170	0.59	263	4	66
Cu^{2+}	73	2.74	2100	6	350
Mg^{2+}	72	2.78	1922	6	320
Fe^{2+}	78	2.56	1920	6	320
Ca^{2+}	100	2.0	1592	6	265
Sr^{2+}	116	1.72	1445	6	241
Ba^{2+}	136	1.47	1304	6	217
Al^{3+}	53	5.66	4660	6	777
Fe^{3+}	65	4.62	4376	6	729

[1] Taken from Atkins (1982), based on 6-coordination; 1 pm = 0.001 nm.
[2] z = cation valency; r = cation radius.
[3] Taken from Burgess (1978).

of the indicator, may account for the limited ability of Na^+- and K^+-montmorillonites to protonate adsorbed Ph_3COH. The amount of Ph_3C^+ ions formed in these systems may thus be assigned to the Brønsted acid activity associated with coordinatively unsaturated aluminum ions exposed at particle edges. By comparison, Li^+, Rb^+, and Cs^+ ions would retain their respective spheres of hydration so that the corresponding forms of montmorillonite are more active in protonating triphenylmethanol than would be expected from their respective ionic potentials.

The question why Ca^{2+}, Sr^{2+}, and Ba^{2+} counterions, which are comparable in size to Na^+ and K^+, fail to key was not adequately addressed by Helsen et al. (1975). We suggest that the answer lies in the relatively high hydration enthalpy of divalent cations (Mackenzie 1964; Burgess 1978; Noller 1982; Izumi et al. 1997). Table 2.7 (column 6) indicates that the bond between a divalent ion and a water molecule in the inner hydration sphere is 3–4 times stronger than for monovalent species. Indeed, in the case of trivalent cations this bond is comparable in strength to that of a single H–H covalent bond (Brown 1954). That multivalent cations can tenaciously hold their coordinated water molecules has been well established. For example, Russell and Farmer (1964) were able to show by infrared spectroscopy that the Mg^{2+} counterions in montmorillonite could individually retain one water molecule against heating at 300°C. Similarly, Gonzalez Garcia et al. (1986) reported that exchangeable Ba^{2+} retained one molecule, and Cr^{3+} two molecules, of water after heating the corresponding forms of montmorillonite to 200°C and 400°C, respectively.

It can therefore be said that (hydrated) multivalent counterions in montmorillonite are sterically hindered from entering the ditrigonal cavities in the basal oxygen plane, unless the mineral has been thoroughly dehydrated (e.g., by heating at 300°C under vacuum). Rather, such cations tend to take up positions approximately midway between two opposing silicate layers (Liu et al. 2011a) as illustrated by Figure 1.16 for interlayer Mg^{2+} ions in vermiculite. Besides being closely related to the ionic potential of the counterion, the number of Brønsted acid sites in montmorillonite tends to decrease after heating the samples at ≥200°C, while Lewis acidity may actually increase (Brown and Rhodes 1997a; Tyagi et al. 2006; Liu et al. 2013b; McCabe and Adams 2013). Mishra and Parida (1998), for example, observed that acid activation substantially increased the activity of montmorillonite in terms of the Brønsted acid-catalyzed dehydration of alcohols and the cracking of cumene.

After heating at 400°C, however, cracking activity was appreciably reduced, while the Lewis acid-catalyzed dehydrogenation of cumene concomitantly increased. Similarly, Cseri et al. (1995) found that cation-exchanged K10 montmorillonites predominantly acted as Lewis acids after calcination at 500°C, while samples that had been dried at 120°C showed appreciable Brønsted acidity.

Any Brønsted acidity that remains after calcination at ≥400°C (when interlayer cation-coordinated water has been completely removed) may be ascribed to structural hydroxyl groups on planar and edge surfaces (Theng 1982), or to re-adsorption of water to incompletely coordinated Al^{3+} ions (Liu et al. 2011b). The results of surface acidity measurements by Liu et al. (2013a) for heated and cation-exchanged montmorillonites (Table 2.10) are in keeping with the aforementioned hypothesis regarding the behavior of interlayer mono- and multivalent counterions in montmorillonite.

A similar picture (Figure 2.6, inset) has emerged from the data of Chaussidon and Calvet (1975) who used the measured ratio of $[Ph_3C^+]/[Ph_3COH]$ to derive the corresponding surface acid strengths by applying Equation 2.17. Thus, as the water content of Ca^{2+}-montmorillonite fell from 9% to 3% (w/w), its acid strength increased from $H_R = -3.1$ to $H_R = -4.9$ corresponding to an increase from 28% to 40% w/w sulfuric acid (Figure 2.1). Assuming that all of the water is associated with the exchangeable cation, the earlier water content values translate to 10 and 3.3 water molecules per Ca^{2+} ion.

This analysis again illustrates that for clay systems with less than 10% w/w water, a small reduction in hydration status can cause a large increase in acid strength. Also noteworthy is that the acid strength of Ca^{2+}-montmorillonite with 6% w/w water ($H_R = -4.4$) is close to that given by the Li^+-exchanged form that has previously been dried a water content of 1%. This observation reflects the relative polarizing power of the two counterions toward their respective coordinated water molecules.

Since H_R is essentially a measure of the carbocation-generating capacity of the surface, its magnitude and range for a given solid acid would correlate with catalytic activity. The validity of this principle has been demonstrated by Damon et al. (1977) for a series of silica-aluminas, and by Jacobs (1977, 1984) in the case of zeolites. Figure 2.7 gives a general, qualitative picture of the relationship between reaction facility (conversion selectivity) and relative acid strength (Jacobs 1984; Corma and Wojciechowski 1985). Thus, solid acids of low strength ($H_R < +0.82$) can only catalyze dehydration reactions, while very strong Brønsted sites ($H_R < -16$) are required for alkane (hydrocarbon) cracking. Solids with an acid strength between these two H_R values can catalyze a variety of isomerization and other reactions. The important point to note here is that the number of possible conversions generally increases, while the fraction of active sites catalyzing a given reaction tends to diminish as the surface becomes more acidic (Jacobs 1984). This observation reflects the relative stability

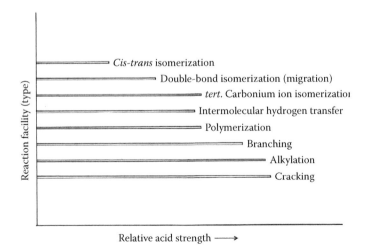

FIGURE 2.7 Diagram showing the relative strength of acid sites required for catalyzing various organic reactions and transformations involving carbocation (carbenium ion) intermediates as described in the text. (Adapted from Corma, A. and Wojciechowski, B.W., *Cat. Rev. Sci. Eng.*, 27, 29–150, 1985.)

of the carbocation intermediates as well as differences in the free energy change between reactants and products and in the reaction kinetics involved (Germain 1969; Poutsma 1976; Bourdillon et al. 1990). Further, the lifetime of the carbocations has a determining influence on the selectivity between isomerization and cracking. The stronger the Brønsted acid sites, the greater the yield of long-living carbocations, and the higher the preference for hydrocracking over isomerization (Jacobs 1984).

It is therefore difficult, in practice, to devise or select a single model reaction for probing the active sites on catalyst surfaces. Nevertheless, by using well-characterized samples and varying the experimental conditions in a systematic manner, a number of hydrocarbon-type conversions have been described that can serve as models for this purpose (Pines 1982; Guisnet 1985). Ballantine et al. (1983) have also pointed out that the number of products, formed in the Brønsted acid-catalyzed organic conversions by H^+/Al^{3+}-montmorillonites, is influenced by the ease with which the carbocation intermediates can be rearranged. High yields are obtained for conversions involving the formation of only one carbocation intermediate.

Solid acid-catalyzed reactions requiring a relatively high acid strength may often be facilitated by increasing the reaction temperature. Besides being accompanied by a series of side reactions, however, such conversions often lead to rapid deactivation of the catalyst as a result of carbon deposition. *Coke* formation, leading to a decrease in pore volume, is especially troublesome when clays and pillared clays are used as catalysts in hydrocarbon cracking (Occelli and Lester 1985; Ballantine 1986; Maes and Vansant 1997; Forzatti and Lietti 1999). In this context, we might mention that carbon deposition and surface blackening also occur when certain clay-organic complexes are heated, even at temperatures that are not sufficiently high to decompose the corresponding free (uncomplexed) organic species (Table 2.8).

TABLE 2.8
Clay-Catalyzed Decomposition of Some Organic Compounds Giving Rise to Carbon Deposition ("Coking") and Surface Blackening

Organic Compound	Clay Mineral	Remarks	Reference (in Chronological Order)
Glycerol	Montmorillonite	Clay-organic complex heated until dry	Walker (1967)
	Vermiculite	Clay immersed in boiling glycerol	Walker (1967)
n-Octacosane	Bentonite	Heated at 275°C	Henderson et al. (1968)
Alkylammonium ions	Hectorite Montmorillonite	Interlayer clay-organic complexes heated at 180°C–350°C. Reaction may involve dehydrogenation and polymerization	Chou and McAtee (1969)
n-Docosanoic acid (Behenic acid)	Ca^{2+}-montmorillonite	Clay-organic complex heated at 250°C. Reaction involves decarboxylation, cracking, and polymerization	Shimoyama and Johns (1971)
n-Hexane, 1-Butanol, 2-Ethoxyethanol (boiling point < 140°C)	Montmorillonite	Clay sample boiled in organic liquid Little, if any, blackening with air-dry mineral Intense blackening with mineral, preheated at 250°C, causing interlayer collapse	Eltantawy and Arnold (1973)
Ethylene glycol, Glycerol, 2,2'-Oxyethanol, n-Dodecane (boiling point > 200°C)	Montmorillonite Illite (air-dry) Kaolinite Halloysite	Clay samples boiled in organic liquid Probable collapse of montmorillonte interlayers	Eltantawy and Arnold (1973)
Ethylene	Cation-exchanged montmorillonites	Hydration to ethanol	Atkins et al. (1983)
Alk-1-enes	Cu^{2+}-montmorillonite	Reaction with interlayer water yielding ethers	Adams et al. (1984)
1-Hexene, Toluene	Al-pillared bentonite	Cracking of organic species at 350°C–550°C	Occelli and Lester (1985)

2.4 LEWIS ACIDITY

As already mentioned, a Lewis acid may be defined as an ion or molecule capable of accepting a *pair* of electrons from a base, acting as an electron-pair donor. In the clay catalysis literature, however, Lewis acids are often used in the wider sense to denote surface sites capable of accepting an electron from a suitable electron donor. In other words, Lewis acids may be identified with electron-accepting sites (Solomon 1968; Forni 1974; Theng 1982; Breen 1991). For clay minerals that have not been heated to the point of dehydroxylation, the principal Lewis acid sites may be identified with ferric ions within the layer structure, and under-coordinated aluminum exposed at clay particle edges.

The presence of such sites has been demonstrated by the ability of clay minerals to effect the following reactions: (a), oxidization of adsorbed organic molecules; and (b), interlayer complex formation with arenes and substituted arenes, often followed by their dimerization, oligomerization, and polymerization. The activity and role of Lewis acids in homogeneous and heterogeneous catalysis have been reviewed by Corma and Garcia (2003), while Balogh and Laszlo (1993) and McCabe and Adams (2013) have listed a range of Lewis acid-catalyzed type reactions involving clay minerals.

A well-known example of reaction (a) is the oxidation of benzidine (Bz) to its blue monovalent radical-cation (Bz$^{+}\bullet$) on adsorption and intercalation by smectites. This transformation involves a single electron transfer from the neutral, colorless diamine to structural Fe^{3+} ions or edge Al^{3+} ions (Solomon et al. 1968; Theng 1971; Furukawa and Brindley 1973; Warren et al. 1986; do Nascimento et al. 2006). Thus, prior reduction of the mineral by treatment with hydrazine, or *masking* the edge surface by sorption of sodium polyphosphate, inhibits color development or markedly reduces the blue color intensity. Mössbauer spectroscopy strongly indicates that the conversion of Bz to Bz$^{+}\bullet$ primarily occurs by electron transfer to Fe^{3+} ions in the octahedral sheet, leading to a marked increase in Fe^{2+}/Fe^{3+} ratio (Tennakoon et al. 1974a, 1974b). Thus, when the amount of adsorbed or intercalated benzidine exceeds 60 cmol/kg, this ratio increases from 0.16 (for the parent montmorillonite) to 0.41 for the corresponding benzidine blue complex. It would therefore appear that electron transfer can occur across the silicate layer boundary as Rozenson and Heller-Kallai (1978) have suggested although …"the pathway by which electrons penetrate the 2:1 layer is still unknown" (Stucki 2006).

On the other hand, McBride (1979) could not detect a significant change in the redox status of structural iron in hectorite with less than 5 cmol benzidine per kg clay although color development was clearly visible. In accord with the earlier suggestion by Solomon et al (1968), only a small fraction of the total structural iron may be involved in the electron transfer process since the adsorbed Bz$^{+}\bullet$ would be stabilized by cation exchange (Tennakoon et al. 1974a), and possibly by π electron interactions between Bz$^{+}\bullet$ and the basal oxygens of the silicate layer (Yariv et al. 1976). We might also add that under acidic conditions (pH $\leq$ 2), or when the blue complex is dried, the radical-cation disproportionates to yield a yellow divalent radical cation plus a colorless benzidinium dication (Theng 1971, 1982; Furukawa and Brindley 1973; Tennakoon and Tricker 1975; Lagaly et al. 2006):

$$\frac{2Bz^{+}\bullet + 2H^{+} \rightarrow Bz^{2+} + BzH_2^{2+}}{\text{(blue)} \qquad \text{(yellow) (colorless)}} \qquad (2.21)$$

Equation 2.21 applies equally well to tetramethylbenzidine (Kovar et al. 1984).

Interestingly, Tricker et al. (1975b) found that even the largely sodium-exchanged Wyoming montmorillonite was able to catalyze the conversion of triphenylamine to N,N,N',N'-tetraphenylbenzidine. The process presumably involved electron transfer to structural ferric ions, giving rise to the triphenylamine radical-cation, followed by dimerization, deprotonation, and the benzidine rearrangement. Alternatively, the conversion could occur by direct coupling of two radical cations with the benzene rings in the *para* position, followed by proton elimination (Figure 2.8). Similarly, structural Fe^{3+} ions in montmorillonite are apparently involved in the oxidation of diaminomaleonitrile (Ferris et al. 1982)

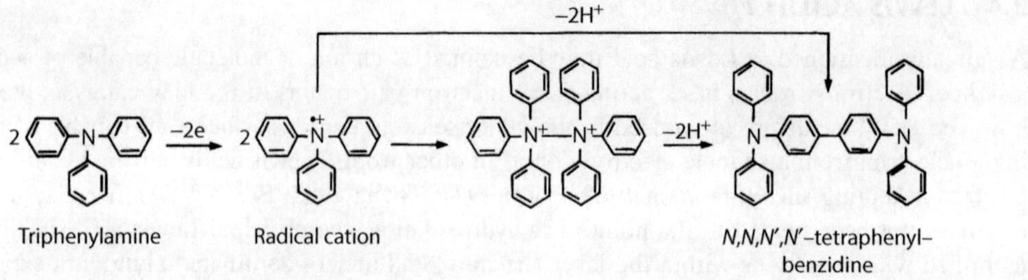

FIGURE 2.8 Diagram showing the two alternative pathways through which triphenylamine may be converted to N,N,N′,N′-tetraphenylbenzidine in the interlayer space of Na⁺-montmorillonite. (Adapted from Tricker, M.J. et al., *Clays Clay Miner.*, 23, 77–82, 1975b.)

and tetrathiafulvalene (Mandair et al. 1987). The same can be said for the oxidative degradation of hydrocortisone by palygorskite in aqueous media. Sepiolite with a closely similar layer structure to, but containing less iron than, palygorskite is relatively less active (Cornejo et al. 1983).

The classic example of reaction (b) is the formation of (colored) complexes when benzene vapor is presented to Cu^{2+}-montmorillonite under dehydrating conditions. Infrared spectroscopy indicates that in the Type I (green) complex the benzene molecule is edge-π-bonded to interlayer Cu^{2+} ion with the retention of both ring planarity and aromaticity. On the other hand, in the Type II (red) complex, the benzene ring is distorted and aromaticity is lost (Doner and Mortland 1969; Mortland and Pinnavaia 1971; Vande Poel et al. 1973; Pinnavaia et al. 1974). Type I complex forms at a higher water content than Type II, while both types are interconvertible by adjusting the hydration status of the system within narrow limits.

Electron spin resonance (Rupert 1973; Pinnavaia et al. 1974; Hall 1980; Boyd and Mortland 1986) and resonance Raman spectroscopy (Soma and Soma 1989) clearly indicate that the formation of Type II complexes involves the transfer of a single electron from the arene (Ar) to the interlayer transition metal ion (M^{n+}), yielding the corresponding radical-cation ($Ar^{+\bullet}$) together with the reduced form of the counterion ($M^{(n-1)+}$),

$$Ar + M^{n+}\text{-smectite} \rightarrow Ar^{+\bullet} + M^{(n-1)+}\text{-smectite} \qquad (2.22)$$

where M represents Cu^{2+}, Fe^{3+}, Ru^{3+}, or VO^{2+}. The infrared spectrum of pyridine adsorbed to Ni^{2+}- and Co^{2+}-exchanged montmorillonites indicates that these counterions can also act as Lewis acid sites or electron acceptors (Breen 1991) although p-xylene apparently fails to form a radical-cation on Co^{2+}-montmorillonite (Johnston et al. 1992). We might interpolate here that the nitration of phenols to the corresponding mononitrated products in the presence of iron(III) nitrate, supported on K10 montmorillonite (cf. Chapter 4), is compatible with a radical-cation pathway (Cornélis and Laszlo 1985).

Equation 2.22 is therefore analogous to the benzidine → benzidine blue transformation, but here it is the exchangeable cations (counterions) that act as electron acceptors rather than structural Fe^{3+} or edge-surface Al^{3+} ions. It is not always easy, however, to differentiate between exchangeable and structural ferric ions. As mentioned earlier, the activity of (anhydrous) Fe^{3+}-montmorillonite in catalyzing the oxidative oligomerization of chloro- and methoxy-phenol (Desjardins et al. 1999) may be ascribed to the ability of exchangeable ferric ions to act as electron acceptors (cf. Table 2.4) as Sawhney et al. (1984) and Yong et al. (1997) have earlier suggested with respect to o-methylphenol and 2,6-dimethylphenol. Fe^{3+}-montmorillonite is also effective in catalyzing the condensation of aromatic aldehydes with dimedone and aniline (Song et al. 2007; Luo et al. 2008). Subsequently, Wallis et al. (2011) used mixed Fe^{3+}/choline montmorillonite for the Lewis acid catalyzed coupling of 2-naphthol and anthrone, and the conjugate addition of indole to methyl vinyl ketone.

Like benzene, symmetrical arenes such as biphenyl, naphthalene, and anthracene (Rupert 1973) as well as thiophene (Cloos et al. 1973) can form both Type I and Type II complexes. On the other hand, substituted benzenes (e.g., mesitylene, phenol) only form Type I complexes (Pinnavaia and Mortland 1971; Fenn and Mortland 1973) although anisole can form both types of complexes (Fenn et al. 1973). Other aromatic compounds that can form radical-cations in the interlayers of transition metal ion-exchanged-montmorillonites, as described by Equation 2.22, include aniline, p-chloroaniline, 3- and 4-chloroanisoles, 4-chlorophenol, N,N-dimethylaniline, 4,4′-disubstituted biphenyls, dibenzo-p-dioxin, pentachlorophenol, N,N,N',N'-tetramethylbenzidine, toluene, p-dimethoxybenzene, and p-xylene (Cloos et al. 1979; Boyd and Mortland, 1985, 1986; Moreale et al. 1985; Soma et al. 1983, 1985; Govindaraj et al. 1987; Soma and Soma 1988; Johnston et al. 1992; Gu et al. 2011).

Such radical-cation species can interact, either mutually or with other intercalated molecules, to yield dimers, trimers, and oligomers as exemplified by triphenylamine (Figure 2.8). Similarly, Fenn et al. (1973) have reported that interlayer anisole in Cu^{2+}-hectorite can form 4,4′-dimethoxybiphenyl, presumably by coupling of two Ph-OCH$_3$$^{+}$• in the para position with the loss of protons. When benzene is also present in the interlayer space, anisole-benzene combinations, and anisole oligomers can form (Mortland and Halloran 1976). Likewise, when 1,1-diphenylethylene is refluxed, in the presence of Cu^{2+}-montmorillonite, the resultant radical-cation dimerizes to give 1-methyl-1,3,3-triphenylindan (Adams et al. 1977) according to Scheme A in Figure 2.9. The indan dimer can also arise by protonation of the monomer (Thomas et al. 1977) as shown in Figure 2.9, Scheme B. Various substituted styrenes such as α-methylstyrene, *trans*-stilbene, anethole, and isohomogenol can similarly form dimers (Tricker et al. 1975a; Adams 1987; Madhavan et al. 2001) as do dibenzo-p-dioxin, pentachlorophenol, and 3- and 4-chloroanisoles (Boyd and Mortland 1985, 1986; Govindaraj et al. 1987), aniline, N,N-dimethylaniline (Soma and Soma 1988), and pyrene (Joseph-Ezra et al. 2014). Likewise, toluene forms numerous oligomers and alkylated oligomers when heated to 160°C in the presence of Cu^{2+}-montmorillonite (Tipton and Gerdom 1992).

The polymerization of benzene, biphenyl, and p-terphenyl, intercalated in transition metal ion-exchanged montmorillonites, to yield poly(p-phenylene) has been reported by Stoessel et al. (1977) and Soma et al. (1984, 1986). Similarly, thiophene and methylthiophenes can yield the corresponding polymers (Soma et al. 1987), while aniline forms polyaniline in the presence of Cu^{2+}-hectorite (Eastman et al. 1996; Eastman and Porter 2000) although p-dimethoxybenzene apparently fails to polymerize (Johnston et al. 1991). Soma and Soma (1989) have pointed out that arene radical-cations are stable when the *para* positions of the ring are occupied by substituents but can undergo polymerization when these positions are open.

Interestingly, arene radical-cations can polymerize in organic solvents, and even in an aqueous environment (Cloos et al. 1979; Moreale et al. 1985; Faguy et al. 1995). For example, when an aqueous aniline solution was percolated through a sand column containing Fe^{3+}-montmorillonite, colored streaks and spots were visible along the column, indicative of type II complex formation. Treating the separated complex with 0.1 M HCl or alkaline $Na_4P_2O_7$ did not change its color nor its carbon content, a behavior not unlike that shown by soil humin or kerogen (Cloos et al. 1981). Likewise, the soil-catalyzed conversion of aniline to yield azobenzene, azoxybenzene, phenazine, anilide, and acetanilide, reported by Pillai et al. (1982), is indicative of an oxidation process involving ferric ions.

More recently, Gupta et al. (2014) obtained polyaniline by intercalating the monomer (from tetrahydrofuran) into an organically modified montmorillonite, containing ferric ions. Fe^{3+}-montmorillonite can also catalyze the Friedel–Crafts sulfonylation of arenes (Choudary et al. 1999) as well as various Diels–Alder cycloaddition reactions (Adams et al. 1987; Laszlo and Moison 1989; Cativiéla et al. 1992). Likewise, Qin et al. (2015) have found Fe^{3+}-montmorillonite to be effective in catalyzing the interlayer oxidative oligomerization of 17β-estradiol in the presence of water. Interestingly, the oligomers formed do not remain in the interlayer space but rapidly settle out from the aqueous phase. The reactivity of Cu^{2+}- and Fe^{3+}-smectites toward aromatic compounds opens… "a vast territory of synthetic organic chemistry…" (Thomas et al. 1977) waiting to be explored and exploited.

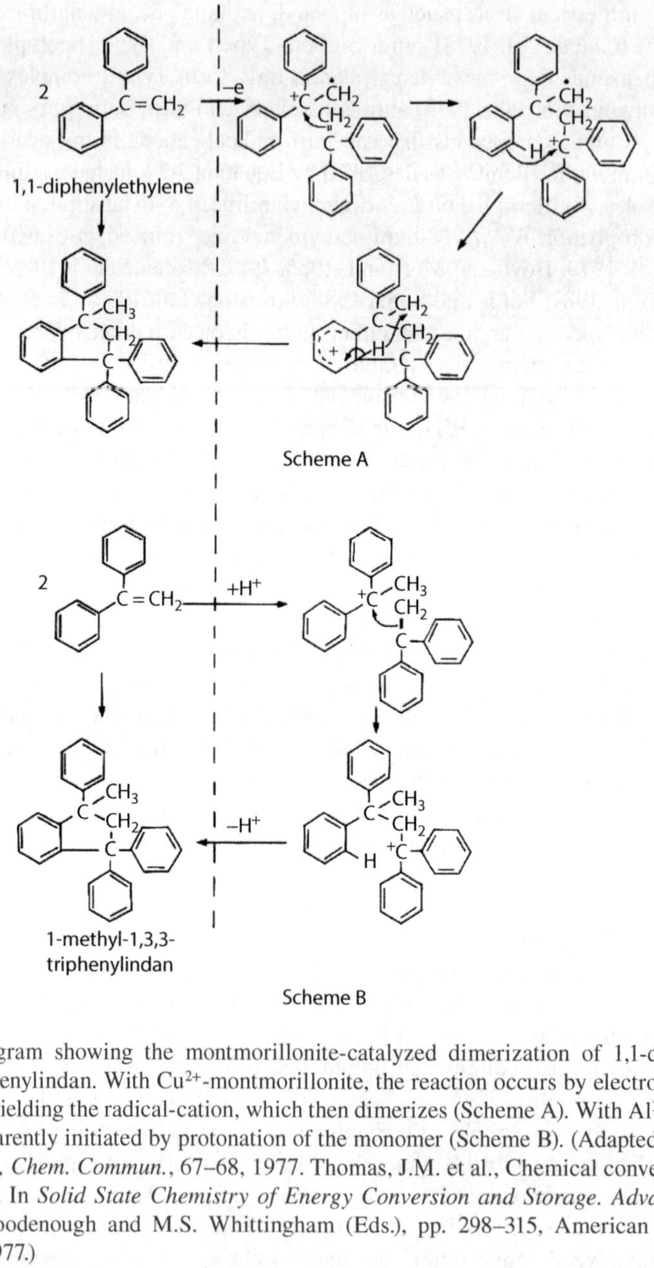

FIGURE 2.9 Diagram showing the montmorillonite-catalyzed dimerization of 1,1-diphenylethylene to 1-methyl-1,3,3-triphenylindan. With Cu^{2+}-montmorillonite, the reaction occurs by electron transfer (to interlayer copper ions), yielding the radical-cation, which then dimerizes (Scheme A). With Al^{3+}-montmorillonite, dimerization is apparently initiated by protonation of the monomer (Scheme B). (Adapted from Adams, J.M. et al., *J. Chem. Soc., Chem. Commun.*, 67–68, 1977. Thomas, J.M. et al., Chemical conversions using sheet-silicate intercalates, In *Solid State Chemistry of Energy Conversion and Storage. Advances in Chemistry Series 163*, J.B. Goodenough and M.S. Whittingham (Eds.), pp. 298–315, American Chemical Society, Washington, DC, 1977.)

It has long been known that styrene can polymerize in the presence of *non*-transition metal ion-exchanged clays (Solomon 1968; Theng and Walker 1970). Here, electron transfer from the adsorbed monomer to (coordinatively unsaturated) edge aluminum ions gives rise to a radical-cation, which rapidly dimerizes. Both the styrene radical-cation and its dimer (dication) can act as the propagating species in the formation of polystyrene (Figure 2.10, Scheme A). Thus, *masking* the particle edge surface by pretreating the clay samples with polyphosphate leads to a marked reduction in catalytic activity. Polymerization is also inhibited in the presence of water (and amines) since these compounds can successfully compete with styrene for Lewis sites at clay particle edges (Solomon and Rosser 1965; Solomon 1968; Theng and Walker 1970). We hasten to add that polystyrene can also form by a Brønsted acid-catalyzed process, initiated by the protonated monomer (carbocation), or the

2.5 BRØNSTED–LEWIS ACID COMBINATION AND SYNERGY

The occurrence in clay minerals of both Brønsted and Lewis acid sites, and their involvement in organic catalysis, have been the subject of many reports. Brown and Rhodes (1997a), for example, found that an acid-treated montmorillonite, exchanged with various polyvalent cations, could mediate the Brønsted acid-catalyzed conversion of α-pinene to camphene as well as the Lewis acid-catalyzed rearrangement of camphene hydrochloride to isobornyl chloride. Brønsted acidity peaked following thermal activation of the clay catalyst at 150°C, while maximum Lewis acidity was measured after heating at 200°C–300°C. That Brønsted and Lewis acidity may operate competitively has been suggested by Frenkel and Heller-Kallai (1983) during the high-temperature transformation of limonene in the presence of Na^+-, Mg^{2+}-, Al^{3+}-, and H^+/Al^{3+}-exchanged montmorillonites. The isomerization-disproportionation of limonene is apparently controlled by Brønsted acidity, which in turn is influenced by the nature of the interlayer cations. On the other hand, the formation of p-cymene by oxidation (of limonene) is a Lewis acid catalyzed reaction. Similarly, Motokura et al. (2007) have proposed that the high activity of Al^{3+}-montmorillonite in catalyzing the α-benzylation of 1,3-dicarbonyl compounds with primary alcohols is due to cooperation between Brønsted and Lewis acid sites in the interlayer space. Synergy between Brønsted and Lewis acidities may also lie behind the high efficiency of Fe^{3+}- and Zn^{2+}-(K10) montmorillonites in catalyzing the Friedel–Crafts acylation of aromatic ethers with acetic anhydride as compared with the Cu^{2+}-, Al^{3+}-, and Co^{2+}-exchanged samples (Choudary et al. 1998).

Since Brønsted and Lewis acidity can coexist (under anhydrous and dehydrating conditions), it is often difficult to say which type of acidity is dominantly involved in promoting a given organic reaction/transformation. For example, the benzidine → benzidine blue reaction is a Lewis acid-catalyzed process involving electron transfer to structural Fe^{3+} ions, while the blue → yellow transformation (Equation 2.21) that occurs on dehydration is mediated by Brønsted acids. As already mentioned, both Brønsted and Lewis acids may be involved in the clay-catalyzed polymerization of styrene. Atkins et al. (1983) have suggested similarly for the hydration of ethylene (to ethanol) in the interlayer space of montmorillonite. Likewise, the reactions of conjugated dienes (Adams and Clapp 1986), and the decarboxylation of long-chain fatty acids to the corresponding n-alkanes that are subsequently cracked into a mixture of petroleum-like hydrocarbons (Shimoyama and Johns 1971; Johns and Shimoyama 1972; Yuan et al. 2013), display the characteristics of both a Brønsted acid- and Lewis acid-driven process (Chapter 8).

Following Jurg and Eisma (1964), Almon and Johns (1976) have investigated the decarboxylation of behenic acid ($C_{21}H_{43}COOH$) in the presence of various clay minerals. In the case of smectites, the rate of decarboxylation at 250°C decreased in the order of nontronite > oxidized montmorillonite > montmorillonite > saponite. With montmorillonite as the catalyst, this rate increased by ~43% on addition of H_2O_2 to the system, and it decreased by nearly 36% when the clay sample had previously been treated with polyphosphate. Almon and Johns (1976) rationalized the results in terms of electron transfer from the fatty acid to structural ferric ions and/or incompletely coordinated Al^{3+} ions, exposed at particle edges, to yield the corresponding acyl radical (RCO•). The latter decomposed to an alkyl radical, which in turn formed an alkane by hydrogenation. Gas/liquid chromatography indicated that the principal n-alkane formed was $C_{21}H_{44}$, which was subsequently *cracked* to give a range of normal and branched (C_{14}–C_{20}) alkanes of shorter chain. The high ratio (~4.5) of branched to normal alkanes, measured under anhydrous conditions, is consistent with a Brønsted acid-catalyzed process, involving the formation of carbocation intermediates (Johns 1979; Lao et al. 1989). A similar mechanism has been suggested by Liu et al. (2013c, 2013d) for the thermal decomposition of 12-aminolauric acid in the interlayer space of montmorillonite, and by Li et al. (1998) for the clay-catalyzed degradation of immature kerogen. The role of clay minerals in the formation, migration, and accumulation of crude oil has been reviewed by Wu et al. (2012).

In this context, we might mention the interesting observation by Barlow et al. (1993) on the benzylation of anisole in the presence of *clayzic* ($ZnCl_2$-impregnated K10 montmorillonite).

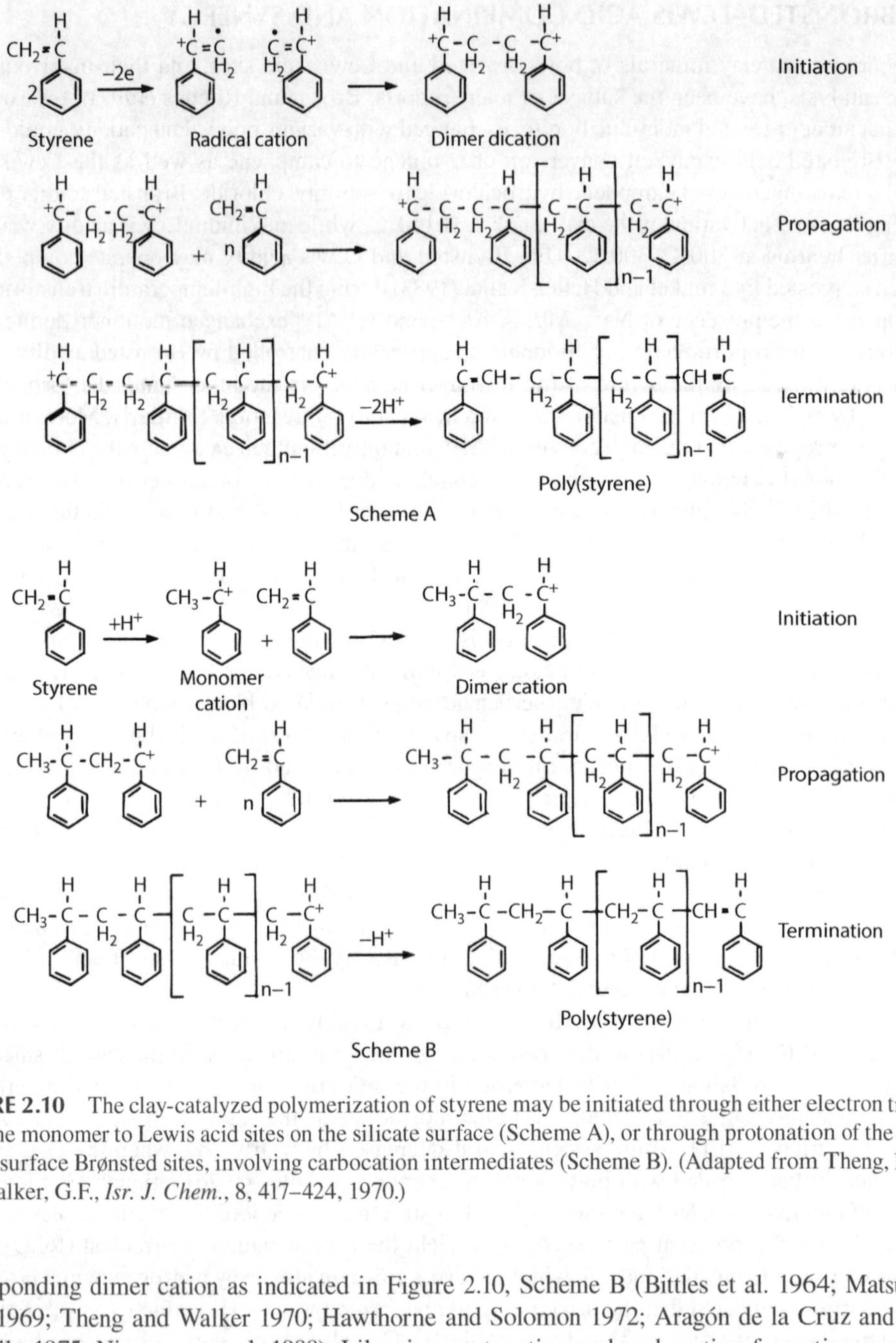

FIGURE 2.10 The clay-catalyzed polymerization of styrene may be initiated through either electron transfer from the monomer to Lewis acid sites on the silicate surface (Scheme A), or through protonation of the monomer at surface Brønsted sites, involving carbocation intermediates (Scheme B). (Adapted from Theng, B.K.G. and Walker, G.F., *Isr. J. Chem.*, 8, 417–424, 1970.)

corresponding dimer cation as indicated in Figure 2.10, Scheme B (Bittles et al. 1964; Matsumoto et al. 1969; Theng and Walker 1970; Hawthorne and Solomon 1972; Aragón de la Cruz and Viton Barbolla 1975; Njopwouo et al. 1988). Likewise, protonation and carbocation formation appear to be involved in the oligomerization and polymerization of indene, divinylbenzene, α-methylstyrene, α-phenylstyrene, 4-vinylpyridine, and various alkenes by acid-washed montmorillonite (Pillai and Ravindranathan 1994; Pillai et al. 1995; Fournaris et al. 2001; Madhavan et al. 2001; Harrane et al. 2002).

Other clay-catalyzed organic conversions involving Lewis acids and supported Lewis acids such as the Diels–Alder and Friedel–Crafts reactions (Laszlo 1987; Balogh and Laszlo 1993; Brown 1994; Clark and Macquarrie 1996; Kawabata et al. 2005; McCabe and Adams 2013) are described in Chapter 5.

Below 40°C the reaction was catalyzed by Brønsted acids, whereas above this temperature both Brønsted and Lewis acid sites were apparently involved in the transformation. Another example of a combined Brønsted and Lewis acid catalyzed reaction is the chlorination by carbon tetrachloride of adamantane to 1-adamantyl chloride in the presence of K10 or $Fe(NO_3)_3$-doped K10 montmorillonite (Laszlo 1986). Similarly, Clark et al. (1989) and Cornélis et al. (1990) reported that the activity of K10 montmorillonite in catalyzing the Friedel–Crafts reaction of benzene with benzyl chloride to form diphenylmethane could be dramatically enhanced by prior impregnation of the clay sample with $ZnCl_2$. Sukumar et al. (1998) have observed similarly for Zn(II)- and Fe(III)-impregnated kaolinite and metakaolinite. The use of *clayzic* and other K10 montmorillonite-supported reagents (Balogh and Laszlo 1993; Brown 1994; Clark et al. 1994; Clark and Macquarrie 1996) as catalysts of various organic reactions and transformations is described in Chapter 4.

The coexistence of Brønsted and Lewis acid sites also offers scope for a synergistic interaction between the two types of acidity. Swift and Black (1974), for example, found that the activity of calcined nickel-substituted synthetic mica-montmorillonite (SMM), described in Chapter 1, increased with nickel content. On the basis of subsequent work by Sohn and Ozaki (1980), we suggest that calcination leads to layer dehydroxylation and the formation of electron-deficient (Lewis acid) sites (**1**), which in turn can enhance the acidity of neighboring structural hydroxyl groups (Brønsted sites) (**2**) through an inductive effect, as depicted by Scheme 2.1.

The conversion of tertiary butylalcohol to isobutylene (and water) over *deammoniated* smectites (Davidtz 1976) according to Equation 2.23,

$$(CH_3)_3C-OH \rightarrow (CH_3)_2C=CH_2 + H_2O \tag{2.23}$$

provides another example of a synergistic interaction between Brønsted and Lewis acid sites. These types of smectites are formed when an NH_4^+-exchanged sample (**1**) is heated at 200°C–300°C, causing the ammonium ions to decompose into (gaseous) ammonia and protons (Jankovič and Komadel 2000). In dioctahedral smectites, the protons from this source can migrate into vacancies in the octahedral sheet (Wright et al. 1972). Alternatively, the protons can attach themselves to surface oxygens in the tetrahedral sheet (**2**), analogous to what Uytterhoeven et al. (1965) have suggested for X-zeolites and Y-zeolites, as shown in Scheme 2.2. In Davidtz' (1976) case, however, the deammoniation process was carried out by electrodialysis.

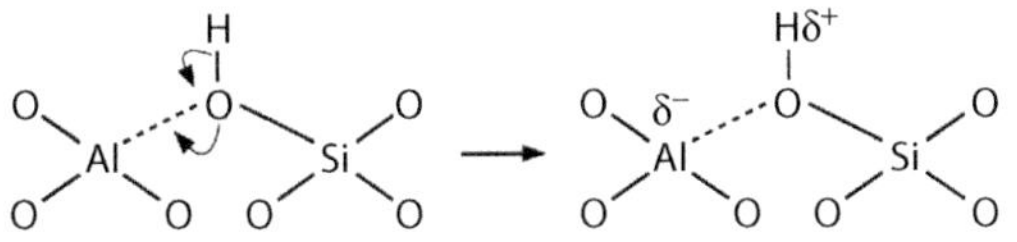

SCHEME 2.1 Possible enhancement of the acidity of structural hydroxyl groups, acting as Brønsted acid sites (**2**), by adjacent electron-deficient aluminums, acting as Lewis acids (**1**), through an inductive effect.

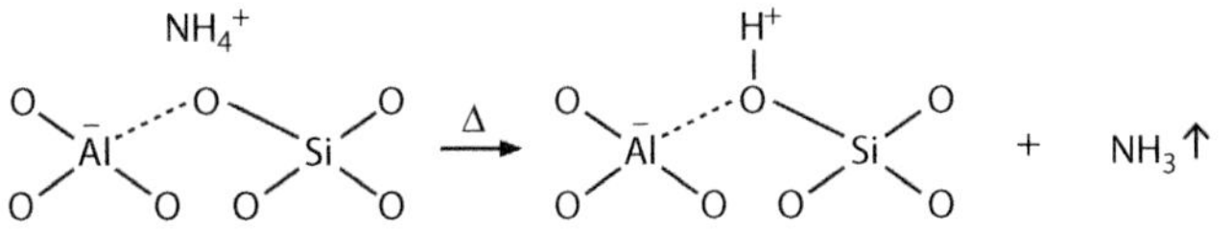

SCHEME 2.2 Attachment of a proton to a surface oxygen atom in the tetrahedral sheet of a smectite (**2**). The proton arises from the "deammoniation" of an NH_4^+-exchanged clay (**1**) by heating at 200°C–300°C with the concomitant evolution of ammonia.

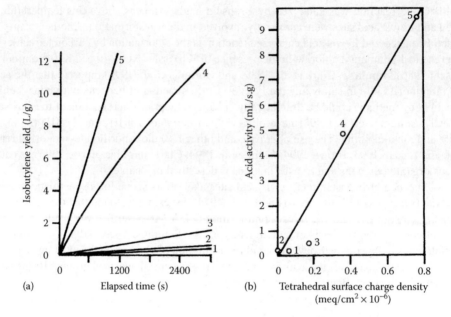

FIGURE 2.11 Conversion of *tert*-butyl alcohol to isobutylene over five deammoniated dioctahedral smectites (1–5) with varying isomorphous substitution in octahedral and tetrahedral positions. Diagram (a) shows rate plots for the formation of isobutylene by dehydration of *tert*-butanol, while diagram (b) shows the relationship between catalytic (*acid*) activity, derived from the slope of the rate plots, and tetrahedral surface charge density of the smectite catalysts. NB: meq = milliequivalent. (Adapted from Davidtz, J.C., *J. Catal.*, 43, 260–263, 1976.)

Using five dioctahedral smectites with varying degrees of isomorphous substitution in octahedral and tetrahedral positions, Davidtz (1976) observed that isobutylene production increased linearly with reaction time, at least over the first 1,200 seconds (Figure 2.11a). Attempts at relating catalytic activity, as represented by the slope of the rate plots, to cation exchange capacity, specific surface area, or octahedral substitution met with little or no success. However, when (acid) activity was plotted against tetrahedral surface charge density, given by the concentration of tetrahedrally coordinated Al per unit surface area, a reasonably good linear relationship was obtained (Figure 2.11b).

In explanation, Davidtz (1976) suggested that the protons, migrating into the octahedral sheet during deammoniation, were shielded or *deactivated*. Infrared spectroscopic analysis of deammoniated dioctahedral smectites by Chourabi and Fripiat (1981), however, would indicate that such protons form silanol groups (Brønsted sites). Since only tetrahedrally coordinated aluminum ions can enhance the acidity of adjacent silanols, the synergistic effect would increase with Al^{3+} for Si^{4+} substitution (Scheme 2.1). The high activity of NH_4^+-montmorillonite in catalyzing the acetylation of 3,4,5-trimethoxybenzaldehyde with acetic anhydride, after heating the mineral at 300°C (Jankovič and Komadel 2000), is consistent with such a concept. By the same token, beidellite is more active (and selective) than montmorillonite in the cracking of cumene (Swarnakar et al. 1996) because the negative surface charge in the former mineral derives from Al^{3+} for Si^{4+} substitution in the tetrahedral sheet. Similarly, Koster et al. (1998) have suggested that tetrahedral substitution lies behind the catalytic activity of montmorillonite in the dimerization of oleic acid.

More direct evidence for a Brønsted–Lewis site synergism comes from the studies by Sieskind et al. (1979) on the clay-catalyzed decomposition of cholestanol. Heating this compound (146°C, 16 h) in the presence of kaolinite and montmorillonite yielded a mixture of steranes and sterenes. In the absence of clay minerals, this transformation can only be realized under *superacidic* conditions as in a solution of toluenesulfonic-acetic acid or $HF-SbF_5$. Similarly, Macedo et al. (1994) have

Surface Acidity and Catalytic Activity

suggested that the activity of an acid-treated (dealuminated), calcined kaolinite in cumene cracking is due to a synergism between Brønsted and Lewis sites. Using *ab initio* methods to model mineral catalysis, Rimola et al. (2005, 2007) found that simultaneous interaction with, and synergy between, neighboring Brønsted and Lewis acid sites (as in feldspar) can dramatically decrease the activation barrier for peptide bond formation.

Interestingly, the clay volatiles (i.e., the gases liberated by heating clay minerals) together with their corresponding condensates, would qualify as super acid catalysts in that they can mediate the cracking of *n*-alkanes under mild conditions (Heller-Kallai 2002). Motokura et al. (2007) have also made the interesting suggestion that the high catalytic activity of Al^{3+}-montmorillonite in the α-benzylation of 1,3-dicarbonyl compounds with primary alcohols is due to the ability of interlayer Al^{3+} cations to act as both a Brønsted and a Lewis acid. As already mentioned, the strong Brønsted acidity of Al^{3+}-montmorillonite arises from the dissociation of cation-coordinated water molecules (Equation 2.20). After dehydration (by heating at >200°C) the Al^{3+} counterion in montmorillonite would also be able to act as an efficient Lewis acid.

2.6 CONCENTRATION AND DISTRIBUTION OF SURFACE ACID SITES

The majority of clay-catalyzed organic conversions take place under conditions where the surface is largely dehydrated. For this reason, the concentration, distribution, and strength of Brønsted and Lewis acid sites at clay mineral surfaces are commonly measured in non-aqueous media using near-dehydrated samples. Of the variety of techniques that have been developed for solid catalysts (Forni 1974; Benesi and Winquist 1978; Jacobs 1984; Anderson and Pratt 1985), only two have been extensively applied to clay mineral systems. These are (1) amine titration in the presence of Hammett indicators to which we have already referred and (2) adsorption and thermal desorption of basic (probe) molecules, in conjunction with Fourier-transform infrared (FTIR) spectroscopic analysis (Breen et al. 1987; Breen 1991; Barzetti et al. 1996; Billingham et al. 1996; Arena et al. 1998; Ahmet and Beytullah 2001; Carrado 2004; Reddy et al. 2007, 2009; Liu et al. 2011b, 2013a, 2013b).

2.6.1 Amine Titration

This method was introduced by Tamele (1950) who titrated a series of synthetic alumina-silicas with *n*-butylamine in the presence of butter yellow. As did Johnson (1955) subsequently, Tamele added the titrant in small increments to the catalyst, suspended in benzene, identifying the end point by the change in color of the adsorbed indicator from the acid (red) to the basic (yellow) form (Table 2.1). Walling (1950), however, was the first to relate the color change of adsorbed indicators to surface acidity in terms of the Hammett function. He also noted that for some catalysts, including an acid-activated clay (*Super-filtrol*), the acid strength in a nonpolar liquid (isooctane) was appreciably lowered when a basic solvent (e.g., acetone) was added to the system. This finding is perhaps hardly surprising since such solvents would complete with the amine for the same acid sites on the catalyst surface (Solomon et al. 1971).

A major disadvantage of the direct titration method is the length of time required for reaching the end point, during which the system may take up moisture from the atmosphere. To get around this problem, Benesi (1957) added *n*-butylamine incrementally to the catalyst (suspended in dry benzene), contained in stoppered vials. The indicator (dissolved in benzene) was added only after the system had been equilibrated by shaking the vials for a specified period (e.g., overnight). By this means, the titers given by a series of indicators could be obtained simultaneously. A variant of Benesi's (1957) method and one that further reduces the time of analysis is to treat the suspension ultrasonically during titration (Bertolacini 1963; Kladnig 1976; Anderson and Pratt 1985; Anderson and Klinowski 1986). This treatment, however, may lead to particle comminution and/or indicator decomposition, and hence should be applied with due caution (Unger et al. 1981; Jacobs 1984).

For a given indicator, the concentration of acid sites, A (mmol/g), may be derived from Equation 2.24 (Unger et al. 1981; Liu et al. 2011b),

$$A = \frac{V.N}{m} \qquad (2.24)$$

where:
V (ml) and N (mmol/g) refer to the volume and concentration of the titrant, respectively
m (g) is the mass of catalyst

By using a series of indicators covering a wide range of pK_a values, a distribution of acid sites may be obtained (Table 2.4). For some indicators, the end point of the titration may be visually assessed with a reasonable degree of accuracy, giving a reproducibility of better than 0.005 mmol/g (Table 2.1). For other indicators, especially the more weakly basic types, the end points should preferably be determined, or at least ascertained, by spectroscopic means if over-titration is to be avoided (Drushel and Sommers 1966; Anderson and Klinowski 1986).

Following this procedure, Frenkel (1974) measured the acid strength distribution of a number of montmorillonites under different conditions of pretreatment, using *n*-butylamine as the titrant and isooctane as the suspending agent. The results for Wyoming montmorillonite are shown in Figure 2.4, which includes those obtained by Benesi (1957), using montmorillonite from the same locality. Benesi's results are therefore comparable with those pertaining to the sodium-exchanged sample used by Frenkel. The two sets of data are clearly discordant, particularly at the lower acid strength end. Following Jacobs (1984), we suggest that Benesi's (1957) sample contained some sorbed water acting as Brønsted sites. Furthermore, the suspending medium (benzene) may associate with exposed hydroxyl groups and Lewis sites at particle edges by π-bonding interactions (Mortland 1970; Theng 1974).

Figure 2.4 also shows that the surface acidity of hydrogen-exchanged (H^+) and acid-activated (H_A^+) montmorillonite (after drying at 130°C) is comparable to that of the Al^{3+}-exchanged sample, while the hydrogen clay that has been dried under ambient conditions (H_C^+) is significantly more acidic than any of the three forms earlier. These observations may be explained in terms of the *autotransformation* of air-dried H^+-montmorillonite to the corresponding mixed H^+/Al^{3+}-exchanged form (Davis et al. 1962; Rhodes and Brown 1995; Komadel et al. 1997; Komadel and Madejová. 2013). This process is further described in Chapter 3.

Another noteworthy feature of Frenkel's (1974) data is that the (total) concentration of acid sites for the aluminum clay and, by extension, for the hydrogen forms, is close to 0.80 mmol/g. This value is a little smaller than the cation exchange capacity (CEC) of the Wyoming montmorillonite used (cf. Table 1.2). Since up to 0.07 mmol/g of the total titratable acidity can be apportioned to particle edges, ca. 0.73 mmol/g may be assigned to external and interlayer basal surfaces. Assuming a ratio of 1:10 for external to interlayer (basal) areas, the acid site concentration associated with the external surface would amount to 0.073 mmol/g. On this basis, the majority of acid sites (0.657 mmol/g) would be associated with interlayer surfaces and expected to be accessible to the titrant. Indeed, the X-ray diffraction analysis of the titrated clays shows that, with the possible exception of Cs^+-montmorillonite, *n*-butylamine is capable of penetrating the interlayer space. That being the case, the low value of titratable acidity for Na^+-montmorillonite could not be ascribed to restricted intercalation of the titrant (from benzene) as Schoonheydt (1991) has suggested; rather, it is an intrinsic property of the clay sample.

The question also arises whether the indicators themselves need to be freely intercalated for interlayer sites to be titrated. For some zeolites, Kladnig (1979) found that the use of Hammett indicators was limited by the pore diameter of the solid catalysts. This finding, however, would apply to the use of indicators in conjunction with amine titration when the amount of indicator added is much smaller than the number of titratable sites. Moreover, protons are capable of migrating and

Surface Acidity and Catalytic Activity

diffusing from *inaccessible* parts of the catalyst structure to sites that are accessible to the indicator (Anderson and Klinowski 1986). In other words, even if a given indicator were excluded from certain pores, it could still convert into its conjugate acid form since only a small fraction of the total sites need to react with the indicator. The dimension of the amine titrant is, therefore, more critical than that of the indicator molecule (Jacobs 1984; Barthomeuf 1985).

Similarly, the intercalatability or otherwise of the indicator does not materially affect the titratability of acid sites in expanding layer silicates, such as montmorillonite. When the titrant has reacted with all accessible sites with an acid strength equal to, or greater than, the pK_a of the (Hammett) indicator, the indicator would be displaced from its location at the silicate surface (Frenkel 1974). On the other hand, steric factors related to the size and shape of the amine titrant used would affect the measurement of acid sites. This situation may be illustrated by the data for kaolinite where neither indicator nor titrant would be able to intercalate. Figure 2.12 shows that the titratable acidity, measured using a series of amines in the presence of Hammett indicators, decreases in the order *n*-butylamine > isobutylamine > *n*-octylamine > *tert*-butylamine > *n*-dodecylamine. In explanation, Solomon et al. (1971) suggested that the *neutralizing* power of amines increased with the length of, or area covered by, the alkyl chain.

These observations, however, may be better rationalized in terms of steric and conformational effects associated with molecular size (occupancy) and chain branching. In the case of kaolinite, there is good evidence to indicate that amines, as well as indicator molecules, are largely adsorbed on particle edges (Lloyd and Conley 1970; Conley and Lloyd 1971; Sivalov et al. 1981, 1983). Depending on the experimental conditions used, the acid sites may be identified with protonated hydroxyl groups, coordinated to edge aluminum ions (cf. Figure 1.8) (Sivalov et al. 1980). The adsorption data of Conley and Lloyd (1971) further indicate that *n*-butylamine can be accommodated by an edge site on a 1:1 basis with the alkyl chain lying flat on the surface. The inset in

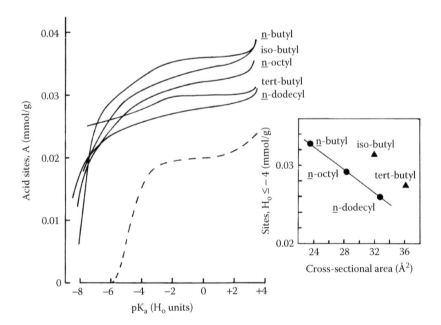

FIGURE 2.12 Concentration and relative strength of acid sites for kaolinite determined by titration (in benzene) with various amines in the presence of Hammett indicators. Solid curves are taken from Solomon et al. (1971), while the broken curve represents the *n*-butylamine titration results obtained by Benesi (1957) for a kaolinite sample dried at 120°C. Inset: Relationship between concentration of acid sites with $H_o \leq -4$ and cross-sectional area of amine titrants. NB: 1 Å = 0.1 nm.

Figure 2.12 shows that for straight-chain amines the decrease in titratable acidity is linearly related to their van der Waals cross-sectional area, which in turn is proportional to the number of carbon atoms in the molecule (Unger et al. 1981; Jacobs 1984). This would suggest that n-octylamine and n-dodecylamine are adsorbed in a similar conformation to n-butylamine. If so, part of the alkyl chain of the former two molecules would extend to, and cover up, an increasing proportion of adjacent sites, making these less accessible to the following titrant molecules. On the other hand, isobutylamine and *tert*-butylamine could react with a greater number of acid sites than is possible with normal (*straight-chain*) molecules of a comparable cross-sectional area. We may therefore infer that *branched-chain* amines adopt a different surface conformation such that the *cover-up* effect (Hendricks 1941) is relatively less pronounced.

Although the particle edges in kaolinite appear to be the principal site of adsorption, the basal surfaces are by no means inert toward amines (Bundy and Harrison 1986). With highly polarizing counterions and under dehydrating conditions, amines adsorbed at these surfaces would also be expected to become protonated (Theng 1982). As in montmorillonite, the acid sites here presumably arise from the dissociation of residual water molecules coordinated to the exchangeable cations (Gribina and Tarasevich 1975). Even under favorable conditions, however, the titratable acidity from this source would be quite small since the CEC of kaolinites rarely exceeds 3.5 cmol$_c$/kg (cf. Table 1.7).

The acid site distribution curve for kaolinite (Figure 2.12) is also very different in shape from their counterparts in montmorillonite (Figure 2.4) in that the majority of the relatively few Brønsted sites in kaolinite are strongly acidic ($H_o < -3$). This might be because the sites in kaolinite are, for the most part, located on particle edges whereas those in montmorillonite are associated with basal and interlayer surfaces. Again, the titratable acidity measured by Benesi (1957) differs from that reported by Solomon et al. (1971). Since the same titrant (n-butylamine) and suspending medium (benzene) were used in both studies, the discrepancy is probably related to differences in particle size and layer-stacking order of the kaolinite samples used (Conley and Lloyd 1971).

Early on, Johnson (1955) reported that the rate of propylene polymerization over a series of silica-aluminas was closely correlated with the concentration of surface acid sites as determined by n-butylamine titration using Hammett indicators. Figure 2.13 shows that such a correlation also

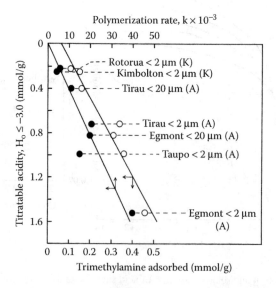

FIGURE 2.13 Relationship between the concentration of strong acid sites ($H_o \leq -3.0$), titratable with n-butylamine, and the rate of propylene polymerization for some soil allophanes (A) and kaolinites (K) of different particle size. The amount of trimethylamine adsorbed as a function of titratable acidity is also given. (Modified from Birrell, K.S., *N. Z. J. Sci.*, 5, 453–462, 1962.)

obtains with soil kaolinite and allophane (Birrell 1962). Further, essentially the same sites appear to be involved in propylene polymerization as in trimethylamine adsorption. The nature and location of Brønsted acid sites in kaolinite have already been referred to. In allophane, these sites may be identified with hydroxyl groups associated with surface defects (perforations) in the spherule wall (cf. Figure 1.24). By analogy with zeolites, it seems probable that these groups are responsible for protonating adsorbed trimethylamine (Jacobs et al. 1972; Earl et al. 1987). Access to surface hydroxyl groups, however, would be more restricted in the case of trimethylamine (diameter ca. 0.64 nm) as compared with *n*-butylamine (diameter ca. 0.43 nm). Accordingly, the amount of trimethylamine adsorbed was appreciably smaller than the concentration of acid sites that were titratable with *n*-butylamine (Pines and Haag 1960). This hypothesis is further supported by the decline in trimethylamine adsorption as the particle size of the clay samples increases. The finite intercept of the adsorption line on the vertical axis would also suggest that a small proportion of trimethylamine was associated with sites other than surface hydroxyls.

In this connection it seems pertinent to mention the work by Sohn and coworkers, as summarized by Sohn (2002), on the dimerization of ethylene and the isomerization of *n*-butene (1-butene) over a synthetic nickel-substituted montmorillonite. They found that catalytic activity over the temperature range at which the mineral had been evacuated (50°C–500°C) ran closely parallel to surface acidity as measured by *n*-butylamine titration in benzene (Figure 2.14). The steep rise in surface acidity for samples previously heated at 50°C–100°C may be ascribed to increased dissociation of water molecules associated with the counterions, while the marked decline in acidity between 100°C and 300°C is due to loss of cation-coordinated water through dehydration. It seems likely, therefore, that

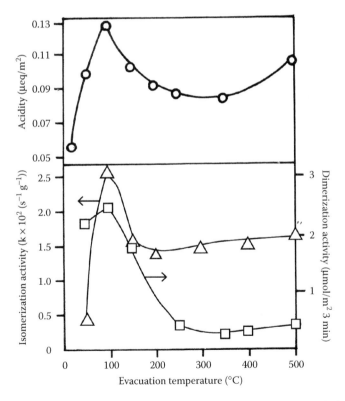

FIGURE 2.14 Surface acidity of a synthetic nickel-substituted montmorillonite in relation to its activity for ethylene dimerization and *n*-butene isomerization as a function of the temperature at which the mineral has previously been evacuated. NB: µeq = microequivalent. (Adapted from Sohn, J.R. and Ozaki, A. *J Catal.*, 61, 29–38, 1980.)

both dimerization and isomerization are catalyzed by Brønsted acid sites, involving carbocation intermediates. The slight increase in surface acidity between 300°C and 500°C is probably due to the formation of Lewis acid sites (e.g., 3- or 5-coordinated aluminum) at the expense of Brønsted sites (Theng 1982; Tennakoon et al. 1986; Macedo et al. 1994).

Despite its limitation, amine titration in a nonaqueous solvent using Hammett indicators is still a widely used method for assessing the surface acidity of solid acids, including clay minerals. Among the variety of amines that have been investigated, n-butylamine is still the preferred choice because of its low volatility, compact size (van der Waals diameter = 0.547 nm), and ease of intercalation into montmorillonite and other smectite species. Shuali et al. (1990) have reported that n-butylamine can also penetrate into the channels of sepiolite and palygorskite, replacing the zeolitic and bound water molecules (cf. Figures 1.20 and 1.21).

2.6.2 Adsorption and Desorption of Basic Probe Molecules

As remarked on earlier, the Hammett and arylmethanol indicator methods, when used in conjunction with amine titration, can yield sensible information about both the strength and concentration of acid sites (Table 2.4). Nevertheless, this methodology has largely been superseded by Fourier-transform infrared (FTIR) spectroscopy together with temperature-programmed desorption (TPD) analysis of adsorbed basic (probe) molecules such as ammonia, pyridine, and amines. The early literature on the application of IR spectroscopy to the characterization of adsorbed molecules at solid surfaces has been reviewed by Little (1966), Hair (1967), and Jacobs (1977).

Mapes and Eischens (1954) were the first to study the chemisorption of ammonia to silica-alumina cracking catalysts using infrared spectroscopy. The IR spectrum of the dehydrated (calcined) material showed strong bands near 3330 and 1640 cm^{-1} due to NH_3 coordinated to Lewis centres. The weak bands near 3120 and 1450 cm^{-1}, assigned to NH_4^+, intensified when water was added to the system, indicating the presence of Brønsted acid sites. Ahmet and Beytullah (2001) have found similarly for acid-activated kaolinite. The assessment of surface acidity of solid catalysts, including clay minerals, by combining ammonia sorption with FTIR spectroscopy has been the topic of many papers and reviews (Mortland et al. 1963; Russell 1965; Forni 1974; Benesi and Winquist 1978; Jacobs 1984; Anderson and Pratt 1985; Rupert et al. 1987).

Temperature-programmed desorption (TPD) analysis of adsorbed ammonia by itself does not appear to discriminate between Brønsted and Lewis acidity, but it can provide information about acid strength distribution (Arena et al. 1998; Okada et al. 2006). Hart and Brown (2004) have pointed out that ammonia can usefully serve as a probe molecule when the surface acid sites are dominantly of the Brønsted type. Ammonia adsorption data, however, are difficult to interpret when the catalyst surface contains a significant amount of Lewis acid sites to which ammonia is strongly bound. Nevertheless, when combined with FTIR spectroscopy, ammonia adsorption can provide a quantitative estimate of the concentration of Brønsted and Lewis sites.

Using this approach, Liu et al. (2013b) have found that the acid sites in the original (unheated) montmorillonite (Mt) are dominantly of the Brønsted type. Figure 2.15 shows that for samples that have been preheated at increasing temperatures (up to 600°C), the concentration of Brønsted sites first increases and then decreases. By contrast, the concentration of Lewis sites tends to increase with preheating temperature. The rise in Brønsted acidity for the sample previously heated at 120°C (Mt-120) may be ascribed to increased dissociation of water molecules coordinated to counterions (Equation 2.20), while the marked decline in the concentration of Brønsted sites for samples preheated at 200°C (Mt-200) and 400°C (Mt-400) is due to interlayer dehydration. Structural dehydroxylation of the calcined material (Mt-600) would account for the formation of Lewis acid sites at the expense of Brønsted sites (Theng 1982).

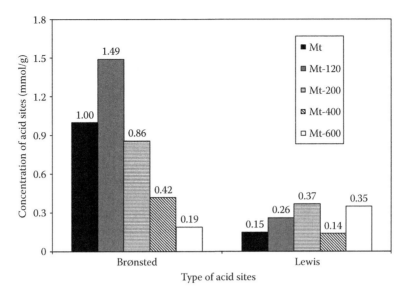

FIGURE 2.15 Concentration of Brønsted and Lewis acid sites, determined by Fourier-transform infrared (FTIR) spectroscopy and temperature-programmed desorption (TPD) of adsorbed ammonia, for monmorillonite (Mt) before and after heating at 120°C (Mt-120), 200°C (Mt-200), 400°C (Mt-400), and 600°C (Mt-600) in a muffle oven for 3 h. (Modified from Liu, D. et al., *Appl. Clay Sci.* 80–81, 407–412, 2013b.)

Earlier, Brown and Rhodes (1997b) measured both the concentration (number) and strength of acid sites of three commercial acid-activated smectites (K10, KSF, Fulcat 22A) by using a combination of thermogravimetry (TG), differential scanning calorimetry (TSC), and ammonia adsorption. The KSF material with the highest number of acid sites of moderate acid strength was the most effective in catalyzing the transesterification reaction between isopropenyl acetate and ethanol, and the addition of methanol to isoamylene. Hart and Brown (2004) used ammonia adsorption microcalorimetry to assess the surface acidity of K5 and K10 montmorillonites. In being able to access acid sites on both external and interlayer surfaces, (anhydrous) ammonia is a suitable probe molecule, especially when Bronsted acidity is dominant. It is perhaps not surprising that the concentration of acid sites (determined by pyridine adsorption) in acid-activated montmorillonite and vermiculite increases with the molarity of the HCl solutions used in the activation process (Ravichandran and Sivasankar 1997).

We might also add that montmorillonites exchanged with ferric ions behave as strong Lewis acids after heating at 500°C–550°C (Cativiéla et al. 1993; Cseri et al. 1995; Brown and Rhodes 1997b). Thus, Fe^{3+}-exchanged K10 montmorrilonite can effectively catalyze the benzylation of aromatics (Cseri et al. 1995; Hart and Brown 2004), and the acetylation of cyclohexanol with acetic anhydride (Shimizu et al. 2008). This material is equally effective in catalyzing the rearrangement of cyclohexadienone-phenol (Chalais et al. 1986), and in the Friedel–Crafts acylation of anisole, mesitylene, and *p*-xylene (Cornélis et al. 1990) as well as in Diels–Alder cycloaddition reactions (Laszlo and Lucchetti 1984; Adams et al. 1987) and the condensation of methyl acrylate with cyclopentadiene (Cativiéla et al. 1993). More recently, Wallis et al. (2006) reported that Fe(III)-treated K10 montmorillonite, both hydrated and anhydrous, could catalyze the oxidative coupling of 2-naphthol although it was not efficient in promoting the same reaction involving most substrates. Wallis et al. (2007) also found that (further) treatment of K10 montmorillonite with HCl led to a significant enhancement of catalytic activity in the tetrahydropyranylation of ethanol, the deacetylation of benzaldehyde, and the esterification of succinic anhydride.

Since its introduction by Parry (1963), pyridine has become the preferred basic probe molecule in determining the Brønsted and Lewis acidity of solid acid catalysts, including clay minerals. As a base, pyridine ($pK_b = 8.8$) is weaker than ammonia ($pK_b = 4.75$), and hence is more selective for strongly acid sites. Furthermore, the principal infrared-active absorption bands of adsorbed pyridine are sufficiently far removed from those due to hydroxyl stretching and bending vibrations. Physically adsorbed pyridine, giving an infrared band near 1575 cm^{-1}, can be removed by evacuation at room temperature, while hydrogen-bonded pyridine is lost by evacuation at ca. 150°C or by thermal desorption at 120°C (Anderson and Pratt 1985; Reddy et al. 2009).

Wright et al. (1972) made an early attempt at recording the IR spectra of pyridine adsorbed to synthetic ammonium mica-montmorillonite (SMM). Pyridine, adsorbed to deammoniated SMM, was completely converted into the pyridinium ion as indicated by the appearance of a band near 1547 cm^{-1}. On the other hand, about equal amounts of pyridinium and Lewis acid-coordinated pyridine (band near 1456 cm^{-1}) were formed with SMM that had been dried at 200°C under vacuum. As for ammonia, the addition of water to the preheated sample caused a marked decrease in the intensity of the 1456 cm^{-1} band, and a concomitant intensification of the 1547 cm^{-1} band (Equation 2.25):

$$\equiv Al:pyr + H_2O \rightleftharpoons \equiv A\overline{l}(OH) + pyrH^+ \quad (2.25)$$

The positions of principal infrared bands for adsorbed ammonia and pyridine together with their relative intensities are listed in Table 2.9. The concentration of Brønsted and Lewis acid sites can be estimated by integrating the respective peak areas, using Beer's law, or an internal standard (Jacobs 1984; Liu et al. 2013a).

The analysis of surface acid sites on clay minerals by adsorption of pyridine (and several amines), in conjunction with FTIR and TPD, and its relationship to catalytic activity, have been reported by

TABLE 2.9
Principal Infrared Absorption Bands for Ammonia and Pyridine Adsorbed to Solid Acid Catalysts, Including Clay Minerals

Surface Site	Ammonia		Pyridine	
	Band Position (cm^{-1})	Assignment	Band Position (cm^{-1})	Assignment
Brønsted acid	1430–1450 (s)	Asym. N-H bending	~1540 (s)	Combination modes,
			~1640 (s)	Involving C-C stretch
Lewis acid	1600–1620 (s)	Asym. N-H bending	1450–1470 (vs)	Combination modes,
			1600–1630 (s)	Including C-C stretch
Hydrogen-bonded	~3320(s), ~3400 (s)	N-H stretching	1440–1450 (vs)	Combination modes,
			1580–1590 (s)	Involving C-C stretch
Brønsted plus Lewis	N.a.		1485–1500 (vs)	Combination modes,
				Involving C-C stretch

Sources: Parry, E.P., *J. Catal.*, 2, 3713–3779, 1963; Forni, L., *Catal. Rev.: Sci. Eng.*, 8, 65–115, 1974; Anderson, J.R. and K.C. Pratt, *Introduction to Characterization and Testing of Catalysts*, Academic Press, Sydney, Australia, 1985; Barzetti, T. et al., *J. Chem. Soc. Faraday Trans.*, 92, 1401–1407, 1996; Jankovič, L. and Komadel, P., *J. Catal.*, 218, 227–233, 2003; Liu, H. et al., *Phys. Chem. Min.*, 40, 479–489, 2013a.

N.a. = not applicable; s = strong; vs = very strong.

TABLE 2.10
Concentration of Acid Sites in the Original Sample of Montmorillonite (Mt) and in Mt After Heating at 120°C, 200°C, 400°C, and 600°C, and After Exchanging the Counterions Initially Present (Ca^{2+}) with Other Cations (Al^{3+}, Mg^{2+}, Li^+, Na^+, K^+)

	Concentration of Acid Sites (mmol/g)		
Sample	Total[1]	Brønsted[2]	Lewis[3]
Mt	0.80	0.26 (33)[4]	0.54 (67)[4]
Mt-120	1.50	0.51 (34)	0.99 (66)
Mt-200	0.95	0.29 (31)	0.66 (69)
Mt-400	0.33	0.10 (30)	0.23 (70)
Mt-600	0.16	0.08 (50)	0.08 (50)
Al^{3+}-Mt	1.20	0.91 (76)	0.29 (24)
Mg^{2+}-Mt	1.00	0.56 (56)	0.44 (44)
Li^+-Mt	0.65	0.27 (42)	0.38 (58)
Na^+-Mt	0.30	0.09 (30)	0.21 (70)
K^+-Mt	0.24	0.02 (8)	0.22 (91)

Source: Adapted from Liu, H. et al., *Phys. Chem. Min.,* 40, 479–489, 2013a.
[1] Determined by titration with *n*-butylamine in the presence of Hammett indicators.
[2] Derived from the infrared band near 1540 cm^{-1} for adsorbed pyridinium.
[3] Obtained by difference between Total and Brønsted acid sites.
[4] Numbers in brackets denote percentages of Total acid sites.

many workers (Breen et al. 1987, 1995; Cseri et al. 1995; Barzetti et al. 1996; Arena et al. 1998; Janković and Komadel 2003; Akçay 2005; Tyagi et al. 2006; Reddy et al. 2007, 2009; Shimizu et al. 2008; Bilgiç et al. 2010).

The study by Liu et al. (2013a) provides a good example of using a combined approach to assess the concentration, strength, and type of acid sites in heated and cation-exchanged montmorillonites. These workers measured the total concentration (and strength) of acid sites by *n*-butylamine titration in the presence of Hammett indicators, while the type of acid sites (Brønsted and Lewis) was assessed by diffuse reflectance Fourier transform infrared (DRIFT) spectroscopy of adsorbed pyridine.

The concentration of Brønsted and Lewis acid sites in the original (unheated) sample, and in the samples after heating at 120°C, 200°C, 400°C, and 600°C (Table 2.10) are quantitatively different from those shown in Figure 2.15 for which ammonia was used as the probe although the total acid site concentration (Brønsted plus Lewis) shows a similar trend.

Interestingly, there are at least twice as many Lewis as Brønsted sites except for the calcined sample (Mt-600). These observations reflect differences in basicity and surface accessibility between pyridine and ammonia. The total concentration of acid sites for different cation-exchanged montmorillonites (Mt) increases in the order: K^+-Mt < Na^+-Mt < Li^+-Mt < Mg^{2+}-Mt < Al^{3+}-Mt. The increase in the concentration (and percentages) of Brønsted acid sites follows the same sequence, running parallel to the ionic potential of the exchangeable cations (Table 2.7) as Frenkel (1974) has found earlier (Figure 2.4). The percentage of Lewis acid sites, on the other hand, decreases in the aforementioned sequence. The range of acid strengths ($-3.0 \leq H_o \leq +4.8$), measured for the different homoionic montmorillonites (Figure 2.16), is also consistent with Frenkel's (1974) observation.

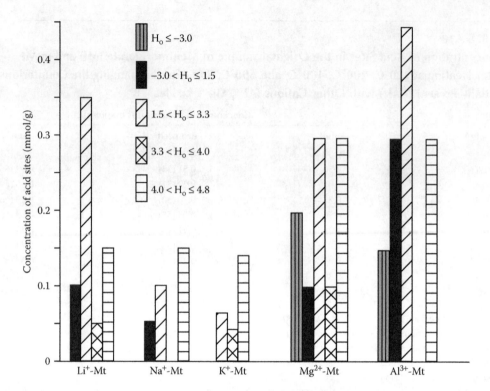

FIGURE 2.16 Concentration and strength of surface acid sites for different cation-exchanged montmorillonites (Mt), determined by n-butylamine titration in petroleum ether. (Modified from Liu, H. et al., *Phys. Chem. Miner.*, 40, 479–489, 2013a.)

REFERENCES

Abdo, S., M.I. Cruz, and J.J. Fripiat. 1980. Metallation-demetallation reaction of tin tetra (4-pyridyl)porphyrin in Na-hectorite. *Clays and Clay Minerals* 28: 125–129.

Abraham, M.H., R.M. Doherty, M.J. Kamlet, and R.W. Taft. 1986. A new look at acids and bases. *Chemistry in Britain* 22: 551–554.

Adams, J.M. 1987. Synthetic organic chemistry using pillared, cation-exchanged and acid-treated montmorillonite catalysts—A review. *Applied Clay Science* 2: 309–342.

Adams, J.M. and T.V. Clapp. 1985. Infrared studies of 1-hexene adsorbed onto Cr^{3+}-exchanged montmorillonite. *Clays and Clay Minerals* 33: 15–20.

Adams, J.M. and T.V. Clapp. 1986. Reactions of the conjugated dienes, butadiene and isoprene, alone and with methanol over ion-exchanged montmorillonites. *Clays and Clay Minerals* 34: 287–294.

Adams, J.M., J.A. Ballantine, S.H. Graham et al. 1979. Selective chemical conversions using sheet silicate intercalates: Low temperature addition of water to 1-alkenes. *Journal of Catalysis* 58: 238–252.

Adams, J.M., A. Bylina, and S.H. Graham. 1981. Shape selectivity in low temperature reactions of C_6-alkenes catalysed by a Cu^{2+}-exchanged montmorillonite. *Clay Minerals* 16: 325–332.

Adams, J.M., K. Martin, and R.W. McCabe. 1987. Catalysis of Diels–Alder cycloaddition reactions by ion-exchanged montmorillonites. In *Proceedings of the International Clay Conference, Denver, 1985* (Eds.) L.G. Schultz, H. van Olphen, and F.A. Mumpton, pp. 324–328. Bloomington, IN: The Clay Minerals Society.

Adams, J.M., T.V. Clapp, D.E. Clement, and P.I. Reid. 1984. Reactions of alk-1-enes over ion-exchanged montmorillonites. *Journal of Molecular Catalysis* 27: 179–194.

Adams, J.M., S.H. Graham, P.I. Reid, and J.M. Thomas. 1977. Chemical conversions using sheet silicates: Ready dimerization of diphenylethylene. *Journal of the Chemical Society, Chemical Communications* 67–68.

Ahmet, T. and A. Beytullah. 2001. Firmly adsorbed ammonia and pyridine species at activated kaolinite surfaces. *Adsorption Science & Technology* 19: 673–679.

Akçay, M. 2005. The surface acidity and characterization of Fe-montmorillonite probed by in situ FT-IR spectroscopy of adsorbed pyridine. *Applied Catalysis A: General* 294: 156–160.

Almon, W.R. and W.D. Johns. 1976. Petroleum-forming reactions: Clay-catalyzed fatty acid decarboxylation. In *Proceedings of the International Clay Conference 1975* (Ed.) S.W. Bailey, pp. 399–409. Wilmette, IL: Applied Publishing.

Anderson, M.W. and J. Klinowski. 1986. Use of Hammett indicators for the study of acidity of zeolite catalysts. *Zeolites* 6: 150–153.

Anderson, J.R. and K.C. Pratt. 1985. *Introduction to Characterization and Testing of Catalysts*. Sydney, Australia: Academic Press.

Aragón de la Cruz, F. and C. Viton Barbolla. 1975. Polymerization of styrene, indene and divinylbenzene in the interlamellar space of montmorillonite. *Advances in Organic Geochemistry, Proceedings of the 7th International Meeting*, pp. 851–856.

Arena, F., R. Dario, and A. Parmaliana. 1998. A characterization study of the surface acidity of solid catalysts by temperature programmed methods. *Applied Catalysis A: General* 170: 127–137.

Atkins, P.W. 1982. *Physical Chemistry*, 2nd ed., p. 750. Oxford, UK: Oxford University Press.

Atkins, M.P., D.J.H. Smith, and D.J. Westlake. 1983. Montmorillonite catalysts for ethylene hydration. *Clay Minerals* 18: 423–429.

Atkinson, D. and G. Curthoys. 1979. The acidity of solid surfaces and its determination by amine titration and adsorption of coloured indicators. *Chemical Society Reviews* 8: 475–497.

Atkinson, D. and G. Curthoys. 1980. Acid-site distribution on faujasite-type zeolites determined by n-butylamine titration. 1. Comparison of the use of Hammett and arylmethanol indicators on zeolite X. *Journal of Physical Chemistry* 84: 1358–1360.

Bailey, G.W., J.L. White, and T. Rothberg. 1968. Adsorption of organic herbicides by montmorillonite: Role of pH and chemical character of adsorbate. *Soil Science Society of America Proceedings* 32: 333–234.

Ballantine, J.A. 1986. The reactions in clays and pillared clays. In *Chemical Reactions in Organic and Inorganic Constrained Systems* (Ed.) R. Setton, pp. 197–212. Dordrecht, the Netherlands: D. Reidel.

Ballantine, J.A., J.H. Purnell, and J.M. Thomas. 1983. Organic reactions in a clay microenvironment. *Clay Minerals* 18: 347–356.

Balogh, M. and P. Laszlo. 1993. *Organic Chemistry Using Clays*. Berlin, Germany: Springer-Verlag.

Barlow, S.L., T.W. Bastock, J.H. Clark, and S.R. Cullen. 1993. Explanation of an unusual substituent effect in the benzylation of anisole and identification of the origin of the active site in Clayzic. *Tetrahedron Letters* 34: 3339–3342.

Barthomeuf, D. 1985. Importance of the acid strength in heterogeneous catalysis. *Studies in Surface Science and Catalysis* 20: 75–89.

Barzetti, T., E. Selli, D. Moscotti, and L. Forni. 1996. Pyridine and ammonia as probes for FTIR analysis of solid acid catalysts. *Journal of the Chemical Society, Faraday Transactions* 92: 1401–1407.

Bender, M.L., R.J. Bergeron, and M. Kamiyama. 1984. *The Bioorganic Chemistry of Enzymatic Catalysis*. New York: John Wiley & Sons.

Benesi, H.A. 1956. Acidity of catalyst surfaces. I. Acid strength from colors of adsorbed indicators. *Journal of the American Chemical Society* 78: 5490–5494.

Benesi, H.A. 1957. Acidity of catalyst surfaces. II. Amine titration using Hammett indicators. *Journal of Physical Chemistry* 61: 970–973.

Benesi, H.A. and B.H.C. Winquist. 1978. Surface acidity of solid catalysts. In *Advances in Catalysis, Vol. 27* (Eds.) D.D. Eley, H. Pines and P.B. Weisz, pp. 97–182. New York: Academic Press.

Bertolacini, R.J. 1963. Acidity distributions in cracking catalysts. *Analytical Chemistry* 35: 599–600.

Bilgiç, C., D. Topaloglu Yazici, and N. Vural. 2010. Characterizing the surface acidity of bentonite by various methods. *Surface and Interface Analysis* 42: 1000–1004.

Billingham, J., C. Breen, and J. Yarwood. 1996. In situ determination of Brønsted/Lewis acidity on cation-exchanged clay mineral surfaces by ATR-IR. *Clay Minerals* 31: 513–522.

Birrell, K.S. 1962. Surface acidity of subsoils derived from volcanic ash deposits. *New Zealand Journal of Science* 5: 453–462.

Bittles, J.A., A.K. Chaudhuri, and S.W. Benson. 1964. Clay catalyzed reactions of olefins. I. Polymerization of styrene. *Journal of Polymer Science Part A: General Papers* 2: 1221–1231.

Bourdillon, G., C. Gueguen, and M. Guisnet. 1990. Characterization of acid catalysts by means of model reactions. I. Acid strength necessary for the catalysis of various hydrocarbon reactions. *Applied Catalysis* 61: 123–139.

Boyd, S.A. and M.M. Mortland. 1985. Dioxin radical formation and polymerization on Cu(II)-smectite. *Nature* 316: 532–535.

Boyd, S.A. and M.M. Mortland. 1986. Radical formation and polymerization of chlorophenols and chloroanisole on copper(II)-smectite. *Environmental Science & Technology* 20: 1056–1058.

Breen, C. 1991. Thermogravimetric and infrared study of the desorption of butylamine, cyclohexylamine and pyridine from Ni- and Co-exchanged montmorillonite. *Clay Minerals* 26: 487–496.

Breen, C., A.T. Deane, and J.J. Flynn. 1987. The acidity of trivalent cation-exchanged montmorillonite. Temperature-programmed desorption and infrared studies of pyridine and *n*-butylamine. *Clay Minerals* 22: 169–178.

Breen, C., J. Madejová, and P. Komadel. 1995. Characterisation of moderately acid-treated, size-fractionated montmorillonites using IR and MAS NMR spectroscopy and thermal analysis. *Journal of Materials Chemistry* 5: 469–474.

Brønsted, J.N. 1923. Einige Bemerkungen über den Begriff der Säuren und Basen. *Recueil des Travaux Chimiques des Pays-Bas* 42: 718–728.

Brown, D.R. 1994. Clays as catalyst and reagent supports. *Geologica Carpathica–Series Clays* 45: 4556.

Brown, G.I. 1954. *A Simple Guide to Modern Valency Theory*. Harlow, UK: Longmans, Green and Co.

Brown, D.R. and C.N. Rhodes. 1997a. Brønsted and Lewis acid catalysis with ion-exchanged clays. *Catalysis Letters* 45: 35–40.

Brown, D.R. and C.N. Rhodes. 1997b. A new technique for measuring surface acidity by ammonia adsorption. *Thermochimica Acta* 294: 33–37.

Brown, C.B. and J.L. White. 1969. Reactions of 12 *s*-triazines with soil clays. *Soil Science Society of America Proceedings* 33: 863–867.

Buchler, J.W. 1975. Static coordination chemistry of metalloporphyrins. In *Porphyrins and Metalloporphyrins* (Ed.) K.M. Smith, pp. 157–231. Amsterdam, the Netherlands: Elsevier.

Bundy, W.M. and J.L. Harrison. 1986. Nature and utility of hexamethylene diamine-produced kaolin floc structure. *Clays and Clay Minerals* 34: 81–86.

Burgess, J. 1978. *Metal Ions in Solution*, pp. 182–183. Chichester, UK: Ellis Horwood.

Busca, G. 2007. Acid catalysts in industrial hydrocarbon chemistry. *Chemical Reviews* 197: 5366–5410.

Cady, S.S. and T.J. Pinnavaia. 1978. Porphyrin intercalation in mica-type silicates. *Inorganic Chemistry* 17: 1501–1507.

Calvet, R. 1975. Dielectric properties of montmorillonites saturated by bivalent cations. *Clays and Clay Minerals* 23: 257–265.

Camejo-Abreu, C., V. Tabernero, M.D. Alba, T. Cuenca, and P. Terreros. 2014. Enhanced activity of clays and its crucial role for the activity in ethylene polymerization. *Journal of Molecular Catalysis A: Chemical* 393: 96–104.

Canesson, P., M.I. Cruz, and H. Van Damme. 1979. X.P.S. study of the interaction of some porphyrins and metalloporphyrins with montmorillonite. In *International Clay Conference 1978. Developments in Sedimentology* 27 (Eds.) M.M. Mortland and V.C. Farmer, pp. 217–225. Amsterdam, the Netherlands: Elsevier.

Carrado, K.A. 2004. Introduction: Clay structure, surface acidity, and catalysis. In *Handbook of Layered Materials* (Eds.) S.M. Auerbach, K.A. Carrado, and P.K. Dutta, pp. 1–37. New York: Marcel Dekker.

Cativiéla, C., F. Figueras, J.M. Fraile et al. 1993. Comparison of the catalytic properties of protonoic zeolites and exchange clays for Diels–Alder synthesis. *Applied Catalysis A: General* 101: 253–267.

Cativiéla, C., J.M. Fraile, J.I. García, J.A. Mayoral, and E. Pires. 1992. Effect of clay calcination on clay-catalysed Diels–Alder reactions of cyclopentadiene with methyl and (−)-methyl acrylates. *Tetrahedron* 48: 6467–6476.

Chalais, S., P. Laszlo, and A. Mathy. 1986. Catalysis of the cyclohexadienone-phenol rearrangement by a Lewis-acidic clay system. *Tetrahedron Letters* 27: 2627–2630.

Chaussidon, J. and R. Calvet. 1975. Catalytic reactions on clay surfaces. *IUPAC Third International Congress of Pesticide Chemistry, Helsinki, 1974*, pp. 230–236. Stuttgart, Germany: Georg Thieme Verlag.

Chizallet, C. and P. Raybaud. 2010. Acidity of amorphous silica-alumina: From coordination promotion of Lewis sites to proton transfer. *ChemPhysChem* 11: 105–108.

Chou, C.C. and J.L. McAtee, Jr. 1969. Thermal decomposition of organo-ammonium compounds exchanged onto montmorillonite and hectorite. *Clays and Clay Minerals* 17: 339–346.

Choudary, B.M., M. Sateesh, M.L. Kantam, and K.V.R. Prasad. 1998. Acylation of aromatic ethers with acid anhydrides in the presence of cation-exchanged clays. *Applied Catalysis A: General* 171: 155–160.

Choudary, B.M., N.S. Chowdari, M.L. Kantam, and R. Kannan. 1999. Fe(III) exchanged montmorillonite: A mild and ecofriendly catalyst for sulfonylation of aromatics. *Tetrahedron Letters* 40: 2859–2862.

Chourabi, B. and J.J. Fripiat. 1981. Determination of tetrahedral substitutions and interlayer surface heterogeneity from vibrational spectra of ammonium in smectites. *Clays and Clay Minerals* 29: 260–268.

Clark, J.H., A.P. Kybett, D.J. Macquarrie, S.J. Barlow, and P. Landon. 1989. Montmorillonite supported transition metal salts as Friedel–Crafts alkylation catalysts. *Journal of the Chemical Society, Chemical Communications* 1353–1354.

Clark, J.H., S.R. Cullen, S.J. Barlow, and T.W. Bastock. 1994. Environmentally friendly chemistry using supported reagent catalysts: Structure-property relationships for clayzic. *Journal of the Chemical Society, Perkin Transcations* 2: 1117–1130.

Clark, J.H. and D.J. Macquarrie. 1996. Environmentally friendly catalytic methods. *Chemical Society Reviews* 25: 303–310.

Clark, J.H. and D.J. Macquarrie. 1996. Environmentally friendly catalytic methods. *Chemical Society Reviews* 25: 303–310.

Cloos, P. 1972. Interactions entre les pesticides et la fraction minérale du sol. *Pédologie* 22: 148–173.

Cloos, P., C. Badot, and A. Herbillon. 1981. Interlayer formation of humin in smectites. *Nature* 289: 391–393.

Cloos, P., A. Moreale, C. Broers, and C. Badot. 1979. Adsorption and oxidation of aniline and p-chloroaniline by montmorillonite. *Clay Minerals* 14: 307–321.

Cloos, P., D. Vande Poel, and J.P. Camerlynck. 1973. Thiophene complexes on montmorillonite saturated with different cations. *Nature Physical Science* 243: 54–55.

Collet, C. and P. Laszlo. 1991. Clay catalysis of the non-aqueous Diels–Alder reaction and the importance of humidity control. *Tetrahedron Letters* 32: 2905–2908.

Conley, R.F. and M.K. Lloyd. 1971. Adsorption studies on kaolinites. II. Adsorption of amines. *Clays and Clay Minerals* 19: 273–282.

Corma, A. and H. García. 1997. Organic reactions catalyzed over solid acids. *Catalysis Today* 38: 257–308.

Corma, A. and H. García. 2003. Lewis acids: from conventional homogeneous to green homogeneous and heterogeneous catalysis. *Chemical Reviews* 103: 4307–4365.

Corma, A. and B.W. Wojciechowski. 1985. The chemistry of catalytic cracking. *Catalysis Reviews: Science and Engineering* 27: 29–150.

Cornejo, J., M.C. Hermosin, J.L. White, J.R. Barnes, and S.L. Hem. 1983. Role of ferric iron in the oxidation of hydrocortisone by sepiolite and palygorskite. *Clays and Clay Minerals* 31: 109–112.

Cornélis, A., A. Gerstmans, P. Laszlo, A. Mathy, and I. Zieba. 1990. Friedel–Crafts acylations with modified clays as catalysts. *Catalysis Letters* 6: 103–110.

Cornélis, A. and P. Laszlo. 1985. Clay-supported copper(II) and iron(III) nitrates: Novel multi-purpose reagents for organic synthesis. *Synthesis* 909–918.

Cornélis, A., P. Laszlo, M.W. Zettler, B. Das, and K.V.N.S. Srinivas. 1999. Montmorillonite K10. *e-EROS Encyclopedia of Reagents for Organic Synthesis*, pp. 3667–3671. Wiley Online Library.

Cruz, M., J.L. White, and J.D. Russell. 1968. Montmorillonite-*s*-triazine interactions. *Israel Journal of Chemistry* 6: 315–323.

Cseri, T., S. Békássy, F. Figueras, and S. Rizner. 1995. Benzylation of aromatics on ion-exchanged clays. *Journal of Molecular Catalysis A: Chemical* 98: 101–107.

Damon, J-P., B. Delmon, and J-M Bonnier. 1977. Acidic properties of silica-alumina gels as a function of chemical composition. Titration and catalytic activity measurements. *Journal of the Chemical Society, Faraday Transactions 1* 73: 372–380.

Dasgupta, S. and B. Török. 2008. Application of clay catalysts in organic synthesis. A review. *Organic Preparations and Procedures International* 40: 1–65.

Davidtz, J.C. 1976. The acid activity of 2:1 layer silicates. *Journal of Catalysis* 43: 260–263.

Davis, L.E., R. Turner, and L.D. Whittig. 1962. Some studies of the auto-transformation of H-bentonite to Al-bentonite. *Soil Science Society of America Proceedings* 26: 441–443.

Deno, N.C., H.E. Berkheimer, W.L. Evans, and H.J. Peterson. 1959. Carbonium ions. VII. An evaluation of the H_R acidity function in aqueous perchloric and nitric acids. *Journal of the American Chemical Society* 81: 23–44.

Deno, N.C., J.J. Jaruzelski, and A. Schriesheim. 1955. Carbonium ions. I. An acidity function (C_0) derived from arylcarbonium ion equilibria. *Journal of the American Chemical Society* 77: 3044–3051.

Desjardins, S., J.A. Landry, and J.P. Farant. 1999. Effects of water and pH on the oxidative oligomerization of chloro and methoxyphenol by a montmorillonite clay. *Journal of Soil Contamination* 8: 175–195.

do Nascimento, G.M., P.S.M. Barbosa, V.R.L. Constantino, and M.L.A. Temperini. 2006. Benzidine oxidation on cationic clay surfaces in aqueous suspension monitored by in situ resonance Raman spectroscopy. *Colloids and Surfaces A: Physicochemical and Engineering Aspects* 289: 39–46.

Doner, H.B. and M.M. Mortland. 1969. Benzene complexes with Cu(II) montmorillonite. *Science* 166: 1406–1407.

Drushel, H.V. and A.L. Sommers. 1966. Catalyst acidity distribution using visible and fluorescent indicators. *Analytical Chemistry* 38: 1723–1731.

Earl, W.L., P.O. Fritz, A.A.V. Gibson, and J.H. Lunsford. 1987. A solid-state NMR study of acid sites in zeolite Y using ammonia and trimethylamine as probe molecules. *Journal of Physical Chemistry* 91: 2091–2095.

Eastman. M.P., M.E. Hagerman, J.L. Attuso, E.D. Bain, and T.L. Porter. 1996. Polymerization of benzene and aniline on Cu(II)-exchanged hectorite clay films: A scanning force microscope study. *Clays and Clay Minerals* 44: 769–773.

Eastman, M.P. and T.L. Porter. 2000. Polymerization of organic monomers and biomolecules on hectorite. In *Polymer-Clay Nanocomposites* (Eds.) T.J. Pinnavaia and G.W. Beall, pp. 65–93. Chichester, UK: John Wiley & Sons.

Eltantawy, I.M. and P.W. Arnold. 1973. Catalytic decomposition of organic molecules by clays. *Nature Physical Science* 244: 144.

Faguy, P.W., R.A. Lucas, and W. Ma. 1995. An FT-IR-ATR spectroscopic study of the spontaneous polymerization of pyrrole in iron-exchanged montmorillonite. *Colloids and Surfaces A: Physicochemical and Engineering Aspects* 105: 105–112.

Farmer, V.C. and M.M. Mortland. 1966. An infrared study of the co-ordination of pyridine and water to exchangeable cations in montmorillonite and saponite. *Journal of the Chemical Society A*, 344–351.

Fenn, D.B. and M.M. Mortland. 1973. Interlamellar metal complexes on layer silicates. II. Phenol complexes in smectites. In *Proceedings International Clay Conference, Madrid* (Ed.) J.M. Serratosa, pp. 591–603. Madrid: División de Ciencias CSIC.

Fenn, D.B., M.M. Mortland, and T.J. Pinnavaia. 1973. The chemisorption of anisole on Cu(II) hectorite. *Clays and Clay Minerals* 21: 315–322.

Ferris, J.P., W.J. Hagan, Jr., K.W. Alwis, and J. McCrea. 1982. Chemical evolution 40. Clay-mediated oxidation of diaminomaleonitrile. *Journal of Molecular Evolution* 18: 304–309.

Forni, L. 1974. Comparison of the methods for the determination of surface acidity of solid catalysts. *Catalysis Reviews: Science and Engineering* 8: 65–115.

Forzatti, P. and L. Lietti. 1999. Catalyst deactivation. *Catalysis Today* 52: 165–181.

Fournaris, K.G., N. Boukos, and D. Petridis. 2001. Aqueous polymerization of protonated 4-vinylpyridine in montmorillonite. *Applied Clay Science* 19: 77–88.

Fowkes, F.M., H.A. Benesi, L.B. Ryland et al. 1960. Clay-catalyzed decomposition of insecticides. *Journal of Agricultural and Food Chemistry* 8: 203–210.

Frenkel, M. 1974. Surface acidity of montmorillonites. *Clays and Clay Minerals* 22: 435–441.

Frenkel, M. 1975. Use and abuse of some Hammett indicators for the determination of surface acidity. *Analytical Chemistry* 47: 598–599.

Frenkel, M. and L. Heller-Kallai. 1983. Interlayer cations as reaction directors in the transformation of limonene on montmorillonite. *Clays and Clay Minerals* 31: 92–96.

Fripiat, J.J. 1968. Surface fields and transformation of adsorbed molecules in soil colloids. *Transactions 9th International Congress of Soil Science, Adelaide* 1: 679–689.

Fripiat, J.J. 1971. Some applications of nuclear magnetic resonance to surface chemistry. *Catalysis Reviews: Science and Engineering* 5: 269–282.

Fripiat, J.J., J. Helsen, and L. Vielvoye. 1964. Formation des radicaux libres sur la surface des montmorillonites. *Bulletin Groupe Français des Argiles* 15: 3–10.

Fripiat, J.J., A.N. Jelli, G. Poncelet, and J. André. 1965. Thermodynamic properties of adsorbed water molecules and electrical conduction in montmorillonites and silicas. *Journal of Physical Chemistry* 69: 2185–2197.

Furukawa, T. and G.W. Brindley 1973. Adsorption and oxidation of benzidine and aniline by montmorillonite and hectorite. *Clays and Clay Minerals* 21: 279–288.

Fusi, P., G.G. Ristori, and A. Malquori. 1980. Montmorillonite-asulam interactions: I. Catalytic decomposition of asulam adsorbed on H- and Al-clay. *Clay Minerals* 15: 147–155.

Germain, J.E. 1969. *Catalytic Conversion of Hydrocarbons*. New York: Academic Press.

Ghosh, A.K. and R.A. Kydd. 1985. Fluorine-promoted catalysts. *Catalysis Reviews: Science and Engineering* 27: 539–589.

Gold, V. and B.W.V. Hawes. 1951. The ionisation of triarylcarbinols in strong acids and the definition of a new acidity function. *Journal of the Chemical Society* 2101–2111.

Goldstein, T.P. 1983. Geocatalytic reactions in formation and maturation of petroleum. *American Association of Petroleum Geologists Bulletin* 67: 152–159.
Gonzalez Garcia, S., J.M. Suarez Cardeso, A. Rodriguez Alonso, and G. Dios Cancela. 1986. Spectrophotometric and magnetic behaviour of montmorillonite complexes. 1. Hydrates and DMSO solvates of Cr(III) samples. *Analitica Quimica* 82: 13–18.
Govindaraj, N., M.M. Mortland, and S.A. Boyd. 1987. Single electron transfer mechanism of oxidative dechlorination of 4-chloroanisole on copper(II)-smectite. *Environmental Science & Technology* 21: 1119–1123.
Graham, J. 1964. Adsorbed water on clays. *Reviews of Pure and Applied Chemistry* 14: 81–90.
Gribina, I.A. and Yu. I. Tarasevich. 1975. On the role of exchangeable cations and bound water in the formation of kaolinite acidity. *Kolloidnyi Zhurnal* 37: 140–143.
Gu, C., C. Liu, C.T. Johnston, B.J. Teppen, H. Li, and S.A. Boyd. 2011. Pentachlorophenol radical cations generated on Fe(III)-montmorillonite initiate octachlorodibenzo-*p*-dioxin formation in clays: Density functional theory and Fourier Transform infrared studies. *Environmental Science & Technology* 45: 1399–1406.
Guisnet, M. 1985. Characterization of acid catalysts by use of model reactions. *Studies in Surface Science and Catalysis* 20: 283–297.
Guisnet, M. 1990. Model reactions for characterizing the acidity of solid catalysts. *Accounts of Chemical Research* 23: 392–398.
Gupta, B., A. Rakesh, A.A. Melvin, A.C. Pandey, and R. Prakash. 2014. In-situ synthesis of polyaniline coated montmorillonite (Mt) clay using Fe^{+3} intercalated Mt as oxidizing agent. *Applied Clay Science* 95: 50–54.
Hair, M.L. 1967. *Infrared Spectroscopy in Surface Chemistry*. New York: Marcel Dekker.
Hall, P.L. 1980. The application of electron spin resonance spectroscopy to studies of clay minerals: II. Interlamellar complexes—Structure, dynamics and reactions. *Clay Minerals* 15: 337–349.
Hammett, L.P. and A.J. Deyrup. 1932. A series of simple basic indicators. I. The acidity function of mixtures of sulfuric and perchloric acids with water. *Journal of the American Chemical Society* 54: 2721–2739.
Harrane, A., R. Meghabar, and M. Belbachir. 2002. A protons exchanged montmorillonite clay as an efficient catalyst for the reaction of isobutylene polymerization. *International Journal of Molecular Sciences* 3: 790–800.
Hart, M.P. and D.R. Brown. 2004. Surface acidities and catalytic activities of acid-activated clays. *Journal of Molecular Catalysis A: Chemical* 212: 315–321.
Harter, R.D. and J.L. Ahlrichs. 1969. Effect of acidity on reactions of organic acids and amines with montmorillonite clay surfaces. *Soil Science Society of America Proceedings* 33: 859–863.
Hattori, H. and Y. Ono. 2015. *Solid Acid Catalysis: From Fundamentals to Applications*. Singapore: Pan Stanford Publishing.
Hawthorne, D.G. and D.H. Solomon. 1972. Catalytic activity of sodium kaolinites. *Clays and Clay Minerals* 20: 75–78.
Heller-Kallai, L. 2002. Clay catalysis in reactions of organic matter. In *Organo-Clay Complexes and Interactions* (Eds.) S. Yariv and H. Cross, pp. 567–613. New York: Marcel Dekker.
Heller-Kallai, L., S. Yariv, and M. Riemer. 1973. Effect of acidity on the sorption of histidine by montmorillonite. In *Proceedings International Clay Conference, Madrid* (Ed.) J.M. Serratosa, pp. 651–662. Madrid: División de Ciencias CSIC.
Helsen, J. 1970. Détermination du type d'acidité et mésure spectrophotométrique de la dissociation de l'eau des montmorillonites par adsorption du triphenylcarbinol. *Bulletin Groupe Français des Argiles* 22: 139–155.
Helsen, J. 1982. Clay minerals as solid acids and their catalytic properties. *Journal of Chemical Education* 59: 1063–1065.
Helsen, J.A., R. Drieskens, and J. Chaussidon. 1975. Position of exchangeable cations in montmorillonites. *Clays and Clay Minerals* 23: 334–335.
Henderson, W., G. Eglinton, P. Simmonds, and J.E. Lovelock. 1968. Thermal alteration as a contributory process to the genesis of petroleum. *Nature* 219: 1012–1016.
Hendricks, S.B. 1941. Base exchange of the clay mineral montmorillonite for organic cations and its dependence upon adsorption due to van der Waals forces. *Journal of Physical Chemistry* 45: 65–81.
Henmi, T. and K. Wada. 1974. Surface acidity of imogolite and allophane. *Clay Minerals* 10: 231–245.
Hirschler, A.E. 1963. The measurement of catalyst acidity using indicators forming stable surface carbonium ions. *Journal of Catalysis* 2: 428–439.

Hirschler, A.E. and A. Schneider. 1961. Acid strength distribution studies of catalyst surfaces. *Journal of Chemical & Engineering Data* 6: 313–318.

Ikhsan, J., M.J. Angove, J.D. Wells, and B.B. Johnson. 2005. Surface complexation modeling of the sorption of 2-, 3-, and 4-aminopyridine by montmorillonite. *Journal of Colloid and Interface Science* 284: 383–392.

Imamura, S., Y. Hayashi, K. Kajiwara, H. Hoshino, and C. Kaito. 1993. Imogolite: A possible new type of shape-selective catalyst. *Industrial & Engineering Chemistry Research* 32: 600–603.

Izumi, Y., K. Urabe, and M. Onaka. 1997. Recent advances in liquid-phase organic reactions using heteropoly-acid and clay. *Catalysis Surveys from Japan* 1: 17–23.

Jacobs, P.A. 1977. *Carboniogenic Activity of Zeolites*. Amsterdam, the Netherlands: Elsevier.

Jacobs, P.A. 1984. The measurement of surface acidity. In *Characterization of Heterogeneous Catalysts* (Ed.) F. Delannay, pp. 367–404. New York: Marcel Dekker.

Jacobs, P.A., B.K.G. Theng, and J.B. Uytterhoeven. 1972. Quantitative infrared spectroscopy of amines in synthetic zeolites X and Y. II. Adsorption of amines on Na-Hydrogen zeolites X and Y. *Journal of Catalysis* 26: 191–201.

Jankovič, L. and P. Komadel. 2000. Catalytic properties of a heated ammonium-saturated dioctahedral smectite. *Collection of Czechoslovak Chemical Communications* 65: 1527–1536.

Jankovič, L. and P. Komadel. 2003. Metal cation-exchanged montmorillonite catalyzed protection of aromatic aldehydes with Ac_2O. *Journal of Catalysis* 218: 227–233.

Johns, W.D. 1979. Clay mineral catalysis and petroleum generation. *Annual Review of Earth and Planetary Sciences* 7: 183–198.

Johns, W.D. and A Shimoyama. 1972. Clay minerals and petroleum-forming reactions during burial and diagenesis. *American Association of Petroleum Geologists Bulletin* 56: 2160–2167.

Johnson, O. 1955. Acidity and polymerization activity of solid acid catalysts. *Journal of Physical Chemistry* 59: 827–831.

Johnston, C.T., T. Tipton, D.A. Stone, C. Erickson, and S.L. Trabue. 1991. Chemisorption of *p*-dimethoxybenzene on copper-montmorillonite. *Langmuir* 7: 289–296.

Johnston, C.T., T. Tipton, S.L. Trabue, C. Erickson, and D.A. Stone. 1992. Vapor-phase sorption of *p*-xylene on Co- and Cu-exchanged SAz-1 montmorillonite. *Environmental Science & Technology* 26: 382–390.

Joseph-Ezra, H., A. Nasser, and U. Mingelgrin. 2014. Surface interactions of pyrene and phenanthrene on Cu-montmorillonite. *Applied Clay Science* 95: 348–356.

Jurg, J.W. and E. Eisma. 1964. Petroleum hydrocarbons: Generation from a fatty acid. *Science* 144: 1451–1452.

Kawabata, T., M. Kato, T. Mizugaki, K. Ebitani, and K. Kaneda. 2005. Monomeric metal aqua complexes in the interlayer space of montmorillonites as strong Lewis acid catalysts for heterogeneous carbon–carbon bond-forming reactions. *Chemistry—A European Journal* 11: 288–297.

Kitayama, Y., M. Kamimura, K-i. Wakui, M. Kanamori, T. Kodama, and J. Abe. 1999. Cyclodehydration of diethylene glycol (DEG) catalyzed by clay mineral sepiolite. *Journal of Molecular Catalysis A: Chemical* 142: 237–245.

Kladnig, W. 1976. Surface acidity of cation exchanged Y-zeolites. *Journal of Physical Chemistry* 80: 262–269.

Kladnig, W.F. 1979. Use of Hammett indicators for acidity measurements in zeolites. *Journal of Physical Chemistry* 83: 765–766.

Komadel, P., M. Janek, J. Madejová, A. Weekes, and C. Breen. 1997. Acidity and catalytic activity of mildly acid-treated Mg-rich montmorillonite and hectorite. *Journal of the Chemical Society, Faraday Transactions* 93: 4207–4210.

Komadel, P. and J. Madejová. 2013. Acid activation of clay minerals. In *Handbook of Clay Science*, 2nd edition. Developments in Clay Science, Vol. 5A. (Eds.) F. Bergaya and G. Lagaly, pp. 385–409. Amsterdam, the Netherlands: Elsevier.

Koster, R.M., M. Bogert, B. de Leeuw, E.K. Poels, and A Bliek. 1998. Active sites in the clay catalysed dimerisation of oleic acid. *Journal of Molecular Catalysis A: Chemical* 134: 159–169.

Kovar, L., R. DellaGuardia, and J.K. Thomas. 1984. Reaction of radical cations of tetramethylbenzidine with colloidal clays. *Journal of Physical Chemistry* 88: 3595–3599.

Lagaly, G., M. Ogawa, and I. Dékány. 2006. Clay mineral organic interactions. In *Handbook of Clay Science*, Developments in Clay Science, Vol. 1 (Eds.) F. Bergaya, B.K.G. Theng, and G. Lagaly, pp. 309–377. Amsterdam, the Netherlands: Elsevier.

Lao, Y., J. Korth, J. Ellis, and P.T. Crisp. 1989. Heterogeneous reactions of 1-pristene catalysed by clays under simulated geological conditions. *Organic Geochemistry* 14: 375–379.

Laszlo, P. 1986. Catalysis of organic reactions by inorganic solids. *Accounts of Chemical Research* 19: 121–127.

Laszlo, P. 1987. Chemical reactions on clays. *Science* 235: 1473–1477.
Laszlo. P. and J. Lucchetti. 1984. Acceleration of the Diels–Alder reaction by clays suspended in organic solvents. *Tetrahedron Letters* 25: 2147–2150.
Laszlo, P. and H. Moison. 1989. Catalysis of Diels–Alder reactions with acrolein as dienophile by iron(III)-doped montmorillonite. *Chemistry Letters* 1031–1034.
Laura, R.D. and P. Cloos. 1975a. Adsorption of ethylene diamine (EDA) on montmorillonite saturated with different cations—III. Na-, K-, and Li-montmorillonite: Ion exchange, protonation, co-ordination and hydrogen bonding. *Clays and Clay Minerals* 23: 61–69.
Laura, R.D. and P. Cloos. 1975b. Adsorption of ethylene diamine (EDA) on montmorillonite saturated with different cations—IV. Al-, Ca-, and Mg-montmorillonite: Ion exchange, protonation, co-ordination and hydrogen bonding. *Clays and Clay Minerals* 23: 343–348.
Lewis, G.N. 1923. *Valence and the Structure of Atoms and Molecules.* New York: Chemical Catalog Company.
Lewis, G.N. and J. Bigeleisen. 1943. Methylene blue and other indicators in general acids. The acidity function. *Journal of the American Chemical Society* 65: 1144–1150.
Li, S., S. Guo, and X. Tan. 1998. Characteristics and kinetics of catalytic degradation of immature kerogen in the presence of mineral and salt. *Organic Geochemistry* 29: 1431–1439.
Little, L.H. 1966. *Infrared Spectra of Adsorbed Species.* New York: Academic Press.
Liu, H., D. Liu, P. Yuan et al. 2013a. Studies on the solid acidity of heated and cation-exchanged montmorillonite using n-butylamine titration in non-aqueous system and diffuse reflectance Fourier transform infrared (DRIFT) spectroscopy. *Physics and Chemistry of Minerals* 40: 479–489.
Liu, X., X. Lu, R. Wang, E.J. Meijer, and H. Zhou. 2011a. Acidities of confined water in interlayer space of clay minerals. *Geochimica et Cosmochimica Acta* 75: 4978–4986.
Liu, D., P. Yuan, H. Liu et al. 2011b. Influence of heating on the solid acidity of montmorillonite: A combined study by DRIFT and Hammett indicators. *Applied Clay Science* 52: 358–363.
Liu, D., P. Yuan, H. Liu et al. 2013b. Quantitative characterization of the solid acidity of montmorillonite using combined FTIR and TDP based on the NH_3 adsorption system. *Applied Clay Science* 80–81: 407–412.
Liu, H., P. Yuan, D. Liu, D. Tan, H. He, and J. Zhu. 2013c. Effects of solid acidity of clay minerals on the thermal decomposition of 12-aminolauric acid. *Journal of Thermal and Analytical Calorimetry* 114: 125–130.
Liu, H., P. Yuan, Qin, H. et al. 2013d. Thermal degradation of organic matter in the interlayer clay-organic complex: A TG-FTIR study on a montmorillonite/12-aminolauric acid system. *Applied Clay Science* 80–81: 398–406.
Lloyd, M.K. and R.F. Conley. 1970. Adsorption studies on kaolinites. *Clays and Clay Minerals* 18: 37–46.
Lunsford, J.H. 1968. Surface interactions of NaY and decationated Y zeolites with nitric oxide as determined by electron paramagnetic resonance spectroscopy. *Journal of Physical Chemistry* 72: 4163–4168.
Luo, H., Y. Kang, H. Nie, and L. Yang. 2008. Fe^{3+}-montmorillonite: An efficient solid catalyst for one-pot synthesis of decahydroacridine derivatives. *Journal of the Chinese Chemical Society* 55: 1280–1285.
Macedo, J.C.D., C.J.A. Mota, S.M.C. de Menezes, and V. Camorim. 1994. NMR and acidity studies of dealuminated metakaolin and their correlation with cumene cracking. *Applied Clay Science* 8: 321–330.
Mackenzie, R.C. 1964. Hydratationseigenschaften von Montmorillonit. *Berichte der deutschen keramischen Gesellschaft* 41: 696–708.
Madhavan, D., M. Murugalakshmi, A. Lalitha, and K. Pitchumani. 2001. Dimerisation of olefins catalysed by K10 montmorillonite clays. *Catalysis Letters* 73: 1–4.
Maes, N. and E.F. Vansant. 1997. Modification of the porosity of pillared clays by carbon deposition. II. Hydrocarbon cracking. *Journal of Porous Materials* 4: 5–15.
Mandair, A-P.S., W.R. McWhinnie, and P. Monsef-Mirzai. 1987. Charge transfer interactions in smectite clays. *Inorganica Chimica Acta* 134: 99–103.
Mapes, J.E. and R.P. Eischens. 1954. The infrared spectra of ammonia chemisorbed on cracking catalysts. *Journal of Physical Chemistry* 58: 1059–1062.
Matsumoto, T., I. Sakai, and M. Arihara. 1969. Polymerization of styrene by acid clay. *Kobunshi Ronbunshu* 26: 378–384.
McBride, M.B. 1979. Reactivity of adsorbed and structural iron in hectorite as indicated by oxidation of benzidine. *Clays and Clay Minerals* 27: 224–230.
McBride, M.B. 1994. *Environmental Chemistry of Soils.* New York: Oxford University Press.
McCabe, R.W. and J.M. Adams. 2013. Clay minerals as catalysts. In *Handbook of Clay Science,* 2nd ed. *Developments in Clay Science, Vol. 5B* (Eds.) F. Bergaya and G. Lagaly, pp. 491–538. Amsterdam, the Netherlands: Elsevier.

Mishra, T. and K. Parida. 1998. Transition metal pillared clay: 4. A comparative study of textural, acidic and catalytic properties of chromia pillared montmorillonite and acid activated montmorillonite. *Applied Catalysis A: General* 166: 123–133.

Moreale, A., P. Cloos, and C. Badot. 1985. Differential behaviour of Fe(III) and Cu(II)-montmorillonite with aniline: I. Suspensions with constant solid:liquid ratio. *Clay Minerals* 20: 29–37.

Mortland, M.M. 1966. Urea complexes with montmorillonite: An infrared absorption study. *Clay Minerals* 6: 143–156.

Mortland, M.M. 1968. Protonation of compounds at clay mineral surfaces. *Transactions 9th International Congress of Soil Science, Adelaide* 1: 691–699.

Mortland, M.M. 1970. Clay-organic complexes and interactions. *Advances in Agronomy* 22: 75–117.

Mortland, M.M. and L.J. Halloran. 1976. Polymerization of aromatic molecules on smectite. *Soil Science Society of America Journal* 40: 367–370.

Mortland, M.M. and T.J. Pinnavaia. 1971. Formation of copper(II)-arene complexes on the interlamellar surfaces of montmorillonite. *Nature Physical Science* 229: 75–77.

Mortland, M.M. and K.V. Raman. 1968. Surface acidity of smectites in relation to hydration, exchangeable cation, and structure. *Clays and Clay Minerals* 16: 193–198.

Mortland, M.M., J.J. Fripiat, J. Chaussidon, and J. Uytterhoeven. 1963. Interaction between ammonia and the expanding lattices of montmorillonite and vermiculite. *Journal of Physical Chemistry* 67: 248–258.

Motokura, K., N. Nakagiri, T. Mizugaki, K. Ebitani, and K. Kaneda. 2007. Nucleophilic substitution reactions of alcohols with use of montmorillonite catalysts as solid Brønsted acids. *Journal of Organic Chemistry* 72: 6006–6015.

Njopwouo, D., G. Roques, and R. Wandji. 1988. A contribution to the study of the catalytic action of clays on the polymerization of styrene: II. Reaction mechanism. *Clay Minerals* 23: 35–43.

Noller, H. 1982. Catalysis from the standpoint of coordination chemistry. *Acta Chimica Academiae Scientiarum Hungaricae* 109: 429–448.

Occelli, M.L. and J.E. Lester. 1985. Nature of active sites and coking reactions in a pillared clay mineral. *Industrial & Engineering Chemistry Product Research Development* 24: 27–32.

Okada, K., N. Arimitsu, Y. Kameshima, A. Nakajima, and K.J.D. MacKenzie. 2006. Solid acidity of 2:1 type clay minerals activated by selective leaching. *Applied Clay Science* 31: 185–193.

Parry, E.P. 1963. An infrared study of pyridine adsorbed on acidic solids. Characterization of surface acidity. *Journal of Catalysis* 2: 3713–79.

Paul, M.A. and F.A. Long. 1957. H_o and related indicator acidity functions. *Chemical Reviews* 57: 1–45.

Pillai, P., C.S. Helling, and J. Dragun. 1982. Soil-catalyzed oxidation of aniline. *Chemosphere* 11: 299–317.

Pillai, S.M. and M. Ravindranathan. 1994. Oligomerization of dec-1-ene over montmorillonite clay catalysts. *Journal of the Chemical Society, Chemical Communications* 1813–1814.

Pillai, S.M., A. Wali, and S. Satish. 1995. Oligomerization of vinylic compounds on montmorillonite clay catalysts. *Reaction Kinetics and Catalysis Letters* 55: 251–257.

Pines, H. 1982. Use of organic probes in detecting active sites in heterogeneous catalysis. *Journal of Catalysis* 78: 1–16.

Pines, H. and W.O. Haag. 1960. Alumina: Catalyst and support. I. Alumina, its intrinsic acidity and catalytic activity. *Journal of the American Chemical Society* 82: 2471–2483.

Pinnavaia, T.J. 1983. Intercalated clay catalysts. *Science* 220: 365–371.

Pinnavaia, T.J. and M.M. Mortland. 1971. Interlamellar metal complexes of layer silicates. I. Copper(II)-arene complexes on montmorillonite. *Journal of Physical Chemistry* 75: 3957–3962.

Pinnavaia, T.J., P.L. Hall, S.S. Cady, and M.M. Mortland. 1974. Aromatic radical cation formation on the intercrystal surfaces of transition metal layer lattice silicates. *Journal of Physical Chemistry* 78: 994–999.

Poinsignon, C., J.M. Cases, and J.J. Fripiat. 1978. Electrical polarization of water molecules adsorbed by smectites. An Infrared study. *Journal of Physical Chemistry* 82: 1855–1860.

Poutsma, M.L. 1976. Mechanistic considerations of hydrocarbon transformations catalyzed by zeolites. In *Zeolite Chemistry and Catalysis. ACS Monograph 171* (Ed.) J.A. Rabo, pp. 437–528. Washington, DC: American Chemical Society.

Qin, C., D. Troya, C. Shang, S. Hildreth, R. Helm, and K. Xia. 2015. Surface catalyzed oxidative oligomerization of 17β-estradiol by Fe^{3+}-saturated montmorillonite. *Environmental Science & Technology* 49: 956–964.

Ravichandran, J. and B. Sivasankar. 1997. Properties and catalytic activity of acid-modified montmorillonite and vermiculite. *Clays and Clay Minerals* 45: 854–858.

Reddy, C.R., G. Nagendrappa, and B.S. Jai Prakash. 2007. Surface acidity study of M^{n+}-montmorillonite clay catalysts by FT-IR spectroscopy: Correlation with esterification activity. *Catalysis Communications* 8: 241–246.

Reddy, C.R., P. Iyengar, G. Nagendrappa, and B.S. Jai Prakash. 2005. Esterification of succinic anhydride to di-(*p*-cresyl) succinate over M^{n+} montmorillonite clay catalysts. *Journal of Molecular Catalysis A: Chemical* 229: 31–37.

Reddy, C.R., Y.S. Bhat, G. Nagendrappa, and B.S. Jai Prakash. 2009. Brønsted and Lewis acidity of modified montmorillonite clay catalysts determined by FT-IR spectroscopy. *Catalysis Today* 141: 157–160.

Rhodes, C.N. and D.R. Brown. 1995. Autotransformation and ageing of acid-treated montmorillonite catalysts: A solid-state ^{27}Al NMR study. *Journal of the Chemical Society, Faraday Transactions* 91: 1031–1035.

Rimola, A., M. Sodupe, and P. Ugliengo. 2007. Aluminosilicate surfaces as promoters for peptide bond formation: An assessment of Bernal's hypothesis by ab initio methods. *Journal of the American Chemical Society* 129: 8333–8344.

Rimola, A., S. Tosone, M. Sodupe, and P. Ugliengo. 2005. Peptide bond formation activated by the interplay of Lewis and Brønsted catalysts. *Chemical Physics Letters* 408: 295–301.

Rozenson, I. and L. Heller-Kallai. 1978. Reduction and oxidation of Fe^{3+} in dioctahedral smectites. III. Oxidation of octahedral iron in montmorillonite. *Clays and Clay Minerals* 26: 88–92.

Ruíz-Guerrero, R., J. Cárdenas, L. Bautista, M. Vargas, E. Vázquez-Labastida, and M. Salmón. 2006. Catalytic synthesis of 1,3,5-triphenylbenzenes, β-methylchalcones, and 2,4,6-triphenyl pyrilium salts, promoted by a super acid trifluoromethane sulfonic clay from acetophenones. *Journal of the Mexican Chemical Society* 50: 114–118.

Rupert, J.P. 1973. Electron spin resonance spectra of interlamellar Cu(II)-arene complexes on montmorillonite. *Journal of Physical Chemistry* 77: 784–790.

Rupert, J.P., W.T. Granquist, and T.J. Pinnavaia. 1987. Catalytic properties of clay minerals. In *Chemistry of Clays and Clay Minerals, Monograph No. 6* (Ed.) A.C.D. Newman, pp. 289–318. London, UK: Mineralogical Society.

Russell, J.D. 1965. Infra-red study of the reactions of ammonia with montmorillonite and saponite. *Transactions of the Faraday Society* 61: 2284–2294.

Russell, J.D., M. Cruz, and J.L. White. 1968a. The adsorption of 3-aminotriazole by montmorillonite. *Journal of Agricultural and Food Chemistry* 16: 21–24.

Russell, J.D., M. Cruz, J.L. White et al. 1968b. Mode of chemical degradation of *s*-triazines by montmorillonite. *Science* 160: 1340–1342.

Russell, J.D. and V.C. Farmer. 1964. Infrared spectroscopic studies of the dehydration of montmorillonite and saponite. *Clay Minerals Bulletin* 5: 443–464.

Salmón, M., M. Pérez-Luna, C. Lopéz-Franco et al. 1997. Catalytic conversion of propylene oxide on a super acid sulfonic clay (SASC) system. *Journal of Molecular Catalysis A: Chemical* 122: 169–174.

Santos, A., A. Rodriguez, J. Gonzales-Garmendia, and J. Barrios. 1970. Espectros infrarrojos de muestras homoiónicas de montmorillonita y vermiculita y sus complejos interlaminares con aminas. *Réunion Hispano-Belga de Minerales de la Arcilla, Madrid* 87–97.

Sawhney, B.L., R.K. Kozloski, P.J. Isaacson, and M.P.N. Gent. 1984. Polymerization of 2,6-dimethylphenol on smectite surfaces. *Clays and Clay Minerals* 32: 108–114.

Schoonheydt, R.A. 1991. Clays: From two to three dimensions. *Studies in Surface Science and Catalysis* 58: 201–239.

Shimizu, K.-i., T. Higuchi, E. Takasugi, T. Hatamachi, T. Kodama, and A. Satsuma. 2008. Characterization of Lewis acidity of cation-exchanged montmorillonite K-10 clay as effective heterogeneous catalyst for acetylation of alcohol. *Journal of Molecular Catalysis A: Chemical* 284: 89–96.

Shimoyama, A. and W.D. Johns. 1971. Catalytic conversion of fatty acids to petroleum-like paraffins and their maturation. *Nature Physical Science* 232: 140–144.

Shuali, U., M. Steinberg, S. Yariv, M. Müller-Vonmoos, G. Kahr, and A. Rub. 1990. Thermal analysis of sepiolite and palygorskite treated with butylamine. *Clay Minerals* 25: 107–119.

Sieskind, O., G. Joly, and P. Albrecht. 1979. Simulation of the geochemical transformations of sterols: Superacid effect of clay minerals. *Geochimica et Cosmochimica Acta* 43: 1675–1679.

Sivalov, E.G., N.G. Vasil'ev, and F.D. Ovcharenko. 1980. Nature of the acidity of kaolinite lateral faces. Effect of Group I exchange metals. *Doklady Akademii Nauk SSSR* 253: 1176–1179.

Sivalov, E.G., N.G. Vasil'ev, and F.D. Ovcharenko. 1981. Investigation of the interaction of *p*-aminoazobenzene with the surface of acid forms of laminar silicates. *Kolloidnyi Zhurnal* 43: 792–794.

Sivalov, E.G., N.G. Vasil'ev, and A. Savkin. 1983. Spectrometric study of acid active centers on a kaolinite surface. *Kinetika i Kataliz* 24: 1510–1513.

Šnirková, S., E. Jóna, L. Lajdová et al. 2009. Ni-exchanged montmorillonite with methyl-, dimethyl- and trimethylamine and their thermal properties. *Journal of Thermal Analysis and Calorimetry* 96: 63–66.

Sohn, J.R. 2002. Catalytic activities of nickel-containing catalysts for ethylene dimerization and butene isomerization and their relationship to acidic properties. *Catalysis Today* 73: 197–209.

Sohn, J.R. and A. Ozaki. 1980. Acidity of nickel silicate and its bearing on the catalytic activity for ethylene dimerization and butene isomerization. *Journal of Catalysis* 61: 29–38.

Solomon, D.H. 1968. Clay minerals as electron acceptors and/or donors in organic reactions. *Clays and Clay Minerals* 16: 13–39.

Solomon, D.H. and D.G. Hawthorne. 1971. The reaction of 2,2-diphenyl-1-picrylhydrazyl with minerals. *Journal of Macromolecular Science: Chemistry* A5: 575–586.

Solomon, D.H. and D.G. Hawthorne. 1983. *Chemistry of Pigments and Fillers*. New York: John Wiley & Sons.

Solomon, D.H. and M.J. Rosser. 1965. Reactions catalyzed by minerals. Part I. Polymerization of styrene. *Journal of Applied Polymer Science* 9: 1261–1271.

Solomon, D.H., B.C. Loft, and J.D. Swift. 1968. Reactions catalysed by minerals–IV. The mechanism of the benzidine blue reaction on silicate minerals. *Clay Minerals* 7: 389–397.

Solomon, D.H., J.D. Swift, and A.J. Murphy. 1971. The acidity of clay minerals in polymerization and related reactions. *Journal of Macromolecular Science: Chemistry* A5: 587–601.

Soma, Y. and M. Soma. 1988. Adsorption of benzidines and anilines on Cu- and Fe-montmorillonites studied by resonance Raman spectroscopy. *Clay Minerals* 23: 1–12.

Soma, Y. and M. Soma. 1989. Chemical reactions of organic compounds on clay surfaces. *Environmental Health Perspectives* 83: 205–214.

Soma, Y., M. Soma, and I. Harada. 1983. Raman spectroscopic evidence of formation of p-dimethoxybenzene cation on Cu- and Ru-montmorillonites. *Chemical Physics Letters* 94: 475–478.

Soma, Y., M. Soma, and I. Harada. 1984. Reactions of aromatic molecules in the interlayer of transition-metal ion-exchanged montmorillonite studied by resonance Raman spectroscopy. 1. Benzene and p-phenylenes. *Journal of Physical Chemistry* 88: 3034–3038.

Soma, Y., M. Soma, and I. Harada. 1985. Reactions of aromatic molecules in the interlayer of transition-metal ion-exchanged montmorillonite studied by resonance Raman spectroscopy. 2. Monosubstituted benzenes and 4,4'-disubstituted biphenyls. *Journal of Physical Chemistry* 89: 738–742.

Soma, Y., M. Soma, and I. Harada. 1986. The oxidative polymerization of aromatic molecules in the interlayer of montmorillonites studied by resonance Raman spectroscopy. *Journal of Contaminant Hydrology* 1: 95–106.

Soma, Y., M. Soma, Y. Furukawa, and I. Harada. 1987. Reactions of thiophene and methylthiophenes in the interlayer of transition-metal ion-exchanged montmorillonite studied by resonance Raman spectroscopy. *Clays and Clay Minerals* 35: 53–59.

Song, G., B. Wang, H. Luo, and L. Yang. 2007. Fe^{3+}-montmorillonite as a cost-effective and recyclable solid acidic catalyst for the synthesis of xanthenediones. *Catalysis Communications* 8: 673–676.

Stoessel, F., J.L. Guth, and R. Wey. 1977. Polymérisation de benzène en polyparaphénylène dans une montmorillonite cuivrique. *Clay Minerals* 12: 255–259.

Stucki, J.W. 2006. Properties and behaviour of iron in clay minerals. In *Handbook of Clay Science, Developments in Clay Science, Vol. 1*, (Eds.) F. Bergaya, B.K.G. Theng, and G. Lagaly, pp. 423–475. Amsterdam, the Netherlands: Elsevier.

Sukumar, R., K.R. Sabu, L.V. Bindu, and M. Lalithambika. 1998. Kaolinite supported metal chlorides as Friedel–Crafts alkylation catalysts. *Studies in Surface Science and Catalysis* 113: 557–562.

Swarnakar, R., K.B. Brandt, and R.A. Kydd. 1996. Catalytic activity of Ti- and Al-pillared montmorillonite and beidellite for cumene cracking and hydrocracking. *Applied Catalysis A: General* 142: 61–71.

Swift, H.E. and E.R. Black. 1974. Superactive nickel-aluminosilicate catalysts for hydroisomerization and hydrocracking of light hydrocarbons. *Industrial & Engineering Chemistry, Product Research and Development* 13: 106–110.

Swoboda, A.R. and G.W. Kunze. 1968. Reactivity of montmorillonite surfaces with weak organic bases. *Soil Science Society of America Proceedings* 32: 806–811.

Tahoun, S. and M.M. Mortland. 1966. Complexes of montmorillonite with primary, secondary, and tertiary amides: I. Protonation of amides on the surface of montmorillonite. *Soil Science* 102: 248–254.

Tamele, M.W. 1950. Chemistry of the surface and the activity of aluminasilica cracking catalyst. *Discussions of the Faraday Society* 8: 270–279.

Tanabe, K. 1981. Solid acid and base catalysts. In *Catalysis–Science and Technology, Vol. 2* (Eds.) J.R. Anderson and M. Boudart, pp. 231–273. Berlin, Germany: Springer-Verlag.

Tarasevich, Yu. I. and V.E. Doroshenko. 1983. Interaction of triphenylcarbinol with montmorillonite. *Kolloidnyi Zhurnal* 45: 1167–1170.

Tarasevich, Yu. I. and V.P. Fedorova. 1979. Electronic spectra of arylcarbinols sorbed on argillaceous minerals. *Kolloidnyi Zhurnal* 41: 812–814.

Telichkun, V. P., Yu. I. Tarasevich, and V.V. Goncharuk. 1977. Electronic spectra of Hammett indicator dyes sorbed by montmorillonite. *Teoritcheskaya Eksperimentalnaya Khimiya* 13: 131–134.

Tennakoon, D.T.B. and M.J. Tricker. 1975. Surface and intercalate chemistry of layered silicates. Part V. Infrared, ultraviolet, and visible spectroscopic studies of benzidine-montmorillonite and related systems. *Journal of the Chemical Society, Dalton Transactions* 1802–1806.

Tennakoon, D.T.B., J.M. Thomas, and M.J. Tricker. 1974b. Surface and intercalate chemistry of layered silicates. Part II. An iron-57 Mössbauer study of the role of lattice-substituted iron in the benzidine blue reaction of montmorillonite. *Journal of the Chemical Society, Dalton Transactions* 2211–2215.

Tennakoon, D.T.B., J.M. Thomas, M.J. Tricker, and J.O. Williams. 1974a. Surface and intercalate chemistry of layered silicates. Part I. General introduction and the uptake of benzidine and related organic molecules by montmorillonite. *Journal of the Chemical Society, Dalton Transactions* 2207–2211.

Tennakoon, D.T.B., J.M. Thomas, W. Jones, T.A. Carpenter, and S. Ramdas. 1986. Characterization of clays and clay-organic systems. *Journal of the Chemical Society, Faraday Transactions 1* 82: 545–562.

Theng, B.K.G. 1971. Mechanisms of formation of colored clay-organic complexes. A review. *Clays and Clay Minerals* 19: 383–390.

Theng, B.K.G. 1974. *The Chemistry of Clay-Organic Reactions*. London, UK: Adam Hilger.

Theng, B.K.G. 1982. Clay-activated organic reactions. In *International Clay Conference 1981. Developments in Sedimentology 35.* (Eds.) H. van Olphen and F. Veniale, pp. 197–238. Amsterdam, the Netherlands: Elsevier.

Theng, B.K.G. and G.F. Walker. 1970. Interactions of clay minerals with organic monomers. *Israel Journal of Chemistry* 8: 417–424.

Thomas, J.M. 1982. Sheet silicate intercalates: New agents for unusual chemical conversions. In *Intercalation Chemistry*, (Eds.) M.S. Whittingham and A.J. Jacobson, pp. 55–99. New York: Academic Press.

Thomas, J.M., J.M. Adams, S.H. Graham, and D.T.B. Tennakoon. 1977. Chemical conversions using sheet-silicate intercalates. In *Solid State Chemistry of Energy Conversion and Storage. Advances in Chemistry Series 163* (Eds.) J.B. Goodenough and M.S. Whittingham, pp. 298–315. Washington, DC: American Chemical Society.

Tipton, T. and L.E. Gerdom. 1992. Polymerization and transalkylation reactions of toluene on Cu(II)-montmorillonite. *Clays and Clay Minerals* 40: 429–435.

Touillaux, R., P. Salvador, C. Vandermeersche, and J.J. Fripiat. 1968. Study of water layers adsorbed on Na and Ca montmorillonite by the pulsed nuclear magnetic resonance technique. *Israel Journal of Chemistry* 6: 337–349.

Tricker, M.J., D.T.B. Tennakoon, J.M. Thomas, and S.H. Graham. 1975a. Novel reactions of hydrocarbon complexes of metal-substituted sheet silicates; thermal dimerisation of *trans*-stilbene. *Nature* 253: 110–111.

Tricker, M.J., D.T.B. Tennakoon, J.M. Thomas, and J. Heald. 1975b. Organic reactions in clay-mineral matrices: Mass-spectrometric study of the conversion of triphenylamine to N,N,N',N'-tetraphenylbenzidine. *Clays and Clay Minerals* 23: 77–82.

Tyagi, B., C.D. Chudasama, and R.V. Jasra. 2006. Characterization of surface acidity of an acid montmorillonite activated with hydrothermal, ultrasonic and microwave techniques. *Applied Clay Science* 31: 16–28.

Unger, K.K., U.R. Kittelmann, and W.K. Kreis. 1981. Examination and standardisation of amine titration in surface acidity measurements of silicas, aluminas and synthetic zeolites. *Journal of Chemical Technology and Biotechnology* 31: 453–469.

Uytterhoeven, J.B., L.G. Christner, and W.K. Hall. 1965. Studies of hydrogen held by solids. VIII. The decationated zeolites. *Journal of Physical Chemistry* 69: 2117–2126.

Van Damme, H., M. Crespin, F. Obrecht, M.I. Cruz, and J.J. Fripiat. 1978. Acid-base and complexation behavior of porphyrins on the intracrystal surface of swelling clays: Meso-tetraphenylporphyrin and meso-tetra (4-pyridyl)porphyrin on montmorillonites. *Journal of Colloid and Interface Science* 66: 43–54.

Vande Poel, D., P. Cloos, J. Helsen, and E. Janini. 1973. Comportement particulier du benzène adsorbé sur la montmorillonite cuivrique. *Bulletin du Groupe français des Argiles* 25: 115–126.

Walker, G.F. 1967. Catalytic decomposition of glycerol by layer silicates. *Clay Minerals* 7: 111–112.

Walling, C. 1950. The acid strength of surfaces. *Journal of the American Chemical Society* 72: 1164–1168.

Wallis, P.J., K.J. Booth, A.F. Patti, and J.L. Scott. 2006. Oxidative coupling revisited: Solvent-free, heterogeneous and in water. *Green Chemistry* 8: 333–337.

Wallis, P.J., W.P. Gates, A.F. Patti, J.L. Scott. 2011. Catalytic activity of choline modified Fe(III) montmorillonite. *Applied Clay Science* 53: 336–340.

Wallis, P.J., W.P. Gates, A.F. Patti, J.L. Scott, and E. Teoh. 2007. Assessing and improving the catalytic activity of K10 montmorillonite. *Green Chemistry* 9: 980–986.

Warren, D.S., A.L. Clark, and R. Perry. 1986. A review of clay-aromatic interactions with a view to their use in hazardous waste disposal. *The Science of the Total Environment* 54: 157–172.

Weil-Malherbe, H. and J. Weiss. 1948. Colour reactions and adsorption of some aluminosilicates. *Journal of the Chemical Society* 2164–2169.

White, J.L. 1976. Protonation and hydrolysis of s-triazines by Ca-montmorillonite as influenced by substitutions at the 2-, 4- and 6-positions. In *Proceedings of the International Clay Conference 1975* (Ed.) S.W. Bailey, pp. 391–398. Wilmette, IL: Applied Publishing.

Wilson, K. and J.H. Clark. 2000. Solid acids and their use as environmentally friendly catalysts in organic synthesis. *Pure and Applied Chemistry* 72: 1313–1319.

Wilson, M.C. and A.K. Galwey. 1976. Reactions of stearic acid, of *n*-dodecanol, and of cyclohexanol on the clay minerals illite, kaolinite and montmorillonite. *Journal de Chimie Physique* 73: 441–446.

Wright, A.C., W.T. Granquist, and J.V. Kennedy. 1972. Catalysis by layer lattice silicates. I. The structure and thermal modification of a synthetic ammonium dioctahedral clay. *Journal of Catalysis* 25: 6580.

Wu, D.Y., N. Matsue, T. Henmi, and N. Yoshinaga. 1992. Surface acidity of pyrophyllite and talc. *Clay Science* 8: 367–379.

Wu, L.M., C.H. Zhou, J. Keeling, D.S. Tong, and W.H. Yu. 2012. Towards an understanding of the role of clay minerals in crude oil formation, migration and accumulation. *Earth-Science Reviews* 115: 373–386.

Yariv, S. and H. Cross. (Eds.), 2002. *Organo-Clay Complexes and Interactions*. New York: Marcel Dekker.

Yariv, S. and L. Heller. 1970. Sorption of cyclohexylamine by montmorillonite. *Israel Journal of Chemistry* 8: 935–945.

Yariv, S., L. Heller, Z. Sofer, and W. Bodenheimer. 1968. Sorption of aniline by montmorillonite. *Israel Journal of Chemistry* 6: 741–756.

Yariv, S., N. Lahav, and M. Lacher. 1976. On the mechanism of staining montmorillonite by benzidine. *Clays and Clay Minerals* 24: 51–52.

Yong, R.N., S. Desjardins, J.P. Farant, and P. Simon. 1997. Influence of pH and exchangeable cation on oxidation of methylphenols by a montmorillonite clay. *Applied Clay Science* 12: 93–110.

Yuan, P., H. Liu, D. Liu, D. Tan, W. Yan, and H. He. 2013. Role of the interlayer space of monrmorillonite in hydrocarbon generation: An experimental study based on high temperature-pressure pyrolysis. *Applied Clay Science* 75–76: 82–91.

3 Surface Activation and Modification

3.1 INTRODUCTION

The catalytic activity of clays and clay minerals, as we have seen, is intimately related to their ability to act as solid acids either in the Brønsted or Lewis sense. We have also mentioned that the surface acidity of clays can be modified, if not controlled, by cation exchange, dehydration, heating, and wetting (e.g., Adams et al. 1983). Thus, clay surfaces can become highly acidic through dehydration (cf. Figures 2.2 and 2.6), while the relative concentration of Brønsted and Lewis acid sites varies with the temperature at which the clay samples have been preheated (cf. Figure 2.15). In order to enhance and entrench surface acidity, however, it is common practice to *activate* the clay samples by treatment with mineral acid solutions; that is, by introducing protons into the interlayer exchange sites. Indeed, a large number and variety of clay-catalyzed organic reactions are carried out using so-called *acid-activated* smectites, notably the commercially available *K-catalysts*, either as such or after impregnation with certain metal salts (Balogh and Laszlo 1993; Chitnis and Sharma 1997a; Clark and Macquarrie 1997; Nikalje et al. 2000; Flessner et al. 2001; Varma 2002; Dasgupta and Török 2008; Fernandes et al. 2012; McCabe and Adams 2013; Kumar et al. 2014). The majority of K-catalysts (Table 3.1), for example, are formed by treatment of montmorillonite with HCl of variable concentrations at 80°C–90°C (De Stefanis and Tomlinson 2006). These materials and other clay-based catalysts may be further modified by heating (calcination), ultrasonic treatment, or microwave irradiation (Varma 2002; Mahmoud et al. 2003; Noyan et al. 2006; Tyagi et al. 2006; Korichi et al. 2009, 2012; Hussin et al. 2011; Heller-Kallai 2013; Franco et al. 2014). The use of thermally modified, acid-activated smectites as catalysts of various organic reactions has been summarized by Balogh and Laszlo (1993) and McCabe and Adams (2013).

Acid-activated bentonite, kaolinite, montmorillonite, and palygorskite have long been used to catalyze some organic reactions on an industrial scale. A well-known example is the *Houdry* process by which crude oil (petroleum) is converted to high-octane gasoline (Houdry et al. 1938; Hettinger 1991). The literature on the role of different clay minerals in the catalytic cracking of petroleum has been summarized by Hansford (1950), Ryland et al. (1960), Thomas (1970), Banerjee and Sen (1974), and Heller-Kallai (2002). By the mid-1960s, however, this particular function of clays was taken over by synthetic zeolites (Barrer 1978; Venuto and Habib 1979; Breck 1980) which, in turn, were supplanted by pillared interlayered clays (PILC) during the early 1980s (Occelli 1987, 1988; Figueras 1988; Lambert and Poncelet 1997; Vicente et al. 2013). Nevertheless, clay minerals and their surface-modified forms continue to serve as catalysts in the petroleum refining industry (Emam 2013). Indeed, more PILC are being used for petroleum cracking than any other solid acid catalysts (Adams 1987; Ding et al. 2001; Vicente et al. 2013) since these materials can give satisfactory yields of liquid hydrocarbons in the gasoline and diesel range (De Stefanis et al. 2013). Other important industrial applications of acid-activated clay and PILC catalysts include the decolorization (bleaching) of edible oils and fats, the cracking of vegetable oils to produce biofuels, and the synthesis of fine chemicals (Anderson and Williams 1962; Siddiqui 1968; Fahn 1973; Theng and Wells 1995; Christidis et al. 1997; Chitnis and Sharma

TABLE 3.1
Some Surface Properties of Commercial K-Catalysts Produced by Süd-Chemie (Germany) and Available from Fluka and Sigma-Aldrich

Catalyst[a]	Specific Surface Area (m²/g)	Average Pore Diameter (nm)	Pore Volume (mL/g)	Cation Exchange Capacity (cmol/kg)	pH (H$_o$)[b]	Acidity[c] Brønsted (mmol/g)	Lewis (mmol/g)
Parent clay	88	4.4	0.097	91			
KSF	9–40	5.0	0.011		1.5 (−8 to −9)	0.59	0.15
KSF/0	117	7.4	0.215		1.3	1.03	0.20
KP10	169	7.1	0.300		1.8	0.49	0.09
K10	229–254	5.6[d]	0.320		4.5 (−5.6 to −5.9)	0.33	0.29[e]
K0	268		0.380		5.5		
KS	322		0.465		3.0	0.45	0.26
Cu^{2+}-K10	236			39		0.13	0.68
Zn^{2+}-K10	213			36		0.16	0.73
Al^{3+}-K10	234		0.293	30			
Fe^{3+}-K10	239			54		0.35	0.34

Sources: Cseri, T. et al., *J. Mol. Catal. A: Chem.*, 98, 101–107, 1995a; Cseri, T. et al., *Appl. Catal. A: Gen.*, 132, 141–155, 1995b; Békássy, S. et al., *Top. Catal.*, 13, 287–290, 2000; Flessner, U. et al., *J. Mol. Catal. A: Chem.*, 168, 247–256, 2001; Vodnár, J. et al., *Appl. Catal. A: Gen.*, 208, 329–334, 2001; Hart, M.P. and Brown, D.R., *J. Mol. Catal. A: Chem.*, 212, 315–321, 2004.

[a] The parent clay is a montmorillonite-rich bentonite from Bavaria, Germany, often referred to as Moosburg montmorillonite in the literature. KSF is obtained by treating the parent material with sulfuric acid at room temperature, while the other K-catalysts are formed by treatment with HCl of variable concentrations at 80°C–90°C (De Stefanis and Tomlinson 2006).

[b] pH values are measured on a 10% suspension of the clay samples in water. Ho is the Hammett acidity function (Laszlo 1987; Villemin et al. 1992).

[c] Determined by adsorption of NH$_3$ and pyridine combined with infrared spectroscopy.

[d] Pore diameters of 6–10 nm were reported by Butruille and Pinnavaia (1992).

[e] Limited Lewis acidity was detected by Clark and Macquarrie (1997).

1997a; Clark and Macquarrie 1997; Falaras et al. 1999, 2000; Ding et al. 2001; Valenzuela Díaz and de Souza Santos 2001; Kloprogge et al. 2005; Centi and Perathoner 2008; Hussin et al. 2011; Önal and Sarikaya 2012; Tong et al. 2014).

Because of their large surface area and pore volume, thermal stability, and high surface acidity, PILC feature prominently in organic catalysis (Table 3.5). Indeed, the term *pillaring* is often mentioned in relation to the formation of catalytically active microporous solids (Figueras 1988; Lambert and Poncelet 1997; Gil et al. 2000, 2008, 2010; Ding et al. 2001; Vicente et al. 2013). As the name suggests, PILC are formed by intercalating metal (hydr)oxides into smectites through a cation exchange process, and converting the intercalated species into the corresponding metal oxide *pillars* by heating at ≥300°C. PILC may therefore be likened to two-dimensional zeolites in terms of structure, thermal stability, and catalytic activity.

It has long been known that the layers of expanding clay minerals, such as smectites and vermiculites, can also be propped apart (*pillared*) by intercalation of compact alkylammonium

Surface Activation and Modification

cations (Barrer and MacLeod 1955), yielding porous sorbents and molecular sieves. The early literature on the synthesis, properties, and applications of such organically modified clay minerals has been summarized by Grim (1968), Theng (1974), and Barrer (1978, 1984, 1989), while the general chemistry of the clay-organic interaction has been described in reviews written or edited by Mortland (1970), Theng (1974, 2012), Rausell-Colom and Serratosa (1987), Yariv and Cross (2002), and Lagaly et al. (2013).

Organically modified clays or *organoclays*, formed by intercalation of quaternary ammonium cations into smectites (de Paiva et al. 2008; He et al. 2014; Zhu et al. 2015), however, are unstable at temperatures above 225°C although the quaternary phosphonium-exchanged varieties do not decompose until close to 300°C (Xie et al. 2001, 2002; Hedley et al. 2007; Kooli and Yan 2013). Organoclays have a large propensity for taking up non-ionic, hydrophobic organic contaminants and pollutants (Jaynes and Boyd 1991; Theng et al. 2008; Yuan et al. 2013; Zhu et al. 2015, 2016) but are essentially inactive as catalysts of organic reactions unless they have previously been treated with mineral acids. Because of their hydrophobic character, however, organically modified clays can usefully serve as phase-transfer and triphase catalysts (Cornélis and Laszlo 1982; Cornélis et al. 1983; Lin and Pinnavaia 1991; Pinnavaia 1995; Akelah et al. 1999; Varma 1999; Yadav and Naik 2000; Ghiaci et al. 2005; Shabestary et al. 2007; Wibowo et al. 2010; McCabe and Adams 2013).

Clay mineral surfaces can also be made hydrophobic by covalent grafting or *silylation* with organosilanes. The preparation and characterization of silylated clay minerals have recently been reviewed by He et al. (2013). Such organically functionalized layer silicates, however, do not play an important role in organic catalysis. On the other hand, the deposition, immobilization, or impregnation of various metal salts, metal-organic complexes, metal nanoparticles, and heteropoly acids on clay mineral surfaces afford a whole new class of catalytically active *heterostructural* materials (Balogh and Laszlo 1993; Brown 1994; Varma 2002; Yadav 2005; McCabe and Adams 2013; Kumar et al. 2014). The preparation, characterization, and activity of clay-supported catalysts and reagents are summarized in Chapter 4.

3.2 ACID ACTIVATION

As mentioned in Chapter 2, the surface of *dry* smectites can be highly acidic in the Brønsted sense due to the dissociation of water molecules that are directly coordinated to (polyvalent) interlayer counterions. The number and strength of surface Brønsted acid sites may be greatly enhanced by replacing the counterions, initially present, with protons. Thus, the conversion of Na^+-montmorillonite into H^+/Al^{3+}-exchanged and K10 montmorillonite by treatment with mineral acids, increases its surface acidity, H_o, from ca. +2.0 to as high as −8, depending on the severity of treatment (Table 2.5).

Proton exchange or *acid activation* is commonly carried out by treating the raw clay samples with a dilute solution of a mineral acid (HCl, H_2SO_4, HNO_3), more often than not under reflux conditions, and removing the excess electrolyte by washing with water. Acid activation can be greatly accelerated by ultrasonic treatment or microwave irradiation (e.g., Tyagi et al. 2006). Alternatively, protons may be introduced by passing a dilute aqueous suspension of the clay mineral through a H^+-exchange resin (Banin and Ravikovitch 1966; Barshad 1969). The amount of hydrogen—more precisely hydronium—ions, occupying exchange positions at the mineral surface, may be estimated directly by conductometric or potentiometric titration with a suitable base (Komadel and Madejová 2013). Alternatively, the introduced protons may be exchanged for other cations and the quantity so displaced measured using standard analytical procedures (Barshad 1960; Davis et al. 1962; Loeppert et al. 1986). Either way, the concentration of hydrogen ions per unit weight of clay might

TABLE 3.2
Decomposition of Montmorillonite (Utah bentonite) Induced by Acid Treatment as Indicated by Changes in Cation Exchange Capacity (CEC) with Treatment Cycle

Number of Cycles[a]	Exchangeable Cations Extracted (cmol/kg)			CEC[c] (cmol/kg)
	H^+	Al^{3+}	Mg^{2+}	
1	8	62	10	80
2	9	60	10	79
3	14	49	9	72
4	15	43	8	66
5	12	34	7	53
Total		248[b]	44	

Source: Coleman, N.T. and Craig, D., *Soil Sci.*, 91, 14–18, 1961.

[a] A cycle consists of leaching the clay sample with 1M HCl, washing with water, storing at 90°C for 24 h, and extracting the cations with 1M KCl.

[b] The total amount of extractable Al^{3+} (after five cycles) corresponds to a 22% decomposition of the layer structure, in accord with the percentage decrease in CEC.

[c] Sum of extractable H^+, Al^{3+}, and Mg^{2+}.

---

be expected to approach the cation exchange capacity (CEC) of the sample being used. This expectation, however, does not accord with practice in that the concentration of exchangeable protons is commonly appreciably smaller than the corresponding CEC (Table 3.2). The generally accepted explanation for this discrepancy is that acid-treated clay minerals are inherently unstable, transforming spontaneously into mixed H^+/Al^{3+}-exchanged systems (Coleman and Craig 1961; Eeckman and Laudelout 1961; Miesserov 1969; Barshad and Foscolos 1970; Vasil'ev and Ovcharenko 1977; Sindhu 1984; Rhodes and Brown 1995).

As a result, the CEC of acid-activated montmorillonites, such as K10, is noticeably smaller than that of the parent clay (Table 3.1). Likewise, Massam and Brown (1998) reported a steady decline in the CEC of a montmorillonite after refluxing in 30 wt % H_2SO_4 for increasing lengths of time. Önal (2007) observed similarly for a bentonite that had been treated with sulfuric acid of increasing concentrations (from 0 to 70 wt % H_2SO_4). Little, if any, reduction in CEC, however, occurred when Ramesh et al. (2012) treated a montmorillonite with the non-protonic *p*-toluene sulfonic acid.

Some of the aluminum ions, occupying exchange sites, may originate from adsorbed hydroxy-aluminum compounds whose presence as surface contaminants is by no means uncommon (e.g., Barnhisel and Bertsch 1989). For the most part, however, Al^{3+} and other polyvalent cations (Mg^{2+}, Fe^{3+}) originate and derive from the layer structure (Heyding et al. 1960; Rhodes and Brown 1995). Being highly mobile, the introduced protons would first attack the edge surface of the clay particles and then diffuse into (and ultimately dissolve) the octahedral sheet of the mineral (Steudel et al. 2009; Komadel and Madejová 2013). This process is reminiscent of the heat-induced migration of Li^+ (and other small counterions such as Cu^{2+} and Ce^{3+}) from exchange sites to octahedral vacancies and ditrigonal depressions, referred to in Chapter 1 (Section 1.3.4). Unlike lithium ions, however, protons can move rather freely by a series of jumps in hydrogen bonding to oxygen ions (Fripiat 1971), inducing Al^{3+} (Mg^{2+}, Fe^{3+}) ions to migrate from octahedral positions to basal and interlayer exchange sites (Glaeser et al. 1961; Maricic et al. 1982).

Rhodes and Brown (1995) used solid-state ^{27}Al NMR spectroscopy, in conjunction with chemical analysis, to investigate the *autotransformation* of acid-treated montmorillonite (under conditions of high humidity). Although structural aluminum and interlayer (hydrated) Al^{3+} have the same chemical shifts, the respective NMR line widths are distinguishable, enabling them to follow the migration of aluminum from the layer structure to interlayer exchange sites. Chemical, X-ray diffraction, infrared, and NMR spectroscopic analyses further indicate that proton diffusion into, and migration through, the octahedral sheet in montmorillonite can lead to Si–O–Al bond rupture (Komadel and Madejová 2013). In structures where the layer charge dominantly resides in the tetrahedral sheet, such as beidellite and vermiculite, the introduced protons can apparently displace aluminum substituting for silicon in tetrahedral positions (Brückman et al. 1976; Vasil'ev et al. 1982; Breen et al. 1995a; Christidis et al. 1997).

The rate of acid dissolution is influenced by the octahedral sheet composition, crystallinity, and particle size of the clay mineral sample, the type and concentration of acid as well as the duration and temperature of acid treatment. As such, clay minerals vary in their resistance and stability to acid attack. Irrespective of mineral species, however, the final product of acid activation consists of a mixture of unaltered layers and a protonated, hydrated silica phase of short-range order, arising from the decomposition of the tetrahedral sheet (Komadel 2016) (cf. Figure 4.1). This is the case with both 1:1 type layer silicates such as kaolinite, halloysite, and chrysotile (Aglietti et al. 1988; Wypych et al. 2005; Panda et al. 2010; Zhang et al. 2012; Rozalen and Huertas 2013) and 2:1 type phyllosilicates such as smectites and vermiculites (Fahn 1973; Komadel et al. 1990; Suquet et al. 1991, 1994; Shinoda et al. 1995a; Vicente-Rodriguez et al. 1996; Madejová et al. 1998; Steudel et al. 2009). The same is true for 2:1 layer-ribbon type structures, such as sepiolite and palygorskite. Containing more octahedrally coordinated Mg^{2+} ions, and having larger channel dimensions, sepiolite is more susceptible to acid decomposition than is palygorskite (Suárez Barrios et al. 1995; Dékány et al. 1999; Komadel and Madejová 2013; Franco et al. 2014). For the same reason, the octahedral sheet of a magnesium-rich palygorskite is more easily dissolved by treatment with HCl as compared to its aluminum-rich counterpart (González et al. 1989).

More often than not, however, the X-ray diffraction patterns of acid-activated clay minerals show a vestige of the characteristic basal reflection, while the corresponding electron micrographs indicate the particle (shape) outline, of the corresponding original (parent) materials (Aglietti et al. 1988; Komadel et al. 1990; Mokaya and Jones 1994; Suquet et al. 1991, 1994; Shinoda et al. 1995a; Suárez Barrios et al. 1995; Wallis et al. 2007; Boudriche et al. 2011; Zhang et al. 2012). Besides causing a reduction in the cation exchange capacity of the clay samples, acid activation leads to a marked increase in specific surface area (SSA) and pore volume (Table 3.1) (González-Pradas et al. 1991; Rhodes and Brown 1995; Suárez Barrios et al. 1995; Breen et al. 1995b; Balci 1999; Hart and Brown 2004; Steudel et al. 2009; Panda et al. 2010; Zhang et al. 2010a; Tomić et al. 2011; Kumar et al. 2013). Shinoda et al. (1995a), for example, found that the SSA of Na^+-montmorillonite rose from 25 to ca. 175 m^2/g, while the concentration of acid sites with $H_o \leq -5.6$ increased from 0 to 0.5 mmol/g, following treatment with 30 wt % H_2SO_4 for 1 h. The reviews by Komadel (2003, 2016) and Komadel and Madejová (2013) provide further information on the acid activation of clay minerals.

In the presence of water, H^+-exchanged clay minerals are inherently unstable, being progressively transformed into the corresponding mixed H^+/Al^{3+} forms with partial dissolution of the layer structure (Table 3.2). Water may also react with, and hence obscure Lewis sites associated with incompletely coordinated aluminum ions, exposed at clay particle edges. The rate of

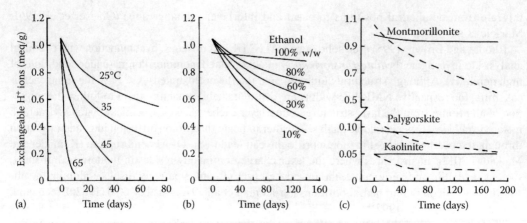

FIGURE 3.1 Diagram showing the rate of autotransformation (*aging*) of different clay minerals as indicated by the change in the concentration of exchangeable H+ ions over time of storage: (a), aqueous H+-montmorillonite suspension (0.3% w/v) showing the effect of storage temperature; (b), H+-montmorillonite suspension (0.3% w/v) showing the effect of increasing proportions of ethanol in the suspending medium; and (c), H+-exchanged montmorillonite, palygorskite, and kaolinite showing the effect of prior drying where the solid lines refer to dehydrated samples, and the broken lines to air-dry samples. (Adapted from Vasil'ev, N.G. and Ovcharenko, F.D., *Russian Chem. Rev.*, 46, 775–788, 1977.)

autotransformation (*aging*) is proportional to the square root of time. Delvaux and Laudelout (1964), for example, observed a 50% transformation in ca. 11 days for a montmorillonite. Figure 3.1a shows that the rate of aging is also influenced by the temperature at which the acid-treated clay suspension is stored, or dried (Mitra and Singh, 1959; Coleman and Craig, 1961; Vasil'ev and Ovcharenko 1977). For aqueous suspensions at 90°C, autotransformation can be completed as quickly as four days (Komadel and Madejová 2013). The aging process can be greatly retarded, if not altogether inhibited, by drying or storing the acid-treated material in ethanol (Figure 3.1b). Interestingly, Wallis et al. (2007) found that the catalytic activity of the commercial (acid-activated) K10 montmorillonite varied among batches of the supplied material, probably as a result of aging during storage. Indeed, further treatment of the K10 samples with HCl increased catalytic activity with respect to the tetrahydropyranilation of ethanol, the deacetylation of benzaldehyde, and the esterification of succinic anhydride. A simple and practical means of preserving the integrity of H+-exchanged clay minerals is to dehydrate the freshly prepared samples (Figure 3.1c), by either evacuating over P_2O_5 at ambient temperature, or freeze-drying and storing the dry clays in a refrigerated sealed container under dry organic liquids, such as benzene (Clark et al. 1997).

Breen et al. (1995a, 1995b, 1997) used the yields of tetrahydropyranyl ether (by reacting 2,3-dihydropyran with methanol) to compare and test the catalytic activity of a range of acid-treated smectites. In this instance, mild acid treatment apparently gave rise to high Brønsted acidity and catalytic activity. More recently, Yadav et al. (2004a) reported that the isomerization of α-pinene, catalyzed by montmorillonites that had been treated with 0.5–2.0 M sulfuric acid, yielded limonene as the main product, while α-terpinene was primarily produced in the presence of severely treated (2.5–4.5 M H_2SO_4) clay samples.

Acid-treated (H+/Al^{3+}-exchanged) smectites, especially the commercially available K10 and KSF montmorillonites, have been used to catalyze a wide range and variety of organic reactions and conversions (Ballantine 1986; Balogh and Laszlo 1993; Kaur and Kishore 2012; McCabe and Adams 2013), the vast majority of which are listed in Table 3.3.

TABLE 3.3
Organic Conversions and Reactions Catalyzed by Cation-Exchanged and Acid-Treated Montmorillonites with Particular Reference to the Commercially Available K-Series Catalysts

Reaction Type (in Alphabetical Order)	Reactant/Substrate	Product	Clay Catalyst	Reference (in Chronological Order)
Acetalization	Carbonyls with trimethyl orthoformate	Dimethyl acetals	K10 montmorillonite	Taylor and Chiang (1977)
	Enol ethers and 1,2-diols	Cyclic acetals	KSF montmorillonite	Vu and Maitte (1979)
	Cyclohexanones, benzaldehydes with methanol	Corresponding acetals	Ce^{3+}-montmorillonite	Tateiwa et al. (1995)
	Aldehydes and ketones with ethylene glycol	Acetals	K10 montmorillonite	Li et al. (1997a)
	Carbonyl compounds with alcohols	Corresponding acetals	Ti^{4+}-montmorillonite	Kawabata et al. (2001)
	Ketones with p-Toluenesulphonic acid	Dimethyl acetals	K10 montmorillonite	Mansilla and Regás (2006)
	Benzaldehyde with methanol	Corresponding acetal	Natural bentonite	Venkatathri (2006)
	Ketones with methanol	Dimethyl acetals	K10 and (Ce^{3+}, Fe^{3+}, Al^{3+})-exchanged K10 montmorillonites	Thomas and Sugunan (2006); Thomas et al. (2011)
	Unsaturated aldehydes with ethylene glycol	2-(R)-dioxolanes	Acid-activated smectite	Besbes et al. (2012)
	Aldehydes with 2-mercaptoethanol	1,3-Oxathiolanes	Sulfonic acid-grafted Na^+-montmorillonite	Shirini et al. (2013)
Acetylation and Acylation	Friedel–Crafts acylation of aromatic compounds (benzene, toluene) with carboxylic acids	Ketones	Cation-exchanged montmorillonites	Chiche et al. (1987)
	Friedel–Crafts acylation of anisole, mesitylene, and p-xylene with benzoyl chloride or benzoic anhydride	Various benzophenones	Fe^{3+}-exchanged and $ZnCl_2$-impregnated K10 montmorillonite;	Cornélis et al. (1990)
	Acetylation of diisobutylene and diisoamylene with acetic anhydride	Corresponding methyl ketones	F-24 montmorillonite	Shah and Sharma (1993)
	Acetylation of benzo-15-crown-5 ether with acetyl chloride	4'-Acetyl-benzo-15-crown-5 ether	Cu^{2+}-exchanged and $CuCl_2$-doped K10 montmorillonite	Cseri et al. (1996)
	Mono-, di-, and tri-saccharides with acetic anhydride	Corresponding acetylated products	K10 montmorillonite	Bhaskar and Loganathan (1998)
	Aromatic ethers with acetic anhydrides	Corresponding acylated products	Cation-exchanged K10 montmorillonites	Choudary et al. (1998)
	Alcohols, thiols, phenols and amines with acetic anhydride	Corresponding acetates	KSF and K10 montmorillonite	Li et al. (1997a); Li and Li (1998)

(*Continued*)

TABLE 3.3 (Continued)
Organic Conversions and Reactions Catalyzed by Cation-Exchanged and Acid-Treated Montmorillonites, with Particular Reference to the Commercially Available K-Series Catalysts

Reaction Type (in Alphabetical Order)	Reactant/Substrate	Product	Clay Catalyst	Reference (in Chronological Order)
Acetylation and Acylation (Continued)	Aldehydes with acetic anhydride	Corresponding acylals (1,1-diacetates)	Fe^{3+}-montmorillonite, Zn-doped montmorillonite	Zhang et al. (1997a); Li et al. (1998a); Nagy et al. (2002)
	Acylation of resorcinol with phenylacetyl chloride	Ketone and ester	K-series catalysts and cation-exchanged forms	Békássy et al. (2000)
	Acetylation of benzo-15-crown-5 ether with acetyl chloride	4′-Acetyl-benzo-15-crown-5 ether	Sn^{2+}-exchanged K10 montmorillonite	Biró et al. (2000)
	N-acetylation of amines with acetic acid	Corresponding acetates	K10 and cation-exchanged K10 montmorillonites	Choudary et al. (2001)
	Acylation of resorcinol with benzoic acid followed by Fries rearrangement of resorcinol monobenzoate	2,4-Dihydroxybenzo-phenone	F-22B and K-series montmorillonites	Bolognini et al. (2004)
	Acylation of 1-methyl-1-cyclohexene and 1-trimethylsilyl-1-alkynes with acyl chlorides	Corresponding acylated products	Cation-exchanged montmorillonite	Nishimura et al. (2004)
	Acylation of sulphonamides with carboxylic acid anhydrides	N-acyl-sulfonamides	Fe^{3+}-exchanged K10 montmorillonite	Singh et al. (2004b)
	Acylation of anisole with acetic anhydride or dodecanoic acid	4-Methoxyaceto-phenone or 1-(4-methoxyphenyl)-1-dodecanone	Ti^{4+}-montmorillonite	Ebitani et al. (2006)
	Glycerol with acetic acid	Mono-, di-, and tri-acetin	K10 montmorillonite	Gonçalves et al. (2008)
	Cyclohexanol with acetic anhydride	Corresponding acetate	Fe^{3+}-exchanged K10 montmorillonite	Shimizu et al. (2008)
	D-glucose, glycerol, D-mannitol, 1,2,3-triazolic derivatives with acetic anhydride	Corresponding acetylated products	K10 and KSF montmorillonite	De Oliveira et al. (2014)
	Glycerol with acetic acid	Diacetylglycerols	La^{3+}-montmorillonite	Mizugaki et al. (2014)
Addition	Interlayer water to 1-alkenes	Corresponding ethers	Cation-exchanged montmorillonite	Adams et al. (1979)
	Alcohols to alkenes	Ethers	Al^{3+}-bentonite	Ballantine et al. (1983, 1984b)
	Thiols to alkenes	Alkyl sulphides	Al^{3+}-bentonite	Ballantine et al. (1983)

(*Continued*)

TABLE 3.3 (Continued)
Organic Conversions and Reactions Catalyzed by Cation-Exchanged and Acid-Treated Montmorillonites, with Particular Reference to the Commercially Available K-Series Catalysts

Reaction Type (in Alphabetical Order)	Reactant/Substrate	Product	Clay Catalyst	Reference (in Chronological Order)
Addition (Continued)	Carboxylic acids to alkenes	Esters	Al^{3+}- and H^+/Al^{3+}-montmorillonite	Ballantine et al. (1983, 1984a)
	Tertiary alcohols to nitriles	Substituted amides	Al^{3+}-montmorillonite	Ballantine (1986)
	Dienes to acrolein	Cyclic olefins	Fe^{3+}-K10 montmorillonite	Laszlo and Moison (1989)
	Oxiranes with carbonyl compounds	1,3-Dioxolanes	Acid-treated bentonite (Tonsil)	Cabrera et al. (1992)
	Aliphatic thiols to styrene	Corresponding sulphides	K10 and cation-exchanged forms	Kannan and Pitchumani (1997)
	Benzyl chlorides to thiols	Benzyl sulfides	Silylated K10 and KSF montmorillonite	Kannan et al. (1999)
	Phenols with formaldehyde	Salicylic aldehydes	KSF montmorillonite with trimethylamine	Bigi et al. (2000)
	Alcohols to nitriles	Secondary amides	Fe^{3+}-K10 montmorillonite	Lakouraj et al. (2000)
	Thiols and thiobenzoic acids to olefins	Dithiocarboxylic esters	K10 montmorillonite	Kanagasabapathy et al. (2001)
	Cyclization of homoallylic alcohols with aldehydes	Tetrahydropyranols	KSF montmorillonite	Yadav et al. (2001)
	Indoles with 3-formylindole	Triindolylmethanes	K10 montmorillonite	Chakrabarty and Sarkar (2002)
	Aryl amines to trimethylsilyl cyanide	α-Aminonitriles	KSF montmorillonite	Yadav et al. (2004c)
	Nucleophilic addition of silyl enolate to aldimines	Amino carbonyl compounds	K10 montmorillonite	Akiyama et al. (2005)
	Organic disulfides to alkenes	vic-Bis(alkyl/aryl-thio)alkanes	Fe^{3+}-montmorillonite	Nishimura et al. (2005)
	Amines to α,β-ethylenic compounds	Mono- and di-addition products	K10 montmorillonite	Joseph et al. (2006)
	Nucleophilic addition of sulphonamides and carboxamides to alkenes	Hydroaminated products	H^+/Al^{3+}-montmorillonite	Motokura et al. (2006b)
	1,3-Dicarbonyl compounds to alkenes and alcohols	Corresponding adducts	H^+/Al^{3+}-montmorillonite	Motokura et al. (2006a)
	Aniline to 1,4-naphthaquinone	2-(R-anilino)-1,4-naphthoquinones	Natural bentonite	Leyva et al. (2008)
	Michael addition of indoles to nitroolefins	3-Substituted indole derivatives	K10 montmorillonite	Chen and Li (2009)
	Imines to ethyl diazoacetate	cis-Aziridines (diastereoselective	K10 montmorillonite	Borkin et al. (2010)
	Isobutanol to epoxidised palm oil olein	Palm-based polyols	K10 montmorillonite	Noor et al. (2013)

(Continued)

TABLE 3.3 (Continued)
Organic Conversions and Reactions Catalyzed by Cation-Exchanged and Acid-Treated Montmorillonites, with Particular Reference to the Commercially Available K-Series Catalysts

Reaction Type (in Alphabetical Order)	Reactant/Substrate	Product	Clay Catalyst	Reference (in Chronological Order)
Addition (Continued)	*tert*-Butyldimethylsilyl cyanide to ketones	Cyanohydrin *tert*-butyldimethylsilyl ethers	Sn^{4+}-montmorillonite	Wang et al. (2012)
	Electrophilic addition of indole to aldehydes	Bis(indoyl) methanes	Fe^{3+}-exchanged K10 montmorillonite	Fekri et al. (2015)
	Nucleophilic addition of indole derivatives with azo-linked aldehydes	Corresponding azo-linked diindolyl methanes	Fe^{3+}-exchanged K10 montmorillonite	Nikpassand et al. (2017)
Alkylation	Benzene with *t*-butylbromide	Mono- and di-substituted products	Cu^{2+}-bentonite	Lotz and Gosselck (1979)
	Benzene with propene or ethylene	Cumene or ethylbenzene	H^+- and Al^{3+}-montmorillonite	Westlake et al. (1985)
	Substituted benzenes and naphthalenes with alcohols and esters	Corresponding alkylated products	Montmorillonite (Wyoming)	Rigby et al. (1986)
	3-Hydroxy-2-methylenealkanoic esters with benzene	Tri-substituted alkenes	K10 montmorillonite	Saib and Foucaud (1987)
	Phenol with cyclohexene	*o*- and *p*-Cyclohexyl phenols	F-24 montmorillonite	Chakrabarti and Sharma (1992a)
	Benzene with primary, secondary, and tertiary alcohols	Various isomeric alkylbenzenes	K10 montmorillonite	Sieskind and Albrecht (1993)
	Friedel–Crafts alkylation of hydroxy and methoxy aromatics with 4-hydroxybutan-2-one	Raspberry ketone and other pharmaceutically active compounds	Cation-exchanged montmorillonite	Tateiwa et al. (1994c)
	Friedel–Crafts alkylation of benzene with (Z,Z)-1,5-cyclooctadiene	Isomers of phenylbicyclo octanes	Cation-exchanged montmorillonite	Tateiwa et al. (1994a)
	Methylation of phenanthrene and alkylphenanthrenes	Corresponding methylated derivatives	Al^{3+}-montmorillonite	Alexander et al. (1995)
	Benzene and toluene with benzyl chloride and benzyl alcohol	*Ortho*- and *para*-isomers of toluene	KSF, KSF/0, K0, KP10, K10, KS montmorillonites; cation-exchanged K10	Cseri et al. (1995a, 1995b)

(*Continued*)

TABLE 3.3 (Continued)
Organic Conversions and Reactions Catalyzed by Cation-Exchanged and Acid-Treated Montmorillonites, with Particular Reference to the Commercially Available K-Series Catalysts

Reaction Type (in Alphabetical Order)	Reactant/Substrate	Product	Clay Catalyst	Reference (in Chronological Order)
Alkylation (Continued)	Diphenylamine with α-methylstyrene and diisobutylene	Mono- and di-alkylated diphenylamines	F-24 montmorillonite	Chitnis and Sharma (1996)
	Methyl salicylate with isoamylene and isobutylene	Methyl [5-tert-amyl]- and methyl [5-tert-butyl] salicylate	F-24 montmorillonite	Chopade and Sharma (1996)
	Aniline with ethanol	N-ethylaniline and N,N′-diethylaniline	K10 and vanadia-impregnated K10 montmorillonite	Narayanan and Deshpande (1996)
	Phenol with aldehydes and ketones	Corresponding bisphenols and alkylphenols	Al^{3+}-montmorillonite	Tateiwa et al. (1996)
	Substituted anilines with phenylacetylene	1,1-diaryl-ethylenes	KSF montmorillonite	Arienti et al. (1997)
	Aniline with α-methylstyrene	Ortho- and para-isomers of aniline	F-24 montmorillonite	Chitnis and Sharma (1997b)
	Benzene with propene	Cumene	Cation-exchanged Al-pillared bentonite	Geatti et al. (1997)
	Benzene with isopropanol	Isopropyl benzene (cumene)	Acid-treated montmorillonite	Ravichandran and Sivasankar (1997)
	Phenol with cyclohexanones	Trans-(4-substituted-cyclohexyl) phenols	Al^{3+}-montmorillonite	Nishimura et al. (2000)
	Benzylation of arenes with benzyl chloride	Mono- and di-benzylated products	$FeCl_3$-impregnated and Fe-pillared K10 montmorillonite	Pai et al. (2000); Singh and Samant (2004)
	Friedel–Crafts alkylation of arenes with alcohols	Corresponding alkyl sulfonates	Fe^{3+}-montmorillonite with sulfonic acid	Choudary et al. (2002)
	Toluene with cyclohexene	Cyclohexyltoluenes	F-24 montmorillonite	Hoefnagel and van Bekkum (2003)
	Diphenylamine with α-methylstyrene	Monocumyldiphenylamine	F-20 montmorillonite	Liu et al. (2004)
	Pyrroles and indoles with cyclic hemiacetals	Di-pyrrolyl and bis-idolyl alkanols	KSF montmorillonite	Yadav et al. (2004b)
	Benzene with benzyl chloride	Diphenylmethane	Zn^{2+}- and Cd^{2+}-montmorillonite	Ahmed and Dutta (2005)
	Benzene, substituted benzenes, biphenyl and naphthalene with benzyl chloride	Diphenylmethane, corresponding isomers, and benzylated derivatives	Acid-treated bentonite	Pushpaletha et al. (2005)

(Continued)

TABLE 3.3 (Continued)
Organic Conversions and Reactions Catalyzed by Cation-Exchanged and Acid-Treated Montmorillonites, with Particular Reference to the Commercially Available K-Series Catalysts

Reaction Type (in Alphabetical Order)	Reactant/Substrate	Product	Clay Catalyst	Reference (in Chronological Order)
Alkylation (Continued)	Friedel–Crafts alkylation of benzene with propylene	Isopropyl benzene (cumene)	Synthetic Al^{3+}-exchanged Zn-saponite	Vogels et al. (2005)
	Friedel–Crafts alkylation of cage enone with pyrrole and thiophene	Aralkyl substituted diones	K10 montmorillonite	James et al. (2007)
	Benzylation of 1,3-dicarbonyl compounds with primary alcohols	Corresponding benzylated products	H^+/Al^{3+}-montmorillonite	Motokura et al. (2007)
	Hydroxyalkylation of p-cresol with formaldehyde	Dihydroxydiaryl methane	Montmorillonite	Garade et al. (2009)
	Friedel–Crafts alkylation of indoles and pyrrole with nitroalkenes	Corresponding adducts	K10 montmorillonite	An et al. (2010)
Allylation	Anisole with benzyl ether	Substituted diphenylmethane	K10 montmorillonite	Nie et al. (2017)
	Allylation of ketones and aldehydes	Corresponding adducts	K10 montmorillonite	Nowrouzi et al. (2009)
	Allylation of α-Aryl alcohols with allyltrimethylsilane	Corresponding adducts	Sn^{4+}-montmorillonite	Wang et al. (2010a)
	Tetramethoxytoluene with (E)-4-chloro-2-methyl-1-phenyl-sulfonyl-2-butene	Toward a synthesis of coenzyme Q_{10} (CoQ_{10})	K10 montmorillonite	Lee et al. (2013)
Allylsilylation	Primary alcohol with allyltrimethylsilane	Corresponding silylated alcohol (and gaseous propene)	K10 and cation-exchanged montmorillonites	Onaka et al. (1993)
	Aromatic and aliphatic alkenes	Corresponding allylsilylated compounds	H^+/Al^{3+}-montmorillonite	Motokura et al. (2010)
	Alkynes with allylsilanes	Corresponding allylsilylated compounds	H^+/Al^{3+}-montmorillonite	Motokura et al. (2011)
Azidation	Benzylic and allylic alcohols with trimethylsilyl azide	Corresponding azidated products	Acidified natural montmorillonite	Tandiary et al. (2015)
Carbonylation	Allylic alcohols with CO in presence of triphenylphosphine	β,γ-Unsaturated acids	Montmorillonite-supported palladium acetate	Naigre and Alper (1996)
	Amines with di-tert-butyl dicarbonate	Corresponding N-tert-butylcarbamates	K10 and KSF montmorillonite	Chankeshwara and Chakraborti (2006)

(Continued)

TABLE 3.3 (Continued)
Organic Conversions and Reactions Catalyzed by Cation-Exchanged and Acid-Treated Montmorillonites, with Particular Reference to the Commercially Available K-Series Catalysts

Reaction Type (in Alphabetical Order)	Reactant/Substrate	Product	Clay Catalyst	Reference (in Chronological Order)
Carbonylation (Continued)	Methanol with carbon monoxide	Acetic acid	Montmorillonite-supported rhodium complexes	Saikia et al. (2016)
Condensation	Acetals with vinyl ethers	α,β-Unsaturated aldehydes	K10 montmorillonite	Fishman et al. (1981)
	Toluene with bromine	o- and p-Methyl-phenylmethanes	Acid-treated bentonite (Tonsil)	Salmón et al. (1990)
	Pyrrole with aldehydes	Porphyrins	K10 montmorillonite	Laszlo and Luchetti (1993)
	Aldol condensation of cyclohexanone	2-(1-cyclohexen-1-yl) cyclohexanone	Acid-treated clay	Shah et al. (1994)
	Diethyl malonate with aromatic aldehydes	Benzylidenemalonate compounds	Acid-treated bentonite (Tonsil)	Delgado et al. (1995)
	Trimethylhydro-quinone with isophytol	α-Tocopherol	Metal-exchanged montmorillonites	Matsui and Yamamoto (1996)
	Arylaldehyde with arylhydroxylamine	α,N-diaryl nitrones	K10 montmorillonite	Venkatachalapathy et al. (1997)
	Aromatic aldehydes with N,N-dimethyl-aniline	Triarylmethanes	K10 montmorillonite	Zhang et al. (1997b)
	Aldehydes with pyrroles	Porphyrins	K10 montmorillonite	Izumi et al. (1998)
	Phenols with ethyl acetoacetate	Coumarins	K10 montmorillonite	Li et al. (1998b)
	Aldehydes and ketones with 2,2-bis (hydroxymethyl) propane-1,3-diol	Diacetals	KSF and K10 montmorillonite	Zhang et al. (1998a)
	Aldehydes with silyl enol ethers	Corresponding aldol compounds	K10 montmorillonite	Loh and Li (1999)
	Methyl diazoacetate with various aldehydes	Corresponding β-Keto esters	Rh^{3+}- and Cu^{2+}-exchanged clays	Phukan et al. (1999)
	Aldehyde with malononitrile then cyclocondensation of benzylidene malenonitrile with α-naphthol	Substituted 2-amino-2-chromenes	KSF montmorillonite	Ballini et al. (2000)
	Tosylation of alcohols and diols with p-toluene sulfonic acid	Tosylated derivatives	K10 and Fe^{3+}-exchanged K10 montmorillonite	Choudary et al. (2000)
	Phenoxyethanol with fluoren-9-one	9,9-bis[4-(2-hydroxyethoxy) phenyl]fluorine	Ti^{4+}-montmorillonite	Ebitani et al. (2000)
	Ethyl diazoacetate with aldehydes	β-Keto esters	K10 montmorillonite	Bandgar et al. (2001)
	1,3-Propanedithiol with aldehydes and ketones	2-Substituted-1,3-dithianes	Acid-activated bentonite (Tonsil Actisil FF)	Miranda et al. (2001)

(Continued)

TABLE 3.3 (Continued)
Organic Conversions and Reactions Catalyzed by Cation-Exchanged and Acid-Treated Montmorillonites, with Particular Reference to the Commercially Available K-Series Catalysts

Reaction Type (in Alphabetical Order)	Reactant/Substrate	Product	Clay Catalyst	Reference (in Chronological Order)
Condensation (Continued)	Indoles with aldehydes, ketones	Diindolylalkanes	K10 montmorillonite	Chakrabarty et al. (2002)
	Aldol condensation of formaldehyde with acetaldehyde	Acrolein and croton-aldehyde	Cation-exchanged montmorillonite	Azzouz et al. (2003)
	2,3-Dihydro-4H-pyran-4-ones with aryl hydrazines	5-Substituted pyrazoles	KSF montmorillonite	Yadav et al. (2004d)
	Allyl chloride with secondary amines	Cyclic allylamines	Acid-treated montmorillonite (Mag-H$^+$)	Hachemaoui and Belbachir (2005)
	Enaminoketones with ethyl acetoacetate	2,3-disubstituted-6-arylpyridines	K10 montmorillonite	Reddy et al. (2005c)
	β-Pinene with paraformaldehyde	Nopol	Zn^{2+}-montmorillonite	Yadav and Jasra (2006)
	Isatoic anhydride and primary amines with aromatic aldehydes	Mono- and di-substituted dihydro-quinazolinones	K10 montmorillonite	Salehi et al. (2006)
	β-Naphthol, aromatic aldehyde, and amines/urea	Amidoalkyl naphthols	K10 montmorillonite	Kantevari et al. (2007)
	Knoevenagel condensation of carbonyls with active methylene derivatives	Coumarin compounds	KSF and K10 montmorillonite	Moussaoui and Salem (2007)
	Dimedone with (unprotected) sugars	9-hydroxyalkyl-3,3,6,6-tetramethyl-3,4,5,6,7,9-hexa-hydro-1H-xanthene-1,8(2H)-diones	Sc^{3+}-montmorillonite	Sato et al. (2007)
	Aromatic aldehydes with dimedone	Xanthenedione derivatives	Fe^{3+}-montmorillonite	Song et al. (2007)
	Aromatic aldehydes, dimedone, and aniline in ethanol	Decahydroacridine derivatives	Fe^{3+}-montmorillonite	Luo et al. (2008)
	1,2-Diamines with 1,2-dicarbonyl compounds	Quinoxaline derivatives	K10 montmorillonite	Huang et al. (2008)
	Aryl and alkyl 1,2-diamines with diketones	Quinoxaline derivatives	K10 montmorillonite	Hasaninejad et al. (2009)
	1,2-Phenylene diamines with aldehydes	Corresponding benzimidazoles	K10-supported ZrOCl$_3$.nH$_2$O	Rostamizadeh et al. (2009)
	o-tert-Butylaniline with paraformaldehyde	4,4'-diamino-3,3'-dibutyl-diphenyl methane	Acid-treated montmorillonite	Liu et al. (2010)
	2,5-Disubstituted-3-cyanoacetyl indoles with substituted orthophenylene diamine	Corresponding benzimidazoles	K10 montmorillonite	Biradar and Sharanbasappa (2011)

(Continued)

TABLE 3.3 (Continued)
Organic Conversions and Reactions Catalyzed by Cation-Exchanged and Acid-Treated Montmorillonites, with Particular Reference to the Commercially Available K-Series Catalysts

Reaction Type (in Alphabetical Order)	Reactant/Substrate	Product	Clay Catalyst	Reference (in Chronological Order)
Condensation (Continued)	Dihydroxyacetone and glyceraldehyde with alcohol	Corresponding alkyl lactates	Sn^{4+}-montmorillonite	Wang et al. (2011a)
	Benzophenone hydrazine with ketones/aldehydes	Azine derivatives	$BiCl_3$-impregnated K10 montmorillonite	Ravi et al. (2012)
Cyclization and Cyclocondensation	Cyclocondensation of aliphatic aldehydes with pyrrole	*meso*-Tetraalkyl-porphyrins	K10 montmorillonite	Onaka et al. (1993); Shinoda et al. (1995b)
	Cyclization of phenyl-thioacetals	Benzothiophene and derivatives	$ZnCl_2$-impregnated K10 montmorillonite	Clark et al. (1995)
	Ethyl diazoacetate with styrene	*cis*-Cyclopropane	Cu^{2+}-K10 montmorillonite	Fraile et al. (1996)
	Stereoselective cyclization of citronellals	Isopulegols	Cation-exchanged montmorillonite	Tateiwa et al. (1997)
	Carbonyl compounds with phenylhydrazine	1,2,3,4-tetrahydro carbazole and indoles	K10 and H^+/Al^{3+}-montmorillonite	Dhakshinamoorthy and Pitchumani (2005)
	Cyclocondensation of aromatic aldehydes with dimedone, ethylacetoacetate and ammonium acetate	Polyhydroquinoline derivatives	K10 montmorillonite	Song et al. (2005)
	Cyclocondensation of *o*-phenylenediamines with ketones	1,5-Benzodiazepine derivatives	Zn^{2+}-exchanged montmorillonite; K10 montmorillonite	Varala et al. (2006); An et al. (2008)
	Anthranilic acid with various amides	Quinazolin-4(3*H*)-ones derivatives	K10 montmorillonite	Roopan et al. (2008)
	Baylis–Hillman adducts (aryl aldehyde, methylvinyl ketone) with arylhydrazine	1,5-diaryl pyrazoles	KSF montmorillonite	Nikpassand et al. (2009)
	Heterocyclization of 5-amino-1-phenyl-1*H*-pyrazole-4-carboxamides with orthoesters	Substituted pyrazolo[3,4-*d*]pyrimidin-4-ones	K10 montmorillonite	Davoodnia et al. (2012)
	Cyclization of acetylenic amines	Pyrrolocoumarins, pyrroloquinolones	K10 montmorillonite	Majumdar et al. (2012)
	Cyclocondensation of aromatic ketones with aromatic aldehydes and ammonium acetate	2,4,6-trisubstituted pyridines	K10 montmorillonite	Kannan and Sreekumar (2013a)
	Cyclocondensation of aldehydes and amines	Tetra-substituted imidazoles	K10-supported titanocene dichloride	Kannan and Sreekumar (2013b)

(*Continued*)

TABLE 3.3 (Continued)
Organic Conversions and Reactions Catalyzed by Cation-Exchanged and Acid-Treated Montmorillonites, with Particular Reference to the Commercially Available K-Series Catalysts

Reaction Type (in Alphabetical Order)	Reactant/Substrate	Product	Clay Catalyst	Reference (in Chronological Order)
Cyclization and Cyclocondensation (Continued)	Intramolecular cyclization of *o*-substituted amides	2-Substituted 3-ethyl-3*H*-imidazo-[4,5-*b*]-pyridines	Al^{3+}-exchanged K10 montmorillonite	Suresh et al. (2013b)
	Intramolecular cyclization of amide to 2-substituted benzoxazole and oxazolopyridine	Benzoxazole and oxazolopyridine derivatives	Al^{3+}-exchanged K10 montmorillonite	Suresh et al. (2013a)
	Iminoalkynes	Substituted isoquinolines	Al^{3+}-exchanged K10 montmorillonite	Jeganathan and Pitchumani (2014)
	Arylhydrazones	1*H*-Indazoles	K10 montmorillonite	Yu et al. (2015)
	Methylcarbonyl and thiourea	2-aminothiazoles	K10 montmorillonite	Safari and Sadeghi (2016)
	Cyclocondensation of 1,2-phenylenediamine with acetone	1,5-benzodiazepine	Fe_2O_3-impregnated montmorillonite	González et al. (2017)
Decomposition	Diaminomaleonitrile	Oxalic acid and hydrogen cyanide	Cation-exchanged bentonites and montmorillonites	Ferris et al. (1979, 1982)
	tert-Butyl acetate	Isobutylene and acetic acid	Cation-exchanged bentonites	Goldstein (1983)
	Cumene hydroperoxide	Phenol and acetone	Acid-treated bentonite	Selvin et al. (2010)
Dehydration	*n*-Hexanol	*n*-Hexene isomers	Natural montmorillonite	Galwey (1969)
	Intermolecular dehydration of primary alcohols	Di(alk-1-yl) ethers	Cation-exchanged montmorillonite	Ballantine et al. (1983, 1984b)
	Secondary and tertiary alcohols	Corresponding alkenes	Cation-exchanged montmorillonite	Ballantine et al. (1983, 1984b)
	Cyclodehydration of non-aromatic diols	Corresponding cyclic ethers	Al^{3+}-montmorillonite	Kotkar and Ghosh (1986)
	n-Pentanol	1,1-dialkyl ether	Al^{3+}- and Fe^{3+}-montmorillonite	Habib et al. (1988)
	2-Propanol	Propylene, di-isopropyl ether	Synthetic hectorite	Suzuki et al. (1988a)
	Tertiary alcohols	Olefins	K10 montmorillonite	Kantam et al. (1993)
	Methanol	C1–C4 hydrocarbons	Acid-activated montmorillonite	Ravichandran and Sivasankar (1997)
	Isopropanol	Isopropyl ether and propene	Acid-activated bentonite	Haffad et al. (1998)
	Isopropanol	Propene	Calcined surfactant-modified K10 montmorillonite	Yao and Kawi (1999)

(*Continued*)

TABLE 3.3 (Continued)
Organic Conversions and Reactions Catalyzed by Cation-Exchanged and Acid-Treated Montmorillonites, with Particular Reference to the Commercially Available K-Series Catalysts

Reaction Type (in Alphabetical Order)	Reactant/Substrate	Product	Clay Catalyst	Reference (in Chronological Order)
Dehydration (Continued)	Glucose	Furfural compounds	Cation-exchanged montmorillonite	Gonzalez and Laird (2006)
	Glycerol	Acrolein	Acid-activated montmorillonite	Zhao et al. (2013)
	Fructose and glucose	5-hydroxymethyl furfural	Cr^{3+}-exchanged K10 montmorillonite	Fang et al. (2014)
	Glucose	5-methoxymethyl furfural; then hydrated to methyl levulinate	Various metal ion-exchanged montmorillnites	Liu et al. (2017)
Dehydrogenation	Cumene	α-Methylstyrene	Fluorinated montmorillonite	Brückman and Haber (1979)
	Methanol	Methyl formate	Cu^{2+}-fluorotetrasilicic mica; Cu^{2+}-laponite	Morikawa et al. (1982); Matsuda et al. (1995)
	Ethanol	Butadiene, acetaldehyde	Synthetic hectorite	Suzuki et al. (1988a)
	Ethylbenzene	Styrene	Al-PILC-supported cobalt nitrate/acetate	González and Moronta (2004)
	Limonene	p-Cymene	Acid-treated bentonite	Fernandes et al. (2007)
	Propane	Propene	Fe-treated montmorillonite	Grygar et al. (2007)
Deprotection and Cleavage	Tetrahydropyranyl ethers	Farnesylhydro-quinone	Acid-treated bentonite (Tonsil)	Cruz-Almanza et al. (1990)
	Acetal and ketal	Corresponding carbonyl compounds	K10 montmorillonite	Gautier et al. (1997)
	1,1-Diacetates	Corresponding aldehydes	KSF and K10 montmorillonite	Li et al. (1997b)
	Dealkylation of organic sulfides	Corresponding disulfides	K10 montmorillonite	Naicker et al. (1998b)
	Desilylation of trimethylsilyl ethers	Corresponding hydroxyl compounds	K10 montmorillonite	Zhang et al. (1998b)
	Tetrahydropyranyl ethers	Alcohols and phenols	K10 and KSF montmorillonite	Li et al. (1999)
	Removal of tert-butyoxycarbonyl from aromatic amines	Corresponding aromatic amines	K10 montmorillonite	Shaikh et al. (2000)
	tert-Butyl transfer from tert-butylphenols	Parent phenol and tert-butyltoluenes	KSF montmorillonite	Bigi et al. (2001)

(Continued)

TABLE 3.3 (Continued)
Organic Conversions and Reactions Catalyzed by Cation-Exchanged and Acid-Treated Montmorillonites, with Particular Reference to the Commercially Available K-Series Catalysts

Reaction Type (in Alphabetical Order)	Reactant/Substrate	Product	Clay Catalyst	Reference (in Chronological Order)
Deprotection and Cleavage (Continued)	Hydrolysis of *tert*-butyl esters	Parent carboxylic acids	KSF montmorillonite	Yadav et al. (2002c)
	Removal of acetonide	Parent amino alcohol	K10 montmorillonite	Shaikh et al. (2004)
	Cleavage of *tert*-butoxycarbonyl (Boc) group in N,N-di-Boc-protected amines and α-amino acids	Corresponding pure compounds	K10 montmorillonite	Hernández et al. (2007)
	Oxidative cleavage of C–C bonds in α-substituted ketones	Corresponding keto-acids and di-acids	K10 montmorillonite	Younssi et al. (2012)
	Deprotection of *p*-Methoxybenzyl-protected esters and ethers	*O*-methylated quinone methides	H^+/Al^{3+}-montmorillonite	Chen et al. (2014)
Diels–Alder reactions and Cycloadditions	1,3 Cyclohexadiene with 4-*t*-Butylphenol	Corresponding cycloadducts	Fe^{3+}-doped K10 montmorillonite	Laszlo and Lucchetti (1984a)
	Cyclopentadiene with methylvinylketone	Ditto	Ditto	Laszlo and Lucchetti (1984b)
	Furans with α,β-unsaturated aldehydes and ketones	Ditto	Ditto	Laszlo and Lucchetti (1984c)
	Addition of acryloylferrocene to 1-phenyl-1,3-butadiene	3-phenyl-4-ferrocenoyl-cyclohexene	Cation-exchanged and Fe^{3+}-K10 montmorillonites	Toma et al. (1987)
	Cyclic dienes with various dienophiles (dimethyl maleate, 1,4-benzoquinone, methyl vinyl ketone)	Corresponding cycloadducts	K10 montmorillonite	Avalos et al. (1998)
	Danishefsky's diene with aldimines	2-substituted 2,3-dihydro-4-pyridones	K10 montmorillonite	Akiyama et al. (2002)
	Arylamines with cyclic enol ethers	Pyrano- and furano-quinolines	KSF montmorillonite	Yadav et al. (2002d)
	Arenecarbaldehydes, arylamines and methylenecyclo-propanes	Quinoline derivatives	KSF montmorillonite	Shao and Shi (2003)
	Cyclopentadiene with trans-2-methylene-1,3-dithiolane 1,3 dioxide	Corresponding Diels–Alder adduct	Fe^{3+}-doped K10 montmorillonite	Gültekin (2004)
	2,3-Dimethyl-1,3-butadiene with benzaldehydes	Corresponding Diels–Alder adducts	K10 montmorillonite	Dintzner et al. (2007)

(*Continued*)

TABLE 3.3 (Continued)
Organic Conversions and Reactions Catalyzed by Cation-Exchanged and Acid-Treated Montmorillonites, with Particular Reference to the Commercially Available K-Series Catalysts

Reaction Type (in Alphabetical Order)	Reactant/Substrate	Product	Clay Catalyst	Reference (in Chronological Order)
Elimination	Ammonia from pairs of primary amines	Secondary amines	Cation-exchanged montmorillonites	Ballantine et al. (1981a)
	Hydrogen sulphide from primary and secondary thiols	Dialkyl sulphides	Cation-exchanged montmorillonites	Ballantine et al. (1981b)
	Trichloroacetamide from reaction of glycosyl trichloroacet-imidates with phenols	Aryl O-glycosides	K10 montmorillonite	Li et al. (2010)
Ene reaction	Olefins with diethyloxomalonate	γ-Lactones	K10 montmorillonite	Roudier and Foucaud (1984)
	Intermolecular carbonyl-ene reaction of α-methylstyrenes with paraformaldehyde	Corresponding homoallylic alcohols	Cation-exchanged montmorillonites	Tateiwa et al. (1997)
Esterification and Transesterification	Acetic acid with cyclohexene, and myristic acid with propylene	Cyclohexyl acetate and isopropyl myristate	F-24 montmorillonite	Chakrabarti and Sharma (1992b, 1992c)
	Transesterification of cyclohexyl acrylate with n-butanol and 2-ethylhexanol	n-Butyl acrylate and 2-ethylhexyl acrylate	F-24 montmorillonite	Saha and Streat (1999)
	Carboxylic acids with methanol or ethanol	Corresponding esters	Fe^{3+}- and Fe^{3+}-K10 montmorillonite	Kantam et al. (2002); Srinivas and Das (2003)
	Carboxylic acids with 3-phenyl-1-propanol	Corresponding esters	Ti^{4+}-montmorillonite	Kawabata et al. (2003)
	Phenylacetic acid with phenols and cresols	Phenylacetates	Al^{3+}-montmorillonite	Reddy et al. (2004)
	Dicarboxylic acids with alcohols and phenols	Corresponding diesters	Cation-exchanged montmorillonites	Reddy et al. (2005a)
	Succinic anhydride with p-cresol	Di-(p-Cresyl) succinate	Al^{3+}- and H^+/Al^{3+}-montmorillonites	Reddy et al. (2005b)
	p-Hydroxy benzoic acids with alcohols	Corresponding esters (parabens)	K10 montmorillonite	Hazarika et al. (2007)
	Stearic and various carboxylic acids with p-cresol	Corresponding esters	H^+/Al^{3+}-bentonite	Vijayakumar et al. (2005, 2009)
	Succinic acid with iso-butanol	Di-(iso-butyl) succinate	Cation-exchanged montmorillonites	Reddy et al. (2007)
	Palm oil with methanol	Corresponding methyl esters (biodiesel)	KSF montmorillonite; KOH-impregnated bentonite	Kansedo et al. (2009); Soetaredjo et al. (2011)
	Fatty acids with short-chain alcohols	Corresponding esters	KSF, KSF/0, KP10, and K10 montmorillonites	Neji et al. (2009)

(*Continued*)

TABLE 3.3 (Continued)
Organic Conversions and Reactions Catalyzed by Cation-Exchanged and Acid-Treated Montmorillonites, with Particular Reference to the Commercially Available K-Series Catalysts

Reaction Type (in Alphabetical Order)	Reactant/Substrate	Product	Clay Catalyst	Reference (in Chronological Order)
Esterification and Transesterification (Continued)	Aliphatic and aromatic ketones with cyanotrimethylsilane	Corresponding cyanohydrin trimethylsilyl esters	Sn^{4+}-montmorillonite	Wang et al. (2009)
	Propionic acid with p-cresol	p-Cresylpropionate	Al^{3+}, H^+, and cation-exchanged montmorillonites	Reddy et al. (2009, 2010)
	Oleic acid with methanol	Corresponding methyl ester	Calcined and acid-activated kaolinite	do Nascimento et al. (2011b)
	Ditto	Ditto	Bentonite modified with acidic ionic liquids	Ghiaci et al. (2011)
	Phenolic acids with hydroxyl compounds	Corresponding esters	Fe^{3+}-montmorillonite	Mahanta and Handique (2011)
	Waste cooking palm oil with methanol	Corresponding methyl esters	Calcined local clay	Olutoye and Hameed (2013)
	Lauric acid with methanol	Methyl laurate	Phosphoric acid-activated montmorillonite	Zatta et al. (2013)
	Soybean oil with methanol	Corresponding methyl esters (biodiesel)	KF-treated smectites	Alves et al. (2014); Silva et al. (2014)
	Decanoic acid with p-cresol	p-Cresyldecanoate	Montmorillonite treated with different organic acids	Venkatesha et al. (2015)
	Stearic, oleic, and palmitic acids with ethylene glycol and glycerol	Corresponding esters	KSF, KSF/0, KP10, and K10 montmorillonites	Chaari et al. (2017)
	Long-chain fatty acids with ethanol	Corresponding ethyl esters	K10 montmorillonite	Kanda et al. (2017)
Etherification and O-alkylation	Epichlorohydrin with alcohols	Monoethers of 3-chloro-1,2 propanediol	K10 montmorillonite	Vu et al. (1982)
	2-Methyl pent-2-ene with primary alcohols	Corresponding ethers	Al^{3+}-montmorillonite	Adams et al. (1983)
	Isobutene, 2-methylpropene, tert-butanol with methanol	Methyl t-Butyl ether (MTBE)	K10, Al^{3+}- and Cu^{2+}-montmorillonites	Bylina et al. (1980); Adams et al. (1982b, 1986)
	Cyclic ketones with thiols	Enolthioethers	KSF montmorillonite	Labiad and Villemin (1989)
	Paraformaldehyde with alcohols	Dialkoxymethanes	Montmorillonite	Deshmukh et al. (1995)
				(Continued)

TABLE 3.3 (Continued)
Organic Conversions and Reactions Catalyzed by Cation-Exchanged and Acid-Treated Montmorillonites, with Particular Reference to the Commercially Available K-Series Catalysts

Reaction Type (in Alphabetical Order)	Reactant/Substrate	Product	Clay Catalyst	Reference (in Chronological Order)
Etherification and O-alkylation (Continued)	Allylic and benzylic alcohols with trimethylorthoacetate	Corresponding ethers	KSF montmorillonite	Kumar et al. (1997)
	N-tosyl aziridines with alcohols	β-Amino ethers	KSF montmorillonite	Yadav et al. (2002b)
	Epoxidized methyl oleate with methanol	β-Hydroxyethers	Nb^{5+}-K10 montmorillonite	Gallo et al. (2006)
	Aryl aldehyde bisulfite adducts with sodium borohydride and hexamethyldisilazine	Corresponding aryl trimethylsilyl ethers	K10 montmorillonite	Khodaei et al. (2006)
	Glycerol with benzyl alcohol	Monobenzyl glycerol ether	K10 montmorillonite	da Silva et al. (2009)
	Aromatic aldehydes with allyltrimethylsilane	Homoallylic silyl ethers	K10 montmorillonite	Dintzner et al. (2009)
Hydrolysis	Phosmet (O,O-dimethyl, S-(N-phthalimidomethyl) dithiophosphate	Phthalimide	Cation-exchanged smectites	Sánchez Camazano and Sánchez Martín (1983)
	s-Triazine	Formamide	Na^+-montmorillonite	Nguyen (1986)
	Aryl-substituted α,β-difluoroallyl alcohols	(Z)-α-fluoro-β-aryl-substituted acrylaldehydes	K10 montmorillonite	Funabiki et al. (1999)
	Diethatyl-ethyl	Diethatyl and dechlorinated ethyl ester derivatives	Cation-exchanged montmorillonites	Liu et al. (2000)
	Carbosulfan	Carbofuran	Cu^{2+}- and Al^{3+}-montmorillonite	Wei et al. (2001)
	4-Nitrophenyl phosphate	4-Nitrophenol and inorganic phosphate	La^{3+}-hectorite	Frey et al. (2003)
	Cellulose	Reducing sugar	Acid-activated montmorillonite	Tong et al. (2013)
Isomerization[a]	2′-Aminochalcones	2-Aryl-1,2,3,4-tetrahydro-4-quinolones	K10 montmorillonite	Varma and Saini (1997)
	α-Pinene	Camphene and limonene	Acid-activated montmorillonite and saponite	Volzone et al. (2001); Beşün et al. (2002); Yadav et al. (2004a)
	cis- and trans-Butene	Corresponding isomers	F24 and F124 montmorillonites	Moronta et al. (2005)

(Continued)

TABLE 3.3 (Continued)
Organic Conversions and Reactions Catalyzed by Cation-Exchanged and Acid-Treated Montmorillonites, with Particular Reference to the Commercially Available K-Series Catalysts

Reaction Type (in Alphabetical Order)	Reactant/Substrate	Product	Clay Catalyst	Reference (in Chronological Order)
Multi-component Reactions	Aldehyde, malononitrile, and α-naphthol	Substituted 2-amino-2-chromenes	KSF montmorillonite	Ballini et al. (2000)
	Aldehyde, β-carbonyl, and urea	Substituted dihydro-pyrimidines	KSF montmorillonite	Bigi et al. (1999); Ballini et al. (2000)
	β-Keto ester, aryl aldehyde, and urea	5-Alkoxycarbonyl-4-aryl-3,4-dihydro-pyrimidin-2(1H)ones	KSF montmorillonite	Lin et al. (2000)
	Aromatic amines, aromatic aldehydes, and cyclopentadiene	Tetrahydroquinolines	Commercial bentonite	Sartori et al. (2001)
	Aryldehydes, ketones, and acetylchloride	β-Acetamido ketones	K10 montmorillonite	Bahulayan et al. (2003)
	Coupling of diazonium-clay complex with phenols, naphthols, and aromatic amines	Sulfanilate azo dyes	Bentonite, K10 montmorillonite	Dabbagh et al. (2007)
	Carbonyl compounds, amines, and trimethylsilyl cyanide	α-Amino nitriles	Sn^{4+}-montmorillonite	Wang et al. (2010b)
	p-Nitrobenzaldehyde with methanol, 3-buten-1-ol, and benzene	Tetrahydro-1-(4-nitrophenyl)-4-phenyl-2H-pyran	K10 montmorillonite	Dintzner et al. (2012)
	β-Keto ester, arylaldehyde, malononitrile, and alcohol	Cyano pyridines	K10 montmorillonite	Reddy et al. (2012)
	Aldehydes and terminal alkynes	3,5-Disubstituted isoxazoles	Cu^{2+}-montmorillonite with NaN_3	Bharate et al. (2013a)
	Amines, aldehydes, 1,3-dicarbonyl compounds, and nitroalkanes	Functionalized pyrroles	K10 and KSF montmorillonites	Bharate et al. (2013b)
	Reaction of 1,2-diamines with phenacyl bromides	2-Substituted quinoxalines	K10 montmorillonite	Jeganathan et al. (2014a)
	Coupling of terminal alkynes, amines, and aqueous formaldehyde	Corresponding propargylamines	Ag^+-K10 montmorillonite	Jeganathan et al. (2014b)
	Aldehyde, amine, and thioglycolic acid	Thiazolidinones	K10 montmorillonite	Sharma et al. (2015)
	Phenyl acetylene/1-hexeyne, aromatic aldehyde, and benzamide/acetamide	N,N'-alkylidene bismamide derivatives	K10 montmorillonite	Lambat et al. (2016)

(Continued)

TABLE 3.3 (Continued)
Organic Conversions and Reactions Catalyzed by Cation-Exchanged and Acid-Treated Montmorillonites, with Particular Reference to the Commercially Available K-Series Catalysts

Reaction Type (in Alphabetical Order)	Reactant/Substrate	Product	Clay Catalyst	Reference (in Chronological Order)
Multi-component Reactions (Continued)	Acetophenones, aldehydes, and 2-sulfanyl-1,3-benzoxazole-5-sulfonamide	1,3-Benzoxazole-5-sulfonamide derivatives	Fe^{3+}-montmorillonite	Vinoda et al. (2016a)
	Acetophenones, aldehydes, and thiourea	1,3-thiazine derivatives	Fe^{3+}-montmorillonite	Vinoda et al. (2016b)
Nitration	Aromatic hydrocarbons	Corresponding nitro derivatives	Natural bentonite/dilute HNO_3	Bahulayan et al. (2002)
	Aniline, naphthalene, fluorine, chrysene, dibenzofluorene	Corresponding nitro derivatives	$Bi(NO_3)_3$-impregnated montmorillonite	Samajdar et al. (2001)
	Phenolic compounds	Nitrated products	Metal-impregnated, HNO_3-treated KSF	Yin and Shi (2005)
	Benzyl alcohol	Benzaldehyde	$Cu(NO_3)_2$-impregnated acid-treated montmorillonite	Massam and Brown (1998)
Oxidation[b]	α-Methylpyrroles	α-Formylpyrroles	Tl^{3+}-montmorillonite	Jackson et al. (1984)
	Epoxidation of allyl alcohols	Corresponding derivatives	Ti-PILC	Choudary et al. (1990)
	Oxidative decarboxylation of isocitric acid	α-Ketoglutaric acid	Na^+-montmorillonite	Naidja and Siffert (1990)
	Selected alcohols, thiophenol	Corresponding aldehydes and ketones	K10 supported potassium ferrate	Delaude et al. (1995)
	Ethylbenzene, diphenylmethane, and benzylalcohol	Acetophenone, benzophenone, and benzaldehyde	Natural bentonite/dilute HNO_3	Bahulayan et al. (2002)
	3-Amino-4-hydroxyphenylarsonic acid	Azobenzene arsenic acid	Ca^{2+}-montmorillonite	Wershaw et al. (2003)
	Alkyl halides	Corresponding aldehydes and ketones	K10 supported iodic acid	Hashemi et al. (2004)
	Olefins and chalcones	Corresponding aldehydes and derivatives	K10 supported iron-salen complex	Dhakshinamoorthy and Pitchumani (2006)
	Aliphatic aldehydes	Corresponding carboxylic acids	KSF montmorillonite	Dintzner et al. (2010)
	Adsorbed acetone, benzene, ethylbenzene	CO_2, H_2O	Al- and Al–Ce-PILC supported Pd	Zuo et al. (2012)
	5-Hydroxymethyl furfural	5-Hydroxymethyl-2-furancarboxylic acid	K10 supported molybdenum acetylacetonate	Zhang et al. (2014)

(Continued)

TABLE 3.3 (Continued)
Organic Conversions and Reactions Catalyzed by Cation-Exchanged and Acid-Treated Montmorillonites, with Particular Reference to the Commercially Available K-Series Catalysts

Reaction Type (in Alphabetical Order)	Reactant/Substrate	Product	Clay Catalyst	Reference (in Chronological Order)
Rearrangement	Cholest-5-ene	Cholest-13(17)-ene	K10 montmorillonite	Sieskind and Albrecht (1985)
	Cyclohexadienone	Corresponding phenol derivatives	Fe^{3+}-K10 montmorillonite	Chalais et al. (1986)
	Substituted allyl phenyl ethers	ortho-Allyl phenols, chromans, and coumarans	KSF and K10 montmorillonites	Dauben et al. (1990)
	Alkyl phenyl ethers	Alkyl phenols	Cation-exchanged montmorillonites	Tateiwa et al. (1994b)
	N-methyl-N-nitrosoaniline	N-methyl-4-nitrosoaniline	KSF and K10 montmorillonites	Kannan et al. (1997)
	Phenyl acetate and phenyl benzoate	Corresponding p-isomers	K10 and cation-exchanged K10 montmorillonite	Venkatachalapathy and Pitchumani (1997)
	N-phenyl-hydroxylamine	p-Nitrosodiphenyl amine	K10 and cation-exchanged K10 montmorillonite	Naicker et al. (1998a)
	Allyl phenyl ether	ortho-Prenyl phenol	K10 montmorillonite	Dintzner et al. (2004)
	D-glycals	Furan diols	KSF montmorillonite	Yadav et al. (2005)
	Alicyclic, aromatic, and aliphatic ketoximes	Corresponding lactams and amides	Ti^{4+}-montmorillonite	Mitsudome et al. (2012b)
	p-Cresyldecanoate	1-(2-hydroxy-5-methylphenyl) decan-1-one	Montmorilonite treated with different organic acids	Venkatesha et al. (2015)
Reduction	Nitroarenes with hydrazine	Corresponding amino compounds	K10 montmorillonite	Han and Jang (1990)
Ring-opening	Epoxidized methyl oleate with methanol	β-Hydroxyethers	Nb^{5+}-K10 montmorillonite	Gallo et al. (2006)
Substitution	Hydroxyl groups of allylic and benzylic alcohols by aniline	Variety of compounds	H^+/Al^{3+}-montmorillonite	Motokura et al. (2007)
	Propargylic alcohols with alkynylsilanes	1,4-diynes	Acid-treated K10 montmorillonite	Wang et al. (2010c)
	Hydroxyls of allylic and benzylic alcohols by cyanide groups of trialkylsilyl cyanide	Corresponding nitriles	Sn^{4+}- and Ti^{4+}-montmorillonites	Wang et al. (2011b)
Vinylation (Heck)	Substituted anilines with vinylacetate	Corresponding methyl cinnamates	Pd^{2+}- and Cu^{2+}-exchanged K10	Waterlot et al. (2000)

[a] The clay-catalyzed isomerization of organic compounds is described in Chapter 6.
[b] Oxidation reactions catalyzed by clay-supported reagents are mentioned in Chapter 4. The clay-catalyzed oxidation of organic species, effected through a Fenton-like or photo-Fenton process in the presence of oxidizing agents, is described in Chapter 7.

3.3 THERMAL ACTIVATION AND RELATED TREATMENTS

3.3.1 HEATING AND CALCINATION

For the sake of convenience, we may distinguish four temperature ranges within which marked changes occur in clay mineral structures (Wolters and Emmerich 2007; Che et al. 2011; Heller-Kallai 2013): (1), below −5°C when part of the surface-associated water is frozen; (2), room temperature to 300°C when dehydration occurs with the loss of water from external particle and interlayer surfaces, and from the hydration shell of exchangeable cations; (3), above 400°C when dehydroxylation takes place with the loss of water from condensation of two structural hydroxyl groups; and (4), between 600°C and 900°C when structural decomposition and recrystallization occur.

In the case of kaolinite, dehydroxylation at 600°C–900°C gives rise to metakaolinite, composed of distorted tetrahedral sheets and altered octahedral sheets. Solid-state ^{27}Al NMR spectroscopy indicates that a large proportion of the aluminum in metakaolinite occurs in the form of five- and four-coordinate species (Rocha and Klinowski 1990; Massiot et al. 1995; Fabbri et al. 2013). Formation of Al^V, at the expense of Al^{VI}, was also observed by Tennakoon et al. (1986b) for montmorillonite that had been dehydroxylated at 700°C. Kaolinite dehydroxylation is a two-step process, involving the interaction of two hydroxyl groups to form a water molecule whereas silanol groups are formed when halloysite is dehydroxylated (Frost and Vassallo 1996). Above 900°C metakaolinite transforms into a glass and spinel phase (Siddique and Klaus 2009; Rouquerol et al. 2013).

The ability of partially dealuminated metakaolinite to catalyze cumene cracking has been attributed to synergism between Brønsted and Lewis acid sites, associated with Al^{IV} and Al^V, respectively (Macedo et al. 1994). The infrared spectrum of adsorbed pyridine indicates that air-dry acid-activated metakaolinite contains both Brønsted and Lewis acid sites. After heating at 200°C, the concentration of Brønsted acid sites declines while that of Lewis sites increases (Belver et al. 2005). Earlier, Sabu et al. (1993, 1999) reported that acid-activated metakaolinite could effectively catalyze the Friedel–Crafts alkylation of benzene with benzyl chloride, and the conversion of benzyl chloride to diphenylmethane. Zhao et al. (2016) have reported that a natural (tubular) halloysite can promote the Friedel–Crafts alkenylation of *p*-xylene with phenylacetylene after calcination at 500°C, and similarly, calcined acid-activated kaolinite can catalyze the (skeletal) isomerization of 1-butene and α-pinene, the dehydration of isopropanol (Perissinotto et al. 1997; Volzone et al. 2005; Lenarda et al. 2007), and the esterification of oleic acid with methanol (do Nascimento et al. 2011a, 2011b).

Although very few clay-catalyzed organic reactions are carried out at low (below zero) temperatures, we should mention that freezing, like drying, has the effect of enhancing the surface acidity of smectites (Heller-Kallai 2013). This finding may be ascribed to the partial removal, by freezing, of water molecules associated with interlayer counterions. Using time-resolved synchroton X-ray diffraction, Svensson and Hansen (2010) have shown that Na$^+$-montmorillonite is capable of retaining a monolayer of water molecules after freezing at −50°C.

Dehydration may also lead to changes in layer stacking and porosity. Hydrated halloysite, for example, can lose its interlayer water, even under ambient conditions of temperature and humidity as a result of which the basal spacing decreases from 1.01 to 0.72 nm (Joussein et al. 2005; cf. Figure 1.10). In the case of montmorillonite, dehydration results in the loss of water from external particle surfaces and intraparticle pores, followed by the stepwise removal of interlayer water (Cases et al. 1992; Heller-Kallai 2013). When monovalent cation-exchanged montmorillonites are heated at ~150°C, interlayer water is lost, and the basal spacing decreases from about 2 to 0.98 nm, indicative of interlayer collapse (cf. Table 1.2). On the other, the interlayer space of montmorillonite, exchanged with polyvalent cations, does not fully collapse even after heating at 300°C. Interestingly, Sarikaya et al. (2000) and Noyan et al. (2006) found that the surface area, pore volume, and total surface acidity of bentonite did not appreciably change on heating between 100°C and 450°C–500°C although the values steeply declined when the sample was heated at 900°C–1200°C.

We might add here that ethers can form directly from alcohols intercalated into polyvalent cation-exchanged montmorillonites (Adams et al. 1981a, 1982a; Habib et al. 1988; Mitsudome et al. 2012a). Such materials can also convert 1-alkenes to di-2,2'-alkylether through addition of interlayer water. Using clay samples containing a single sheet of interlayer water, the reaction was completed in ~3 h, whereas ~150 h was required with montmorillonite having two interlayer water sheets. Ether formation essentially ceases when all the interlayer water is consumed, after which the ether decomposes into alk-1-ene, alk-2-ene, and alkan-2-ol (Ballantine et al. 1983; Adams et al. 1984; Adams 1987). Furthermore, the low-temperature conversions of C_6 alkenes (over hydrated Cu^{2+} montmorillonite) show shape selectivity in that the reaction products are influenced by the branching and double-bond position in the alkene substrates (Adams et al. 1981b).

Atkins et al. (1983) also found that the interlayer conversion of ethylene and water to ethanol was greatly inhibited when the montmorillonite structure had collapsed through calcination. Similarly, the Brønsted acid-catalyzed formation of phenylacetates (from phenols and phenylacetic acid) in Al^{3+}-montmorillonite is inhibited following the removal of interlayer water (Reddy et al. 2004). Interestingly, the addition of water to dehydrated Fe^{3+}-montmorillonite can partly restore the clay-catalyzed hydration of interlayer styrene derivatives (Ajjou et al. 1997). Similarly, Motokura et al. (2012) have observed that water addition to a proton-exchanged montmorillonite can dramatically accelerate the allylsilylation of aromatic and aliphatic alkenes, possibly by stabilizing the active disilyl cation intermediate. Earlier, Ballantine et al. (1985) reported that hydrated Al^{3+}-montmorillonite (containing ca. 12% interlayer water) could catalyze the addition of water to 2-methylpropene to give *tert*-butanol. Following dehydration, however, the same material turned into an effective catalyst for the oligomerization of 2-methylpropene.

Another interesting finding relates to the variation in surface acidity of a commercial acid-activated montmorillonite (Fulcat 40) after heating from 75°C to 350°C. Brown and Rhodes (1997), for example, found that the rate constant for the Brønsted-acid catalyzed rearrangement of α-pinene to camphene was highest at 150°C, coincident with the development of maximum Brønsted acidity. The same temperature maximum was observed by Yadav et al. (2004a) for acid-activated montmorillonite, and by Noyan et al. (2006) for a natural montmorillonite, while Liu et al. (2011) found a slightly lower value (120°C). The Fulcat 40 sample, used by Brown and Rhodes (1997), shows maximum Lewis acidity after heating at 250°C–300°C as indicated by the rate constant of the Lewis acid catalyzed rearrangement of camphene to isobornyl chloride.

In this context, we might also mention the work by Cseri et al. (1995b) on the Friedel–Crafts alkylation of toluene over a series of commercially available, acid-activated K catalysts. They found that the rate of benzyl alcohol alkylation was related to Brønsted acidity whereas Lewis acid sites were involved when benzyl chloride was used. Cseri et al. (1995a) further noted that the acidity of cation-exchanged K10 montmorillonites, after drying at 120°C, was dominantly of the Brønsted type but became almost pure Lewis after heating at 500°C. In order to minimize the activity of Brønsted acid sites, Cativiela et al. (1995) also used K10 montmorillonite that had previously been calcined at 550°C to catalyze the Friedel–Crafts alkylation of anisole with dienes. The high efficiency of Fe^{3+}- and Zn^{2+}-exchanged K10 montmorillonite in promoting the Friedel–Crafts acylation of aromatic ethers with acid anhydrides is ascribed to a synergy between Brønsted and Lewis acidities (Choudary et al. 1998).

Complex structural and textural changes occur when sepiolite and palygorskite are heated at increasing temperatures, as indicated in Figure 1.22 with respect to specific surface areas. We have also seen (cf. Section 1.3.8) that the loss of water molecules from sepiolite (between 300°C and 400°C) causes the ribbons to rotate and the structure to fold. With acid-activated sepiolite, however, heating (even up to 550°C) does not lead to particle folding but rather to an open-channel structure (Valentin et al. 2007). Dehydroxylation of sepiolite (between 680°C and 900°C) eventually gives rise to clinoenstatite, while palygorskite becomes X-ray amorphous at 700°C (Brigatti et al. 2013). Using controlled rate thermal analysis and differential scanning calorimetry, Frost and Ding (2003) were able to identify three dehydration and five dehydroxylation steps.

The DTA and TG curves of allophane and imogolite show a large endotherm (weight loss) between 50°C and 300°C due to the removal of physisorbed water. For allophane, the endotherm peak temperature varies from 150°C to 185°C, depending on the Al/Si ratio of the mineral, and the type of counterion (Henmi 1980). On further heating, allophane loses weight continuously (through dehydration and dehydroxylation) until the appearance of an exotherm at 900°C–1000°C, signaling the formation of mullite (Henmi 1980; Wada 1989; MacKenzie et al. 1991). The curves for imogolite resemble those for allophane except for the appearance of a small, distinct endotherm at ca 400°C due to structural dehydroxylation (Yoshinaga and Aomine 1962; Wada 1989).

The removal of adsorbed (*molecular*) water at 300°C from a synthetic imogolite makes the inner silanol groups—which are as acidic as those of amorphous silica—accessible to such molecules as ammonia and methanol. Filimonova et al. (2016) have suggested that Brønsted acid sites on allophane are identifiable with aluminol (Al–OH) and silanol (Si–OH) groups at defect sites on spherule surfaces, while undercoordinated Al^{3+} ions can act as Lewis acids. Removal of surface-adsorbed water and dehydroxylation would therefore enhance Lewis acidity. Imogolite loses its nanotube structure when heated at 500°C. The resultant lamellar phase is much more active in catalyzing the conversion of phenol (to *o*-cresol) than the material that has been dehydrated by heating at 300°C (Bonelli et al. 2009). Kang et al. (2010) also observed that imogolite-like single-walled aluminosilicate nanotubes were fully dehydrated by heating at 250°C (under vacuum), while dehydroxylation at >300°C led to partial disordering of the wall structure.

3.3.2 Microwave and Ultrasound Irradiation

Microwave irradiation has been widely used to assist and promote clay-catalyzed organic reactions. Being transparent to the reaction vessel, microwaves can heat the reactants directly. As such, a more uniform heating rate can be attained within a much shorter time than is possible with conventional heating (Varma et al. 1997; Gajare et al. 2000; Ramesh et al. 2015). More importantly, microwave-assisted organic reactions can proceed in the presence of very little solvent, opening a pathway to performing environmentally benign (*green*) synthetic chemistry. Indeed, the majority of microwave-activated organic conversions, in the presence or absence of clays, take place under solvent-free conditions (Varma 1999, 2002; Loupy 2004; Dasgupta and Török 2008; Dastan et al. 2012). Nor does microwave irradiation appear to have a detrimental effect on clay mineral structure and crystallinity until a high power ($\geq$650 W) is reached (Korichi et al. 2012; Al Bakain et al. 2014) although Pillai et al. (2013) have observed a slight weakening of the basal reflection in the X-ray diffraction pattern of a montmorillonite after irradiation with microwaves. Microwave activation, however, can alter the textural properties of clay minerals, but the changes that occur are similar in extent to those induced by conventional thermal treatment (Gajare et al. 2000). Franco et al. (2014), for example, reported that the specific surface area (SSA) of an acid-treated sepiolite, irradiated with microwaves for just a few minutes, was similar to the value measured for the same sample that had been conventionally heated for two days. In the case of trioctahedral smectites (saponite, stevensite), however, Franco et al. (2016) observed appreciable dissolution of the octahedral sheet, formation of X-ray amorphous silica, and a large increase in SSA after only 20 min of microwave treatment. The extensive literature on microwave-assisted organic reactions, in the absence and presence of clay catalysts, has been reviewed by Caddick (1995), Strauss and Trainor (1995), Varma (1999, 2002), Lidström et al. (2001), Loupy (2004), Man and Shahidan (2007), Polshettiwar and Varma (2008), and McCabe and Adams (2013).

Ramesh et al. (2010) have reported that Al^{3+}-montmorillonite can effectively catalyze the Brønsted acid esterification of butyl alcohol with different alkanoic acids under microwave irradiation, and that the yield of esters is solvent-dependent. Montmorillonite, treated with p-toluene sulfonic acid, is as efficient as the Al^{3+}-exchanged sample in catalyzing the solvent-free esterification of benzoic acid with p-cresol (Ramesh et al. 2012). Likewise, Naskar et al. (2010) used microwave irradiation to synthesize 2-(1′,3′-dihydro-1H-[2,3′]biindolyl-2′-ylidne)-indan-1,3-diones/bis-indolylindane-1,3-diones from ninhydrin and 3-substituted/unsubstituted indoles under solvent-free conditions in the presence of K10 montmorillonite. More recently, de Oliveira et al. (2013) used an acid-activated metakaolinite to promote the esterification of oleic acid with methanol under microwave irradiation (400 W for 15 min).

Like microwave irradiation, ultrasonic activation can enhance the rate of organic reactions with minimal use of solvents, and often under solvent-free conditions. The structural integrity of montmorillonite is not appreciably affected by (low-frequency) ultrasound irradiation although the treatment can cause a marked decrease in average particle diameter. On the other hand, ultrasound treatment of palygorskite (in suspension) can lead to particle aggregation (Novikova et al. 2016). The combination of sonochemistry and clay science, including the use of ultrasound in organic adsorption and catalysis by clay minerals, has been described in a recent review by Chatel et al. (2016).

Chtourou et al. (2010) used commercial acid-activated montmorillonites (KSF, KSF/0, K10, KP10) to synthesize *trans*-chalcones from aryl methyl ketones and aryl aldehydes under ultrasound irradiation and solvent-free conditions. Using a similar approach, Mohammadpoor-Baltork et al. (2010) obtained a variety of 2-aryl-5,6-dihydro-4H-1,3-oxazines from arylnitriles and 3-amino-1-propanol, in the presence of K10 and KSF montmorillonites.

Table 3.4 lists a number of clay-catalyzed organic reactions and conversions, assisted by either microwave irradiation or ultrasonication. In the majority of cases, commercial (acid-activated) K10 and KSF montmorillonites have served as catalysts.

TABLE 3.4
Organic Conversions and Syntheses Catalyzed by Cation-Exchanged Montmorillonites, KSF and K10 Montmorillonites, and Metal Salt Impregnated K-Catalysts Assisted by Microwave or Ultrasound Irradiation Under Solvent-Free Conditions

Microwave-Assisted Reaction	Clay Catalyst	Reference (in Chronological Order)
Conversion of pinacol to pinacolone	Di- and tri-valent cation-exchanged montmorllonites	Gutierrez et al. (1989)
Synthesis of indoles from phenylhydrazine and ketones	KSF montmorillonite	Villemin et al. (1989)
Condensation of tetronic acid, 3-methyl-1-phenyl-5-pyrazolone, and barbituric acid with aldehydes	KSF montmorillonite	Villemin and Labiad (1990a, 1990b, 1990c)
Synthesis of acetal derivatives of L-galactono-1,4-lactone	KSF montmorillonite	Csiba et al. (1993)
Formation of anhydrides from carboxylic acids and isopropenyl acetate	KSF montmorillonite	Villemin et al. (1993)
Synthesis of 5-nitrofuraldehyde and 5-nitrofurfurylidene	KSF montmorillonite and K10-supported $ZnCl_2$	Villemin and Martin (1994)
Synthesis of N-substituted imidazoles from imidazole and ethyl acrylate	Li^+- and Cs^+-montmorillonites	Martin-Aranda et al. (1997)
Oxidation of alcohols to carbonyl compounds	K10 montmorillonite-supported $Fe(NO_3)_3$ (clayfen)	Varma and Dahiya (1997)
Synthesis of imines and enamines from amines and aldehydes/ketones	K10 montmorillonite	Varma et al. (1997)
Oxidative deprotection of tetrahydropyranyl ethers to the corresponding carbonyl compounds	K10-supported $Fe(NO_3)_3$ (clayfen)	Heravi et al. (1999)
Synthesis of 2-(2-pyridyl)indole derivatives	K10-supported $ZnCl_2$ (clayzic)	Lipińska et al. (1999)
Synthesis of imidazo [1,2-α] annulated pyridines, pyrazines, and pyrimidenes	K10 montmorillonite	Varma and Kumar (1999)
Formation of N-sulfonylimines from aldehydes and sulfonamides	K10 montmorillonite	Vass et al. (1999)
Synthesis of spiro [$3H$-indole-3,3'-[$3H$-1,2,4]triazol]-2(1H) ones from 3-arylimino-2H-indol-2-ones and thiosemicarbazide	K10 montmorillonite	Dandia et al. (2000)
Isomerization of natural furfuran lignans	KSF montmorillonite	Das et al. (2000)
Selective deprotection of allyl esters	K10 montmorillonite	Gajare et al. (2000)
Synthesis of 3-substituted benzofurans from phenoxy acetophenones	KSF montmorillonite	Meshram et al. (2000)
Condensation of malonic acid and aromatic aldehydes	Heat activated bentonite, K10 and KSF montmorillonites	Loupy et al. (2001)
Isomerization of acetates	K10 montmorillonite	Shanmugam and Singh (2001)
Synthesis of 2,3-unsaturated O-glycosides from tri-O-acetyl-D-glucal and alcohols	K10 montmorillonite	de Oliveira et al. (2002)
Synthesis of warfarin acetals from warfarin with methanol/ethanol	K10 montmorillonite	Krstić et al. (2002)
Synthesis of elvirol, curcuphenol, and sesquichamaenol	K10 montmorillonite	Singh et al. (2002)

(Continued)

TABLE 3.4 (Continued)
Organic Conversions and Syntheses Catalyzed by Cation-Exchanged Montmorillonites, KSF and K10 Montmorillonites, and Metal Salt Impregnated K-Catalysts, Assisted by Microwave or Ultrasound Irradiation Under Solvent-Free Conditions

Microwave-Assisted Reaction	Clay Catalyst	Reference (in Chronological Order)
Synthesis of 2H-benz[e]-1,3-oxazin-2-ones by cyclohydrazination of salicylaldehyde semicarbazones	K10 montmorillonite	Yadav et al. (2002a)
Nitration of β-lactams	Montmorillonite-supported Bi(NO$_3$)$_3$	Bamik et al. (2003)
Hydroxylation of phenol	Fe/Al-pillared smectites	Letaief et al. (2003)
Synthesis of nitriles from aldehydes and hydroxylamine HCl	K10 and KSF montmorillonites	Dewan et al. (2004)
Transformation of 2-tert-butylphenol	K10 montmorillonite	Kurfürstová and Hájek (2004)
Synthesis of methylene-dioxyprecocene from sasamol/phenols and 3-methyl-2-butenal	K$^+$-exchanged K10 montmorillonite	Dintzner et al. (2005)
Synthesis of cis-3a,4,7,7a-tetrahydroisoindole-1,3-diones and cyclic diimides from aromatic and aliphatic diamines with cis-1,2,3,6-tetrahydro phthalic anhydride and maleic anhydride	K10 montmorillonite	Habibi and Marvi (2005)
Synthesis of bithiazole derivatives from thioamides or thiourea and α-halo ketones	K10 montmorillonite	Hashemi et al. (2005)
Synthesis of thiobarbituric acid and barbituric acid derivatives	K10 montmorillonite	Kidwai et al. (2005)
Synthesis of achiral and chiral diamines from N-tosylaziridines and amines	K10 montmorillonite	Nadir and Singh (2005)
Synthesis of isoxazolyl triazinethiones and isoxazolyl oxadiazinethiones	K10 montmorillonite	Rajanarendar et al. (2005)
Synthesis of thieno[2,3-d]pyrimidines from corresponding aromatic and heterocyclic carboxylid acids	K10 montmorillonite	Kidwai et al. (2006)
Synthesis of nitriles from aldehydes	KSF and K10 montmorillonite	Li et al. (2006)
Synthesis of dihydropyrimidones	Zr-pillared montmorillonite	Singh et al. (2006)
Synthesis of trifluoromethylated imines from 1,1,1-trifluoroacetophenone and amines	K10 montmorillonite	Abid et al. (2007)
Synthesis of 1,4-dioxo-3,4-dihydrophthalazine-2(1H)-carboxamides/carbothioamides from phthalic anhydrides and semicarbazide/thiosemicarbazide	K10 montmorillonite	Habibi and Marvi (2007)
Oxidative coupling of substituted benzylamines with anilines and aliphatic amines	K10 montmorillonite	Landge et al. (2007)

(Continued)

TABLE 3.4 (Continued)
Organic Conversions and Syntheses Catalyzed by Cation-Exchanged Montmorillonites, KSF and K10 Montmorillonites, and Metal Salt Impregnated K-Catalysts, Assisted by Microwave or Ultrasound Irradiation Under Solvent-Free Conditions

Microwave-Assisted Reaction	Clay Catalyst	Reference (in Chronological Order)
Cyclization of 3-allyl-4-hydroxy coumarins and pyrenes with N-bromosuccinimide	K10 montmorillonite	Majumdar et al. (2007)
Synthesis of condensed benzo[N,N]-hetrocycles from phenylenediamines and aldehydes/ketones	K10 montmorillonite	Landge and Török (2008)
Synthesis of substituted isobenzofuran-1(3H)-ones from phthalaldehydic acid and ketones	K10 montmorillonite	Landge et al. (2008)
Synthesis of β-keto esters by reacting methyl diazoacetate with aldehydes	K10 and KSF montmorillonites	Marvi and Habibi (2008)
Condensation of aromatic and aliphatic aldehydes with β-naphthol	K10 montmorillonite	Sharifi et al. (2008)
Synthesis of dihydropyrano-xanthones from hydroxyxanthone precursors	K10 montmorillonite	Castanheiro et al. (2009)
Synthesis of 1-acyl-3-aryl-imidazolines from N-acyl-N′-arylethylenediamines with formaldehyde	K10 montmorillonite	Caterina et al. (2009)
Synthesis of quinolones from anilines and cinnamaldehydes	K10 montmorillonite	De Paolis et al. (2009)
Synthesis of indan-1,3-diones from phthalic anhydrides and diethylmalonate	KSF montmorillonite	Marvi and Giahi (2009)
Synthesis of 2-(1′,3′-dihydro-1H-[2,3′]biindolyl-2′-ylidene)-indan-1,3-dione/bisindolylindane-1,3-diones from ninhydrin and 3-substituted/unsubstituted indoles	K10 montmorillonite	Naskar et al. (2010)
Esterification of butyl alcohol with different alkanoic acids	Al^{3+}-montmorillonite	Ramesh et al. (2010)
Oxidative coupling of amines to imines	K10 montmorillonite	Atanassova et al. (2011)
Synthesis of 5-substituted 1-H-tetrazoles from aryl and benzyl nitriles and sodium azide	K10 montmorillonite	Marvi et al. (2011)
Synthesis of octahydroxanthenes by condensation of dimedone with aryl aldehydes	Cr-pillared montmorillonite	Kar et al. (2013)
Synthesis of carvenone by isomerization of 1,2-limonene oxide	Natural montmorillonite	Thi Nguyen et al. (2013)
Synthesis of aryl and heteroaryl *trans*-chalcones from substituted acetophenones and benzaldehydes	KSF montmorillonite	Rocchi et al. (2014)
Synthesis of 2-aryl substituted 1,3,4-oxadiazoles and 1,2,4-oxadiazole derivatives from acid hydrazides and trimethyl orthoformate	Al^{3+}-K10 montmorillonite	Suresh et al. (2014)

(Continued)

TABLE 3.4 (Continued)
Organic Conversions and Syntheses Catalyzed by Cation-Exchanged Montmorillonites, KSF and K10 Montmorillonites, and Metal Salt Impregnated K-Catalysts, Assisted by Microwave or Ultrasound Irradiation Under Solvent-Free Conditions

Microwave-Assisted Reaction	Clay Catalyst	Reference (in Chronological Order)
Synthesis of benzanilides	Montmorillonite doped with Pd nanoparticles	Dar et al. (2015)
Synthesis of diethyl 1-(4-aryl)-4-phenyl-1H-pyrrole-2,3-dicarboxylates	K10 montmorillonite	Kabeer et al. (2015)
Aldol condensation between cyclohexanone and aldehyde, and amide synthesis from aromatic acid and aniline	Montmorillonite treated with p-toluene sulfonic acid or HCl	Ramesh et al. (2015)

Ultrasound-Assisted Reaction	Clay Catalyst	Reference (in Chronological Order)
Aromatization of Hantzsch 1,4-dihydropyridines	K10 montmorillonite-supported $Cu(NO_3)_2$ (claycop)	Maquestiau et al. (1991)
Glucosylation of butanol and dodecanol	KSF/0 montmorillonite	Brochette et al. (1997)
Stereoselective isomerization of Baylis–Hillman acetates	K10 montmorillonite	Shanmugam and Singh (2001)
Condensation of imidazole with ethyl acrylate	Li^+- and Cs^+-montmorillonites	Martin-Aranda et al. (2002)
Friedel–Crafts acylation of 2-methoxynaphthalene with acetic anhydride	K10 montmorillonite and Filtrol-24	Yadav and Rahuman (2003)
Synthesis of arylmethylene-malononitriles by condensation of malononitrile with aromatic aldehydes	K10 montmorillonite-supported $ZnCl_2$ (clayzic)	Li et al. (2004)
Aromatization of *trans*- and *cis*-pyrazolines	K10 montmorillonite-supported $Cu(NO_3)_2$ (claycop)	Mallouk et al. (2004)
Knoevenagel condensation of malononitrile with carbonylic compounds	Li^+- and Cs^+-montmorillonites	Martin-Aranda et al. (2005)
Synthesis of 5-substituted 1-H-tetrazoles from nitriles and sodium azide	K10 montmorillonite	Chermahini et al. (2010)
Synthesis of 3-aryl-3-hydroxy-2-(1H-indol-3-yl)-1-phenyl-1-propanone via cleavage of epoxides with indole	K10 montmorillonite-supported $ZnCl_2$ (clayzic)	Li et al. (2010)
Synthesis of substituted oxindoles	KSF montmorillonite	Dandia et al. (2011)
Synthesis of 2-substituted benzothiazoles from aromatic aldehydes and *o*-aminothiophenol	K10 montmorillonite-supported $FeCl_3$	Chen et al. (2013)
Synthesis of 1,4-disubstituted 1,2,3-triazoles	Cu(II)-doped KSF montmorillonite	Dar et al. (2013)
Synthesis of 5,5-disubstituted hydantoins from aldehyde/ketone, KCN, and ammonium carbonate	K10 montmorillonite	Safari and Javadian (2013)
Synthesis of β-phosphonomalononitriles	KSF and K10 montmorillonites	Dar et al. (2014)
N,N-dibenzylation of anilines	Cu(II)-doped KSF montmorillonite	Dar et al. (2015)
Synthesis of 2-aryl substituted *N*-(4-oxo-1,2-dihydroquinazolin-3(4H)-yl)aryl or alkylamide derivatives	K10 montmorillonite	Rani et al. (2016)

3.4 PILLARED INTERLAYERED CLAYS AND POROUS CLAY HETEROSTRUCTURES

The synthesis of pillared interlayered clays (PILC) dates back to the late 1970s (e.g., Vaughan and Lussier 1980), in response to the demand for cracking catalysts with a larger pore size than that found in zeolites. For a historical account of research and development of PILC, the papers by Vaughan (1988) and Gil et al. (2008) make interesting reading. Since then a large volume of literature has accumulated on the preparation, characterization, and catalytic applications of PILC, sometimes also referred to as *cross-linked smectites* (CLS), and *pillared layered solids* (PLS). Here we will outline the synthesis of PILC, and summarize their catalytic activity. The reviews by Figueras (1988), Lambert and Poncelet (1997), Ohtsuka (1997), Kloprogge (1998), Gil et al. (2000, 2008, 2010, 2011), Ding et al. (2001), Cool and Vansant (2004), Bergaya et al. (2006), and Vicente et al. (2013) should be consulted for more details.

As already remarked on, the first step in preparing PILC consists of intercalating a *pillaring agent* into 2:1 type phyllosilicates, commonly swelling Na^+-smectites, through a cation exchange process. As such, the cation exchange capacity (CEC) of the parent clay material is a key factor controlling pillar density and interpillar distance. Although pillar density might be expected to vary from interlayer to interlayer within a given particle because of the heterogeneous layer charge distribution, referred to in Chapter 1, this expectation does not accord with reality (Pinnavaia 1983). The application of microwave irradiation or ultrasonication during this step can markedly speed up the intercalation process as well as allowing concentrated clay suspensions to be used. The resultant PILC also show improved textural properties and thermal stability, while their acidity and catalytic activity are largely unaffected (Fetter et al. 1996, 1997; Katdare et al. 1999, 2000; Awate et al. 2001; Singh et al. 2004a; Gyftopoulou et al. 2005; Pérez-Zurita et al. 2005; Olaya et al. 2009; Fetter and Bosch 2010; Tomul 2011).

The most widely used pillaring agent is the oligomeric aluminum polyoxo cation, denoted as the *Keggin Al_{13}* ion, or simply Al_{13} *cation* (Brindley and Sempels 1977; Lahav et al. 1978; Vaughan and Lussier 1980; Pinnavaia et al. 1984; Plee et al. 1985). Being composed of twelve Al octahedra, arranged in four groups of three octahedra around one central Al tetrahedron, the Al_{13} cation has a structural formula of $[Al_{13}O_4(OH)_{24}(H_2O)_{12}]^{7+}$. This pillaring agent may be obtained by one of three means: (1), hydrolysis of 0.1–0.001 M $AlCl_3$ solutions with NaOH at an Al/OH ratio of 2.0–2.2; (2), dissolving Al powder in an $AlCl_3$ solution to give aluminum chlorhydrol (ACH), which is commercially available; and (3), electrolysis of an $AlCl_3$ solution. A recent thermodynamic analysis by Wang et al. (2016) indicates that the (monolayer) intercalation of Al_{13} is spontaneous and endothermic.

The second step in the pillaring process is to remove excess ions from the clay-Al_{13} complex by repeated washing-centrifugation, or by dialysis, in order to improve product stability. The washed (or dialyzed) material is then dried (in air) or freeze-dried, giving a characteristic basal spacing of ca. 1.80 nm, and an interlayer separation (distance) of ca. 0.8 nm. The third and final step consists of calcining (heating) the washed complex at 400°C–500°C, converting the Al_{13} cation to an Al oxide pillar or alumina cluster without significantly reducing the basal spacing. As a result, the (dehydroxylated) pillar is anchored to the silicate layer by covalent bonding (Tennakoon et al. 1986a) as is evidently the case in tetrahedrally substituted smectites, such as beidellite and saponite (Plee et al. 1985, 1987; Malla and Komarneni 1993; Lambert and Poncelet 1997). The synthesis of PILC is diagrammatically shown in Figure 3.2.

The polyoxo cations of Cr, Fe. Ga, Si, Ti, and Zr as well as mixed Al–Cr, Al–Fe, Al–Ga, Al–Si, and Al–Zr cations have also been used as pillaring agents. The isostructural Ga_{13} cation has received particular attention as an alternative to Al_{13} (Bradley and Kydd 1993a, 1993b; Brandt and Kydd 1997a). Similarly, Ti-PILC have attracted much interest because of their photocatalytic activity and potential applications in environmental remediation (Sterte 1986; Yamanaka et al. 1987; Ding et al. 2001; Kaneko et al., 2001; Shimizu et al. 2002; Liu et al. 2006; Romero et al. 2006; Centi and Perathoner 2008; Ding et al. 2008; Zhou et al. 2010).

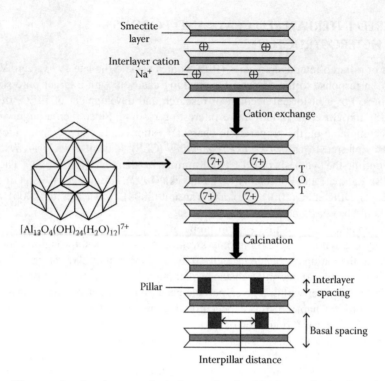

FIGURE 3.2 Diagram showing the two principal steps involved in the synthesis of an alumina-pillared interlayered smectite: exchanging the interlayer counterions (Na$^+$) on the clay sample with the Al$_{13}$ pillaring agent (Keggin cation), followed by calcination of the smectite-Al$_{13}$ intercalate. The basal spacing of the pillared product is ca. 1.8 nm, giving an interlayer spacing (separation) of ca. 0.84 nm. The interpillar distance controls the porosity of the pillared clay. T = tetrahedral sheet; O = octahedral sheet. (Modified from Pinnavaia, T.J., Clay catalysts: Opportunities for use in improving environmental quality, in *Clays: Controlling the Environment. Proceedings of the 10th International Clay Conference, Adelaide, Australia, 1993*, G.J. Churchman, R.W. Fitzpatrick, and R.A. Eggleton (Eds.), CSIRO Publishing, Melbourne, Australia, pp. 3–8, 1995.)

An interesting variant is the material formed by intercalation of imogolite nanotubes into Na$^+$-montmorillonite, giving a basal spacing of ~3.4 nm, which is retained after heating at 400°C. The interactions of imogolite with interlayer montmorillonite surfaces apparently involve H-bonding and electrostatic attraction, rather than through covalent bonding as in conventional PILC (Johnson et al. 1988).

The intercalation of a non-ionic surfactant (Michot and Pinnavaia 1992), polyvinyl alcohol (Suzuki et al. 1988b,1991), or polyethylene glycol-400 (Gao et al. 2014) in conjunction with the Al$_{13}$ pillaring agent can appreciably reduce the amount of water required for PILC synthesis as well as increase the basal spacing and surface area of the product. Furthermore, the pore structure and dimensions can be *tuned* by introducing surfactants in the pillaring process (Zhu and Lu 2001; Zhu et al. 2002). Using Laponite and alkyl polyether surfactants, Zhu and Lu (2001) obtained materials with pores of 2–10 nm in diameter. By intercalating Al$_{13}$ in the presence of a non-ionic surfactant (Igepal CO-720), González et al. (2009) could similarly prepare a disordered mesoporous Al-PILC (Figure 3.3), capable of catalyzing the isomerization of 1-butene to yield *cis-* and *trans-*butene.

Al-pillared smectites, commonly denoted as Al-PILC, have surface areas of 250–350 m^2/g and micropore volumes ranging from 0.06 to 0.13 cm^3/g. A more important property for catalytic purposes is the surface acidity of the materials. Fourier-transform infrared spectroscopy (FTIR) of adsorbed bases, notably pyridine, and temperature programmed desorption of ammonia (cf. Chapter 2), have indicated the presence in Al-PILC of both Brønsted and Lewis acid sites although little, if any,

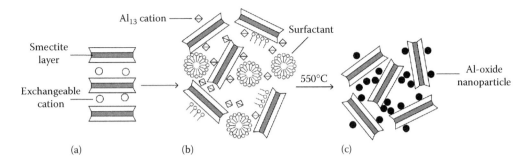

FIGURE 3.3 Diagram showing the formation of a disordered (synthetic) smectite-type clay by exchanging the interlayer inorganic cations (a) with Al_{13} cations in the presence of a non-ionic surfactant (b), and calcining the exfoliated clay-Al_{13}-surfactant complex to yield a mesoporous material impregnated with aluminum oxide nanoparticles (c). (Modified from González, E. et al., *Clays Clay Miner.*, 57, 383–391, 2009.)

Brønsted acidity was detected by Reddy et al. (2009). Lewis acid sites are associated with coordinatively unsaturated Al in the alumina pillars, and their number (concentration) may exceed that of Brønsted sites (Schutz et al. 1987; Figueras 1988; Gil et al. 2000). Indeed, Lewis acid sites appear to predominate in Al-PILC that have been heated at >300°C under vacuum (Occelli and Tindwa 1983; Occelli and Lester 1985; Occelli 1986).

However, FTIR analysis of pyridine adsorbed on Al-pillared montmorillonite, synthetic mica-montmorillonite, and rectorite by Bagshaw and Cooney (1993) indicates that both Brønsted and Lewis acid sites are associated with the pillars. Protons that form either during the pillaring step when the Al^{3+} species polymerizes, or at the calcination step when the hydrated pillars are converted into their corresponding oxides, are at the origin of Brønsted acidity (e.g., Tennakoon et al. 1986a). In beidellite and saponite where the layer charge arises from tetrahedral substitution of Al^{3+} for Si^{4+}, proton attack of Si–O–Al linkages in the tetrahedral sheet gives rise to Brønsted sites identifiable with Si–OH–Al groups (Schutz et al. 1987; Chevalier et al. 1994a, 1994b; Kloprogge and Frost, 1999). In investigating the dehydration and dehydrogenation of C_{1-3} alcohols, catalyzed by alumina-pillared montmorillonite, beidellite, and saponite, Raimondo et al. (1998) observed that the variation of catalytic activity with temperature went through a crossover point. They suggested that at this *iso-acidic point* the Brønsted and Lewis acids are equal in strength. In any case, the combination of high acidity and porosity lies behind the efficiency of Al-PILC (and their surfactant-modified forms) in catalyzing many organic reactions, including the isomerization of 1-butene (Moronta et al. 2008; González et al. 2009).

In this context, it seems pertinent to mention the earlier work by Schutz et al. (1987) who compared the acidity and catalytic activity of Al_{13}-pillared beidellite (PB) with those of Al_{13}-pillared montmorillonite (PM). Besides having a higher (Brønsted) acidity than PM, the PB material is more selective as to the isomerization *n*-decane. This might be because PB has a relatively large (interpillar) pore volume, allowing a higher amount of branched isomers to be produced in, and diffuse out of, the interlayer space as compared with PM. Moronta et al. (1996) have also noted that Pt-impregnated pillared saponites are more efficient than their Al oxide pillared counterparts in catalyzing the hydroisomerization of heptane, presumably because the former materials have a relatively high concentration of Brønsted acid sites, identifiable with tetrahedral Si–OH–Al groups.

According to Molina et al. (2011), an average of 30 papers per annum were published on organic catalysis by PILC during the first decade of the twenty-first century. As a result, a large volume of literature has accumulated on this topic as the reviews by Szostak and Ingram (1995), Lambert and Poncelet (1997), Ding et al. (2001), and Cool and Vansant (2004) would indicate.

Table 3.5 lists a number of reaction types, catalyzed by pillared interlayered smectites, exclusive of oxidative processes. The application of PILC to the oxidative- and photo-degradation of organic

TABLE 3.5
Organic Conversions and Reactions Catalyzed by Clay Minerals Pillared with Various Polyoxo Metal (Hydroxymetal) Cations, Commonly Referred to as Pillared Interlayered Clays (PILC)

Reaction Type (in Alphabetical Order)	Example (Substrate)	Product	PILC Catalyst[a]	Important Parameter	Reference (in Chronological Order)
Acetalization	Carbonyl compounds with methanol	Corresponding acetals	AlFe-PILC	Brønsted and Lewis acidity	Cramarossa et al. (1997)
	Acetone with ethylene glycol	2,2-dimethyl-1,3-dioxolane	ZrAl-pillared bentonite	Brønsted acidity	Mnasri and Frini-Srasra (2012, 2013)
Acylation	Hexanol and benzylalcohol with acetic acid	Hexylacetate and benzylacetate	Fe-, Cr-, Cr-Fe-, Fe-Cr-PILC	Lewis acidity	Akçay (2004)
Addition	Methanol to propylene oxide	2-Methoxy-1-propanol	Organo-zirconium pillared bentonite and taeniolite	Brønsted acidity	Permana et al. (2004)
	Ditto	Ditto	Al-PILC	Ratio of Lewis/Brønsted acicity	Timofeeva and Khankhasaeva (2009)
Alcoholysis	Epoxidized methyl oleate with methanol	β-Hydroxyethers	Nb-PILC	Brønsted acidity	Gallo et al. (2006)
	Propylene oxide with methanol	Propylene glycol methyl ether	Al-PILC and ZrAl-PILC	Brønsted acid sites	Timofeeva et al. (2011, 2014)
Alkylation	Ethylation of toluene	Ethyltoluene isomers	Al-pillared bentonite	Brønsted acidity	Occelli et al. (1985)
	Toluene alkylation with methanol	Xylene isomers trimethylbenzene	Al-PILC, Al-pillared saponite,	Pore structure	Urabe et al. (1986)
	Ditto	Xylene isomers	Al-pillared fluor-tetrasilicic mica	La^{3+} impregnation	Sakurai et al. (1988)
	Ditto	Xylene isomers trimethylbenzene	Al-PILC	Interpillar spacing	Horio et al. (1991)
	Ditto	Ditto	Al- and GaAl-PILC	Acid site concentration	Benito et al. (1999)
	Ditto	Ditto	Al-PILC, Zr-PILC, AlZr-PILC	Brønsted acidity	Vasques et al. (2005)
	Ditto	Ditto	Cr-pillared KSF montmorillonite	Selectivity to *p*-xylene	Binitha and Sugunan (2008)
	Ditto	Ditto	Al-PILC, Zr-PILC		Fernandes et al. (2008)

(*Continued*)

TABLE 3.5 (Continued)
Organic Conversions and Reactions Catalyzed by Clay Minerals Pillared with Various Polyoxo Metal (Hydroxymetal) Cations, Commonly Referred to as Pillared Interlayered Clays (PILC)

Reaction Type (in Alphabetical Order)	Example (Substrate)	Product	PILC Catalyst[a]	Important Parameter	Reference (in Chronological Order)
Alkylation (Continued)	Propene alkylation of biphenyl	*Meta-* and *para-*alkylated isomers	Al-PILC	Brønsted acidity	Butruille and Pinnavaia (1992)
	Friedel–Crafts alkylation of aromatic hydrocarbons	Alkylated products	Fe-PILC, Zr-pillared K10 montmorillonite	Lewis acidity	Choudary et al. (1997); Singh and Samant (2004)
	Benzene alkylation with propene	Cumene	La^{3+}- and Al^{3+}-exchanged Al-bentonite	Lewis acidity	Geatti et al. (1997)
	Benzene alkylation with ethylene	Ethylbenzene	Al-PILC, Ga–Al-PILC	Brønsted acidity	Lenarda et al. (1999)
	Ethylation of benzene with ethanol	Ethylbenzene	Ti-PILC	Brønsted and Lewis acidity	Binitha and Sugunan (2006)
	Benzene alkylation with 1-dodecene	Linear mono-alkylbenzenes	Fe-PILC	Lewis acidity	Farzaneh et al. (2007)
	Benzene alkylation with 1-hexene, 1-octene, and 1-dodecene	Monoalkylated products	Zr-PILC	Surface acidity, alkene chain length	Guerra et al. (2008)
	Benzene alkylation with 1-octene	Mono-alkyl benzene and 2-phenyloctane	Ti-pillared KSF montmorillonite	Brønsted acidity	Narayanan and Sugunan (2008)
	Benzene alkylation with 1-decene, and 1-dodecene	Mono-alkyl benzenes	Al-PILC, acid-activated Al-PILC	Surface acidity, benzene/ olefin ratio	Faghihian and Mohammadi (2013, 2014)
	Toluene benzylation with benzyl alcohol	*Ortho* and *para* isomers of methyl diphenyl methane	Al-PILC, Ce-exchanged Al-PILC	Surface acidity	Narayanan and Deshpande (2000)
	Aniline ethylation with ethanol	*N*-ethylaniline, *N*,*N*'- diethylaniline, toluidines	Al-PILC, Ce-exchanged Al-PILC	Surface acidity	Narayanan and Deshpande (2000)
	Alkylation of phenol with methanol	Anisole, cresols, polyalkylated phenols	Zr-PILC	Brønsted and Lewis acidity	Mishra and Rao (2004)

(*Continued*)

TABLE 3.5 (Continued)
Organic Conversions and Reactions Catalyzed by Clay Minerals Pillared with Various Polyoxo Metal (Hydroxymetal) Cations, Commonly Referred to as Pillared Interlayered Clays (PILC)

Reaction Type (in Alphabetical Order)	Example (Substrate)	Product	PILC Catalyst[a]	Important Parameter	Reference (in Chronological Order)
Alkylation (Continued)	Ethylation of benzene with ethanol	Ethylbenzene	Ti-PILC	Brønsted and Lewis acidity pore size	Binitha and Sugunan (2006)
	tert-Butylation of phenol	p-tert-Butylphenol, 2,4-di-tert Butylphenol	Fe-PILC with and without transition metal cations (V, Co, Ni)	Lewis acidity	Kurian and Sugunan (2006)
Aromatization	C_3 and C_4 hydrocarbons	Benzene and xylene	ZnAl-PILC	Weak acidity	Liu et al. (1999)
Cleavage	O-methyl bond in anisole	Alkylated phenols	Al-pillared bentonite	Microporosity	Carrado et al. (1990)
Condensation	Acetone	Diacetone alcohol and mesityl oxide	Al-PILC	Ratio of Lewis/Brønsted acidity	Timofeeva and Khankhasaeva (2009)
	Cumene	Benzene (and α-methylstyrene by dehydrogenation)	Al-PILC, Al-pillared saponite	Acidity and pore structure	Kikuchi and Matsuda (1991)
	Ditto	Ditto	Cr-PILC, Cr-pillared vermiculite	Brønsted acidity	Vijayakumar et al. (1994)
	Ditto	Ditto	Al-pillared acid-activated clay,	Brønsted acidity	Mokaya and Jones (1995b)
	Ditto	Ditto	TiAl-PILC and TiAl-pillared beidellite	Surface acidity, tetrahedral charge	Swarnakar et al. (1996)
	Ditto	Ditto	FeCr-PILC	Fe/Cr ratio	Mishra and Parida (1998b)
Cracking	n-Heptane	Iso-butane, propene	CeAl-PILC	Brønsted acidity	González et al. (1992)
	Heavy oil/petroleum	Olefins, gasoline	Al-PILC (400°C in vacuo)	Lewis acidity	Occelli (1983)
	Ditto	Ditto	Al-pillared saponites		Chevalier et al. (1994a)
	Ditto	Ditto	Al-PILC, AlGa-PILC	Pore structure	González et al. (1999)
	Polyethylene	Liquid hydrocarbons	Al-PILC	Surface acidity	Manos et al. (2001, 2002)

(Continued)

TABLE 3.5 (Continued)
Organic Conversions and Reactions Catalyzed by Clay Minerals Pillared with Various Polyoxo Metal (Hydroxymetal) Cations, Commonly Referred to as Pillared Interlayered Clays (PILC)

Reaction Type (in Alphabetical Order)	Example (Substrate)	Product	PILC Catalyst[a]	Important Parameter	Reference (in Chronological Order)
Dehydration	Methanol	C_2–C_4 olefins	Zr-PILC	Surface area	Burch and Warburton (1986)
	1-Butanol	Butene isomers	Ti-PILC; pillared acid-activated clay (PAAC)	Surface acidity; Brønsted acidity	Del Castillo and Grange (1993); Mokaya and Jones (1995b)
	Pentan-1-ol	Pentenes	Al-pillared and acid-activated Al-PILC	Brønsted acidity and pillar density	Jones and Purnell (1994); Bovey and Jones (1995)
	Glucose	Formic and 4-oxopentanoic acids	Fe-PILC, Cr-PILC, Al-PILC	Acidity and pore size	Louvranij and Rorrer (1994, 1997)
	1-Phenylethanol	3-oxa-2,4-diphenylpentane	Ti-PILC	Accessibility of strong acid sites	Gil et al. (1996)
	Methanol	Dimethyl ether	Al-PILC	Brønsted acidity	Hashimoto et al. (1996)
	2-Propanol and methanol	Propene and hydrocarbons	Cr-PILC; Al-, Ga, AlGa-pillared saponite	Brønsted acidity	Mishra and Parida (1998a); Vicente et al. (2009)
	Iso-Propanol	Propene	Al-PILC, Al-pillared saponite	Brønsted and Lewis acidity	Trombetta et al. (2000)
	C_2–C_8 alcohols	Olefins	FeAl-PILC, FeAl-pillared beidellite	Lewis acids on pillars	De Stefanis et al. (2003)
Dehydrogenation	Cyclohexane	Benzene	Cr-PILC	Large interlayer separation	Tzou and Pinnavaia (1988)
	Cumene	α-Methylstyrene	Ga-PILC, GaAl-PILC,	Lewis site density	Bradley and Kydd (1993a); Brandt and Kydd (1997b)
	Ditto	Ditto	TiAl-PILC and TiAl-pillared beidellite	Surface acidity	Swarnakar et al. (1996)

(Continued)

TABLE 3.5 (Continued)
Organic Conversions and Reactions Catalyzed by Clay Minerals Pillared with Various Polyoxo Metal (Hydroxymetal) Cations, Commonly Referred to as Pillared Interlayered Clays (PILC)

Reaction Type (in Alphabetical Order)	Example (Substrate)	Product	PILC Catalyst[a]	Important Parameter	Reference (in Chronological Order)
Dehydrogenation (Continued)	Ditto	Ditto	Cr-PILC; FeCr-PILC	Lewis acidity and Fe/Cr ratio	Mishra and Parida (1998a, 1998b)
	Ethylbenzene	Styrene	AlCr-pillared saponites	Chromium content	Vicente et al. (2002)
	Ditto	Ditto	Fe-pillared smectite		Huerta et al. (2003)
	Ditto	Ditto	Co-impregnated Fe-pillared clay	Cobalt content	Gonzalez and Moronta (2004)
	Ditto	Ditto	Al-pillared Fe-rich saponite	Acid strength	Gandia et al. (2005)
	Ditto	Ditto	Co/Mo-impregnated Al-PILC	Surface acidity and metal reduction	Moronta et al. (2006)
Dehydrogenation (oxidative)	Propane	Propene (propylene)	V-impregnated Zr-PILC and CrAl-PILC	Microporosity, reaction temperature	Bahranowski et al. (2000) and De León et al. (2014)
Disproportionation	1,2,4-Trimethyl-benzene	1,2,4,5-Tetramethyl-benzene and *o*-xylene	Al-PILC and Zr-PILC	Brønsted acidity and interlayer distance	Kikuchi et al. (1984, 1985)
	Ditto	Ditto	Al-PILC and Al-pillared saponite	Lewis (and Brønsted) acidity, spilt-over hydrogen	Matsuda et al. (1988a, 1988b, 1995)
	Ditto	1,2,4,5-Tetramethyl-benzene	Al-PILC, Al-pillared beidellite, Al-pillared synthetic mica-montmorillonite	Isomerization of alkyl benzenes	Kojima et al. (1991)
	Ditto	1,2,4,5-Tetramethyl-benzene and *o*-xylene	LaAl-PILC; CrAl-PILC	La(II) content; Cr/Al ratio	Zhao et al. (1993, 1995)
	Toluene	Benzene and *p*-xylene	Zr-PILC, Cr-PILC, Zr- and Cr-pillared hectorite	Lewis acidity	Auer and Hofmann (1993)
	o-Xylene	Toluene and trimethylbenzene	Al-PILC, Cr-PILC	Lewis and Brønsted acidity	Krajčovič et al. (1995)

(*Continued*)

TABLE 3.5 (Continued)
Organic Conversions and Reactions Catalyzed by Clay Minerals Pillared with Various Polyoxo Metal (Hydroxymetal) Cations, Commonly Referred to as Pillared Interlayered Clays (PILC)

Reaction Type (in Alphabetical Order)	Example (Substrate)	Product	PILC Catalyst[a]	Important Parameter	Reference (in Chronological Order)
Disproportionation (Continued)	Ethylbenzene	*m*-Diethylbenzene	Al-PILC, Zr-PILC	Surface acidity	Jerónimo et al. (2007)
Esterification	*n*-Hexanol and benzylalcohol with acetic acid	Hexylacetate and benzylacetate	Fe-PILC, Cr-PILC, CrFe-PILC, and FeCr-PILC	Lewis acidity	Akçay (2004)
	2-Methoxy- and 2-ethoxy ethanol with acetic acid	2-Methoxyethanol acetate and 2-ethoxyethanol acetate	Al-PILC	Lewis acidity	Wang and Li (2000, 2004)
	n-Butanol with acetic acid (solvent-free)	Butylacetate	Dodecatungsto-phosphoric acid-impregnated Ti-PILC		Varadwaj and Parida (2011)
	Ethanol with acetic acid	Ethylacetate	Ti-PILC	Brønsted acidity	Peter et al. (2012)
Hydrocracking and Hydroisomerization	*n*-Pentane	Single-branched isomers	Al-pillared Ni-substituted mica montmorillonite		Gaaf et al. (1983)
	Ditto	Ditto	Pt- and Re-impregnated Al-PILC	Brønsted acidity?	Parulekar and Hightower (1987)
	n-Hexane	Single-branched isomers	Pt- and Re-impregnated Al-PILC	Brønsted acidity?	Parulekar and Hightower (1987)
	n-Decane	C_4–C_6 hydrocarbons	Fe- and Cr-exchanged Al-PILC	Location of Fe^{3+} and Cr^{3+}	Skoularikis et al. (1988)
	Ditto	Ditto	AlGa-pillared beidellite and montmorillonite	Tetrahedral substituition	Molina et al. (1994)
	Ditto	Isomers	AlZr-, AlHf-, and AlCe-vermiculite	Interlayer separation and porosity	Campos et al. (2008)
	n-Heptane	Isomers	Pt- and Re-impregnated Al-PILC	Brønsted acidity?	Parulekar and Hightower (1987)
	Ditto	Ditto	Al-PILC and AlGa-PILC	Surface area	Pesquera et al. (1995)
	Ditto	Ditto	Pt-impregnated Al-, AlZr-, Zr-pillared montmorillonites, and saponites	Brønsted acidity, tetrahedral substitution, and Pt hydrogenolysis	Moreno et al. (1996, 1999)

(*Continued*)

TABLE 3.5 (Continued)
Organic Conversions and Reactions Catalyzed by Clay Minerals Pillared with Various Polyoxo Metal (Hydroxymetal) Cations, Commonly Referred to as Pillared Interlayered Clays (PILC)

Reaction Type (in Alphabetical Order)	Example (Substrate)	Product	PILC Catalyst[a]	Important Parameter	Reference (in Chronological Order)
Hydrocracking and Hydroisomerization (Continued)	Ditto	Ditto	AlCe-PILC	Surface area and porosity	Hernando et al. (2001)
	Ditto	Ditto	Pd-impregnated sulfated Zr-PILC	Metal number/acid site ratio	Bouchenafa-Saib et al. (2004)
	Ditto	Ditto	Al-, AlZr-, Zr-, AlSi-PILC	Addition of Zr or Si increases acidity and selectivity	Molina et al. (2005)
	n-Octane	Isomers	Al-PILC and Al-pillared saponite	Surface acidity	Kooli et al. (2008)
	Ditto	Ditto	Al-PILC and Pt-doped Al-PILC	Brønsted acidity	Doblin et al. (1991, 1994)
	Coal extract and coal-derived liquids	Decrease in material fraction with boiling points > 450°C	Pt/Al-pillared vermiculite and phlogopite	Interpillar porosity	del Rey-Perez-Caballero et al. (2000)
			Sn-, Cr-, and Al-pillared montmorillonite and laponite	Surface acidity and microporosity	Bodman et al. (2002) and Gyftopoulou et al. (2005)
Hydrogenation	Cyclohexene	Cyclohexane	Pd-impregnated Al-PILC	Particle size	Szűcs et al. (1998)
	Benzene, xylene, and mesitylene	Cyclohexane, C_8 cycloalkane	Pt/Pd-impregnated Al-PILC	Pore size and Pt/Pd loading	Liu et al. (1999)
	Benzene	Cyclohexane	La/Ni-impregnated Al-PILC, Ni-PILC, Ni-impregnated Si-PILC	Pore size, metal loading, calcination temperature, and particle size microporosity	Louloudi and Papayannakos (1998, 2000, 2016)
	Ditto	Ditto	Ni-impregnated Al-pillared saponite	Ni loading and location of acid sites	Louloudi et al. (2003)
	Quinoline	Decahydro-quinoline	Rh-exchanged MgAlCe-PILC	Rh particle size (≤2 nm)	Campanati et al. (2002)

(Continued)

TABLE 3.5 (Continued)
Organic Conversions and Reactions Catalyzed by Clay Minerals Pillared with Various Polyoxo Metal (Hydroxymetal) Cations, Commonly Referred to as Pillared Interlayered Clays (PILC)

Reaction Type (in Alphabetical Order)	Example (Substrate)	Product	PILC Catalyst[a]	Important Parameter	Reference (in Chronological Order)
Hydrogenation (Continued)	Phenyl alkyl acetylenes	cis-Alkene	Pd-impregnated Al-PILC	Reactant/Pd ratio and stereoselectivity	Marín-Astorga et al. (2005)
	Dimethyl adipate	Monomethyl ester, caprolactone, and 1,6-hexanediol, methyl caproate	Pt- and Ru-impregnated Al-PILC	Surface acidity	Figueiredo et al. (2008, 2009)
Hydroxylation	Phenol	Dihydroxy-benzenes	Ti-PILC	Brønsted acidity and solvent type	Del Castillo et al. (1996)
	Ditto	Ditto	Cu-doped Al-PILC	Adsorption and activation of phenol	Bahranowski et al. (1998)
	Ditto	Ditto	Zr-PILC	Brønsted acidity	Awate et al. (2001)
	Ditto	Ditto	FeAl-pillared smectites	Brønsted acidity and redox centers	Letaïef et al. (2003)
	Ditto	Ditto	Fe-PILC exchanged with La, Ce, Th	Phenol and solvent concentration	Kurian et al. (2012)
	Ditto	Ditto	Fe-grafted Si-pillared montmorillonite	Tetrahedrally coordinated Fe(III)	Yang et al. (2013)
	Benzene	Phenol	Cu-doped Al-PILC	Benzene/H_2O_2 mole ratio	Pan et al. (2008)
	Ditto	Ditto	Cu-doped Si-PILC	Copper attached to silica pillar	Yue et al. (2014)
Isomerization	m-Xylene	Isomers	Al-PILC	Shape selectivity	Mori and Suzuki (1989)
	α-Acetylenic alcohols	α,β-Ethylenic carbonyl compounds	V-PILC	Shape selectivity	Choudary et al. (1990b)
	1-Butene	cis- and trans-Butene	Al- and Fe-PILC	Surface acidity	Béres et al. (1995)
	Ditto	Ditto	Surfactant-modified and Al-pillared smectite	Surface acid sites of moderate strength	González et al. (2009)

(Continued)

TABLE 3.5 (Continued)
Organic Conversions and Reactions Catalyzed by Clay Minerals Pillared with Various Polyoxo Metal (Hydroxymetal) Cations, Commonly Referred to as Pillared Interlayered Clays (PILC)

Reaction Type (in Alphabetical Order)	Example (Substrate)	Product	PILC Catalyst[a]	Important Parameter	Reference (in Chronological Order)
Isomerization (Continued)	Ditto	*iso*-Butene	Mono- and bi-pillared montmorillonite and saponite	Brønsted and Lewis acid sites	Trombetta et al. (2000)
	Ditto	Ditto	Al- and Fe-PILC	Surface acidity and surface area	Moronta et al. (2008)
	Longifolene	*iso*-Longifolene	Ru-doped Al-PILC	Weak acid sites	Singh et al. (2007)
	n-Hexane	Mono- and di-branched isomers	Sulfated Zr-pillared bentonite		Radwan et al. (2009)
Oxidation	Alcohols	Carbonyl compounds	Cr-PILC	In the presence of *tert*-butyl hydroperoxide	Choudary et al. (1990a)
	Methanol	Formaldehyde	MoO$_3$-impregnated Al-PILC	Brønsted acidity	Klissurski et al. (1996)
	Toluene	Carbon dioxide (and phenol?)	Fe$_2$O$_3$-impregnated Ti-PILC	Lewis and Brønsted acicity	Liang et al. (2016)
Sulfonylation	Arenes	Corresponding sulfones	Fe-PILC	Lewis acidity (Fe^{3+})	Singh et al. (2004c)

[a] Unless otherwise specified, the parent clay mineral of PILC is montmorillonite.

compounds, including wet hydrogen peroxide oxidations through a Fenton-like reaction (Garrido-Ramírez et al. 2010; Herney-Ramírez and Madeira 2010; Herney-Ramírez et al. 2010; Perathoner and Centi 2010; Vicente et al. 2010), is described in Chapter 7.

The surface acidity, pore volume, average pore diameter, and often the surface area of pillared clays may be enhanced by using acid-activated smectites as the starting (host) materials (Mokaya and Jones 1995b; Bieseki et al. 2013). Pillared acid-activated clays (PAAC) have both microporosity and mesoporosity. The development of micropores (at the expense of mesopores) is dependent on the level of acid treatment and pillaring conditions (Bovey and Jones 1995; Mokaya and Jones 1995a). The results for benzene alkylation and 1-pentanol dehydration indicate that the combination of acid activation and pillaring make for highly efficient solid acid catalysts, being superior to Al^{3+}-exchanged and K10 montmorillonite (Mokaya and Jones 1994). The high Brønsted acidity of PAAC, relative to the corresponding non-activated (conventional) PILC, is also indicated by their increased efficiency in catalyzing 1-butanol dehydration and cumene hydrocracking (Mokaya and Jones 1995b). Interestingly, Tennakoon et al. (1987) have found Al-PILC to be inactive in n-pentanol dehydration at 200°C, probably because the protons are immobilized within the silicate layers, and hence unavailable for catalysis. Catalytic activity, however, is greatly enhanced at 250°C when the liberated protons can diffuse to the substrate.

Galarneau et al. (1995) have described the synthesis of *porous clay heterostructures* (PCH) in which the contiguous layers of a swelling smectite (Li^+-fluorohectorite) are propped apart by an open-framework silica. The formation of PCH involves the intercalation of a surfactant cation such as hexadecyltrimethylammonium by a cation exchange reaction. After washing the smectite-HDTMA complex, a neutral cosurfactant (e.g., dodecylamine or hexadecylamine) is introduced, which reacts with the cationic surfactant to form a micellar template in the interlayer space. A silica source, such as tetraethylorthosilicate (TEOS), is then added and allowed to hydrolyze and condense around the outer surface of the micelle. The hydrolysis of interlayer TEOS can also be induced by ammonia (Li et al. 2009). Removal of the interlayer micelles by calcination at 550°C–600°C (or by solvent extraction) produces a porous solid with 1.4–2.2 nm wide pores, capable of catalyzing the selective dehydration of 2-methylbut-3-yn-2-ol to 2-methylbut-3-yn-1-ene (Galarneau et al. 1997). Figure 3.4 shows the principal steps involved in the preparation of PCH.

Using a synthetic saponite as the parent clay mineral, Polverejan et al. (2000a, 2000b) were able to obtain PCH with a basal spacing of ~3.4 nm, a pore volume of ~0.45 cm^3/g, a BET surface area of ~850 m^2/g, and a total acidity of 0.64–0.77 mmol/g (determined by temperature-programmed desorption of cyclohexylamine). The synthesized PCH could catalyze the condensed phase Friedel–Crafts alkylation of the bulky 2,4-di-*tert*-butylphenol with cinnamyl alcohol to yield 6,8-di-*tert*-butyl-2,3-dihydro[4*H*]benzopyran. By comparison, dealkylated 4-*tert* butylphenol was the main product formed over H$^+$-saponite and K10 montmorillonite because of diffusion and accessibility

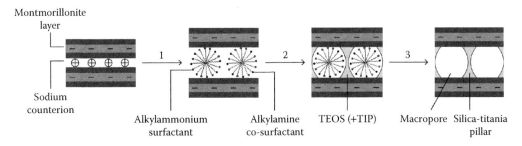

FIGURE 3.4 Diagram showing the steps involved in the synthesis of a porous clay heterostructure (PCH): (1), replacing the interlayer sodium counterions with alkylammonium cations and adding melted alkylamine co-surfactant; (2), introducing a mixture of tetraethylorthosilicate (TEOS) and titanium isopropoxide (TIP); and (3), burning off the interlayer surfactants by calcination to yield a macroporous silica-titania-pillared material. (Adapted from Chmielarz, L. et al., *J. Solid State Chem.*, 182, 1094–1104, 2009.)

problems. More recently, Zhang et al. (2010b) synthesized a PCH with a surface area of 554 m^2/g and a total pore volume of 0.626 cm^3/g for the hydrogenation of benzene.

Similar silica-pillared montmorillonites have been prepared by Ahenach et al. (2000) using aminopropyltrioxysilane as the silica source, while Ruiz-Hitzky and Aranda (2014) used an alcohol, instead of a neutral amine, to expand the clay interlayers prior to introducing the silicon alkoxide. The surface acidity (largely of the Brønsted type) of PCH may be enhanced by using acid-activated smectites as host minerals (Pichowicz and Mokaya 2001; Kooli et al. 2006), or by incorporating titania into the silica pillars (Chmielarz et al. 2009) as illustrated in Figure 3.4.

3.5 ORGANICALLY MODIFIED CLAY MINERALS

The vast majority of organically modified clay minerals, used in organic catalysis, are essentially formed by one of two processes: (1) intercalation of cationic organic compounds and (2) surface grafting of organosilanes (*silylation*) and other organic species.

3.5.1 ORGANOCLAYS AND RELATED MATERIALS

The propensity of various clay minerals for adsorbing cationic organic compounds has been known for nearly a century although it was not until the early 1950s that the formation and properties of organically modified smectites began to be systematically investigated (Weiss 1963; Theng 1974; Barrer 1978; Lagaly et al. 2013). Although cation exchange is the principal mechanism involved in the intercalation of alkylammonium ions into smectites (and vermiculites), van der Waals interactions (between alkyl chain and silicate surface) and entropy effects can make an appreciable contribution to the overall adsorption energy (Theng et al. 1967; Vansant and Uytterhoeven 1972).

Interlayer complexes of smectites with short-chain, compact alkylammonium ions such as tetramethylammonium (TMA), commonly denoted as type I organoclays, have basal spacings of ca. 1.5 nm with interlayer separations of ca. 0.55 nm. As mentioned earlier, the organic cations can act as *pillars*, creating an interpillar free space (*pore*) into which various guest molecules can be accommodated. Thus type I organoclays are highly porous having specific surface areas of 100–300 m^2/g.

In terms of organic catalysis, however, montmorillonites pillared with the protonated cage compound 1,4,-diazabicyclo[2,2,2]octane (Dabco), also known as triethylene diamine, have received more attention than type I organoclays since Dabco-exchanged materials have a high thermal stability and can resist swelling in organic solvents (Shabtai et al. 1977). On the basis of X-ray diffraction and chemical analysis, Mortland and Berkheiser (1976) have suggested that the Dabco^{2+} ion can simultaneously adsorb to negatively charged (exchange) sites on opposing interlayer surfaces, giving a basal spacing of 1.42 nm. A slightly larger value of 1.48 nm was measured by Shabtai et al. (1977) for the interlayer Dabco-complex in which the di-protonated cation adopted a perpendicular orientation with respect to the silicate layer.

The interlayer structure of smectite complexes with mono-protonated (Dabco^{1+}) and di-protonated (Dabco^{2+}) species has been systematically investigated by Van Leemput et al. (1983). Figure 3.5a shows that the *edge-to-edge* interpillar separation (λ) for three montmorillonites (WB, MM, OM) and a hectorite (H) is inversely related to the mean interlayer charge deficit (i.e., the charge per formula unit, x) (cf. Table 1.1). Shabtai et al. (1977) reported a λ value of 0.6 nm for monmorillonite, while Slade et al. (1989) measured a value of ~0.3 nm for a high-charge vermiculite. As might be expected, the interpillar distance for complexes with Dabco^{1+} is about half as large as that for Dabco^{2+}-exchanged materials. For a given smectite sample, λ is also influenced by the interlayer orientation of the organic cation. Figure 3.5b shows that both λ and interlayer separation (Δ) are larger when the interlayer Dabco^{2+} (cross-sectional area ~0.28 nm^2) is oriented perpendicularly to the silicate surface (a) than when this cation adopts a parallel orientation with a cross-sectional area of ~0.40 nm^2 (b). For the montmorillonite from Moosburg (MM), λ_1 is 0.89 nm and λ_2 is 0.78 nm (Stul et al. 1983).

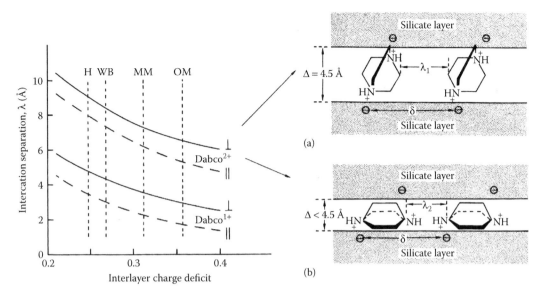

FIGURE 3.5 Diagram showing the relationship between mean interlayer charge deficit (density) and intercation (interpillar) separation, λ, for complexes of hectorite (H), Wyoming bentonite (WB), Moosburg montmorillonite (MM), and Otay montmorillonite (OT) with 1,4-diazobicyclo(2,2,2)-octane (Dabco) cations in either the mono-protonated (Dabco^{1+}) or di-protonated (Dabco^{2+}) form. Solid curves refer to complexes in which the interlayer Dabco cations adopt a perpendicular orientation with respect to the silicate layers, shown in (a), while the broken curves refer to complexes where the organic cations assume a parallel orientation, shown in (b). (Modified from Van Leemput, L. et al., *Clays Clay Miner.*, 31, 261–268, 1983.)

Besides being more active in mediating the esterification of carboxylic acids than type I organoclays, Dabco^{2+}-exchanged montmorillonite (Mt) can promote shape-selective organic conversions (Kikuchi and Matsuda 1988). An example is provided by comparing the rates of acetic acid esterification with isomeric C_5 and C_6 alcohols (Shabtai et al. 1977), summarized in Table 3.6. The high esterification activity of H$^+$/Al^{3+}-Mt may be ascribed to the accessibility of its proton-rich interlayers to the reactants. The interpillar pore space of Dabco-Mt is also accessible but to different extents depending on the shape of the starting alcohols. The lack of Brønsted acidity, and the high occupancy of the interlayer space by dodecylammonium ions (DA) would account for the low overall esterification activity of DA-Mt.

The ability of Dabco^{2+}-exchanged montmorillonite to catalyze the conversion of acetonitrile to acetamide has been reported by Mortland and Berkheiser (1976). Similarly, Stul et al. (1983) have found Zn-Dabco-montmorillonite to be highly efficient in catalyzing the isomerization of *cis*-2-butene to *trans*-2-butene and 1-butene (at 150°C) whereas TMA$^+$-montmorillonite is comparatively inactive. We should mention that the free, neutral Dabco base can serve as a homogeneous catalyst for various organic conversions, including the coupling of functionalized aldehydes to acrylic esters (Hoffmann and Rabe 1983), and that of acrylonitrile to α-keto esters (Basavaiah et al. 1987). This compound can also catalyze the Baylis–Hillman reaction between aromatic aldehydes and activated olefins (Basavaiah et al. 2003; Chandrasekhar et al. 2004; de Souza et al. 2008), and the synthesis of quinoxalines by reacting 1,2-diamines with phenacyl bromides (Meshram et al. 2010). More recently, Meshram et al. (2012a, 2012b) used Dabco to promote the synthesis of pyrrole by reacting phenacyl bromides, pentane-2,4-dione and amine, and that of benzofurans from phenacyl halides and *o*-hydroxy benzaldehyde. Whether or not such reactions can be efficiently and selectively catalyzed by clay-intercalated Dabco awaits investigation.

TABLE 3.6
Relative Rates of Esterification[a] of Acetic Acid with Isomeric C_5 and C_6 Alcohols Catalyzed by Acid-Activated Montmorillonite (H/Al-Mt), Dabco^{2+}-Montmorillonite (Dabco-Mt), and *n*-Dodecylammonium-Montmorillonite (DA-Mt)

Alcohol	Catalyst		
	H/Al-Mt	Dabco-Mt	DA-Mt
Isomeric C_5 alcohols			
n-pentanol	1.80	1.00	0.07
3-methylbutan-1-ol	1.71	0.08	0.08
2,1-dimethylpropan-1-ol	1.67	<0.02	0.07
Primary C_6 alcohols			
n-hexanol	1.72	1.00	0.05
2-methylpentan-1-ol	1.58	0.65	0.05
Secondary C_6 alcohols			
hexan-2-ol	0.96	0.40	<0.04
2-methylpentan-2-ol	0.85	0.08	<0.04

Source: Shabtai, J. et al., Synthesis and catalytic properties of a 1,4-diaza-bicyclo[2,2,2]octane-montmorillonite system—A novel type of molecular sieve, in *Proceedings of the 6th International Congress of Catalysis, 1976*, G.C. Bond, P.B. Wells, and F.C. Tompkins (Eds.), The Chemical Society, London, UK, pp. 660–667, 1977.

[a] For esterification with C_5 alcohols: conversion of acetic acid into corresponding ester/conversion of acetic acid into *n*-pentyl acetate in the presence of Dabco-Mt. For esterification with C_6 alcohols: conversion of acetic acid into corresponding ester/conversion of acetic acid into *n*-hexyl acetate in the presence of Dabco-Mt. See Shabtai et al. (1977) for other experimental details.

Type II organoclays are formed by intercalation of long-chain quaternary alkylammonium ions, such as hexadecyltrimethylammonium (HDTMA), through a cation exchange mechanism. Such cationic surfactants can adopt various arrangements in the interlayer space, giving rise to different basal spacings (Figure 3.6), in dependence of alkyl chain length and smectite layer charge density (Lagaly 1986; Jaynes and Boyd 1991; Lagaly et al. 2013). The basal spacing of the organoclays also increases with surfactant cation loading (He et al. 2010). Since the interlayer space here is largely occupied by the surfactant, type II organoclays have a low porosity, while their surface area is one order of magnitude smaller than that of the type I variety. More often than not, this type of organoclay is less efficient in catalyzing a given organic conversion than the corresponding parent material (e.g., Naicker et al. 1998b).

Commercial type II organoclays, such as Cloisite 20A and Cloisite 30B, are available from Southern Clay Products, now a part of BYK Additives. For more information on the preparation and properties of organoclays together with their environmental and industrial applications, than can be included here, the reader is referred to the reviews by Xu et al. (1997), de Paiva et al. (2008), Theng et al. (2008), Theng (2012), Yuan et al. (2013), He et al. (2014), and Zhu et al. (2015).

As already remarked on, organically modified smectites are for the most part catalytically inactive unless the intercalated organic species is a catalyst in its own right. Even then, the products of the catalytic conversion may differ between the homogeneous and heterogeneous systems. For example, the hydrogenation of 1-hexene by the rhodium triphenylphosphine cation [Rh(PPh$_3$)$_2$]$^+$ in homogeneous solution yields hexane and 2-hexene with a ratio of ca. 65 to 35. When intercalated into hectorite, however, the Rh(PPh$_3$)$_2$$^+$ cation can (under certain experimental conditions) quantitatively convert 1-hexene to hexane. Likewise, no isomerization occurs during the catalytic hydrogenation of 1-hexene in methanol by the intercalated complex with Rh(PPh$_3$)$_n$$^+$ although

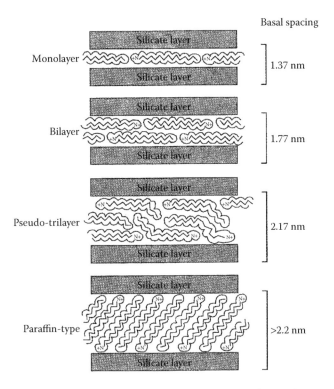

FIGURE 3.6 Possible arrangements of long-chain quaternary ammonium cations, such as hexadecyltrimethylammonium (cetyltrimethylammonium), in the interlayer space of smectites. The basal spacing of the resultant (type II) organoclays is indicated. (Adapted from Theng, B.K.G. et al., Organically modified clays for pollution uptake and environmental protection, in *Soil Mineral-Microbe-Organic Interactions*, Q. Huang, P.M. Huang, and A. Violante (Eds.), Springer-Verlag, Berlin, Germany, pp. 145–174, 2008; Lagaly, G. et al., Clay mineral-organic interactions, in *Handbook of Clay Science*, 2nd ed. *Developments in Clay Science*, F. Bergaya and G. Lagaly (Eds.), Elsevier, Amsterdam, the Netherlands, Vol. 5A, pp. 435–505, 2013.)

appreciable polymerization is observed in homogeneous solution (Pinnavaia et al. 1979; Pinnavaia 1983). The hectorite-intercalated Rh(NBD)(dppe)$^+$ cation, where NBD is norbornadiene and dppe represents 1,2-bis(diphenylphosphino)ethane, is also effective in catalyzing the hydrogenation of 1,3-butadienes and 1-hexene (Raythatha and Pinnavaia 1981, 1983). Similarly, the [Rh(COD)(PPh$_3$)$_2$]$^+$ and [Ir(COD)(PPh$_3$)$_2$]$^+$ cations, where COD denotes 1,5-cyclooctadiene, intercalated into a synthetic fluorotetrasilicic mica (Miyazaki et al. 1985) and montmorillonite (Chin et al., 1993), respectively, can catalyze the hydrogenation of various olefins. Mitsudome et al. (2008) have made the interesting observation that the interlayer complex of montmorillonite with the chiral organocatalyst (5S)-2,2,3-trimethyl-5-phenylmethyl-4-imidazolidinone cation can effectively catalyze the asymmetric Diels–Alder reaction of cyclohexene with acrolein, affording a 80% yield with a high enantioselectivity for the endo cycloaddition product. Ghiaci et al. (2011) have also noted that a bentonite, intercalated with acidic ionic liquids, such as 3,3′-(butane-1,6-diyl)bis(6-sulfo-1-(4-sulfobenzyl)-1H-benzimidazolium hydrogensulfate, is capable of catalyzing the esterification of fatty acids (e.g., oleic acid) with methanol. As we will see in Chapter 4, clay minerals can host and support many different kinds of catalytically active reagents capable of converting and transforming a wide range of adsorbed organic species.

We hasten to mention that organoclays can become catalytically active after treatment with mineral acids through which protons are introduced into the interlayer space (Pálkova et al. 2013). Thus, TMA$^+$-smectites are inactive in converting α-pinene into camphene (and limonene) but can catalyze the reaction after treatment with HCl (Breen and Moronta 1999; Komadel and

Madejová 2013). Similarly, Moronta et al. (2002) found that TMA$^+$-exchanged montmorillonites, after acid activation, were effective in catalyzing the isomerization (at 300°C) of 1-butene to *cis*- and *trans*-2-butene. Interestingly, the samples that were acid treated before being intercalated with TMA$^+$ were less active, presumably because the interlayer protons could largely be replaced by TMA$^+$ ions. Breen and Moronta (2000) also observed that the efficiency of mixed Al^{3+}/TMA$^+$-montmorillonites in transforming α-pinene markedly declined when more than 20% of the exchange complex was occupied by TMA ions due to restricted access of the reactant to active (Al^{3+}) sites. Likewise, the partial replacement of Fe^{3+} counterions in montmorillonite by choline cations had a negative effect on the conjugate addition of indole with methyl vinyl ketone because of decreased accessibility of interlayer Fe^{3+} acting as Lewis acid sites (Wallis et al. 2011). Interestingly, a bentonite-polystyrene complex that has been treated with chlorosulfonic acid can serve as an efficient catalyst in the Biginelli reaction of benzaldehyde, ethyl acetoacetate, and urea to yield 3,4-dihydropyrimidin-2(1H)-ones under solvent-free conditions (Kalbasi et al. 2012). Similarly, an interlayer complex of tetrabutylammonium periodate with acid-activated (K10) montmorillonite could catalyze the oxidation of benzyl alcohols to the corresponding carbonyl compounds (Venkatachalapathy et al. 1999).

We should also add that organically modified clays, acting as phase-transfer catalysts, can actually assist some organic conversions. In other words, the partial/full occupancy of the interlayer space by organic cations/surfactants has the effect of promoting, rather than suppressing, catalytic activity. This effect may be explained in terms of the provision by the organoclay of a hydrophobic (solid) phase, into which the reagents (usually contained in two immiscible liquid phases) can be transferred and allowed to react (Cornélis and Laszlo 1982; Wibowo et al. 2010). For example, Varma and coworkers (Varma and Kumar 1998; Varma and Naicker 1998; Varma et al. 1999a, 1999b, 1999d) used type 1 organoclays together with surfactant and 18-crown-6 ether intercalated smectites as phase transfer catalysts to synthesize α-azidoketones from α-tosyloxyketones and sodium azide, alkyl azides from alkyl bromides and sodium azide, and α-azidoketones from alkyl halides and α-tosyloxyketones as well as benzyl cyanides, thiocyanides, and alcohols from the corresponding benzyl halides and sodium cyanide, thiocyanide and hydroxide, respectively. Earlier, Cornélis et al. (1983) used a commercial type II organoclay (Tixogel VP) as a phase transfer catalyst in the preparation of formaldehyde acetals by reacting alcohols with dihalomethanes in the presence of 50% aqueous NaOH. Likewise, Yadav and Naik (2000) obtained benzoic anhydride from benzoyl chloride and sodium benzoate under alkaline conditions at 30°C, using a partially HDTMA$^+$-exchanged acid-activated bentonite (Tonsil). More recently, Sedaghat et al. (2014) were able to synthesize 3,4-dihydropyrano[c]chromene derivatives by a one-pot reaction of 4-hydrocoumarin, aromatic aldehydes, and malononitrile in aqueous ethanol in the presence of a similar organoclay material.

Intercalation of tetraphenylphosphonium bromide into K10 montmorillonite gives rise to a material capable of catalyzing the synthesis of stilbenes from aryl halides and styrene (Varma et al. 1999c) as well as the cross-coupling of aryl halides with arylboronic acids (Varma and Naicker 1999). The partial replacement of Fe^{3+} counterions in montmorillonite by HDTMA$^+$ ions also yields a material with improved catalytic activity for the oxidative coupling of 2-naphthol (Wallis et al. 2010). Similarly, the presence of a cationic polymer (*magnafloc*) in the interlayers of acid-activated bentonites has a positive effect on the isomerization of α-pinene (Breen and Watson 1998).

Besides being highly hydrophobic, type II organoclays are more resistant to acid attack than the corresponding TMA$^+$-exchanged counterparts. Madejová et al. (2009), for example, noted that treatment of TMA$^+$-(SAz-1) montmorillonite with 6 M HCl (2 h at 80°C) yielded a protonated silica phase whereas the HDTMA$^+$-exchanged sample remained essentially intact. They suggested that the bulky HDTMA ions could fully cover the external and interlayer silicate surfaces, preventing protons from attacking the layer structure. Similarly, Pálkova et al. (2011) have explained the stability against acid treatment of tetrabutylammonium-exchanged montmorillonite in terms of interlayer occupation by the organic cations. Interestingly, complexes formed by intercalation of TMA into previously acid-activated montmorillonite are much more effective in catalyzing the isomerization of 1-butene (to *cis-* and *trans*-2-butene) than the corresponding TMA$^+$-exchanged samples followed by acid activation (Moronta et al. 2002).

3.5.2 Surface Grafting and Silylation

Although organically grafted clay minerals do not feature prominently as catalysts of organic reactions, we will briefly describe their synthesis and properties for the sake of completeness. Research into the covalent grafting of organic functional groups to clay mineral surfaces (*functionalization*) dates back to the 1940s when Berger (1941) reported to have methylated the silanol groups of H-montmorillonite using diazomethane, as indicated by Equation 3.1:

$$[\text{surface}]-\text{OH} + \text{CH}_2\text{N}_2 \rightarrow [\text{surface}]-\text{OCH}_3 + \text{N}_2\uparrow \quad (3.1)$$

Berger's claim and the results of early attempts at preparing various *organic derivatives* of clay minerals (Deuel et al. 1950; Deuel 1952; Spencer and Gieseking 1952), however, have been called into question (Brown et al. 1952; Greenland and Russell 1955) on the grounds that montmorillonite has a low content of exposed $\equiv$Si–OH groups, and diazomethane is prone to forming a polymer in the interlayer space (Bart et al. 1979). Franco et al. (1978) also made the interesting observation that treatment of NH$_4^+$-bentonite with diazomethane led to the methylation of exchangeable ammonium and the formation of tetramethylammonium ion in the interlayer space.

On the other hand, such minerals as sepiolite and palygorskite, which are relatively rich in structural hydroxyl goups (cf. Figures 1.20 and 1.21), can be readily methylated by diazomethane (Hermosín and Cornejo 1986) or *functionalized* by grafting with organochlorosilanes (Ruiz-Hitzky and Fripiat 1976). Similarly, imogolite lends itself to being silylated with organosilanes as do Laponite and acid-treated chrysotile (Johnson and Pinnavaia 1990, 1991; Herrera et al. 2005; Mendelovici and Frost 2005; Wheeler et al. 2005; Zanzottera et al. 2012).

As is the case with organoclays, there has been a resurgence of interest in synthesizing covalent silane-grafted clay minerals, over the past two decades, due to the requirement for hydrophobized silicate nanoparticles in preparing polymer-clay nanocomposites (Manias et al. 2001; Le Pluart et al. 2002; Kim et al. 2003; Chen et al. 2005; Zhang et al. 2006; Choi et al. 2009; Silva et al. 2011; Ianchis et al. 2012; Theng 2012; Huskić et al. 2013; Tao et al. 2014; Daitx et al. 2015; Kotal and Bhowmick 2015; Romanzini et al. 2015).

More often than not, the grafting process (*silylation*) involves hydrolysis of the organosilane coupling agent, R-SiX$_3$ (Equation 3.2), followed by condensation of the hydrolysis product, R-Si(OH)$_3$, with a surface silanol group (Equation 3.3)

$$\text{R-SiX}_3 + 3\text{H}_2\text{O} \rightarrow \text{R-Si(OH)}_3 + 3\text{HX} \quad (3.2)$$

$$[\text{surface}]-\text{OH} + \text{R-Si(OH)}_3 \rightarrow [\text{surface}]-\text{O}-\text{SiR(OH)}_2 + \text{H}_2\text{O} \quad (3.3)$$

where:
R-SiX$_3$ represents a silane coupling agent
X is an alkoxy group (–OCH$_3$ or –OC$_2$H$_5$)
R denotes a functional group, such as –CH$_2$NH$_2$, γ-methacryloxypropyl, γ-aminopropyl, glycidylpropyl, octyldimethyl, phenyldimehyl, and vinyldimethyl

Among these, γ-aminopropyl triethoxysilane (APTES) with the structural formula of NH$_2$(CH$_2$)$_3$Si(OC$_2$H$_5$)$_3$, is perhaps the most widely used compound for the silylation of different clay mineral species, including *kaolinite* (Tonlé et al. 2007; Avila et al. 2010; Guerra et al. 2012; Yang et al. 2012; Tao et al. 2014), *halloysite* (Yuan et al. 2008; Daitx et al. 2015), *fluorohectorite* (He et al.

2005), *imogolite* (Johnson and Pinnavaia 1990; Qi et al. 2008), *Laponite* (Wheeler et al. 2005; Pereira et al. 2007), *montmorillonite* (He et al. 2005; Shanmugharaj et al. 2006; Piscitelli et al. 2010; Su et al. 2012; Bertuoli et al. 2014), acid-activated (K10 and KSF) montmorillonite (Kannan et al. 1999; Pereira et al. 2007; Varadwaj et al. 2013), Al_{13}-pillared montmorillonite (Qin et al. 2010), *saponite* (Avila et al. 2010; Tao et al. 2016), and *vermiculite* (Alves et al. 2013). We should recall, however, that the grafting of APTES to the inner surface hydroxyls of kaolinite (cf. Figure 1.7) invariably requires prior interlayer expansion of the mineral particles by intercalation of small polar molecules, notably dimethylsulfoxide (de Faria et al. 2012; He et al. 2013; Araújo et al. 2014). Organosilanes other than APTES, and kindred organic compounds have also been used for the silylation of pristine clay minerals, acid-activated smectites, pillared interlayered clays, and organoclays (Herrera et al. 2005; Zhu et al. 2007; Piscitelli et al. 2010; Paul et al. 2011; de Faria et al. 2012; Ianchis et al. 2012; Mansoori et al. 2012; Tajbakhsh et al. 2014; Romanzini et al. 2015).

In a recent review on clay mineral silylation, He et al. (2013) have pointed out that the success of silane grafting is dependent on such factors as surface hydroxyl density, functionality and structure of the silane coupling agent, solvent polarity, and reaction temperature. The use of nonpolar solvents (e.g., toluene, cyclohexane) is conducive to intercalation, hydrolysis, and condensation of the silane molecules into polysiloxane oligomers (Su et al. 2013). On intercalation, these oligomers can form covalent bonds with opposing interlayer surfaces, increasing the basal spacing from 1.48 nm (for the starting Ca^{2+}-montmorillonite) to 2.09 nm (Su et al. 2012). As a result, individual layers in a given particle are *locked* together, as is the case with pillared interlayered clays (cf. Figure 3.2), preventing further interlayer expansion when cetyltrimethylammonium bromide (CTAB) is introduced (Figure 3.7). The *locking* of contiguous silicate layers would also inhibit particle exfoliation during the preparation of polymer-clay nanocomposites, and hence adversely affect the properties of the composite products (Theng 2012). On the other hand, the porous organosilane-clay complex can usefully serve as sorbents and molecular sieves of organic contaminants and pollutants (He et al. 2014).

Fraile et al. (1997) have suggested that silylation (with trimethylchlorosilane) of Zn(II)- and Fe(III)-exchanged K10 montmorillonites had the effect of diminishing, if not eliminating, Lewis acid sites. As a result, the catalytic activity of the silylated sample for the Diels–Alder reaction of (–)-methyl acrylate with cyclopentadiene was greatly reduced. Rehydration of the silylated clay, however, could restore Lewis acidity as well as generate Brønsted acid sites. Moraes et al. (2011) have also found that grafting a bentonite with propyl sulfonic acid groups can enhance the Brønsted acid-catalyzed esterification of 1-propanol with acetic acid. Kannan et al. (1996) used

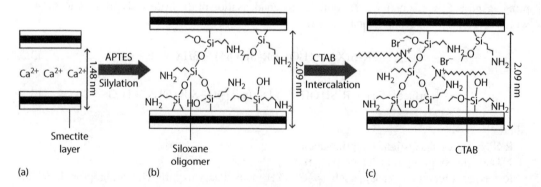

FIGURE 3.7 Diagram showing the silylation of Ca^{2+}-montmorillonite with a basal spacing of 1.48 nm (a) with 3-aminopropyl triethoxysilane (APTES). The intercalated silane and its (siloxane) oligomer are grafted (covalently bonded) to opposing silicate layers, yielding a *locked* complex with a basal spacing of 2.09 nm (b). Because of the locking effect, intercalation of cetyltrimethylammonium bromide (CTAB) does not increase the basal spacing of the silylated material (c). (Adapted from Su, L. et al., *Mater. Chem. Phys.*, 136, 292–295, 2012.)

APTES-grafted montmorillonite to obtain organic sulfides from thiols. Similarly, K10 montmorillonite, grafted with APTES and N-(2-amino ethyl)-3-amino propyl trimethoxysilane, can promote the Knoevenagel condensation of benzaldeyde with diethyl malonate, yielding cinnamic acid as the main product (Varadwaj et al. 2013). Tajbakhsh et al. (2014) also noted that a silylated organoclay, after sulfonation with propane sulfone, was an efficient catalyst for the synthesis of quinoxaline derivatives by condensation of 1,2-diamines with 1,2-diketones in ethanol at room temperature.

Organically functionalized kaolinites have also featured as catalysts. For example, kaolinite grafted with Fe(III) picolinate and Fe(III) dipicolinate can effectively catalyze the epoxidation of *cis*-cyclooctene to *cis*-cyclooctenoxide, and the oxidation of cyclohexane to cyclohexanol and cyclohexanone in the presence of hydrogen peroxide (de Faria et al. 2012). Araújo et al. (2014) used grafted complexes of kaolinite with divalent heavy metal (Co, Mn, Ni)-dipicolinate to catalyze similar organic conversions. Silylation of imogolite with APTES, followed by attachment of osmium tetroxide to the amine groups of the grafted mineral, gives rise to a material capable of catalyzing the achiral hydroxylation of olefins (Qi et al. 2008).

REFERENCES

Abid, M., M. Savolainen, S. Landge et al. 2007. Synthesis of trifluoromethyl-imines by solid acid/superacid catalyzed microwave assisted approach. *Journal of Fluorine Chemistry* 128: 587–594.

Adams, J.M. 1987. Synthetic organic chemistry using pillared, cation-exchanged and acid-treated montmorillonite catalysts—A review. *Applied Clay Science* 2: 309–342.

Adams, J.M., A. Bylina, and S.H. Graham. 1981b. Shape selectivity in low temperature reactions of C_6-alkenes catalysed by a Cu^{2+}-exchanged montmorillonite. *Clay Minerals* 16: 325–332.

Adams, J.M., A. Bylina, and S.H. Graham. 1982a. Conversion of 1-hexene to di-2-hexyl ether using a Cu^{2+}-smectite catalyst. *Journal of Catalysis* 75: 190–195.

Adams, J.M., J.A. Ballantine, S.H. Graham et al. 1979. Selective chemical conversions using sheet silicate intercalates: Low-temperature addition of water to 1-alkenes. *Journal of Catalysis* 58: 238–252.

Adams, J.M., D.E. Clement, and S.H. Graham. 1981a. Low-temperature reaction of alcohols to form t-butyl ethers using clay catalysts. *Journal of Chemical Research (S)* 254–255.

Adams, J.M., D.E. Clement, and S.H. Graham. 1982b. Synthesis of methyl-t-butyl ether from methanol and isobutene using a clay catalyst. *Clays and Clay Minerals* 30: 129–134.

Adams, J.M., K. Martin, R.W. McCabe, and S. Murray. 1986. Methyl *t*-butyl ether (MTBE) production: A comparison of montmorillonite-derived catalysts with an ion-exchange resin. *Clays and Clay Minerals* 34: 597–603.

Adams, J.M., T.V. Clapp, and D.E. Clement. 1983. Catalysis by montmorillonites. *Clay Minerals* 18: 411–421.

Adams, J.M., T.V. Clapp, D.E. Clement, and P.I. Reid. 1984. Reactions of alk-1-enes over ion-exchanged montmorillonites. *Journal of Molecular Catalysis* 27: 179–194.

Aglietti, E.P., J.M. Porto Lopez, and E. Pereira. 1988. Structural alterations in kaolinite by acid treatment. *Applied Clay Science* 3: 155–163.

Ahenach, J., P. Cool, R.E.N. Impens, and E.F. Vansant. 2000. Silica-pillared clay derivatives using aminopropyltriethoxysilane. *Journal of Porous Materials* 7: 475–481.

Ahmed, O.S. and D.K. Dutta. 2005. Friedel–Crafts benzylation of benzene using Zn and Cd ions exchanged clay composites. *Journal of Molecular Catalysis A: General* 229: 227–231.

Ajjou, A.N., D. Harouna, C. Detellier, and H. Alper. 1997. Cation-exchanged montmorillonite catalyzed hydration of styrene derivatives. *Journal of Molecular Catalysis A: Chemical* 126: 55–60.

Akçay, M. 2004. The catalytic acylation of alcohols with acetic acid by using Lewis acid character pillared clays. *Applied Catalysis A: General* 269: 157–160.

Akelah, A., M.A. El-Borai, M.F. Abd El-Aal, A. Rehab, and M.S. Abou-Zeid. 1999. New catalytic systems based on intercalated polymer-montmorillonite supports. *Macromolecular Chemistry and Physics* 200: 955–963.

Akiyama, T., K. Matsuda, and K. Fuchibe. 2002. Montmorillonite K10-catalyzed aza Diels–Alder reaction of Danishefsky's diene with aldimines, generated in situ from aliphatic aldehydes and amine, in aqueous media. *Synlett* 1898–1900.

Akiyama, T., K. Matsuda, and K. Fuchibe. 2005. Montmorillonite K10 catalyzed nucleophilic addition reaction to aldimines in water. *Synthesis* 15: 2606–2608.

Al Bakain, R.Z., Y.S. Al-Degs, A.A. Issa, S. Abdul Jawad, K.A. Abu Safieh, and M.A. Al-Ghouti. 2014. Activation of kaolin with minimum solvent consumption by microwave heating. *Clay Minerals* 49: 667–681.

Alexander, R., T.P. Bastow, S.J. Fisher, and R.I. Kagi. 1995. Geosynthesis of organic compounds: II. Methylation of phenanthrene and alkylphenanthrenes. *Geochimica et Cosmochimica Acta* 59: 4259–4266.

Alves, A.P.M., M.G. Fonseca, and A.F. Wanderley. 2013. Inorganic-organic hybrids originating from organosilane anchored onto leached vermiculite. *Materials Research* 16: 1–8.

Alves, H.J., A.M. da Rocha, M.R. Monteiro et al. 2014. Treatment of clay with KF: New solid catalysts for biodiesel production. *Applied Clay Science* 91–92: 98–104.

An, L-T., F-Q. Ding, J-P. Zou, and X-H. Lu. 2008. Montmorillonite K10: An efficient catalyst for solvent-free synthesis of 1,5-benzodiazepine derivatives. *Synthetic Communications* 38: 1259–1267.

An, L-T., L-L. Zhang, J-P. Zou, and G-L. Zhang. 2010. Montmorillonite K10: Catalyst for Friedel–Crafts alkylation of indoles and pyrrole with nitroalkenes under solventless conditions. *Synthetic Communications* 40: 1978–1984.

Anderson, A.J.C. and P.N. Williams. 1962. *Refining of Oils and Fats for Edible Purposes, 2nd ed.* New York: Pergamon Press.

Araújo, F.R., J.G. Baptista, L. Marçal et al. 2014. Versatile heterogeneous dipicolinate complexes grafted into kaolinite: Catalytic oxidation of hydrocarbons and degradation of dyes. *Catalysis Today* 227: 105–115.

Arienti, A., F. Bigi, R. Maggi et al. 1997. Regioselective electrophilic alkylation of anilines with phenylacetylene in the presence of montmorillonite KSF. *Tetrahedron* 53: 3795–3804.

Atanassova, V., K. Ganno, A. Kulkarni et al. 2011. Mechanistic study on the oxidative coupling of amines to imines on K-10 montmorillonite. *Applied Clay Science* 53: 220–226.

Atkins, M.P., D.J.H. Smith, and D.J. Westlake. 1983. Montmorillonite catalysts for ethylene hydration. *Clay Minerals* 18: 423–429.

Auer, H. and H. Hofmann. 1993. Pillared clays: Characterization of acidity and catalytic properties and comparison with some zeolites. *Applied Catalysis A: General* 97: 23–38.

Avalos, M., R. Babiano, J.L. Bravo et al. 1998. Cycloadditions with clays and alumina without solvents. *Tetrahedron Letters* 39: 2013–2016.

Avila, L.R., E.H. de Faria, K.J. Ciuffi et al. 2010. New synthesis strategies for effective functionalization of kaolinite and saponite with silylating agents. *Journal of Colloid and Interface Science* 341: 186–193.

Awate, S.V., S.B. Waghmode, K.R. Patil, M.S. Agashe, and P.N. Joshi. 2001. Influence of preparation parameters on characteristics of zirconia-pillared clay using ultrasonic technique and its catalytic performance in phenol hydroxylation reaction. *Korean Journal of Chemical Engineering* 18: 257–262.

Azzouz, A., D. Messad, D. Nistor. C. Catrinescu, A. Zvolinschi, and S. Asaftei. 2003. Vapor phase aldol condensation over fully io-exchanged montmorillonite-rich catalysts. *Applied Catalysis A: General* 241: 1–13.

Bagshaw, S.A. and R.P. Cooney. 1993. FTIR surface site analysis of pillared clays using pyridine probe species. *Chemistry of Materials* 5: 1101–1109.

Bahranowski, K., M. Gasior, A. Kielski et al. 1998. Physico-chemical characterization and catalytic properties of copper-doped alumina-pillared montmorillonites. *Clays and Clay Minerals* 46: 98–102.

Bahranowski, K., R. Grabowski, B. Grzybowska et al. 2000. Synthesis and physicochemical properties of vanadium-doped zirconia-pillared montmorillonites in relation to oxidative dehydrogenation of propane. *Topics in Catalysis* 11/12: 255–261.

Bahulayan, D., S.K. Das, and J. Iqbal. 2003. Montmorillonite K10 clay: an efficient catalyst for the one-pot stereoselective synthesis of β-acetamido ketones. *Journal of Organic Chemistry* 68: 5735–5738.

Bahulayan, D., G. Narayan, V. Sreekumar, and M. Lalithambika. 2002. Natural bentonite clay/dilute HNO_3 (40%)—A mild, efficient, and reusable catalyst/reagent system for selective mononitration and benzylic oxidations. *Synthetic Communications* 32: 3565–3574.

Balci, S. 1999. Effect of heating and acid pre-treatment on pore size distribution of sepiolite. *Clay Minerals* 34: 647–655.

Ballantine, J.A. 1986. The reactions in clays and pillared clays. In *Chemical Reactions in Organic and Inorganic Constrained Systems*, (Ed.) R. Setton, pp. 197–212. Dordrecht, the Netherlands: D. Reidel.

Ballantine, J.A., H. Purnell, M. Rayanakorn, J.M. Thomas, and K.J. Williams. 1981a. Chemical conversions using sheet silicates: Novel intermolecular elimination of ammonia from amines. *Journal of the Chemical Society, Chemical Communications* 1: 9–10.

Ballantine, J.A., J.H. Purnell, and J.M. Thomas. 1983. Organic reactions in a clay microenvironment. *Clay Minerals* 18: 347–356.

Ballantine, J.A., M. Davies, I. Patel et al. 1984b. Organic reactions catalysed by sheet silicates: Ether formation by the intermolecular dehydration of alcohols and by addition of alcohols to alkenes. *Journal of Molecular Catalysis* 26: 37–56.

Ballantine, J.A., M. Davies, R.M. O'Neil et al. 1984a. Organic reactions catalysed by sheet silicates: Ester production by the direct addition of carboxylic acids to alkenes. *Journal of Molecular Catalysis* 26: 57–77.

Ballantine, J.A., R.P. Galvin, R.M. O'Neil, H. Purnell, M. Rayanakorn, and J.M. Thomas. 1981b. Chemical conversions using sheet silicates: Novel intermolecular elimination of hydrogen sulphide from thiols. *Journal of the Chemical Society, Chemical Communications* 695–696.

Ballantine, J.A., W. Jones, J.H. Purnell, D.T.B. Tennakoon, and J.M. Thomas. 1985. The influence of interlayer water on clay catalysis. Interlamellar conversion of 2-methylpropene. *Chemistry Letters* 14: 763–766.

Ballini, R., F. Bigi, M.L. Conforti et al. 2000. Multicomponent reactions under clay catalysis. *Catalysis Today* 60: 305–309.

Balogh, M. and P. Laszlo. 1993. *Organic Chemistry Using Clays*. Berlin, Germany: Springer-Verlag.

Bandgar, B.P., S.S. Pandit, and V.S. Sadavarte. 2001. Montmorillonite K-10 catalyzed synthesis of β-keto esters: Condensation of ethyl diazoacetate with aldehydes under mild conditions. *Green Chemistry* 3: 247–249.

Banerjee, B.K. and M.K. Sen. 1974. Application of clay minerals in catalyst industry. *Indian Society of Soil Science Bulletin* 9: 247–253.

Banik, B.K., S. Samajdar, I. Banik, S.S. Ng, and J. Hann. 2003. Montmorillonite impregnated with bismuth nitrate: Microwave-assisted facile nitration of β-lactams. *Heterocycles* 61: 97–100.

Banin, A. and S. Ravikovitch. 1966. Kinetics of reactions in the conversion of Na- or Ca-saturated clay to H-Al clay. *Clays and Clay Minerals* 14: 193–204.

Barnhisel, R.I. and P.M. Bertsch. 1989. Chlorites and hydroxy interlayered vermiculite and smectite. In *Minerals in Soil Environments, 2nd ed*, (Eds.) J.B. Dixon and S.B. Weed, pp. 729–788. Madison, WI: Soil Science Society of America.

Barrer, R.M. 1978. *Zeolites and Clay Minerals as Sorbents and Molecular Sieves*. London, UK: Academic Press.

Barrer, R.M. 1984. Sorption and molecular sieve properties of clays and their importance as catalysts. *Philosophical Transactions of the Royal Society. London, A* 311: 333–352.

Barrer, R.M. 1989. Shape-selective sorbents based on clay minerals: A review. *Clays and Clay Minerals* 37: 385–395.

Barrer, R.M. and MacLeod, D.M. 1955. Activation of montmorillonite by ion exchange and sorption complexes of tetra-alkylammonium montmorillonites. *Transactions of the Faraday Society* 51: 1290–1300.

Barshad, I. 1960. The effect of the total chemical composition and crystal structure of soil minerals on the nature of the exchangeable cations in acidified clays and in natural occurring acid soils. *Transactions 7th International Congress of Soil Science, Madison, WI* 2: 435–444.

Barshad, I. 1969. Preparation of hydrogen-saturated montmorillonites. *Soil Science* 108: 38–42.

Barshad, I., and A.E. Foscolos. 1970. Factors affecting the rate of the interchange reaction of adsorbed H^+ on the 2:1 clay minerals. *Soil Science* 110: 52–60.

Bart, J.C., F. Cariati, L. Erre, C. Gessa, G. Micera, and P. Piu. 1979. Formation of polymeric species in the interlayer of bentonite. *Clays and Clay Minerals* 27: 429–432.

Basavaiah, D., A.J. Rao, and T. Satyanarayana. 2003. Recent advances in the Bayliss-Hillman reaction and applications. *Chemical Reviews* 103: 811–891.

Basavaiah, D., T.K. Bharathi, and V.V.L. Gowriswari. 1987. DABCO-catalyzed coupling of α-keto esters with acrylonitrile and methyl acrylate. *Tetrahedron Letters* 28: 4351–4352.

Békássy, S., J. Farkas, B. Ágai, and F. Figueras. 2000. Selectivity of *C*- versus *O*-acylation of diphenols by clay catalysts. I. Acylation of resorcinol with phenylacetyl chloride. *Topics in Catalysis* 13: 287–290.

Belver, C., C. Breen, F. Clegg, C.E. Fernandes, and M.A. Vicente. 2005. A variable-temperature diffuse reflectance infrared Fourier Transform spectroscopy study of the binding of water and pyridine to the surface of acid-activated metakaolin. *Langmuir* 21: 2129–2136.

Benito, I., A. del Riego, M. Martínez, C. Blanco, C. Pesquera, and F. González. 1999. Toluene methylation on Al_{13}- and $GaAl_{12}$-pillared clay catalysts. *Applied Catalysis A. General* 180: 175–182.

Béres, A., I. Hannus, and I. Kiricsi. 1995. Acid-base testing of catalysts using 1-butene isomerization as test reaction. *Reaction Kinetics and Catalysis Letters* 56: 55–61.

Bergaya, F., A. Aouad, and T. Mandalia. 2006. Pillared clays and clay minerals. In *Handbook of Clay Science. Developments in Clay Science, Vol 1*, (Eds.) F. Bergaya, B.K.G. Theng, and G. Lagaly, pp. 393–421. Amsterdam, the Netherlands: Elsevier.

Berger, G. 1941. The structure of montmorillonite: Preliminary communication on the ability of clays and clay minerals to be methylated. *Chemische Weekblad* 38: 42–43.

Bertuoli, P.T., D. Piazza, L.C. Scienza, and A.J. Zattera. 2014. Preparation and characterization of montmorillonite modified with 3-aminopropyltriethoxysilane. *Applied Clay Science* 87: 46–51.

Besbes, N., D. Hadji, A. Mostefai, A. Rahmouni, M.L. Efrit, and E. Srasra. 2012. Experimental and theoretical investigation on the catalytic acetalization of aldehydes over Tunisian acid activated clay: A mechanistic study. *Journal de la Societé Chimique de Tunisie* 14: 39–46.

Beşün, N., F. Özkan, and G. Gündüz. 2002. Alpha-pinene isomerization on acid-treated clays. *Applied Catalysis A: General* 224: 285–297.

Bharate, J.B., R. Sharma, S. Aravinda et al. 2013b. Montmorillonite clay catalyzed synthesis of functionalized pyrroles through domino four-component coupling of amines, aldehydes, 1,3-dicarbonyl compounds and nitroalkanes. *RSC Advances* 3: 21736–21742.

Bharate, S.B., A.K. Padala, B.A. Dar, R.R. Yadav, and B. Singh. 2013a. Montmorillonite clay Cu(II) catalyzed domino one-pot multicomponent synthesis of 3,5-disubstituted isoxazoles. *Tetrahedron Letters* 54: 3558–3561.

Bhaskar, P.M. and D. Loganathan. 1998. Per-O-acetylation of sugars catalysed by montmorillonite K-10. *Tetrahedron Letters* 39: 2215–2218.

Bieseki, L., H. Treichel, A.S. Araujo, and S.B.C. Pergher. 2013. Porous materials obtained by acid treatment processing followed by pillaring of montmorillonite clays. *Applied Clay Science* 85: 46–52.

Bigi, F., M.L. Conforti, R. Maggi, A. Mazzacani, and G. Sartori. 2001. Montmorillonite KSF-catalysed regioselective *trans-tert*-butylation of *tert*-bytylphenols. *Tetrahedron Letters* 42: 6543–6545.

Bigi, F., M.L. Conforti, R. Maggi, and G. Sartori. 2000. Trialkylamine controlled phenol-formaldehyde reaction over clay catalysts: Selective and environmentally benign synthesis of salicylic aldehydes. *Tetrahedron* 56: 2709–2712.

Bigi, F., S. Carloni, B. Frullanti, R. Maggi, and G. Sartori. 1999. A revision of the Biginelli reaction under solid acid catalysis. Solvent-free synthesis of dihydropyrimidines over montmorillonite KSF. *Tetrahedron Letters* 40: 3465–3468.

Binitha, N.N. and S. Sugunan. 2006. Preparation, characterization and catalytic activity of titania pillared montmorillonite clays. *Microporous and Mesoporous Materials* 93: 82–89.

Binitha, N.N. and S. Sugunan. 2008. Shape selective toluene methylation over chromia pillared montmorillonites. *Catalysis Communications* 9: 2376–2380.

Biradar, J.S. and B. Sharanbasappa. 2011. MK-10 clay-catalyzed synthesis of 2-(2′,5′-disubstituted-1′H-indol-3′-yl)-1H-benzo[d]imidazoles under conventional and microwave irradiation. *Synthetic Communications* 41: 885–890.

Biró, K., S. Békássy, B. Ágai, and F. Figueras. 2000. Heterogeneous catalysis for the acetylation of benzo crown ethers. *Journal of Molecular Catalysis A: Chemical* 151: 179–184.

Bodman, S.D., W.R. McWhinnie, V. Begon et al. 2002. Metal-ion pillared clays as hydrocracking catalysts (I): Catalyst preparation and assessment of performance at short contact times. *Fuel* 81: 449–459.

Bolognini, M., F. Cavai, M. Cimini et al. 2004. An environmentally friendly synthesis of 1,4-dihydroxybenzophenone by the single-step O-mono-benzylation of 1,3-dihydrobenzene (resorcinol) and Fries rearrangement of intermediate resorcinol monobenzoate: The activity of acid-treated montmorillonite clay catalysts. *Comptes Rendus Chimie* 7: 143–150.

Bonelli, B., I. Bottero, N. Ballarini, S. Passeri, F. Cavani, and E. Garrone. 2009. IR spectroscopic and catalytic characterization of the acidity of imogolite-based systems. *Journal of Catalysis* 264: 15–30.

Borkin, D., A. Carlson, and B. Török. 2010. K-10-catalyzed highly diastereoselective synthesis of aziridines. *Synlett* 745–748.

Bouchenafa-Saïb, N., R. Issaadi, and P. Grange. 2004. Hydroconversion of *n*-heptane: A comparative srudy of catalytic properties of Pd/sulfated Zr-pillared montmorillonite, Pd/sulfated zirconia and Pd/γ-alumina. *Applied Catalysis A; General* 259: 9–15.

Boudriche, L., R. Calvet, B. Hamdi, and H. Balard. 2011. Effect of acid treatment on surface properties evolution of attapulgite clay: An application of inverse gas chromatography. *Colloids and Surfaces A: Physicochemical and Engineering Aspects* 392: 45–54.

Bovey, J. and W. Jones. 1995. Characterisation of Al-pillared acid activated clay catalysts. *Journal of Materials Chemistry* 5: 2027–2035.

Bradley, S.M. and R.A. Kydd. 1993a. Ga_{13}, Al_{13}, $GaAl_{12}$, and chromium-pillared montmorillonites: Acidity and reactivity for cumene conversion. *Journal of Catalysis* 141: 239–249.

Bradley, S.M. and R.A. Kydd. 1993b. A comparison of the catalytic activities of Ga_{13}-, Al_{13}-, $GaAl_{12}$-, and chromium-pillared interlayered clay minerals and Ga-H-ZSM-5 zeolite in the dehydrocyclodimerization of propane. *Journal of Catalysis* 142: 448–454.

Brandt, K.B. and R.A. Kydd. 1997a. Characterization of synthetic microporous pillared beidellite of high thermal stability. *Chemistry of Materials* 9: 567–572.

Brandt, K.B. and R.A. Kydd. 1997b. The effect of framework substitution and pillar composition on the cracking activities of montmorillonite and beidellite. *Applied Catalysis A: General* 165: 327–333.

Breck, D.W. 1980. Potential uses of natural and synthetic zeolites in industry. In *The Properties and Applications of Zeolites*, (Ed.) R.P. Townsend, pp. 391–422. London, UK: The Chemical Society.

Breen, C. and A. Moronta. 1999. Influence of layer charge on the catalytic activity of mildly acid-activated tetramethylammonium-exchanged bentonites. *Journal of Physical Chemistry B* 103: 5675–5680.

Breen, C. and A. Moronta. 2000. Characterization and catalytic activity of aluminum- and aluminum/tetramethylammonium-exchanged bentonites. *Journal of Physical Chemistry B* 104: 2702–2708.

Breen, C. and R. Watson. 1998. Acid-activated organoclays: Preparation, characterisation and catalytic activity of polycation-treated bentonites. *Applied Clay Science* 12: 479–494.

Breen, C., F.D. Zahoor, J. Madejová, and P. Komadel. 1997. Characterisation and catalytic activity of acid treated, size fractionated smectites. *Journal of Physical Chemistry B* 101: 5324–5331.

Breen, C., J. Madejová, and P. Komadel. 1995a. Correlation of catalytic activity with infra-red, ^{29}Si MAS NMR and acidity data for HCl-treated fine fractions of montmorillonites. *Applied Clay Science* 10: 219–230.

Breen, C., J. Madejová, and P. Komadel. 1995b. Characterisation of moderately acid-treated, size-fractionated montmorillonites using IR and MAS NMR spectroscopy and thermal analysis. *Journal of Materials Chemistry* 5: 469–474.

Brigatti, M.F., E. Galán, and B.K.G. Theng. 2013. Structure and mineralogy of clay minerals. In *Handbook of Clay Science, 2nd ed. Developments in Clay Science, Vol. 5A*. (Eds.) F. Bergaya and G. Lagaly, pp. 21–81. Amsterdam, the Netherlands: Elsevier.

Brindley, G.W. and R.E. Sempels. 1977. Preparation and properties of some hydroxyl-aluminium beidellites. *Clay Minerals* 12: 229–237.

Brochette, S., G. Descotes, A. Bouchu, Y. Queneau, N. Monnier, and C. Pétrier. 1997. Effect of ultrasound on KSF/0 mediated glycosylations. *Journal of Molecular Catalysis A: Chemical* 123: 123–130.

Brown, D.R. 1994. Clays as catalyst and reagent supports. *Geologica Carpathica, Series Clays* 45: 45–56.

Brown, D.R. and C.N. Rhodes. 1997. Brønsted and Lewis acid catalysis with ion-exchanged clays. *Catalysis Letters* 45: 35–40.

Brown, G., R. Greene-Kelly, and K. Norrish. 1952. Organic derivatives of montmorillonite. *Nature* 169: 756–757.

Brückman, K., and J. Haber. 1979. Active centers in fluorinated montmorillonites. *Reaction Kinetics and Catalysis Letters* 10: 173–177.

Brückman, K., J. Fijal, J. Haber et al. 1976. Influence of different activation methods on the catalytic properties of montmorillonite. *Mineralogia Polonica* 7: 5–13.

Burch, R. and C.I. Warburton. 1986. Zr-containing pillared interlayer clays II. Catalytic activity for the conversion of methanol into hydrocarbons. *Journal of Catalysis* 97: 511–515.

Butruille, J-R. and T.J. Pinnavaia. 1992. Propene alkylation of liquid phase biphenyl catalyzed by alumina pillared clay catalysts. *Catalysis Today* 14: 141–154.

Bylina, A., J.M. Adams, S.H. Graham, and J.M. Thomas. 1980. Chemical conversions using sheet silicates: Simple method for producing methyl t-butyl ether. *Journal of the Chemical Society, Chemical Communications* 21: 1003–1004.

Cabrera, A., D. Vázquez, L. Velasco, M. Salmón, and J.L. Arias. 1992. 1,3-Dioxolane formation with a montmorillonite-type clay catalyst. *Journal of Molecular Catalysis* 75: 101–107.

Caddick, S. 1995. Microwave assisted organic reactions. *Tetrahedron* 51: 10403–10432.

Campanati. M., M. Casagrande, I. Fagiolino et al. 2002. Mild hydrogenation of quinoline 2. A novel Rh-containing pillared layered clay catalyst. *Journal of Molecular Catalysis A: Chemical* 184: 267–272.

Campos, A., B. Gagea, S. Moreno, P. Jacobs, and R. Molina. 2008. Decane hydroconversion with Al-Zr, Al-Hf, Al-Ce-pillared vermiculites. *Applied Catalysis A: General* 345: 112–118.

Carrado, K.A., R. Hayatsu, R.E. Botto, and R.E. Winans. 1990. Reactivity of anisoles on clay and pillared clay surfaces. *Clays and Clay Minerals* 38: 250–256.

Cases, J.M., I. Berend, G. Besson et al. 1992. Mechanism of adsorption and desorption of water vapor by homoionic montmorillonite. 1. The sodium-exchanged form. *Langmuir* 8: 2730–2739.

Castanheiro, R.A.P., M.M.M. Pinto, S.M.M. Cravo, D.C.G.A. Pinto, A.M.S. Silva, and A. Kijjoa. 2009. Improved methodologies for synthesis of prenylated xanthones by microwave irradiation and combination of heterogeneous catalysis (K10 clay) with microwave irradiation. *Tetrahedron* 65: 3848–3857.

Caterina, M.C., M.V. Corona, I. Perillo, and A. Salerno. 2009. An efficient synthesis of 1-acyl-3-arylimidazolines catalyzed by montmorillonite K-10 clay under microwave irradiation. *Heterocycles* 78: 771–781.

Cativiela, C., J.I. García, M. García-Matres et al. 1995. Clay-catalyzed Friedel–Crafts alkylation of anisole with dienes. *Applied Catalysis A: General* 123: 273–287.

Centi, G. and S. Perathoner. 2008. Catalysis by layered materials: A review. *Microporous and Mesoporous Materials* 107: 3–15.

Chaari, A., S.B. Nejl, and M.H. Frikha. 2017. Fatty acid esterification with polyols over acidic montmorillonite. *Journal of Oleo Science* 66: 455–461.

Chakrabarti, A. and M.M. Sharma. 1992a. Alkylation of phenol with cyclohexene catalysed by cationic ion-exchange resins and acid-treated clay: O- versus C-alkylation. *Reactive Polymers* 17: 331–340.

Chakrabarti, A. and M.M. Sharma. 1992b. Cyclohexanol from cyclohexene via cyclohexyl acetate: Catalysis by ion-exchange resin and acid-treated clay. *Reactive Polymers* 18: 107–115.

Chakrabarti, A. and M.M. Sharma. 1992c. Anhydrous esterification of myristic acid with propylene: Ion exchange resin and acid-treated clay as catalyst. *Journal of the American Oil Chemists Society* 69: 1251–1253.

Chakrabarty, M. and S. Sarkar. 2002. Novel clay-mediated, tandem addition-elimination-(Michael) addition reactions of indoles with 3-formylindole: An eco-friendly route to symmetrical and unsymmetrical triidolylmethanes. *Tetrahedron Letters* 43: 1351–1353.

Chakrabarty, M., N. Ghosh, R. Basak, and Y. Harigaya. 2002. Dry reaction of indoles with carbonyl compounds on montmorillonite K10 clay: A mild, expedient synthesis of diindolylalkanes and vibrindole A. *Tetrahedron Letters* 43: 4075–4078.

Chalais, S., P. Laszlo, and A. Mathy. 1986. Catalysis of the cyclohexadienone-phenol rearrangement by a Lewis-acidic clay system. *Tetrahedron Letters* 27: 2627–2630.

Chandrasekhar, S., Ch. Narsihmulu, B. Saritha, and S.S. Sultana. 2004. Poly(ethyleneglycol) (PEG): A rapid and recyclable reaction medium for the DABCO-catalyzed Baylis-Hillman reaction. *Tetrahedron Letters* 45: 5865–5867.

Chankeshwara, S.V. and A.K. Chakraborti. 2006. Montmorillonite K10 and montmorillonite KSF as new and reusable catalysts for conversion of amines to *N-tert*-butylcarbamates. *Journal of Molecular Catalysis A: Chemical* 253: 198–202.

Chatel, G., L. Novikova, and S. Petit. 2016. How efficiently combine sonochemistry and clay science? *Applied Clay Science* 119: 193–201.

Che, C., T.D. Glotch, D.L. Bish, J.R. Michalski, and W. Xu. 2011. Spectroscopic study of the dehydration and/or dehydroxylation of phyllosilicate and zeolite minerals. *Journal of Geophysical Research: Planets* 116: 1–23.

Chen, D., C. Xu, J. Deng et al. 2014. Proton-exchanged montmorillonite-mediated reactions of methoxybenzyl esters and ethers. *Tetrahedron* 70: 1975–1983.

Chen, G-F., H-M. Jia, L-Y. Zhang, B-H. Chen, and J-T. Li. 2013. An efficient synthesis of 2-substituted benzothiazoles in the presence of $FeCl_3$/montmorillonite K-10 under microwave irradiation. *Ultrasonics Sonochemistry* 20: 627–632.

Chen, G-X., J.B. Choi, and J.S. Yoon. 2005. The role of functional group on the exfoliation of clay in poly (L-lactide). *Macromolecular Rapid Communications* 26: 183–187.

Chen, W-Y. and X-S Li. 2009. Montmorillonite K10-catalyzed Michael addition of indoles to nitroolefins under solvent-free conditions. *Synthetic Communications* 39: 2014–2021.

Chermahini, A.N., A. Teimouri, F. Momenbeik et al. 2010. Clay-catalyzed synthesis of 5-substituted 1-*H*-tetrazoles. *Journal of Heterocyclic Chemistry* 47: 913–922.

Chevalier, S., R. Franck, H. Suquet, J-F. Lambert, and D. Barthomeuf. 1994b. Al-pillared saponites. Part 1—IR studies. *Journal of the Chemical Society, Faraday Transactions* 90: 667–674.

Chevalier, S., R. Franck, J-F. Lambert, D. Barthomeuf, and H. Suquet. 1994a. Characterization of the porous structure and cracking activity of Al-pillared saponites. *Applied Catalysis A: General* 110: 153–165.

Chiche, B., A. Finiels, C. Gauthier, P. Geneste, J. Graille, and D. Pioch. 1987. Acylation over cation-exchanged montmorillonite. *Journal of Molecular Catalysis* 42: 229–235.

Chin, C.S., B. Lee, I. Yoo, and T-H. Kwon. 1993. Synthesis, reactions and catalytic activities of iridium complexes intercalated into montmorillonite. *Journal of the Chemical Society, Dalton Transactions* 581–585.

Chitnis, S.R. and M.M. Sharma. 1996. Alkylation of diphenylamine with methylstyrene and diisobutylene using acid-treated clay catalysts. *Journal of Catalysis* 160: 84–94.

Chitnis, S.R. and M.M. Sharma. 1997a. Industrial applications of acid-treated clays as catalysts. *Reactive and Functional Polymers* 32: 93–115.

Chitnis, S.R. and M.M. Sharma. 1997b. Alkylation of aniline with α-methylstyrene and separation of close boiling aromatic amines through reaction with α-methylstyrene, using acid-treated clay catalysts. *Reactive & Functional Polymers* 33: 1–12.

Chmielarz, L., B. Gil, P. Kuśtrowski, Z. Piwowarska, B. Dudek, and M. Michalik. 2009. Montmorillonite-based porous heterostructures (PCHs) intercalated with silica-titania pillars—Synthesis and characterization. *Journal of Solid State Chemistry* 182: 1094–1104.

Choi, Y.Y., S.H. Lee, and S.H. Ryu. 2009. Effect of silane functionalization of montmorillonite on epoxy/montmorillonite nanocomposite. *Polymer Bulletin* 63: 47–55.

Chopade, S.P. and M.M. Sharma. 1996. Alkylation of methyl salicylate with isoamylene and isobutylene: Ion-exchange resins versus acid-treated clay catalysts. *Reactive & Functional Polymers* 28: 253–261.

Choudary, B.M., A.D. Prasad, and V.L.K. Valli. 1990b. Shape-selective isomerisation of α-acetylenic alcohols to α,β-ethylenic carbonyl compounds by vanadium-pillared montmorillonite catalyst. *Tetrahedron Letters* 31: 7521–7522.

Choudary, B.M., A. Durgaprasad, and V.L.K. Valli. 1990a. Selective oxidation of alcohols by chromia-pillared montmorillonite catalyst. *Tetrahedron Letters* 31: 5785–5788.

Choudary, B.M., B.P.C. Rao, N.S. Chowdari, and M.L. Kantam. 2002. Fe^{3+}-montmorillonite: A bifunctional catalyst for one pot Friedel–Crafts alkylation of arenes with alcohols. *Catalysis Communications* 3: 363–367.

Choudary, B.M., M. Sateesh, M.L. Kantam, and K.V.R. Prasad. 1998. Acylation of aromatic ethers with acid anhydrides in the presence of cation-exchanged clays. *Applied Catalysis A: General* 171: 155–160.

Choudary, B.M., M.L. Kantam, M. Sateesh, K.K. Rao, and P.L. Santhi. 1997. Iron pillared clays—Efficient catalysts for Friedel–Crafts reactions. *Applied Catalysis A: General* 149: 257–264.

Choudary, B.M., N.S. Chowdari, and M.L. Kantam. 2000. Montmorillonite clay catalyzed tosylation of alcohols and selective monotosylation of diols with *p*-toluenesulfonic acid: An enviro-economic route. *Tetrahedron* 56: 7291–7298.

Choudary, B.M., V. Bhaskar, M.L. Kantam, K.K. Rao, and K.V. Raghavan. 2001. Acylation of amines with carboxylic acids: The atom economic protocal catalysed by Fe(III)-montmorillonite. *Catalysis Letters* 74: 207–211.

Christidis, G.E., P.W. Scott, and A.C. Dunham. 1997. Acid activation and bleaching capacity of bentonites from the islands of Milos and Chios, Aegean, Greece. *Applied Clay Science* 12: 329–347.

Chtourou, M., R. Abdelhédi, M. H. Frikha, and M. Trabelsi. 2010. Solvent free synthesis of 1,3-diaryl-2-propenones catalysed by commercial acid-clays under ultrasound irradiation. *Ultrasonics Sonochemistry* 17: 246–249.

Clark, J.H. and D.J. Macquarrie. 1997. Heterogeneous catalysis in liquid phase transformations of importance in the industrial preparation of fine chemicals. *Organic Process Research & Development* 1: 149–162.

Clark, J.H., A.J. Butterworth, S.J. Tavener et al. 1997. Environmentally friendly chemistry using supported reagent catalysts: Chemically-modified mesoporous solid catalysts. *Journal of Chemical Technology and Biotechnology* 68: 367–376.

Clark, P.D., A. Kirk, and J.G.K. Yee. 1995. An improved synthesis of benzo[*b*]thiophene and its derivatives using modified montmorillonite catalysts. *Journal of Organic Chemistry* 60: 1936–1938.

Coleman, N.T. and D. Craig. 1961. The spontaneous alteration of hydrogen clay. *Soil Science* 91: 14–18.

Cool, P. and E.F. Vansant. 2004. Pillared clays and porous clay heterostructures. In *Handbook of Layered Materials*, (Eds.) S.M. Auerbach, K.A. Carrado, and P.K. Dutta, 261–311. New York: Marcel Dekker.

Cornélis, A. and P. Laszlo. 1982. Clay-supported reagents; II. Quaternary ammonium-exchanged montmorillonite as catalyst in the phase-transfer preparation of symmetrical formaldehyde acetals. *Synthesis* 2: 162–163.

Cornélis, A., A. Gerstmans, P. Laszlo, A. Mathy, and I. Zieba. 1990. Friedel–Crafts acylations with modified clays as catalysts. *Catalysis Letters* 6: 103–110.

Cornélis, A., P. Laszlo, and P. Pennetreau. 1983. Some organic syntheses with clay-supported reagents. *Clay Minerals* 18: 437–445.

Cramarossa, M.R., L. Forti, and F. Ghelfi. 1997. Acetals by AlFe-pillared montmorillonite catalysis. *Tetrahedron* 53: 15889–15894.

Cruz-Almanza, R., F.J. Pérez-Flores, and M. Avila. 1990. Deprotection of tetrahydropyranyl ethers with a Mexican bentonite, synthesis of farnesylhydroquinone. *Synthetic Communications* 20: 1125–1131.

Cseri, T., S. Békássy, F. Figueras, and S. Rizner. 1995a. Benzylation of aromatics on ion-exchanged clays. *Journal of Molecular Catalysis A: Chemical* 98: 101–107.

Cseri, T., S. Békássy, F. Figueras, E. Cseke, L-C. de Menorval, and R. Dutartre. 1995b. Characterization of clay-based K catalysts and their application in Friedel–Crafts alkylation of aromatics. *Applied Catalysis A: General* 132: 141–155.

Cseri, T., S. Békássy, Z. Bódás, B. Ágai, and F. Figueras. 1996. Acetylation of B15C5 crown ether on Cu modified clay catalysts. *Tetrahedron Letters* 37: 1473–1476.

Csiba, M., J. Cleophax, A. Loupy, J. Malthete, and S.D. Gero. 1993. Liquid crystalline 5,6-O-acetals of L-galactono-1,4-lactone prepared by a microwave irradiation on montmorillonite. *Tetrahedron Letters* 34: 1787–1790.

da Silva, C.R.B., V.L.C. Gonçalves, E.R. Lachter, and C.J.A. Mota. 2009. Etherification of glycerol with benzyl alcohol catalysed by solid acids. *Journal of the Brazilian Chemical Society* 20: 1–5.

Dabbagh, H.A., A. Teimouri, and A.N. Chermahini. 2007. Green and efficient diazotization and diazo coupling reactions on clays. *Dyes and Pigments* 73: 239–244.

Daitx, T.S., L.N. Carli, J.S. Crespo, and R.S. Mauler. 2015. Effects of the organic modification of different clay minerals and their application in biodegradable polymer nanocomposites of PHBV. *Applied Clay Science* 115: 157–164.

Dandia, A., D.S. Bhati, A.K. Jain, and G.N. Sharma. 2011. Ultrasound promoted clay catalyzed efficient and one pot synthesis of substituted oxindoles. *Ultrasonics Sonochemistry* 18: 1143–1147.

Dandia, A., H. Sachdeva, and R.D. Singh. 2000. Montmorillonite catalysed synthesis of novel spiro[3H-indole-3,3'-[3H-1,2,4]triazol]-2(1H) ones in dry media under microwave irradiation. *Journal of Chemical Research (S)* 6: 272–275.

Dar, B.A., A. Bhowmik, A. Sharma et al. 2013. Ultrasound promoted efficient and green protocol for the expeditious synthesis of 1,4 disubstituted 1,2,3-triazoles using Cu(II) doped clay as catalyst. *Applied Clay Science* 80–81: 351–357.

Dar, B.A., N. Pandey, S. Singh et al. 2014. Heterogeneous reusable catalyst, ultrasound energy, and no solvent: A quick and green recipe for one-pot synthesis of β-phosphonomalononitriles at room temperature. *Tetrahedron Letters* 55: 623–628.

Dar, B.A., N. Pandey, S. Singh, P. Kumar, M. Farooqui, and B. Singh. 2015. Solvent-free, scalable and expeditious synthesis of benzanilides under microwave irradiation using clay doped with palladium nanoparticles as a recyclable and efficient catalyst. *Green Chemistry Letters and Reviews* 8: 1–8.

Dar, B.A., V. Shrivastava, A. Bhowmik, M.A. Wagay, and B. Singh. 2015. An expeditious N,N-dibenzylation of anilines under ultrasonic irradiation conditions using low loading Cu(II)-clay heterogeneous catalyst. *Tetrahedron Letters* 56: 136–141.

Das, B., P. Madhusudhan, and B. Venkataiah. 2000. Clay catalysed convenient isomerization of natural furofuran lignans under microwave irradiation. *Synthetic Communications* 30: 4001–4006.

Dasgupta, S. and B. Török. 2008. Application of clay catalysts in organic synthesis. A review. *Organic Preparations and Procedures International* 40: 1–65.

Dastan, A., A. Kulkarni, and B. Török. 2012. Environmentally benign synthesis of heterocyclic compounds by combined microwave-assisted heterogeneous catalytic approaches. *Green Chemsitry* 14: 17–37.

Dauben, W.G., J.M. Cogen, and V. Behar. 1990. Clay catalyzed rearrangement of substituted allyl phenyl ethers: Synthesis of ortho-allyl phenols, chromans and coumarans. *Tetrahedron Letters* 31: 3241–3244.

Davis, L.E., R. Turner, and L.D. Whittig. 1962. Some studies of the auto-transformation of H-bentonite to Al-bentonite. *Soil Science Society of America Proceedings* 26: 441–443.

Davoodnia, A., R. Moloudi, N. Tavakoli-Hoseini, and M. Shaker. 2012. Montmorillonite K10 as an efficient and reusable catalyst for the synthesis of substituted pyrazolo[3,4-d]pyrimidin-4-ones under solvent-free conditions. *Asian Journal of Chemistry* 24: 2195–2198.

de Faria, E.H., G.P. Ricci, L. Marçal et al. 2012. Green and selective oxidation reactions catalyzed by kaolinite covalently grafted with Fe(III) pyridine-carboxylate complexes. *Catalysis Today* 187: 135–149.

De León, M.A., C. De Los Santos, L. Latrónica et al. 2014. High catalytic activity at low temperature in oxidative dehydrogenation of propane with Cr-Al pillared clay. *Chemical Engineering Journal* 241: 336–343.

de Oliveira, A.D.N., L.R. da Silva Costa, L.H. de Oliveira Pires et al. 2013. Microwave-assisted preparation of a new esterification catalyst from wasted *flint* kaolin. *Fuel* 103: 626–631.

de Oliveira, R.N., A.D.L. Xavier, B.M. Guimarães et al. 2014. Combining clays and ultrasound irradiation for an O-acetylation reaction of N-glucopyranosyl and other molecules. *Journal of the Chilean Chemical Society* 59: 2610–2614.

de Oliveira, R.N., J.R. de Freitas Filho, and R.M. Srivastava. 2002. Microwave-induced synthesis of 2,3-unsaturated O-glycosides under solvent-free conditions. *Tetrahedron Letters* 43: 2141–2143.

de Paiva, L.B., A.R. Morales, and F.R. Valenzuela Díaz. 2008. Organoclays: Properties, preparation and applications. *Applied Clay Science* 42: 8–24.

De Paolis, O., L. Teixeira, and B. Török. 2009. Synthesis of quinolines by a solid acid-catalyzed microwave-assisted domino cyclization-aromatization approach. *Tetrahedron Letters* 50: 2939–2942.

de Souza, R.O.M.A., V.L.P. Pereira, P.M. Esteves, and M.L.A.A. Vasconcellos. 2008. The Morita-Baylis-Hillman reaction in aqueous-organic solvent system. *Tetrahedron Letters* 49: 5902–5905.

De Stefanis, A. and A.A.G. Tomlinson. 2006. Towards designing pillared clays for catalysis. *Catalysis Today* 114: 126–141.

De Stefanis, A., G. Perez, and A.A.G. Tomlinson, 2003. PLS vs zeolites as sorbents and catalysts, Part 7. The role of oxide pillars and aluminosilicate sheets in alcohol dehydration on PILCs. *Reaction Kinetics and Catalysis Letters* 79: 227–232.

De Stefanis, A., P. Cafarelli, F. Gallese, E. Borsella, A. Nana, and G. Perez. 2013. Catalytic pyrolysis of polyethylene: A comparison between pillared and restructured clays. *Journal of Analytical and Applied Pyrolysis* 104: 479–484.

Dékány, I., L. Turi, A. Fonseca, and J.B. Nagy. 1999. The structure of acid treated sepiolite: Small-angle X-ray scattering and multi MAS-NMR investigations. *Applied Clay Science* 14: 141–160.

Del Castillo, H.L. and P. Grange. 1993. Preparation and catalytic activity of titanium pillared montmorillonite. *Applied Catalysis A: General* 103: 23–34.

Del Castillo, H.L., A. Gil, and P. Grange. 1996. Hydroxylation of phenol on titanium pillared montmorillonite. *Clays and Clay Minerals* 44: 706–709.

del Rey-Perez-Caballero, F.J., M.L. Sanchez-Henao, and G. Poncelet. 2000. Hydroisomerization of octane on Pt/Al-pillared vermiculite and phlogopite, and comparison with zeolites. *Studies in Surface Science and Catalysis* 130: 2417–2422.

Delaude, L., P. Laszlo, and P. Lehance. 1995. Oxidation of organic substrates with potassium ferrate (VI) in the presence of the K10 montmorillonite. *Tetrahedron Letters* 36: 8505–8508.

Delgado, F., J. Tamariz, G. Zepeda, M. Landa, R. Miranda, and J. García. 1995. Knoevenagel condensation catalyzed by a Mexican bentonite using infrared irradiation. *Synthetic Communications* 25: 753–759.

Delvaux, L. and H. Laudelout. 1964. Catalyse hétérogène de la decomposition de l'ester diazoacetique en suspension aqueuse d'argile hydrogène. *Journal de Chimie Physique* 61: 1153–1161.

Deshmukh, A.R.A.S., V.K. Gumaste, and B.M. Bhawal. 1995. Montmorillonite, an efficient catalyst for the preparation of dialkoxymethanes. *Synthetic Communications* 25: 3939–3944.

Deuel, H. 1952. Organic derivatives of clay minerals. *Clay Minerals Bulletin* 1: 205–214.

Deuel, H., G. Huber, and R. Iberg. 1950. Organische derivate von tonmineralien. *Helvetica Chimica Acta* 33: 1229–1232.

Dewan, S.K., R. Singh, and A. Kumar. 2004. One-pot synthesis of nitriles from aldehydes and hydroxylamine hydrochloride over silica gel, montmorillonite K-10, and KSF catalysts in dry media under microwave irradiation. *Synthetic Communications* 34: 2025–2029.

Dhakshinamoorthy, A. and K. Pitchumani. 2005. Facile clay-induced Fischer indole synthesis: A new approach to synthesis of 1,2,3,4-tetrahydrocarbazole and indoles. *Applied Catalysis A: General* 292: 305–311.

Dhakshinamoorthy, A. and K. Pitchumani. 2006. Clay-anchored non-heme iron-salen complex catalyzed cleavage of C=C bond in aqueous medium. *Tetrahedron* 62: 9911–9918.

Ding, X., T. An, G. Li et al. 2008. Preparation and characterization of hydrophobic TiO_2 pillared clay: The effect of acid hydrolysis catalyst and doped Pt amount on photocatalytic activity. *Journal of Colloid and Interface Science* 320: 501–507.

Ding, Z., J.T. Kloprogge, R.L. Frost, G.Q. Lu, and H.Y. Zhu. 2001. Porous clays and pillared clays-based catalysts. Part 2. A review of the catalytic and molecular sieve applications. *Journal of Porous Materials* 8: 273–293.

Dintzner, M.R., A.J. Little, M. Pacilli et al. 2007. Montmorillonite clay-catalyzed hetero-Diels–Alder reaction of 2,3,-dimethyl-1,3-butadiene with benzaldehydes. *Tetrahedron Letters* 48: 1577–1579.

Dintzner, M.R., J.J. Maresh, C.R. Kinzie, A.F. Arena, and T. Speltz. 2012. A research-based undergraduate organic laboratory project: Investigation of a one-pot, multicomponent, environmentally friendly Prins–Friedel–Crafts-type reaction. *Journal of Chemical Education* 89: 265–267.

Dintzner, M.R., K.M. Morse, K.M. McClelland, and D.M. Coligado. 2004. Investigation of the montmorillonite clay-catalyzed [1,3] shift reaction of 3-methyl-2-butenyl phenyl ether. *Tetrahedron Letters* 45: 79–81.

Dintzner, M.R., T.W. Lyons, M.H. Akroush, P. Wucka, and A.T. Rzepka. 2005. A quick, clean and green synthesis of methylenedioxyprecocene and other chromenes over basic montmorillonite K10 clay. *Synlett* 5: 785–788.

Dintzner, M.R., Y.A. Mondjinou, and B. Unger. 2009. Montmorillonite K10 clay-catalyzed synthesis of homoallylic silyl ethers: An efficient and environmentally friendly Hosomi-Sakurai reaction. *Tetrahedron Letters* 50: 6630–6641.

Dintzner, M.R., Y.A. Mondjinou, and D.J. Pileggi. 2010. Montmorillonite clay catalyzed cyclotrimerization and oxidation of aliphatic aldehydes, *Tetrahedron Letters* 51: 826–827.

do Nascimento, L.A.S., L.M.Z. Tito, R.S. Angélica, C.E.F. da Costa, J.R. Zamian, and G.N. da Rocha Filho. 2011b. Esterification of oleic acid over solid acid catalysts prepared from Amazon *flint* kaolin. *Applied Catalysis B: Environmental* 101: 495–503.

do Nascimento, L.A.S., R.S. Angélica, C.E.F. da Costa, J.R. Zamian, and G.N. da Rocha Filho. 2011a. Comparative study between catalysts for esterification prepared from kaolins. *Applied Clay Science* 51: 267–273.

Doblin, C., J.F. Mathews, and T.W. Turney. 1991. Hydrocracking and isomerization of *n*-octane and 2,2,4-trimethylpentane over a platinum/alumina-pillared clay. *Applied Catalysis* 70: 197–212.

Doblin, C., J.F. Mathews, and T.W. Turney. 1994. Shape selective cracking of *n*-octane and 2,2,4-trimethylpentane over an alumina pillared clay. *Catalysis Letters* 23: 151–160.

Ebitani, K., M. Kato, K. Motokura, T. Mizugaki, and K. Kaneda. 2006. Highly efficient heterogeneous acylations of aromatic compounds with acid anhydrides and carboxylic acids by montmorillonite-enwrapped titanium as a solid acid catalyst. *Research on Chemical Intermediates* 32: 305–315.

Ebitani, K., T. Kawabata, K. Nagashima, T. Mizugaki, and K. Kaneda. 2000. Simple and clean synthesis of 9,9-bis[4-(2-hydroxyethoxy)phenyl]fluorine from the aromatic alkylation of phenoxyethanol with fluoren-9-one catalysed by titanium cation-exchanged montmorillonite. *Green Chemistry* 2: 1571–1660.

Eeckman, J.P. and H. Laudelout. 1961. Chemical stability of hydrogen montmorillonite suspensions. *Kolloid Zeitschrift* 178: 99–107.

Emam, E.A. 2013. Clays as catalysts in petroleum refining industry. *ARPN Journal of Science and Technology* 3: 356–375.

Fabbri, B., S. Gualtieri, and C. Leonardi. 2013. Modifications induced by the thermal treatment of kaolin and determination of reactivity of metakaolin. *Applied Clay Science* 73: 2–10.

Faghihian, H. and M.H. Mohammadi. 2013. Surface properties of pillared acid-activated bentonite as catalyst for selective production of linear alkylbenzene. *Applied Surface Science* 264: 492–499.

Faghihian, H. and M.H. Mohammadi. 2014. Acid activation effect on the catalytic performance of Al-pillared bentonite in alkylation of benzene with olefins. *Applied Clay Science* 93–94: 1–7.

Fahn, R. 1973. Einfluβ der Struktur und der Morphologie von Bleicherden auf die Bleichwirkung bei Ölen und Fetten. *Fette, Seifen, Anstrichmittel* 75: 77–82.

Falaras, P., F. Lezou, G. Seiragakis, and D. Petrakis. 2000. Bleaching properties of alumina-pillared acid-activated montmorillonite. *Clays and Clay Minerals* 48: 549–556.

Falaras, P., I. Kovanis, F. Lezou, and G. Seiragakis. 1999. Cottonseed oil bleaching by acid-activated montmorillonite. *Clay Minerals* 34: 221–232.

Fang, Z., B. Liu, J. Luo, Y. Ren, and Z. Zhang. 2014. Efficient conversion of carbohydrates into 5-hydroxymethylfurfural catalyzed by the chromium-exchanged montmorillonite K-10 clay. *Biomass & Bioenergy* 60: 171–177.

Farzaneh, F., F. Zarkesh, and M. Ghandi. 2007. Utilization of metal-pillared montmorillonite as a remarkable solid acid catalyst for the synthesis of linear monoalkylbenzenes. *Journal of Science, Islamic Republic of Iran* 18: 135–138.

Fekri, L.Z., M. Nikpassand, and M. Kohansal. 2015. Fe^{3+}-montmorillonite K10, as effective, eco-friendly, an reusable catalyst for the synthesis of bis(1*H*-indol-3-yl)methanes under grinding condition. *RussianJournal of General Chemistry* 85: 2861–2866.

Fernandes, C., C. Catrinescu, P. Castilho, P.A. Russo, M.R. Carrott, and C. Breen. 2007. Catalytic conversion of limonene over acid activated Serra de Dentro (SD) bentonite. *Applied Catalysis A: General* 318: 108–120.

Fernandes, C.I., C.D. Nunes, and P.D. Vaz. 2012. Clays in organic synthesis—Preparation and catalytic applications. *Current Organic Synthesis* 9: 670–694.

Fernandes, S., A. Martins, J. Pires, A.P. Carvalho, and H. Vasques. 2008. Catalytic characterization of pillared clays through toluene methylation reaction. *Reaction Kinetics and Catalysis Letters* 95: 373–378.

Ferris, J.P., E.H. Edelson, N.M. Mount, and A.E. Sullivan. 1979. The effect of clays on the oligomerization of HCN. *Journal of Molecular Evolution* 13: 317–330.

Ferris, J.P., W.J. Hagan, K.W. Alwis, and J. McCrea. 1982. Chemical evolution 40. Clay-mediated oxidation of diaminomaleonitrile. *Journal of Molecular Evolution* 18: 304–309.

Fetter, G., G. Heredia, A.M. Maubert, and P. Bosch. 1996. Synthesis of Al-intercalated montmorillonites using microwave irradiation. *Journal of Materials Chemistry* 6: 1857–1858.

Fetter, G., G. Heredia, L.A. Velázquez, A.M. Maubert, and P. Bosch. 1997. Synthesis of aluminium-pillared motmroillonites using highly concentrated clay suspensions. *Applied Catalysis A: General* 162: 41–45.

Fetter, G. and P. Bosch. 2010. Microwave effect on clay pillaring. In: *Pillared Clays and Related Catalysts*, (Eds.) A. Gil, S.A. Korili, R. Trujillano, and M.A. Vicente, pp. 1–21. New York: Springer.

Figueiredo, F.C.A., E. Jordão, and W.A. Carvalho. 2008. Adipic ester hydrogenation catalyzed by platinum supported in alumina, titania and pillared clays. *Applied Catalysis A: General* 351: 259–266.

Figueiredo, F.C.A., E. Jordão, R. Landers, and W.A. Carvalho. 2009. Acidity control of ruthenium pillared clay and its application as a catalyst in hydrogenation reactions. *Applied Catalysis A: General* 371: 131–141.

Figueras, F. 1988. Pillared clays as catalysts. *Catalysis Reviews—Science and Engineering* 30: 457–499.

Filimonova, S., S. Kaufhold, F.E. Wagner, W. Häusler, and I. Kögel-Knabner. 2016. The role of allophane nanostructure and Fe oxide speciation for hosting soil organic matter in an allophanic Andosol. *Geochimica et Cosmochimica Acta* 180: 284–302.

Fishman, D., J.T. Klug, and A. Shani. 1981. α,β-Unsaturated aldehydes; montmorillonite clay K-10. An effective catalyst for the preparation of unsaturated aldehydes via condensation of acetals with vinyl ethers. *Synthesis* 2: 137–138.

Flessner, U., D.L. Jones, J. Rozière et al. 2001. A study of the surface acidity of acid-treated montmorillonite clay catalysts. *Journal of Molecular Catalysis A: Chemical* 168: 247–256.

Fraile, J.M., J.I. García, and J.A. Mayoral. 1996. Cyclopropanation reactions catalysed by copper(II)-exchanged clays and zeolites. Influence of the catalyst on the selectivity. *Chemical Communications* 11: 1319–1320.

Fraile, J.M., J.I. García, D. Gracia, J.A. Mayoral, T. Tarnai, and F. Figueras. 1997. Contribution of different mechanisms and different active sites to the clay-catalyzed Diels–Alder reactions. *Journal of Molecular Catalysis A: Chemical* 121: 97–102.

Franco, F., M. Pozo, J.A. Cecilia, M. Benítez-Guerrero, and M. Lorente. 2016. Effectiveness of microwave assisted acid treatment on dioctahedral and trioctahedral smectites. The influence of octahcdral composition. *Applied Clay Science* 120: 70–80.

Franco, F., M. Pozo, J.A. Cecilia, M. Benítez-Guerrero, E. Pozo, and J.A. Martín Rubí. 2014. Microwave assisted acid treatment of sepiolite: The role of composition and 'crystallinity'. *Applied Clay Science* 102: 15–27.

Franco, M.A., C. Gessa, and F. Cariati. 1978. Identification of tetramethylammonium ion in methylated NH_4-bentonite. *Clays and Clay Minerals* 26: 73–75.

Frey, S.T., B.M. Hutchins, B.J. Anderson, T.K. Schreiber, and M.E. Hagerman. 2003. Catalytic hydrolysis of 4-nitrophenyl phosphate by lanthanum(III)-hectorite. *Langmuir* 19: 2188–2192.

Fripiat, J.J. 1971. Some applications of nuclear magnetic resonance to surface chemistry. *Catalysis Reviews: Science and Engineering* 5: 269–282.

Frost, R.L. and A.M. Vassallo. 1996. The dehydroxylation of the kaolinite clay minerals using infrared emission spectroscopy. *Clays and Clay Minerals* 44: 635–651.

Frost, R.L. and Z. Ding. 2003. Controlled rate thermal analysis and differential scanning calorimetry of sepiolites and palygorskites. *Thermochimica Acta* 397: 119–128.

Funabiki, K., E. Murata, Y. Fukushima, M. Matsui, and K. Shibata. 1999. Montmorillonite K10 (clay) catalyzed hydrolysis of aryl-substituted α,β-difluorallyl alcohols leading to (Z)-α-fluoro-β-aryl-substitutee acrylaldehydes. *Tetrahedron* 55: 4637–4642.

Gaaf, J., R. van Santen, A. Knoester, and B. van Wingerden. 1983. Synthesis and catalytic properties of pillared nickel substituted mica montmorillonite clays. *Journal of the Chemical Society, Chemical Communications* 655–657.

Gajare, A.S., N.S. Shaikh, B.K. Bonde, and V.H. Deshpande. 2000. Microwave accelerated selective and facile deprotection of allyl esters catalysed by montmorillonite K-10. *Journal of the Chemical Society, Perkin Transactions* 1: 639–640.

Galarneau, A., A. Barodawalla, and T.J. Pinnavaia. 1995. Porous clay heterostructures formed by gallery-templated synthesis. *Nature* 374: 529–531.

Galarneau, A., A. Barodawalla, and T.J. Pinnavaia.1997. Porous clay heterostructures (PCH) as acid catalysts. *Chemical Communications* 17: 1661–1662.

Gallo, J.M.R., S. Teixeira, and U. Schuchardt. 2006. Synthesis and characterization of niobium modified montmorillonite and its use in the acid-catalyzed synthesis of β-hydroxyethers. *Applied Catalysis A: General* 311: 199–203.

Galwey, A.K. 1969. Reactions of alcohols adsorbed on montmorillonite and the role of minerals in petroleum genesis. *Journal of the Chemical Society D: Chemical Communications* 11: 577–578.

Gandia, L.M., A. Gil, M.A. Vicente, and C. Belver. 2005. Dehydrogenation of ethylbenzene on alumina-pillared Fe-rich saponites. *Catalysis Letters* 101: 229–234.

Gao, Y., W. Li., H. Sun et al. 2014. A facile in situ pillaring method—The synthesis of Al-pillared montmorillonite. *Applied Clay Science* 88–89: 228–232.

Garade, A.C., V.R. Mate, and C.V. Rode. 2009. Montmorillonite for selective hydroxyalkylation of *p*-cresol. *Applied Clay Science* 43: 113–117.

Garrido-Ramírez, E.G., B.K.G. Theng, and M.L. Mora. 2010. Clays and oxide minerals as catalysts in Fenton-like reactions. *Applied Clay Science* 47: 182–192.

Gautier, E.C.L., A.E. Graham, A. McKillop, S.P. Standen, and R.J.K. Taylor. 1997. Acetal and ketal deprotection using montmorillonite K10: The first synthesis of syn 4,8 dioxatricyclo[5.1.0.0^{3,5}]-2,6-octanedione. *Tetrahedron Letters* 38: 1881–1884.

Geatti, A., M. Lenarda, L. Storaro, R. Ganzerla, and M. Perissinotto. 1997. Soild acid catalysts from clays: Cumene synthesis by benzene alkylation with propene catalyzed by cation exchanged aluminum pillared clays. *Journal of Molecular Catalysis A: Chemical* 121: 111–118.

Ghiaci, M., B. Aghabarari, and A. Gil. 2011. Production of biodiesel by esterification of natural fatty acids over modified organoclay catalysts. *Fuel* 90: 3382–3389.

Ghiaci, M., M.E. Sedaghat, R.J. Kalbasi, and A. Abbaspur. 2005. Applications of surfactant-modified clays to synthetic organic chemistry. *Tetrahedron* 61: 5529–5534.

Gil, A., H.L. Del Castillo, J. Masson, J. Court, and P. Grange. 1996. Selective dehydration of 1-phenylethanol to 3-oxa-2,4-diphenylpentane on titanium pillared montmorillonite. *Journal of Molecular Catalysis A: Chemical* 107: 185–190.

Gil, A., L.M. Gandia, and M.A. Vicente. 2000. Recent advances in the synthesis and catalytic applications of pillared clays. *Catalysis Reviews* 42: 145–212.

Gil, A., S.A. Korili, and M.A. Vicente. 2008. Recent advances in the control and characterization of the porous structure of pillared clay catalysts. *Catalysis Reviews* 50: 153–221.

Gil, A., S.A. Korili, R. Trujillano, and M.A. Vicente (Eds.). 2010. *Pillared Clays and Related Catalysts*. New York: Springer.

Gil, A., S.A. Korili, R. Trujillano, and M.A. Vicente. 2011. A review on characterization of pillared clays by specific techniques. *Applied Clay Science* 53: 97–105.

Glaeser, R., I. Mantin, and J. Méring. 1961. Études sur l'acidité de la montmorillonite. *International Geological Congress, XXI Session, Norden, 1960*, pp. 28–34.

Goldstein, T.P. 1983. Geocatalytic reactions in formation and maturation of petroleum. *American Association of Petroleum Geologists Bulletin* 67: 152–159.

Gonçalves, V.L.C., B.P. Pinto, J.C. Silva, and C.J.A. Mota. 2008. Acetylation of glycerol catalyzed by different solid acids. *Catalysis Today* 133–135: 673–677.

González, B., R. Trujillano, M.A. Vicente et al. 2017. Two synthesis approaches of Fe-containing intercalated montmorillonites: Differences as acid catalysts for the synthesis of 1,5-benzodeazepine from 1,2-phenylenediamine and acetone. *Applied Clay Science* 146: 388–396.

González, E. and A. Moronta. 2004. The dehydrogenation of ethylbenzene to styrene catalyzed by a natural and an Al-pillared clays impregnated with cobalt compounds: A comparative study. *Applied Catalysis A: General* 258: 99–105.

González, E., D. Rodriguez, L. Huerta, and A. Moronta. 2009. Isomerization of 1-butene catalyzed by surfactant-modified Al_2O_3-pillared clays. *Clays and Clay Minerals* 57: 383–391.

González, F., C. Pesquera, C. Blanco, I. Benitáo, S. Mendioroz, and J.A. Pajares. 1989. Structural and textural evolution of Al- and Mg-rich palygorskite. I. Under acid treatment. *Applied Clay Science* 4: 373–388.

González, F., C. Pesquera, I. Benito, E. Herrero, C. Poncio, and S. Casuscelli. 1999. Pillared clays: Catalytic evaluation in heavy oil cracking using a microactivity test. *Applied Catalysis A: General* 181: 71–76.

González, F., C. Pesquera, I. Benito, S. Mendioroz, and G. Poncelet. 1992. High conversion and selectivity for cracking of *n*-heptane on cerium-aluminium montmorillonite catalysts. *Journal of the Chemical Society, Chemical Communications* 491–493.

Gonzalez, J.M. and D.A. Laird. 2006. Smectite-catalyzed dehydration of glucose. *Clays and Clay Minerals* 54: 38–44.

González-Pradas, E., E. Villafranca-Sánchez, M. Villafranca-Sánchez, F. del del Rey-Bueno, A. Valverde-García, and A. García-Rodríguez. 1991. Evolution of surface properties in a bentonite as a function of acid and heat treatments. *Journal of Chemical Technology and Biotechnology* 52: 211–218.

Greenland, D.J. and E.W. Russell. 1955. Organo-clay derivatives and the origin of the negative charge on clay particles. *Transactions of the Faraday Society* 51: 1300–1307.

Grim, R.E. 1968. *Clay Mineralogy, 2nd ed.* New York: McGraw-Hill.

Grygar, T., D. Hradil, P. Bezdička, B. Doušova, L. Čapek, and O. Schneeweiss. 2007. Fe(III)-modified montmorillonite and bentonite: Synthesis, chemical and UV-Vis spectral characterization, arsenic sorption, and catalysis of oxidative dehydrogenation of propane. *Clays and Clay Minerals* 55: 165–176.

Guerra, D.L., S.P. Oliveira, R.A.S. Silva, E.M. Silva, and A.C. Batista. 2012. Dielectric properties of organo-functionalized kaolinite clay and application in adsorption mercury cation. *Ceramic International* 38: 1687–1696.

Guerra, S.R., L.M.O.C. Merat, R.A.S. San Gil, and L.C. Dieguez. 2008. Alkylation of benzene with olefins in the presence of zirconium-pillared clays. *Catalysis Today* 133–135: 223–230.

Gültekin, Z. 2004. Iron(III)-doped montmorillonite catalysis of alkenes bearing sulphoxide groups in Diels–Alder reactions. *Clay Minerals* 39: 345–348.

Gutierrez, E., A. Loupy, G. Bram, and E. Ruiz-Hitzky. 1989. Inorganic solids in 'dry media'. An efficient way for developing microwave irradiation activated organic reactions. *Tetrahedron Letters* 30: 945–948.

Gyftopoulou, M.E., M. Millan, A.V. Bridgwater, D. Dugwell, R. Kandiyoti, and J.A. Hriljac. 2005. Pillared clays as catalysts for hydrocracking of heavy liquid fuels. *Applied Catalysis A: General* 282: 205–214.

Habib, A.M., A.A. Saafan, A.K. Abou-Seif, and M.A. Salem. 1988. Catalytic conversion of *n*-pentanol to 1,1-dialkyl ether by aluminium and iron exchanged montmorillonite catalyst. *Colloids and Surfaces* 29: 337–341.

Habibi, D. and O. Marvi. 2005. Montmorillonite K-10 supported one-pot synthesis of some symmetric diimides and 3a,4,7,7a-tetrahydroisoindole-1,3-dione derivatives under solvent-free conditions using microwaves. *Journal of the Serbian Chemical Society* 70: 579–583.

Habibi, D. and O. Marvi. 2007. Montmorillonite KSF clay as an efficient catalyst for the synthesis of 1,4-dioxo-3,4-dihydrophthalazine-2(1*H*)-carboxamides and -carbothioamides under solvent-free conditions using microwave irradiation. *Catalysis Communications* 8: 127–130.

Hachemaoui, A. and M. Belbachir. 2005. Montmorillonite clay-catalysed synthesis of cyclic allylamines. *Mendeleev Communications* 15: 124–125.

Haffad, D., A. Chambellan, and J.C. Lavalley. 1998. Characterisation of acid-treated bentonite. Reactivity, FTIR study and ^{27}Al MAS NMR. *Catalysis Letters* 54: 227–233.

Han, B.H. and D.G. Jang. 1990. Montmorillonite catalyzed reduction of nitroarenes with hydrazine. *Tetrahedron Letters* 31: 1181–1182.

Hansford, R.C. 1950. Chemical concepts of catalytic cracking. In *Advances in Catalysis, Vol. IV*, (Eds.) W.G. Frankenburg, E.K. Rideal, and V.I. Komarewsky, pp. 1–30. New York: Academic Press.

Hart, M.P. and D.R. Brown. 2004. Surface acidities and catalytic activities of acid-activated clays. *Journal of Molecular Catalysis A: Chemical* 212: 315–321.

Hasaninejad, A., A. Zare, M. Shekouhy, and A.R. Moosavi-Zare. 2009. Bentonite clay K-10 as an efficient reagent for the synthesis of quinoxaline derivatives at room temperature. *E-Journal of Chemistry* 6(S1): S247–S253.

Hashemi, M.M., A. Asadollahi, and R. Mostaghim. 2005. Microwave-assisted synthesis of bithiazole derivatives under solvent-free conditions. *Russian Journal of Organic Chemistry* 41: 623–624.

Hashemi, M.M., A. Rahimi, and Y. Ahmadibeni. 2004. Microwave-expedited synthesis of aromatic aldehydes and ketones from alkyl halides without solvent using wet montmorillonite K10 supported iodic acid as oxidant. *Acta Chimica Slovenica* 51: 333–336.

Hashimoto, K., Y. Hanada, Y. Minami, and Y. Kera. 1996. Conversion of methanol to dimethyl ether and formaldehyde over alumina intercalated in a montmorillonite. *Applied Catalysis A: General* 141: 57–69.

Hazarika, M.K., R. Parajuli, and P. Phukan. 2007. Synthesis of parabens using montmorillonite K10 clay as catalyst: A green protocol. *Indian Journal of Chemical Technology* 14: 104–106.

He, H., J. Duchet, J. Galy, and J-F. Gérard. 2005. Grafting of swelling clay minerals with 3-aminopropyl-triethoxysilane. *Journal of Colloid and Interface Science* 288: 171–176.

He, H., L. Ma, J. Zhu, P. Yuan, and Y. Qing. 2010. Organoclays prepared from montmorillonites with different cation exchange capacity and surfactant configuration. *Applied Clay Science* 48: 67–72.

He, H., L. Ma, J. Zhu, R.L. Frost, B.K.G. Theng, and F. Bergaya. 2014. Synthesis of organoclays. A critical review and some unresolved issues. *Applied Clay Science* 100: 22–28.

He, H., Q. Tao, J. Zhu, P. Yuan, W. Shen, and S. Yang. 2013. Silylation of clay mineral surfaces. *Applied Clay Science* 71: 15–20.

Hedley, C.B., G. Yuan, and B.K.G. Theng. 2007. Thermal analysis of montmorillonite modified with quaternary phosphonium and ammonium surfactants. *Applied Clay Science* 35: 180–188.

Heller-Kallai, L. 2002. Clay catalysis in reactions of organic matter. In *Organo-Clay Complexes and Interactions*, (Eds.) S. Yariv and H. Cross, pp. 567–613. New York: Marcel Dekker.

Heller-Kallai, L. 2013. Thermally modified clay minerals. In *Handbook of Clay Science, 2nd ed. Developments in Clay Science, Vol. 5A*. (Eds.) F. Bergaya and G. Lagaly, pp. 411–433. Amsterdam, the Netherlands: Elsevier.

Henmi, T. 1980. Effect of SiO_2/Al_2O_3 ratio on the thermal reactions of allophane. *Clays and Clay Minerals* 28: 92–96.

Heravi, M.M., D. Ajami, M.M. Mojtahedi, and M. Ghassemzadeh. 1999. A convenient oxidative deprotection of tetrahydropyranyl ethers with iron(III) nitrate and clay under microwave irradiation in solvent free conditions. *Tetrahedron Letters* 40: 561–562.

Hermosín, M.C. and J. Cornejo. 1986. Methylation of sepiolite and palygorskite with diazomethane. *Clays and Clay Minerals* 34: 591–596.

Hernández, J.N., F.R. Pinacho Crisótomo, T. Martín, and V.S. Martín. 2007. A practical method for selective cleavage of a *tert*-butoxycarbamoyl *N*-protective group from *N,N*-diproteced α-amino acid derivatives using montmorillonite K-10. *European Journal of Organic Chemistry* 30: 5050–5058.

Hernando, M.J., C. Pesquera, C. Blanco, and F. González. 2001. Synthesis, characterization, and catalytic properties of pillared montmorillonite with aluminum/cerium polyoxycations. *Chemistry of Materials* 13: 2154–2159.

Herney-Ramírez, J. and L.M. Madeira. 2010. Use of pillared clay-based catalysts for wastewater treatment through Fenton-like process. In: *Pillared Clays and Related Catalysts*, (Eds.) A. Gil, S.A. Korili, R. Trujillano, and M.A. Vicente, pp. 129–165. New York: Springer.

Herney-Ramírez, J., M.A. Vicente, and L.M. Madeira. 2010. Heterogeneous photo-Fenton oxidation with pillared clay-based catalysts for wastewater treatment: A review. *Applied Catalysis B: Environmental* 98: 10–26.

Herrera, N.N., J-M. Letoffe, J-P. Reymond, and E. Bourgeat-Lami. 2005. Silylation of laponite clay particles with monofuntional and trifunctional vinyl alkoxysilanes. *Journal of Materials Chemistry* 15: 863–871.

Hettinger, J.P. 1991. Contribution to catalytic cracking in the petroleum industry. *Applied Clay Science* 5: 445.

Heyding, R.D., R. Ironside, A.R. Norris, and R.Y. Prysiazniuk. 1960. Acid activation of montmorillonite. *Canadian Journal of Chemistry* 38: 1003–1016.

Hoefnagel, A.J. and H. van Bekkum. 2003. Selective alkylation of methylbenzenes with cyclohexene catalyzed by solid acids. *Catalysis Letters* 85: 7–11.

Hoffmann, H.M. and J. Rabe. 1983. Preparation of 2-(1-hydroxyalkyl)acrylic esters; simple three-step synthesis of mikanecic acid. *Angewandte Chemie International Edition in English* 22: 795–796.

Horio, M., K. Suzuki, H. Masuda, and T. Mori. 1991. Alkylation of toluene with methanol on alumina-pillared montmorillonite. Suppression of deactivation by control of the lateral spacings of pillars. *Applied Catalysis* 72: 109–118.

Houdry, E., W.F. Burt, A.E. Pew, and W.A. Peters, Jr. 1938. Catalytic processing by the Houdry process. *National Petroleum News* 48: R570–R580.

Huang, T-K., R. Wang, L. Shi, and X-X. Lu. 2008. Montmorillonite K-10: An efficient and reusable catalyst for the synthesis of quinoxaline derivatives in water. *Catalysis Communications* 9: 1143–1147.

Huerta, L., A. Meyer, and E. Choren. 2003. Synthesis, characterization and catalytic application for ethylbenzene dehydrogenation of an iron pillared clay. *Microporous and Mesoporous Materials* 57: 219–227.

Huskić, M., M. Žigon, and M. Ivanković. 2013. Comparison of the properties of clay polymer nanocomposites prepared by montmorillonite modified by silane and quaternary ammonium salts. *Applied Clay Science* 85: 109–115.

Hussin, F., M.K. Aroua, and W.M.A.W. Daud. 2011. Textural characteristics, surface chemistry and activation of bleaching earth: A review. *Chemical Engineering Journal* 170: 90–116.

Ianchis, R., M.C. Corobea, D. Donescu et al. 2012. Advanced functionalization of organoclay nanoparticles by silylation and their polystyrene nanocomposites obtained by miniemulsion polymerization. *Journal of Nanoparticle Research* 14: 1233–1245.

Izumi, Y., K. Urabe, and M. Onaka. 1998. Advances in liquid-phase organic reactions using heteropolyacid and clay. *Microporous and Mesoporous Materials* 21: 227–233.

Jackson, A.H., K.R.N. Rao, N.S. Ooi, and E. Adelakun. 1984. Reactions on solid supports. Part I: Novel preparation of α-formylpyrroles from α-methylpyrroles by oxidation with thallium(III) nitrate on clay. *Tetrahedron Letters* 25: 6049–6050.

James, B., E. Suresh, and M.S. Nair. 2007. Friedel–Crafts alkylation of a cage enone: Synthesis of aralkyl substituted tetracyclo[5.3.1.0[2,6].0[4,8]]undeca-9,11-diones and the formation of fascinating novel cage compounds with pyrrole and thiophene using montmorillonite K-10. *Tetrahedron Letters* 48: 6059–6063.

Jaynes, W.F. and S.A. Boyd. 1991. Clay mineral type and organic-compound sorption by hexadecyltrimethyl-ammonium-exchanged clays. *Soil Science Society of America Journal* 55: 43–48.

Jeganathan, M. and K. Pitchumani. 2014. Synthesis of substituted isoquinolines *via* iminoalkyne cyclization using Ag(I) exchanged K10-montmorillonite clay as a reusable catalyst. *RSC Advances* 4: 38491–38497.

Jeganathan, M., A. Dhakshinamoorty, and K. Pitchumani. 2014a. One-pot synthesis of 2-substituted quinoxalines using K10-montmorillonite as heterogeneous catalyst. *Tetrahedron Letters* 55: 1616–1620.

Jeganathan, M., A. Dhakshinamoorty, and K. Pitchumani. 2014b. One-pot synthesis of propargylamines using Ag(I)-exchanged K10 montmorillonite clay as reusable catalyst in water. *ACS Sustainable Chemistry & Engineering* 2: 781–787.

Jerónimo, D., J.M. Guil, B.M. Corbella et al. 2007. Acidity characterization of pillared clays through microcalorimetric measurements and catalytic ethylbenzene test reaction. *Applied Catalysis A: General* 330: 89–95.

Johnson, I.D., T.A. Werpy, and T.J. Pinnavaia. 1988. Tubular silicate-layered silicate intercalation compounds: A new family of pillared clays. *Journal of the American Chemical Society* 110: 8545–8547.

Johnson, L.M. and T.J. Pinnavaia. 1990. Silylation of tubular aluminosilicate polymer (imogolite) by reaction with hydrolyzed (γ-aminopropyl)triethoxysilane. *Langmuir* 6: 307–311.

Johnson, L.M. and T.J. Pinnavaia. 1991. Hydrolysis of (γ-aminopropyl)triethoxysilane-silylated imogolite and formation of a silylated tubular silicate-layered silicate nanocomposite. *Langmuir* 7: 2636–2641.

Jones, J.R. and J.H. Purnell. 1994. The catalytic dehydration of pentan-1-ol by alumina pillared Texas montmorillonites of differing pillar density. *Catalysis Letters* 28: 283–289.

Joseph, T., G.V. Shanbhag, D.P. Sawant, and S.B. Halligudi. 2006. Chemoselective *anti*-Markovnikov hydroamination of α,β-ethylenic compounds with amines using montmorillonite clay. *Journal of Molecular Catalysis A: Chemical* 250: 210–217.

Joussein, E., S. Petit, J. Churchman, B. Theng, D. Righi, and B. Delvaux. 2005. Halloysite clay minerals—A review. *Clay Minerals* 40: 383–426.

Kabeer, S.A., N.M. Reddy, G. Mohan, B. Mahesh, and C.S. Reddy. 2015. Montmorillonite K-10: An efficient and reusable catalyst for the one-pot multicomponent microwave synthesis of diethyl 1-(4-aryl)-4-phenyl-1*H*-pyrrole-2,3-dicarboxylates. *Der Pharma Chemica* 7: 504–509.

Kalbasi, R.J., A.R. Massah, and B. Daneshvarnejad. 2012. Preparation and characterization of bentonite/PS-SO$_3$H nanocomposites as an efficient acid catalyst for the Biginelli reaction. *Applied Clay Science* 55: 1–9.

Kanagasabapathy, S., A. Sudalai, and B.C. Benicewicz. 2001. Montmorillonite K10-catalyzed regioselective addition of thiols and thiobenzoic acids onto olefins: An efficient synthesis of dithiocarboxylic esters. *Tetrahedron Letters* 42: 3791–3794.

Kanda, L.R.S., M.L. Corazza, L. Zatta, and F. Wypych. 2017. Kinetics evaluation of the ethyl esterification of long chain fatty acids using commercial *montmorillonite K10* as catalyst. *Fuel* 193: 265–274.

Kaneko, T., H. Shimotsuma, M. Kajikawa, T. Hatamachi. T. Kodama, and Y. Kitayama. 2001. Synthesis and photocatalytic activity of titania pillared clays. *Journal of Porous Materials* 8: 295–301.

Kang, D-Y., J. Zang, E.R. Wright, A.L. McCanna, C.W. Jones, and S. Nair. 2010. Dehydration, dehydroxylation, and rehydroxylation of single-walled aluminosilicate nanotubes. *ACS Nano* 4: 4897–4907.

Kannan, P. and K. Pitchumani. 1997. Clay-catalysed radical addition of aliphatic thiols to styrene. *Catalysis Letters* 45: 271–273.

Kannan, P., H.S. Banu, and K. Pitchumani. 1999. Syntheses of organic benzyl sulphides from thiols using a modified clay catalyst. *Proceedings of the Indian Academy of Sciences (Chemical Sciences)* 111: 555–561.

Kannan, P., K. Pitchumani, S. Rajagopal, and C. Srinivasan. 1996. Synthesis of organic sulphides from thiols using montmorillonite-3-aminopropyltriethoxysilane as a new catalyst. *Chemical Communications* 3: 369–370.

Kannan, P., K. Pitchumani, S. Rajagopal, and C. Srinivasan. 1997. Sheet silicate catalysed demethylation and Fischer-Hepp rearrangement of *N*-methyl-*N*-nitrosoaniline. *Journal of Molecular Catalysis A: Chemical* 118: 189–193.

Kannan, V. and K. Sreekumar. 2013a. Montmorillonite K10 clay catalyzed one pot synthesis of 2,4,6-tri substituted pyridine under solvent free condition. *Modern Research in Catalysis* 2: 42–46.

Kannan, V. and K. Sreekumar. 2013b. Clay-supported titanium catalyst for the solvent free synthesis of tetra-substituted imidazoles and benzimidazoles. *Journal of Molecular Catalysis A: Chemical* 376: 34–39.

Kansedo, J., K.T. Lee, and S. Bhatia. 2009. Biodiesel production from palm oil via heterogeneous transesterification. *Biomass and Bioenergy* 33: 271–276.

Kantam, M.L., P.L. Santhi, and M.F. Siddiqui. 1993. Montmorillonite catalyzed dehydration of tertiary alcohols to olefins. *Tetrahedron Letters* 34: 1185–1186.

Kantam, M.L., V. Bhaskar, and B.M. Choudary. 2002. Direct condensation of carboxylic acids with alcohols: The atom economic protocol catalysed by Fe^{3+}-montmorillonite. *Catalysis Letters* 78: 185–188.

Kantevari, S., S.V.N. Vuppalapati, and L. Nagarapu. 2007. Montmorillonite K10 catalyzed efficient synthesis of amidoalkyl naphthols under solvent free conditions. *Catalysis Communications* 8: 1857–1862.

Kar, P., S. Samantaray, and B.G. Mishra. 2013. Catalytic application of chromia-pillared montmorillonite towards environmentally benign synthesis of octahydroxanthenes. *Reaction Kinetics Mechanisms and Catalysis* 108: 241–251.

Katdare, S.P., V. Ramaswamy, and A.V. Ramaswamy. 1999. Ultrasonication: A competitive method of intercalation for the preparation of alumina-pillared montmorillonite catalyst. *Catalysis Today* 49: 313–320.

Katdare, S.P., V. Ramaswamy, and A.V. Ramaswamy. 2000. Factors affecting the preparation of alumina pillared montmorillonite employing ultrasonics. *Microporous and Mesoporous Materials* 37: 329–336.

Kaur, N. and D. Kishore. 2012. Montmorillonite: An efficient, heterogeneous and green catalyst for organic synthesis. *Journal of Chemical and Pharmaceutical Research* 4: 991–1015.

Kawabata, T., T. Mizugaki, K. Ebitani, and K. Kaneda. 2001. Highly efficient heterogeneous acetalization of carbonyl compounds catalyzed by a titanium cation-exchanged montmorillonite. *Tetrahedron Letters* 42: 8329–8332.

Kawabata, T., T. Mizugaki, K. Ebitani, and K. Kaneda. 2003. Highly efficient esterification of carboxylic acids with alcohols by montmorillonite-enwrapped titanium as a heterogeneous acid catalyst. *Tetrahedron Letters* 44: 9205–9208.

Khodaei, M., A.R. Khosropour, and J. Abbasi. 2006. A novel and highly efficient method for the direct conversion of aryl aldehyde bisulfite adducts to their aryl trimethylsilyl ethers in a one-pot manner catalyzed by monmorillonite K-10. *Phosphorus, Sulfur, and Silicon* 181: 93–97.

Kidwai, M., R. Thakur, and R. Mohan. 2005. Ecofriendly synthesis of novel antifungal (thio)barbituric acid derivatives. *Acta Chimica Slovenica* 52: 88–92.

Kidwai, M., V. Bansal, and R. Thakur. 2006. Solid supported synthesis of new thieno[2,3-d]pyrimidines. *Journal of Sulfur Chemistry* 27: 57–63.

Kikuchi, E. and T. Matsuda. 1988. Shape selective acid catalysis by pillared clays. *Catalysis Today* 2: 297–307.

Kikuchi, E. and T. Matsuda. 1991. Influence of pore structure on the catalytic behavior of clay compounds. *Studies in Surface Science and Catalysis* 60: 377–384.

Kikuchi, E., T. Matsuda, H. Fujiki, and Y. Morita. 1984. Disproportionation of 1,2,4-trimethylbenzene over montmorillonite pillared by aluminium oxide. *Applied Catalysis* 11: 331–340.

Kikuchi, E., T. Matsuda, J. Ueda, and Y. Morita. 1985. Conversion of trimethylbenzenes over montmorillonites pillared by aluminium and zirconium oxides. *Applied Catalysis* 16: 401–410.

Kim, J-T., D-Y. Lee, T-S. Oh, and D-H. Lee. 2003. Characteristics of nitrile-butadiene rubber layered silicate nanocomposites with silane coupling agent. *Journal of Applied Polymer Science* 89: 2633–2640.

Klissurski, D., D. Petridis, N. Abadzhieva, and K. Hadjiivanov. 1996. MoO_3 supported on montmorillonite type pillared clays: Characterization, surface acidity and catalytic properties towards the oxidation of methanol. *Applied Clay Science* 10: 451–459.

Kloprogge, J.T. 1998. Synthesis of smectites and porous pillared clay catalysts: A review. *Journal of Porous Materials* 5: 5–41.

Kloprogge, J.T. and R.L. Frost. 1999. Infrared emission spectroscopy of Al-pillared beidellite. *Applied Clay Science* 15: 431–445.

Kloprogge, J.T., L.V. Duong, and R.L. Frost. 2005. A review of the synthesis and characterisation of pillared clays and related porous materials for cracking of vegetable oils to produce biofuels. *Environmental Geology* 47: 967–981.

Kojima, M., R. Hartford, and C.T. O'Connor. 1991. The effect of pillaring montmorillonite and beidellite on the conversion of trimethyl benzenes. *Journal of Catalysis* 128: 487–498.

Komadel, P. 2003. Chemically modified smectites. *Clay Minerals* 38: 127–138.

Komadel, P. 2016. Acid activated clays: Materials in continuous demand. *Applied Clay Science* 131: 84–99.

Komadel, P. and J. Madejová. 2013. Acid activation of clay minerals. In *Handbook of Clay Science*, 2nd ed. *Developments in Clay Science, Vol. 5A*. (Eds.) F. Bergaya and G. Lagaly, pp. 385–409. Amsterdam, the Netherlands: Elsevier.

Komadel, P., D. Schmidt, J. Madejová, and B. Čičel. 1990. Alteration of smectites by treatments with hydrochloric acid and sodium carbonate solutions. *Applied Clay Science* 5: 113–122.

Kooli, F. and L. Yan. 2013. Chemical and thermal properties of organoclays derived from highly stable bentonite in sulfuric acid. *Applied Clay Science* 83–84: 349–356.

Kooli, F., P.C. Hian, Q. Weirong, S.F. Alshahateet, and F. Chen. 2006. Effect of the acid-activated clays on the properties of porous clay heterostructures. *Journal of Porous Materials* 13: 319–324.

Kooli, F., Y. Liu, S.F. Alshateet, P. Siril, and R. Brown. 2008. Effect of pillared clays on the hydroisomerization of *n*-heptane. *Catalysis Today* 131: 244–249.

Korichi, S., A. Elias, A. Mefti, and A. Bensmaili. 2012. The effect of microwave irradiation and conventional acid activation on the textural properties of smectite: Comparative study. *Applied Clay Science* 59–60: 76–83.

Korichi, S., A. Elias, and A. Mefti. 2009. Characterization of smectite after acid activation with microwave irradiation. *Applied Clay Science* 42: 432–438.

Kotal, M. and A.K. Bhowmick. 2015. Polymer nanocomposites from modified clays: Recent advances and challenges. *Progress in Polymer Science* 51: 127–187.

Kotkar, D., and P.K. Ghosh. 1986. Cyclodehydration of non-aromatic diols using aluminium(III)-exchanged montmorillonite as a solid Brönsted acid catalyst. *Journal of the Chemical Society, Chemical Communications* 9: 650–651.

Krajčovič, J., P. Hudec, and F. Grejták. 1995. Catalytic properties of pillared montmorillonites. *Reaction Kinetics and Catalysis Letters* 54: 87–97.

Krstić, I.J., S. Sukdolak, and S. Solujić. 2002. An efficient synthesis of warfarin acetals on montmorillonite clay K-10 with microwaves. *Journal of the Serbian Chemical Society* 67: 325–329.

Kumar, B.S., A. Dhakshinamoorthy, and K. Pitchumani. 2014. K10 montmorillonite clays as environmentally benign catalysts for organic reactions. *Catalysis Science & Technology* 4: 2378–2396.

Kumar, H.M.S., B.V.S. Reddy, P.K. Mohanty, and J.S. Yadav. 1997. Clay catalyzed highly selective O-alkylation of primary alcohols with orthoesters. *Tetrahedron Letters* 38: 3619–3622.

Kumar, S., Panda, A.K., and R.K. Singh. 2013. Preparation and characterization of acids and alkali treated kaolin clay. *Bulletin of Chemical Reaction Engineering & Catalysis* 8: 61–69.

Kurfürstová, J. and M. Hájek. 2004. Microwave-induced catalytic transformation of 2-*tert*-butylphenol at low temperatures. *Research on Chemical Intermediates* 30: 673–681.

Kurian, M. and S. Sugunan. 2006. *tert*-Butylation of phenol catalysed by metal exchanged iron pillared montmorillonites. *Catalysis Communications* 7: 417–421.

Kurian, M., M. Joy, and D. Raj. 2012. Hydroxylation of phenol over rare earth exchanged iron pillared montmorillonites. *Journal of Porous Materials* 19: 633–640.

Labiad, B. and D. Villemin. 1989. Clay catalysis: A simple and efficient synthesis of enolthioethers from cyclic ketones. *Synthesis* 2: 143–144.

Lagaly, G. 1986. Interactions of alkylamines with different types of layered compounds. *Solid State Ionics* 22: 43–51.

Lagaly, G., M. Ogawa, and I. Dékány. 2013. Clay mineral-organic interactions. In *Handbook of Clay Science*, 2nd ed. *Developments in Clay Science, Vol. 5A*. (Eds.) F. Bergaya and G. Lagaly, pp. 435–505. Amsterdam, the Netherlands: Elsevier.

Lahav, N., U. Shani, and J. Shabtai. 1978. Cross-linked smectites. I. Synthesis and properties of hydroxyl-aluminum-montmorillonite. *Clays and Clay Minerals* 26: 107–115.

Lakouraj, M.M., B. Movassagh, and J. Fasihi. 2000. Fe^{3+}-montmorillonite K10: An efficient catalyst for selective amidation of alcohols with nitriles under non-aqueous condition. *Synthetic Communications* 30: 821–827.

Lambat, T.L., S.S. Deo, F.S. Inam, T.B. Deshmukh, and A.R. Bhat. 2016. Montmorillonite K10: An efficient organo heterogeneous catalyst for one-pot synthesis of new *N,N'*-alkylidene bisamide derivatives under solvent free condition. *Karbala International Journal of Modern Science* 2: 63–68.

Lambert, J-F. and G. Poncelet. 1997. Acidity in pillared clays: Origin and catalytic manifestations. *Topics in Catalysis* 4: 43–56.

Landge, S.M. and B. Török. 2008. Synthesis of condensed benzo[*N,N*]-hetrocycles by microwave-assisted solid acid catalysis. *Catalysis Letters* 122: 338–343.

Landge, S.M., M. Berryman, and B. Török. 2008. Microwave-assisted solid acid-catalyzed one-pot synthesis of isobenzofuran-1(3*H*)-ones. *Tetrahedron Letters* 49: 4505–4508.

Landge, S.M., V. Atanassova, M. Thimmaiah, and B. Török. 2007. Microwave-assisted oxidative coupling of amines to imines on solid acid catalysts. *Tetrahedron Letters* 48: 5161–5164.

Laszlo, P. 1987. Chemical reactions on clays. *Science* 235: 1473–1477.

Laszlo, P. and H. Moison. 1989. Catalysis of Diels–Alder reactions with acrolein as dienophile by iron(III)-doped montmorillonite. *Chemistry Letters* 18(6): 1031–1034.

Laszlo, P. and J. Luchetti. 1984a. Catalysis of the Diels–Alder reaction in the presence of clays. *Tetrahedron Letters* 25: 1567–1570.

Laszlo, P. and J. Luchetti. 1984b. Acceleration of the Diels–Alder reaction of clays suspended in organic solvents. *Tetrahedron Letters* 25: 2147–2150.

Laszlo, P. and J. Luchetti. 1984c. Easy formation of Diels–Alder cycloadducts between furans and α,β-unsaturated aldehydes and ketones at normal pressure. *Tetrahedron Letters* 25: 4387–4388.

Laszlo, P. and J. Luchetti. 1993. Porphyrin synthesis using clays. Taking advantage of statistical product distributions. *Chemistry Letters* 22(3): 449–452.

Le Pluart, L., J. Duchet, H. Sautereau, and J.F. Gérard. 2002. Surface modifications of montmorillonite for tailored interfaces in nanocomposites. *Journal of Adhesion* 78. 645–662.

Lee, S.W., H.B. Lee, B.C. Kim, K. Sadaiah, K. Lee, and H. Shin. 2013. A large scale formal synthesis of CoQ$_{10}$: Highly stereoselective Friedel–Crafts allylation reaction of tetramethoxytoluene with (*E*)-4-chloro-2-methyl-1-phenylsulfonyl-2-butene in the presence of montmorillonite K-10. *Bulletin of the Korean Chemical Society* 34: 1257–1259.

Lenarda, M., L. Storaro, A. Talon, E. Moretti, and P. Riello. 2007. Solid acid catalysts from clays: Preparation of mesoporous catalysts by chemical activation of metakaolin under acid conditions. *Journal of Colloid and Interface Science* 311: 537–543.

Lenarda, M., L. Storaro, G. Pellegrini, L. Piovesan, and R. Ganzerla. 1999. Solid acid catalysts from clays. Part 3: Benzene alkylation with ethylene catalyzed by aluminum and aluminum gallium pillared bentonites. *Journal of Molecular Catalysis A: Chemical* 145: 237–244.

Letaïef. S., B. Casal, P. Aranda, M.A. Martín-Luengo, and E. Ruiz-Hitzky. 2003. Fe-containing pillared clays as catalysts for phenol hydroxylation. *Applied Clay Science* 22: 263–277.

Leyva, E., L.I. López, E. Moctezuma, and H. de Lasa. 2008. A bentonitic clay assisted method for the preparation of 2-(R-anilino)-1,4-naphthoquinones. *Topics in Catalysis* 49: 281–287.

Li, A-X., T-S. Li, and T-H. Ding. 1997a. Montmorillonite K-10 and KSF as remarkable acetylation catalysts. *Chemical Communications* 15: 1389–1390.

Li, B., H. Mao, X. Li, W. Ma, and Z. Liu. 2009. Synthesis of mesoporous silica-pillared clay by intragallery ammonia-catalyzed hydrolysis of tetraethoxysilane using quaternary ammonium surfactants as gallery templates. *Journal of Colloid and Interface Science* 336: 244–249.

Li, J-T., C-Y. Xing, and T-S. Li. 2004. An efficient and environmentally friendly method for synthesis of arylmethylenemalenonitrile catalyzed by montmorillonite K10-ZnCl$_2$ under ultrasound irradiation. *Journal of Chemical Technology and Biotechnology* 79: 1275–1278.

Li, J-T., M-X. Sun, and Y. Yin. 2010. Ultrasound promoted efficient method for the cleavage of 3-aryl-2,3-epoxyl-1-phenyl-1-propanone with indole. *Ultrasonics Sonochemistry* 17: 359–362.

Li, L-J., Y-X. Song, Y-S. Gao, Y.F. Li, and J-F. Zhang. 2006. Solvent-free synthesis of nitriles from aldehydes catalysed by KF/Al$_2$O$_3$, montmorillonite KSF and K10. *E-Journal of Chemistry* 3: 164–168.

Li, T.S., Z-H. Zhang, and T-S. Jin. 1999. Montmorillonite clay catalysis IX: A mild and efficient method for removal of tetrahydropyranyl ethers. *Synthetic Communications* 29: 181–188.

Li, T-S. and A-X. Li. 1998. Montmorillonite clay catalysis. Part 10. K-10 and KSF-catalysed acylation of alcohols, phenols, thiols and amines: Scope and limitation. *Journal of the Chemical Society, Perkin Transactions 1* 12: 1913–1918.

Li, T-S., S-H. Li, J-T. Li, and H-Z. Li. 1997a. Montmorillonite clay catalysis. Part 2. An efficient and convenient procedure for the preparation of acetals catalysed by montmorillonite K-10. *Journal of Chemical Research (S)* 1: 26–27.

Li, T-S., Z-H. Zhang, F. Yang, and C-G. Fu. 1998b. Montmorillonite clay catalysis. Part 7. An environmentally friendly procedure for the synthesis of coumarins *via* Pechmann condensation of phenols with ethyl acetoacetate. *Journal of Chemical Research (S)* 1: 38–39.

Li, T-S., Z-H. Zhang, and C-G. Fu. 1997b. Montmorillonite clay catalysis V: An efficient and facile procedure for deprotection of 1,1-diacetates. *Tetrahedron Letters* 38: 3285–3288.

Li, T-S., Z-H. Zhang, and Y-J. Gao. 1998a. A rapid preparation of acylals of aldehydes catalysed by Fe^{3+}-montmorillonite. *Synthetic Communications* 28: 4665–4671.

Li, X., X. Yan, Z. Li, H. Chen, and P. Zhang. 2010. Catalysis and deactivation of montmorillonite K10 in the aryl O-glycosylation of glycosyl trichloroacetoimidates. *Frontiers of Chemical Engineering in China* 4: 342–347.

Liang, X., F. Qi, P. Liu et al. 2016. Performance of Ti-pillared montmorillonite supported Fe catalysts for toluene oxidation: The effect of Fe on catalytic activity. *Applied Clay Science* 132–133: 96–104.

Lidström, P., J. Tierney, B. Wathey, and J. Westman. 2001. Microwave assisted organic synthesis—A review. *Tetrahedron* 57: 9225–9283.

Lin, C-L. and T.J. Pinnavaia. 1991. Organo-clay assemblies for triphase catalysis. *Chemistry of Materials* 3: 213–215.

Lin, H., J. Ding, X. Chen, and Z. Zhang. 2000. An efficient synthesis of 5-alkoxycarbonyl-4-aryl-3,4-dihydropyrimidin-2(1H)-ones catalyzed by KSF montmorillonite. *Molecules* 5: 1240–1243.

Lipińska, T., E. Guibé-jampel, A. Petit, and A. Loupy. 1999. 2-(2-pyridyl)indole derivatives preparation via Fischer reaction on montmorillonite K10/zinc chloride under microwave irradiation. *Synthetic Communications* 29: 1349–1354.

Liu Y.L., S.S. Kim, and T.J. Pinnavaia. 2004. Mesostructured aluminosilicate alkylation catalysts for the production of aromatic amine antioxidants. *Journal of Catalysis* 225: 381–387.

Liu, D., P. Yuan, H. Liu et al. 2011. Influence of heating on the solid acidity of montmorillonite: A combined study by DRIFT and Hammett indicators. *Applied Clay Science* 52: 358–363.

Liu, J., B-B. Yang, X-Q. Wang, C-L. Liu, R-Z. Yang, and W-S Dong. 2017. Glucose conversion to methyl levulinate catalyzed by metal ion-exchanged montmorillonites. *Applied Clay Science* 141: 118–124.

Liu, S., J-H. Yang, and J-H. Choy. 2006. Microporous SiO_2-TiO_2 nanosols pillared montmorillonite for photocatalytic decomposition of methyl orange. *Journal of Photochemistry and Photobiology A: Chemistry* 179: 75–80.

Liu, W., J. Gan, S.K. Papiernik, and S.R. Yates. 2000. Sorption and catalytic hydrolysis of diethatyl-ethyl on homoionic clays. *Journal of Agricultural and Food Chemistry* 48: 1935–1940.

Liu, W-Q., L. Zhao, G-D. Sun, and E-Z. Min. 1999. Saturation of aromatics and aromatization of C_3 and C_4 hydrocarbons over metal loaded pillared clay catalysts. *Catalysis Today* 51: 135–140.

Liu, Z., Q. Xu, X. Peng, D. Li, and X. Wang. 2010. Confinement effect as a tool for selectivity orientation in heterogeneous synthesis of 4,4′-diamino-3,3′-dibutyl-diphenyl methane over montmorillonite catalysts. *Journal of Molecular Catalysis A: Chemical* 325: 55–59.

Loeppert, R.H., L.W. Zelazny, and B.G. Volk. 1986. Acidic properties of montmorillonite in selected solvents. *Clays and Clay Minerals* 34: 87–92.

Loh, T-P. and X-R. Li. 1999. Clay montmorillonite K10 catalyzed aldol-type reaction of aldehydes with silyl enol ethers in water. *Tetrahedron* 55: 10789–10802.

Lotz, W.W. and J. Gosselck. 1979. Friedel–Crafts-Alkylierung mit Cu^{2+}-Bentonit. *Zeitschrift für Naturforschung* 34b: 121–122.

Louloudi, A. and N. Papayannakos. 1998. Hydrogenation of benzene on La-Ni and clay supported La-Ni catalysts. *Applied Catalysis A: General* 175: 21–31.

Louloudi, A. and N. Papayannakos. 2000. Hydrogenation of benzene on Ni/Al-pillared montmorillonite catalysts. *Applied Catalysis A: General* 204: 167–176.

Louloudi, A. and N. Papayannakos. 2016. Performace of Ni/Si-pillared clay catalytic extrudates for benzene hydrogenation reaction. *Applied Clay Science* 123: 47–55.

Louloudi, A., J. Michalopoulos, N-H. Gangas, and N. Papayannakos. 2003. Hydrogenation of benzene on Ni/Al-pillared saponite catalysts. *Applied Catalysis A: General* 242: 41–49.

Loupy, A. 2004. Solvent-free microwave organic synthesis as an efficient procedure for green chemistry. *Comptes Rendus Chimie* 7: 103–112.

Loupy, A., S-J. Song, S-M. Sohn, Y-M. Lee, and T-W. Kwon. 2001. Solvent-free bentonite-catalyzed condensation of malonic acid and aromatic aldehydes under microwave irradiation. *Journal of the Chemical Society, Perkin Transactions 1* 10: 1220–1222.

Lourvanij, K. and G.L. Rorrer. 1994. Dehydration of glucose to organic acids in microporous pillared clay catalysts. *Applied Catalysis A: General* 109: 147–165.

Lourvanij, K. and G.L. Rorrer. 1997. Reaction rates for the partial dehydration of glucose to organic acids in solid-acid, molecular-sieving catalyst powders. *Journal of Chemical Technology and Biotechnology* 69: 35–44.

Luo, H., Y. Kang, H. Nie, and L. Yang. 2008. Fe^{3+}-montmorillonite: An efficient solid catalyst for one-pot synthesis of decahydroacridine derivatives. *Journal of the Chinese Chemical Society* 55: 1280–1285.

Macedo, J.C.D., C.J.A. Mota, S.M.C. de Menezes, and V. Camorim. 1994. NMR and acidity studies of dealuminated metakaolin and their correlation with cumene cracking. *Applied Clay Science* 8: 321–330.

MacKenzie, K.J.D., M.E. Bowden, and R.H. Meinhold. 1991. The structure and thermal transformations of allophanes studied by ^{29}Si and ^{27}Al high resolution solid-state NMR. *Clays and Clay Minerals* 39: 337–346.

Madejová, J., H. Pálková, M. Pentrák, and P. Komadel. 2009. Near-infrared spectroscopic analysis of acid-treated organo-clays. *Clays and Clay Minerals* 57: 392–403.

Madejová, J., J. Bujdák, M. Janek, and P. Komadel. 1998. Comparative FT-IR study of structural modifications during acid treatment of dioctahedral smectites and hectorite. *Spectrochimica Acta Part A* 54: 1397–1406.

Mahanta, D. and J.G. Handique. 2011. Highly efficient esterification of phenolic acid over metal exchanged montmorillonite clay. *Asian Journal of Science and Technology* 4: 51–58.

Mahmoud, S., A. Hammoudeh, and M. Al-Noaimi. 2003. Pretreatment effects on the catalytic activity of Jordanian bentonite. *Clays and Clay Minerals* 51: 52–57.

Majumdar, K.C., H. Rahaman, and B. Roy. 2007. Solid-state regioselective cyclization initiated by the electrophilic attack on the double bond by N-bromosuccinimide on montmorillonite K10 clay support under microwave irradiation. *Synthetic Communications* 37: 1477–1484.

Majumdar, K.C., S. Ponra, and T. Ghosh. 2012. Montmorillonite K10 catalyzed, microwave-assisted cyclization of acetylenic amines: An efficient synthesis of pyrrolocoumarins and pyrroloquinolones. *Synthesis* 44: 2079–2083.

Malla, P.B. and S. Komarneni. 1993. Properties and characterization of Al_2O_3 and SiO_2-TiO_2 pillared saponite. *Clays and Clay Minerals* 41: 472–483.

Mallouk, S., K. Bougrin, H. Doua, R. Benhida, and M. Soufiaoui. 2004. Ultrasound-accelerated aromatisation of *trans*- and *cis*-pyrazolines under heterogeneous conditions using claycop. *Tetrahedron Letters* 45: 4143–4148.

Man, A.K. and R. Shahidan. 2007. Microwave-assisted chemical reactions. *Journal of Macromolecular Science, Part A: Pure and Applied Chemistry* 44: 651–657.

Manias, E., A. Touny, L. Wu, K. Strawhecker, B. Lu, and T.C. Chung. 2001. Polypropylene/montmorillonite nanocomposites. Review of the synthetic routes and materials properties. *Chemistry of Materials* 13: 3516–3523.

Manos, G., I.Y. Yusof, N. Papayannakos, and N.H. Gangas. 2001. Catalytic cracking of polyethylene over clay catalysts. Comparison with ultrastable Y zeolite. *Industrial & Engineering Chemistry Research* 40: 2220–2225.

Manos, G., I.Y. Yusof, N.H. Gangas, and N. Papayannakos. 2002. Tertiary recycling of polyethylene to hydrocarbon fuel by catalytic cracking over aluminum pillared clays. *Energy Fuels* 16: 485–489.

Mansilla, H. and D. Regás. 2006. Simple method for the preparation of dimethyl acetals from ketones with montmorillonite K10 and *p*-toluenesulfonic acid. *Synthetic Communications* 36: 2195–2201.

Mansoori, Y., K. Roojaei, M.R. Zamanloo, and G. Imanzadeh. 2012. Polymer-clay nanocomposites via chemical grafting of polyacrylonitrile onto cloisite 20A. *Bulletin of Materials Science* 35: 1063–1070.

Maquestiau, A., A. Mayence, and J-J. Vanden Eynde. 1991. Ultrasound-promoted aromatization of Hantzsch 1,4-dihydropyridines by clay-supported cupric nitrate. *Tetrahedron Letters* 32: 3839–3840.

Maricic, M., M. Tecilazic-Stevanovic, T. Janackovic, and L. Kostic-Gvozdenovic. 1982. Migration of octahedral cations from the crystal lattice to an exchangeable position in the H-form of montmorillonite. *Bulletin de la Société Chimique Beograd* 47: 625–629.

Marín-Astorga, N., G. Alvez-Manoli, and P. Reyes. 2005. Stereoselective hydrogenation of phenyl alkyl acetylenes on pillared clays supported palladium catalysts. *Journal of Molecular Catalysis A: Chemical* 226: 81–88.

Martin-Aranda, R.M., E. Ortega-Cantero, M.L. Rojas-Cervantes, M.A. Vicente-Rodríguez, and M.A. Bañares-Muñoz. 2002. Sonocatalysis and basic clays. Michael addition between imidazole and ethyl acrylate. *Catalysis Letters* 84: 201–204.

Martin-Aranda, R.M., E. Ortega-Cantero, M.L. Rojas-Cervantes, M.A. Vicente-Rodríguez, and M.A. Bañares-Muñoz. 2005. Ultrasound-activated Knoevenagel condensation of malenonitrile with carbonylic compounds catalysed by alkaline-doped saponites. *Journal of Chemical Technology and Biotechnology* 80: 234–238.

Martín-Aranda, R.M., M.A. Vicente-Rodríguez, J.M. López-Pestaña et al. 1997. Application of basic clays in microwave activated Michael additions: Preparation of N-substituted imidazoles. *Journal of Molecular Catalysis A: Chemical* 124: 115–121.

Marvi, O. and D. Habibi. 2008. Microwave-induced solvent-free synthesis of β-keto esters using montmorillonite KSF and K10 clays as efficient and recyclable heterogeneous solid acids. *Chinese Journal of Chemistry* 26: 522–524.

Marvi, O. and M. Giahi. 2009. Montmorillonite KSF clay as novel and recyclable heterogeneous catalyst for the microwave mediated synthesis of indan-1,3-diones. *Bulletin of the Korean Chemical Society* 30: 2918–2920.

Marvi, O., A. Alizadeh, and S. Zarrabi. 2011. Montmorillonite K-10 clay as an efficient reusable heterogeneous catalyst for the solvent-free microwave mediated synthesis of 5-substituted 1*H*-tetrazoles. *Bulletin of the Korean Chemical Society* 32: 4001–4004.

Massam, J. and D.R. Brown. 1998. Acid-treated montmorillonite supports for copper(II)nitrate oxidation reagents. *Applied Catalysis A: General* 172: 259–264.

Massiot, D., P. Dion, J.F. Alcover, and F. Bergaya. 1995. ^{27}Al and ^{29}Si MAS NMR study of kaolinite thermal decomposition by controlled rate thermal analysis. *Journal of the American Ceramic Society* 78: 2940–2944.

Matsuda, T., H. Seki, and E. Kikuchi. 1995. The effect of spiltover hydrogen on the stabilization of catalytic activities of Y-type zeolite and pillared montmorillonite for the disproportionation of 1,2,4-trimethylbenzene. *Journal of Catalysis* 154: 41–46.

Matsuda, T., K. Yogo, C. Pantawong, and E. Kikuchi. 1995. Catalytic properties of copper-exchanged clays for the dehydrogenation of methanol to methyl formate. *Applied Catalysis A: General* 126: 177–186.

Matsuda, T., M. Asanuma, and E. Kikuchi. 1988a. Effect of high-temperature treatment on the activity of montmorillonite pillared by alumina in the conversion of 1,2,4-trimethylbenzene. *Applied Catalysis* 38: 289–299.

Matsuda, T., N. Nagashima, and E. Kikuchi. 1988b. Physical and catalytic properties of smectite clays pillared by alumina in disproportionation of 1,2,4-trimethylbenzene. *Applied Catalysis* 45: 171–182.

Matsui, M. and H. Yamamoto. 1996. Metal ion-exchanged montmorillonites as practical and useful catalysts for the synthesis of α-tocopherol. *Bulletin of the Chemical Society of Japan* 69: 137–139.

McCabe, R.W. and J.M. Adams. 2013. Clay minerals as catalysts. In *Handbook of Clay Science, 2nd ed. Developments in Clay Science, Vol. 5B* (Eds.) F. Bergaya and G. Lagaly, pp. 491–538. Amsterdam, the Netherlands: Elsevier.

Mendelovici, E. and R.L. Frost. 2005. Pioneer studies on HCl and silylation treatments of chrysotile. *Journal of Colloid and Interface Science* 289: 597–599.

Meshram, H.M., B.C. Reddy, B.R.V. Prasad, P.R. Goud, G.S. Kumar, and R.N. Kumar. 2012b. DABCO-promoted efficient and convenient synthesis of benzofurans. *Synthetic Communications* 42: 1669–1676.

Meshram, H.M., G.S. Kumar, P. Ramesh, and B.C. Reddy. 2010. A mild and convenient synthesis of quinoxalines via cyclization-oxidation process using DABCO as catalyst. *Tetrahedron Letters* 51: 2580–2585.

Meshram, H.M., K.C. Sekhar, Y.S.S. Ganesh, and J.S. Yadav. 2000. Clay catalyzed facile cyclohydration under microwave: Synthesis of 3-substituted benzofurans. *Synlett* 9: 1273–1274.

Meshram, H.M., V.M. Bangade, B.C. Reddy, G.S. Kumar, and P.B. Thakur. 2012a. DABCO promoted and efficient and convenient synthesis of pyrrole inaqueous medium. *International Journal of Organic Chemistry* 2: 159–165.

Michot, L.J. and T.J. Pinnavaia. 1992. Improved synthesis of alumina-pillared montmorillonite by surfacatant modification. *Chemistry of Materials* 4: 1433–1437.

Miesserov, K.G. 1969. Nature of active sites of aluminosilicate catalysts. *Journal of Catalysis* 13: 169–186.

Miranda, R., R. Osnaya, R. Garduño, F. Delgado, C. Alvarez, and M. Salmon. 2001. A general alternative to obtain S.S-acetals using TAFF, a bentonitic clay, as the catalyst. *Synthetic Communications* 31: 1587–1597.

Mishra, B.G. and G.R. Rao. 2004. Physicochemical and catalytic properties of Zr-pillared montmorillonite with varying pillar density. *Microporous and Mesoporous Materials* 70: 43–50.

Mishra, T. and K. Parida. 1998a. Transition metal pillared clay: 4. A comparative study of textural, acidic and catalytic properties of chromia pillared montmorillonite and acid activated montmorillonite. *Applied Catalysis A: General* 166: 123–133.

Mishra, T. and K. Parida. 1998b. Transition metal pillared clay: 5. Synthesis, characterisation and catalytic activity of iron-chromium mixed oxide pillared montmorillonite. *Applied Catalysis A: General* 174: 91–98.

Mitra, R.P. and H. Singh. 1959. On the strong acid character of montmorillonite clays and its disappearance on ageing. *Naturwissenschaften* 46: 319–320.

Mitsudome, T., K. Nose, T. Mizugaki, K. Jitsukawa, and K. Kaneda. 2008. Reusable montmorillonite-entrapped organocatalyst for asymmetric Diels–Alder reaction. *Tetrahedron Letters* 49: 5464–5466.

Mitsudome, T., T. Matsuno, S. Sueoka, T. Mizugaki, K. Jitsukawa, and K. Kaneda. 2012a. Direct synthesis of unsymmetrical ethers from alcohols by titanium cation-exchanged montmorillonite. *Green Chemistry* 14: 610–613.

Mitsudome, T., T. Matsuno, S. Sueoka, T. Mizugaki, K. Jitsukawa, and K. Kaneda. 2012b. Titanium cation-exchanged montmorillonite as an active heterogeneous catalyst for the Beckman rearrangement under mild reaction conditions. *Tetrahedron Letters* 53: 5211–5214.

Miyazaki, T., A. Tsuboi, H. Urata et al. 1985. Hydrogenation of olefins with cationic rhodium complex intercalated in fluoro tetrasilicic mica. *Chemistry Letters* 14: 793–796.

Mizugaki, T., R. Arundhathi, T. Mitsudome, K. Jitsukawa, and K. Kaneda. 2014. Highly efficient and selective transformations of glycerol using reusable heterogeneous catalysts. *ACS Sustainable Chemistry & Engineering* 2: 574–578.

Mnasri, S. and N. Frini-Srasra. 2012. Influence of aluminium incorporation in the preparation of zirconia-pillared clay and catalytic performance in the acetalization reaction. *Clay Minerals* 47: 453–463.

Mnasri, S. and N. Frini-Srasra. 2013. Evolution of Brönsted and Lewis acidity of single and mixed pillared bentonite. *Infrared Physics & Technology* 58: 15–20.

Mohammadpoor-Baltork, I., M. Moghadam, S. Tangestaninejad, V. Mirkhani, and Z. Eskandari. 2010. Ultrasound promoted selective synthesis of 2-aryl-5,6-dihydro-4H-1,3-oxazines catalyzed by K-10 and KSF montmorillonite clays: A practical procedure under mild and solvent-free conditions. *Ultrasonics Sonochemistry* 17: 857–862.

Mokaya, R. and W. Jones. 1994. Pillared acid-activated clay catalysts. *Journal of the Chemical Society, Chemical Communications* 8: 929–930.

Mokaya, R. and W. Jones. 1995a. The microstructure of alumina pillared acid-activated clays. *Journal of Porous Materials* 1: 97–110.

Mokaya, R. and W. Jones. 1995b. Pillared clays and pillared acid-activated clays: A comparative study of physical, acidic, and catalytic properties. *Journal of Catalysis* 153: 76–85.

Molina, C.B., J.A. Casas, A.H. Pizarro, and J.J. Rodriguez. 2011. Pillared clays as green chemistry catalysts: Application to wastewater treatment. In *Clay: Types, Properties and Uses*, (Eds.) J.P. Humphrey and D.E. Boyd, pp. 435–474. New York: Nova Science Publishers.

Molina, M.F., R. Molina, and S. Moreno. 2005. Hydroconversion of heptane over a Colombian montmorillonite modified with mixed pillars of Al-Zr and Al-Si. *Catalysis Today* 107–108: 426–430.

Molina, R., S. Moreno, A. Vieira-Coelho, J.A. Martens, P.A. Jacobs, and G. Poncelet. 1994. Hydroisomerisation-hydrocracking of decane over Al- and Ga-pillared clays. *Journal of Catalysis* 148: 304–314.

Moraes, D.S., R.S. Angélica, C.E.F. Costa, G.N. Rocha Filho, and J.R. Zamian. 2011. Bentonite functionalized with propyl sulfonic acid groups used as catalyst in esterification reactions. *Applied Clay Science* 51: 209–213.

Moreno, S., R.S. Kou, and G. Poncelet. 1996. Hydroconversion of heptane over Pt/Al-pillared montmorillonites and saponites. *Journal of Catalysis* 162: 198–208.

Moreno, S., R.S. Kou, R. Molina, and G. Poncelet. 1999. Al-, Al, Zr-, and Zr-pillared montmorillonites and saponites: Preparation, characterization, and catalytic activity in heptane hydroconversion. *Journal of Catalysis* 182: 174–185.

Mori, T. and K. Suzuki. 1989. Shape selective property of alumina-pillared montmorillonite with different lateral distances in m-xylene conversion. *Chemistry Letters* 18(12): 2165–2168.

Morikawa, Y., K. Takagi, Y. Moro-oka, and Y. Ikawa. 1982. Cu-fluor tetra silisic mica. A novel effective catalyst for the dehydrogenation of methanol to form methyl formate. *Chemistry Letters* 11: 1805–1808.

Moronta, A., J. Luengo, Y. Ramírez, J. Quiñónez, E. González, and J. Sánchez. 2005. Isomerization of *cis*-2-butene and *trans*-2-butene catalysed by acid- and ion-exchanged smectite-type clays. *Applied Clay Science* 29: 117–123.

Moronta, A., M.E. Troconis, E. Gonzaléz et al. 2006. Dehydrogenation of ethylbenzene to styrene catalyzed by Co, Mo and CoMo catalysts supported on natural and aluminium-pillared clays. Effect of the metal reduction. *Applied Catalysis A: General* 310: 199–204.

Moronta, A., T. Oberto, G. Carruyo et al. 2008. Isomerization of 1-butene catalyzed by ion-exchanged, pillared and ion-exchanged/pillared clays. *Applied Catalysis A: General* 334: 173–178.

Moronta, A., V. Ferrer, J. Quero, G. Arteaga, and E. Choren. 2002. Influence of preparation method on the catalytic properties of acid-activated tetramethylammonium-exchanged clays. *Applied Catalysis* 230: 127–135.

Moronta, S., R.S. Kou, and G. Poncelet. 1996. Hydroconversion of heptane over Pt/Al-pillared montmroillonites and saponites. *Journal of Catalysis* 162: 198–208.

Mortland, M.M. 1970. Clay-organic complexes and interactions. *Advances in Agronomy* 22: 75–117.

Mortland, M.M. and V. Berkheiser. 1976. Triethylene diamine-clay complexes as matrices for adsorption and catalytic reactions. *Clays and Clay Minerals* 24: 60–63.

Motokura, K., N. Fujita, K. Mori, Nakagiri, T. Mizugaki, K. Ebitani, and K. Kaneda. 2006a. Brønsted acid mediated heterogeneous addition reaction of 1,3-dicarbonyl compounds to alkenes and alcohols. *Angewandte Chemie* 118: 2667–2671.

Motokura, K., N. Nakagiri, K. Mori et al. 2006b. Efficient C–N bond formations catalyzed by a proton-exchanged montmorillonite as a heterogeneous Brønsted acid. *Organic Letters* 8: 4617–4620.

Motokura, K., N. Nakagiri, T. Mizugaki, K. Ebitani, and K. Kaneda. 2007. Nucleophilic substitution reactions of alcohols with use of montmorillonite catalysts as solid Brønsted acids. *Journal of Organic Chemistry* 72: 6006–6015.

Motokura, K., S. Matsunaga, A. Miyaji, T. Yashima, and T. Baba. 2011. Solvent-induced selectivity switching: Intermolecular allylsilylation, arylsilylation, and silylation of alkynes over montmorillonite catalyst. *Tetrahedron Letters* 52: 6687–6692.

Motokura, K., S. Matsunaga, A. Miyaji, Y. Sakamoto, and T. Baba. 2010. Heterogeneous allylsilylation of aromatic and aliphatic alkenes catalyzed by proton-exchanged montmorillonite. *Organic Letters* 12: 1508–1511.

Motokura, K., S. Matsunaga, H. Noda, A. Miyaji, and T. Baba. 2012. Water-accelerated allylsilylation of alkenes using a proton-exchanged montmorillonite catalyst. *ACS Catalysis* 2: 1942–1946.

Moussaoui, Y. and R.B. Salem. 2007. Catalyzed Knoevenagel reactions on inorganic solid supports: Application to the synthesis of coumarin compounds. *Comptes Rendus Chimie* 10: 1162–1169.

Nadir, U.K. and A. Singh. 2005. Microwave-induced clay-catalyzed ring opening of *N*-tosylaziridines: A green approach to achiral and chiral diamines. *Tetrahedron Letters* 46: 2083–2086.

Nagy, N.M., M.A. Jakab, J. Kónya, and S. Antus. 2002. Convenient preparation of 1,1-diacetates from aromatic aldehydes catalyzed by zinc-montmorillonite. *Applied Clay Science* 21: 213–216.

Naicker, K.P., A. Lalitha, K. Pitchumani, and C. Srinivasan. 1998b. Clay-catalysed dealkylation of organic sulfides. *Catalysis Letters* 56: 237–239.

Naicker, K.P., K. Pitchumani, and R.S. Varma. 1998a. The catalytic influence of clays on Bamberger rearrangement: Unexpected formation of *p*-nitrosodiphenyl amine from *N*-phenylhydroxylamine. *Catalysis Letters* 54: 165–167.

Naidja, A. and S. Siffert. 1990. Oxidative decarboxylation of isocitric acid in the presence of montmorillonite. *Clay Minerals* 25: 27–37.

Naigre, R. and H. Alper. 1996. Palladium clay catalyzed regio- and stereospecific synthesis of β,γ-unsaturated acids by the carbonylation of allylic alcohols. *Journal of Molecular Catalysis A: Chemical* 111: 11–15.

Narayanan, B.N. and S. Sugunan. 2008. Alkylation of benzene with 1-octene over titania pillared montmorillonite. *Reaction Kinetics and Catalysis Letters* 94: 77–83.

Narayanan, S. and K. Deshpande. 1996. A comparative aniline alkylation activity of montmorillonite and vanadia-montmorillonite with silica and vanadia-silica. *Applied Catalysis A: General* 135: 125–135.

Narayanan, S. and K. Deshpande. 2000. Alumina pillared montmorillonite: chacterization and catalysis of toluene benzylation and aniline ethylation. *Applied Catalysis A: General* 193: 17–27.

Naskar, S., P. Paira, R. Paira et al. 2010. Montmorillonite K-10 clay catalyzed solvent-free synthesis of bis-idolylindane-1,3-dione, 2-(1′,3′-dihydro-1*H*[2,3′]biindolyl-2′-ylidene)-indan-1,3-dione and bisindolylindeno[1,2-*b*]quinoxaline under microwave irradiation. *Tetrahedron* 66: 5196–5203.

Neji, S.B., M. Trabelsi, and M.H. Frikha. 2009. Esterification of fatty acids with short-chain alcohols over commercial acid clays in a semi-continuous reactor. *Energies* 2: 1107–1117.

Nguyen, T.T. 1986. Infrared spectroscopic study of the formamide-Na-montmorillonite complex. Conversion of s-triazine to formamide. *Clays and Clay Minerals* 34: 521–528.

Nie, G., X. Zhang, P. Han et al. 2017. Lignin-derived multi-cyclic high density biofuel by alkylation and hydrogenated intramolecular cyclization. *Chemical Engineering Science* 158: 64–69.

Nikalje, M.D., P. Phukan, and A. Sudalai. 2000. Recent advances in clay-catalyzed organic transformations. *Organic Preparations and Procedures International* 32: 1–40.

Nikpassand, M., L.Z. Fekri, and M.R. Mousavi. 2017. Synthesis of azo-linked diindolyl methanes using Fe^{3+}-montmorillonite K10 under solvent-free condition. *Journal of Taibah University for Science* 11: 151–158.

Nikpassand, M., M. Mamaghani, K. Tabatabaeian, and M.K. Abiazi. 2009. KSF: An efficient catalyst for the regioselective synthesis of 1,5-diaryl pyrazoles using Baylis-Hillman adducts. *Molecular Diversity* 13: 389–393.

Nishimura, T., S. Ohtaka, A. Kimura et al. 2000. Metal cation-exchanged montmorillonite (M^{n+}-mont)-catalyzed reductive alkylation of phenol and 1-naphthol with cyclohexanones. *Applied Catalysis A: General* 194–195: 415–425.

Nishimura, T., S. Ohtaka, K. Hashimoto et al. 2004. Metal cation-exchanged montmorillonite (M^{n+}-mont)-catalyzed Friedel–Crafts acylation of 1-methyl-1-cyclohexene and 1-trimethylsilyl-1-alkynes. *Bulletin of the Chemical Society of Japan* 77: 1765–1766.

Nishimura, T., T. Yoshinaka, and S. Uemura. 2005. Metal cation-exchanged montmorillonite-catalyzed addition of organic disulfides to alkenes. *Bulletin of the Chemical Society of Japan* 78: 1138–1141.

Noor, N.M., T.N.M.T. Ismail, Y.S. Kian, and H.A. Hassan. 2013. Synthesis of palm-based polyols: effect of K10 montmorillonite catalyst. *Journal of Oil Palm Research* 25: 92–99.

Novikova, L., P. Ayrault, C. Fontaine, G. Chatel, F. Jérôme, and L. Belchinskaya. 2016. Effect of low frequency ultrasound on the surface properties of natural aluminosilicates. *Ultrasonics Sonochemistry* 31: 598–609.

Nowrouzi, F., A.N. Thadani, and R.A. Batey. 2009. Allylation and crotylation of ketones and aldehydes using potassium organotrifluoroborate salts under Lewis acid and montmorillonite K10 catalyzed conditions *Organic Letters* 11: 2631–2634.

Noyan, H., M. Önal, and Y. Sarikaya. 2006. The effect of heating on the surface area, porosity and surface acidity of a bentonite. *Clays and Clay Minerals* 54: 375–381.

Occelli, M.L. 1983. Catalytic cracking with an interlayered clay. A two-dimensional molecular sieve. *Industrial & Engineering Chemistry Product Research Development* 22: 553–559.

Occelli, M.L. 1986. New routes to the preparation of pillared montmorillonite catalysts. *Journal of Molecular Catalysis* 35: 377–389.

Occelli, M.L. 1987. Surface and catalytic properties of some pillared clays. In *Proceedings of the International Clay Conference, Denver, 1985*, (Eds.) L.G. Schultz, H. van Olphen, and F.A. Mumpton, pp. 319–323. Bloomington, IN: The Clay Minerals Society.

Occelli, M.L. 1988. Surface properties and cracking activity of delaminated clay catalysts. *Catalysis Today* 2: 339–355.

Occelli, M.L. and J.E. Lester. 1985. Nature of active sites and coking reactions in a pillared clay mineral. *Industrial and Engineering Chemistry Product Research Development* 24: 27–32.

Occelli, M.L. and R.M. Tindwa. 1983. Physicochemical properties of montmorillonite interlayered with cationic oxyaluminum pillars. *Clays and Clay Minerals* 31: 22–28.

Occelli, M.L., R.A. Innes, F.S.S. Hwu, and J.W. Hightower. 1985. Sorption and catalysis on sodium-montmorillonite interlayered with aluminum oxide clusters. *Applied Catalysis* 14: 69–82.

Ohtsuka, K. 1997. Preparation and properties of two-dimensional microporous pillared interlayered solids. *Chemistry of Materials* 9: 2039–2050.

Olaya, A., G. Blanco, S. Bernal, S. Moreno, and R. Molina. 2009. Synthesis of pillared clays with A-Fe and Al-Fe-Ce starting from concentrated suspensions of clay using microwaves or ultrasound, and their catalytic activity in the phenol oxidation reaction. *Applied Catalysis B: Environmental* 93: 56–65.

Olutoye, M.A. and B.H. Hameed. 2013. A highly active clay-based catalyst for the synthesis of fatty acid methyl ester from waste cooking palm oil. *Applied Catalysis A: General* 450: 57–62.

Onaka, M., T. Shinoda, Y. Izumi, and E. Nolen. 1993. Porphyrin synthesis in clay nanospaces. *Chemistry Letters* 22: 117–120.

Onaka, M., Y. Hosokawa, K. Higuchi, and Y. Izumi. 1993. Acidity comparison between ion-exchanged clay montmorillonites by using silylation of alcohol. *Tetrahedron Letters* 34: 1171–1172.

Önal, M. 2007. Swelling and cation exchange capacity relationship for samples obtained from a bentonite by acid activations and heat treatments. *Applied Clay Science* 37: 74–80.

Önal, M. and Y. Sarikaya. 2012. Maximum bleaching of vegetable oils by acid-activated bentonite: Influence of nanopore radius. *Adsorption Science & Technology* 30: 97–104.

Pai, S.G., A.R. Bajpai, A.B. Deshpande, and S.D. Samant. 2000. Benzylation of arenes in the presence of montmorillonite K10 modified using aqueous and acetonitrile solutions of $FeCl_3$. *Journal of Molecular Catalysis A: Chemical* 156: 233–243.

Pálkova, H., L. Jankovič, M. Zimowska, and J. Madejová. 2011. Alterations of the surface and morphology of tetraalkyl-ammonium modified montmorillonites upon acid treatment. *Journal of Colloid and Interface Science* 363: 213–222.

Pálkova, H., V. Hronský, L. Jankovič, and J. Madejová. 2013. The effect of acid treatment on the structure and surface acidity of tetraalkylammonium montmorillonites. *Journal of Colloid and Interface Science* 395: 166–175.

Pan, J., C. Wang, S. Guo, J. Li, and Z. Yang. 2008. Cu supported over Al-pillared interlayer clays catalysts for direct hydroxylation of benzene to phenol. *Catalysis Communications* 9: 176–181.

Panda, A.K., B.G. Mishra, D.K. Mishra, and R.K. Singh. 2010. Effect of sulphuric acid treatment on the physico-chemical characteristics of kaolin clay. *Colloids and Surfaces A: Physicochemical and Engineering Aspects* 363: 98–104.

Parulekar, V.N. and J.W. Hightower. 1987. Hydroisomerization of *n*-paraffins on a platinum-rhenium/pillared clay mineral catalyst. *Applied Catalysis* 35: 249–262.

Paul, B., W.N. Martens, and R.L. Frost. 2011. Organosilane grafted acid-activated beidellite clay for the removal of non-ionic alachlor and anionic imazaquin. *Applied Surface Science* 257: 5552–5558.

Perathoner, S. and G. Centi. 2010. Catalytic wastewater treatment using pillared clays. In: *Pillared Clays and Related Catalysts*, (Eds.) A. Gil, S.A. Korili, R. Trujillano, and M.A. Vicente, pp. 167–200. New York: Springer.

Pereira, C., S. Patrício, A.R. Silva et al. 2007. Copper acetylacetonate anchored onto amine-functionalised clays. *Journal of Colloid and Interface Science* 316: 570–579.

Pérez-Zurita, M.J., G.J. Pérez-Quintana, A.J. Hasblady et al. 2005. Synthesis of Al-PILC assisted by ultrasound: Reducing the intercalation time and the amount of synthesis water. *Clays and Clay Minerals* 53: 528–535.

Perissinotto, M., M. Lenarda, L. Storaro, and R. Ganzerla. 1997. Solid acid catalysts from clays: Acid leached metakaolin as isopropanol dehydration and 1-butene isomerization catalyst. *Journal of Molecular Catalysis A: Chemical* 121: 103–109.

Permana, Y., S. Shimazu, N. Ichikuni, and T. Uematsu. 2004. Selective synthesis of primary methoxypropanol using clay supported tris(2,4-pentanedionato)zirconium(IV). *Journal of Molecular Catalysis A: Chemical* 221: 141–144.

Pesquera, C., F. González, M.J. Hernando, C. Blanco, and I. Benito. 1995. Selectivity in the conversion of *n*-heptane on an Al-PILC modified with Ga. *Reaction Kinetics and Catalysis Letters* 55: 267–274.

Peter, O.I., O. Chidi, and M.A. Iheanacho. 2012. The preparation and application of environmentally benign titanium pillared clay catalyst for esterification of ethanol and acetic acid. *American Chemical Science Journal* 2: 45–59.

Phukan, P., J.M. Mohan, and A. Sudalai. 1999. Reaction of methyl diazoacetate with aldehydes, amines, thiols, alcohols and acids over transition metal-exchanged clays. *Journal of the Chemical Society: Perkin Transactions 1* 24: 3685–3689.

Pichowicz, M. and R. Mokoya. 2001. Porous clay heterostructures with enhanced acidity obtained from acid-activated clays. *Chemical Communications* 20: 2100–2101.

Pillai, S.K., S.S. Ray, M. Scriba, J. Bandyopadhyay, M.P. Roux-van der Merwe, and J. Badenhorst. 2013. Microwave assisted green synthesis and characterization of silver/montmorillonite heterostructures with improved antimicrobial properties. *Applied Clay Science* 83–84: 315–321.

Pinnavaia, T.J. 1983. Intercalated clay catalysts. *Science* 220: 365–371.

Pinnavaia, T.J. 1995. Clay catalysts: Opportunities for use in improving environmental quality. In *Clays: Controlling the Environment. Proceedings of the 10th International Clay Conference, Adelaide, Australia, 1993*, (Eds.) G.J. Churchman, R.W. Fitzpatrick, and R.A. Eggleton, pp. 3–8. Melbourne, Australia: CSIRO Publishing.

Pinnavaia, T.J., M-S. Tzou, S.D. Landau, and R.H. Raythatha. 1984. On the pillaring and delamination of smectite clay catalysts by polyoxo cations of aluminum. *Journal of Molecular Catalysis* 27: 195–212.

Pinnavaia, T.J., R. Raythatha, J.G-S. Lee, L.J. Halloran, and J.F. Hoffman. 1979. Intercalation of catalytically active metal complexes in mica-type silicates. Rhodium hydrogenation catalysts. *Journal of the American Chemical Society* 101: 6891–6897.

Piscitelli, F., P. Posocco, R. Toth et al. 2010. Sodium montmorillonite silylation: Unexpected effect of the aminosilane chain length. *Journal of Colloid and Interface Science* 351: 108–115.

Plee, D., F. Borg, L. Gatineau, and J.J. Fripiat. 1985. High resolution solid-state ^{27}Al and ^{29}Si nuclear magnetic resonance study of pillared clays. *Journal of the American Chemical Society* 107: 2362–2369.

Plee, D., L. Gatineau, and J.J. Fripiat. 1987. Pillaring processes of smectites with and without tetrahedral substitution. *Clays and Clay Minerals* 35: 81–88.

Polshettiwar, V. and Varma, R.S. 2008. Microwave-assisted organic synthesis and transformations using benign reaction media. *Accounts of Chemical Research* 41: 629–639.

Polverejan, M., T.R. Pauly, and T.J. Pinnavaia. 2000b. Acidic porous clay heterostructures (PCH): Intragallery assembly of mesoporous silica in synthetic saponite clays. *Chemistry of Materials* 12: 2698–2704.

Polverejan, M., Y. Liu, and T.J. Pinnavaia. 2000a. Mesostructured clay catalysts: A new porous clay heterostructure (PCH) derived from synthetic saponite. *Studies in Surface Science and Catalysis* 129: 401–408.

Pushpaletha, P., S. Rugmini, and M. Lalithambika. 2005. Correlation between surface properties and catalytic activity of clay catalysts. *Applied Clay Science* 30: 141–153.

Qi, X., H. Yoon, S-H. Lee, J. Yoon, and S-J. Kim. 2008. Surface-modified imogolite by 3-APS-OsO_4 complex: Synthesis, characterization and its application in the dihydroxylation of olefins. *Journal of Industrial and Engineering Chemistry* 14: 136–141.

Qin, Z., P. Yuan, J. Zhu, H. He, D. Liu, and S. Yang. 2010. Influences of thermal pretreatment temperature and solvent on the organosilane modification of Al_{13}-intercalated/Al-pillared montmorillonite. *Applied Clay Science* 50: 546–553.

Radwan, D., L. Saad, S. Mikhail, and S.A. Selim. 2009. Catalytic evaluation of sulfated zirconia pillared clay in *n*-hexane transformation. *Journal of Applied Sciences Research* 5: 2332–2342.

Raimondo, M., A. De Stefanis, G. Perez, and A.A.G. Tomlinson. 1998. PLS vs. zeolites as sorbents and catalysts. 5. Evidence for Brønsted/Lewis acid crossover and high acididty in conversion of C_{1-3} alcohols in some alumina-pillared smectite clays. *Applied Catalysis A: General* 171: 85–97.

Rajanarendar, E., K. Ramu, D. Karunakar, and P. Ramesh. 2005. Microwave-assisted synthesis of new isoxazolyl triazinethiones and isoxazolyl oxadiazinethiones in dry media. *Journal of Heterocyclic Chemistry* 42: 711–715.

Ramesh, S., B.S. Jai Prakash, and Y.S. Bhat. 2010. Enhancing Brønsted acid site activity of ion exchanged montmorillonite by microwave irradiation for ester synthesis. *Applied Clay Science* 48: 159–163.

Ramesh, S., B.S. Jai Prakash, and Y.S. Bhat. 2012. Highly active and selective C-alkylation of *p*-cresol with cyclohexanol using *p*-TSA treated clays under solvent-free microwave irradiation. *Applied Catalysis A: General* 413–414: 157–162.

Ramesh, S., B.S. Jai Prakash, and Y.S. Bhat. 2015. Nanoporous montmorillonite catalyzed condensation reactions under microwave irradiation: A green approach. *Current Organocatalysis* 2: 1–7.

Ramesh, S., Y.S. Bhat, and B.S. Jai Prakash. 2012. Microwave-activated p-TSA dealuminated montmorillonite—A new material with improved catalytic activity. *Clay Minerals* 47: 231–242.

Rani, C.S., N. Suresh, M.V.B. Rao, and M. Pal. 2016. Montmorillonite K10 catalyzed one-pot synthesis of 2-aryl substituted *N*-(4-oxo-1,2-dihydroquinazolin-3(4*H*)-yl)aryl or alkylamide derivatives under ultrasound irradiation. *Arabian Journal of Chemistry*, http://dx.doi.org/10.1016/j.arabjc.2016.02.014.

Rausell-Colom, J.A. and J.M. Serratosa. 1987. Reactions of clays with organic substances. In *Chemistry of Clays and Clay Minerals*, (Ed.) A.C.D. Newman, pp. 371–422. London, UK: Mineralogical Society.

Ravi, K., B. Krishnakumar, and M. Swaminathan. 2012. An efficient protocol for the green and solvent-free synthesis of azine derivatives at room temperature using $BiCl_3$-loaded montmorillonite K10 as a new recyclable heterogeneous catalyst. *ISRN Organic Chemistry*; doi:10.5402/595868.

Ravichandran, J. and B. Sivasankar. 1997. Properties and catalytic activity of acid-modified montmorillonite and vermiculite. *Clays and Clay Minerals* 45: 854–858.

Raythatha, R. and T.J. Pinnavaia. 1981. Hydrogenation of 1,3-butadienes with a rhodium complex-layered silicate intercalation catalyst. *Journal of Organometallic Chemistry* 218: 115–122.

Raythatha, R. and T.J. Pinnavaia. 1983. Clay intercalation catalysts interlayered with rhodium phosphine complexes. Surface effects on the hydrogenation and isomerization of 1-hexene. *Journal of Catalysis* 80: 47–55.

Reddy, C.R., B. Vijayakumar, P. Iyengar, G. Nagendrappa, and B.S. Jai Prakash. 2004. Synthesis of phenylacetates using aluminium-exchanged montmorillonite clay catalyst, *Journal of Molecular Catalysis A: Chemical* 223: 117–122.

Reddy, C.R., G. Nagendrappa, and B.S. Jai Prakash. 2007. Surface acidity study of M^{n+}-montmorillonite clay catalysts by FT-IR spectroscopy: Correlation with esterification activity. *Catalysis Communications* 8: 241–246.

Reddy, C.R., P. Iyengar, G. Nagendrappa, and B.S. Jai Prakash. 2005a. Esterification of dicarboxylic acids to diesters over M^{n+}-montmorillonite clay catalysts. *Catalysis Letters* 101: 87–91.

Reddy, C.R., P. Iyengar, G. Nagendrappa, and B.S. Jai Prakash. 2005b. Esterification of succinic anhydride to di-(*p*-cresyl) succinate over M^{n+}-montmorillonite clay catalysts. *Journal of Molecular Catalysis: Chemical* 229: 31–37.

Reddy, C.R., S. Ramesh, Y.S. Bhat, G. Nagendrappa, and B.S. Jai Prakash. 2010. Synthesis of *p*-cresylpropionate over M^{n+}-monmorillonite catalysts: Aspect of catalyst solvent interactions by DRIFTS study. *Reaction Kinetics Mechanisms and Catalysis* 100: 289–300.

Reddy, C.R., Y.S. Bhat, G. Nagendrappa, and B.S. Jai Prakash. 2009. Brønsted and Lewis acidity of modified montmorillonite clay catalysts determined by FT-IR spectroscopy. *Catalysis Today* 141: 157–160.

Reddy, G.J., D. Latha, C. Thirupathaiah, and S.K. Rao. 2005c. A facile synthesis of 2,3-disubstituted-6-arylpyridines from enaminones using montmorillonite K10 as solid acid support. *Tetrahedron Letters* 46: 301–302.

Reddy, T.R., G.R. Reddy, L.S. Reddy et al. 2012. Montmorillonite K-10 mediated green synthesis of cyano pyridines: Their evaluation as potential inhibitors of PDE4. *European Journal of Medicinal Chemistry* 48: 265–274.

Rhodes, C.N. and D.R. Brown. 1995. Autotransformation and ageing of acid-treated montmorillonite catalysts: A solid-state ^{27}Al NMR study. *Journal of the Chemical Society: Faraday Transactions* 91: 1031–1035.

Rigby, D., T.D. Gilbert, and J.W. Smith. 1986. The synthesis of alkyl aromatic hydrocarbons and its geochemical implications. *Organic Geochemistry* 9: 255–264.

Rocchi, D., J.F. González, and J.C. Menéndez. 2014. Montmorillonite clay-promoted, solvent-free cross-aldol condensation under focused microwave irradiation. *Molecules* 19: 7317–7326.

Rocha, J. and J. Klinowski. 1990. ^{29}Si and ^{27}Al magic-angle-spinning NMR studies of the thermal transformation of kaolinite. *Physics and Chemistry of Minerals* 17: 179–186.

Romanzini, D., V. Piroli, A. Frache, A.J. Zattera, and S. C. Amico. 2015. Sodium montmorillonite modified with methacryloxy and vinylsilanes: Influence of silylation on the morphology of clay/unsaturated polyester nanocomposites. *Applied Clay Science* 114: 550–557.

Romero, A., F. Dorado, I. Asencio, P.B. García, and J.L. Valverde. 2006. Ti-pillared clays: Synthesis and general characterization. *Clays and Clay Minerals* 54: 737–747.

Roopan, S.M., T. Maiyalagan, and F.N. Khan. 2008. Solven-free syntheses of some quinazolin-4(3*H*)-ones derivatives. *Canadian Journal of Chemistry* 86: 1019–1025.

Rostamizadeh, S., A.M. Amani, R. Aryan, H.R. Ghaieni, and L. Norouzi. 2009. Very fast and efficient synthesis of some novel substituted 2-arylbenzimidazoles in water using $ZrOCl_2.nH_2O$ on montmorillonite K10 as catalyst. *Monatshefte für Chemie* 140: 547–552.

Roudier, J-F. and A. Foucaud. 1984. Clay catalyzed ene-reaction. Synthesis of γ-lactones. *Tetrahedron Letters* 25: 4375–4378.

Rouquerol, R., J. Rouquerol, and P. Llewellyn. 2013. Thermal analysis. In *Handbook of Clay Science, 2nd ed. Developments in Clay Science, Vol. 5B*. (Eds.) F. Bergaya and G. Lagaly, pp. 361–379. Amsterdam, the Netherlands: Elsevier.

Rozalen, M. and F.J. Huertas. 2013. Comparative effect of chrysotile leaching in nitric, sulfuric and oxalic acids at room temperature. *Chemical Geology* 352: 134–142.

Ruiz-Hitzky, E. and J.J. Fripiat. 1976. Organomineral derivatives obtained by reacting organochlorosilanes with the surface of silicates in organic solvents. *Clays and Clay Minerals* 24: 25–30.

Ruiz-Hitzky, E. and P. Aranda. 2014. Novel architectures in porous materials based on clays. *Journal of Sol-Gel Science and Technology* 70: 307–316.

Ryland, L.B., M.W. Tamele, and J.N. Wilson. 1960. Cracking catalysts. In *Catalysis, Vol. VII*, (Ed.) P.H. Emmett, pp. 1–91. New York: Reinhold.

Sabu, K.R., R. Sukumar, and M. Lalithambika. 1993. Acidic properties and catalytic activity of natural kaolinitic clays for Friedel–Crafts alkylation. *Bulletin of the Chemical Society of Japan* 66: 3535–3541.

Sabu, K.R., R. Sukumar, R. Rekha, and M. Lalithambika. 1999. A comparative study on H_2SO_4, HNO_3 and $HClO_4$ treated metakaolinite of a natural kaolinite as Friedel–Crafts alkylation catalyst. *Catalysis Today* 49: 321–326.

Safari, J. and L. Javadian. 2013. Montmorillonite K-10 as a catalyst in the synthesis of 5,5-disubstituted hydantoins under ultrasound irradiation. *Journal of Chemical Science* 125: 981–987.

Safari, J. and M. Sadeghi. 2016. Montmorillonite K10: An effective catalyst for synthesis of 2-aminothiazoles. *Research on Chemical Intermediates* 42: 8175–8183.

Saha, B. and M. Streat. 1999. Transesterification of cyclohexyl acrylate with *n*-butanol and 2-ethylhexanol: Acid-treated clay, ion exchange resins and tetrabutyl titanate as catalysts. *Reactive & Functional Polymers* 40: 13–27.

Saib, D. and A. Foucaud. 1987. Allylic phenylation of 3-hydroxy-2-methylenealkanoic esters with benzene and K10 montmorillonite. *Journal of Chemical Research (S)* 11: 372–373.

Saikia, P.K., P.P. Sarmah, B.J. Borah, L. Saikia, and D.K. Dutta. 2016. Functionalized montmorillonite supported rhodium complexes: Efficient catalysts for carbonylation of methanol. *Journal of Molecular Catalysis A: Chemical* 412: 27–33.

Sakurai, H., K. Urabe, and Y. Izumi. 1988. New acidic pillared clay catalysts prepared from fluor-tetrasilicic mica. *Journal of the Chemical Society, Chemical Communications* 23: 1519–1520.

Salehi, P., M. Dabiri, M. Baghbanzadeh, and M. Bahramnejad. 2006. One-pot, three-component synthesis of 2,3-dihydro-4(1*H*)-quinazolinones by montmorillonite K-10 as an efficient and reusable catalyst. *Synthetic Communications* 36: 2287–2292.

Salmón, M., E. Angeles, and R. Miranda. 1990. Bromine/bentonite earth system, promoter of phenylmethanes from toluene. *Journal of the Chemical Society, Chemical Communications* 17: 1188–1190.

Samajdar, S., F.F. Becker, and B.K. Banik. 2001. Montmorillonite impregnated with bismuth nitrate: A versatile reagent for the synthesis of nitro compounds of biological significance. *Arkivoc* 8: 27–33.

Sánchez Camazano, M.S. and M.J. Sánchez Martín. 1983. Montmorillonite-catalyzed hydrolysis of phosmet. *Soil Science* 136: 89–93.

Sarikaya, Y., M. Önal, B. Baran, and T. Alemdaroğlu. 2000. The effect of thermal treatment on some of the physicochemical properties of a bentonite. *Clays and Clay Minerals* 48: 557–562.

Sartori, G., F. Bigi, R. Maggi, A. Mazzacani, and G. Oppici. 2001. Clay/water mixtures—A heterogeneous and ecologically efficient catalyst for the three-component stereoselective synthesis of tetrahydroquinolines. *European Journal of Organic Chemistry* 13: 2513–2518.

Sato, S., Y. Naito, and K. Aoki. 2007. Scandium cation exchanged montmorillonite catalyzed direct C-glycosylation of a 1,3-diketone, dimedone, with unprotected sugars in aqueous solution. *Carbohydrate Research* 342: 913–918.

Schutz, A., D. Plee, F. Borg, P. Jacobs, G. Poncelet, and J.J. Fripiat. 1987. Acidity and catalytic properties of pillared montmorillonite and beidellite. In *Proceedings of the International Clay Conference, Denver, 1985*, (Eds.) L.G. Schultz, H. van Olphen, and F.A. Mumpton, pp. 305–310. Bloomington, IN: The Clay Minerals Society.

Sedaghat, M.E., M.R. Booshehri, M.R. Nazarifar, and F. Farhadi. 2014. Surfactant modified bentonite (CTMAB-bentonite) as a solid heterogeneous catalyst for the rapid synthesis of 3,4-dihydropyrano[c] chromene derivatives. *Applied Clay Science* 95: 55–59.

Selvin, R., H-L. Hsu, P. Aneesh, S-H. Chen, and L.H. Li. 2010. Preparation of acid-modified bentonite for selective decomposition of cumene hydroperoxide into phenol and acetone. *Reaction Kinetics Mechanisms and Catalysis* 100: 197–204.

Shabestary, N., S. Khazaeli, D. Dutko, and B.L. Cutts. 2007. Clay-supported quaternary ammonium and phosphonium cations in triphase catalysis and the effect of cosolvent in catalytic activity. *Scientia Iranica* 14: 297–302.

Shabtai, J., N. Frydman, and R. Lazar. 1977. Synthesis and catalytic properties of a 1,4-diaza-bicyclo[2,2,2] octane-montmorillonite system—A novel type of molecular sieve. In *Proceedings of the 6th International Congress of Catalysis, 1976*, (Eds.) G.C. Bond, P.B. Wells, and F.C. Tompkins, pp. 660–667. London, UK: The Chemical Society.

Shah, N.F. and M.M. Sharma. 1993. A convenient method for the preparation of methyl ketones: Acetylation of diisobutylene and diisoamylene with acetic anhydride in the presence of cation exchange resin and acid-treated clay as catalyst. *Reactive Polymers* 20: 47–56.

Shah, N.F., M.S. Bhagwat, and M.M. Sharma. 1994. Cross-dimerization of α-methylstyrene with isoamylene and aldol condensation of cyclohexanone using a cation-exchange resin and acid-treated clay catalysts. *Reactive Polymers* 22: 19–34.

Shaikh, N.S., A.S. Gajare, V.H. Deshpande, and A.V. Bedekar. 2000. A mild procedure for the clay catalyzed selective removal of the tert-butoxycarbonyl protecting group from aromatic amines. *Tetrahedron Letters* 41: 385–387.

Shaikh, N.S., S.S. Bhor, A.S. Gajare, V.H. Deshpande, and R.D. Wakharkar. 2004. Mild and facile procedure for clay-catalyzed acetonide protection and deprotection of *N*(Boc)-amino alcohols and protection of 1,2-diols. *Tetrahedron Letters* 45: 5395–5398.

Shanmugam, P. and P.R. Singh. 2001. Montmorillonite K10 clay-microwave assisted isomerisation of acetates of the Baylis-Hillman adducts: A facile method of stereoselective synthesis of (*E*)-trisubstituted alkenes. *Synlett* 8: 1314–1316.

Shanmugharaj, A.M., K.Y. Rhee, and S.H. Ryu. 2006. Influence of dispersing medium on grafting of aminopropyltriethoxysilane in swelling clay materials. *Journal of Colloid and Interface Science* 298: 854–859.

Shao, L-X. and M. Shi. 2003. Montmorillonite KSF-catalyzed one-pot, three-component, aza-Diels–Alder reactions of methylenecyclopropanes with arenecarbaldehydes and arylamines. *Advanced Synthesis & Catalysis* 345: 963–966.

Sharifi, A., M.S. Abaee, A. Tavakkoli, M. Mirzaei, and A. Zolfaghari. 2008. Facile montmorillonite K-10-supported synthesis of xanthene derivatives under microwave and thermal conditions. *Synthetic Communications* 38: 2958–2966.

Sharma, G.V.R., B. Devi, K.S. Reddy, M.V. Reddy, A.K. Kondapi, and C. Bhaskar. 2015. Montmorillonite K10 catalyzed multi component reactions (MCR): Synthesis of novel thiazolidinones as anticancer agents. *Heterocyclic Communications* 21: 187–190.

Shimizu, K-I., T. Higuchi, E. Takasugi, T. Hatamachi, T. Kodama, and A. Satsuma. 2008. Characterization of Lewis acidity of cation-exchanged montmorillonite K-10 clay as effective heterogeneous catalyst for acetylation of alcohol. *Journal of Molecular Catalysis A: Chemical* 284: 89–96.

Shimizu, K-I., T. Kaneko, T. Fujishima, T. Kodama, H. Yoshida, and Y. Kitayama. 2002. Selective oxidation of liquid hydrocarbons over photoirradiated TiO_2 pillared clays. *Applied Catalysis A; General* 225: 185–191.

Shinoda, T., M. Onaka, and Y. Izumi. 1995a. Proposed models of mesopore structures in sulfuric acid-treated montmorillonites and K10. *Chemistry Letters* 24: 495–496.

Shinoda, T., M. Onaka, and Y. Izumi. 1995b. The reason why K10 is an effective promoter for *meso*-tretraalkylporphyrin synthesis. *Chemistry Letters* 24: 493–494.

Shirini, F., S.V. Atghia, and M. Mamaghani. 2013. Sulfonic acid-functionalized ordered nanoporous Na^+-montmorillonite as an efficient, eco-benign, and water-tolerant nanoreactor for chemoselective oxathio-acetalization of aldehydes. *International Nano Letters* 3: 3.

Siddique, R. and J. Klaus. 2009. Influence of metakaolin on the properties of mortar and concrete: A review. *Applied Clay Science* 43: 392–400.

Siddiqui, M.K.H. 1968. *Bleaching Earths*. Oxford, UK: Pergamon Press.

Sieskind, O. and P. Albrecht. 1985. Efficient synthesis of rearranged choles-13(17)-enes catalysed by montmorillonite-clay. *Tetrahedron Letters* 26: 2135–2136.

Sieskind, O. and P. Albrecht. 1993. Synthesis of alkylbenzenes by Friedel–Crafts reactions catalysed by K10 montmorillonite. *Tetrahedron Letters* 34: 1197–1200.

Silva, A.A., K. Dahmouche, and B.G. Soares. 2011. Nanostructure and dynamic mechanical properties of silane-functionalized montmorillonite/epoxy nanocomposites. *Applied Clay Science* 54: 151–158.

Silva, L.C.A., E.A. Silva, M.R. Monteiro, C. Silva, J.G. Teleken, and H.J. Alves. 2014. Effect of the chemical composition of smectites used in KF/clay catalysts on soybean oil transesterification into methyl esters. *Applied Clay Science* 102: 121–127.

Sindhu, P.S. 1984. The acid character of attapulgite in terms of its structure. *Clay Research* 3: 75–80.

Singh, B., J. Patial, P. Sharma, S.G. Agarwal, G.N. Qazi, and S. Maity. 2007. Influence of acidity of montmorillonite and modified montmorillonite caly minerals for the conversion of longifolene to isolongifolene. *Journal of Molecular Catalysis A: Chemical* 266: 215–220.

Singh, D.U. and S.D. Samant. 2004. Comparative study of benzylation of benzene using benzyl chloride in the presence of pillared bentonite; ion-exchanged and pillaring solution impregnated montmorillonite K10. *Journal of Molecular Catalysis A: Chemical* 223: 111–116.

Singh, D.U., P.R. Singh, and S.D. Samant. 2004b. Fe-exchanged montmorillonite K10—The first heterogeneous catalyst for acylation of sulfonamides with carboxylic acid anhydrides. *Tetrahedron Letters* 45: 4805–4807.

Singh, D.U., P.R. Singh, and S.D. Samant. 2004c. Fe-pillared bentonite—An efficient catalyst for sulfonylation of arenes using aryl and alkyl sulfonyl chlorides. *Tetrahedron Letters* 45: 9079–9082.

Singh, V., A. Khurana, I. Kaur, V. Sapehiyia, G.L. Kad, and J. Singh. 2002. Microwave-assisted facile synthesis of elvirol, curcuphenol and sesquichamaenol using montmorillonite K-10 in dry media. *Journal of the Chemical Society, Perkin Transactions* 1 15: 1766–1768.

Singh, V., V. Sapehiyia, and G.L. Kad. 2004a. Ultrasound and microwave activated preparation of ZrO_2-pillared clay composite: Catalytic activity for selective, solventless acylation of 1, *n*-diols. *Journal of Molecular Catalysis A: Chemical* 210: 119–124.

Singh, V., V. Sapehiyia, V. Srivastava, and S. Kaur. 2006. ZrO_2-pillared clay: An efficient catalyst for solventless synthesis of biologically active multifunctional dihydropyrimidinones. *Catalysis Communications* 7: 571–578.

Skoularikis, N.D., R.W. Coughlin, A. Kostapapas, K. Carrado, and S.L. Suib. 1988. Catalytic performance of iron(III) and chromium (III) exchanged pillared clays. *Applied Catalysis* 39: 61–76.

Slade, P.G., P.K. Schultz, and E.R.T. Tiekink. 1989. Structure of a 1.4-diazabicyclo[2,2,2] octane-vermiculite intercalate. *Clays and Clay Minerals* 37: 81–88.

Soetaredjo, F.E., A. Ayucitra, S. Ismadji, and A.L. Maukar. 2011. KOH/bentonite catalysts for transesterification of palm oil to biodiesel. *Applied Clay Science* 53: 341–346.

Song, G., B. Wang, H. Luo, and L. Yang. 2007. Fe^{3+}-montmorillonite as a cost-effective and recyclable solid acidic catalyst for the synthesis of xanthenediones. *Catalysis Communications* 8: 673–676.

Song, G., B. Wang, X. Wu, Y. Kang, and L. Yang. 2005. Montmorillonite K10 clay: An effective solid catalyst for one-pot synthesis of polyhydroquinoline derivatives. *Synthetic Communications* 35: 2875–2880.

Spencer, W.F. and J.E. Gieseking. 1952. Organic derivatives of montmorillonite. *Journal of Physical Chemistry* 56: 751–753.

Srinivas, K.V.N.S. and B. Das. 2003. A highly convenient, efficient, and selective process for preparation of esters and amides from carboxylic acids using Fe^{3+}-K10 montmorillonite clay. *Journal of Organic Chemistry* 68: 1165–1167.

Sterte, J. 1986. Synthesis and properties of titanium oxide cross-linked montmorillonite. *Clays and Clay Minerals* 34: 658–664.

Steudel, A., L.F. Batenburg, H.R. Fischer, P.G. Weidler, and K. Emmerich. 2009. Alteration of swelling clay minerals by acid activation. *Applied Clay Science* 44: 105–115.

Strauss, C.R. and R.W. Trainor. 1995. Invited review. Developments in microwave-assisted organic chemistry. *Australian Journal of Chemistry* 48: 1665–1692.

Stul, M.S., L. Van Leemput, and J.B. Uytterhoeven. 1983. Sorption and isomerization of normal olefins on cross-linked and simply exchanged montmorillonites. *Clays and Clay Minerals* 31: 158–159.

Su, L., Q. Tao, H. He, J. Zhu, and P. Yuan. 2012. Locking effect: A novel insight in the silylation of montmorillonite surfaces. *Materials Chemistry and Physics* 136: 292–295.

Su, L., Q. Tao, H. He, J. Zhu, P. Yuan, and R. Zhu. 2013. Silylation of montmorillonite surfaces: Dependence on solvent nature. *Journal of Colloid and Interface Science* 391: 16–20.

Suárez Barrios, M., L.V. Flores González, M.A. Vicente Rodríguez, and J.M. Martín Pozas. 1995. Acid activation of a palygorskite with HCl: Development of physico-chemical, textural and surface properties. *Applied Clay Science* 10: 247–258.

Suquet, H., R. Franck, J-F. Lambert, F. Elsass, C. Marcilli, and S. Chevalier. 1994. Catalytic properties of two pre-cracking matrices: A leached vermiculite and a Al-pillared saponite. *Applied Clay Science* 8: 349–364.

Suquet, H., S. Chevalier, C. Marcilly, and D. Barthomeuf. 1991. Preparation of porous materials by chemical activation of the Llano vermiculite. *Clay Minerals* 26: 49–60.

Suresh, D., A. Dhakshinamoorthy, and K. Pitchumani. 2013a. A green route for the synthesis of 2-substituted benzoxazole derivatives catalyzed Al^{3+}-exchanged K10 clay. *Tetrahedron Letters* 54: 6415–6419.

Suresh, D., A. Dhakshinamoorthy, K. Kanagaraj, and K. Pitchumani. 2013b. Synthesis of 2-substituted 3-ethyl-3*H*-imidazo[4,5-*b*]pyridines catalyzed by Al^{3+}-exchanged K10 clay as solid acids. *Tetrahedron Letters* 54: 6479–6484.

Suresh, D., K. Kanagaraj, and K. Pitchumani. 2014. Microwave promoted one-pot synthesis of 2-aryl substituted 1,3,4-oxadiazoles and 1,2,4-oxadiazole derivatives using Al^{3+}-K10 clay as a heterogeneous catalyst. *Tetrahedron Letters* 55: 3678–3682.

Suzuki, E., S. Idemura, and Y. Ono. 1988a. Catalytic conversion of 2-propanol and ethanol over synthetic hectorite and its analogues. *Applied Clay Science* 3: 123–134.

Suzuki, K., M. Horio, H. Matsuda, and T. Mori. 1991. Control of the distance between the silicate layers of hectorite by pillaring with alumina in the presence of polyvinyl alcohol. *Journal of the Chemical Society, Chemical Communications* 13: 873–874.

Suzuki, K., T. Mori, K. Kawase, H. Sakami, and S. Iida. 1988b. Preparation of delaminated clay having a narrow micropore distribution in the presence of hydroxyaluminum cations and polyvinyl alcohol. *Clays and Clay Minerals* 36: 142–152.

Svensson, P.D. and S. Hansen. 2010. Freezing and thawing of montmorillonite—A time-resolved synchrotron X-ray diffraction study. *Applied Clay Science* 49: 127–134.

Swarnakar, R., K.B. Brandt, and R.A. Kydd. 1996. Catalytic activity of Ti- and Al-pillared montmorillonite and beidellite for cumene cracking and hydrocracking. *Applied Catalysis A: General* 142: 51–71.

Szostak, R. and C. Ingram. 1995. Pillared layered structures (PLS): From microporous to nano-phase materials. In *Catalysis by Microporous Materials*, (Eds.) H.K. Beyer, H.G. Karge, I. Kiricsi, and J.B. Nagy, pp. 13–38. Amsterdam, the Netherlands: Elsevier.

Szücs, A., Z. Király, F. Berger, and I. Dékány. 1998. Preparation and hydrogen sorption of Pd nanoparticles on Al_2O_3 pillared clays. *Colloids and Surfaces A: Physicochemical and Engineering Aspects* 139: 109–118.

Tajbakhsh, M., M. Bazzar, S.F. Ramzanian, and M. Tajbakhsh. 2014. Sulfonated nanoclay minerals as a recyclable eco-friendly catalyst for the synthesis of quinoxaline derivatives in green media. *Applied Clay Science* 88–89: 178–185.

Tandiary, M.A., Y. Masui, and M. Onaka. 2015. A combination of trimethylsilyl chloride and hydrous natural montmorillonite clay: An efficient solid acid catalyst for the azidation of benzylic and allylic alcohols with trimethylsilyl azide. *RSC Advances* 5: 15736–15739.

Tao, Q., L. Su, R.L. Frost, H. He, and B.K.G. Theng. 2014. Effect of functionalized kaolinite on the curing kinetics of cycloaliphatic epoxy/anhydride system. *Applied Clay Science* 95: 317–322.

Tao, Q., Y. Fang, T. Li et al. 2016. Silylation of saponite with 3-aminopropyltriethoxysilane. *Applied Clay Science* 132–133: 133–139.

Tateiwa, J-I., A. Kimura, M. Takasuka, and S. Uemura. 1997. Metal cation-exchanged montmorillonite (M^{n+}-mont)-catalysed carbonyl-ene reactions. *Journal of the Chemical Society, Perkin Transactions 1* 15: 2169–2174.

Tateiwa, J-I., E. Hayama, T. Nishimura, and S. Uemura. 1996. Al^{3+}-exchanged montmorillonite as an effective solid catalyst for selective synthesis of alkylphenols and bisphenols. *Chemistry Letters* 25(1): 59–60.

Tateiwa, J-I., H. Horiuchi, and S. Uemura. 1995. Ce^{3+}-exchanged montmorillonite (Ce^{3+}-mont) as a useful substrate-selective acetalization catalyst. *Journal of Organic Chemistry* 60: 4039–4043.

Tateiwa, J-I., H. Horiuchi, K. Hashimoto, T. Yamauchi, and S. Uemura. 1994c. Cation-exchanged montmorillonite-catalyzed facile Friedel–Crafts alkylation of hydroxyl and methoxy aromatics with 4-hydroxybutan-2-one to produce raspberry ketone and some pharmaceutically active compounds. *Journal of Organic Chemistry* 59: 5901–5904.

Tateiwa, J-I., H. Horiuchi, M. Suama, and S. Uemura. 1994a. Cation-exchanged montmorillonite-catalyzed Friedel–Crafts transannular alkylation with (Z,Z)-1,5-cyclooctadiene. *Bulletin of the Chemical Society of Japan* 67: 2883–2885.

Tateiwa, J-I., T. Nishimura, H. Horiuchi, and S. Uemura. 1994b. Rearrangement of alkyl phenyl ethers to alkylphenols in the presence of cation-exchanged montmorillonite (M^{n+}-mont). *Journal of the Chemical Society, Perkin Transactions 1* 23: 3367–3371.

Taylor, E.C. and C-S. Chang. 1977. Trimethyl orthoformate absorbed on the montmorillonite clay K-10: An effective reagent for acetal formation. *Synthesis* 7: 467.

Tennakoon, D.T.B., J.M. Thomas, W. Jones, T.A. Carpenter, and S. Ramdas. 1986b. Characterization of clays and clay-organic systems. *Journal of the Chemical Society, Faraday Transactions 1* 82: 545–562.

Tennakoon, D.T.B., W. Jones, and J.M. Thomas. 1986a. Structural aspects of metal-oxide-pillared sheet silicates. *Journal of the Chemical Society, Faraday Transactions 1* 82: 3081–3095.

Tennakoon, D.T.B., W. Jones, J.M. Thomas, J.H. Ballantine, and J.H. Purnell. 1987. Characterization of clay and pillared clay catalysts. *Soild State Ionics* 24: 205–212.

Theng, B.K.G. 1974. *The Chemistry of Clay-Organic Reactions*. London, UK: Adam Hilger.

Theng, B.K.G. 2012. *Formation and Properties of Clay-Polymer Complexes, 2nd ed*. Amsterdam, the Netherlands: Elsevier.

Theng, B.K.G. and N. Wells. 1995. Assessing the capacity of some New Zealand clays for decolorizing vegetable oil and butter. *Applied Clay Science* 9: 321–326.

Theng, B.K.G., D.J. Greenland, and J.P. Quirk. 1967. Adsorption of alkylammonium cations by montmorillonite. *Clay Minerals* 7: 1–17.

Theng, B.K.G., G.J. Churchman, W.P. Gates, and G. Yuan. 2008. Organically modified clays for pollution uptake and environmental protection. In *Soil Mineral-Microbe-Organic Interactions*, (Eds.) Q. Huang, P.M. Huang, and A. Violante, pp. 145–174. Berlin, Germany: Springer-Verlag.

Thi Nguyen, T-T., D-K. Nguyen Chau, F. Duus, and T.N. Le. 2013. Green synthesis of carvenone by montmorillonite-catalyzed isomerization of 1,2-limonene oxide. *International Journal of Organic Chemistry* 3: 206–209.

Thomas, B. and S. Sugunan. 2006. Acetalization of ketones on K-10 clay and rare earth exchanged HFAU-Y zeolites: A mild and facile procedure for the synthesis of dimethylacetals. *Journal of Porous Materials* 13: 99–106.

Thomas, B., V.G. Ramu, S. Gopinath et al. 2011. Catalytic acetalization of carbonyl compounds over cation (Ce^{3+}, Fe^{3+} and Al^{3+}) exchanged montmorillonites and Ce^{3+}-exchanged Y zeolites. *Applied Clay Science* 53: 227–235.

Thomas, C.L. 1970. *Catalytic Processes and Proven Catalysts*. New York: Academic Press.

Timofeeva, M.N. and S.Ts. Khankhasaeva. 2009. Regulating the physicochemical and catalytic properties of layered aluminosilicates. *Kinetics and Catalysis* 50: 57–64.

Timofeeva, M.N., V.N. Panchenko, A. Gil, Y.A. Chesalov, T.P. Sorokina, and V.A. Likholobov. 2011. Synthesis of propylene glycol methyl ether from methanol and propylene oxide over alumina-pillared clays. *Applied Catalysis B: Environmental* 102: 433–440.

Timofeeva, M.N., V.N. Panchenko, M.M. Matrosova et al. 2014. Factors affecting the catalytic performance of Zr, Al-pillared clays in the synthesis of propylene glycol methyl ether. *Industrial & Engineering Chemistry Research* 53: 13565–13574.

Toma, S., P. Elečko, J. Gažová, and E. Solčániová. 1987. Diels–Alder reaction of acryloylferrocene with 1-phenyl-1,3-butadiene catalysed by homoionic forms of montmorillonite. *Collection of Czechoslovak Chemical Communications* 52: 391–394.

Tomić, Z.P., V.P. Logar, B.M. Babic, J.R. Rogan, and P. Makreski. 2011. Comparison of structural, textural and thermal characteristics of pure and acid treated bentonites from Aleksinac and Petrovac (Serbia). *Spectrochimica Acta Pert A: Molecular and Biomolecular Spectroscopy* 82: 389–395.

Tomul, F. 2011. Effect of ultrasound on the structural and textural properties of copper-impregnated cerium-modified zirconium-pillared bentonite. *Applied Surface Science* 258: 1836–1848.

Tong, D.S., X. Xia, X.P. Luo et al. 2013. Catalytic hydrolysis of cellulose to reducing sugar over acid-activated montmorillonite catalysts. *Applied Clay Science* 74: 147–153.

Tong, D.S., Y.M. Zheng, W.H. Yu, L.M. Wu, and C.H. Zhou. 2014. Catalytic cracking of rosin over acid-activated montmorillonite catalysts. *Applied Clay Science* 100: 123–138.

Tonlé, I.K., T. Diaco, E. Ngameni, and C. Detellier. 2007. Nanohybrid kaolinite-based materials obtained from the interlayer grafting of 3-aminopropyltriethoxysilane and their potential use as electrochemical sensors. *Chemistry of Materials* 19: 6629–6636.

Trombetta, M., G. Busca, M. Lenarda et al. 2000. Solid acid catalysts from clays. Evaluation of surface acidity of mono- and bi-pillared smectites by FT-IR spectroscopy measurements, NH_3-TDP and catalytic tests. *Applied Catalysis A: General* 193: 55–69.

Tyagi, B., C.D. Chudasama, and R.V. Jasra. 2006. Characterization of surface acidity of an acid montmorillonite activated with hydrothermal, ultrasonic and microwave techniques. *Applied Clay Science* 31: 16–28.

Tzou, M.S. and T.J. Pinnavaia. 1988. Chromia pillared clays. *Catalysis Today* 2: 243–259.

Urabe, K., H. Sakurai, and Y. Izumi. 1986. Pillared synthetic saponite as an efficient alkylation catalyst. *Journal of the Chemical Society, Chemical Communications* 14: 1074–1076.

Valentin, J.L., M.A. López-Manchado, A. Rodríguez, P. Posadas, and L. Ibarra. 2007. Novel anhydrous unfolded structure by heating of acid pre-treated sepiolite. *Applied Clay Science* 36: 245–255.

Valenzuela Díaz, F.R. and P. de Souza Santos. 2001. Studies on the acid activation of Brazilian smectite clays. *Quimica Nova* 24: 1–17.

Van Leemput, L., M.S. Stul, A. Maes, J.B. Uytterhoeven, and A Cremers. 1983. Surface properties of smectites exchanged with mono-and biprotonated 1,4, diazobicyclo(2,2,2)-octane. *Clays and Clay Minerals* 31: 261–268.

Vansant, E.F. and J.B. Uytterhoeven. 1972. Thermodynamics of the exchange of *n*-alkylammonium ions on Na-montmorillonite. *Clays and Clay Minerals* 20: 47–54.

Varadwaj, G.B.B. and K.M. Parida. 2011. Facile synthesis of dodecatungstophosphoric acid @ TiO_2 pillared montmorillonite and its effectual exploitation towards solvent-free esterification of acetic acid with *n*-butanol. *Catalysis Letters* 141: 1476–1482.

Varadwaj, G.B.B., S. Rana, and K.M. Parida. 2013. Amine functionalized K10 montmorillonite: A solid acid-base catalyst for the Knoevenagel condensation reaction. *Dalton Transactions* 42: 5122–5129.

Varala, R., R. Enugala, and S.R. Adapa. 2006. Zinc montmorillonite as a reusable heterogeneous catalyst for the synthesis of 2,3-dihydro-1*H*-1,5-benzodiazepine derivatives. *Arkivoc* 13: 171–177.

Varma, R.S. 1999. Solvent-free organic syntheses using supported reagents and microwave irradiation. *Green Chemistry* 1: 43–55.

Varma, R.S. 2002. Clay and clay-supported reagents in organic synthesis. *Tetrahedron* 58: 1235–1255.

Varma, R.S. and D. Kumar. 1998. Surfactant pillared clays as phase-transfer catalysts: A facile synthesis of α-azidoketones from α-tosyloxyketones and sodium azide. *Catalysis Letters* 53: 225–227.

Varma, R.S. and D. Kumar. 1999. Microwave-accelerated three-component condensation reaction on clay: Sovent-free synthesis of imidazo[1,2-α] annulated pyridines, pyrazines and pyrimidines. *Tetrahedron Letters* 40: 7665–7669.

Varma, R.S. and R. Dahiya. 1997. Microwave-assisted oxidation of alcohols under solvent-free conditions using clayfen. *Tetrahedron Letters* 38: 2043–2044.

Varma, R.S. and R.K. Saini. 1997. Microwave-assisted isomerization of 2′-aminochalcones on clay: An easy route to 2-aryl-1,2,3,4-tetrahydro-4-quinolones. *Synlett* 7: 857–858.

Varma, R.S. and K.P. Naicker. 1998. Surfactant pillared clays in phase-transfer catalysis: A new route to alkyl azides from alkyl bromides and sodium azide. *Tetrahedron Letters* 39: 2915–2918.

Varma, R.S. and K.P. Naicker. 1999. Palladium chloride/tetraphenylphosphonium bromide intercalated clay: New catalyst for cross-coupling of aryl halides with arylboronic acids. *Tetrahedron Letters* 40: 439–442.

Varma, R.S. K. Pitchumani, and K.P. Naicker. 1999d. Triphasic catalyst systems based on surfactant/clay composites—Facile synthesis of cyano, thiocyano and hydroxyl compounds using a triphasic catalyst. *Green Chemistry* 1: 95–97.

Varma, R.S., K.P. Naicker, and D. Kumar. 1999b. Can ultrasound substitute for a phase-transfer catalyst? Triphase catalysis and sonochemical acceleration in nucleophilic substitution of alkyl halides and α-tosyloxyketones: Synthesis of alkyl azides and α-azidoketones. *Journal of Molecular Catalysis A: Chemical* 149: 153–160.

Varma, R.S., K.P. Naicker, and J. Aschberger. 1999a. A facile preparation of alkylazides from alkyl bromides and sodium azide using 18-crown-6 ether doped clay. *Synthetic Communications* 29: 2823–2830.

Varma, R.S., K.P. Naicker, and P.J. Liesen. 1999c. Palladium chloride and tetraphenylphosphonium bromide intercalated clay as a new catalyst for the Heck reaction. *Tetrahedron Letters* 40: 2075–2078.

Varma, R.S., R. Dahiya, and S. Kumar. 1997. Clay catalyzed synthesis of imines and enamines under solvent-free conditions using microwave irradiation. *Tetrahedron Letters* 38: 2039–2042.

Vasil'ev, N.G. and F.D. Ovcharenko. 1977. The chemistry of the surfaces of the acid forms of natural layer silicates. *Russian Chemical Reviews* 46: 775–788.

Vasil'ev, N.G., M.A. Buntova, and B.P. Tsemko. 1982. Nature of the acidity of soil minerals. *Pochvovedeniye* 6: 145–149.

Vasques, H., A. Miranda, A. Martins et al. 2005. Toluene methylation over pillared clays with Al, Zr and Al/Zr. *Studies in Surface Science and Catalysis* 158, Part B: 1469–1476.

Vass, A., J. Dudás, and R.S. Varma. 1999. Solvent-free synthesis of *N*-sulfonylimines using microwave irradiation. *Tetrahedron Letters* 40: 4951–4954.

Vaughan, D.E.W. 1988. Pillared clays—A historical perspective. *Catalysis Today* 2: 187–198.

Vaughan, D.E.W. and R.J. Lussier. 1980. Preparation of molecular sieves based on pillared interlayered clays (PILC). In *Proceedings 5th International Conference on Zeolites, Naples, 1980*, (Ed.) L.V.C. Rees, pp. 94–101. London, UK: Heyden.

Venkatachalapathy, C. and K. Pitchumani. 1997. Fries rearrangement of esters in montmorillonite clays: Steric control and selectivity. *Tetrahedron* 53: 17171–17176.

Venkatachalapathy, C., M. Rajarajan, H.S. Banu, and K. Pitchumani. 1999. Clay-supported tetrabutylammonium periodate as a versatile oxidant for alcohols and sulfides. *Tetrahedron* 55: 4071–4076.

Venkatachalapathy, C., M. Thirumalaikumar, S. Muthusubramanian, K. Pitchumani, and S. Sivasubramanian. 1997. A clean clay catalysed synthesis of α-*N*-diarylnitrones. *Synthetic Communications* 27: 4041–4047.

Venkatathri, N. 2006. Characterization and catalytic properties of a naturally occurring clay, bentonite. *Bulletin of the Catalysis Society of India* 5: 61–72.

Venkatesha, N.J., B.S. Jai Prakash, and Y.S. Bhat. 2015. The active site accessibility aspect of montmorillonite for ketone yield in ester rearrangement. *Catalysis Science & Technology* 5: 1629–1637.

Venuto, P.B. and E.T. Habib, Jr. 1979. *Fluid Catalytic Cracking with Zeolite Catalysts*. New York: Marcel Dekker.

Vicente, M.A., A. Gil, and F. Bergaya. 2013. Pillared clays and clay minerals. In *Handbook of Clay Science, 2nd ed. Developments in Clay Science, Vol. 5A*. (Eds.) F. Bergaya and G. Lagaly, pp. 523–557. Amsterdam, the Netherlands: Elsevier.

Vicente, M.A., A. Meyer, E. González, M.A. Bañares-Muñoz, L.M. Gandía, and A. Gil. 2002. Dehydrogenation of ethylbenzene on alumina-chromia pillared saponites. *Catalysis Letters* 78: 99–103.

Vicente, M.A., C. Belver, M. Sychev, R. Prihod'ko, and A. Gil. 2009. Relationship between the surface properties and the catalytic performance of Al-, Ga-, and AlGa-pillared saponites. *Industrial & Engineering Chemistry Research* 48: 406–414.

Vicente, M.A., R. Trujillano, K.J. Ciuffi, E.J. Nassar, S.A. Korili, and A. Gil. 2010. Pillared clay catalysts in green oxidation reactions. In: *Pillared Clays and Related Catalysts*, (Eds.) A. Gil, S.A. Korili, R. Trujillano, and M.A. Vicente, pp. 301–318. New York: Springer.

Vicente-Rodriguez, M.A., M. Suarez, M.A. Bañares-Muñoz, and J. d. D. Lopez-Gonzalez. 1996. Comparative FT-IR study of the removal of octahedral cations and structural modifications during acid treatment of several silicates. *Spectrochimica Acta Part A* 52: 1685–1694.

Vijayakumar, B., G. Nagendrappa, and B.S. Jai Prakash. 2009. Acid activated Indian bentonite, an efficient catalyst for esterification of carboxylic acids. *Catalysis Letters* 128: 183–189.

Vijayakumar, B., P. Iyengar, G. Nagendrappa, and B.S. Jai Prakash. 2005. Direct esterification of carboxylic acids with *p*-cresol catalysed by acid activated Indian bentonite. *Indian Journal of Chemistry* 44B: 1950–1953.

Vijayakumar, S., C. Vijaya, K. Rengaraj, and B. Sivasankar. 1994. Surface acidity-cracking activity correlation in chromium pillared clays. *Bulletin of the Chemical Society of Japan* 67: 3107–3111.

Villemin, D. and B. Labiad. 1990a. Clay catalysis: Dry condensation of tetronic acid with aldehydes under microwave irradiation. Synthesis of 3-(arylmethylene)-2,4-(3*H*, 5*H*) furandiones. *Synthetic Communications* 20: 3207–3212.

Villemin, D. and B. Labiad. 1990b. Clay catalysis: Dry condensation of 3-methyl-1-phenyl-5-pyrazolone with aldehydes under microwave irradiation. *Synthetic Communications* 20: 3213–3218.

Villemin, D. and B. Labiad. 1990c. Clay catalysis: Dry condensation of barbituric acid with aldehydes under microwave irradiation. *Synthetic Communications* 20: 3333–3337.

Villemin, D. and B. Martin. 1994. Clay catalysis: An easy synthesis of 5-nitrofuraldehyde and 5-nitrofurfurylidene derivatives under microwave irradiation. *Journal of Chemical Research (S)* 25: 146–147.

Villemin, D., B. Labiad, and A. Loupy. 1993. Clay catalysis: A convenient and rapid formation of anhydride from carboxylic acid and isopropenyl acetate under microwave irradiation. *Synthetic Communications* 23: 419–424.

Villemin, D., B. Labiad, and M. Hammadi. 1992. Formation of dithianes and dithiolanes catalysed by clay. An unexpected functional selectivity. *Journal of the Chemical Society, Chemical Communications* 17: 1192–1193.

Villemin, D., B. Labiad, and Y. Ouhilal. 1989. One-pot synthesis of indoles catalysed by montmorillonite under microwave irradiation. *Chemistry & Industry* 21(10): 607–608.

Vinoda, B.M., Y.D. Bodke, M. Vinuth, M.A. Sindhe, T. Venkatesh, and S. Telkar. 2016a. One pot synthesis, antimicrobial and *in silico* molecular docking study of 1,3-benzoxazole-5-sulfonamide derivatives. *Organic Chemistry Current Research* 5: 163. doi:10.4172/2161-0401.1000163.

Vinoda, B.M., Y.D. Bodke, M. Vinuth, R. Kenchappa, S. Telkar, and S. Patil. 2016b. Synthesis of some novel 1,3-thiazine derivatives using Fe(III)-montmorillonite catalysts as potent antimicrobial agents. *Inventi Rapid: Med Chem* 2016(2): 1–7.

Vodnár, J., J. Farkas, and S. Békássy. 2001. Catalytic decomposition of 1,4-diisoprpylbenzene dihydroperoxide on montmorillonite-type catalysts. *Applied Catalysis A: General* 208: 329–334.

Vogels, R.J.M.J., J.T. Kloprogge, and J.W. Geus. 2005. Catalytic activity of synthetic saponite clays: Effects of tetrahedral and octahedral composition. *Journal of Catalysis* 231: 443–452.

Volzone, C., O. Masini, N.A. Comelli, L.M. Grzona, E.N. Ponzi, and M.I. Ponzi. 2001. Production of camphene and limonene from pinene over acid di- and trioctahedral smectite clays. *Applied Catalysis A: General* 214: 213–218.

Volzone, C., O. Masini, N.A. Comelli, L.M. Grzona, E.N. Ponzi, and M.I. Ponzi. 2005. α-Pinene conversion by modified-kaolinitic clay. *Materials Chemistry and Physics* 93: 296–300.

Vu, M.T. and P. Maitte. 1979. Réaction des éthers d'énols sur les diols en présence de montmorillonite: Une synthèse commode d'acétals cycliques. *Bulletin de la Société Chimique de France* 5–6: 264–265.

Vu, M.T., H. Petit, and P. Maitte. 1982. Alcoolyse de l'epichlorhydrine du glycerol sur catalyseur solide. Synthèse commode d'éthers du chloro-3 propane diol-1,2. *Bulletin des Sociétés Chimiques Belges* 91: 261–262.

Wada, K. 1989. Allophane and imogolite. In *Minerals in Soil Environments, 2nd ed*, (Eds.) J.B. Dixon and S.B. Weed, pp. 1051–1087. Madison, WI: Soil Science Society of America.

Wallis, P.J., A.L. Chaffee, W.P. Gates, A.F. Patti, and J.L. Scott. 2010. Partial exchange of Fe(III) montmorillonite with hexadecyltrimethylammonium cation increases catalytic activity for hydrophobic substrates. *Langmuir* 26: 4258–4265.

Wallis, P.J., W.P. Gates, A.F. Patti, and J.L. Scott. 2011. Catalytic activity of choline modified Fe(III) montmorillonite. *Applied Clay Science* 53: 336–340.

Wallis, P.J., W.P. Gates, A.F. Patti, J.L. Scott, and E. Teoh. 2007. Assessing and improving the catalytic activity of K-10 montmorillonite. *Green Chemistry* 9: 980–986.

Wang, G., X. Su, Y. Hua et al. 2016. Kinetics and thermodynamic analysis of the adsorption of hydroxy-Al cations by montmorillonite. *Applied Clay Science* 129: 79–87.

Wang, J., Y. Masui, and M. Onaka. 2010a. Direct allylation of α-aryl alcohols with allyltrimethylsilane catalyzed by heterogeneous tin ion-exchanged montmorillonite. *Tetrahedron Letters* 51: 3300–3303.

Wang, J., Y. Masui, and M. Onaka. 2010b. Synthesis of α-amino nitriles from carbonyl compounds, amines, and trimethylsilyl cyanide: Comparison between catalyst-free conditions and the presence of tin ion-exchanged montmorillonite. *European Journal of Organic Chemistry* 9: 1763–1771.

Wang, J., Y. Masui, and M. Onaka. 2011a. Conversion of triose sugars with alcohols to alkyl lactates catalyzed by Brønsted acid tin ion-exchanged montmorillonite. *Applied Catalysis B: Environmental* 107: 135–139.

Wang, J., Y. Masui, and M. Onaka. 2011b. Direct synthesis of nitriles from alcohols with trialkylsilyl cyanide using Brønsted acid montmorillonite catalysts. *ACS Catalysis* 1: 446–454.

Wang, J., Y. Masui, K. Watanabe, and M. Onaka. 2009. Highly efficient cyanosilylation of sterically bulky ketones catalyzed by tin ion-exchanged montmorillonite. *Advanced Synthesis & Catalysis* 351: 553–557.

Wang, J., Y. Masui, T. Hattori, and M. Onaka. 2012. Rapid additions of bulky *tert*-butyldimethylsilyl cyanide to hindered ketones promoted by heterogeneous tin ion-exchanged montmorillonite catalyst. *Tetrahedron Letters* 53: 1978–1981.

Wang, T., R-D. Ma, L. Liu, and Z-P. Zhan. 2010c. Solvent-free solid acid-catalyzed nucleophilic substitution of propargylic alcohols: A green approach for the synthesis of 1,4-diynes. *Green Chemistry* 12: 1576–1579.

Wang, Y. and W. Li. 2000. Kinetics of acetic acid esterification with 2-methoxyethanol over a pillared clay catalyst. *Reaction Kinetics and Catalysis Letters* 69: 169–176.

Wang, Y. and W. Li. 2004. Kinetics of acetic acid esterification with 2-ethoxyethanol over an Al-pillared catalyst. *Reaction Kinetics and Catalysis Letters* 83: 195–203.

Waterlot, C., D. Couturier, and B. Rigo. 2000. Montmorillonite-palladium-copper catalyzed cross-coupling of methyl acrylate with arylamines. *Tetrahedron Letters* 41: 317–319.

Wei, J., G. Furrer, S. Kaufmann, and R. Schulin. 2001. Influence of clay minerals on the hydrolysis of carbamate pesticides. *Environmental Science & Technology* 35: 2226–2232.

Weiss, A. 1963. Mica-type layer silicates with alkylammonium ions. *Clays and Clay Minerals* 10: 191–224.

Wershaw, R.L., D.W. Rutherford, C.E. Rostad et al. 2003. Mass spectrometric identification of an azobenzene derivative produced by smectite-catalyzed conversion of 3-amino-4-hydroxyphenylarsonic acid. *Talanta* 59: 1219–1226.

Westlake, D.J., M.P. Atkins, and R. Gregory. 1985. The use of layered clays for the production of petrochemicals. *Acta Physica et Chemica* 31: 301–307.

Wheeler, P.A., J. Wang, J. Baker, and L.J. Mathias. 2005. Synthesis and characterization of covalently functionalized Laponite clay. *Chemistry of Materials* 17: 3012–3018.

Wibowo, T.Y., A.Z. Abdullah, and R. Zakaria. 2010. Organo-montmorillonites as catalysts for selective synthesis of glycerol monolaurate. *Applied Clay Science* 50: 280–281.

Wolters, F. and K. Emmerich. 2007. Thermal reactions of smectites–Relation of dehydroxylation temperature to octahedral structure. *Thermochimica Acta* 462: 80–88.

Wypych, F., L.B. Adad, N. Mattoso, A.A.S. Marangon, and W.H. Schreiner. 2005. Synthesis and characterization of disordered layered silica obtained by selective leaching of octahedral sheets from chrysotile and phlogopite structures. *Journal of Colloid and Interface Science* 283: 107–112.

Xie, W., R. Xie, W-P. Pan et al. 2002. Thermal stability of quaternary phosphonium modified montmorillonites. *Chemistry of Materials* 14: 4837–4845.

Xie, W., Z. Gao, W-P. Pan, D. Hunter, A. Singh, and R. Vaia. 2001. Thermal degradation chemistry of alkyl quaternary ammonium montmorillonite. *Chemistry of Materials* 13: 2979–2990.

Xu, S., G. Sheng, and S.A. Boyd. 1997. Use of organoclays in pollution abatement. *Advances in Agronomy* 59: 25–62.

Yadav, G.D. 2005. Synergism of clay and heteropoly acids as nano-catalysts for the development of green processes with potential industrial applications. *Catalysis Surveys from Asia* 9: 117–137.

Yadav, G.D. and M.S.M.M. Rahuman. 2003. Synergism of ultrasound and solid acids in intensification of Friedel–Crafts acylation of 2-methoxynaphthalene with acetic anhydride. *Ultrasonics Sonochemistry* 10: 135–138.

Yadav, G.D. and S.S. Naik. 2000. Clay-supported liquid-liquid-solid phase transfer catalysis: Synthesis of benzoic anhydride. *Organic Process Research & Development* 4: 141–146.

Yadav, J.S., B.V.S. Reddy, and A.V. Madhavi. 2005. Montmorillonite KSF clay as novel and recyclable heterogeneous solid acid for the conversion of D-glycals into furan diol. *Journal of Molecular Catalysis A: Chemical* 226: 213–214.

Yadav, J.S., B.V.S. Reddy, and G. Satheesh. 2004b. Montmorillonite clay catalyzed alkylation of pyrroles and indoles with cyclic hemi-acetals. *Tetrahedron Letters* 45: 3673–3676.

Yadav, J.S., B.V.S. Reddy, B. Eeshwaraiah, and M. Srinivas. 2004c. Montmorillonite KSF clay catalyzed one-pot synthesis of α-aminonitriles. *Tetrahedron* 60: 1767–1771.

Yadav, J.S., B.V.S. Reddy, E. Balanarsaiah, and S. Raghavendra. 2002b. Montmorillonite clay catalyzed cleavage of aziridines with alcohols. *Tetrahedron Letters* 43: 5105–5107.

Yadav, J.S., B.V.S. Reddy, G.M. Kumar, and Ch. V.S.R. Murthy. 2001. Montmorillonite clay catalyzed in situ Prins-type cyclisation reaction. *Tetrahedron Letters* 42: 89–91.

Yadav, J.S., B.V.S. Reddy, K.S. Rao, and K. Harikishan. 2002c. Montmorillonite clay: A novel reagent for the chemoselective hydrolysis of *t*-butyl esters. *Synlett* 5: 826–828.

Yadav, J.S., B.V.S. Reddy, M. Srinivas, A. Prabhakar, and B. Jagadeesh. 2004d. Montmorillonite KSF clay-promoted synthesis of enantiomerically pure 5-substituted pyrrazoles from 2,3-dihydro-4*H*-pyran-4-ones. *Tetrahedron Letters* 45: 6033–6036.

Yadav, J.S., B.V.S. Reddy, S. Sadasiv, and P.S.R. Reddy. 2002d. Montmorillonite clay-catalyzed [4+2] cycloaddition reactions: A facile synthesis of pyrano- and furanoquinolines. *Tetrahedron Letters* 43: 3853–3856.

Yadav, L.D.S., S. Singh, and A. Singh. 2002a. Novel clay-catalysed cyclisation of salicylaldehyde semicarbazones to 2H-benz[e]-1,3-oxazin-2-ones under microwave irradiation. *Tetrahedron Letters* 43: 8551–8553.

Yadav, M.K. and R.V. Jasra. 2006. Synthesis of nopol from β-pinene using $ZnCl_2$ impregnated Indian montmorillonite. *Catalysis Communications* 7: 889–895.

Yadav, M.K., C.D. Chudasama, and R.V. Jasra. 2004a. Isomerisation of α-pinene using modified montmorillonite clays. *Journal of Molecular Catalysis A: Chemical* 216: 51–59.

Yamanaka, S., T. Nishihara, M. Hattori, and Y. Suzuki. 1987. Preparation and properties of titania pillared clay. *Minerals Chemistry and Physics* 17: 87–101.

Yang, S., G. Liang, A. Gu, and H. Mao. 2013. Synthesis of mesoporous iron-incorporated silica-pillared clay and catalytic performance for phenol hydroxylation. *Applied Surface Science* 285P: 721–726.

Yang, S-Q., P. Yuan, H-P. He et al. 2012. Effect of reaction temperature on grafting of γ-aminopropyl triethoxysilane (APTES) onto kaolinite. *Applied Clay Science* 62–63: 8–14.

Yao, Y.Z. and S. Kawi. 1999. Surfactant-treated K10 montmorillonite: A high-surface-area clay catalyst. *Journal of Porous Materials* 6: 77–85.

Yariv, S. and H. Cross (Eds.). 2002. *Organo-Clay Complexes and Interactions*. New York: Marcel Dekker.

Yin, W-P. and M. Shi. 2005. Nitration of phenolic compounds by metal-modified montmorillonite KSF. *Tetrahedron* 61: 10861–10867.

Yoshinaga, N. and S. Aomine. 1962. Imogolite in some Ando soils. *Soil Science and Plant Nutrition* 8: 22–29.

Younssi, I.E., T. Rhadfi, A. Atlamsani, J-P. Quisefit, F. Herbst, and K. Draoui. 2012. K-10 montmorillonite: An efficient and reusable catalyst for the aerobic C–C bond cleavage of substituted α-ketones. *Journal of Molecular Catalysis A: Chemical* 363–364: 437–445.

Yu, J., J.W. Lim, S.Y. Kim, J. Kim, and J.N. Kim. 2015. An efficient transition-metal-free synthesis of 1H-indazoles from arylhydrazones with montmorillonite K-10 under O_2 atmosphere. *Tetrahedron Letters* 56: 1432–1436.

Yuan, G.D., B.K.G. Theng, G.J. Churchman, and W.P. Gates. 2013. Clays and clay minerals for pollution control. In *Handbook of Clay Science, 2nd edition. Developments in Clay Science, Vol. 5B*. (Eds.) F. Bergaya and G. Lagaly, pp. 587–644. Amsterdam, the Netherlands: Elsevier.

Yuan, P., P.D. Southon, Z. Liu et al. 2008. Functionalization of halloysite clay nanotubes by grafting with γ-aminopropyltriethoxysilane. *Journal of Physical Chemistry C* 112: 15742–15751.

Yue, L., B. Li, Z. Ren, N. Wu, and X. Li. 2014. Facile synthesis of copper-doped mesoporous silica pillared clay (Cu-MSPC) as a high-performance catalyst for hydroxylation of benzene to phenol. *Chemistry Letters* 43: 1473–1475.

Zanzottera, C., A. Vicente, E. Celasco, C. Fernandez, E. Garrone, and B. Bonelli. 2012. Physico-chemical properties of imogolite nanotubes functionalized on both external and internal surfaces. *Journal of Physical Chemistry C* 116: 7499–7506.

Zatta, L., L.P. Ramos, and F. Wypych. 2013. Acid-activated montmorillonites as heterogeneous catalysts for the esterification of lauric acid with methanol. *Applied Clay Science* 80–81: 236–244.

Zhang, A., R. Zhang, N. Zhang, S. Hong, and M. Zhang. 2010b. Synthesis of new NiO-SiO_2-sol pillared montmorillonite and its catalytic activity in the hydrogenation of benzene. *Kinetics and Catalysis* 51: 710–713.

Zhang, A-B., P. Li, H-Y. Zhang et al. 2012. Effects of acid treatment on the physico-chemical and pore characteristics of halloysite. *Colloids and Surfaces A: Physicochemical and Engineering Aspects* 396: 182–188.

Zhang, J., Q. Wang, H. Chen, and A. Wang. 2010a. XRF and nitrogen adsorption studies of acid-activated palygorskite. *Clay Minerals* 45: 145–156.

Zhang, J., R.K. Gupta, and C.A. Wilkie. 2006. Controlled silylation of montmorillonite and its polyethylene nanocomposites. *Polymer* 47: 4537–4543.

Zhang, Z., B. Liu, K. Lu, J. Sun, and K. Deng. 2014. Aerobic oxidation of biomass derived 5-hydroxymethylfurfural into 5-hydroxymethyl-2-furancarboxylic acid catalyzed by a montmorillonite K-10 clay immobilized molybdenum acetylacetonate complex. *Green Chemistry* 16: 2762–2770.

Zhang, Z-H., P. Yang, T-S. Li, and C-G. Fu. 1997b. Montmorillonite clay catalysis VI: Synthesis of triarylmethanes via Baeyer condensation of aromatic aldehydes with N,N-dimethylaniline catalysed by montmorillonite K-10. *Synthetic Communications* 27: 3823–3828.

Zhang, Z-H., T-S. Li, and C-G. Fu. 1997a. Montmorillonite clay catalysis. Part 4. An efficient and convenient procedure for preparation of 1,1-diacetates from aldehydes. *Journal of Chemical Research (S)* 5: 174–175.

Zhang, Z-H., T-S. Li, F. Yang, and C-G. Fu. 1998b. Montmorillonite clay catalysis XI: Protection and deprotection of hydroxyl group by formation and cleavage of trimethylsilyl ethers catalysed by montmorillonite K-10. *Synthetic Communications* 28: 3105–3114.

Zhang, Z-H., T-S. Li, T-S. Jin, and J-T. Li. 1998a. Montmorillonite clay catalysis. Part 12. An efficient and practical procedure for synthesis of diacetals from 2,2-bis(hydroxymethyl) propane-1,3-diol with carbonyl compounds. *Journal of Chemical Research (S)* 10: 640–641.

Zhao, D., Y. Yang, and X. Guo. 1993. Preparation and characterization of lanthanum-doped pillared clays. *Materials Research Bulletin* 28: 939–949.

Zhao, D., Y. Yang, and X. Guo. 1995. Synthesis and characterization of hydroxyl-CrAl pillared clays. *Zeolites* 15: 58–66.

Zhao, H., C.H. Zhou, L.M. Wu et al. 2013. Catalytic dehydration of glycerol to acrolein over sulfuric acid-activated montmorillonite catalysts. *Applied Clay Science* 74: 154–162.

Zhao, Z., J. Ran, Y. Jiao, W. Li, and B. Miao. 2016. Modified natural halloysite nanotube solely employed as an efficient and low-cost solid acid catalyst for alpha-styrene production via direct alkenylation. *Applied Catalysis A: General* 513: 1–8.

Zhou, C-H., D. Tong, and X. Li. 2010. Synthetic hectorite: Preparation, pillaring and applications in catalysis. In: *Pillared Clays and Related Catalysts*, (Eds.) A. Gil, S.A. Korili, R. Trujillano, and M.A. Vicente, pp. 67–97. New York: Springer.

Zhu, H.Y. and G.Q. Lu. 2001. Engineering the structures of nanoporous clays with micelles of alkyl polyether surfactants. *Langmuir* 17: 588–594.

Zhu, H.Y., Z. Ding, C.Q. Lu, and G.Q. Lu. 2002. Molecular engineered porous clays using surfactants. *Applied Clay Science* 20: 165–175.

Zhu, L., S. Tian, J. Zhu, and Y. Shi. 2007. Silylated pillared clay (SPILC): A novel bentonite-based inorgano-organo composite sorbent synthesized by integration of pillaring and silylation. *Journal of Colloid and Interface Science* 315: 191–199.

Zhu, R., Q. Chen, Q. Zhou, Y. Xi, J. Zhu, and H. He. 2016. Adsorbents based on montmorillonite for contaminant removal from water: A review. *Applied Clay Science* 123: 239–258.

Zhu, R., Q. Zhou, J. Zhu, Y. Xi, and H. He. 2015. Organoclays as sorbents of hydrophobic organic contaminants: Sorptive characteristics and approaches to enhancing sorption capacity. *Clays and Clay Minerals* 63: 199–221.

Zuo, S., F. Liu, R. Zhou, and C. Qi. 2012. Adsorption/desorption and catalytic oxidation of VOCs on montmorillonite and pillared clays. *Catalysis Communications* 22: 1–5.

4 Organic Catalysis by Clay-Supported Reagents

4.1 INTRODUCTION

The use of supported reagents in organic synthesis dates back to the late 1960s. In their review on this topic, McKillop and Young (1979a, 1979b) listed several solids that have usefully served as support materials, including K10 and KSF montmorillonites. It was not until the 1980s, however, that clay minerals—often after acid activation and/or pillaring—began to be widely investigated and used as catalyst supports. Besides being easily available, inexpensive, noncorrosive, and nontoxic, clay-based supports are recyclable, paving the way for eco-friendly (*green*) chemistry (Clark et al. 1997a; Varma 1999; Yadav 2005; Nagendrappa 2011; Zhou 2011). Moreover, organic catalysis by clay-supported reagents can often proceed under mild conditions showing high yields and substrate selectivity (McKillop and Young 1979a; Laszlo 1998).

As its title suggests, this chapter focuses on the catalytic activity of the supported reagents rather than the clay mineral host. This is not to say, however, that the mineral component is only passively involved in the catalytic process. Indeed, the attributes that make layer silicates effective as heterogeneous catalysts such as layer structure, reduced dimensionality of reaction space, inherent surface acidity, extensive surface area, and abundance of mesopores are important to their functioning as solid supports (Cornélis and Laszlo 1980, 1994; Cornélis et al. 1983a, 1983b, 1990; Laszlo 1986; Clark and Macquarrie 1996; McCabe and Adams 2013). In terms of catalysis, supported reagents offer several advantages over their solution counterparts by way of restricted reactant diffusion, substrate activation/stabilization, selective reaction pathway promotion, and ease of work-up through the immobilization of by-products or toxic chemicals (Cornélis and Laszlo 1985). Moreover, impregnation or intercalation into clay minerals can stabilize the reagent against leaching and decomposition.

There are essentially five classes of clay-supported reagents for organic synthesis: (1), metal salts; (2), metal oxides and metal sulfides; (3), metal nanoparticles; (4), organic compounds and metal-organic complexes; and (5), heteropoly acids. The clay mineral support is said to be *impregnated* or *doped* with the reagent in question—although the term *doping*, when used in relation to metal cations, often denotes intercalation through a cation exchange process, while the reagent may be *adsorbed*, *encapsulated*, *entrapped*, *immobilized*, or *intercalated* by the clay mineral.

4.2 CLAY-SUPPORTED METAL SALTS

An early example of an organic reaction, catalyzed by a clay-supported metal salt, is the oxidative rearrangement of alkyl aryl ketones by thallium(III) nitrate (TTN) on K10 montmorillonite, prepared by stirring the clay mineral with a solution of TTN in a mixture of methanol and trimethyl orthoformate and evaporating to dryness (Taylor et al. 1976). Thus, TTN/K10 can rapidly and smoothly convert propiophenone into methyl α-methyl phenylacetate in 98% yield whereas TTN by itself (in refluxing acidic methanol) yields a mixture of methyl α-methyl phenylacetate (45%) and α-methoxypropiophenone (32%). Similarly, montmorillonite-supported TTN can efficiently transform β-methoxycarbonylmethyl pyrroles to the α-formyl pyrrole rather than the expected α-methyl pyrrole (Jackson et al. 1984). Taylor et al. (1976) have also noted that tetralone converts into more than ten products by TTN in methanol but only yields a 1:1 mixture of methylindane-1-carboxylate and 2-methoxy-1-tetralone in the presence of TTN/K10. Likewise, TTN/K10 can promote the formation of methyl indan-1-carboxylates (by ring contraction) and/or 2-methoxy-1-tetralones (by α-oxidation) from a series of 1-tetralones (Ferraz et al. 2001).

Research into organic catalysis by clay-supported metal salts took off in the late 1980s with the synthesis of K10-supported zinc chloride, ferric nitrate, cupric nitrate, and ammonium nitrate, and their use as catalysts for a wide range and variety of organic conversions (Cornélis and Laszlo 1985; Laszlo 1986; Balogh and Laszlo 1993; Clark et al. 1994b; Clark and Macquarrie 1996; Varma 2002). This development, in turn, has led to the manufacture and commercialization of environmentally friendly clay-supported catalysts by the name of *Envirocats*, available from Contract Chemicals, UK (Bandgar et al. 2000a, 2000b, 2000c; Lee and Ko 2004; Ghatpande and Mahajan 2005; Shaikh et al. 2011).

4.2.1 Clay-Supported Zinc(II) Chloride and Other Metal Chlorides

The preparation of K10-supported $ZnCl_2$ (*clayzic*) together with its surface properties and use as a Friedel–Crafts alkylation catalyst, has been the topic of several pioneering investigations by Laszlo's group in Liège, Belgium (Laszlo 1998) and by Clark and coworkers in York, England (Clark et al. 1994). The activity of clayzic is appreciably higher than the sum of its constituent parts, indicative of a synergistic effect between K10 montmorillonite and $ZnCl_2$ both of which are weak Lewis acids (Rhodes et al. 1991; Clark et al. 1994). Impregnation of K10 with $ZnCl_2$ (e.g., from an acetonitrile solution), followed by thermal activation of the product, yields a highly efficient and selective catalyst for Friedel–Crafts alkylations and acylations (Cornélis et al. 1990; Cornélis and Laszlo 1994).

When Clark et al. (1989) reacted benzene (**1**) with benzyl chloride (at room temperature) in the presence of their clayzic preparation, all of the benzyl chloride (**2**) was consumed within 15 min giving diphenylmethane (**3**) in 80% isolated yield, according to Scheme 4.1. Izumi et al. (1997a), however, observed only a 45% conversion of benzyl chloride and a modest 32% yield of diphenylmethane. Zn^{2+}-exchanged K10 can also give full conversion of benzyl chloride but yields only 60% diphenylmethane in 90 min. under reflux, similar to what Laszlo and Mathy (1987) have found earlier and what Izumi et al. (1997a) have reported for Zn^{2+}-nontronite.

Interestingly, K10-supported $AlCl_3$ and $FeCl_3$ are relatively ineffective, giving ≤6% conversion in 60 min. On the other hand, Natekar and Samant (1995) reported that the catalytic activity of $AlCl_3$ and $FeCl_3$, supported on a bentonite from India, was comparable to that shown by the $ZnCl_2$-impregnated clay sample. No details, however, were given about the clay host or about the surface properties of the bentonite-supported catalysts. Pinho et al. (1995) also noted that chrysotile-supported $ZnCl_2$, $CoCl_2$, $FeCl_3$, $NiCl_2$, $CrCl_3$ gave diphenylmethane in 60%–75% yield, while the acylation of anisole with acetyl chloride was markedly selective for *para*-acetylanisole. Likewise, Phukan et al. (2003) observed that both the Zn^{2+}-exchanged montmorillonite (from Wyoming) and its $ZnCl_2$-impregnated counterpart gave comparable yields of diphenylmethane (Scheme 4.1).

Clark et al. (1989) showed that benzyl alcohol reacted much more slowly (with benzene) than did benzyl chloride, while Laszlo (1998) reported that (at 20°C) toluene was not alkylated in the presence of clayzic and an equimolar mixture of benzyl alcohol and benzyl chloride. Aromatic hydrocarbon substrates can apparently act as their own solvents. Thus, the presence of benzene as

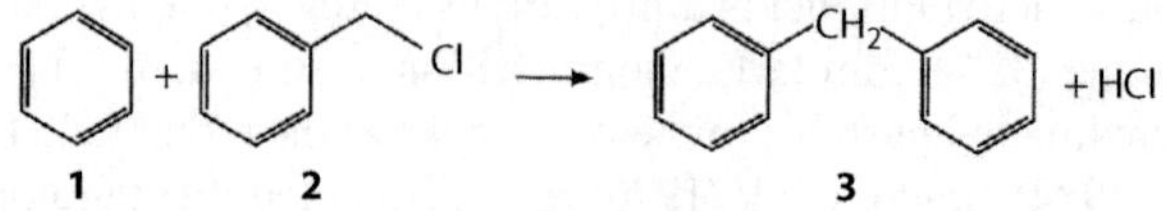

SCHEME 4.1 Reaction of benzene (**1**) with benzyl chloride (**2**) to yield diphenylmethane (**3**) in the presence of K10-supported $ZnCl_2$ ("clayzic"). (After Clark, J.H. et al., *J. Chem. Soc., Chem. Commun.*, 1353–1354, 1989.)

a co-reactant significantly enhances the alkylation (by benzyl chloride) of toluene, mesitylene, or p-xylene (Cornélis et al. 1991a). Further, in a mixture of mesitylene and toluene, the relative reactivity toward benzyl chloride is *inverted* in that mesitylene is favored over toluene whereas the converse applies when the two substrates are separately benzylated (Cornélis et al. 1991b). Reactivity inversion in the benzylation of toluene by benzyl chloride and benzyl alcohol may partly be ascribed to preferential adsorption (to the solid catalyst) of benzyl alcohol (Yadav and Thorat 1996).

Similarly, Clark et al. (1994a) observed that thiophene could be selectively benzylated at the 2-position in >90% yield using clayzic whereas benzylation of benzo[b]thiophene yielded the 3-benzyl isomer (Clark and Mesher 1995). Besides being effective in catalyzing the cyclization of phenylthioacetals to yield benzo[b]thiophene and its derivatives (Clark et al. 1995), clayzic can promote the thioalkylation of thiophenes and benzo[b]thiophenes (Clark et al. 1996).

Since impregnation with $ZnCl_2$ would add Lewis acid sites to K10 montmorillonite, the overall activity of the resultant clayzic material would be enhanced. The question remains why clayzic is so much more active and selective (as an alkylation catalyst) in comparison with bulk (unsupported) $ZnCl_2$ and Zn^{2+}-exchanged K10. In response, Clark and coworkers (Barlow et al. 1994; Clark et al. 1994b; Clark and Macquarrie 1996) and Rhodes and Brown (1992, 1993) have proposed that acid treatment of montmorillonite causes partial dissolution of the octahedral sheet, and the release of Al^{3+} (and other octahedrally coordinated cations), giving rise to a mesoporous solid with pores of 6–10 nm in diameter together with a short-range order (*amorphous*) silica phase (cf. Chapter 3). The extent of octahedral sheet dissolution, together with the increase in specific surface area and mesoporosity in terms of both pore diameter and pore volume, is dependent on the severity of acid treatment (Rhodes and Brown 1993).

The addition of $ZnCl_2$ (up to ten times its cation exchange capacity) to the acid-activated material leads to the replacement of Al^{3+} (and protons) by Zn^{2+} followed by entry into, and occupancy of, the mesopores by $ZnCl_2$ (Figure 4.1). The deposition of $ZnCl_2$, even at a low loading of 1 mmol/g, would render a large proportion of Brønsted acid sites, initially present in the K10 support, inaccessible to the substrate (Massam and Brown 1995). Excess $ZnCl_2$ would adsorb on external K10 particle surfaces and behave much like the bulk salt in showing low catalytic activity. A similar scheme has been suggested by Sukumar et al. (1998) for acid-treated metakaolinite impregnated with various metal chlorides.

Thermal analysis has indicated that metal nitrates, supported on K10 montmorillonite, are present in the form of their respective crystalline hydrates, rather than as the acetone solvates (Cseri et al. 1996b). Since these reagents tend to decompose at a lower temperature as compared with the unsupported (bulk) counterparts, solvent removal by evaporation (during catalyst preparation) should be carried out at room temperature or below 35°C. Interestingly, metal nitrates that decompose at a relatively low temperature are generally very effective in catalyzing the oxidation of alcohols and nitration of phenols.

Barlow et al. (1993) have proposed that zinc ions occupying the mesopores of K10 (Shinoda et al. 1995) can act as Lewis acid sites in the Friedel–Crafts alkylation of aromatic substrates with benzyl chloride in the presence of clayzic (dried at 280°C). Coordination of the polar anisole with these sites would account for the low rate of its benzylation as compared with benzene. On the other hand, both Cornélis et al. (1993a) and Massam and Brown (1995) have suggested that alkylation with benzyl alcohol in the presence of clayzic (dried at 120°C) involves participation of Brønsted acid sites (protons) arising from dissociation of Zn^{2+}-coordinated water molecules (cf. Chapter 2). Such a mechanism, however, would not be operative with clayzic that has been dehydrated (by thermal activation) at 280°C (Clark et al. 1994). Indeed, Choudhary and Mantri (2002) have proposed that thermal activation leads to the formation of new Lewis acid sites (–O–Zn–Cl) by reaction of $ZnCl_2$ with surface hydroxyl groups of K10 and the evolution of HCl. Similarly, Clark et al. (1997b) have suggested that the activity of K10-supported $MnCl_2$ for the thiolation by dimethyldisulfide of several aromatic compounds involves coordination of the substrate to Mn(II). Choudhary and Jha (2008) went so far as to suggest that the reaction of anhydrous $GaCl_3$ (or $AlCl_3$) with the surface

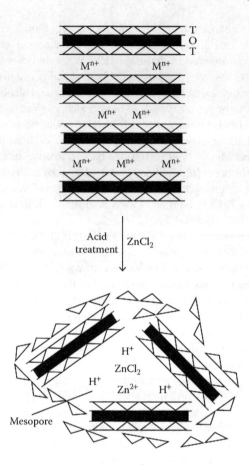

FIGURE 4.1 Diagram showing the preparation of *clayzic* by depositing $ZnCl_2$ in the mesopores of an acid-activated smectite, such as K10 montmorillonite. Prior acid treatment of the clay mineral leads to the partial replacement of exchangeable counterions by protons and the formation of a short-range order (*amorphous*) silica phase, depicted here by the dispersed open triangles. T = tetrahedral sheet; O = octahedral sheet; M^{n+} = polyvalent counterions. (Adapted from Clark, J.H. and Macquarrie, D.J., *Chem. Soc. Rev.*, 25, 303–310, 1996.)

hydroxyl groups of K10 (in dry carbon tetrachloride under reflux) gave rise to catalytically active, moisture-stable $(-O-)_2-Ga/Al$ sites. If so, these sites would be largely confined to the clay particle edge surface, consistent with the earlier suggestion by Cornélis et al. (1993b).

Clayzic is also an efficient Lewis acid catalyst for the *green* synthesis of benzimidazoles and quinoxalines from carbonyl compounds and *o*-phenylenediamine in aqueous media at room temperature (Dhakshinamoorthy et al. 2011). The proposed mechanism involves both carbonyl coordination to Zn^{2+} and protonation of the diamine. We might also add that K10-supported $ZnBr_2$ (after thermal activation at 200°C) can swiftly and selectively catalyze the *para*-bromination of bromobenzene. Interestingly, $ZnBr_2$ by itself does not catalyze the reaction, while K10-supported zinc acetate and zinc sulfate give very low conversions and selectivities because of poor dispersion in the clay matrix (Ross et al. 1998).

The presence of (hydrated) zinc ions in its mesoporous structure (Figure 4.1) makes clayzic more *polar* than the parent montmorillonite. As a result, polar substrates are more easily accommodated into the clayzic structure as compared with their nonpolar counterparts. Thus, in a mixture of alkyl alcohol and alkyl halide, the former compound is the first to react although the alcohol is less reactive than the corresponding halide. Further, the activity of clayzic in the benzylation of arylhalides decreases in the sequence: bromobenzene > chlorobenzene > fluorobenzene paralleling the polarizability of the substrates whereas the reverse order is observed under homogeneous conditions. By promoting the partitioning of alkylbenzenes into mesopores, dehydration also increases the catalytic performance of clayzic (Clark and Macquarrie 1996).

K10 montmorillonite-supported metal chloride reagents, other than $ZnCl_2$, have featured as catalysts in a variety of organic reactions. Early on, Chalais et al. (1985) reported that $FeCl_3$-impregnated K10 montmorillonite could promote the selective chlorination and arylation of adamantine. With carbon tetrachloride as the solvent, 1-chloro- and 1,3-dichloroadamantanes were formed, while monosubstituted 1-adamantyl derivatives or 1,3-disubstituted adamantanes were produced when benzene was used as the solvent.

In assessing the catalytic activity of some metal chlorides supported on K10 montmorillonite, for the reaction of benzene with benzyl chloride (Scheme 4.1), Clark et al. (1989) found that the yield of diphenylmethane decreased in the order $ZnCl_2 > NiCl_2 > CuCl_2 > CdCl_2$. Interestingly, Choudhary and Jana (2002) observed the following sequence for diphenylmethane selectivity: $InCl_3$/K10 > $GaCl_3$/K10 > $FeCl_3$/K10 > $ZnCl_2$/K10. The same order was also reported by Choudhary et al. (2001a) for the selective esterification of *tert*-butanol by acetic anhydride to yield *tert*-butyl acetate (and acetic acid) as represented by Equation 4.1:

$$(CH_3)_3COH + (CH_3CO)_2O \rightarrow CH_3COOC(CH_3)_3 + CH_3COOH \tag{4.1}$$

Further, the activity of $InCl_3$/K10 in the benzylation of benzene and substituted benzenes decreases in the order benzene > toluene > mesitylene > anisole being the reverse to that observed for the Friedel–Crafts reaction under homogeneous conditions (Choudhary and Jana 2002). The observed sequence of selectivity reflects the ease with which these compounds can enter the mesopores of the clay mineral support (Figure 4.1), coordinate to the transition metal ions, and react with benzyl chloride. K10-supported $InCl_3$ (or $GaCl_3$) is equally effective in catalyzing the acylation of benzene, aromatic alcohols, and phenols by acyl chloride, even in the presence of moisture (Choudhary et al. 2001b, 2004).

$CuCl_2$/K10 is a useful catalyst for the acetylation of benzo-15-crown-5 ether (Cseri et al. 1996a), while $CuBr_2$/K10 can promote the solvent-free coupling of aldehydes, amines, and alkynes to yield 2,4-disubstituted quinolines (Yadav et al. 2003c). Similarly, $H_2[PtCl_6]$/K10 can promote the selective hydrogenation of C = O groups of α,β-unsaturated aldehydes (Szöllősi et al. 1999), while $ZrOCl_2.8H_2O$/K10 can catalyze the conjugate addition of amines to α,β-unsaturated alkenes at room temperature under solvent-free conditions (Hashemi et al. 2006). Liu et al. (2009) found $SbCl_3$/K10 to be efficient in catalyzing the Friedel–Crafts alkylation of indoles and pyrroles with epoxides, giving good yields of the corresponding C-alkylated derivatives with high regioselectivity. More recently, Ravi et al. (2012) made the interesting observation that K10-supported $BiCl_3$ could effectively promote the synthesis of azine derivatives from benzophenone hydrazone and ketones/aldehydes by simply grinding the clay catalyst with the reactants.

In this context, we might mention the FTIR and NMR spectroscopic studies by Asseid et al. (1992) on some K10-supported metal fluorides. They suggested that CuF_2 could attack the tetrahedral sheet of K10 montmorillonite to from a $CuSiF_6$ complex whereas ZnF_2 tended to attack the octahedral sheet, giving rise to AlF_3 and AlF_4^- complexes, causing the replacement of octahedrally coordinated Al^{3+} by Zn^{2+}. Accordingly, ZnF_2/K10 was more efficient in catalyzing the reaction of benzene with benzyl chloride (Scheme 4.1) than the CuF_2/K10 and CdF_2/K10 samples.

Envirocat EPZ10, the commercially available counterpart of clayzic, was used by Bandgar and Kasture (2000b) to catalyze the Friedel–Crafts arylmethylation of phenol and anisole with various benzyl halides to yield the corresponding *para*-arylmethylated products. Envirocat EPZ10, activated by azeotrope drying with toluene, is also effective in catalyzing the Biginelli-type synthesis (cf. Chapter 5) of 3,4-dihydropyrimidin-2(1*H*)-ones from aliphatic aldehydes, ethyl acetoacetate, and urea (Lee and Ko 2004). This commercial catalyst can also promote the synthesis (in water) of

1,8-dioxo-octahydroxanthenes from aldehydes and dimedone or cyclohexane-1,3-dione (Pore et al. 2010) as well as the formation of 3-acetoacetylcoumarins from salicylaldehydes and 4-hydroxy-6-methyl-2*H*-pyran-2-one (Shaikh et al. 2011).

EPZ10 and other Envirocats such as EPZG (clay-supported ferric chloride) and EPIC (clay-supported polyphosphoric acid) can also catalyze the Friedel–Crafts acylation of substituted benzenes with acyl chloride or acid anhydride under microwave irradiation (Bandgar and Kasture 2000b). Similarly, Ghatpande and Mahajan (2005) used Envirocat EPZG to synthesize various benzophenones by reacting substituted benzenes with benzyl chlorides. Envirocat EPZG could also promote the conversion of aldoximes into nitriles (Bandgar et al. 1995), the selective acetalization of aldehydes and ketones (Bandgar and Gaikwad 1998), the transesterification of β-keto esters with various alcohols (Bandgar et al. 2001), and the synthesis of *N*-benzyl substituted amides from benzyl alcohols and substituted nitriles (Veverková and Toma 2005). This commercial catalyst is equally efficient in mediating the synthesis of imines (from aldehydes and primary amines) and enamines (from ketones and secondary amines) as well as the benzoylation of various benzene, naphthalene, and thiophene derivatives (with benzoyl chloride or benzoic anhydride) under solvent-free conditions and microwave irradiation (Varma and Dahiya 1997a; Veverková et al. 2000).

4.2.2 Clay-Supported Iron(III) Nitrate and Other Metal Nitrates

We have summarized the activity of K10-supported thallium(III) nitrate in catalyzing various organic conversions and reactions and will now describe the use of other transition metal nitrate reagents in organic catalysis, starting with K10-supported ferric nitrate (*clayfen*). This reagent may be obtained by dissolving $Fe(NO_3)_3 \cdot 9H_2O$ in acetone, adding the K10 material, and removing the solvent from the suspension by rotary evaporation under reduced pressure. Only freshly prepared samples should be used as clayfen slowly decomposes with the emission of nitrous fumes. Clayfen also loses activity when left in air at room temperature. The versatility of clayfen as a catalyst is related to its ability to serve as a source of nitrosonium (NO^+) ions, while the ferric ions can act as potent electron acceptors (Lewis acids) (Cornélis et al. 1983a; Cornélis and Laszlo 1985, 1994).

Many of the early clayfen-catalyzed reactions are concerned with the oxidation and nitration of organic substrates. An example of an oxidation reaction is the conversion of alcohols into aldehydes and ketones, involving nitrous esters as intermediates (Cornélis and Laszlo 1980; Cornélis et al. 1982). The conversion may be represented by Equation 4.2 where R_1 denotes a phenyl group and R_2 is a methyl group. The NO formed is then oxidized in air to NO_2.

$$3R_1R_2CHOH + 2H^+ + 2NO_3^- \rightarrow 3R_1R_2CO + 2NO + 4H_2O \qquad (4.2)$$

Clayfen is similarly effective in converting alcohols into carbonyl compounds (Varma and Dahiya 1997b) and oxidizing sulfides to sulfoxides (Varma and Dahiya 1998a) under solvent-free conditions and with the assistance of microwave irradiation.

Cornélis et al. (1983a, 1983c) have also found clayfen to be efficient in catalyzing the oxidative coupling of thiols (RSH) into symmetrical disulfides (RSSR). An example is shown in Scheme 4.2 with

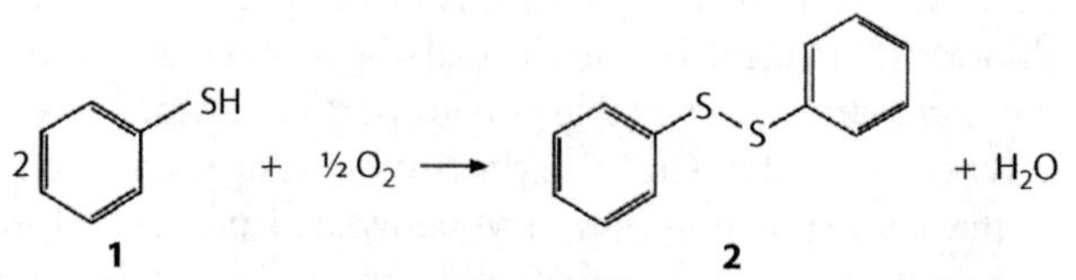

SCHEME 4.2 Oxidative coupling of thiophenol (**1**) to give diphenyl disulfide (**2**) in the presence of K10-supported ferric nitrate ("clayfen"). (After Cornélis, A. et al., *Tetrahedron Lett.*, 24, 3103–3106, 1983c.)

respect to the conversion of thiophenol (**1**) to diphenyl disulfide (**2**), involving thionitrite (RSNO) as an intermediate, according to Equations 4.3 through 4.5.

$$RSH + NO^+ \rightarrow RSNO + H^+ \tag{4.3}$$

$$RSNO \rightarrow RS\cdot + NO \tag{4.4}$$

$$2RS\cdot \rightarrow RSSR \tag{4.5}$$

Diffuse reflectance FT-infrared spectroscopy has shown that nitrosonium (and nitrate) ions are formed when transition metal nitrates are deposited on K10 montmorillonite (Srinivasan and Ganguly 1991). Thus, clayfen can efficiently catalyze the mono-nitration of various phenols with a high degree of regioselectivity as shown by the data in Table 4.1 (Cornélis et al. 1983b, 1984). Clayfen is similarly effective in promoting the synthesis of β-nitrostyrenes from styrene and its substituted derivatives under solvent-free conditions (Varma et al. 1998).

Clayfen is also an efficient catalyst for the cleavage of thioacetals (Balogh et al. 1984a; Varma and Saini 1997), *N,N'*-dimethylhydrazones (Laszlo and Polla 1984a), and oximes (Heravi et al. 2000), giving rise to the corresponding carbonyl compounds. In addition, clayfen can convert hydrazines into azides by nitrosation in dichloromethane (Laszlo and Polla 1984b) and promote the aromatization of 1,4-dihydropyridines (Balogh et al. 1984b).

Many of the aforementioned conversions can also be carried out using K10-supported cupric nitrate (*claycop*) and ammonium nitrate (*clayan*). Claycop is prepared by adding K10 to a solution of $Cu(NO_3)_2 \cdot 3H_2O$ in acetone (Balogh et al. 1984b), while clayan is obtained by suspending the clay in an aqueous solution of NH_4NO_3 (Meshram et al. 1998a). The suspensions are then placed in a rotary evaporator, and the solvent is removed under reduced pressure at ~50°C. Dry powders of claycop and clayan can be stored for at least 30 days without loss of activity.

TABLE 4.1
Clayfen-Catalyzed Nitration of Phenolic Compounds

		Yield (%)	
Phenolic Compound	**Reaction Conditions**	**ortho-Product**	**para-Product**
Phenol	Tetrahydrofuran or ether, 20 h	39	41
4-Methylphenol	Ether, 20 h	58	
3-Methylphenol	Ether, 20 h	20[a]	34
4-Chlorophenol	Ether, 20 h	88	
4-Fluorophenol	Tetrahydrofuran, 20 h or toluene, 5 h	69	
4-*tert*-Butylphenol	Tetrahydrofuran, 3 h or toluene, 2 h	92	
3-Hydroxyphenol	Ether, 2 h		50[b]
4-Hydroxybenzaldehyde	Toluene, 72 h	93	
4-Hydroxybenzonitrile	Toluene, 48 h	88	
2-Hydroxynaphthalene	Tetrahydrofuran, 2 h	63[c]	
Estrone	Toluene, 24 h	55[d,e]	

Source: Cornélis, A. and Laszlo, P., *Synthesis*, 909–918, 1985.

[a] 5-Methyl-2-nitro product
[b] 2-Hydroxy-4-nitro product
[c] 2-Hydroxy-1-nitro product
[d] 3-Hydroxy-2-nitro product
[e] No reaction in tetrahydrofuran

Like clayfen, claycop can catalyze the oxidation of aromatic compounds. Massam and Brown (1998), for example, were able to relate the activity of a claycop-like material in converting benzyl alcohol into benzaldehyde to the duration of acid treatment of the montmorillonite support and the severity of thermal activation. They suggested that Brønsted acid sites (protons) on the acid-treated montmorillonite were involved in the oxidation process, at least with respect to the clay-supported cupric nitrate reagent that had been dried at 120°C. Varma and Dahiya (1998b) also found claycop to be efficient in oxidizing a variety of organic molecules under solvent-free conditions and microwave irradiation. Earlier, Imamura et al. (1996) reported that copper-impregnated imogolite could serve as a shape-selective catalyst for the isomerization of 1-butene, and the decomposition of *tert*-butylhydroperoxide and 1,1-bis(*tert*-butyldioxy)cyclododecane. Similarly, copper-doped Al-pillared montmorillonite could mediate the oxidation of toluene and xylene (Bahranowski et al. 1999), and convert benzene into phenol (Pan et al. 2008), in the presence of hydrogen peroxide.

As we might expect, claycop is an effective catalyst for the selective nitration of aromatic compounds. Thus, Laszlo and Pennetreau (1987) found improved *para*-nitration of halobenzenes by claycop (in the presence of acetic anhydride), and Gigante et al. (1995) reported the preferential *ortho*-nitration of aromatic compounds apart from anisole. Another striking example of selective catalysis is the nitration of aromatic olefins in the presence of a catalytic amount of 2,2,6,6-tetra methylpiperidine-1-yl)oxyl (TEMPO) (Begari et al. 2014). Besides giving high product yields, the reaction showed an exclusive preference for the *E*-isomer. Thus, styrene (**1**) can be converted into [(*E*)-2-nitrovinyl]benzene (**2**) in 95% yield using dioxane as a solvent (Scheme 4.3). Claycop is similarly effective in nitrating *para*-substituted styrenes to the corresponding *E*-isomers whereas clayan (and clayfen) give low yields and produced polymeric compounds (Varma et al. 1998).

KSF montmorillonite-supported $Bi(NO_3)_3$ is an efficient catalyst for the regiospecific nitration of polycyclic aromatic compounds, estrone, and β-lactams (Samajdar et al. 2001a) as well as for the oxidation of some benzylic and allylic alcohols (Samajdar et al. 2001b). Similarly, Ravi and Tewari (2012) have reported that K10 montmorillonite, impregnated with $Bi(NO_3)_3$, can effectively promote the nitration of various pyrrazoles. González and Moronta (2004) impregnated $Co(NO_3)_2$ into a natural clay (presumably smectite) and its Al-pillared derivative, and used the supported reagents to promote the dehydrogenation of ethylbenzene to styrene. Impregnation of $Ni(NO_3)_2$ with or without $Ce(NO_3)_3$ or $Pr(NO_3)_3$ into K10, a bentonite, a montmorillonite, and a smectite together with their respective Al-pillared interlayered and polymer-modified clays, followed by calcination at 500°C (in air), yields reagents capable of catalyzing carbon dioxide reforming of methane to produce synthesis gas (Wang et al. 1998; Daza et al. 2009; Gamba et al. 2011).

Balogh et al. (1984a) used claycop to catalyze the cleavage of thioacetals into the corresponding aldehydes and ketones. Similarly, ketoximes can be cleaved into carbonyl compounds using a bentonite-supported copper(II) nitrate reagent (Sanabria et al. 1995), and dithiocarbamates decompose into isothiocyanates in the presence of claycop with benzene as a solvent (Meshram et al. 1997a).

Meshram et al. (2003) used K10-supported ammonium nitrate (clayan) for the mono-nitration of substituted arenes in the presence of perchloric acid. They suggested that the nitrosonium ion (from the supported reagent) combined with $HClO_4$ to give nitrosyl perchlorate. Reaction of $NOClO_4$ with the various arene substrates yielded nitroso arenes which, on oxidation, gave the corresponding nitro arenes. Like clayfen, clayan is an efficient reagent for the oxidative coupling of thiols into disulfides

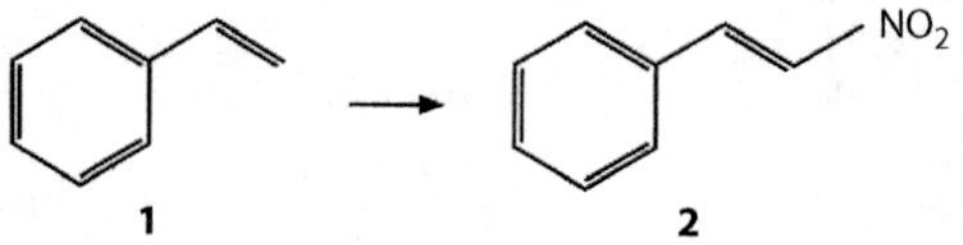

SCHEME 4.3 Quantitative conversion of styrene (**1**) to [(*E*)-2-nitrovinyl]benzene (**2**), catalyzed by K10-supported cupric nitrate ("claycop") in the presence of 2,2,6,6-tetramethylpiperidine-1-yl)oxyl (TEMPO). (After Begari, E. et al., *Synlett*, 25, 1997–2000, 2014.)

(cf. Scheme 4.2) in dry media (Meshram et al. 2000). Clayan is also effective in catalyzing the dethioacetalization of thioacetals and dithianes (Meshram et al. 1997b; 1999a), and the cleavage of electron-rich oximes, phenylhydrazones, and tosylhydrazones to yield the corresponding carbonyls (Meshram et al. 1998a, 1998b). Similarly, semicarbozones and phenylhydrazones can be cleaved to the corresponding ketones using K10-supported ammonium persulfate (Varma and Meshram 1997). Clayan can also catalyze the deprotection of tetrahydropyranyl ethers, acetals, and acetonides (Meshram et al. 1999b) as well as 4-methoxyphenyl)-methyl ethers (Yadav et al. 1998). For the most part, the earlier conversions take place under solvent-free conditions using microwave irradiation.

4.3 CLAY-SUPPORTED METAL OXIDES AND METAL SULFIDES

Clay-supported metal oxides are commonly prepared by mixing an aqueous suspension of the clay mineral, or its organically modified form, with a sol of the metal oxide/salt, and autoclaving the mixture—more commonly by washing the *heterocoagulated* clay-metal oxide/salt complex, and calcining or drying the washed product (Körösi et al. 2004; Zhu et al. 2005; Hur et al. 2006; Kun et al. 2006; Han et al. 2010; Khaorapapong et al. 2011, 2015; Khumchoo et al. 2015). As such, these materials are structurally similar to pillared interlayered clays (PILC) described in the previous chapter. However, apart from their use as photocatalysts and nanocatalysts in Fenton-like reactions (Garrido-Ramírez et al. 2010; Navalon et al. 2010; Liu and Zhang 2014), or as adsorbents of chemical contaminants in water (Zhu et al. 2016), only a small proportion of the accumulated literature on clay-supported metal oxides or *metal oxide-clay nanocomposites* relates to their capacity for catalyzing organic reactions and conversions.

An early reference to using clay-supported metal oxides in organic catalysis is that by Alvarez et al. (1991) who reported that MnO_2/bentonite could catalyze the solvent-free oxidation of various 4-aryl-1,4-dihydropyridines under microwave irradiation. Subsequently, Gandía et al. (2000) prepared a series of manganese oxide reagents, supported on saponite and montmorillonite and their respective Al- and Zr-pillared clays. The performance of these catalysts, in terms of the complete oxidation of acetone (at 337°C–387°C), increased in the order Al-pillared clays < unpillared clays < Zr-pillared clays, while catalyst stability was higher with the unpillared than the pillared clays (Gandía et al. 2002; Gil et al. 2006). More recently, Fang et al. (2016) used montmorillonite-supported MnO_2 nanoparticles to catalyze the degradation of bisphenol A.

Hectorite- and kaolinite-supported TiO_2 (anatase) has been shown to be an efficient catalyst for the degradation of toluene and D-limonene under ultraviolet irradiation (Kibanova et al. 2009). Likewise, He et al. (2011) reported that VO_x-WO_y/TiO_2 particles, supported on palygorskite, could promote the decomposition of adsorbed 1,2-dichlorobenzene. More recently, Belaidi et al. (2015) observed that vanadia supported on chromia-pillared acid-activated bentonite could catalyze the epoxidation of cyclohexene in the presence of *tert*-butyl hydroperoxide.

Montmorillonite- and hectorite-supported SnO_2 nanoparticles can efficiently promote the photodegradation of salicylic acid (Körösi et al. 2004). More recently, Patil et al. (2015) used a Bi_2O_3-montmorillonite nanocomposite as a photocatalyst for the degradation of Congo red, while Akkari et al. (2016) chose montmorillonite-supported ZnO to catalyze the photodecomposition of methylene blue, under visible or UV light irradiation.

Narayanan and Deshpande (1996) prepared vanadia/K10 montmorillonite by mixing the acid treated clay sample with ammonium metavanadate in aqueous oxalic acid, evaporating the suspension, oven-drying at 120°C, and calcining at 450°C. The resultant clay-supported reagent was then used for the alkylation of aniline to yield *N*-ethyl- and *N,N*′-diethyl-aniline. Kurian and Sugunan (2003, 2004) refluxed Fe-, Al- and mixed Fe-Al pillared clays as well as K10 and KSF montmorillonites with the ammonium salt of Mo, V, and Cr in acetone to obtain the respective clay-supported metal oxide catalysts for the selective benzylation of *o*-xylene with benzyl chloride to yield the monoalkylated product.

Using a similar approach, Gao and Xu (2006) obtained clay-supported vanadium oxide catalysts for the selective hydroxylation of benzene to phenol in the presence of H_2O_2. Subsequently,

Qi et al. (2008) used a complex of 3-APS-OsO$_4$ with synthetic imogolite, where APS denotes 3-aminopropyltriethoxysilane, to promote the dihydroxylation of olefins. Similarly, Dar et al. (2013) observed that Cu(OH)$_x$, supported on KSF montmorillonite, could catalyze the *ipso*-hydroxylation of arylboronic acids, while Masui et al. (2014) found montmorillonite, impregnated with Sn(OH)$_4$ nanoparticles, to be an efficient acid catalyst for various organic reactions involving carbon-carbon bond formation.

Nogueira et al. (2011) impregnated iron oxide into montmorillonite and used the product to catalyze the oxidation of toluene. Similarly, Virkutyte and Varma (2014) found that K10-supported Fe$_3$O$_4$ could promote the degradation of dichlorphenol in the presence of various oxidizing agents such as peroxymonosulfate and hydrogen peroxide. More recently, Saikia et al. (2016) reported that Fe$_3$O$_4$ nanoparticles, generated *in situ* within the porous structure of acid-activated montmorillonite, could effectively catalyze the Baeyer–Villiger oxidation (cf. Chapter 5) of various cyclic and aromatic ketones in the presence of hydrogen peroxide. The superior magnetic properties of highly crystalline Fe$_3$O$_4$ nanoparticles in the interlayer space of montmorillonite were ascribed by Son et al. (2010) to anisotropic magnetic interactions among the two-dimensionally distributed nanoparticles. Montmorillonite-supported Fe$_3$O$_4$ nanoparticles are generally associated with interlayer surfaces with only a small proportion being found in mesopores and on external surfaces of the clay heterostructure (Hur et al. 2006; Yuan et al. 2009).

Nanosize metal sulfides have attracted much interest because of their optical and semi-conducting properties. Such particles may be stabilized by impregnation and intercalation into montmorillonite and related layer silicates including hectorite, laponite, rectorite, and saponite. Clay-supported metal sulfides may be synthesized by one of three means: (1), mixing the clay mineral with a solution of the thiourea complex of the metal, and autoclaving the mixture (Han et al. 2005, 2006, 2008; Xiao et al. 2007; Boukhatem et al. 2013); (2), mixing preformed metal sulfide nanoparticles—obtained by reacting the acetate salt of the metal and sodium sulfate in the presence of hexadecyltrimethylammonium bromide—with the clay mineral (Praus et al. 2011); and (3), a solid-solid reaction between the metal cation-exchanged montmorillonites and sodium sulfide (Khaorapapong et al. 2009, 2010; Kabilaphat et al. 2015). Clay-supported CdS can promote the photodegradation of Rhodamine (B and 6G) and methylene blue (under light or UV irradiation) as well as the photoreduction of carbon dioxide. In this regard, its activity is superior to that of either pure (unsupported) CdS or the pristine (raw) clay sample (Xiao et al. 2007; Praus et al. 2011; Nascimento et al. 2012; Boukhatem et al. 2013).

4.4 CLAY-SUPPORTED METAL NANOPARTICLES

The basic approach to preparing clay-supported metal nanoparticles for organic catalysis is to bring the clay mineral sample, often after acid treatment or pillaring, into contact with a suitable metal salt (or metal-organic complex), followed by reduction of the intercalated or surface-adsorbed metal compound to yield the corresponding zero-valent metal nanoparticles. Figure 4.2 illustrates the

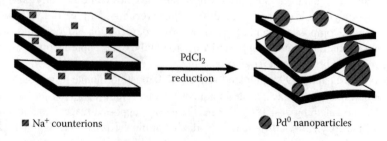

FIGURE 4.2 Diagram showing the impregnation of Pd^0 nanoparticles into montmorillonite by adding a solution of PdCl$_2$ (pH 4–5) to a suspension of Na$^+$-montmorillonite in ethanol-water, replacing the Na$^+$ counterions with Pd^{2+}, and converting the latter into Pd^0 by *in situ* reduction with ethanol. (Modified from Papp, S. et al., *Solid State Ionics*, 141–142, 169–176, 2001a; Papp, S. and Dékány, I., *Colloid Polym. Sci.*, 281, 727–737, 2003.)

process for montmorillonite-supported Pd^0. The metal *precursor* may be impregnated into, or deposited (precipitated) on, the clay mineral support from solution or the vapor phase, while (chemical) reduction may be carried out using a polyol, hydrogen, alkali metal borohydride, UV irradiation, or such organic compounds as ascorbic acid, ethanol, hydrazine, and tetrahydrofuran (Papp et al. 2008; Varadwaj and Parida 2013; Dutta et al. 2015).

Early on, Roy and coworkers (Ravindranathan et al. 1990; Malla et al. 1991, 1992) prepared a copper metal-montmorillonite complex (Cu^0/montmorillonite) by intercalating hydroxy copper acetate into Na^+-montmorillonite, replacing the Na^+ counterions with Cu^{2+}, suspending the intercalate in ethylene glycol (*polyol*) which acted as both solvent and reducing agent, and refluxing the suspension (at 195°C) in an argon atmosphere. X-ray diffraction and transmission electron microscopy indicate the presence of copper metal clusters (0.4–0.5 nm diameter) in the interlayer space, and large metal aggregates on external particle surfaces. Sarmah and Dutta (2012) used a similar approach to place Ru^0 nanoparticles (~5 nm diameter) on an acid-activated montmorillonite for the hydrogenation (reduction) of substituted nitrobenzenes to the corresponding anilines.

Using a cationic gold precursor in the form of $Au(en)_2Cl_2$ (en = ethylene diamine), and sodium borohydride as the reductant, Zhu et al. (2009) were able to obtain montmorillonite- and sepiolite-supported gold nanoparticles. Subsequently, Borah et al. (2011) chose $Cu(CH_3COO)_2$ as the precursor, and $NaBH_4$ as the reducing agent, to prepare Cu^0 nanoparticles on an acid-activated bentonite. A similar approach was also used to impregnate montmorillonite, acid-activated montmorillonite, and laponite with various metal nanoparticles, including Pd^0 (Crocker et al. 1993); Ni^0 and Ag^0 (Ayyappan et al. 1996); Au^0 and Ag^0 (Aihara et al. 1998); Co^0, Ni^0, and Zn^0 (Ahmed and Dutta 2003a, 2003b; Dutta et al. 2011); and Ru^0 (Sarmah and Dutta 2012).

Earlier, Anderson et al. (1997) obtained Ni^0/sepiolite catalysts by suspending the clay material in a solution of nickel nitrate at pH 3.5, heating the suspension at 90°C, precipitating the metal precursor with urea, and exposing the precipitate to a flow of H_2 in argon. Although the bentonite- and palygorskite-supported Ni^0 catalysts prepared by this means showed comparable activity in the hydrogenation of linoleic acid, their respective selectivity to oleic acid was markedly different (Anderson et al. 1993). Tetrasilicic mica has also been used as support of palladium nanoparticles (Sivakumar et al. 2001; Divakar et al. 2008a).

Another widely used method is to intercalate the metal cation (e.g., Pd^{2+}) from its salt (e.g., H_2PdCl_4) into montmorillonite in an ethanol-water binary liquid mixture, and allow it to be reduced *in situ* by the interlayer ethanol to yield the corresponding metal nanoparticles (Papp et al. 2001a, 2008). Similarly, Szöllősi et al. (2001) prepared montmorillonite-supported Pt^0 catalysts by intercalating $[Pt(NH_3)_4](NO_3)_2$ and then reducing the metal precursor with H_2 gas or $NaBH_4$ in solution. Li et al. (2010) introduced Fe^0 nanoparticles into the interlayer space of an organo-montmorillonite by dispersing the organoclay in a solution of $FeCl_2$ and adding $NaBH_4$ to the dispersion. By mixing palygorskite with iron and nickel nitrates, and reducing with H_2, Liu et al. (2013) obtained a bimetallic Fe^0Ni^0/palygorskite nanocomposite for the catalytic cracking of benzene. Subsequently, Masui et al. (2014) used montmorillonite-supported $Sn(OH)_4$ nanoparticles, formed by treating the clay with $SnCl_4.5H_2O$, to catalyze the cyanotrimethylsilylation of benzophenone with cyanotrimethylsilane.

A simplified version of the method is to exchange the interlayer cations in smectite with Fe^{3+} ions, which are then reduced to yield zero-valent iron (Gu et al. 2010). Similarly, Dai et al. (2010) immobilized Ru^0 nanoparticles in montmorillonite by treating the Ru^{3+}-exchanged mineral with hydrogen at 240°C. The resultant Ru^0/montmorillonite was then used to catalyze the methanolysis of ammonia borane. Jia and Wang (2012) and Jia et al. (2012) intercalated Fe^0 nanoparticles into an organo-montmorillonite, and Pd^0/Fe^0 into a smectite, and used the products to mediate the dechlorination of 2,4-dichlorophenol and pentachlorophenol, while Luo et al. (2013) chose a Fe^0/rectorite for the decoloration of Orange II. Yang and Kim (2013) immobilized Au^0 nanoparticles on laponite by reducing an acidified Au(III) chloride solution with $NaBH_4$ in the presence of the mineral.

Earlier, Török et al. (1999) prepared Pt^0/smectites, containing cinchonidine as a chiral modifier, by impregnating the clay samples with H_2PtCl_6, reducing the precursor metal ions with ethanol, and intercalating the organic compound, for the enantioselective hydrogenation of ethyl pyruvate and 2-methyl-2-pentenoic acid. A decade later, Pan et al. (2009) used $H_2PtCl_6.6H_2O$ as the precursor, and supercritical methanol as the reducing agent, to obtain a Pt^0/montmorillonite catalyst for the hydrogenation of nitrobenzene. Sometime later, Varade and Haraguchi (2013) reported that a synthetic hectorite (Laponite XLG) and a natural Na^+-montmorillonite (Kunipia F) could serve as an *in situ* reducing agent of intercalated Pt ions for the formation of thermally stable Pt^0/clay catalysts. Pillai et al. (2013) could similarly obtain silver nanoparticles on montmorillonite in the absence of any reducing agent by using microwave irradiation.

Nanoparticles of palladium and silver can be introduced into kaolinite when the interlayer space of the mineral has previously been expanded by intercalation of dimethylsulfoxide (Patakfalvi et al. 2003; Patakfalvi and Dékány 2004; Papp et al. 2008). Similarly, prior expansion of the smectite interlayers by pillaring with alumina, or intercalation of water-soluble polymers (e.g., polyvinyl pyrrolidone), can promote metal nanoparticle formation and stability (Szücs et al. 1998, 2000; Papp and Dékány 2003). Thus, impregnation of Al-, Ce/Al-, and Zr-pillared montmorillonites with cobalt nanoparticles gives rise to clay-supported catalysts capable of mediating the Fischer–Tropsch reaction (Su et al. 2009; Ahmad et al. 2013). Ni^0/Zr-pillared laponite has also been used by Hao et al. (2003) to catalyze the CO_2 reforming of methane to syngas.

A potentially useful innovation in preparing clay-supported metal nanoparticles is to intercalate the metal precursors into organically modified clays (*organoclays*) (He et al. 2014) or clay-polymer complexes (Theng 2012). For example, Király et al. (1996) added a solution of Pd(II) acetate to a suspension of hexadecylammonium $(HDAM)^+$-montmorillonite (in ethanol-toluene), followed by *in situ* reduction of the precursor, to form an interlayer complex of montmorillonite with Pd^0 nanoparticles. The activity of the resultant nanocomposite in catalyzing the hydrogenation of 1-octene and styrene is related to the interlayer expansion of the complex with Pd-HDAM in various organic solvents (Mastalir et al. 1997). A similar approach was adopted by several investigators to prepare nanoparticles of Pd^0 and Pt^0 (Papp and Dékány 2002; Sivakumar et al. 2004; Divakar et al. 2008b; Wang et al. 2009), Fe^0 (Li et al. 2010; Wu et al. 2012), and Ru^0 (Zhou et al. 2013) in montmorillonite and fluorotetrasilicic mica whose interlayers have previously been expanded and *hydrophobized* by intercalation of alkylpyridinium or long-chain quaternary ammonium ions. Likewise, Mastalir et al. (2002) were able to synthesize catalytically active clay-supported platinum nanoparticles by impregnating H_2PtCl_6 into dihydrocinchonidine-exchanged montmorillonite (and hectorite), and reducing the metal precursor with $NaBH_4$.

Patel et al. (2008) used montmorillonite exchanged with $[(CH_3)_2(HT)_2N]^+$ cations (where HT denotes hydrogenated tallow) to intercalate $[Pd(NH_3)_4]Cl_2$ and $[Rh(NH_3)_6]Cl_3$ which, on reduction with $NaBH_4$, yielded interlayer Pd^0/clay and Rh^0/clay nanocomposites. Similarly, Zhang et al. (2009) used the commercial equivalent of the same organoclay (Cloisite 20A, Southern Clay Products, USA) and platinum(II) acetylacetonate as the metal precursor to obtain a Pt^0/montmorillonite nanocomposite that was highly stable against mechanical agitation and high temperature treatment.

A variant approach to preparing clay-supported metal (Au, Cu, Pd, Pt, Ru) catalysts is to mix the metal salt precursors with an aqueous solution of a cationic surfactant such as cetyltrimethylammonium (CTA) or tetradecyltrimethylammonium bromide, reduce the metal cations with $NaBH_4$ or hydrazine, and then add a dispersion of the clay mineral (montmorillonite, hectorite) to the mixture (Mastalir et al. 2000, 2001, 2004; Manikandan et al. 2007a, 2008; Dhanagopal et al. 2010; Teixeira-Neto et al. 2015; Mekewi et al. 2016). Such clay-intercalated nanosize noble metals can efficiently catalyze the hydrogenation and reduction of various organic compounds. The various types of organic reactions that can be catalyzed by clay-supported metal nanoparticles are summarized in Table 4.2.

A simple method of preparing Au^0/clay nanocomposites is to intercalate preformed gold nanoparticles into Na^+- and CTA^+-exchanged montmorillonite, using ultrasonic irradiation (Belova et al.

TABLE 4.2
Organic Conversions and Reactions Catalyzed by Clay-Supported Metal Nanoparticles

Reaction Type (in Alphabetical Order)	Substrate	Product	Clay-Supported Catalyst	Reference (in Chronological Order)
Condensation	Aromatic aldehyde, dimedone, ethyl-acetoacetate, and ammonium acetate	Polyhydroquinoline	Ni^0 on acid-activated montmorillonite	Saikia et al. (2012)
Coupling	Iodobenzene with alkenes	Corresponding arylated products	Pd^0 on organo-montmorillonite	Ghiaci et al. (2012)
	Aryl halides with alkenes and alkynes	Corresponding derivatives	Pd^0 on acid-activated montmorillonite	Borah and Dutta (2013)
	Organoboronic acids with aryl halides	Cross-coupled products	Pd^0 on acid-activated montmorillonite	Borah et al. (2014a)
	Aryl bromides with arylboronic acids	Cross-coupled products	Pd^0 on K10 montmorillonite	Gogoi et al. (2014)
	Aldehyde, amine, and alkyne	Propargylamines	Cu^0 and Ag^0 on acid-activated montmorillonite	Borah et al. (2014b); Borah and Das (2016)
Cycloaddition	Anilines with benzoylchloride	Benzanilides	Pd^0 on KSF montmorillonite	Dar et al. (2015)
	Azides and terminal alkynes	1,2,3-Triazoles	Cu^0 on acid-activated montmorillonite	Borah et al. (2011)
Debromination	Decabromodiphenyl ether	Di-, tri-, tetra-, and penta-bromo-diphenyl ethers	Fe^0 on montmorillonite and organo-montmorillonite	Yu et al. (2012); Pang et al. (2014)
Dehydrogenation	Ethylbenzene	Styrene	Pt^0 and Mo^0 on synthetic smectite	Morán et al. (2007)
Hydrodeoxygenation	Phenolic compounds	Cycloalkanes	Ru^0 on montmorillonite	Xu et al. (2014)
Hydrogenation	1-Hexene	Hexane	Rh^0 on montmorillonite and palygorskite	Herrero et al. (1989)
	Linoleic acid	Stearic acid; Oleic acid	Ni^0 on bentonite and palygorskite	Anderson et al. (1993)
	Ethylene	Ethane	Pt^0 on synthetic smectite-like minerals	Arai et al. (1996)
	Ethyl pyruvate; 2-Methyl-2-pentenoic acid	Ethyl lactate; Methyl pentanoic acid [enantioselective for (R)-product]	Smectite-supported Pt^0 and Pd^0 containing cinchonidine	Török et al. (1999); Mastalir et al. (2002)
	Cinnamaldehyde	Cinnamyl alcohol	Pt^0 on K10 montmorillonite, montmorillonite, and hectorite	Szöllösi et al. (1998, 1999); Hájek et al. (2006); Manikandan et al. (2007a, 2007b)
	1-Phenyl-1-butyne; 1-Phenyl-1-pentyne	1-Phenyl-1-butene; 1-Phenyl-*cis*-1-pentene	Pd^0 on organo-montmorillonite	Mastalir et al. (2000, 2001)

(Continued)

TABLE 4.2 (Continued)
Organic Conversions and Reactions Catalyzed by Clay-Supported Metal Nanoparticles

Reaction Type (in Alphabetical Order)	Substrate	Product	Clay-Supported Catalyst	Reference (in Chronological Order)
Hydrogenation (Continued)	Crotonaldehyde	Crotyl alcohol (main product)	Pt^0 on K10 montmorillonite	Kun et al. (2001)
	Phenylacetylene; 4-Octyne	Styrene; 4-Octene (selective to *cis*-alkene)	Pd^0 on organo-montmorillonite	Mastalir et al. (2004)
	Phenyl alkyl acetylenes	Corresponding alkene (high selectivity for *cis* isomer)	Pd^0 on Al-pillared montmorillonite	Marín-Astorga et al. (2005)
	Benzene/cyclo-hexene	Cyclohexane	Ru^0 on montmorillonite and palygorskite	Miao et al. (2006, 2007)
	α,β-Unsaturated aldehydes	Corresponding unsaturated alcohols	Pd^0 on vermiculite and tetrasilicic mica; Pt^0 and Ru^0 on montmorillonite and hectorite	Divakar et al. (2007, 2008a); Dhanagopal et al. (2010)
	Aromatic alkenes and alkynes	Corresponding alkanes and alkenes	Ni^0 on K10 montmorillonite	Dhakshinamoorthy and Pitchumani (2008)
	Benzaldehyde	Benzyl alcohol	Pd^0 on bentonite	Divakar et al. (2008b)
	Citral	Geraniol and nerol	Pt^0 and Ru^0 on montmorillonite	Manikandan et al. (2008)
	Aromatic compounds (benzene, toluene, xylenes, naphthalene, anthracene)	Corresponding cyclohexanes, hydronaphthalene, hydroanthracene	Rh^0 on montmorillonite	Sidhpuria et al. (2009)
	Substituted nitrobenzenes	Corresponding anilines	Ru^0 on acid-activated montmorillonite	Sarmah and Dutta (2012)
	Acetophenone	1-Phenylethanol	Ni^0 on acid-activated montmorillonite	Dutta et al. (2011)
	Benzene	Cyclohexene and cyclohexane	Ru^0 on soil smectite and acid-activated montmorillonite	Shawkataly et al. (2012); Chen et al. (2016)
	Monoterpenes (e.g., limonene, geraniol)	Corresponding products (e.g., *p*-menthene, citronellol)	Rh^0 on K10 montmorillonite	Agarwal and Ganguli (2013)
	Halonitrobenzene	Haloaniline	Ni^0 on acid-activated montmorillonite	Dutta and Dutta (2014)
	Aromatic carbonyl compounds	Corresponding alcohols	Rh^0 on acid-activated montmorillonite	Sarmah and Dutta (2014)

(*Continued*)

TABLE 4.2 (Continued)
Organic Conversions and Reactions Catalyzed by Clay-Supported Metal Nanoparticles

Reaction Type (in Alphabetical Order)	Substrate	Product	Clay-Supported Catalyst	Reference (in Chronological Order)
Hydrogenation (Continued)	Quinoline	Tetrahydro- and decahydro-quinoline	Rh^0 on montmorillonite	Zhou et al. (2013)
Hydrogenolysis	Glycerol	1,2-Propanediol	Ru^0–Cu^0 bimetallic bentonite	Jiang et al. (2009)
Hydroisomerization	Hexanes (2-methyl-pentane, n-hexane, cyclohexane)	Corresponding isomers and cracked products	Pd^0 and Pt^0 on Al-pillared montmorillonite	Issaadi et al. (2001); Zakarina et al. (2008)
Hydroxylation	Arylboronic acids	Phenols	Ag^0 on K10 montmorillonite	Begum et al. (2015)
Oxidation	Ethanol	Carbon dioxide, water	Co^0 and Cu^0 on smectite and Al-pillared smectite	Pérez et al. (2014)
	Acetone, methyl-ethylketone	Carbon dioxide, water	Pt^0 on Al-pillared smectite	Gil et al. (2001)
	Phenol	Carbon dioxide, tar	Fe^0 on montmorillonite	Zhou et al. (2006)
	Benzyl alcohol	Benzaldehyde	Pd^0 on organo-montmorillonite	Wang et al. (2009)
	Ditto	Ditto	Au^0 and Pd^0 on montmorillonite	Teixeira-Neto et al. (2015)
	Ditto	Ditto	Au^0 on halloysite	Philip et al. (2017)
	p-Xylene	Terephthalic acid	Co and Mn nanoparticles on organo-bentonite	Ghiaci et al. (2012)
Reduction	Nitrobenzene	Aniline	Pt^0 and Fe^0 on montmorillonite	Pan et al. (2009); Gu et al. (2010) Jia et al. (2011)
	4-Nitrophenol	Aminophenol	Ag^0 on halloysite; Ag^0 and Au^0 on montmorillonite	Liu and Zhao (2009); Praus et al. (2013); Ammari and Chenouf (2015)
Substitution (allylic)	Allyl methyl carbonate with ethylacetoacetate	2-Acetyl-4-pentenoic acid ethyl ester	Pd^0 on montmorillonite	Mitsudome et al. (2007)

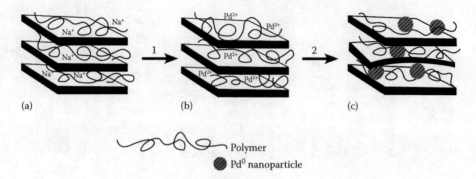

FIGURE 4.3 Same as for Figure 4.2 but using an interlayer complex of Na$^+$-montmorillonite with a polymer (e.g., poly(N-vinyl-2-pyrrolidone) as the starting material (a), adding a PdCl$_2$ solution (Step 1) to replace the Na$^+$ counterions with Pd^{2+} (b), and reducing the Pd^{2+} cations to Pd^0 (Step 2) to yield a clay-polymer complex containing Pd^0 nanoparticles (c). (Modified from Papp, S. et al., *Applied Clay Science*, 19, 155–172, 2001b; Papp, S. and Dékány, I., *Colloid Polym. Sci.*, 281, 727–737, 2003.)

2008, 2009a). Alternatively, the gold nanoparticles can be introduced by ultrasonic replacement of polyethylene glycol from the interlayer space of montmorillonite (Belova et al. 2009b).

Earlier, Papp et al. (2001b) prepared Pd^0 nanoparticles on montmorillonite by intercalating Pd^{2+} ions (from a solution of PdCl$_2$) into the clay-polymer complex and reducing with ethanol (Figure 4.3). Likewise, Chen and Kuo (2006) successfully synthesized clay-supported gold nanoparticles by introducing HAuCl$_4$ from solution into an interlayer montmorillonite-polyethylenimine complex and allowing the gold ions to be reduced *in situ*. More recently, Philip et al. (2017) capped gold nanoparticles with polyethylenimine and allowed the positively charged metal-polymer assemblage to coat the external surfaces of a tubular halloysite. The *decorated* halloysite particles were efficient in promoting the conversion of benzyl alcohol to benzaldehyde. When the capping polymer was removed by calcination, the *naked* gold nanoparticles could catalyze the oxidation of benzyl alcohol to benzoic acid.

In this context, we should also mention the use of room-temperature ionic liquids, such as 1,1,3,3-tetramethylguanidium trifluoroacetate and 1,1,3,3-tetramethylguanidinium lactate, to immobilize metal nanoparticles on montmorillonite and palygorskite (*attapulgite*). The procedure involves exchanging the inorganic counterions on the clay mineral surface with the ionic liquid, intercalating the desired metal ions (from its salt), and then reducing them (H$_2$, 300°C) to the corresponding zero-valent species (Miao et al. 2006, 2007; Jiang et al. 2009a; Xu et al. 2014). Thus, Ratti et al. (2010) were able to intercalate a sulfonic acid functionalized ionic liquid into montmorillonite, and used the modified clay as a catalyst for the transesterification of β-ketoesters with various alcohols. More recently, Martínez et al. (2014) immobilized palladium nanoparticles in 1-butyl-3-methylimidazolium hexafluorophosphate ([bmim][PF$_6$]), which, after impregnation into laponite, could catalyze the Mizoroki–Heck cross-coupling of aryl halides with olefins under solvent-free conditions (cf. Chapter 5).

4.5 CLAY-SUPPORTED ORGANIC AND METAL-ORGANIC REAGENTS

An early reference to the use of clay-organic complexes as catalysts is that by Taylor and Chiang (1977) who reported that trimethyl orthoformate supported on K10 montmorillonite can effectively convert carbonyl compounds into the corresponding acetals. Subsequently, Heravi and Ajami (1998) used bis(trimethylsilyl) chromate on K10 montmorillonite to promote the oxidative deprotection of tetrahydropyranyl ethers to the corresponding carbonyl compounds, while Varma and Naicker (1998) were able to convert arylaldehydes into nitriles using hydroxylamine hydrochloride on K10 montmorillonite under microwave irradiation and solvent-free conditions. Similarly, K10-supported tetrabytulammonium periodate is efficient in catalyzing the oxidation of benzylic alcohols to the corresponding aldehydes and of aryl sulfides to the corresponding sulfoxides (Venkatachalapathy et al. 1999). Subsequently, Kalbasi et al. (2012) prepared a bentonite-polystyrene nanocomposite by *in situ* polymerization of styrene. Treatment with chlorosulfonic acid turned the sample into an efficient catalyst for the Biginelli reaction of aldehydes with ethyl acetoacetate and urea to yield 3,4-dihydropyrimidin-2(1H)-ones (cf. Chapter 5).

Pinnavaia and coworkers (Pinnavaia and Welty 1975; Quayle and Pinnavaia 1979; Pinnavaia et al. 1979; Farzaneh and Pinnavaia 1983; Pinnavaia 1983) did much of the pioneering work on the preparation of clay-supported metal-organic reagents by intercalating $Rh(PPh_3)_2^+$, $RhH_2Cl(P-P^+)_x$, or $Rh(COD)(PPh_3)_2^+$, where PPh_3 stands for triphenylphosphine, $(P-P^+)$ denotes $[(C_6H_5)_2P-(CH_2)_2P(C_6H_5)_2(CH_2C_6H_5)^+]$, and COD is 1,5-cyclooctadiene, into Na^+-hectorite. The resultant intercalated reagents were then used to catalyze the hydrogenation and hydroformylation of 1-hexene as well as the reduction of alkynes to *cis* alkenes in the presence of methanol. The interlayer complex of fluorotetrasilicic mica with $[Rh(COD)(PPh_3)_2]^+$ was also efficient in promoting the hydrogenation of 4-vinylcyclohexene (Miyazaki et al. 1985). Similarly, Herrero et al. (1989, 1991) used a cationic rhodium complex, $[Rh(nbd)-(Me_2CO)_x]ClO_4^-$ (where "nbd" denotes 2,5-norbornadiene) as the metal precursor, and hydrogen as the reducing agent, to prepare montmorillonite- and palygorskite-supported rhodium catalysts for the hydrogenation of 1-hexene. Crocker and Herold (1991) also found that the interlayer complex of K10 montmorillonite with $[Pd(PPh_3)_3(NCMe)]^{2+}$ could catalyze the carbomethoxylation of ethylene to methyl propionate with >90% selectivity, while K10-supported $[Ir(COD)(PPh_3)_2]X$, where X stands for PF_6 or BF_4, was able to promote the hydrogenation of imines (Margalef-Català et al. 1999). Various rhodium-phosphine complexes, supported on smectites, can also catalyze the asymmetric hydrogenation of substituted acrylic acids (Mazzei et al. 1980), α,β-unsaturated carboxylic acid esters (Shimazu et al. 1996) and itaconates (Sento et al. 1999) as well as the hydrogenation of benzene to cyclohexane (Halligudi et al. 1992). Clay-catalyzed enantioselective organic conversions are described in Chapter 7.

Montmorillonite-supported $PdCl_2$ complexes with chlorodiphenylphosphine and bipyridine are similarly efficient in catalyzing the selective hydrogenation of alkenes, alkynes, and styrene (Choudary and Bharati 1987; Choudary et al. 1985, 1991), and the conversion of organic azides to amines (Sharma and Chandrasekhar 1989). Likewise, the intercalation of $PdCl_2$ and tetraphenylphosphonium bromide into K10 yielded a catalytically active material for the cross-coupling of aryl halides with arylboronic acids (Varma and Naicker 1999), and the Heck reaction between aryl halides and styrene to yield *trans*-stilbenes (Varma et al. 1999). Similar reactions have been reported by Poyatos et al. (2003) using complexes of Pd(II) with $[[CHNCH-(CH_2Ph-NO_2)_2][Br]_2]$supported on K10 montmorillonite.

The preparation and catalytic activity of clay-supported metalloporphyrins and Schiff-base complexes have attracted much interest as these reagents can serve as models of biomimetic oxidation catalysts. For example, natural porphyrins can be stabilized (against oxidative decomposition) by adsorption to a clay mineral support to yield a complex mimicking the polypeptide envelope that protects the active site of cytochrome P450 enzymes (Carrado and Winans 1990; Bedioui 1995; Crestini et al. 2004; Gupta et al. 2009). We might mention that K10 montmorillonite can catalyze the formation of alkylporphyrins from aliphatic aldehydes and pyrroles (Izumi et al. 1997b).

Barloy et al. (1990, 1992) used the montmorillonite-intercalated tetracationic $Mn(TMPyP)^{4+}$ complex, where TMPyP denotes tetrakis(4-*N*-methyl-4-pyridiniumyl)porphyrin, for the epoxidation of alkenes in the presence of iodosylbenzene (PhIO) as oxidant. This clay-supported reagent could also catalyze the hydroxylation of alkanes, while showing a marked preference for small, linear molecules to the more bulky substrates. Similar results were obtained by Martinez-Lorente et al. (1996) using a K10-supported Mn(III) porphyrin complex and H_2O_2 as the oxidant, while Machado et al. (2002) and Nakagaki et al. (2006) reported the preferential formation of cyclohexanol to cyclohexanone during the oxidation of cyclohexane by Fe(III) porphyrin complexes immobilized on montmorillonite and chrysotile. Montmorillonite-supported $Mn(TMPyP)^{4+}$ was also efficient in catalyzing the oxidation of mono- and di-meric lignin model compounds (Crestini et al. 2004), and 2,4,6-trichlorophenol could be oxidatively degraded by a montmorillonite-hemin complex (Xiong et al. 2014). Intercalation of a Mn-porphyrin complex into a previously silanized montmorillonite and kaolinite gave rise to biomimetic catalysts for the enhanced oxidation of catechol by H_2O_2 (Nuzzo and Piccolo 2013). Bizaia et al.

(2009) prepared an interlayer complex of kaolinite with [*meso*-tetrakis(pentafluorophenyl)porphinato] iron(III) by first expanding the clay layers with dimethylsulfoxide (DMSO) and then displacing the intercalated DMSO with ethanolamine. The resultant kaolinite-porphyrin complex was highly effective in catalyzing the epoxidation of cyclooctene as well as in converting cyclohexane into cyclohexanone. More recently, Jondi et al. (2016) used the tetra(4-pyridylporphyrinato-Mn(III)) complex, intercalated into natural clays, to catalyze the hydrosilylation reaction of 1-octene and triethoxysilane to yield the linear tri(oxy)silyl-1-octene as the sole product. Bouhlel et al. (1993) reported that impregnation of nickel acetylacetonate into K10 montmorillonite gave a complex capable of promoting the epoxidation of olefins by molecular oxygen in the presence of isobutyraldehyde (as a *sacrificial* reductant). Pereira et al. (2008) used laponite and K10 montmorillonite, before and after grafting with 3-aminopropyl triethoxysilane (APTES), as supports of vanadyl acetylacetonate for the epoxidation of geraniol, and *tert*-butyl hydroperoxide as oxidant. A K10-supported molybdenum acetylacetonate complex was used by Farias et al. (2011) for the epoxidation of soybean and castor oils, and by Zhang et al. (2014) for the aerobic oxidation of 5-hydroxymethylfurfural to 5-hydroxymethyl-2-furancarboxylic acid. Likewise, Kameyama et al. (2006) were able to obtain 1,2-epoxycyclohexane from cyclohexene using montmorillonite-intercalated cobalt porphyrins. Interestingly, intercalation into a synthetic fluorohectorite made Co(TMPyP)$^{4+}$ less active than the unsupported (*free*) counterpart in the oxidation of 2,6-di-*tert*-butylphenol by dioxygen, apparently because the cobalt porphyrin complex adopted such a surface orientation as to restrict substrate access (Chibwe et al. 1996; Dias et al. 2000).

Dixit and Srinivasan (1988) investigated the catalytic activity of a kaolinite-supported Mn(III)-Schiff base complex for the epoxidation of olefins using *tert*-butylhydroperoxide (TBHP) as the oxidant. They suggested that the decomposition of TBHP was suppressed by adsorption to the supported catalyst. Schiff base ligands are typically prepared by condensing primary amines with aldehydes to yield imines, which can coordinate to transition metal ions via the lone pair electrons of nitrogen (Gupta and Sutar 2008). Two common Schiff base complexes/ligands are *N,N'*-bis(salicylidene)-ethylenediamine (*salen*) (Figure 4.4a) and *N,N'*-bis(salicylidene)-1,2-phenyldiamine (*salophen*) (Fraile et al. 1998; Kadwa et al. 2014). The use of chiral ligands, such as (*R, R*)-(−)-*N,N'*-bis-(3,5-di-*tert*-butylsalicylidene)-1,2-cyclohexanediamine (*salhd*) (Figure 4.4b), opens the way to the enantioselective conversion of substrates (Fraile et al. 1998; Kureshy et al. 2003, 2004; Kuźniarska-Biernacka et al. 2009). The different metal (e.g., cobalt, iron, manganese)-Schiff base complexes may be immobilized or intercalated into the clay support by either direct cation exchange or treatment of the metal-exchanged clays with a solution of the ligand. The use of metal-salen complexes in catalyzing the oxygenation of organic sulfur and nitrogen compounds has been reviewed by Venkataramanan et al. (2005).

Metal-salen and related reagents supported on clay minerals can efficiently catalyze the epoxidation of alkenes in the presence of various terminal oxidants, such as hydrogen peroxide, iodosylbenzene, sodium hypochlorite, and sodium periodate (Table 4.3). We might add that

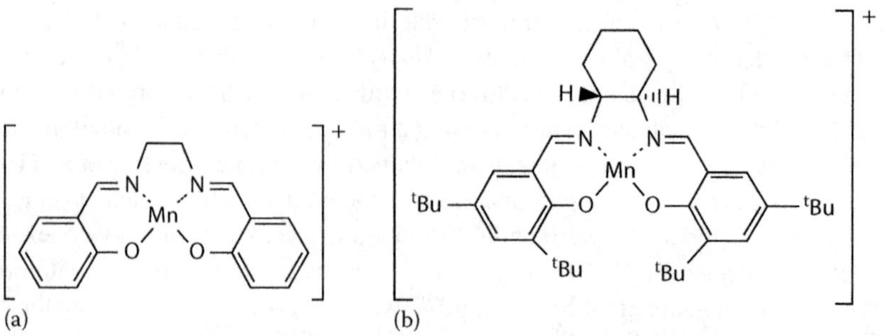

FIGURE 4.4 Structure of an Mn complex with a non-chiral salen ligand, *N,N'*-bis(salicylidene)-ethylenediamine (a), and with a chiral ligand, (*R, R*)-(−)-*N,N'*-bis-(3,5-di-*tert*-butylsalicylidene)-1,2-cyclohexanediamine (b). (After Fraile, J.M. et al., *J. Mol. Catal. A Chem.*, 136, 47–57, 1998.)

TABLE 4.3
Organic Conversions and Reactions Catalyzed by Clay-Supported Metal-Schiff Base (Ligand) Complexes

Reaction Type (in Alphabetical Order)	Substrate	Product	Metal-Ligand Complex on Clay	Reference
Epoxidation	Chromenes	Corresponding epoxides	Mn(III)-salen on montmorillonite	Kureshy et al. (2003, 2004)
	Cyclohexene	Cyclohexene epoxide	Mn(III)-salen on montmorillonite and laponite	Fraile et al. (1998)
	Ditto	Ditto	Mn(III)- and Ni(II)-salen on montmorillonite	Chatterjee and Mitra (1999)
	Ditto	Ditto	Cu(II)-diaza dioxa macrocyclic ligand on K10 montmorillonite	Salavati-Niasari et al. (2007)
	Ditto	Ditto	Co(II)-salophen on montmorillonite	Jiang et al. (2009)
	Ditto	Ditto	Mn(III)-salen and salhd on Al-PILC[a]	Garcia et al. (2016)
	Cyclooctene	Corresponding epoxide	Mn(III)-salen containing phosphonium groups on montmorillonite	Bahramian et al. (2006)
	1,2-Dihydro-naphthalene	Corresponding epoxide	Mn(III)-salen on montmorillonite and laponite	Fraile et al. (1998)
	1-Hexene	Corresponding epoxide	Mn(III)- and Ni(II)-salen on montmorillonite	Chatterjee and Mitra (1999)
	Indene	Corresponding epoxide	Mn(III)-salen on montmorillonite	Kureshy et al. (2003, 2004)

(Continued)

TABLE 4.3 (*Continued*)
Organic Conversions and Reactions Catalyzed by Clay-Supported Metal-Schiff Base (Ligand) Complexes

Reaction Type (in Alphabetical Order)	Substrate	Product	Metal-Ligand Complex on Clay	Reference
Epoxidation (Continued)	α-Methylstyrene	Corresponding epoxide	Mn(III)-salen on organo-bentonite and PCH[b]	Kuźniarska-Biernacka et al. (2009, 2010)
	Styrene	Corresponding epoxide	Mn(III)-salen on montmorillonite	Kureshy et al. (2003)
	Ditto	Ditto	Mn(III)-salen on Al-PILC[a] and organically modified Al-PILC[a]	Kuźniarska-Biernacka et al. (2004, 2007, 2009, 2010); Cardoso et al. (2005); Das et al. (2006);
	Ditto	Ditto	Mn(III)-salen on laponite	Kuźniarska-Biernacka et al. (2005)
Hydrogenation	Aromatic nitro compounds	Corresponding hydrogenated products	Pt(II)-Schiff base complex on montmorillonite	Parida et al. (2011)
Hydroxylation	Adamantane, cyclohexane, cyclooctane, 1,2,3,4-tetrahydro-naphthalene	Corresponding alcohols and ketones	Mn(III)-salen containing phosphonium groups on montmorillonite	Bahramian et al. (2006)
Oxidation	Cyclohexene	Cyclohexene- oxide, 2-cyclo-hexenol, 2-cyclo hexenone	Mn(III)-salen on montmorillonite and laponite	Fraile et al. (1998)
	Chalcones, olefins	Corresponding ketones and aldehydes	Fe(III)-salen on K10 montmorillonite	Dhakshinamoorthy and Pitchumani (2006)
	n-Octane	Mixture of octanal, octanols, and octanones	Fe(III)-salen and related Schiff base complexes on montmorillonite	Kadwa et al. (2014)

[a] PILC = Alumina pillared interlayered clay
[b] PCH = Porous clay heterostructure (cf. Chapter 3)

montmorillonite-supported iodosylbenzene can serve as an efficient catalyst for the oxidation of alkyl, aryl, and diaryl sulfides to the corresponding sulfoxides (Kannan et al. 1997).

4.6 CLAY-SUPPORTED HETEROPOLYACIDS

Of the hundred or so heteropoly acids (HPA) that are known to exist, only those with a Keggin structure have featured widely as acid and oxidation (redox) catalysts because of their availability and stability against dehydration at 150°C–200°C (Misono et al. 2000; Timofeeva 2003). In the Keggin HPA, the heteropoly anion has the general formula $[XM_{12}O_{40}]^{n-}$ where X represents P^{5+}, Si^{4+}, and M denotes W^{6+}, Mo^{6+} while the central XO_4 tetrahedron is surrounded by 12 edge- and corner-sharing MO_6 octahedra (Kozhevnikov 2009). In terms of clay-supported HPA, dodeca-tungstophosphoric acid (DTP) or 12-tungstophosphoric acid with the formula $H_3PW_{12}O_{40}$ has received the greatest attention (Yadav 2005; Pacula et al. 2014). With a Hammett acidity function (H_o) of <-8.2 (cf. Figure 2.1), solid DTP may be regarded as a superacid of the Brønsted type (Timofeeva 2003).

Partial replacement of the protons in DTP with Cs^+, as in $Cs_{2.5}H_{0.5}PW_{12}O_{40}$, increases catalytic activity and stability against heating and leaching. Being insoluble in most liquids, the cesium-substituted HPA can be easily recycled and serve as efficient catalysts in a variety of organic conversions such as the Friedel–Crafts alkylation of benzene with benzyl chloride, the benzo-ylation of *p*-xylene, the Fries rearrangement of aryl esters, the esterification of hexanoic acid with methanol, the transesterifcation of ethyl propanoate/hexanoate with methanol, the Beckmann rearrangement of oximes, and the cycloaddition of epichlorohydrin with acetone (Izumi et al. 1995; Kozhevnikova et al. 2003; Alsalme et al. 2008; Shiju et al. 2009; Kozhevnikov 2009; Yadav and Surve 2013).

Other clay-supported HPA reagents with a Keggin-like structure include dodecatungstosilicic acid (DTS), $H_4SiW_{12}O_{40}$, dodecamolybdophosphoric acid (DMP), $H_3PMo_{12}O_{40}$, and dodecamolyb-dosilicic acid (DMS), $H_4SiMo_{12}O_{40}$. Joshi and Narasimhan (1989, 1993), for example, have shown that unsupported DTS (and DMP) can catalyze various organic reactions, including the acetalization and ketalization of aldehydes and ketones, the transfer hydrogenation of nitro compounds in the presence of $NaBH_4$, the reduction of nitro compounds to ketones in the presence of aldehydes, and the conversion of aldehydes to 1,2-diacetates. A related HPA is 11-molybdo-vanado-phosphoric acid, $H_4PVMo_{11}O_{40}$. When impregnated into an acid-activated bentonite, this reagent can catalyze the epoxidation of cyclohexene (Boudjema et al. 2015).

Impregnation of HPA into montmorillonite or K10 is conventionally done using the *incipient wetness technique*; adding an aqueous or methanolic solution of HPA to the previously dehydrated clay, stirring the mixture (or mixing the resultant paste with a spatula), and drying the product at 110°C–120°C, and then calcining at 260°C–300°C. Clay-supported $Cs_{2.5}H_{0.5}PW_{12}O_{40}$ may be obtained by mixing the mineral with aqueous CsCl solution, drying/calcining, and adding a methanolic solution of DTP. A 20% w/w HPA loading of the clay support appears to be optimum for many catalyzed organic conversions (Yadav and Doshi 2000; Yadav 2005; Pacula et al. 2014). X-ray diffraction analysis of $H_3PW_{12}O_{40}$-impregnated montmorillonite by Pacula et al. (2014) gives no indication of reagent intercalation. In the case of K10 montmorillonite and related acid-activated smectites, the DTP molecules presumably occupy mesopores of the mineral as depicted in Figure 4.1 for $ZnCl_2$ (Marme et al. 1998). Garade et al. (2010b) have further shown that the ratio of Brønsted to Lewis acid sites in K10 montmorillonite increases with DTP loading. Impregnation of 20% DTP into a commercial acid-activated bentonite (*tonsil*) by Altiokka et al. (2014) produced an efficient catalyst for the esterification of formic acid with *n*-butyl alcohol. The range and variety of organic reactions catalyzed by clay-supported heteropoly acids are summarized in Table 4.4.

TABLE 4.4
Organic Conversions and Reactions Catalyzed by Clay-Supported Heteropoly Acids (HPA), with Particular Reference to Dodecatungstophosphoric Acid (DTP) and Cesium-Exchanged DTP ($Cs_{2.5}H_{0.5}PW_{12}O_{40}$)

Reaction Type (in Alphabetical Order)	Substrate	Product	Clay-Supported HPA	Reference (in Chronological Order)
Acetalization	n-Octanal with methanol	Corresponding acetal	DTP on K10 montmorillonite	Yadav and Pujari (1999)
Acylation	1,3-Dibenzyloxy-benzene with acetic anhydride	3,5-Dibenzyl-oxyacetophenone	Cs-DTP on K10 montmorillonite	Yadav and Badure (2008)
	Ferrocene with acyl chlorides	Corresponding acylferrocenes	PTFMSS[a] on Ca^{2+}-bentorite	Hu and Li (2004)
Alkylation	Hydroquinone with methyl-tert-butyl-ether and tert-butanol	2-tert-Butyl-hydroquinone and 2,5-di-tert-butyl-hydroquinone	DTP on K10 montmorillonite	Yadav and Doshi (2000)
	Catechol, resorcinol, and anisole with methyl-tert-butyl ether	tert-Butylated dihydroxy and alkoxy benzenes	DTP on K10 montmorillonite	Yadav et al. (2001)
	Phenol with methyl-tert-butyl-ether and tert-butanol	C-alkylated products	DTP on K10 montmorillonite	Yadav and Doshi (2002)
	Aniline with methyl-tert-butyl-ether and tert-butanol	Selective for C-alkylated products	DTP on K10 montmorillonite	Yadav and Doshi (2003)
	Phenol with cyclohexene	Selective for O-alkylated cyclohexyl phenyl ether	DTP and Cs-DTP on K10 montmorillonite	Yadav and Kumar (2005)
	Ethylbenzene with ethanol	Diethylbenzene	DTP on K10 montmorillonite	Bokade and Yadav (2008)
	Aniline with methanol	Selective for N-methyl aniline	DTP on K10 montmorillonite	Nehate and Bokade (2009)
	Xylenes with isopropyl alcohol	Isopropyl xylenes (dimethyl cumenes)	DTP and Cs-DTP on K10 montmorillonite	Yadav and Kamble (2009)
	Anisole with cyclohexene	2- and 4-cyclohexyl anisole	Cs-DTP on K10 montmorillonite	Yadav and More (2011)
Benzoylation	Anisole with benzoyl chloride	4-Methoxybenzo-phenone	Cs-DTP on K10 montmorillonite	Yadav et al. (2003a)
	p-Xylene and xylene isomers	2,5-Dimethyl-benzophenone and dimethyl-benzophenone isomers	Cs-DTP on K10 montmorillonite	Yadav et al. (2003b, 2004a)
	Diphenyloxide with benzoic anhydride	Mono-acylated product	Al-DTP on K10 montmorillonite	Tiwari and Yadav (2016)
Condensation	Phenol and acetone	Bisphenol-A	DTP on K10 montmorillonite	Yadav and Kirthivasan (1997)
	1,2-phenylene-diamines and ketones	1,5-benzo-diazepines	DTP and DMP[b] on KSF and K10 montmorillonite	Fazaeli and Aliyan (2007)
	β-Naphthol with arylaldehydes	Various aryl-14H-dibenzo[a,j] xanthenes	DMP[b] on Al-PILC[c]	Naik et al. (2009)

(Continued)

TABLE 4.4 (Continued)
Organic Conversions and Reactions Catalyzed by Clay-Supported Heteropoly Acids (HPA), with Particular Reference to Dodecatungstophosphoric Acid (DTP) and Cesium-Exchanged DTP ($Cs_{2.5}H_{0.5}PW_{12}O_{40}$)

Reaction Type (in Alphabetical Order)	Substrate	Product	Clay-Supported HPA	Reference (in Chronological Order)
Decomposition	Cumene hydroperoxide	Phenol and acetone	Cs-DTP on K10 montmorillonite	Yadav and Asthana (2003)
Dehydration	Methanol	Dimethyl ether	DTP on acid-activated montmorillonite	Marme et al. (1998)
	Ethanol	Ethylene and diethyl ether	DTP on K10 montmorillonite	Bokade and Yadav (2011)
Disproportionation	Ethylbenzene	p-Diethylbenzene (and benzene)	DTP on K10 montmorillonite	Bokade and Yadav (2012)
Epoxidation	Cyclohexene	Cyclohexene epoxide	11-Molybdo-vanado-phosphoric acid on acid-treated bentonite	Boudjema et al. (2015)
Esterification	Transesterification of vegetable oil with alcohols	Bio-diesel (methyl esters)	DTP on K10 montmorillonite	Bokade and Yadav (2007, 2009)
	Phenol with benzoic acid (followed by Fries rearrangement)	4-Hydroxy-benzophenone	Cs-DTP on K10 montmorillonite	Yadav and George (2008)
	Acetic acid with n-butanol	n-Butylacetate	DTP on acid-treated montmorillonite; DTP on Ti-PILC[c]	Bhorodwaj et al. (2009); Varadwaj and Parida (2011)
	Acetic acid with primary, secondary and tertiary butanol	Corresponding butylacetates	DTP on acid-activated montmorillonite	Bhorodwaj and Dutta (2010, 2011)
Etherification	tert-Butylalcohol with methanol	Methyl tert-butyl ether	DTP on K10 montmorillonite	Yadav and Kirthivasan (1995)
	Phenylethyl alcohol with alkanols	Corresponding phenylethyl ethers	DTP on K10 montmorillonite	Yadav and Bokade (1996); Bokade (2001)
Hydroxyalkylation	Hydroquinone with methanol	Hydroquinone monomethyl ether	DTP on K10 montmorillonite	Yadav et al. (2005)
	p-Cresol with formaldehyde	2,2'-methylenebis (4-methylphenol)	DTP on bentonite	Garade et al. (2010a)
	Phenol with formaldehyde	Bisphenol F	DTP on K10 montmorillonite	Garade et al. (2010b)
	Phenol with formaldehyde	Bisphenol F	DTP on acid-treated bentonite	Liu et al. (2015)
Isopropylation	Benzene with isopropanol	Cumene	Cs-DTP on K10 montmorillonite	Yadav et al. (2004b)
	Phenol with diisopropyl ether	2,6-Diisopropyl-phenol	Cs-DTP on K10 montmorillonite	Yadav and Salgaonkar (2005)
Nitration	Arenes with nitric acid	Corresponding nitrated compounds	Phosphoric acid on K10 montmorillonite	Bharadwaj et al. (2014)
Oxidation	Methyl mandelate (in the presence of H_2O_2)	Methyl phenyl glyoxylate	Cs-DTP on K10 montmorillonite	Yadav and Bhagat (2004)
	2-Propanol	Acetone	Cs-DTP on K10 montmorillonite	Rožić et al. (2011)

[a] PTFMSS = Polytrifluormethanesulfosiloxane
[b] DMP = Dodecamolybdophosphoric acid
[c] PILC = Pillared interlayered clay (cf. Chapter 3)

REFERENCES

Agarwal, S. and J.N. Ganguli. 2013. Selective hydrogenation of monoterpenes on rhodium(0) nanoparticles stabilized in montmorillonite K-10 clay. *Journal of Molecular Catalysis A: Chemical* 372: 44–50.

Ahmad, N., S.T. Hussain, B. Muhammad, N. Ali, S.M. Abbas, and Z. Ali. 2013. Zr-pillared montmorillonite for Fischer–Tropsch synthesis. *Progess in Natural Science: Materials International* 23: 374–381.

Ahmed, O.S. and D.K. Dutta. 2003a. Generation of metal nanoparticles on montmorillonite K10 and their characterization. *Langmuir* 19: 5540–5541.

Ahmed, O.S. and D.K. Dutta. 2003b. In situ generation of metal clusters in interlamellar spacing of montmorillonite clay and their thermal behaviour. *Thermochimica Acta* 395: 209–216.

Aihara, N., K. Torigoe, and K. Esumi. 1998. Preparation and characterization of gold and silver nanoparticles in layered laponite suspensions. *Langmuir* 14: 4945–4949.

Akkari, M., P. Aranda, H. Ben Rhaiem, A. Ben Haj Amara, and E. Ruiz-Hitzky. 2016. ZnO/clay nanoarchitectures: Synthesis, characterization and evaluation as photocatalysts. *Applied Clay Science* 131: 131–139.

Alsalme, A., E.F. Kozhevnikova, and I.V. Kozhevnikov. 2008. Heteropolyacids as catalysts for liquid-phase esterification and transesterification. *Applied Catalysis A: General* 349: 170–176.

Altiokka, M.R., E. Akbay, and Z. Him. 2014. Impregnation of 12-tungstophosphoric acid on tonsil: An effective catalyst for esterification of formic acid with *n*-butyl alcohol and kinetic modeling. *Journal of Molecular Catalysis A: Chemical* 385: 18–25.

Alvarez, C., F. Delgado, O. García, S. Medina, and C. Márquez. 1991. MnO_2/bentonite: A new reactive for the oxidation of Hantzsch's dihydropyridines using microwave irradiation, in the absence of solvent (I). *Synthetic Communications* 21: 619–624.

Ammari, F. and M. Chenouf. 2015. Synthesis of gold nanoparticles stabilized in Na-montmorillonite for nitrophenol reduction. *International Journal of Chemical, Molecular, Nuclear, Materials and Metallurgical Engineering* 9: 858–861.

Anderson, J.A., S.E. Falconer, and M. Galán-Fereres. 1997. Ni/sepioliete hydrogenation catalysts Part 1: Precursor-support interaction and nature of exposed metal surfaces. *Spectrochimica Acta Part A* 53: 2627–2639.

Anderson, J.A., M.T. Rodrigo, L. Daza, and S. Mendioroz. 1993. Influence of the support in the selectivity of Ni/clay catalysts for vegetable oil hydrogenation. *Langmuir* 9: 2485–2490.

Arai, M., S.-L. Guo, M. Shirai, Y. Nishiyama, and K. Torii. 1996. The catalytic acitivity of platinum-loaded porous smectite-like clay minerals containing different divalent cations for butane hydrogenolysis and ethylene hydrogenation. *Journal of Catalysis* 161: 704–712.

Asseid, F.M., J.M. Miller, and J.H. Clark. 1992. FT-IR and ^{29}Si, ^{27}Al, and ^{19}F MAS NMR studies of the adsorption of CdF_2, ZnF_2, and CuF_2 onto montmorillonite K10; activity towards Friedel–Crafts alkylation. *Canadian Journal of Chemistry* 7: 2398–2404.

Ayyappan, S., G.N. Subbanna, R.S. Gopalan, and C.N.R. Rao. 1996. Nanoparticles of nickel and silver produced by the polyol reduction of the metal salts intercalated in montmorillonite. *Solid State Ionics* 84: 271–281.

Bahramian, B., V. Mirkhani, M. Moghadam, and S. Tangestaninejad. 2006. Manganese(III) salen immobilized on montmorillonite as biomimetic alkene epoxidation and alkane hydroxylation catalyst with sodium periodate. *Catalysis Communications* 7: 289–296.

Bahranowski, K., M. Gasior, A. Kielski et al. 1999. Copper-doped alumina-pillared montmorillonites as catalysts for oxidation of toluene and xylenes with hydrogen peroxide. *Clay Minerals* 34: 79–87.

Balogh, M., A. Cornélis, and P. Laszlo. 1984a. Cleavage of thioacetals by clay-supported metal nitrates. *Tetrahedron Letters* 25: 3313–3316.

Balogh, M., I. Hermecz, Z. Mészáros, and P. Laszlo. 1984b. Aromatization of 1,4-dihydropyridines by clay-supported metal nitrates. *Helvetica Chimica Acta* 67: 2270–2272.

Balogh, M. and P. Laszlo. 1993. *Organic Chemistry Using Clays*. Berlin, Germany: Springer-Verlag.

Bandgar, B.P. and N.B. Gaikwad. 1998. Solid supported reactions and reagents XIV [1]: Envirocat EPZG® as a novel catalyst for selective acetalization of aldehydes and ketones. *Monatshefte für Chemie* 129: 719–722.

Bandgar, B.P., S.R. Jagtap, S.B. Ghodeshwar, and P.P. Wadgaonkar. 1995. Envirocat EPZG®, a new catalyst for the conversion of aldoximes into nitriles. *Synthetic Communications* 25: 2993–2998.

Bandgar, B.P. and S.P. Kasture. 2000a. Environmentally friendly arylmethylation of aromatics with benzyl halides using Envirocat EPZ10 as the catalyst. *Monatshefte für Chemie* 131: 913–915.

Bandgar, B.P. and S.P. Kasture. 2000b. Envirocats, as novel solid-supported catalysts for Friedel–Crafts acylation. *Journal of the Chinese Chemical Society* 47: 1243–1246.

Bandgar, B.P., S.P. Kasture, K. Tidke, and S.S. Makone. 2000c. Envirocat EPZG catalysis: An efficient and rapid procedure for the deprotection of 1,1-diacetates without solvent. *Green Chemistry* 2: 152–153.

Bandgar, B.P., L.S. Uppalla, and V.S. Sadavarte. 2001. Envirocat EPZG and natural clay as efficient catalysts for transesterification of β-keto esters. *Green Chemistry* 3: 39–41.

Barlow, S.J., T.W. Bastock, J.H. Clark, and S.R. Cullen. 1993. Explanation of an unusual substitutent effect in the benzylation of anisole and identification of the origin of the active sites in clayzic. *Tetrahedron Letters* 34: 3339–3342.

Barlow, S.J., T.W. Bastock, J.H. Clark, and S.R. Cullen. 1994. Activation of clayzic and its effect on the relative rates of benzylation of aromatic substrates. *Journal of the Chemical Society, Perkin Transactions 2*: 411–414.

Barloy, L., P. Battioni, and D. Mansuy. 1990. Manganese porphyrins supported on montmorillonite as hydrocarbon mono-oxygenation catalysts: Particular efficacy for linear alkane hydroxylation. *Journal of the Chemical Society, Chemical Communications* 19: 1365–1367.

Barloy, L., J.P. Lallier, P. Battioni et al. 1992. Manganese porphyrins adsorbed or intercalated in different mineral matrices: Preparation and compared properties as catalysts for alkene and alkane oxidation. *New Journal of Chemistry* 16: 71–80.

Bedioui, F. 1995. Zeolite-encapsulated and clay-intercalated metal porphyrin, phthalocyanine and Schiff-base complexes as models for biomimetic oxidation catalyst: An overview. *Coordination Chemistry Reviews* 144: 39–68.

Begari, E., C. Singh, U. Nookaraju, and P. Kumar. 2014. Clay-supported copper nitrate (claycop): A mild reagent for the selective nitration of aromatic olefins. *Synlett* 25: 1997–2000.

Begum, T., A. Gogoi, P.K. Gogoi, and U. Bora. 2015. Catalysis by mont K-10 supported silver nanoparticles: A rapid and green protocol for the efficient *ipso*-hydroxylation of arylboronic acids. *Tetrahedron Latters* 56: 95–97.

Belaidi, N., S. Bedrane, A. Choukchou-Braham, and R. Bachir. 2015. Novel vanadium-chromium-bentonite green catalysts for cyclohexene epoxidation. *Applied Clay Science* 107: 14–20.

Belova, V., H. Möhwald, and D.G. Shchukin. 2008. Sonochemical intercalation of preformed gold nanoparticles into multilayered clays. *Langmuir* 24: 9747–9753.

Belova, V., H. Möhwald, and D.G. Shchukin. 2009a. Ultrasonic intercalation of gold nanoparticles into a clay matrix in the presence of surface-active materials. Part II: Negative sodium dodecylsulfate and positive cetyltrimethylammonium bromide. *Journal of Physical Chemistry C* 113: 6751–6760.

Belova, V., D.V. Andreeva, H. Möhwald, and D.G. Shchukin. 2009b. Ultrasonic intercalation of gold nanoparticles into clay matrix in the presence of surface-active materials. Part I: Neutral polyethylene glycol. *Journal of Physical Chemistry C* 113: 5381–5389.

Bharadwaj, S.K., P.K. Boruah, and P.K. Gogoi. 2014. Phosphoric acid modified montmorillonite clay: A new heterogeneous catalyst for nitration of arenes. *Catalysis Communications* 57: 124–128.

Bhorodwaj, S.K. and D.K. Dutta. 2010. Heteropoly acid supported modified montmorillonite clay: An effective catalyst for the esterification of acetic acid with sec-butanol. *Applied Catalysis A: General* 378: 221–226.

Bhorodwaj, S.K. and D.K. Dutta. 2011. Activated clay supported heteropoly acid catalysts for esterification of acetic acid with butanol. *Applied Clay Science* 53: 347–352.

Bhorodwaj, S.K., M.G. Pathak, and D.K. Dutta. 2009. Esterification of acetic acid with *n*-butanol using heteropolyacid supported modified clay catalyst. *Catalysis Letters* 133: 185–191.

Bizaia, N., E.H. de Faria, G.P. Ricci et al. 2009. Porphyrin-kaolinite as efficient catalyst for oxidation reactions. *ACS Applied Materials & Interfaces* 1: 2667–2678.

Bokade, V.V. 2001. Effect of etherifying species on O-alkylation of phenylethyl alcohol to perfumery alkyl ethers. *Chemical Engineering Research and Design* 79: 625–630.

Bokade, V.V. and G.D. Yadav. 2007. Synthesis of bio-diesel and bio-lubricant by transesterification of vegetable oil with lower and higher alcohols over heteropolyacids supported by clay (K-10). *Process Safety and Environmental Protection* 85(B5): 372–377.

Bokade, V.V. and G.D. Yadav. 2008. Heteropolyacid supported on acidic clay: A novel efficient catalyst for alkylation of ethylbenzene with dilute ethanol to diethylbenzene in presence of C_8 aromatics. *Journal of Molecular Catalysis A: Chemical* 285: 155–161.

Bokade, V.V. and G.D. Yadav. 2009. Transesterification of edible and non-edible vegetable oils with alcohols over heteropolyacids supported on acid-treated clay. *Industrial & Engineering Chemistry Research* 48: 9408–9415.

Bokade, V.V. and G.D. Yadav. 2011. Heteropolyacid supported on montmorillonite catalyst for dehydration of dilute bio-ethanol. *Applied Clay Science* 53: 263–271.

Bokade, V.V. and G.D. Yadav. 2012. Dodecatungstophosphoric acid supported on acidic clay catalyst for disproportionation of ethylbenzene in presence of C_8 aromatics. *Industrial & Engineering Chemistry Research* 51: 1209–1217.

Borah, S.J. and D.K. Das. 2016. Modified montmorillonite clay stabilized silver nanoparticles: An active heterogeneous catalytic system for the synthesis of propargylamines. *Catalysis Letters* 146: 656–665.

Borah, B.J. and D.K. Dutta. 2013. In situ stabilization of Pd^0-nanoparticles into the nanopores of modified montmorillonite: Efficient heterogeneous catalysts for Heck and Sonogashira coupling reactions. *Journal of Molecular Catalysis A: Chemical* 366: 202–209.

Borah, B.J., S.J. Borah, K. Saikia, and D.K. Dutta. 2014a. Efficient Suzuki-Miyaura coupling reaction in water: Stabilized $Pd°$-montmorillonite clay composites catalyzed reaction. *Applied Catalysis A: General* 469: 350–356.

Borah, B.J., S.J. Borah, L. Saikia, and D.K. Dutta. 2014b. Efficient three-component coupling reactions catalyzed by $Cu°$-nanoparticles stabilized on modified montmorillonite. *Catalysis Science & Technology* 4: 1047–1054.

Borah, B.J., D. Dutta, P.P., Saikia, N.C. Barua, and D.K. Dutta. 2011. Stabilization of $Cu°$-nanoparticles into the nanopores of modified montmorillonite: An implication on the catalytic approach for "Click" reaction between azides and terminal alkynes. *Green Chemistry* 13: 3453–3460.

Boudjema, S., E. Vispe, A. Choukchou-Braham, J.A. Majoral, R. Bachir, and J.M. Fraile. 2015. Preparation and characterization of activated montmorillonite clay supported 11-molybdo-vanado-phosphoric acid for cyclohexene oxidation. *RSC Advances* 5: 6853–6863.

Bouhlel, E., P. Laszlo, M. Levart, M.-T. Montaufier, and G.P. Singh. 1993. Epoxidation of olefins by molecular oxygen with clay-impregnated nickel catalysts. *Tetrahedron Letters* 34: 1123–1126.

Boukhatem, H., L. Djouadi, N. Abdelaziz, and H. Khalaf. 2013. Synthesis, characterization and photocatalytic activity of CdS-montmorillonite nanocomposites. *Applied Clay Science* 72: 44–48.

Cardoso, B., J. Pires, A.P. Carvalho et al. 2005. Mn(III) *salen* complex immobilised into pillared clays by in situ and simultaneous pillaring/encapsulation procedures. Application in the heterogeneous epoxidation of styrene. *Microporous and Mesoporous Materials* 86: 295–302.

Carrado, K.A. and R.E. Winans. 1990. Interactions of water-soluble porphyrins and metalloporphyrins with smectite clay surfaces. *Chemistry of Materials* 2: 328–335.

Chalais, S., A. Cornélis, A. Gerstmans et al. 1985. Direct clay-catalyzed Friedel–Crafts arylation and chlorination of the hydrocarbon adamantane. *Helvetica Chimica Acta* 68: 1196–1203.

Chatterjee, D. and A. Mitra. 1999. Olefin epoxidation catalysed by Schiff-base complexes of Mn and Ni in heterogenised-homogeneous systems. *Journal of Molecular Catalysis A: Chemical* 144: 363–367.

Chen, C.-C. and P.-L. Kuo. 2006. Gold nanoparticles prepared using polyethylenimine adsorbed onto montmorillonite. *Journal of Colloid and Interface Science* 293: 101–107.

Chen, D., M. Huang, S. He et al. 2016. Ru-MOF enwrapped by montmorillonite for catalysing benzene hydrogenation. *Applied Clay Science* 119: 109–115.

Chibwe, M., L. Ukraineczyk, S.A. Boyd, and T.J. Pinnavaia. 1996. Catalytic properties of biomimetic metallomacrocycles intercalated in layered double hydroxides and smectite clay: The importance of edge-site access. *Journal of Molecular Catalysis A: Chemical* 113: 249–256.

Choudary, B.M. and P. Bharati. 1987. Synthesis of interlamellar montmorillonite-bipyridine palladium(ii) catalysts: The first examples of chelation in smectite clay. *Journal of the Chemical Society, Chemical Communications* 1505–1506.

Choudhary, V.R. and S.K. Jana. 2002. Benzylation of benzene and substituted benzenes by benzyl chloride over $InCl_3$, $GaCl_3$, $FeCl_3$ and $ZnCl_2$ supported on clays and Si-MCM-41. *Journal of Molecular Catalysis A: Chemical* 180: 267–276.

Choudhary, V.R. and R. Jha. 2008. $GaAlCl_x$-grafted Mont. K-10 clay: Highly active and stable solid catalyst for the Friedel–Crafts type benzylation and acylation reactions. *Catalysis Communications* 9: 1101–1105.

Choudhary, V.R. and K. Mantri. 2002. Thermal activation of a clayzic catalyst useful for Friedel–Crafts reactions: HCl evolved with creation of active sites in different thermal treatments to $ZnCl_2$/Mont-K10. *Catalysis Letters* 81: 163–168.

Choudhary, V.R., S.K. Jana, and N.S. Patil. 2001b. Acylation of benzene over clay and mesoporous Si-MCM-41 supported $InCl_3$, $GaCl_3$ and $ZnCl_2$ catalysts. *Catalysis Letters* 76: 235–239.

Choudary, B.M., K.R. Kumar, and M.L. Kantam. 1991. Synthesis and catalytic activity in selective hydrogenation of palladium complexes anchored in montmorillonite. *Journal of Catalysis* 130: 41–51.

Choudary, V.R., K. Mantri, and S.K. Jana. 2001a. Selective esterification of tert-butanol by acetic acid anhydride over clay supported $InCl_3$, $GaCl_3$, $FeCl_3$ and $InCl_2$ catalysts. *Catalysis Communications* 2: 57–61.

Choudhary, V.R., K.Y. Patil, and S.K. Jana. 2004. Acylation of aromatic alcohols and phenols over $InCl_3$/montmorillonite K-10 catalysts. *Journal of Chemical Science* 116: 175–177.

Choudary, B.M., K.R. Kumar, Z. Jamil, and G. Thyagarajan. 1985. A novel 'anchored' palladium(II) phosphinated montmorillonite: The first example in the interlamellars of smectite clay. *Journal of the Chemical Society, Chemical Communications* 931–932.

Clark, J.H., A.J. Butterworth, S.J. Tavener et al. 1997a. Environmentally friendly chemistry using supported reagent catalysts: Chemically-modified mesoporous solid catalysts. *Journal of Chemical Technology and Biotechnology* 68: 367–376.

Clark, J.H., S.R. Cullen, S.J. Barlow, and T.W. Bastock. 1994b. Environmentally friendly chemistry using supported reagent catalysts: Structure-property relationships for clayzic. *Journal of the Chemical Society, Perkin Transactions* 2: 1117–1130.

Clark, J.H., A.P. Kybett, D.J. Macquarrie, S.J. Barlow, and P. Landon. 1989. Montmorillonite supported transition metal salts as Friedel–Crafts alkylation catalysts. *Journal of the Chemical Society, Chemical Communications* 1353–1354.

Clark, J.H. and D.J. Macquarrie. 1996. Environmentally friendly catalytic methods. *Chemical Society Reviews* 25: 303–310.

Clark, P.D. and S.T. Mesher. 1995. Benzylation of benzo[*b*]thiophene using $ZnCl_2$-modified montmorillonite clay. *Phosphorus, Sulfur, and Silicon* 105: 157–162.

Clark, P.D., A. Kirk, and R.A. Kydd. 1994a. Benzylation of thiophene using zinc and iron chloride modified montmorillonite clay. *Catalysis Letters* 25: 163–168.

Clark, P.D., A. Kirk, and J.G.K. Yee. 1995. An improved synthesis of benzo[*b*]thiophene and its derivatives using modified montmorillonite clay catalysts. *Journal of Organic Chemistry* 60: 1936–1938.

Clark, P.D., S.T.E. Mesher, and A. Primak. 1996. Clay-catalyzed thioalkylation of thiophenes and benzo[*b*] thiophenes. *Phosphorus, Sulfur, and Silicon* 114: 99–108.

Clark, P.D., S.T.E. Mesher, A. Primak, and H. Yao. 1997b. C–S bond formation in aromatic substrates using Mn(II)-promoted montmorillonite clays. *Catalysis Letters* 48: 79–82.

Cornélis, A. and P. Laszlo. 1980. Oxidation of alcohols by clay-supported iron(III) nitrate. A new efficient oxidizing agent. *Synthesis* 849–850.

Cornélis, A. and P. Laszlo. 1985. Clay-supported copper(II) and iron(III) nitrates: Novel multi-purpose reagents for organic synthesis. *Synthesis* 909–918.

Cornélis, A. and P. Laszlo. 1994. Molding clays into efficient catalysts. *Synlett* 155–161.

Cornélis, A., P. Laszlo, and P. Pennetreau. 1983a. Some organic synthesis with clay-supported reagents. *Clay Minerals* 18: 437–445.

Cornélis, A., P. Laszlo, and P. Pennetreau. 1983b. Nitration of estrone into 2-nitroestrone by clay-supported ferric nitrate. *Journal of Organic Chemistry* 48: 4771–4772.

Cornélis, A., P. Laszlo, and P. Pennetreau. 1984. Nitration of phenols by clay-supported ferric nitrate. *Bulletin des Sociétés Chimiques Belges* 93: 961–972.

Cornélis, A., P.-Y. Herze, and P. Laszlo. 1982. Clay-supported reagents III. The intermediacy of nitrous esters in the oxidation of alcohols by clay-supported ferric nitrate. *Tetrahedron Letters* 23: 5035–5038.

Cornélis, A., P. Laszlo, and S.-F. Wang. 1993a. On the transition state for "clayzic"-catalyzed Friedel–Crafts reactions upon anisole. *Tetrahedron Letters* 34: 3849–3852.

Cornélis, A., P. Laszlo, and S.-F. Wang. 1993b. Side-product inhibition of the catalyst in electrophilic aromatic substitutions and Friedel–Crafts reactions. *Catalysis Letters* 17: 63–69.

Cornélis, A., N. Depaye, A. Gerstmans, and P. Laszlo. 1983c. Clay-supported reagents IV. A novel coupling of thiols into disulphides, via thionitrite intermediates using a clay-supported nitrosation reagent. *Tetrahedron Letters* 24: 3103–3106.

Cornélis, A., C. Dony, P. Laszlo, and K.M. Nsunda. 1991a. Synergistic acceleration of reaction having clay-based catalysts. *Tetrahedron Letters* 32: 1423–1424.

Cornélis, A., C. Dony, P. Laszlo, and K.M. Nsunda. 1991b. Inversion of the relative reactivities of mesitylene and toluene in clay-catalyzed Friedel–Crafts alkylations. *Tetrahedron Letters* 32: 2901–2902.

Cornélis, A., A. Gerstmans, P. Laszlo, A. Mathy, and I. Zieba. 1990. Friedel–Crafts acylations with modified clays as catalysts. *Catalysis Letters* 6: 103–110.

Crestini, C., A. Pastorini, and P. Tagliatesta. 2004. Metalloporphyrins immobilized on montmorillonite as biomimetic catalysts in the oxidation of lignin model compounds. *Journal of Molecular Catalysis A: Chemical* 208: 195–202.

Crocker, M., J.G. Buglass, and R.H.M. Herold. 1993. Synthesis and characterization of palladium crystallites intercalated in montmorillonite. *Chemistry of Materials* 5: 105–109.

Crocker, M. and R.H.M. Herold. 1991. Carbomethoxylation of ethylene catalysed by Pd(II) complexes intercalated in smectite clay. *Journal of Molecular Catalysis* 70: 209–216.

Cseri, T., S. Békássy, Z. Bódás, B. Agai, and F. Figueras. 1996a. Acetylation of B15C5 crown ether on Cu modified clay catalysts. *Tetrahedron Letters* 37: 1473–1476.

Cseri, T., S. Békássy, G. Kenessey, G. Liptay, and F. Figueras. 1996b. Characterization of metal nitrates and clay supported metal nitrates by thermal analysis. *Thermochimica Acta* 288: 137–154.

Dai, H.-B., X.-D. Kang, and P. Wang. 2010. Ruthenium nanoparticles immobilized in montmorillonite used as catalyst for methanolysis of ammonia borane. *International Journal of Hydrogen Energy* 35: 10317–10323.

Dar, B.A., P. Bhatti, A.P. Singh et al. 2013. Clay entrapped $Cu(OH)_x$ as an efficient heterogeneous catalyst for ipso-hydroxylation of arylboronic acids. *Applied Catalysis A: General* 466: 60–67.

Dar, B.A., N. Pandey, S. Singh, P. Kumar, M. Farooqui, and B. Singh. 2015. Solvent-free, scalable and expeditious synthesis of benzanilides under microwave irradiation using clay doped with palladium nanoparticles as a recyclable and efficient catalyst. *Green Chemistry Letters and Reviews* 8: 1–8.

Das, P., I. Kuźniarska-Biernacka, A.R. Silva, A.P. Carvalho, J. Pires, and C. Freire. 2006. Encapsulation of chiral Mn(III) *salen* complexes into aluminium pillared clays: Application as heterogeneous catalysts in the epoxidation of styrene. *Journal of Molecular Catalysis A: Chemical* 248: 135–143.

Daza, C.E., A. Kiennemann, S. Moreno, and R. Molina. 2009. Dry reforming of methane using Ni-Ce catalysts supported on a modified mineral clay. *Applied Catalysis A: General* 364: 65–74.

Dhakshinamoorthy, A. and K. Pitchumani. 2006. Clay-anchored non-heme iron-salen complex catalyzed cleavage of C=C bond in aqueous medium. *Tetrahedron* 62: 9911–9918.

Dhakshinamoorthy, A. and K. Pitchumani. 2008. Clay entrapped nickel nanoparticles as efficient and recyclable catalysts for hydrogenation of olefins. *Tetrahedron Letters* 49: 1818–1823.

Dhakshinamoorthy, A., K. Kanagaraj, and K. Pitchumani. 2011. Zn^{2+}-K10 clay (clayzic) as an efficient water-tolerant, solid acid catalyst for the synthesis of benzimidazoles and quinoxalines at room temperature. *Tetrahedron Letters* 52: 69–73.

Dhanagopal, M., D. Divakar, R.A. Valentine, M. Ramalinga Viswanathan, and S. Thiripuranthagan. 2010. Nanosized noble metals intercalated in clay as catalysts for selective hydrogenation. *Chinese Journal of Catalysis* 31: 1200–1208.

Dias, P.M., D.L.A. de Faria, and V.R.L. Constantino. 2000. Sprectroscopic studies on the interaction of tetramethylpyridylporphyrins and cationic clays. *Journal of Inclusion Phenomena and Macrocyclic Chemistry* 38: 251–266.

Divakar, D., D. Manikandan, G. Kalidos, and T. Sivakumar. 2008b. Hydrogenation of benzaldehyde over palladium intercalated bentonite catalysts: Kinetic studies. *Catalysis Letters* 125: 277–282.

Divakar, D., D. Manikandan, V. Rupa, E.L. Preethi, R. Chandrasekar, and T. Sivakumar. 2007. Palladium-nanoparticle intercalated vermiculite for selective hydrogenation of α,β-unsaturated aldehydes. *Journal of Chemical Technology and Biotechnology* 82: 253–258.

Divakar, D., D. Manikandan, and T. Sivakumar. 2008a. Tetra silicic mica—A synthetic support for nanoparticle generation and catalytic applications. *Catalysis Communications* 9: 2433–2436.

Dixit, P.S. and K. Srinivasan. 1988. The effect of clay-support on the catalytic epoxidation activity of a manganese(III)-Schiff base complex. *Inorganic Chemistry* 27: 4507–4509.

Dutta, D., B.J. Borah, L. Saikia, M.G. Pathak, P. Sengupta, and D.K. Dutta. 2011. Synthesis and catalytic activity of Ni-acid activated montmorillonite nanoparticles. *Applied Clay Science* 53: 650–656.

Dutta, D.K., B.J. Borah, and P.P. Sarmah. 2015. Recent advances in metal nanoparticles stabilization into nanopores of montmorillonite and their catalytic applications for fine chemicals synthesis. *Catalysis Reviews: Science and Engineering* 57: 257–305.

Dutta, D. and D.K. Dutta. 2014. Selective and efficient hydrogenation of halonitrobenzene catalyzed by clay supported Nio-nanoparticles. *Applied Catalysis A: General* 487: 158–164.

Fang, L., R. Hong, J. Gao, and C. Gu. 2016. Degradation of bisphenol A by nano-sized manganese dioxide synthesized using montmorillonite as templates. *Applied Clay Science* 132–133: 155–160.

Farias, M., M. Martinelli, and G.K. Rolim. 2011. Immobilized molybdenum acetylacetonate complex on montmorillonite K-10 as catalysts for epoxidation of vegetable oils. *Applied Catalysis A: General* 403: 119–127.

Farzaneh, F. and T.J. Pinnavaia. 1983. Metal complex catalysts interlayered in smectite clay. Hydroformylation of 1-hexene with rhodium complexes ion exchanged into hectorite. *Inorganic Chemistry* 22: 2216–2220.

Fazaeli, R. and H. Aliyan. 2007. Clay (KSF and K10)-supported heteropoly acids: Friendly, efficient, reusable and heterogeneous catalysts for high yield synthesis of 1,5-dibenzodiazepine derivatives both in solution and under solvent-free conditions. *Applied Catalysis A: General* 331: 78–83.

Ferraz, H.M.C., L.F. Silva Jr., A.M. Aguilar, and T.O. Vieira. 2001. The reaction of 1-tetralones with thallium trinitrate supported on clay: Ring contraction *vs* α-oxidation. *Journal of the Brazilian Chemical Society* 12: 680–684.

Fraile, J.M., J.I. García, J. Massam, and J.A. Mayoral. 1998. Clay-supported non-chiral and chiral Mn(salen) complexes as catalysts for olefin epoxidation. *Journal of Molecular Catalysis A: Chemical* 136: 47–57.

Gamba, O., S. Moreno, and R. Molina. 2011. Catalytic performance of Ni-Pr supported on delaminated clay in the dry reforming of methane. *International Journal of Hydrogen Energy* 36: 1540–1550.

Gandía, L.M., M.A. Vicente, and A. Gil. 2000. Preparation and characterization of manganese oxide catalysts supported on alumina and zirconia-pillared clays. *Applied Catalysis A: General* 196: 281–292.

Gandía, L.M., M.A. Vicente, and A. Gil. 2002. Complete oxidation of acetone over manganese oxide catalysts supported on alumina- and zirconia-pillared clays. *Applied Catalysis B: Environmental* 38: 295–307.

Gao, X. and J. Xu. 2006. A new application of clay-supported vanadium oxide catalyst to selective hydroxylation of benzene to phenol. *Applied Clay Science* 33: 1–6.

Garade, A.C., V.S. Kshirsagar, A. Jha, and C.V. Rode. 2010a. Structure-activity studies of dodecatungstophosphoric acid impregnated bentonite clay catalyst in hydroxyalkylation of *p*-cresol. *Catalysis Communications* 11: 942–945.

Garade, A.C., V.S. Kshirsagar, R.B. Mane, A.A. Ghalwadkar, U.D. Joshi, and C.V. Rode. 2010b. Acidity tuning of montmorillonite K10 by impregnation with dodecatungstophosphoric acid and hydroxyalkylation of phenol. *Applied Clay Science* 48: 164–170.

Garcia, A.M., V. Moreno, S.X. Delgado et al. 2016. Encapsulation of SALEN- and SALDH-Mn(III) complexes in an Al-pillared clay for bicarbonate-assisted catalytic epoxidation of cyclohexene. *Journal of Molecular Catalysis A: Chemical* 416: 10–19.

Garrido-Ramírez, E.G., B.K.G. Theng, and M.L. Mora. 2010. Clays and oxide minerals as catalysts in Fenton-like reactions. *Applied Clay Science* 47: 182–192.

Ghatpande, S. and S. Mahajan. 2005. Synthesis of arylketones using Envirocat EPZG catalyst. *Indian Journal of Chemistry* 44B: 188–192.

Ghiaci, M., M. Mostajeran, and A. Gil. 2012. Synthesis and characterization of Co-Mn nanoparticles immobilized on a modified bentonite and its application for oxidation of *p*-xylene to terephthalic acid. *Industrial & Engineering Chemistry Research* 51: 15821–15831.

Gigante, B., A.O. Prazeres, M.J. Marcelo-Curto, A. Cornélis, and P. Laszlo. 1995. Mild and selective nitration by "claycop". *Journal of Organic Chemistry* 60: 3445–3447.

Gil, A., M.A. Vicente, and S.A. Korili. 2006. Effect of the nature and structure of pillared clays in the catalytic behaviour of supported manganese oxide. *Catalysis Today* 112: 117–120.

Gil, A., M.A. Vicente, J.-F. Lambert, and L.M. Gandía. 2001. Platinum catalysts supported on Al-pillared clays. Application to the catalytic combustion of acetone and methyl-ethyl-ketone. *Catalysis Today* 68: 41–51.

Gogoi, A., S.J. Chutia, P.K. Gogoi, and U. Bora. 2014. A highly efficient heterogeneous montmorillonite K-10-supported palladium catalyst for Suzuki-Miyaura cross-coupling reaction in aqueous medium. *Applied Organometallic Chemistry* 28: 839–844.

González, E. and A. Moronta. 2004. The dehydrogenation of ethylbenzene to styrene catalyzed by a natural and an Al-pillared clays impregnated with cobalt compounds: A comparative study. *Applied Catalysis A: General* 258: 99–105.

Gu, C., H. Jia, H. Li, B.J. Teppen, and S.A. Boyd. 2010. Synthesis of highly reactive subnano-sized zero-valent iron using smectite clay templates. *Environmental Science & Technology* 44: 4258–4263.

Gupta, K.C. and A.K. Sutar. 2008. Catalytic activities of Schiff base transition metal complexes. *Coordination Chemistry Reviews* 252: 1420–1450.

Gupta, K.C., A.K. Sutar, and C.-C. Lin. 2009. Polymer-supported Schiff base complexes in oxidation reactions. *Coordination Chemistry Reviews* 253: 1926–1946.

Hájek, J., P. Kačer, V. Hulínský, L. Červený, and D. Yu Murzin. 2006. High-selectivity hydrogenation of cinnamaldehyde over platinum supported on aluminosilicates. *Research on Chemical Intermediates* 32: 795–816.

Halligudi, S.B., H.C. Bajaj, K.N. Bhatt, and M. Krishnaratnam. 1992. Hydrogenation of benzene to cyclohexane catalyzed by rhodium(I) complex supported on montmorillonite clay. *Reaction Kinetics and Catalysis Letters* 48: 547–552.

Han, Y.-S., J.-W. Lee, and S.-M. Park. 2010. Preparation and water adsorption property of a mesoporous TiO_2-montmorillonite nanocomposite. *Journal of Ceramic Processing Research* 11: 602–605.

Han, Z., H. Zhu, S.R. Bulcock, and S.P. Ringer. 2005. One-step synthesis and structural features of CdS/montmorillonite nanocomposites. *Journal of Physical Chemistry B* 109: 2673–2678.

Han, Z., H. Zhu, J. Shi, and G.Q. Lu. 2006. A straightforward wet-chemical route to the nanocomposites of general layered clays and metal sulfides. *Materials Letters* 60: 2309–2311.

Han, Z., H. Zhu, K.R. Ratinac, S.P. Ringer, J. Shi, and J. Liu. 2008. Nanocomposites of layered clays and cadmium sulfide: Similarities and differences in formation, structure and properties. *Microporous and Mesoporous Materials* 108: 168–182.

Hao, Z., H.Y. Zhu, and G.Q. Lu. 2003. Zr-laponite pillared clay-based nickel catalysts for methane reforming with carbon dioxide. *Applied Catalysis A: General* 242: 275–286.

Hashemi, M.M., B. Eftekhari-Sis, A. Abdollahifar, and B. Khalili. 2006. $ZrOCl_2.8H_2O$ on montmorillonite K10 accelerated conjugate addition of amines to α,β-unsaturated alkenes under solvent-free conditions. *Tetrahedron* 62: 672–677.

He, H., L. Ma, J. Zhu, R.L. Frost, B.K.G. Theng, and F. Bergaya. 2014. Synthesis of organoclays: A critical review and some unresolved issues. *Applied Clay Science* 100: 22–28.

He, X., A. Tang, H. Yang, and J. Ouyang. 2011. Synthesis and catalytic activity of doped TiO_2-palygorskite composites. *Applied Clay Science* 53: 80–84.

Heravi, M.M. and D. Ajami. 1998. Oxidative deprotection of tetrahydropyranyl ethers to carbonyl compounds with montmorillonite K-10 supported bis(trimethylsilyl)chromate under non-aqueous conditions. *Journal of Chemical Research (S)* 718–719.

Heravi, M.M., D. Ajami, and M.M. Mojtahedi. 2000. Regeneration of carbonyl compounds from oximes on clayfen under conventional heating and microwave irradiation. *Journal of Chemical Research (S)* 126–127.

Herrero, J., C. Blanco, and L.A. Oro. 1989. Preparation of rhodium-pyllosilicate catalysts without leaching in liquid-phase 1-hexene hydrogenation. *Applied Organometallic Chemistry* 3: 553–555.

Herrero, J., J.A. Pajares, and C. Blanco. 1991. Surface acidity of palygorskite-supported rhodium catalysts. *Clays and Clay Minerals* 39: 651–657.

Hu, R.-J. and B.-G. Li. 2004. Novel solid acid catalyst, bentonite-supported polytrifluoro-methanesulfosiloxane for Friedel–Crafts acylation of ferrocene. *Catalysis Letters* 98: 43–47.

Hur, S.G., T.W. Kim, S.-J. Hwang, S.-H. Hwang, J.H. Yang, and J.-H. Choy. 2006. Heterostructured nanohybrid of zinc oxide-montmorillonite clay. *Journal of Physical Chemistry* 110: 1599–1604.

Imamura, S., T. Kokubu, T. Yanashita, Y. Okamoto, K. Kajiwara, and H. Kanai. 1996. Shape-selective copper-loaded imogolite catalyst. *Journal of Catalysis* 160: 137–139.

Issaadi, R., F. Garin, C.E. Chitour, and G. Maire. 2001. Catalytic behaviour of combined palladium-acid catalysts: Use of Al and Zr-pillared montmorillonite as supports Part I. Reactivity of linear, branched and cyclic hexane hydrocarbons. *Applied Catalysis A: General* 207: 323–332.

Izumi, Y., M. Ogawa, and K. Urabe. 1995. Alkali metal salts and ammonium salts of Keggin-type heteropolyacids as solid acid catalysts for liquid-phase Friedel–Crafts reactions. *Applied Catalysis A: General* 132: 127–140.

Izumi, Y., K. Urabe, and M. Onaka. 1997a. Development of catalyst materials for acid-catalyzed reactions in the liquid phase. *Catalysis Today* 35: 183–188.

Izumi, Y., K. Urabe, and M. Onaka. 1997b. Recent advances in liquid-phase organic reactions using heteropolyacid and clay. *Catalysis Surveys from Japan* 1: 17–23.

Jackson, A.H., K.R.N. Rao, N.S. Ooi, and E. Adelakun. 1984. Reactions on solid supports. Part I. Novel preparation of α-formyl pyrroles from α-methyl pyrroles by oxidation with thallium(III) nitrate on clay. *Tetrahedron Letters* 25: 6049–6050.

Jia, H. and C. Wang. 2012. Adsorption and dechlorination of 2,4-dichlorophenol (2,4-DCP) on a multifunctional organo-smectite templated zero-valent iron composite. *Chemical Engineering Journal* 191: 202–209.

Jia, H., C. Gu, H. Li et al. 2012. Effect of groundwater geochemistry on pentachlorophenol remediation by smectite-templated nanosized Pd^0/Fe^0. *Environmental Science and Pollution Research* 19: 3498–3505.

Jia, H., C. Gu, S.A. Boyd et al. 2011. Comparison of reactivity of nanoscaled zero-valent iron formed on clay surfaces. *Soil Science Society of America Journal* 75: 357–364.

Jiang, J., K. Ma, Y. Zheng, S. Cai, R. Li, and J. Ma. 2009b. Cobalt salophen complex immobilized into montmorillonite as catalyst for the epoxidation of cyclohexene by air. *Applied Clay Science* 45: 117–122.

Jiang, T., Y. Zhou, S. Liang, H. Liu, and B. Han. 2009a. Hydrogenolysis of glycerol catalyzed by Ru-Cu bimetallic catalysts supported on clay with the aid of ionic liquids. *Green Chemistry* 11: 1000–1006.

Jondi, W., A. Zyoud, W. Mansour, A.Q. Hussein, and H.S. Hilal. 2016. Highly active and selective catalysts for olefin hydrosilylation reactions using metalloporphyrins intercalated in natural clays. *Reaction Chemistry & Engineering* 1: 194–203.

Joshi, M.V. and C.S. Narasimhan. 1989. Catalysis by heteropolyacids: Some new aspects. *Journal of Catalysis* 120: 282–286.

Joshi, M.V. and C.S. Narasimhan. 1993. Facile conversion of aldehydes to 1,1-diacetates catalysed by ZSM-5 and tungstosilicic acid. *Journal of Catalysis* 141: 308–310.

Kabilaphat, J., N. Khaorapapong, K. Saito, and M. Ogawa. 2015. Preparation of metal sulfide mixtures in montmorillonite by solid-solid reactions. *Applied Clay Science* 115: 248–253.

Kadwa, E., M.D. Bala, and H.B. Friedrich. 2014. Characterisation and application of montmorillonite-supported Fe Schiff base complexes as catalysts for the oxidation of *n*-octane. *Applied Clay Science* 95: 340–347.

Kalbasi, R.J., A.R. Massah, and B. Daneshvarnejad. 2012. Preparation and characterization of bentonite/PS-SO_3H nanocomposites as an efficient acid catalyst for the Biginelli reaction. *Applied Clay Science* 55: 1–9.

Kameyama, H., F. Narumi, T. Hattori, and H. Kameyama. 2006. Oxidation of cyclohexene with molecular oxygen catalyzed by cobalt porphyrin complexes immobilized on montmorillonite. *Journal of Molecular Catalysis A: Chemical* 258: 172–177.

Kannan, P., R. Sevvel, S. Rajagopal, K. Pitchumani, and C. Srinivasan. 1997. Oxidation of organic sulphides with clay-supported iodosylbenzene as oxygen donor. *Tetrahedron* 53: 7635–7640.

Khaorapapong, N., N. Khumchoo, and M. Ogawa. 2011. Preparation of zinc oxide-montmorillonite hybrids. *Materials Letters* 65: 657–660.

Khaorapapong, N., N. Khumchoo, and M. Ogawa. 2015. Preparation of copper oxide in smectites. *Applied Clay Science* 104: 238–244.

Khaorapapong, N., A. Ontam, and M. Ogawa. 2010. Formation of ZnS and CdS in the interlayer spaces of montmorillonite. *Applied Clay Science* 50: 19–24.

Khaorapapong, N., A. Ontam, J. Khemprasit, and M. Ogawa. 2009. Formation of MnS- and NiS-montmorillonites by solid-solid reactions. *Applied Clay Science* 43: 238–242.

Khumchoo, N., N. Khaorapapong, and M. Ogawa. 2015. Formation of zinc oxide particles in cetyltrimethylammonium-smectites. *Applied Clay Science* 105–106: 236–242.

Kibanova, D., M. Trejo, H. Destaillats, and J. Cervini-Silva. 2009. Synthesis of hectorite-TiO_2 and kaolinit-TiO_2 nanocomposites with photocatalytic activity for the degradation of model air pollutants. *Applied Clay Science* 42: 563–568.

Király, Z., I. Dékány, A. Mastalir, and M. Bartók. 1996. In situ generation of palladium nanoparticles in smectite clays. *Journal of Catalysis* 161: 401–408.

Körösi, L., J. Németh, and I. Dékány. 2004. Structural and photooxidation properties of SnO_2/layer silicate nanocomposites. *Applied Clay Science* 27: 29–40.

Kozhevnikov, I.V. 2009. Heterogeneous acid catalysis by heteropoly acids: Approaches to catalyst deactivation. *Journal of Molecular Catalysis A: Chemical* 305: 104–111.

Kozhevnikova, E.F., J. Quartararo, and I.V. Kozhevnikov. 2003. Fries rearrangement of aryl esters catalysed by heteropoly acid. *Applied Catalysis A: General* 245: 69–78.

Kun, R., K. Mogyorósi, and I. Dékány. 2006. Synthesis and structural and photocatalytic properties of TiO_2/montmorillonite nanocomposites. *Applied Clay Science* 32: 99–110.

Kun, R., G. Szöllösi, and M. Bartók. 2001. Crotonaldehyde hydrogenation over clay-supported platinum catalysts. *Journal of Molecular Catalysis A: General* 169: 235–246.

Kureshy, R.I., N.H. Khan, S.H.R. Abdi, I. Ahmad, S. Singh, and R.V. Jasra. 2003. Immobilization of dicationic Mn(III) salen in the interlayers of montmorillonite clay for enantioselective epoxidation of nonfunctional alkenes. *Catalysis Letters* 91: 207–210.

Kureshy, R.I., N.H. Khan, S.H.R. Abdi, I. Ahmad, S. Singh, and R.V. Jasra. 2004. Dicationic chiral Mn(III) salen complex exchanged in the interlayers of montmorillonite clay: A heterogeneous enantioselective catalyst for epoxidation of nonfunctionalized alkenes. *Journal of Catalysis* 221: 234–240.

Kurian, M. and S. Sugunan. 2003. Liquid phase benzylation of *o*-xylene over pillared clays. *Indian Journal of Chemistry* 42A: 2480–2486.

Kurian, M. and S. Sugunan 2004. Selective benzylation of *o*-xylene over transition metal doped montmorillonites. *Reaction Kinetics and Catalysis Letters* 81: 57–64.

Kuźniarska-Biernacka, I., C. Pereira, A.P. Carvalho, J. Pires, and C. Freire. 2009. K10-montmorillonite as support for a cationic manganese(III)-*salen* complex. *Journal of the Brazilian Chemical Society* 20: 1320–1326.

Kuźniarska-Biernacka, I., A.R. Silva, A.P. Carvalho, J. Pires, and C. Freire. 2005. Organo-laponites as novel mesoporous supports for manganese(III) *salen* catalysts. *Langmuir* 21: 10825–10834.

Kuźniarska-Biernacka, I., A.R. Silva, A.P. Carvalho, J. Pires, and C. Freire. 2007. Direct immobilization *versus* covalent attachment of a Mn(III)*salen* complex onto an Al-pillared clay and influence in the catalytic epoxidation of styrene. *Journal of Molecular Catalysis A: Chemical* 278: 82–91.

Kuźniarska-Biernacka, I., A.R. Silva, A.P. Carvalho, J. Pires, and C. Freire. 2010. Anchoring of chiral manganese(III) salen complex onto organoclay and porous clay heterostructure and catalytic activity in alkene epoxidation. *Catalysis Letters* 134: 63–71.

Kuźniarska-Biernacka, I., A.R. Silva, R. Ferreira et al. 2004. Epoxidation of styrene by a manganese(III) salen complex encapsulated in an aluminium pillared clay. *New Journal of Chemistry* 28: 853–858.

Laszlo, P. 1986. Catalysis of organic reactions by inorganic solids. *Accounts of Chemical Research* 19: 121–127.

Laszlo, P. 1998. Heterogeneous catalysis of organic reactions. *Journal of Physical Organic Chemistry* 11: 356–361.

Laszlo, P. and A. Mathy. 1987. Catalysis of Friedel–Crafts alkylation by a montmorillonite doped with transition metal cations. *Helvetica Chimica Acta* 70: 577–586.

Laszlo, P. and P. Pennetreau. 1987. Vastly improved pare preference in the nitration of halobenzenes. *Journal of Organic Chemistry* 52: 2407–2410.

Laszlo, P. and E. Polla. 1984a. Conversion of N,N-dimethylhydrazones to carbonyl compounds by clay-supported ferric nitrate. *Tetrahedron Letters* 25: 3309–3312.

Laszlo, P. and E. Polla. 1984b. Efficient conversion of hydrazines to azides with clay-supported ferric nitrate. *Tetrahedron Letters* 25: 3701–3704.

Lee, K.-Y. and K.-Y. Ko. 2004. Envirocat EPZ10: A recyclable solid acid catalyst for the synthesis of Biginelli-type 3,4-dihydropyrimidin-2(1H)-ones. *Bulletin of the Korean Chemical Society* 25: 1929–1931.

Li, S., P. Wu, H. Li et al. 2010. Synthesis and characterization of organo-montmorillinite supported iron nanoparticles. *Applied Clay Science* 50: 330–336.

Liu, H., T. Chen, D. Chang et al. 2013. Characterization and catalytic performance of Fe_3Ni_8/palygorskite for catalytic cracking of benzene. *Applied Clay Science* 74: 135–140.

Liu, Y.-H., Q.-S. Liu, and Z.-H. Zhang. 2009. An efficient Friedel–Crafts alkylation of nitrogen heterocycles catalyzed by antimony trichloride/montmorillonite K-10. *Tetrahedron Letters* 50: 916–921.

Liu, R, X. Xia, X. Niu et al. 2015. 12-Phosphotungstic acid immobilized on activated-bentonite as an efficient heterogeneous catalyst for the hydroxyalkylation of phenol. *Applied Clay Science* 105–106: 71–77.

Liu, J. and G. Zhang. 2014. Recent advances in synthesis and applications of clay-based photocatalysts: A review. *Physical Chemistry and Chemical Physics* 16: 8178–8192.

Liu, P. and M. Zhao. 2009. Silver nanoparticles supported on halloysite nanotubes catalyzed reduction of 4-nitrophenol (4-NP). *Applied Surface Science* 255: 3989–3993.

Luo, S., P. Qin, J. Shao, L. Peng. Q. Zeng, and J.-D. Gu. 2013. Synthesis of reactive nanoscale zero valent iron using rectorite supports and it application for Orange II removal. *Chemical Engineering Journal* 223: 1–7.

Machado, A.M., F. Wypych, S.M. Drechsel, and S. Nakagaki. 2002. Study of the catalytic behavior of montmorillonite/iron(III) and Mn(III) cationic porphyrins. *Journal of Colloid and Interface Science* 254: 158–164.

Malla, P.B., P. Ravindranathan, S. Komarneni, E. Breval, and R. Roy. 1992. Reduction of copper acetate hydroxide hydrate interlayers in montmorillonite by a polyol process. A new approach in the preparation of metal-supported catalysts. *Journal of Materials Chemistry* 2: 559–565.

Malla, P.B., P. Ravindranathan, S. Komarneni, and R. Roy. 1991. Intercalation of copper metal clusters in montmorillonite. *Nature* 351: 555–557.

Manikandan, D., D. Divakar, and T. Sivakumar. 2007a. Utilization of clay minerals for developing Pt nanoparticles and their catalytic activity in the selective hydrogenation of cinnamaldehyde. *Catalysis Communications* 8: 1781–1786.

Manikandan, D., D. Divakar, and T. Sivakumar. 2008. Selective hydrogenation of citral over noble metals intercalated montmorillonite catalysts. *Catalysis Letters* 123: 107–114.

Manikandan, D., D. Divakar, A.V. Rupa, S. Revathi, M.E.L Preethi, and T. Sivakumar. 2007b. Synthesis of platinum nanoparticles in montmorillonite and their catalytic behaviour. *Applied Clay Science* 37: 193–200.

Margalef-Català, R., P. Salagre, E. Fernández, and C. Claver. 1999. High efficiency and reusability of iridium complexes adsorbed in montmorillonite clay on catalytic hydrogenation of imines. *Catalysis Letters* 60: 121–123.

Marín-Astorga, N., G. Alvez-Manoli, and P. Reyes. 2005. Stereoselective hydrogenation of phenyl alkyl acetylenes on pillared clays supported palladium catalysts. *Journal of Molecular Catalysis A: Chemical* 226: 81–88.

Marme, F., G. Coudurier, and J.C. Védrine. 1998. Acid-type catalytic properties of heteropolyacid $H_3PW_{12}O_{40}$ supported on various porous silica-based materials. *Microporous and Mesoporous Materials* 22: 151–163.

Martínez, A.V., J.A. Mayoral, and J.I. García. 2014. Pd nanoparticles immobilized in [bmim][PF_6] supported on laponite clay as highly recyclable catalysts for the Mizoroki-Heck reaction. *Applied Catalysis A: General* 472: 21–28.

Martinez-Lorente, M.A., P. Battioni, W. Kleemiss, J.F. Bartoli, and D. Mansuy. 1996. Manganese porphyrins covalently bound to silica and montmorillonite K10 as efficient catalysts for alkene and alkane oxidation by hydrogen peroxide. *Journal of Molecular Catalysis A: Chemical* 113: 343–353.

Massam, J. and D.R. Brown. 1995. The roles of Brønsted and Lewis surface acid sites in acid-treated montmorillonite supported $ZnCl_2$ alkylation catalysts. *Catalysis Letters* 35: 335–343.

Massam, J. and D.R. Brown. 1998. Acid-treated montmorillonite supports for copper(II)nitrate oxidation reagents. *Applied Catalysis A: General* 172: 259–264.

Mastalir, A., Z. Király, and F. Berger. 2004. Comparative study of size-quantified Pd-montmorillonite catalysts in liquid-phase semihydrogenations of alkynes. *Applied Catalysis A: General* 269: 161–168.

Mastalir, A., Z. Király, G. Szöllősi, and M. Bartók. 2000. Preparation of organophilic Pd-montmorillonite, an efficient catalyst in alkyne semihydrogenation. *Journal of Catalysis* 194: 146–152.

Mastalir, A., Z. Király, G. Szöllősi, and M. Bartók. 2001. Stereoselective hydrogenation of 1-phenyl-1-pentyne over low-loaded Pd-montmorillonite catalysts. *Applied Catalysis A: General* 213: 133–140.

Mastalir, A., G. Szöllősi, Z. Király, and Zs. Rázga. 2002. Preparation and characterization of platinum nanoparticles immobilized in dihydrocinchonidine-modified montmorillonite and hectorite. *Applied Clay Science* 22: 9–16.

Mastalir, A., F. Notheisz, Z. Király, M. Bartók, and I. Dékány. 1997. Novel clay intercalated metal catalysts: A study of the hydrogenation of styrene and 1-octene on clay intercalated Pd catalysts. In *Heterogeneous Catalysis and Fine Chemicals IV* (Eds.) H.U. Blaser, A. Baiker, and R. Prins, pp. 477–484. Amsterdam, the Netherlands: Elsevier.

Masui, Y., J. Wang, K. Teramura, T. Kogure, T. Tanaka, and M. Onaka. 2014. Unique structural characteristics of tin hydroxide nanoparticles-embedded montmorillonite (Sn-mont) demonstrating efficient acid catalysis for various organic reactions. *Microporous and Mesoporous Materials* 198: 129–138.

Mazzei, M., W. Marconi, and M. Riocci. 1980. Asymmetric hydrogenation of substituted acrylic acids by Rh'-aminephosphine chiral complex supported on mineral clays. *Journal of Molecular Catalysis* 9: 381–387.

McCabe, R.W. and J.M. Adams. 2013. Clay minerals as catalysts. In *Handbook of Clay Science, 2nd edition. Developments in Clay Science, Vol. 5B* (Eds.) F. Bergaya and G. Lagaly, pp. 491–538. Amsterdam, the Netherlands: Elsevier.

McKillop, A. and D.W. Young. 1979a. Organic synthesis using supported reagents—Part I. *Synthesis* 401–422.

McKillop, A. and D.W. Young. 1979b. Organic synthesis using supported reagents—Part II. *Synthesis* 481–500.

Mekewi, M.A., A.S. Darwish, M.S. Amin, G. Eshaq, and H.A. Bourazan. 2016. Copper nanoparticles supported onto montmorillonite clays as efficient catalyst for methylene blue dye degradation. *Egyptian Journal of Petroleum* 25: 269–279.

Meshram, H.M., A. Bandyopadhyay, G.S. Reddy, and J.S. Yadav. 2000. Microwave thermolysis I: A convenient and rapid coupling of thiols using "clayan" in solvent-free condition. *Synthetic Communications* 30: 701–706.

Meshram, H.M., S. Dale, and J.S. Yadav. 1997a. A general synthesis of isothiocyanates from dithiocarbamates using claycop. *Tetrahedron Letters* 38: 8743–8744.

Meshram, H.M., Y.S.S. Ganesh, A.V. Madhavi, B. Eshwaraiah, J.S. Yadav, and D. Gunasekar. 2003. Clay supported ammonium nitrate "clayan": A new reagent for selective nitration of arenes. *Synthetic Communications* 33: 2497–2503.

Meshram, H.M., G.S. Reddy, D. Srinivas, and J.S. Yadav. 1998a. Clay supported ammonium nitrate "clayan": A mild and highly selective reagent for the deoximation of electron rich oximes. *Synthetic Communications* 28: 2593–2600.

Meshram, H.M., G.S. Reddy, G. Sumitra, and J.S. Yadav. 1999a. Microwave thermolysis VI: A rapid and general method for dethioacetalization using "clayan" in dry media. *Synthetic Communications* 29: 1113–1119.

Meshram, H.M., G.S. Reddy, and J.S. Yadav. 1997b. Clay supported ammonium nitrate "clayan": A mild and eco-friendly reagent for dethioacetalization. *Tetrahedron Letters* 38: 8891–8894.

Meshram, H.M., G. Sumitra, G.S. Reddy, Y.S.S. Ganesh, and J.S. Yadav. 1999b. Microwave thermolysis V: A rapid and selective method for the cleavage of THP ethers, acetals and acetonides using clay supported ammonium nitrate "clayan" in dry media. *Synthetic Communications* 29: 2807–2815.

Meshram, H.M., D. Srinivas, G.S. Reddy, and J.S. Yadav. 1998b. Clay supported ammonium nitrate "clayan": A rapid and convenient regeneration of carbonyls in dry media. *Synthetic Communications* 28: 4401–4408.

Miao, S., Z. Liu, B. Han et al. 2006. Ru nanoparticles immobilized on montmorillonite by ionic liquids: A highly efficient heterogeneous catalyst for the hydrogenation of benzene. *Angewandte Chemie International Edition* 45: 266–269.

Miao, S., Z. Liu, Z. Zhang et al. 2007. Ionic liquid-assisted immobilization of Rh on attapulgite and its application in cyclohexene hydrogenation. *Journal of Physical Chemistry C* 111: 2185–2190.

Misono, M., I. Ono, G. Koyano, and A. Aoshima. 2000. Heteropolyacids. Versatile green catalysts usable in a variety of reaction media. *Pure and Applied Chemistry* 72: 1305–1311.

Mitsudome, T., K. Nose, K. Mori et al. 2007. Montmorillonite-entrapped sub-nanoordered Pd clusters as a heterogeneous catalyst for allylic substitution reactions. *Angewandte Chemie* 119: 3352–3354.

Miyazaki, T., A. Tsuboi, H. Urata et al. 1985. Hydrogenation of olefins with cationic rhodium complex intercalated in fluoro tetrasilicic mica. *Chemistry Letters* 793–796.

Morán, C., E. González, J. Sánchez, R. Solano, G. Carruyo, and A. Moronta. 2007. Dehydrogenation of ethylbenzene to styrene using Pt, Mo, and Pt-Mo catalysts supported on clay nanocomposites. *Journal of Colloid and Interface Science* 315: 164–169.

Nagendrappa, G. 2011. Organic synthesis using clay and clay-supported catalysts. *Applied Clay Science* 53: 106–138.

Naik, M.A., S. Samantaray, B.G. Mishra, and A. Dubey. 2009. Phosphomolybdic acid dispersed in Al-pillared clay (PMA/Al-PILC) as heterogeneous catalysts for the benign synthesis of benzoxanthenes under solvent free conditions. *Reaction Kinetics and Catalysis Letters* 98: 125–131.

Nakagaki, S., K.A.D.F. Castro, G.S. Machado, M. Halma, S.M. Drechsel, and F. Wypych. 2006. Catalytic activity in oxidation reactions of anionic iron(III) porphyrins immobilized on raw and grafted chrysotile. *Journal of the Brazilian Chemical Society* 17: 1672–1678.

Narayanan, S. and K. Deshpande. 1996. A comparative aniline alkylation activity of montmorillonite and vanadia-montmorillonite with silica and vanadia-silica. *Applied Catalysis A: General* 135: 125–135.

Nascimento, C.C., G.R.S. Andrade, E.C. Neves et al. 2012. Nanocomposites of CdS nanocrystals with montmorillonite functionalized with thiourea derivatives and their use in photocatalysis. *Journal of Physical Chemistry C* 116: 21992–22000.

Natekar, R.S. and S.D. Samant. 1995. Catalysis of Friedel–Crafts alkylation using virgin and Lewis acid doped bentonite k-10. *Indian Journal of Chemistry* 34B: 257–260.

Navalon, S., M. Alvaro, and H. Garcia. 2010. Heterogeneous Fenton catalysts based on clays, silicas and zeolites. *Applied Catalysis B: Environmental* 99: 1–26.

Nehate, M. and V.V. Bokade. 2009. Selective *N*-alkylation of aniline with methanol over a heteropolyacid on montmorillonite K10. *Applied Clay Science* 44: 255–258.

Nogueira, F.G.E., J.H. Lopes, A.C. Silva, R.M. Lago, J.D. Fabris, and L.C.A. Oliveira. 2011. Catalysts based on clay and iron oxide for oxidation of toluene. *Applied Clay Science* 51: 385–389.

Nuzzo, A. and A. Piccolo. 2013. Enhanced catechol oxidation by heterogeneous biomimetic catalysts immobilized on clay minerals. *Journal of Molecular Catalysis A: Chemical* 371: 8–14.

Pacula, A., K. Pamin, A. Zięba et al. 2014. Physicochemical and catalytic properties of montmorillonite modified with 12-tungstophosphoric acid. *Applied Clay Science* 95: 220–231.

Pan, J., J. Liu, S. Guo, and Z. Yang. 2009. Preparation of platinum/montmorillonite nanocomposites in supercritical methanol and their application in the hydrogenation of nitrobenzene. *Catalysis Letters* 131: 179–183.

Pan, J., C. Wang, S. Guo, J. Li, and Z. Yang. 2008. Cu supported over Al-pillared interlayer clays catalysts for direct hydroxylation of benzene to phenol. *Catalysis Communications* 9: 176–181.

Pang, Z., M. Yan, X. Jia, Z. Wang, and J. Chen. 2014. Debromination of decabromodiphenyl ether by organo-montmorillonite-supported nanscale zero-valent iron: Preparation, characterization and influence factors. *Journal of Environmental Sciences* 26: 483–491.

Papp, S. and I. Dékány. 2002. Growth of Pd nanopartilces on layer silicates hydrophobized with alkyl chains in ethanol-tetrahydrofuran mixtures. *Colloid and Polymer Science* 280: 956–962.

Papp, S. and I. Dékány. 2003. Stabilization of palladium nanoparticles by polymers and layer silicates. *Colloid and Polymer Science* 281: 727–737.

Papp, S., Patakfalvi, R. and I. Dékány. 2008. Metal nanoparticle formation on layer silicate lamellae. *Colloid and Polymer Science* 286: 3–14.

Papp, S., A. Szücs, and I. Dékány. 2001a. Colloid synthesis of monodisperse Pd nanoparticles in layered silicates. *Solid State Ionics* 141–142: 169–176.

Papp, S., A. Szücs, and I. Dékány. 2001b. Preparation of Pd^0 nanoparticles stabilized by polymers and layered silicate. *Applied Clay Science* 19: 155–172.

Parida, K., G.B.B. Varadwaj, S. Sahu, and P.C. Sahoo. 2011. Schiff base Pt(II) complex intercalated montmorillonite: A robust catalyst for hydrogenation of aromatic nitro compounds at room temperature. *Industrial & Engineering Chemistry Research* 50: 7849–7856.

Patakfalvi, R. and I. Dékány. 2004. Synthesis and intercalation of silver nanoparticles in kaolinite/DMSO complexes. *Applied Clay Science* 25: 149–159.

Patakfalvi, R., A. Oszkó, and I. Dékány. 2003. Synthesis and characterization of silver nanoparticle/kaolinite composites. *Colloids and Surfaces A: Physicochemical and Engineering Aspects* 220: 45–54.

Patel, H.A., H.C. Bajaj, and R.V. Jasra. 2008. Synthesis of Pd and Rh metal nanoparticles in the interlayer space of organically modified montmorillonite. *Journal of Nanoparticle Research* 10: 625–632.

Patil, S.P., V.S. Shrivastava, G.H. Sonawane, and S.H. Sonawane. 2015. Synthesis of novel Bi_2O_3-montmorillonite nanocomposite with enhanced photocatalytic performance in dye degradation. *Journal of Environmental Chemical Engineering* 3: 2597–2603.

Pereira, C., A.R. Silva, A.P. Carvalho, J. Pires, and C. Freire. 2008. Vanadyl acetylacetonate anchored onto amine-functionalised clays and catalytic activity in the epoxidation of geraniol. *Journal of Molecular Catalysis A: Chemical* 283: 5–14.

Pérez, A., M. Montes, R. Molina, and S. Moreno. 2014. Modified clays as catalysts for the catalytic oxidation of ethanol. *Applied Clay Science* 95: 18–24.

Philip, A., J. Lihavainen, M. Keinänen, and T.T. Pakkanen. 2017. Gold nanoparticle-decorated halloysite nanotubes–Selective catalysts for benzyl alcohol oxidation. *Applied Clay Science* 143: 80–88.

Phukan, A., J.N. Ganguli, and D.K. Dutta. 2003. $ZnCl_2$-Zn^{2+}-montmorillonite composite: Efficient solid acid catalysts for benzylation of benzene. *Journal of Molecular Catalysis A: Chemical* 202: 279–287.

Pillai, S.K., S.S. Ray, M. Scriba, J. Bandyopadhyay, M.P. Roux-van der Merwe, and J. Badenhorst. 2013. Microwave-assisted green synthesis and characterization of silver/montmorillonite heterostructures with improved antimicrobial properties. *Applied Clay Science* 83–84: 315–321.

Pinho, R.O., J.A.R. Rodrigues, P.J.S. Moran, and I. Joekes. 1995. Chrysotile-supported transition metal salts as Friedel–Crafts catalysts. *Journal of the Brazilian Chemical Society* 6: 373–376.

Pinnavaia, T.J. 1983. Intercalated clay catalysts. *Science* 220: 365–371.

Pinnavaia, T.J., R. Raythatha, J.G.-S. Lee, L.J. Halloran, and J.F. Hoffman. 1979. Intercalation of catalytically active metal complexes in mica-type silicates. Rhodium hydrogenation catalysts. *Journal of the American Chemical Society* 101: 6891–6897.

Pinnavaia, T.J. and P.K. Welty. 1975. Catalytic hydrogenation of 1-hexene by rhodium complexes in the intercrystal space of a swelling layer lattice silicate. *Journal of the American Chemical Society* 97: 3819–3820.

Pore, D.M., T.S. Shaikh, N.G. Patil, S.B. Dongare, and U.V. Desai. 2010. Envirocat EPZ-10: A solid acid catalyst for the synthesis of 1,8-dioxo-octahydroxanthenes in aqueous medium. *Synthetic Communications* 40: 2215–2219.

Poyatos, M., F. Márquez, E. peris, C. Claver, and E. Fernández. 2003. Preparation of a new clay-immobilized highly stable palladium catalyst and its efficient recyclability in the Heck reaction. *New Journal of Chemistry* 27: 425–431.

Praus, P., O. Kozák, K. Kočí, A. Panáček, and R. Dvorský. 2011. CdS nanoparticles deposited on montmorillonite: Preparation, characterization and application for photoreduction of carbon dioxide. *Journal of Colloid and Interface Science* 360: 574–579.

Praus, P., M. Turicová, M. Karlíková, L. Kvítek, and R. Dvorský. 2013. Nanocomposite of montmorillonite and silver nanoparticles: Characterization and application in catalytic reduction of 4-nitrophenol. *Materials Chemistry and Physics* 140: 493–498.

Qi, X., H. Yoon, S.-H. Lee, J. Yoon, and S.-J. Kim. 2008. Surface-modified imogolite by 3-APS-OsO_4 complex: Synthesis, characterization and its application in the dihydroxylation of olefins. *Journal of Industrial and Engineering Chemistry* 14: 136–141.

Quayle, W.H. and T.J. Pinnavaia. 1979. Utilization of a cationic ligand for the intercalation of catalytically active rhodium complexes in swelling layer-lattice silicates. *Inorganic Chemistry* 18: 2840–2847.

Ratti, R., S. Kaur, M. Vaultier, and V. Singh. 2010. Preparation, characterization and catalytic activity of MMT-clay exchanged sulphonic acid functionalized ionic liquid for transesterification of β-ketoesters. *Catalysis Communications* 11: 503–507.

Ravi, K., B. Krishnakumar, and M. Swaminathan. 2012. An efficient protocol for the green and solvent-free synthesis of azine derivatives at room temperature using $BiCl_3$-loaded montmorillonite K10 as a new recyclable heterogeneous catalyst. *ISRN Organic Chemistry* volume 2012, 9 pages; doi:10.5402/2012/595868.

Ravi, P. and S.P. Tewari. 2012. Facile and environmentally friendly synthesis of nitropyrazoles using montmorillonite K-10 impregnated with bismuth nitrate. *Catalysis Communications* 19: 37–41.

Ravindranathan, P., P.B. Malla, S. Komarneni, and R. Roy. 1990. Preparation of metal supported montmorillonite catalyst: A new approach. *Catalysis Letters* 6: 401–408.

Rhodes, C.N. and D.R. Brown. 1992. Structural characterisation and optimisation of acid-treated montmorillonite and high-porosity silica supports for $ZnCl_2$ alkylation catalysts. *Journal of the Chemical Society, Faraday Transactions* 88: 2269–2274.

Rhodes, C.N. and D.R. Brown. 1993. Surface properties and porosities of silica and acid-treated montmorillonite catalyst supports: Influence on activities of supported $ZnCl_2$ alkylation catalysts. *Journal of the Chemical Society, Faraday Transactions* 89: 1387–1391.

Rhodes, C.N., M. Franks, G.M.B. Parkes, and D.R. Brown. 1991. The effect of acid treatment on the activity of clay supports for $ZnCl_2$ alkylation catalysts. *Journal of the Chemical Society, Chemical Communications* 804–807.

Ross, J.C., J.H. Clark, D.J. Macquarrie, S.J. Barlow, and T.W. Bastock. 1998. The use of supported zinc bromide for the fast and selective bromination of aromatic substrates. *Organic Process Research & Development* 2: 245–249.

Rožić, L., B. Grbić, N. Radić et al. 2011. Mesoporous 12-tungstophosphoric acid/activated bentonite catalysts for oxidation of 2-propanol. *Applied Clay Science* 53: 151–156.

Saikia, L., D. Dutta, and D.K. Dutta. 2012. Efficient clay supported Ni^o nanoparticles as heterogeneous catalyst for solvent-free synthesis of Hantzsch polyhydroquinoline. *Catalysis Communications* 19: 1–4.

Saikia, P.K., P.P. Sarmah, B.J. Borah, L. Saikia, K. Saikiah, and D.K. Dutta. 2016. Stabilized Fe_3O_4 magnetic nanoparticles into nanopores of modified montmorillonite clay: A highly efficient catalyst for the Baeyer–Villiger oxidation under solvent free conditions. *Green Chemistry* 18: 2843–2850.

Salavati-Niasari, M., E. Zamani, and M. Bazarganipour. 2007. Epoxidation of cyclohexene with K10-montmorillonite and Schiff-base macrocyclic copper complexes. *Applied Clay Science* 38: 9–16.

Samajdar, S., F.F. Becker, and B.K. Banik. 2001a. Montmorillonite impregnated with bismuth nitrate: A versatile reagent for the synthesis of nitro compounds of biological significance. *Arkivoc* 27–33.

Samajdar, S., F.F. Becker, and B.K. Banik. 2001b. Surface-mediated highly efficient oxidation of alcohols by bismuth nitrate. *Synthetic Communications* 31: 2691–2695.

Sanabria, R., P. Castaneda, R. Miranda, A. Tobón, F. Delgado, and L. Velasco. 1995. Oxidative cleavage of ketoximes with bentonic clay-supported copper(II) nitrate. *Organic Preparations and Procedures International: The New Journal for Organic Synthesis* 27: 480–482.

Sarmah, P.P. and D.K. Dutta. 2012. Chemoselective reduction of a nitro group through transfer hydrogenation catalysed by Ru^0-nanoparticles stabilized on modified montmorillonite clay. *Green Chemistry* 14: 1086–1093.

Sarmah, P.P. and D.K. Dutta. 2014. Stabilized Ru^0-nanoparticles-montmorillonite clay composite: Synthesis and catalytic transfer hydrogenation reaction. *Applied Catalysis A: General* 470: 355–360.

Sento, T., S. Shimazu, N. Ichikuni, and T. Uematsu. 1999. Asymmetric hydrogenation of itaconates by hectorite-intercalated Rh-DIOP complex. *Journal of Molecular Catalysis A: Chemical* 137: 263–267.

Shaikh, T.S., K.A. Undale, D.S. Gaikwad, and D.M. Pore. 2011. Envirocat EPZ-10: An efficient catalyst for the synthesis of 3-acetoacetylcoumarins. *Comptes Rendus Chimie* 14: 987–990.

Sharma, G.V.M. and S. Chandrasekhar. 1989. Selective hydrogenation of organic azides to amines by interlamellar montmorillonite-diphenylphosphine palladium catalyst. *Synthetic Communications* 19: 3289–3293.

Shawkataly, O.B., R. Jothiramalingam, F. Adam, T. Radhika, T.M. Tsao, and M.K. Wang. 2012. Ru-nanoparticle deposition on naturally available clay and rice husk biomass materials—Benzene hydrogenation catalysis and synthetic strategies for green catalyst development. *Catalysis Science & Technology* 2: 538–540.

Shiju, N.R., H.M. Williams, and D.R. Brown. 2009. Cs exchanged phosphotungstic acid as an efficient catalyst for liquid-phase Beckmann rearrangement of oximes. *Applied Catalysis B: Environmental* 90: 451–457.

Shimazu, S., K. Ro., T. Sento, N. Ichikuni, and T. Uematsu. 1996. Asymmetric hydrogenation of α,β-unsaturated carboxylic acid esters by rhodium(I)-phosphine complexes supported on smectites. *Journal of Molecular Catalysis A: Chemical* 107: 297–303.

Shinoda, T., M. Onaka, and Y. Izumi. 1995. Proposed models of mesopore structures in sulfuric acid-treated montmorillonites and K10. *Chemistry Letters* 24: 495–496.

Sidhpuria, K.B., H.A. Patel, P.A. Parikh, P. Bahadur, H.C. Bajaj, and R.V. Jasra. 2009. Rhodium nanoparticles intercalated into montmorillonite for hydrogenation of aromatic compounds in the presence of thiophene. *Applied Clay Science* 42: 386–390.

Sivakumar, T., T. Krithiga, K. Shanthi, T. Mori, J. Kubo, and Y. Morikawa. 2004. Noble metals intercalated/supported mica catalyst—Synthesis and characterization. *Journal of Molecular Catalysis A: Chemical* 223: 185–194.

Sivakumar, T., T. Mori, J. Kubo, and Y. Morikawa. 2001. Selective hydrogenation of 2-methylbenzaldehyde using palladium particles generated in situ in surfactant exchanged fluorotetrasilicic mica. *Chemistry Letters* 30: 860–861.

Son, Y.-H., J.-K. Lee, Y. Soong, D. Martello, and M. Chyu. 2010. Structure-property correlation in iron oxide nanoparticle-clay hybrid materials. *Chemistry of Materials* 22: 2226–2232.

Srinivasan, S.K. and S. Ganguly. 1991. FT-IR spectroscopic studies of metal nitrates supported on a modified montmorillonite clay. *Catalysis Letters* 10: 279–288.

Su, H., S. Zeng, H. Dong, Y. Du, Y. Zhang, and R. Hu. 2009. Pillared montmorillonite supported cobalt catalysts for the Fischer–Tropsch reaction. *Applied Clay Science* 46: 325–329.

Sukumar, R., K.R. Sabu, L.V. Bindu, and M. Lalithambika. 1998. Kaolinite supported metal chlorides as Friedel–Crafts alkylation catalysts. *Studies in Surface Science and Catalysis* 113: 557–562.

Szöllősi, G., I. Kun, A. Mastalir, M. Bartók, and I. Dékány. 2001. Preparation, characterization and application of platinum catalysts immobilized on clays. *Solid State Ionics* 141–142: 273–278.

Szöllősi, G., B. Török, L. Baranyi, and M. Bartók. 1998. Chemoselective hydrogenation of cinnamaldehyde to cinnamyl alcohol over Pt/K-10 catalyst. *Journal of Catalysis* 179: 619–623.

Szöllősi, G., I. Kun, B. Török, and M. Bartók. 1999. Chemoselective hydrogenation of the C=O group in unsaturated aldehydes over clay-supported platinum catalysts. *Studies in Surface Science and Catalysis* 125: 539–546.

Szücs, A., F. Berger, and I. Dékány. 2000. Preparation and structural properties of Pd nanoparticles in layered silicate. *Colloids and Surfaces A: Physicochemical and Engineering Aspects* 174: 387–402.

Szücs, A., Z. Király, F. Berger, and I. Dékány. 1998. Preparation and hydrogen sorption of Pd nanoparticles on Al_2O_3 pillared clays. *Colloids and Surfaces A: Physicochemical and Engineering Aspects* 139: 109–118.

Taylor, E.C. and C.-S. Chiang. 1977. Trimethyl orthoformate absorbed on the montmorillonite clay K-10; an effective reagent for acetal formation. *Synthesis* 467.

Taylor, E.C., C.-S. Chiang, A. McKillop, and J.F. White. 1976. Thallium in organic synthesis. 44. Oxidative rearrangements via oxythallation with thallium(III) nitrate supported on clay. *Journal of the American Chemical Society* 98: 6750–6752.

Teixeira-Neto, A.A., B. Faceto, L.S. Siebra, and E. Teixeira-Neto. 2015. Effect of the synthesis method on the properties and catalytic activities of Au/Pd nanoparticles supported on organophilic clay. *Applied Clay Science* 116–117: 175–181.

Theng, B.K.G. 2012. *Formation and Properties of Clay-Polymer Complexes, 2nd edition*. Amsterdam, the Netherlands: Elsevier.

Timofeeva, M.N. 2003. Acid catalysis by heteropoly acids. *Applied Catalysis A: General* 256: 19–35.

Tiwari, M.S. and G.D. Yadav. 2016. Novel aluminium exchanged dodecatungstophosphoric acid supported on K-10 clay as catalyst: Benzoylation of diphenyloxide with benzoic anhydride. *RSC Advances* 6: 49091–49100.

Török, B., K. Balázsik, I. Kun, G. Szöllösi, G. Szakonyi, and M. Bartók. 1999. Homogeneous and heterogeneous asymmetric reactions. Part 13. Clay-supported noble metal catalysts in enantioselective hydrogenations. *Studies in Surface Science and Catalysis* 125: 515–522.

Varade, D. and K. Haraguchi. 2013. Synthesis of higly active and thermally stable nanostructured Pt/clay materials by clay-mediated reduction. *Langmuir* 29: 1977–1984.

Varadwaj, G.B.B. and K.M. Parida. 2011. Facile synthesis of dodecatungstophosphoric acid @ TiO_2 pillared montmorillonite and its effectual exploitation towards solvent free esterification of acetic acid with n-butanol. *Catalysis Letters* 141: 1476–1483.

Varadwaj, G.B.B. and K.M. Parida. 2013. Montmorillonite supported metal nanoparticles: An update on syntheses and applications. *RSC Advances* 3: 13583–13593.

Varma, R.S. 1999. Solvent-free organic syntheses. *Green Chemistry* 43–55.

Varma, R.S. 2002. Clay and clay-supported reagents in organic synthesis. *Tetrahedron* 58: 1235–1255.

Varma, R.S. and R. Dahiya. 1997a. Microwave-assisted oxidation of alcohols under solvent-free conditions using clayfen. *Tetrahedron Letters* 38: 2043–2044.

Varma, R.S. and R. Dahiya. 1997b. Microwave-assisted facile synthesis of imines and enamines using Envirocat EPZG® as a catalyst. *Synlett* 1245–1246.

Varma, R.S. and R. Dahiya. 1998a. Microwave thermolysis with clayfen: Solvent-free oxidation of sulfides to sulfoxides. *Synthetic Communications* 28: 4087–4095.

Varma, R.S. and R. Dahiya. 1998b. Copper(II) nitrate on clay (claycop)–hydrogen peroxide: Selective and solvent-free oxidation using microwaves. *Tetrahedron Letters* 39: 1307–1308.

Varma, R.S. and H.M. Meshram. 1997. Solid state cleavage of semicarbazones and phenylhydrazones with ammonium persulfate-clay using microwave or ultrasonic irradiation. *Tetrahedron Letters* 38: 7973–7976.

Varma, R.S., and K.P. Naicker. 1998. Hydroxylamine on clay: A direct synthesis of nitriles from aromatic aldehydes using microwaves under solvent-free conditions. *Molecules Online* 2: 94–96.

Varma, R.S., and K.P. Naicker. 1999. Palladium chloride/tetraphenylphosphonium bromide intercalated clay: New catalyst for cross-coupling of aryl halides with arylboronic acids. *Tetrahedron Letters* 40: 439–442.

Varma, R.S. and R.K. Saini. 1997. Solid state dethioacetalyzation using clayfen. *Tetrahedron Letters* 38: 2623–2624.

Varma, R.S., K.P. Naicker, and P.J. Liesen. 1998. Selective nitration of styrenes with clayfen and clayan: A solvent-free synthesis of β-nitrostyrenes. *Tetrahedron Letters* 39: 3977–3980.

Varma, R.S., K.P. Naicker, and P.J. Liesen. 1999. Palladium chloride and tetraphenylphosphonium bromide intercalated clay as a new catalyst for the Heck reaction. *Tetrahedron Letters* 40: 2075–2078.

Venkatachalapathy, C., M. Rajarajan, H.S. Banu, and K. Pitchumani. 1999. Clay-supported tetrabutylammonium periodate as a versatile oxidant for alcohols and sulfides. *Tetrahedron* 55: 4071–4076.

Venkataramanan, N.S., G. Kuppuraj, and S. Rajagopal. 2005. Metal-salen complexes as efficient catalysts for the oxygenation of heteroatom containing organic compounds—Synthetic and mechanistic aspects. *Coordination Chemistry Reviews* 249: 1249–1268.

Veverková, E. and S. Toma. 2005. Microwave-assisted method for conversion of alcohols into *N*-substituted amides using Envirocat EPZG® as a catalyst. *Chemical Papers* 59: 8–10.

Veverková, E., B. Gotov, R. Mitterpach, and S. Toma. 2000. Benzoylation of arenes using Envirocat EPZG® catalyst and microwave irradiation. *Collection of Czechoslovak Chemical Communications* 65: 644–650.

Virkutyte, J. and R.S. Varma. 2014. Eco-friendly magnetic iron oxide-pillared montmorillonite for advanced catalytic degradation of dichlorophenol. *ACS Sustainable Chemistry & Engineering* 2: 1545–1550.

Wang, H., S.-H. Deng, Z.-R. Shen, J.-G. Wang, D.-T. Ding, and T.-H. Chen. 2009. Facile preparation of Pd/organoclay catalysts with high performance in solvent-free aerobic selective oxidation of benzyl alcohol. *Green Chemistry* 11: 1499–1502.

Wang, S., H.Y. Zhu, and G.Q. (Max) Lu. 1998. Preparation, characterization, and catalytic properties of clay-based nickel catalysts for methane reforming. *Journal of Colloid and Interface Science* 204: 128–134.

Wu, P., S. Li, L. Ju et al. 2012. Mechanism of the reduction of hexavalent chromium by organo-montmorillonite supported iron nanoparticles. *Journal of Hazardous Materials* 219–220: 283–288.

Xiao, J., T. Peng, D. Ke, L. Zan, and Z. Peng. 2007. Synthesis, characterization of CdS/rectorite nanocomposites and its photocatalytic activity. *Physics and Chemistry of Minerals* 34: 275–285.

Xiong, J., C. Hang, J. Gao, Y. Guo, and C. Gu. 2014. A novel biomimetic catalyst templated by montmorillonite clay for degradation of 2,4,6-trichlorophenol. *Chemical Engineering Journal* 254: 276–282.

Xu, H., K. Wang, H. Zhang, L. Hao, J. Xu, and Z. Liu. 2014. Ionic liquid modified montmorillonite-supported Ru nanoparticles: Highly efficient heterogeneous catalysts for the hydrodeoxygenation of phenolic compounds to cycloalkanes. *Catalysis Science & Technology* 4: 2658–2663.

Yadav, G.D. 2005. Synergism of clay and heteropoly acids as nano-catalysts for the development of green processes with potential industrial applications. *Catalysis Surveys from Asia* 9: 117–137.

Yadav, G.D. and N.S. Asthana. 2003. Selective decomposition of cumene hydroperoxide into phenol and acetone by a novel cesium substituted heteropolyacid on clay. *Applied Catalysis A: General* 244: 341–357.

Yadav, G.D. and O.V. Badure. 2008. Selective acylation of 1,3-dibenzyloxybenzene to 3,5-dibenzyloxyacetophenone over cesium modified dodecatungstophosphoric acid (DTP) on clay. *Applied Catalysis A: General* 348: 16–25.

Yadav, G.D. and R.D. Bhagat. 2004. Synthesis of methyl phenyl glyoxylate via clean oxidation of methyl mandelate over a nanocatalyst based on heteropolyacid supported on clay. *Organic Process Research & Development* 8: 879–882.

Yadav, G.D. and V.V. Bokade. 1996. Novelties of heteropoly acid supported on clay: Etherification of phenethyl alcohol with alkanols. *Applied Catalysis A: General* 147: 299–323.

Yadav, G.D. and N.S. Doshi. 2000. Alkylation of hydroquinone with methyl-*tert*-butyl-ether and *tert*-butanol. *Catalysis Today* 69: 263–273.

Yadav, G.D. and N.S. Doshi. 2002. Alkylation of phenols with methyl-*tert*-butyl-ether and *tert*-butanol over solid acids: Efficacies of clay-based catalysts. *Applied Catalysis A: General* 236: 129–147.

Yadav, G.D. and N.S. Doshi. 2003. Alkylation of aniline with methyl-*tert*-butyl-ether and *tert*-butanol over solid acids: Product distribution and kinetics. *Journal of Molecular Catalysis A: Chemical* 194: 195–209.

Yadav, G.D. and G. George. 2008. Single step synthesis of 4-hydroxybenzophenone via esterification and Fries rearrangement: Novelty of cesium substituted heteropoly acid supported on clay. *Journal of Molecular Catalysis A: Chemical* 292: 54–61.

Yadav, G.D. and S.B. Kamble. 2009. Alkylation of xylenes with isopropyl alcohol over acidic clay supported catalysts: Efficacy of 20% w/w $Cs_{2.5}H_{0.5}PW_{12}O_{40}$/K-10 clay. *Industrial & Engineering Chemistry Research* 48: 9383–9393.

Yadav, G.D. and N. Kirthivasan. 1995. Single-pot synthesis of methyl *tert*-butyl ether from *tert*-butylalcohol and methanol: Dodecatungstophosphoric acid supported on clay as an efficient catalyst. *Journal of the Chemical Society, Chemical Communications* 203–204.

Yadav, G.D. and N. Kirthivasan. 1997. Synthesis of bisphenol-A: Comparison of efficacy of ion exchange resin catalysts vis-à-vis heteropolyacid supported on clay and kinetic modelling. *Applied Catalysis A: General* 154: 29–53.

Yadav, G.D. and P. Kumar. 2005. Alkylation of phenol with cyclohexene over solid acids: Insight in selectivity of O- versus C-alkylation. *Applied Catalysis A: General* 286: 61–70.

Yadav, G.D. and S.R. More. 2011. Green alkylation of anisole with cyclohexene over 20% cesium modified heteropoly acid on K-10 acidic montmorillonite clay. *Applied Clay Science* 53: 254–262.

Yadav, G.D. and A.A. Pujari. 1999. Kinetics of acetalization of perfumery aldehydes with alkanols over solid acid catalysts. *Canadian Journal of Chemical Engineering* 77: 489–496.

Yadav, G.D. and S.S. Salgaonkar. 2005. Selectivity engineering of 2,6-diisopropylphenol in isopropylation of phenol over $Cs_{2.5}H_{0.5}PW_{12}O_{40}$/K-10 clay. *Industrial & Engineering Chemistry Research* 44: 1706–1715.

Yadav, G.D. and P.S. Surve. 2013. Atom economical green synthesis of chloromethyl-1,3-dioxolanes from epichlorohydrin using supported heteropolyacids. *Industrial & Engineering Chemistry Research* 52: 6129–6137.

Yadav, G.D. and T.S. Thorat. 1996. Role of benzyl ether in the inversion of reactivities in Friedel–Crafts benzylation of toluene by benzyl chloride and benzyl alcohol. *Tetrahedron Letters* 37: 5405–5408.

Yadav, G.D., N.S. Asthana, and V.S. Kamble. 2003a. Cesium-substituted dodecatungstophosphoric acid on K-10 clay for benzoylation of anisole with benzoyl chloride. *Journal of Catalysis* 217: 88–99.

Yadav, G.D., N.S. Asthana, and V.S. Kamble. 2003b. Friedel–Crafts benzoylation of *p*-xylene over clay supported catalysts: Novelty of cesium substituted dodecatungstophosphoric acid on K-10 clay. *Applied Catalysis A: General* 240: 53–69.

Yadav, G.D., N.S. Asthana, and S.S. Salgaonkar. 2004a. Regio-selective benzoylation of xylenes over caesium modified heteropolyacid supported on K-10 clay. *Clean Technologies and Environmental Policy* 6: 105–113.

Yadav, G.D., S.A.R.K. Deshmukh, and N.S. Asthana. 2005. Synthesis of hydroquinone monomethyl ether from hydroquinone and methanol over heteropolyacids supported on clay: Kinetics and mechanism. *Industrial & Engineering Chemistry Research* 44: 7969–7977.

Yadav, G.D., P.K. Goel, and A.V. Joshi. 2001. Alkylation of dihydroxybenzenes and anisole with methyl-*tert*-butyl ether over solid acid catalysts. *Green Chemistry* 3: 92–99.

Yadav, G.D., S.S. Salgaonkar, and N.S. Asthana. 2004b. Selectivity engineering in isopropylation of benzene to cumene over cesium substituted dodecatungstophosphoric acid on K-10 clay. *Applied Catalysis A: General* 265: 153–159.

Yadav, J.S., H.M. Meshram, G.S. Reddy, and G. Sumithra. 1998. Microwave thermolysis IV: Selective deprotection of MPM ethers using clay supported ammonium nitrate "clayan" in dry media. *Tetrahedron Letters* 39: 3043–3046.

Yadav, G.D., B.V.S. Reddy, R.S. Rao, V. Naveenkumar, and K. Nagaiah. 2003c. Microwave-assisted one-pot synthesis of 2,4,-disubstituted quinolines under solvent-free conditions. *Synthesis* 1610–1614.

Yang, S.-K. and Y. Kim. 2013. Nanogold particles produced by NaBH4 reduction of gold salt in the presence of laponite sol. *Bulletin of the Korean Chemical Society* 34: 363–364.

Yu, K., C. Gu, S.A. Boyd et al. 2012. Rapid and extensive debromination of decabromodiphenyl ether by smectite clay-remplated subnanoscale zero-valent iron. *Environmental Science & Technology* 46: 8969–8975.

Yuan, P., M. Fan, D. Yang et al. 2009. Montmorillonite-supported magnetite nanoparticles for the removal of hexavalent chromium [Cr(VI)] from aqueous solutions. *Journal of Hazardous Materials* 166: 821–829.

Zakarina, N.A., L.D. Volkova, A.K. Akurpekova et al. 2008. Isomerization of *n*-hexane on platinum, palladium, and nickel catalysts deposited on columnar montmorillonite. *Petroleum Chemistry* 48: 186–192.

Zhang, W., M.K.S. Li, R. Wang, P.-L. Yue, and P. Gao. 2009. Preparation of stable exfoliated Pt-clay nanocatalyst. *Langmuir* 25: 8226–8234.

Zhang, Z., B. Liu, K. Lu, J. Sun, and K. Deng. 2014. Aerobic oxidation of biomass derived 5-hydroxymethyl-furfural into 5-hydroxymethyl-2-furancarboxylic acid catalyzed by a montmorillonite K-10 clay immobilized molybdenum acetylacetonate complex. *Green Chemistry* 16: 2762–2770.

Zhou, C.H. 2011. An overview on strategies towards clay-based designer catalysts for green and sustainable catalysis. *Applied Clay Science* 53: 87–96.

Zhou, L., X. Qi, X. Jiang, Y. Zhou, H. Fu, and H. Chen. 2013. Organophilic worm-like ruthenium nanoparticles catalysts by the modification of CTAB on montmorillonite supports. *Journal of Colloid and Interface Science* 392: 201–205.

Zhou, C.H., D.S. Tong, M. Bao, Z.X. Du, Z.H. Ge, and X.N. Li. 2006. Generation and characterization of catalytic nanocomposite materials of highly isolated iron nanoparticles dispersed in clays. *Topics in Catalysis* 39: 213–219.

Zhu, R., Q. Chen, Q. Zhou, Y. Xi., J. Zhu, and H. He. 2016. Adsorbents based on montmorillonite for contaminant removal from water: A review. *Applied Clay Science* 123: 239–258.

Zhu, H.Y., J.-Y. Li, J.-C. Zhao, and G.J. Churchman. 2005. Photocatalysts prepared from layered clays and titanium hydrate for degradation of organic pollutants in water. *Applied Clay Science* 28: 79–88.

Zhu, L., S. Letaief, Y. Liu, F. Gervais, and C. Detellier. 2009. Clay mineral supported gold nanoparticles. *Applied Clay Science* 43: 439–446.

5 Clay Mineral Catalysis of *Name* Reactions

5.1 INTRODUCTION

Clay minerals, especially montmorillonites and their various surface-modified forms, have featured as catalysts of many *name* reactions—that is, organic chemical conversions that are named after their respective discoverer(s). For a comprehensive listing of these reactions, their underlying mechanisms, and biographical sketches of the corresponding discoverer(s), we refer to the 2006 book by J.J. Li. However, no more than 15% of the listed name reactions have been carried out in the presence of clay minerals or their surface-modified forms. By bringing the reactants together in an environment of reduced dimensionality, clay minerals can and do increase reaction rates, yields, and product selectivity.

Two prominent examples of clay-catalyzed name reactions are the Diels–Alder and Friedel–Crafts conversions (Laszlo 1986; Cornélis et al. 1990; Balogh and Laszlo 1993; Adams et al. 1994), some of which have already been mentioned in the preceding chapters and listed in Table 3.3. It goes without saying that these and other name reactions can proceed under homogeneous (solution) conditions without the assistance of mineral catalysts. However, when such reactions are carried out in the presence of clay minerals, especially acid-activated montmorillonites, the rate of conversion is usually significantly enhanced as do product yield and selectivity (Fernandes et al. 2012; Kaur and Kishore 2012). Following Li (2006), the following clay-catalyzed *name* reactions are summarized in alphabetical order.

5.2 ALDER–ENE REACTION

The Alder–ene, or simply ene reaction, refers to the addition of an *enophile* (i.e., a compound containing a multiple bond, such as C=C, C=O, C=N, N=N) to an alkene bearing an allelic hydrogen (the *ene*).

Roudier and Foucaud (1984, 1986) reported that K10 montmorillonite could act as a (Lewis acid) catalyst for the ene reaction between dioxymalonate (the enophile) and various alkenes to yield γ-lactones. We might also mention that the reaction of cyclooctene-5-carboxylic acid over some polyvalent cation-exchanged clays gave a mixture of internal lactone products (Adams et al. 1982). Miles et al. (2005) used clay-supported Pt^0, obtained by treating a potter's clay with a H_2PtCl_6 solution, and reducing the metal precursor with H_2 gas to catalyze the reduction of 2-furylhydrazone (**1**) to 2-methylene-2,3-dihydrofuran (**2**). As a highly reactive ene, the latter compound could then combine with an aldehyde (**3**) to yield the corresponding alcohol (**4**) according to Scheme 5.1.

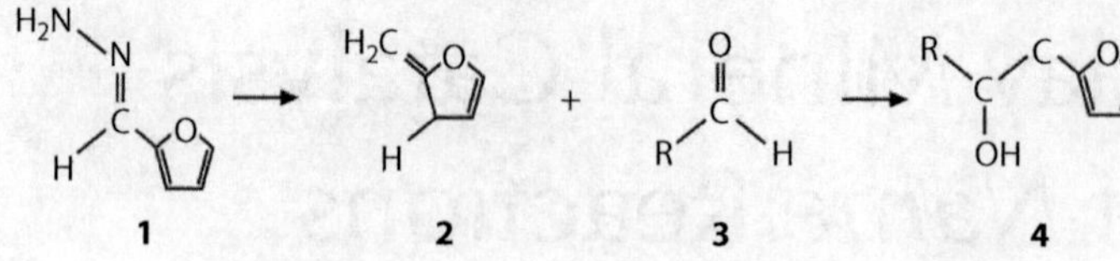

SCHEME 5.1 Reduction of 2-furylhydrazone (**1**) to 2-methylene-2,3-dihydrofuran (**2**) in the presence of clay-supported Pt^0. Compound **2** then combines with an aldehyde (**3**) to afford the corresponding alcohol (**4**). (After Miles, W.H. et al., *J. Org. Chem.*, 70, 2862–2865, 2005.)

5.3 BAEYER–VILLIGER CONDENSATION AND OXIDATION

Using K10 montmorillonite as catalyst, Zhang et al. (1997) synthesized triarylmethanes (**3**) via the Baeyer–Villiger (BV) condensation of aromatic aldehydes (**1**) with *N,N*-dimethylaniline (**2**) under solvent-free conditions (Scheme 5.2). The reaction was promoted by the presence of electron-withdrawing substituents (R), such as a nitro group in the 2, 3, or 4 position on the aromatic ring. Thus with 2-nitrobenzene as substrate, an isolated yield of 96% was recorded within 5 h. On the other hand, with aldehydes having electron-donating substituents (e.g., R = 4-CH_3) the yield dropped to 81% within 18 h. Similarly, Shanmugam and Varma (2001) were able to synthesize heteroaryldiarylmethane leuco bases in good yields by the BV condensation of heterocyclic aldehydes with *N,N*-dimethylaniline at 100°C, using K10 montmorillonite as catalyst.

Lei et al. (2006, 2007) used Sn^{2+}-exchanged palygorskite and montmorillonite, while Hara et al. (2012) chose Sn^{2+}-fluorotaeniolite (mica) to catalyze the BV oxidation of ketones with hydrogen peroxide to yield the corresponding lactones or esters. Similarly, mixed Al-Fe pillared interlayered clays (PILC) (cf. Chapter 3), containing different amounts of iron, can promote the BV oxidation of cyclohexanone to caprolactone at room temperature, using benzaldehyde and oxygen as oxidants (Belaroui et al. 2010; Belaroui and Bengueddach 2012). Catalytic activity correlates with the Fe content (1–3 mmol/g clay) and accessible surface area of the PILC.

de Faria et al. (2012) have reported the conversion of cyclohexanone to caprolactone, using H_2O_2 as oxidizing agent and benzonitrile as solvent, in the presence of kaolinite that have been grafted with Fe(III) picolinate and dipicolinate complexes. More recently, Saikia et al. (2016) used Fe_3O_4 (magnetic) nanoparticles, deposited in the nanopores of an acid-treated montmorillonite, to catalyze the BV oxidation of various cyclic and aromatic ketones in the presence of H_2O_2 at room temperature and under solvent-free conditions.

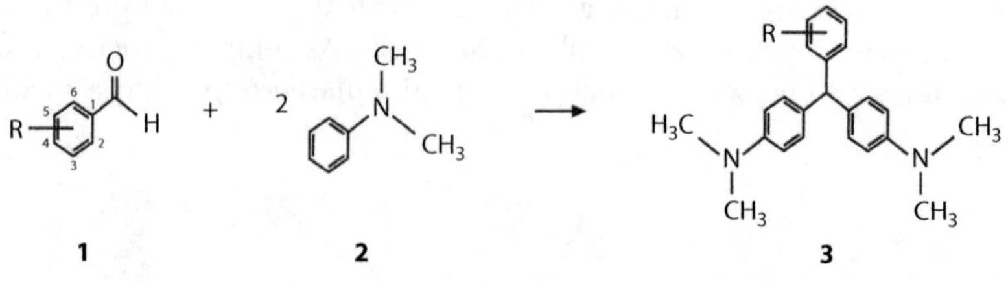

SCHEME 5.2 Baeyer–Villiger condensation of aromatic aldehydes (**1**) with *N,N*-dimethylaniline (**2**) to yield triarylmethanes (**3**) under solvent-free conditions, using K10 montmorillonite as catalyst. (After Zhang, Z-H. et al., *Syn. Commun.*, 27, 3823–3828, 1997.)

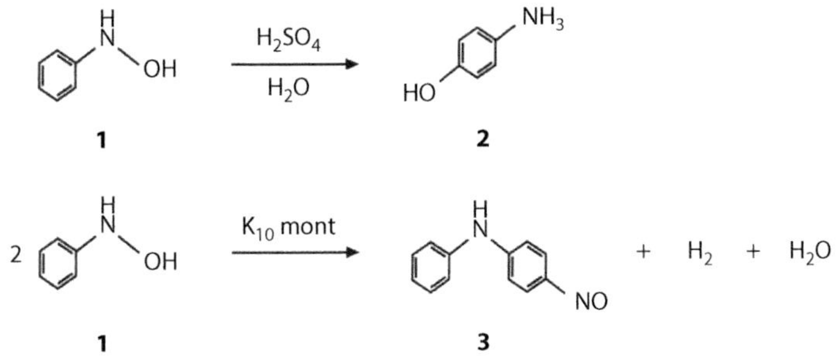

SCHEME 5.3 The acid-catalyzed Bamberger rearrangement of *N*-phenylhydroxylamine (**1**) to 4-aminophenol (**2**) under homogeneous conditions. In the presence of K10 montmorillonite, compound **1** is converted to *p*-nitrosodiphenyl amine (**3**). (After Naicker, K.P. et al., *Catal. Lett.*, 54, 165–167, 1998.)

5.4 BAMBERGER REARRANGEMENT

This reaction refers to the acid-catalyzed rearrangement of *N*-phenylhydroxylamine (**1**) to 4-aminophenol (**2**) as shown in Scheme 5.3. In the presence of K10 montmorillonite, however, *N*-phenylhydroxylamine is *unexpectedly* converted into *p*-nitrosodiphenyl amine (**3**), a reaction that proceeds somewhat faster in the presence of cation-exchanged forms of K10 (Naicker et al. 1998). It would appear that confinement of the substrate to the mesopore structure of K10 prevents the O-protonated form of (**1**) to lose its water and yield the aminophenol product (**2**).

5.5 BAYLIS–HILLMAN REACTION

The Baylis–Hillman (BH), or Morita–Baylis–Hillman, reaction is a carbon-carbon forming reaction, involving the coupling of a carbon electrophile with an electron-poor alkene, containing an electron-withdrawing group (EWG) in the presence of a tertiary amine catalyst. The general reaction is shown in Scheme 5.4 where R = alkyl, aryl, heteroaryl; X = O, NCOOR, NSO$_2$Ph; and EWG = COR, CO$_2$R, CHO, CN, PO(OEt)$_2$, SO$_2$R, SO$_3$R.

As already remarked on in Chapter 3, 1,4,-diazabicyclo[2,2,2]octane (Dabco) is commonly used to catalyze the BH reaction under homogeneous conditions (Basavaiah et al. 2003; Lee et al. 2005; Li 2006; Lima-Junior and Vasconcellos 2012). In this context, we would also mention the *aza*-BH reaction of an electron-deficient alkene, notably an α,β-unsaturated carbonyl compound, with an imine in the presence of a nucleophile.

SCHEME 5.4 Baylis–Hillman (BH) reaction between a carbon electrophile and an alkene, containing an electron-withdrawing group (EWG), to yield the corresponding BH adduct in the presence of a tertiary amine catalyst. (Adapted from Li, J.J., *Name Reactions: A Collection of Detailed Reaction Mechanisms*, 3rd ed., Springer-Verlag, Berlin, Germany, 2006.)

Clay minerals have featured as catalysts for the transformation of the BH adducts (Scheme 5.4), rather than for the BH reaction *per se*. In a series of papers, Shanmugam and Rajasingh (2001, 2002, 2004) have described the activity of K10 montmorillonite in catalyzing the isomerization of the acetates of BH adducts under microwave irradiation. The reaction is highly stereoselective, yielding the corresponding (*E*)-substituted alkenes. K10 montmorillonite was similarly efficient in promoting the solvent-free one-pot protection isomerization of the BH adducts using trimethyl orthoformate and various alcohols as well as the synthesis of substituted 1-aryl indenes from β-phenyl-substituted adducts (Shanmugam and Rajasingh 2005).

Treatment of the BH adducts, derived from acrylate esters (e.g., 3-hydroxy-2-methylene alkanoates) with KSF-supported NaI and NaBr under microwave irradiation, affords the corresponding (*Z*)-allyl iodides and bromides in high yields, while no such transformation occurs in the absence of the montmorillonite catalyst. Interestingly, when the adducts from acrylonitrile, namely, 3-hydroxy-2-methylene alkanenitriles are treated under the same conditions, the corresponding (*E*)- and (*Z*)-allyl iodides and bromides are formed (Yadav et al. 2001a). More recently, Kim et al. (2009) synthesized *N*-tosyl *aza*-BH adducts by reacting the adducts with tosylamide in the presence of K10 monmorillonite. Up to 26% Z-isomers were obtained for adducts with an ester moiety, and 71% for adducts with a nitrile moiety. KSF montmorillonite is similarly efficient in catalyzing the regiosynthesis of 1,5,-diaryl pyrazoles from BH adducts (of aryl aldehyde and methylvinyl ketone) and arylhydrazine (Nikpassand et al. 2009).

5.6 BECKMANN REARRANGEMENT

This reaction refers to the acid-catalyzed rearrangement of an oxime to an amide, while cyclic oximes yield lactams.

Meshram (1990) used the commercially available (acid-activated) KSF montmorillonite (cf. Table 3.1) to promote the Beckmann rearrangement (BR) of various ketoximes. Subsequently, Bosch et al. (1995) reported that K10 montmorillonite could catalyze the same transformation in dry media under microwave irradiation. K10-supported $FeCl_3$ can also mediate the BR of substituted diaryl ketoximes (Pai et al. 1997a). K10 montmorillonite is equally efficient in promoting the vapor-phase BR of salicylaldoxime to benzoxazole, probably involving Brønsted acid sites on the catalyst (Thomas and Sugunan 2006).

More recently, Mitsudome et al. (2012) used Ti^{4+}-exchanged montmorillonite to catalyze the liquid-phase BR of aromatic, aliphatic, and alicyclic ketoximes under mild reaction conditions (90°C in benzonitrile). An added advantage was the ease with which the catalyst could be separated by filtration and reused without significant loss of activity. Another example is the quantitative conversion of cyclododecanone oxime (**1**) into ω-laurolactam (**2**), a monomer of Nylon 12 (Scheme 5.5). Both the extent of conversion and product yield decrease in the order Ti^{4+}-mont > Al^{3+}-mont > Zr^{4+}-mont > Sc^{3+}-mont. This observation is consistent with a Brønsted acid-mediated process since this type of acidity is positively correlated with the ionic potential (valency/radius ratio) of the counterions (cf. Chapter 2).

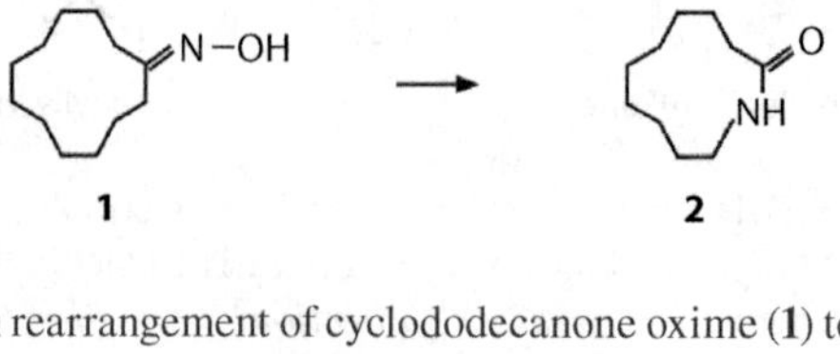

SCHEME 5.5 The Beckmann rearrangement of cyclododecanone oxime (**1**) to ω-laurolactam (**2**) in the presence of Ti^{4+}-exchanged montmorillonite. (After Mitsudome, T. et al., *Tetrahed. Lett.*, 53, 5211–5214, 2012.)

5.7 BIGINELLI REACTION

The Biginelli reaction or Biginelli pyrimidone synthesis, is an acid-catalyzed multicomponent condensation reaction involving an aromatic aldehyde, such as benzaldehyde (**1**), ethyl acetoacetate (**2**), and urea (**3**) to yield 3,4-dihydropyrimidin-2(1*H*)-one (**4**) as shown in Scheme 5.6. The homogeneous reaction is commonly carried out by refluxing the components in ethanol with HCl as catalyst. The mechanism, pharmacological applications, and scope of the Biginelli reaction have been reviewed by Suresh and Sandhu (2012).

Using KSF montmorillonite, Bigi et al. (1999b) synthesized compound (**4**) by heating (**1**), (**2**), and (**3**) at 100°C in water or toluene, and even in the absence of any solvent. Indeed, they were able to obtain 3,4-dihydropyrimidin-2(1*H*)-one in 82% yield and with 98% selectivity by heating the three components at 130°C for 48 h under solvent-free conditions. Similar results were obtained by substituting different aromatic and aliphatic aldehydes for benzyl aldehyde (Scheme 5.6; compound **1**), and β-dicarbonyl compounds for ethyl acetoacetate (Scheme 5.6; compound **2**).

Lin et al. (2000) also found KSF montmorillonite to be highly efficient in promoting the Biginelli reaction of β-keto esters, aryl aldehydes, and urea in methanol to yield 5-alkoxycarbonyl-4-aryl-3,4-dihydropyrimidin-2(1*H*)-ones. Subsequently, Osnaya et al. (2003) used a commercial bentonite (TAFF) and infrared radiation to promote the Biginelli reaction of various benzaldehydes, ethyl acetoacetate, and urea/thiourea. More recently, Sayyahi (2013) and Sayyahi et al. (2013) described the Biginelli synthesis of octahydroquinazolinone and 3,4-dihydropyrimidin-2(1*H*)-one derivatives in water, using a cetyltrimethylammonium-exchanged montmorillonite as a catalyst. It seems likely that the reaction took place in the interlayer space of the organoclay although no X-ray diffraction analysis was carried out to support this possibility.

Salmón et al. (2001) obtained dihydropyrimidones from different aldehydes, ethyl acetoacetate, and urea or thiourea in the presence of a commercial bentonitic clay under infrared irradiation and solvent-free conditions. Sowmiya et al. (2007) have reported similarly, using nanoparticles of sulfated tin oxide impregnated into Al-PILC as the catalyst. Darehkordi et al. (2012) also found acid-activated montmorillonite and Ti-PILC to be efficient in catalyzing the Biginelli synthesis of dihydropyrimidones. The same applies to K10-supported $ZnCl_2$, $GaCl_3$, $InCl_3$, $AlCl_3$, and $FeCl_3$ reagents (Choudhary et al. 2003), and montmorillonite-supported $SmCl_3$ (Li and Bao 2003). Likewise, Lee and Ko (2004) observed that Envirocat EPZ10, the commercially available K10-supported $ZnCl_2$, could promote the Biginelli pyrimidone synthesis in toluene (Centi and Perathoner 2008). Subsequently, Kalbasi et al. (2012) used a bentonite-sulfonated polystyrene nanocomposite as an acid catalyst for the Biginelli synthesis of 3,4-dihydropyrimidin-2(1*H*)-ones/thiones in water. Slimi et al. (2012) used KSF montmorillonite for the same conversion, while Dar et al. (2013) used KSF-supported $AlCl_3$ and $FeCl_3$ under solvent-free conditions.

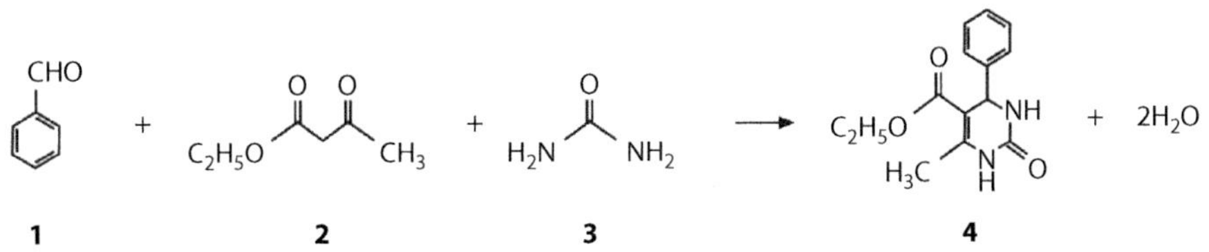

SCHEME 5.6 Biginelli reaction of benzaldehyde (**1**) with ethylacetoacetate (**2**) and urea (**3**) at 100°C to yield 3,4-dihydropyrimidin-2(1*H*)-one (**4**), catalyzed by KSF montmorillonite. (After Bigi et al. 1999b.)

5.8 DIELS–ALDER REACTION

The Diels–Alder (DA) reaction is a Lewis acid-catalyzed [4 + 2]-cycloaddition reaction between a (conjugated) *diene* and a (substituted) alkene, termed the *dienophile*, to yield a (substituted) cyclohexene adduct. The diene can be either open-chain or cyclic and may contain a variety of substituents. In the *normal-demand* DA reaction (Scheme 5.7a), the dienophile has an electron-withdrawing group (EWG) conjugated to the alkene, while in the *inverse-demand* reaction, the dienophile is conjugated to an electron-donating group (EDG) such as amine, ether, or phenol (Scheme 5.7b). In the hetero-DA reaction, either the diene or the dienophile contains a heteroatom such as nitrogen or oxygen (Jørgensen 2000). When imines are used as the dienophile, the reaction is referred to as aza-Diels–Alder (Scheme 5.7c). As the following examples would indicate, DA reactions are also stereoselective.

An early reference to a montmorillonite-catalyzed DA reaction relates to the dimerization of oleic acid (den Otter 1970). As McCabe and Adams (2013) have pointed out, however, the reaction may not be of the Diels–Alder type since high temperatures are used to dehydrogenate the substrate prior to its dimerization. Using five smectites, Čičel et al. (1992) found that dimer yield was influenced by the nature of the exchangeable counterions, while Yang et al. (2013) obtained close to 72% conversion using an organically modified montmorillonite.

Laszlo and coworkers (Laszlo and Luchetti 1984a, 1984b, 1984c; Laszlo and Moison 1989) did the early systematic investigations into some DA reactions using K10-supported ferric nitrate (*clayfen*) (Chapter 4) or Fe^{3+}-exchanged K10 montmorillonite. Thus, when 1,3,-cyclohexadiene was co-dissolved with 4-*tert*-butylphenol (in dichloromethane) in the presence of clayfen at 0°C, they were able to obtain 77% isolated yield of the corresponding adduct in less than 1 h (Laszlo and Luchetti 1984a). They then used Fe^{3+}-K10 montmorillonite to catalyze the DA reaction of cyclopentadiene with methylvinyl ketone (Scheme 5.8) and found an almost quantitative conversion to the cycloadduct with an *endo:exo* ratio of 21:1 or an *endo* preference of 21 (Laszlo and Luchetti 1984b). Fe^{3+}-K10 montmorillonite is similarly effective in catalyzing the cycloaddition of furan and dimethyl-2,5-furan (as dienes) with methylvinyl ketone and acrolein (as dienophiles) (Laszlo Luchetti 1984c), and acrolein can serve as a dienophile in the DA reaction with butadiene, isoprene, cyclopentadiene, cyclohexadiene, 1-methoxy-cyclohexadiene, and anthracene (Laszlo and Moison 1989).

Avalos et al. (1998a, 1998b) have reported that unmodified K10 montmorillonite can catalyze the DA reaction of cyclopentadiene, cyclohexadiene, and furan with α,β-unsaturated carbonyl compounds (as dienophiles) in the absence of any organic solvent. Similar results were obtained by Adams et al. (1994), using transition metal cation-exchanged montmorillonites at room temperature

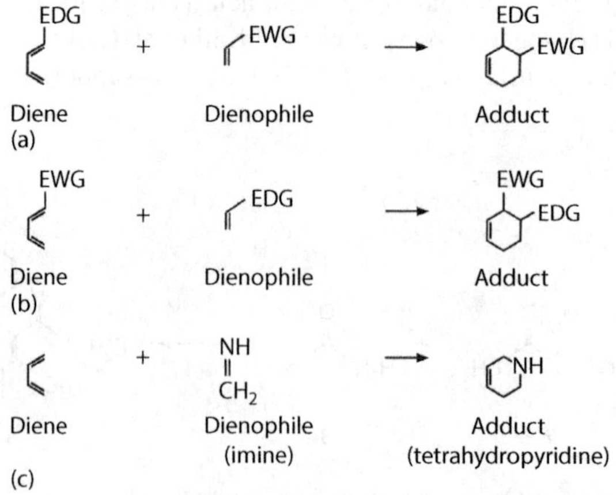

SCHEME 5.7 (a) "Normal-demand" Diels–Alder reaction between a diene and a dienophile to give the corresponding adduct; (b) "inverse-demand" Diels–Alder reaction; and (c), "aza" Diels–Alder reaction. EDG = electron-donating group; EWG = electron-withdrawing group.

and a variety of solvents. Kamath and Samant (1999) and Kamath et al. (2000) have compared the activity of Al^{3+}-, Zn^{2+}-, and Fe^{3+}-K10 montmorillonites, Filtrol-24, bentonite, and pyrophyllite in catalyzing the DA reaction of 4,6-bis(4-methoxyphenyl)-pyran-2(H)-one and 4-(4-methoxyphenyl)-6-methyl-pyran-2(H)-one (as dienes) with naphtoquinone and N-phenylmaleimide (as dienophiles). Filtrol-24 shows the highest activity, while pyrophyllite is the least efficient. The activity of the different cation-exchanged montmorillonites decreases in the order $Fe^{3+} > Zn^{2+} > Al^{3+}$, and the reverse to that is observed for the corresponding homogeneous Lewis acid-catalyzed reactions.

Adams and Clapp (1986) have reported that Ni^{2+}- and Cr^{3+}-montmorillonites can catalyze the DA dimerization of butadiene to yield 4-vinylcyclohexene as essentially the sole product. With isoprene as the conjugated diene, however, the dimers are partly formed by a cycloaddition reaction and partly through a reaction involving carbocation intermediates. The formation of radical cations has been corroborated by electron paramagnetic resonance spectroscopy (Fraile et al. 1997). Following Laszlo and Luchetti (1984b), Adams et al. (1987a, 1987b) have compared the activity of various cation-exchanged montmorillonites in catalyzing the DA reaction between cyclopentadiene (CPD) and methylvinyl ketone (MVK) in dichloromethane (at 22°C) according to Scheme 5.8.

Both Fe^{3+}-montmorillonite and the Cr^{3+}-exchanged sample gave high yields (91%–93%) of cycloadducts with an *endo:exo* ratio of 7:1 and 9:1, respectively. A slightly higher yield (97%) of adduct with an *endo:exo* ratio of 9:1 was obtained with Fe^{3+}-exchanged K10 montmorillonite, while the reaction over Al^{3+}-montmorillonite or in the absence of clay (control) yielded only 17%–21% of cycloadducts. These observations may be explained in terms of a one-electron transfer from the dienophile to the interlayer cation, in the case of Fe^{3+}- and Cr^{3+}-montmorillonites, or to Fe^{3+} ions in the mesopores of Fe^{3+}-exchanged K10 as described in Chapter 2 (cf. Equation 2.22). Thus, Cr^{3+}-vermiculite with a higher layer charge and more restricted interlayer expansion as compared to montmorillonite gives adducts with an *endo:exo* ratio of 2.5:1, indicative of enhanced selection for the less bulky isomer (Adams et al. 1994). In this context, we might also mention that impregnation of a chloroaluminate ionic liquid into an organo-montmorillonite produces an effective catalyst for the conversion of *endo*-tetrahydrodicyclopentadiene into the corresponding *exo*-isomer (Huang et al. 2010).

Endo:exo ratios also tend to increase with the temperature at which the clay samples have been preheated (Cativiéla et al. 1993). López et al. (2007), for example, reported a 99% yield of cycloadduct with an *endo:exo* ratio of 94:6 for the room temperature reaction of cyclopentadiene (CPD) with methylvinyl ketone (MVK), using K10 montmorillonite as catalyst and 1-hexyl-3-methylimidazolium tetrafluoroborate as the (ionic liquid) solvent. Fe^{3+}-exchanged K10 in the presence of butylated hydroxytoluene is also an efficient catalyst for the DA reaction of cyclopentadiene and furan with alkenes bearing sulfoxide groups (e.g., trans-2-methylene-1,3-dithiolane 1,3 dioxide) as dienophiles (Gültekín 2004).

The relative inactivity of Al^{3+}-montmorillonite for the reaction between CPD and MVK may be ascribed to *shielding* of the interlayer Al^{3+} by coordinated water molecules, which apparently resist azeotropic removal with toluene. We would also suggest that interlayer Al^{3+} can form a polyhydroxy complex of the type $[(Al)(OH)_x(H_2O)_{6-x}]^{(3-x)+}$ (Theng and Scharpenseel 1976; Ballantine et al. 1985). Interestingly, Toma et al. (1987) found that the activity of different M^{n+}-exchanged

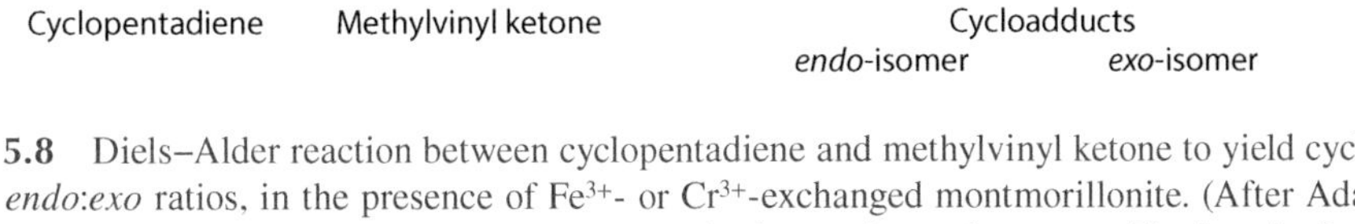

SCHEME 5.8 Diels–Alder reaction between cyclopentadiene and methylvinyl ketone to yield cycloadducts with high *endo:exo* ratios, in the presence of Fe^{3+}- or Cr^{3+}-exchanged montmorillonite. (After Adams, J.M. et al., Catalysis of Diels–Alder cycloaddition reactions by ion-exchanged montmorillonites, in *Proceedings of the International Clay Conference, Denver, 1985* (Eds.) L.G. Schultz, H. van Olphen, and F.A. Mumpton, The Clay Minerals Society, Bloomington, IN , pp. 324–328, 1987a; Adams, J.M. et al., *J. Inclus. Phenom.*, 5, 663–674, 1987b.)

montmorillonites in the Diels–Alder reaction of 1-phenyl-1,3-butadiene with acryloylferrocene (to yield 3-phenyl-4-ferrocenoylcyclohexene) decreased in the order $Al^{3+} > Cr^{3+} > Co^{2+} > Fe^{3+} > Cu^{2+}$. The clay catalysts used, however, were previously dried at 105°C–120°C (24 h), and 4-*tert*-butylphenol was added as a cocatalyst. By the same token, the activity of Zn^{2+}-K10 montmorillonite for the DA reaction between cyclopentadiene and methyl acrylate can be enhanced by prior calcination of the mineral (at 550°C), resulting in layer dehydroxylation and the formation of electron-deficient (Lewis acid) sites as described in Chapter 2 (Cativiéla et al. 1992a). Likewise, Kawabata et al. (2005) have proposed that the DA reaction between 1,3-cyclohexadiene and 3-buten-2-one to form 2-acetylbicyclo[2.2.2]oct-5-ene, in the presence of Cu^{2+}-montmorillonite that has been preheated (under vacuum at 100°C for 2 h), is a Lewis acid-catalyzed conversion. As Figueras et al. (1994) have pointed out, however, diene polymerization may compete with DA cycloaddition. The ease of cycloaddition is also influenced by the organic solvent used. In the presence of water-miscible solvents, this type of reaction gives low yields and shows *endo* preference (Cativiéla et al. 1991, 1992b).

Mitsudome et al. (2008) used an organoclay, obtained by intercalating a chiral organic compound (5S-2,2,3-trimethyl-5-phenyl-methyl-4-imidazolidinone hydrochloride) into a Na^+-montmorillonite, to catalyze the DA reaction between various dienes and dienophiles. For the reaction of cyclohexene with acrolein (in aqueous acetonitrile at room temperature), they reported 80% yield of cycloadduct with an *endo:exo* ratio of 93:7, even after four cycles of catalyst usage. We suggest that the high and sustained efficiency of the reaction is related to the ease with which the reactants can enter the clay interlayer space that has been expanded (from 0.24 to 0.9 nm) by intercalation of the organic reagent (Figure 5.1). Besides having a reduced dimensionality, the interlayer space provides a reactant-compatible hydrophobic nanoenvironment that is conducive to catalysis.

An example of a clay-catalyzed hetero-Diels–Alder reaction is the cycloaddition of 2,3-dimethyl-1,3-butadiene with benzaldehydes (e.g., *o*-anisaldehyde) in carbon tetrachloride (25°C) to yield dihydropyrans in the presence of K10 montmorillonite (Dintzner et al. 2007). Less substituted dienes (e.g., isoprene) give rise to oligomeric rather than cycloaddition products. Preheating the clay at 200°C (1 h) increases Lewis acidity and catalytic activity but suppresses diene dimerization. Earlier, Chiba et al. (1999) were able to synthesize benzodihydropyrans through an intermolecular hetero-DA reaction of *o*-quinomethanes (generated *in situ* by dehydration of *o*-hydroxybenzyl alcohols) with alkenes, using K10 montmorillonite as the catalyst.

Yadav et al. (2002) used KSF montmorillonite to catalyze the synthesis of pyrano- and furano-quinolines from aryl amines and cyclic enol ethers. Yadav et al. (2004b) also succeeded in preparing hexahydro-1*H*-pyrrolo[3,2-c]quinoline derivatives by a hetero-DA reaction between endocyclic enecarbamates and aryl amines. In both instances, good yields of products with moderate to high diastereoselectivity were recorded. Similarly, Ramesh and Raghunathan (2009) were able to synthesize polycyclic pyrano[2,3,4-*kl*]xanthene derivatives by a hetero-DA reaction of 2-(cyclohex-2-enyloxy) benzaldehyde with symmetrical 1,3-diones in the presence of K10 montmorillonite and with the assistance of microwave irradiation.

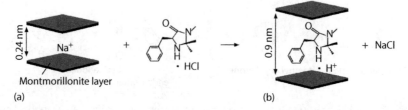

FIGURE 5.1 Diagram showing the reaction of 5S-2,2,3-trimethyl-5-phenyl-methyl-4-imidazolidinone hydrochloride with Na^+-montmorillonite, resulting in the replacement of the interlayer Na^+ counterions by the (chiral) organic cations, and increasing the interlayer spacing from 0.24 nm (a) to 0.9 nm (b). The resultant clay-organic complex can catalyze the Diels–Alder reaction between dienes and dienophiles. (Modified from Mitsudome, T. et al., *Tetrahedron Lett.*, 49, 5464–5466, 2008.)

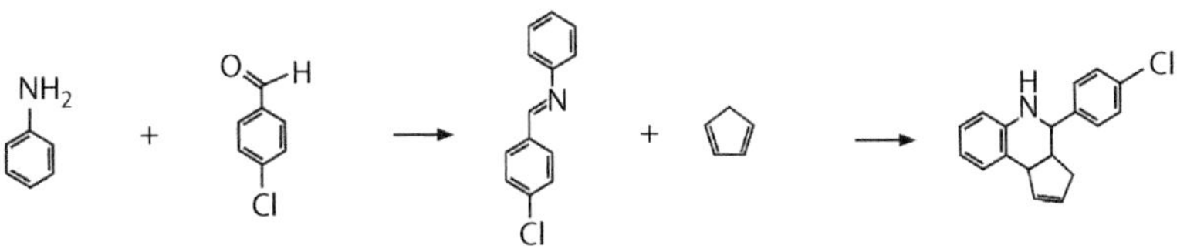

SCHEME 5.9 Aza-Diels–Alder cycloaddition of an imine (**3**) with cyclopentadiene (**4**) to yield tetrahydroquinoline (**5**), in the presence of a commercially available H⁺-bentonite. The imine (intermediate) is obtained by reacting aniline (**1**) with *p*-chlorobenzaldehyde (**2**). (After Sartori, G. et al., *Europ. J. Org. Chem.*, 2513–2518, 2001.)

Sartori et al. (2001) used a commercially available H⁺-exchanged bentonite to promote the aza-Diels–Alder cycloaddition of the imine (**3**), formed by the reaction between aniline (**1**) and *p*-chlorobenzaldehyde (**2**), with cyclopentadiene (**4**) to yield tetrahydroquinoline (**5**) according to Scheme 5.9. Zhu et al. (2009) were able to synthesize tetrahydroquinoline *skeletons* by an aza-DA reaction of methylenecyclopropanes with ethyl (arylimino) acetates, using K10 montmorillonite as the catalyst. Likewise, Akiyama et al. (2002) could obtain 2-substituted 2,3-dihydro-4-pyridones by the aza-DA reaction of Danishefsky's diene (*trans*-1-methoxy-3-trimethylsilyloxy-1,3-butadiene) with aldimines, formed *in situ* from aliphatic aldehydes and *p*-anisidine. KSF montmorillonite is similarly efficient in catalyzing the three-component aza DA reaction (in acetonitrile) of methylenecyclopropane with arenecarbaldehydes and arylamines to form quinolone derivatives with a spirocyclopropyl ring in 90%–100% isolated yield (Shao and Shi 2003).

5.9 FERRIER REARRANGEMENT

The Ferrier (glycal allylic) rearrangement is a Lewis acid catalyzed reaction, combining a nucleophilic substitution reaction with an allylic shift in a glycal (2,3-unsaturated glycoside). An early example of a K10 montmorillonite-catalyzed Ferrier rearrangement (FR) is the glycosidation of various alcohols with 3,4-di-*O*-acetyl-L-rhamnal and 3,4,6-tri-*O*-acetyl-D-glucal in dichloromethane, and the *C*-glycosidation of glycals (Toshima et al. 1995, 1996). Similarly, Shanmugasundaram et al. (2002) were able to synthesize alkyl/aryl 2,3-dideoxy-D-*threo*-hex-2-enopyranosides (**3**) of high α-selectivity by reacting 3,4,6-tri-*O*-acetyl-D-galactal (**1**) with alcohols/phenols (**2**) in chlorobenzene, in the presence of K10 montmorillonite under microwave irradiation as shown in Scheme 5.10 where R¹ = H; R² = OAc; Ac = acetyl. Concurrently, de Oliveira et al. (2002) reported on the microwave-assisted, solvent-free synthesis of unsaturated glycosides from tri-*O*-acetyl-D-glucal and an alcohol, also with K10 montmorillonite as catalyst.

Kumaran et al. (2011) have described the FR of 2-*C*-hydroxymethyl-D-glycals, 3,4,6-tri-*O*-alkyl-D-glycals, and 3,4(dihydro-2*H*-pyran-5-yl)methanol in the presence of K10 montmorillonite. The reaction of the first-mentioned compound with phenols yields pyranol[2,3-*b*]benzopyrans, while that of 2-*C*-hydroxymethyl-D-galactal with 2,6-dimethylphenol leads to a domino transformation, yielding 4-(5′,6′-dihydro-4*H*-pyran-3′-ylmethyl)-2,6,dimethylphenol. Another interesting finding is the ability of

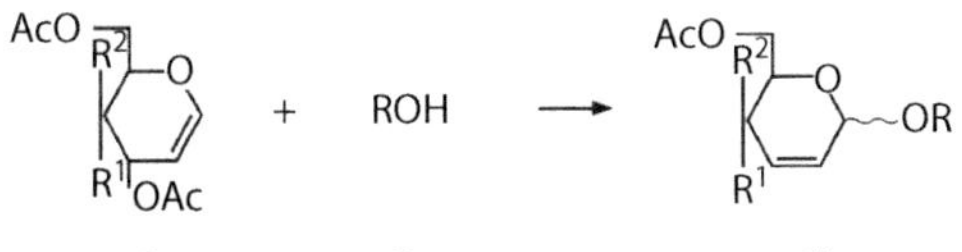

SCHEME 5.10 Ferrier reaction of 3,4,6-tri-*O*-acetyl-D-galactal (**1**) with alcohols/phenols (**2**) to yield alkyl/aryl 2,3-dideoxy-D-*threo*-hex-2-enopyranosides (**3**) in the presence of K10 montmorillonite under microwave irradiation. (After Shanmugasundaram, B. et al., *Tetrahedron Lett.*, 43, 6795–6798, 2002.)

silver-impregnated montmorillonite to catalyze the intramolecular Ferrier reaction of 6-hydroxy glycals (in chloroform at 50°C) to afford 1,6-anhydro disaccharides in 80%–85% yields (Sharma et al. 1994). We might add that bismuth(III) trifluoromethane sulfonate [Bi(OTf)$_3$], in association with K10 montmorillonite, is an efficient catalyst for the FR of a 4,5-oxazoline derivative of sialic acid (Ikeda et al. 2008). In a recent review on the Ferrier rearrangement, Gómez et al. (2013) also mentioned the activity of K10- and KSF-supported dodecatungstophosphoric acid (cf. Chapter 4) in catalyzing the reaction of 3,4,6-tri-*O*-acetyl-D-glucal with benzyl alcohol to yield α-glucoside with complete stereoselectivity.

5.10 FISCHER GLYCOSIDATION/GLYCOSYLATION

This reaction refers to the acid-catalyzed synthesis of a glycoside by reacting an aldose or ketose with an alcohol.

Shanmugam and Nair (1996) found that K10 montmorillonite could promote the glycosidation of 1-*O*-acetyl-2,3-dideoxy-dl-pent-2-enopyrano-4-ulose with various alcohols. Subsequently, Brochette et al. (1997) reported that KSF/0 montmorillonite (cf. Table 3.1) was an efficient catalyst for the Fischer glycosylation of glucose with butanol and dodecanol. The Brønsted acid-catalyzed reaction was enhanced by ultrasonication, which in the case of dodecanol, led to the formation of 1,6-polyglucose.

5.11 FISCHER–HEPP REARRANGEMENT

The Fischer–Hepp rearrangement refers to the conversion of an aromatic *N*-nitroso or nitrosamine to a carbon nitroso compound.

Kannan et al. (1997) have described the ability of K10, KSF, and polyvalent cation-exchanged K10 montmorillonites to promote the Fischer–Hepp rearrangement of *N*-methyl-*N*-nitrosoaniline to *N*-methyl-4-nitrosoaniline, during which *N*-methylaniline and aniline were also formed.

5.12 FISCHER INDOLE SYNTHESIS

The Fischer Indole synthesis refers to the cyclization of arylhydrazones to indoles.

An early example provided by Villemin et al. (1989) is the one-pot synthesis of indoles by reacting phenylhydrazine with various ketones, using microwave irradiation and KSF montmorillonite as catalyst. Adopting a similar approach but using K10-supported ZnCl$_2$ (*clayzic*), Lipińska et al. (1999) were able to obtain 2-(2-oyridyl)indole derivatives from 2-acetylpyridine phenylhydrazone. Microwave irradiation was also used by Janković and Komadel (2003) to assist the synthesis of substituted indoles in the presence of montmorillonite.

Dhakshinamoorthy and Pitchumani (2005) used the Fischer indole synthesis in methanol to prepare 1,2,3,4-tetrahydrocarbazole and substituted indoles (**3**) of high purity by reacting phenylhydrazine (**1**) with various carbonyl compounds (**2**) where R^1 = H and R^2 = H, CH$_3$, C$_2$H$_5$, C$_6$H$_5$, *p*-ClC$_6$H$_4$, *p*-NO$_2$C$_6$H$_4$, or *p*-CH$_3$C$_6$H$_4$, in the presence of K10 montmorillonite (Scheme 5.11). More recently, Chakraborty et al. (2011) reported that KSF montmorillonite could catalyze the solvent-free cyclization of some cyclohexane-1,2-dione-1-phenylhydrazones to the corresponding 1-keto-1,2,3,4-tetrahydrocarbazoles under microwave irradiation.

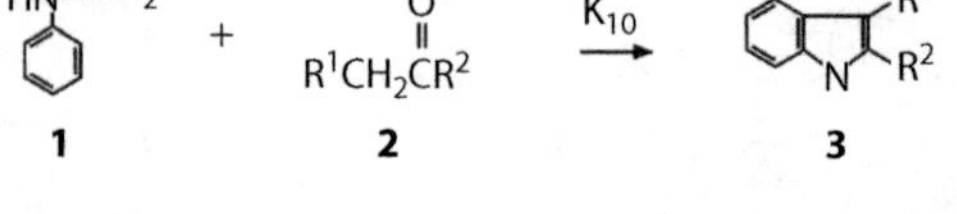

SCHEME 5.11 Fischer indole synthesis of substituted indoles (**3**) by reacting phenylhydrazine (**1**) with various carbonyl compounds (**2**), using K10 montmorillonite as catalyst. (After Dhakshinamoorthy, A. and Pitchumani, K., *Appl. Catal. A: Gen.*, 292, 305–311, 2005.)

5.13 FISCHER–TROPSCH SYNTHESIS

The Fischer–Tropsch (FT) synthesis is a process by which a mixture of carbon monoxide (CO) and hydrogen (H_2), commonly referred to as synthesis gas (*syngas*), is converted into liquid hydrocarbons (Dry 2002) according to the following reaction:

$$(2n + 1)H_2 + nCO \rightarrow C_nH_{2n+2} + nH_2O \tag{5.1}$$

In this respect, Giannelis et al. (1988) and Pinnavaia et al. (1989) did some of the early work on FT catalysis using clusters of ruthenium nanoparticles, embedded in the interlayer space of an alumina pillared montmorillonite (Al-PILC) as described in Chapter 3. In promoting the hydrogenation of CO, the Ru-Al-PILC catalyst showed a marked selectivity toward branched alkanes due to the rearrangement of carbocations formed by protonation of FT terminal alkenes. Subsequently, Rightor et al. (1991) prepared an iron oxide-pillared montmorillonite (Fe-PILC) with an interlayer spacing of 1.5–1.9 nm for the FT synthesis of hydrocarbons (at 270°C). More recently, Upadhyay and Srivastava (2015) used montmorillonite-intercalated ruthenium nanoparticles to catalyze the hydrogenation of alkenes under solvent-free conditions.

Su et al. (2009) were able to obtain C_5^+-selective FT catalysts by wet impregnation of Al- and mixed Ce/Al-PILC with a solution of $Co(NO_3)_3.6H_2O$. The activity of the catalysts could be further enhanced by adsorption of cerium and ruthenium ions to the pillared clays. Using a similar approach, Hao et al. (2010, 2013) and Wang et al. (2011b) prepared a series of Co-loaded cation-exchanged montmorillonites as well as Al-, Si-, and Zr-PILC for the FT synthesis, using a fixed-bed reactor at 1 MPa, 235°C, and $H_2/CO = 2$ (Equation 5.1). The catalyst with a 20% w/w Co loading was efficient in converting CO and more so, following cerium incorporation (Ahmad et al. 2015). At the same time, the selectivity for $C_{4/5}$–C_{12} significantly increased, while that for $C_{21/22}^+$ was lowered, indicative of an acid-catalyzed cracking of long-chain FT hydrocarbons.

Under similar experimental conditions, Ahmad et al. (2013) observed that cobalt nanoparticles supported on Zr-PILC could selectively form C_2–C_{12} hydrocarbons by the FT synthesis, while the selectivity for CH_4 and C_{21} hydrocarbons decreased. Hao et al. (2012) used Co supported on a porous clay heterostructure (cf. Figure 3.4) for controlling the product distribution of FT synthesis, while Zhao et al. (2013) found that KOH-activated Ca^{2+}-montmorillonite, impregnated with Co (10 wt%), was an efficient catalyst. Liu et al. (2014) further showed that Al-PILC, impregnated with MoS_2, could catalyze the formation of mixed (C_1–C_6) alcohols from syngas at 260°C–300°C and 8 MPa. The efficiency of CO conversion and the selectivity for alcohols increased by incorporating 3% w/w K, with or without Ni, into the catalyst. Zhao et al. (2015a) could obtain close to 46% C_5–C_{12} selectivity using an acid-activated montmorillonite loaded with 20% w/w cobalt and 0.1% w/w ruthenium. An even higher (C_5–C_{20}) selectivity was observed with a cobalt-impregnated mixed C/Al-pillared montmorillonite (Zhao et al. 2015b).

5.14 FRIEDEL–CRAFTS REACTION

The Friedel–Crafts (FC) reaction refers to the attachment of an acyl group (*acylation*) or an alkyl group (*alkylation*) to an aromatic ring in the presence of a Lewis acid catalyst. A number of Friedel–Crafts acylations (with acyl chloride, acid anhydride, carboxylic acids) and alkylation reactions (with alkyl halides or alkyl alcohols) have already been described in Chapter 4 and listed in Table 3.3. We might also mention Friedel–Crafts sulfonylation, which refers to the reaction of arenes with sulfonyl halides or sulfonic acid anhydrides to afford the corresponding sulfones.

An early reference to a clay-catalyzed FC alkylation is that by Lotz and Gosselck (1979) who reacted *tert*-butyl bromide with benzene in the presence of Cu^{2+}-bentonite to yield mono- and di-*tert*-butyl benzene. The ratio of mono- to di-alkylated product was influenced by the hydration status of the catalyst, decreasing from 9.36 to 0.78 as the water content increased from

<2 to 22.8% w/w. Subsequently, Chiche et al. (1987) reported that cation-exchanged montmorillonites could effectively catalyze the FC acylation of benzene, toluene, and xylene with carboxylic acids to give the corresponding ketones.

Cation-exchanged clays were also used by Cseri et al. (1995a) to promote the benzylation of aromatic compounds, and by Choudary et al. (1998) to catalyze the acylation of aromatic ethers with acid anhydride. Choudary et al. (2001) subsequently found Fe^{3+}-montmorillonite to be an efficient and selective catalyst for the acylation of amines with acetic acid as did Shinde et al. (2004) for the *tert*-butylation of phenols with *tert*-butyl alcohol.

More recently, Lee et al. (2013) reported that K10 montmorillonite was highly effective in the stereoselective FC alkylation of tetramethoxytoluene (**1**) with (*E*)-4-chloro-2-methyl-1-phenylsulfonyl-2-butene (**2**) according to Scheme 5.12. Thus, in the presence of K10 clay (containing 6.9% w/w water), the reaction of **1** with **2** in dichloroethane at 85 °C for 18 h gave a 74.1% yield of product (**3**) with an *E*:*Z* ratio of 20.4:1. Another interesting K10-catalyzed reaction is the FC alkylation of a polycyclic caged enone with pyrrole and thiophene to yield novel cage compounds (James et al. 2007).

Using transition metal-exchanged K10 montmorillonites as catalysts, Laszlo and Mathy (1987) similarly observed good conversion rates and yields for the FC alkylation of benzene with halides (alcohols, olefins) together with the preferential formation of the mono-alkylated products. Monoalkylation of benzene with primary, secondary, and tertiary alcohols has also been reported by Sieskind and Albrecht (1993). K10 and KSF montmorillonites and their cation-exchanged forms are also effective in promoting the acylation of alcohols, amines, phenols, and thiols with acetyl chloride and benzoyl chloride (Li and Li 1998) as well as the FC reaction of arenes with cholesterol to yield arylcholestenes (Li et al. 1998a). Subsequently, Manju and Sankaran (2004) observed that the benzylation of *o*-xylene, in the presence of transition metal-impregnated montmorillonites, showed a high selectivity for the monoalkylated product.

Although catalytic activity is controlled by Lewis acidity, especially in the case of acid-treated montmorillonites that have been heated at high temperatures (~500°C), Brønsted acidity can affect the rate of alkylation with samples dried at ~100°C (Cseri et al. 1995b). During the alkylation of anisole with various dienes, using cation-exchanged K10 montmorillonites, Cativiéla et al. (1995) similarly found that Brønsted acidity promoted diene polymerization as well as FC dealkylation. Both reactions, however, could be suppressed when the clay catalysts had previously been calcined. Likewise, both Brønsted and Lewis acid sites appeared to be involved in the FC alkylation of aniline with ethanol in the presence of K10 and vanadia-impregnated K10 montmorillonite (Narayanan and Deshpande 1996). They suggested that Brønsted sites on the catalyst controlled the formation of carbocations from ethanol, while Lewis acidity promoted the electrophilic substitution of aniline protons for alkyl carbocations. Similarly, Choudary et al. (1999, 2000) have ascribed the high activity of Fe^{3+}-exchanged K10 montmorillonite for the FC sulfonylation of arenes to the coexistence of Brønsted and Lewis acid sites in the catalyst. The catalytic activity for the alkylation of arenes with alcohols could be further enhanced by the addition of small amounts of *p*-toluenesulfonic or methanesulfonic acid (Choudary et al. 2002). Like the K10 material, montmorillonite treated with *p*-toluenesulfonic acid is an efficient catalyst for the microwave-assisted acylation of *p*-cresol with decanoic acid (Venkatesha et al. 2014).

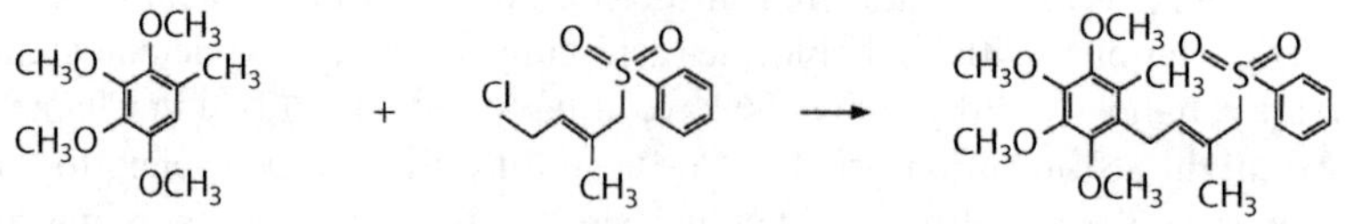

SCHEME 5.12 Stereoselective Friedel–Crafts alkylation of tetramethoxytoluene (**1**) with (*E*)-4-chloro-2-methyl-1-phenylsulfonyl-2-butene (**2**) to yield compound (**3**) with a high *E*:*Z* ratio, in the presence of K10 montmorillonite. (After Lee, S.W. et al., *Bull. Korean Chem. Soc.*, 34, 1257–1259, 2013.)

Tateiwa et al. (1994) made the interesting observation that hydroxyl and methoxy aromatics could be regioselectively C-alkylated with 4-hydroxybutan-2-one in the presence of Al^{3+}-, Fe^{3+}-, Zn^{2+}-, and Zr^{4+}-exchanged montmorillonites. Arienti et al. (1997) have reported similarly for the alkylation of anilines with phenylacetylene to afford 1,1-diarylethylenes, using KSF montmorillonite. Ebitani et al. (2006) also found Ti^{4+}-montmorillonite to be highly efficient in catalyzing the acylation of aromatic compounds with acid anhydrides and carboxylic acids, obtaining 97% yield of 1-(4-methoxyphenyl)-1-dodecanone from the reaction of anisole with dodecanoic acid.

The use of K10-supported $ZnCl_2$ (*clayzic*) as an alkylation catalyst has been described in some detail in Chapter 4. Clayzic-catalyzed FC reactions have been shown to display such interesting features as synergistic acceleration (when two substrates are run in competition), reactivity inversion, and unexpected selectivity for isomer formation (Barlow et al. 1993; Cornélis et al 1993; Yadav et al. 1993; Clark et al. 1994; Pai et al. 1997b; Ballini et al. 2006). It is also worth recalling that clayzic is a much stronger Lewis acid catalyst than (free) zinc chloride, presumably because the reagent in clayzic is dispersed in the mesopores of the acid-treated mineral (Figure 4.1) and hence is readily accessible to the substrate (Clark and Macquarrie 1996). This suggestion is consistent with the observation that K10-supported $FeCl_3$ is superior to its Fe^{3+}-exchanged counterpart in promoting the benzylation of arenes by benzyl chloride (Pai et al. 2000). Easy accessibility to acid sites also lies behind the high efficiency of pillared montmorillonites in catalyzing FC reactions (Choudary et al. 1997; Kurian and Sugunan, 2003, 2005, 2006; Singh and Samant 2004). Like clayzic, K10-supported $SbCl_3$ is efficient in the FC alkylation of indoles and pyrroles with epoxides to give the corresponding C-alkylated derivatives with high yields and regioselectivity (Liu et al. 2009), while K10-supported $Cs_{2.5}H_{0.5}PW_{12}O_{40}$ (cf. Chapter 4) can catalyze the alkylation of anisole with cyclohexene to give 2- and 4-cyclohexylanisoles (Yadav and More 2011).

5.15 FRIEDLÄNDER SYNTHESIS

The Friedländer (quinoline) synthesis refers to the condensation of an aromatic 2-amino aldehyde (or ketone) with another aldehyde (or ketone), containing a reactive α-methylene group (Li 2006; Marco-Contelles et al. 2009).

An early example was provided by Sabitha et al. (1999) who prepared polycyclic quinoline derivatives by a Friedländer synthesis using a clay catalyst and with the assistance of microwave irradiation. Mogilaiah and Sakram (2006) used K10 montmorillonite to catalyze the Friedländer condensation of 2-aminonicotinaldehyde with various carbonyl compounds to yield 1,8-naphthyridines. More recently, Azimi and Abbaspour-Gilandeh (2014) prepared quinolines by reacting 2-aminoarylketones, such as 2-aminobenzophenone (**1**) with 1,3-diketones, such as acetylacetone (**2**) to yield the corresponding quinoline (**3**) using Li+-montmorillonite under solvent-free conditions, according to Scheme 5.13. Likewise, Teimouri and Chermahini (2016) were able to synthesize various poly-substituted quinolines by the Friedländer condensation of 2-aminoarylketones with carbonyl compounds and β-keto esters in the presence of K10 montmorillonite.

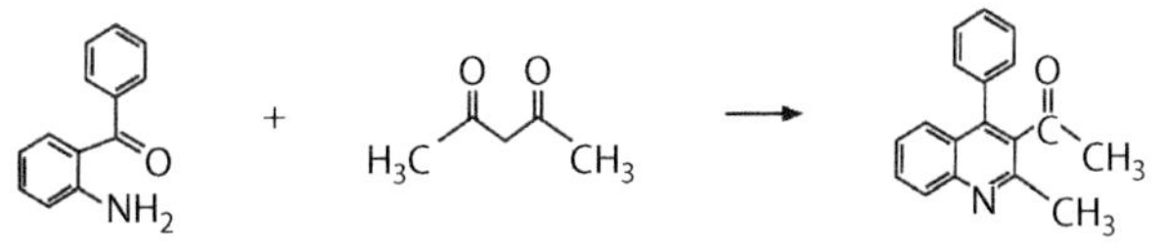

SCHEME 5.13 Friedländer synthesis of a quinoline (**3**) by reacting 2-aminobenzophenone (**1**) with acetylacetone (**2**) in the presence of Li+-montmorillonite under solvent-free conditions. (After Azimi, S.C. and Abbaspour-Gilandeh, E., *J. Nanostr.*, 4, 335–346, 2014.)

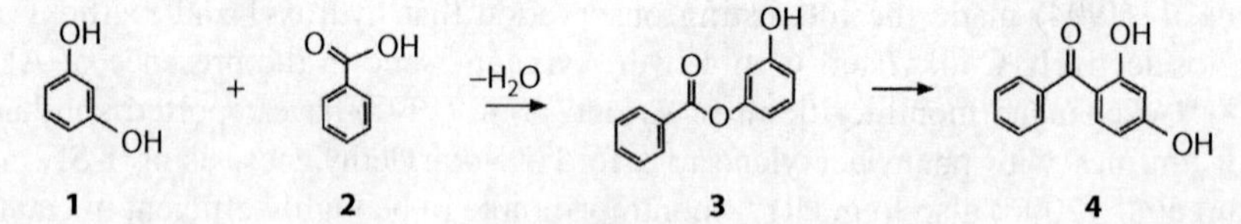

SCHEME 5.14 Synthesis of resorcinol monobenzoate (**3**) by O-benzoylation of resorcinol (**1**) with benzoic acid (**2**) catalyzed by K-series montmorillonites. Consecutive Fries rearrangement of (**3**) then yields 2,4-dihydroxybenzophenone (**4**). (After Bolognini, M. et al., *Comp. Rendus Chimie,* 7, 143–150, 2004.)

5.16 FRIES REARRANGEMENT

The Fries rearrangement refers to the Lewis acid catalyzed conversion of phenolic esters (and lactams) into hydroxyaryl ketones.

Pitchumani and Pandian (1990) gave an early example of a montmorillonite-catalyzed Fries rearrangement (FR) of phenyl toluene-*p*-sulfonate to give 2- and 4-hydroxyphenyl-*p*-tolyl sulfones, the former product being selectively formed. Similarly, K10 and cation (Na^+, H^+, Al^{3+})-exchanged montmorillonites could promote the FR of phenyl and naphthyl esters. They noted that the corresponding 2-isomers were selectively formed, especially in the case of the bulky naphthyl esters (Venkatachalapathy and Pitchumani 1997).

Békássy et al. (2000) subsequently reported that the (ring) acylation of resorcinol with phenylacetyl in the presence of cation-exchanged K10 montmorillonite produced the corresponding ester, which then underwent a Fries rearrangement to yield 1-(2,4-dihydroxyphenyl)-2-phenylethanone. Using the K-series of montmorillonites as catalysts, Bolognini et al. (2004) similarly prepared resorcinol monobenzoate (**3**) by O-benzoylation of resorcinol (**1**) with benzoic acid (**2**) and then obtained 2,4-dihydroxybenzophenone (**4**) through the consecutive FR of the benzoate ester (Scheme 5.14). Besides being correlated with the aluminum content of the minerals, the extent of conversion steadily decreased with the temperature at which the catalysts had previously been heated (calcined). They therefore suggested that catalytic activity was controlled by Brønsted acidity, arising from the dissociation of water molecules associated with Al^{3+} ions (cf. Chapter 2).

The esterification of phenol with benzoic acid and the subsequent FR of the ester to give hydroxybenzophenones (in a one-pot synthesis under solvent-free conditions) could also be achieved using K10-supported $Cs_{2.5}H_{0.5}PW_{12}O_{40}$ (Yadav and George 2008). More recently, Venkatesha et al. (2014, 2015) used montmorillonite that had been treated with *p*-toluenesulfonic acid to promote the O-acylation of *p*-cresol with decanoic acid, followed by the Fries rearrangement of the ester produced (*p*-cresyldecanoate) to the corresponding ketone, namely, 1-(2-hydroxy-5-methylphenyl)decan-1-one. The yield of ketone was related to the surface acidity of the catalyst as well as the accessibility of the *micropores*—in reality, mesopores—to the reactants.

5.17 HANTZSCH DIHYDROPYRIDINE SYNTHESIS

This reaction refers to a multicomponent condensation of an aldehyde, a β-keto ester, and a nitrogen donor, such as ammonia.

Zonouz and coworkers (Zonouz and Hosseini 2008; Zonouz and Sahranavard 2010), for example, obtain 1,4-dihydropyridine derivatives (**3**) in high yields by the Hantzsch synthesis of various aryl aldehydes (**1**) with ethyl acetoacetate (**2**) and ammonium acetate (or ammonia) in ethanol (or water), according to Scheme 5.15, using K10 montmorillonite as the catalyst. K10-supported Ni^0 nanoparticles are similarly effective in catalyzing the solvent-free synthesis of Hantzsch polyhydroquinoline from the condensation of aldehydes, dimedone, ethyl acetoacetate, and ammonium acetate at room temperature (Saikia et al. 2012). Earlier, Maquestiau et al. (1991) reported that K10-supported cupric nitrate or *claycop* (cf. Chapter 4) was an efficient catalyst for the aromatization of Hantzsch 1,4-dihydropyridines under ultrasonic irradiation.

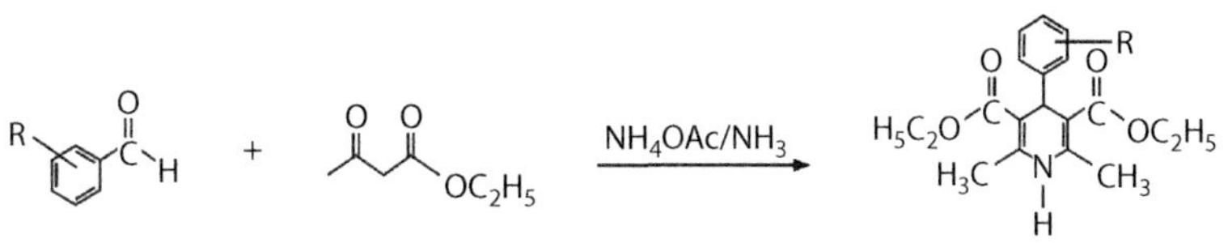

SCHEME 5.15 Hantzsch synthesis of 1,4-dihydropyridine derivatives (**3**) by reacting aryl aldehydes (**1**) with ethyl acetoacetate (**2**) in the presence of ammonium acetate (or ammonia), catalyzed by K10 montmorillonite. (After Zonouz, A.M. and Hosseini, S.B., *Synthetic Commun.*, 38, 290–296, 2008; Zonouz, A.M. and Sahranavard, N., *E-J. Chem.*, 7, S372–S376, 2010.)

5.18 HECK REACTION

The Heck or Mizoroki–Heck reaction refers to a palladium-catalyzed coupling of an unsaturated halide or triflate, such as an aryl, benzyl, vinyl compound, with an alkene to give a substituted alkene.

Using a mixed Pd^{2+}/Cu^{2+}-exchanged K10 montmorillonite as a catalyst, Ramchandani et al. (1997), for example, obtained (*E*)-cinnamates by the Heck reaction of aryl halides with acrylates and (*E*)-stilbenes through the coupling of aryl halides with styrenes. Subsequently, Varma et al. (1999) used K10-intercalated palladium chloride and tetraphenylphosphonium bromide to synthesize *trans*-stilbenes from various aryl halides and styrene. Similarly, Waterlot et al. (2000) were able to obtain methyl cinnamates (without the formation of stilbenes) by the C–C coupling of substituted anilines with methyl acrylate.

Sepiolite-supported $PdCl_2$ can catalyze the Heck reaction of halobenzenes with styrene (Corma et al. 2004). More recently, Firouzabadi et al. (2011) reported that Pd^0 nanoparticles, supported on an aminopropyl-functionalized clay, could effectively promote the Mizoroki–Heck (MH) reaction of iodo- and bromo-arenes with *n*-butyl acrylate and styrene. Similarly, Pd^0 nanoparticles (<10 nm diameter), impregnated into the mesopores of an acid-treated montmorillonite, can promote the Heck vinylation of aryl halides with alkenes (Borah and Dutta 2013). An example of this is shown in Scheme 5.16 for the reaction of iodobenzene (**1**) with methyl acrylate (**2**) to yield methyl cinnamate (**3**) with high *trans* selectivity. The 92% isolated yield of **3**, obtained during the first run, decreases to only 86% by the third run (Dutta et al. 2015). The same reaction can also be catalyzed by Pd^0 nanoparticles immobilized on sepiolite (Tao et al. 2009). Another promising approach is to incorporate the Pd^0 nanoparticles into the interlayer space of a montmorillonite-chitosan complex (Theng 2012), giving both reactants and products easy access to the catalyst (Zeng et al. 2016).

Martínez et al. (2014) used the ionic liquid, 1-butyl-3-methylimidazolium hexafluorophosphate ([bmim][PF_6]), to immobilize palladium nanoparticles on Laponite, obtaining an efficient catalyst for the MH cross-coupling of aryl halides with olefins under solvent-free conditions. In the microwave-assisted reaction between iodobenzene and butyl acrylate, for example, the catalyst could be reused up to 50 times without significant loss of activity (Martínez et al. 2015a, 2015b). Poyatos et al. (2003) have also reported that bis-carbene-pincer complexes of palladium, immobilized on

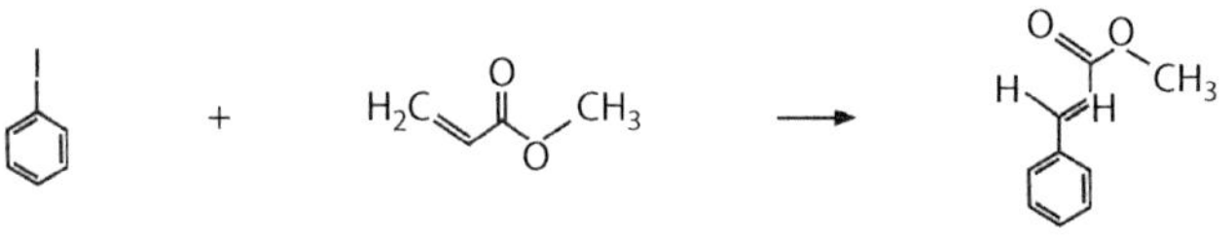

SCHEME 5.16 Mizoroki–Heck reaction of iodobenzene (**1**) with methyl acrylate (**2**) to yield methyl cinnamate (**3**), catalyzed by acid-activated montmorillonite impregnated with Pd^0 nanoparticles. (After Dutta, D.K. et al., *Catal. Rev.: Sci. Eng.*, 57, 257–305, 2015.)

K10 montmorillonite, can serve as a recyclable catalyst for the Heck reaction. The catalyst formed by intercalating an ammonium-tagged oxime carbapalladacycle into montmorillonite can similarly retain its activity over many reaction cycles (Singh et al. 2011). More recently, Firouzabadi et al. (2014) obtained a stable catalyst for the MH reaction by incorporating palladium nanoparticles into a clay-associated phosphinite-functionalized ionic liquid.

5.19 KNOEVENAGEL CONDENSATION

The Knoevenagel condensation refers to the coupling of a carbonyl compound to an activated methylene compound followed by the elimination of water to yield an α,β-unsaturated ketone, and it is commonly catalyzed by amines.

Early on, Rao and Choudary (1991) used a montmorillonite, modified with silylpropylethylene diamine, as a Knoevenagel condensation catalyst. Delgado et al. (1995) subsequently found that a natural (montmorillonite-rich) bentonite could catalyze the Knoevenagel condensation (KC) of aromatic aldehydes with diethyl malonate to yield benzylidenemalonate derivatives when assisted by infrared irradiation. Similarly, Obrador et al. (1998) used a commercial bentonite-based bleaching earth (Tonsil Actisil FF) and infrared irradiation to promote the solvent-free KC of aromatic aldehydes with activated methylene compounds. More recently, Kamble et al. (2013) used a bleaching earth (pH 12.5) as a catalyst, and polyethylene glycol-400 as a solvent, for the KC of aromatic aldehydes with 4-chlorophenyl acetonitrile (followed by the Michael addition of pyrazolone to the condensation product) to yield pyranopyrazole derivatives.

KSF montmorillonite was used by Bigi et al. (1999a) to catalyze the KC of salicylic aldehydes with malonic acid to give coumarin-3-carboxylic acids in >90% yield and selectivity. Similarly, Moussaoui and Salem (2007) were able to synthesize coumarin-3-carboxylic esters in high yields by the KC of aldehydes with various active methylene compounds in the presence of KSF montmorillonite. In investigating the KSF-catalyzed, microwave-assisted, and solvent-free cross-aldol condensation of aromatic aldehydes with ketones, Rocchi et al. (2014) observed the involvement of a Knoevenagel reaction between benzaldehyde and acetylacetone. More recently, Krishnakumar et al. (2015) obtained 4-heteroarylidene-N-arylhomophthalimides by refluxing various heterocyclic aldehydes and active methylene compounds in ethanol in the presence of KSF montmorillonite.

Ramesh and Raghunathan (2009) reported that montmorillonite K10 could promote the synthesis of polycyclic pyrano[2,3,4-kl]xanthene derivatives by the KC of 2-(cyclohex-2-enyloxy)benzaldehyde with symmetrical 1,3-diones when assisted by microwave irradiation. Deshmukh et al. (1999) obtained diethyl 2,2,2-trichloroethylidene propanedioate by reacting diethyl malonate with chloral in the presence of K10 montmorillonite and acetic acid. The K10-catalyzed Knoevenagel condensation of ninhydrin with malenonitrile to produce 2-dicyanomethylene indane-1,3-dione was reported by Chakrabarty et al. (2009).

Heravi et al. (1999) also used K10 montmorillonite to promote the solvent-free Knoevenagel condensation of carbonyl compounds with malononitrile or ethylcyanoacetate under microwave irradiation. Assisted by ultrasound, K10-supported $ZnCl_2$ (*clayzic*) is especially efficient in catalyzing the KC of aromatic aldehydes with malononitrile in anhydrous methanol to give arylmethylene malononitriles in 87%–98% yield within a few minutes (Li et al. 2004). Also noteworthy are the efficiency and reusability of *amine-functionalized*—more precisely, silylated—K10 montmorillonite in promoting the solvent-free Knoevenagel reaction of benzaldehyde with diethyl malonate to yield cinnamic acid (Varadwaj et al. 2013a).

We should also mention that layer silicates other than montmorillonites can catalyze or act as a support for reagents that can promote the Knoevenagel condensation reaction. Martín-Aranda et al. (2005), for example, reported that Li^+- and Cs^+-exchanged saponites could promote the condensation of benzaldehyde (**1**) with malononitrile (**2**) under ultrasonic irradiation to afford the corresponding α,β-unsaturated benzylidene malononitrile (**3**), according to Scheme 5.17. Similar condensation reactions can be catalyzed by kaolinite (Heravi et al. 2009) as well as by zirconium

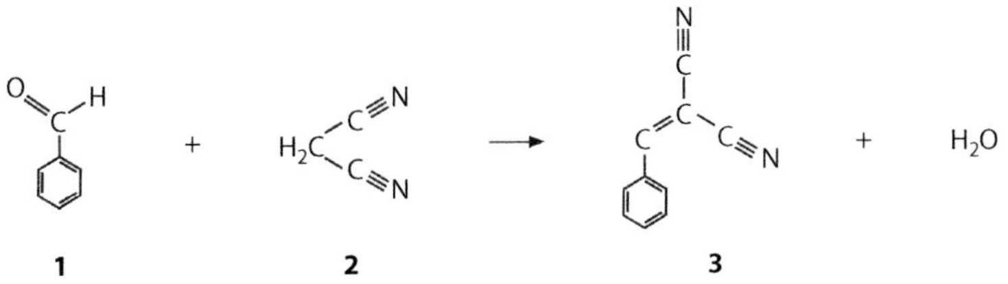

SCHEME 5.17 Knoevenagel condensation of benzaldehyde (**1**) with malononitrile (**2**) to afford the corresponding α,β-unsaturated benzylidene malononitrile (**3**) in the presence of Li$^+$- and Cs$^+$-exchanged saponites under ultrasonic irradiation. (After Martín-Aranda, R.M. et al., *J. Chem. Technol. Biotechnol.*, 80, 234–238, 2005.)

oxide nanoparticles, deposited on the external surface of sepiolite fibers (Letaïef et al. 2011). More recently, Sadjadi et al. (2017) used dodecatungstophosphoric acid, supported on silylated halloysite nanotubes, to promote the synthesis of pyrazolopyranopyrimidine derivatives via the four-component domino reaction of barbituric acid, hydrazine hydrate, ethyl acetoacetate, and benzaldehyde, involving a Knoevenagel condensation step.

5.20 MANNICH REACTION

The Mannich reaction describes the three-component amino alkylation from amine, formaldehyde, and a compound with an acidic methylene group. The final product, a β-amino-carbonyl compound, is referred to as a Mannich base.

Singh et al. (2009) reported the formation of isatin-3-thiosemicarbazones and *N*-Mannich bases by reacting isatins with thiosemicarbazide and their thiosemicarbazone with a secondary amine in the presence of K10 montmorillonite under microwave irradiation. The K10-catalyzed synthesis of various Mannich bases has been summarized by Kaur and Kishore (2012).

Using phosphomolybdic acid, impregnated into a sulfate-grafted Zr-PILC, Samantaray et al. (2011) successfully synthesized β-amino-carbonyl compounds through the Mannich reaction in aqueous media. A sulfonated polystyrene-montmorillonite nanocomposite was similarly efficient in promoting the solvent-free Mannich reaction of anilines, aromatic aldehydes, and cyclohexanone, giving the corresponding β-amino-carbonyl compounds with high (85%–95%) yields and diastereoselectivity (Massah et al. 2012). More recently, Gómez-Sanz et al. (2017) compared the activity of a saponite and some commercial acid-treated montmorillonites (K10, K30, KSF) in catalyzing the synthesis of β-amino ketones. Because of its high porosity and surface acidity, the K10 sample was the most efficient catalyst for the Mannich reaction between preformed imine, *N*-benzylidene aniline, and cyclohexanone.

5.21 MARKOVNIKOV ADDITION RULE

The Markovnikov addition rule states that when a hydrogen halide (HX) adds to an asymmetric alkene, the hydrogen (H) part attaches to the carbon atom containing the least number of alkyl substituents, while the halide (X) part attaches to the carbon atom with more alkyl substituents. In the anti-Markovnikov reaction, on the other hand, the halide adds to the less substituted carbon atom.

Some time ago, Delaude and Laszlo (1991) made the interesting observation that in the absence of a (solid) catalyst, HCl reacted with 1-methylcyclohexene to give the anti-Markovnikov adduct, *trans* 1-chloro-2-methyl cyclohexane, as the major product. In the presence of K10 montmorillonite, however, the reaction was highly regioselective, giving the Markovnikov adduct, 1-chloro-1-methylcyclohexane, in 90% isolated yield. Similarly, Kanagasabapathy et al. (2001) have reported that in the presence of K10, thiols (**1**) can add regioselectively to olefins, such as styrene (**2**), in a Markovnikov manner to yield the corresponding thioesters (**3**). On the other hand, the respective

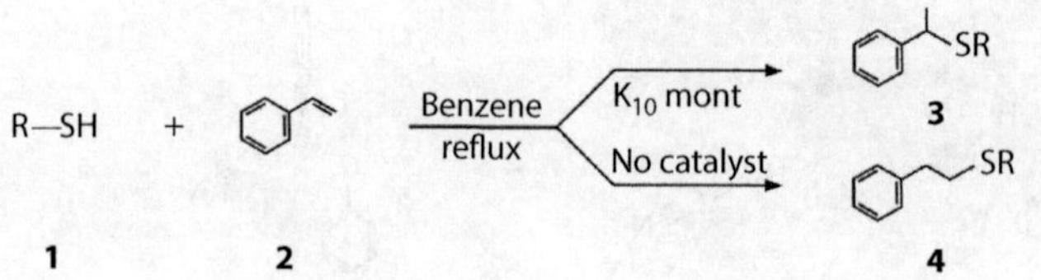

SCHEME 5.18 Markovnikov addition of thiols (1) to styrene (2) to yield the corresponding thioesters (3), catalyzed by K10 montmorillonite. In the absence of a catalyst, the anti-Markovnikov product (4) obtains. (After Kanagasabapathy, S. et al., *Tetrahedron Lett.*, 42, 3791–3794, 2001.)

anti-Markovnikov products (4) are formed in the absence of a catalyst, according to Scheme 5.18 where R denotes phenyl or benzoyl. Further, when the aforementioned esters are treated with Lawesson's reagent, the corresponding dithiocarboxylic esters are formed in high yield. More recently, Lanke et al. (2012) reacted styrene with arenes in the presence of K10 montmorillonite to afford the corresponding 1,1-diarylalkanes as the Markovnikov adducts.

Curiously, Kannan and Pitchumani (1997) found that aliphatic thiols (RSH) where R = butane, hexane, octane, could add to styrene by an anti-Markovnikov reaction when the reactants were heated in a water bath (95°C) for 1 h in the presence of K10 montmorillonite and its cation-exchanged forms. In explanation, they suggested that the reaction involved the formation of RS free radicals. Using K10 montmorillonite as a catalyst, Joseph et al. (2006) similarly observed that the reaction of anilines and substituted anilines with α,β-ethylenic compounds yielded the corresponding anti-Markovnikov adducts with almost 100% selectivity. Thus, aniline reacted with ethyl acrylate (in toluene at 368 K) to give exclusively the anti-Markovnikov mono-addition product, *N*-[2-ethoxycarbonyl)ethyl] aniline. On the other hand, only the Markovnikov addition products (e.g., phenyl-(1-phenylethylidene) amine)) were obtained by Shandbhag and coworkers (Shanbhag and Halligudi 2004; Joseph et al. 2005; Shanbhag et al. 2006) during the hydroamination of alkynes (e.g., phenyl acetylene) with aromatic amines (e.g., aniline) in the presence of Cu^{2+}- and Zn^{2+}-exchanged K10 montmorillonite (cf. Scheme 6.5). They suggested that the exchangeable cations acted as Lewis acid sites in concert with the inherent Brønsted acidity of K10.

5.22 MICHAEL ADDITION

The Michael addition describes the (conjugate) addition of a carbon nucleophile to an α,β-unsaturated carbonyl compound.

The use of Al^{3+}-montmorillonite as an efficient catalyst for the Michael addition (MA) of silyl ketene acetals to α,β-unsaturated esters (enoates) has been reported by Kawai et al. (1987, 1988b), and reviewed by Izumi and Onaka (1992). Iqbal et al. (1988) and Poupaert et al. (1999) used K10 montmorillonite to promote the MA of indoles with α,β-unsaturated carbonyl compounds, such as methylvinyl ketone. Under microwave or ultrasound irradiation, even Li^+- and Cs^+-exchanged montmorillonites can catalyze the solvent-free Michael addition of imidazole (1) to ethyl acrylate (2) to yield the corresponding *N*-substituted imidazole (3), according to Scheme 5.19 (Martín-Aranda et al. 1997, 2002).

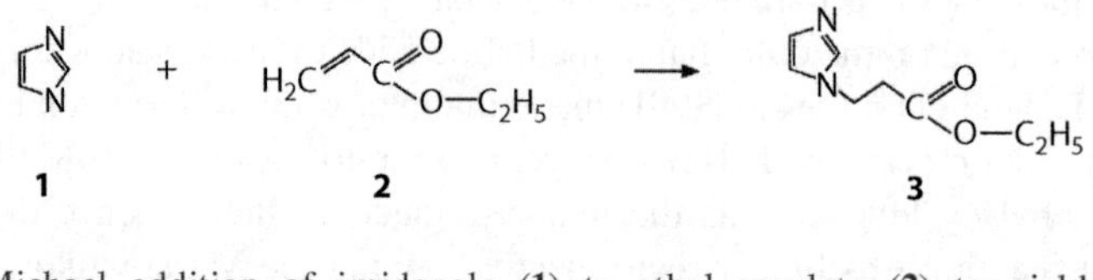

SCHEME 5.19 Michael addition of imidazole (1) to ethyl acrylate (2) to yield the corresponding *N*-substituted imidazole (3) in the presence of Li^+- and Cs^+-exchanged montmorillonites under solvent-free conditions, aided by microwave or ultrasound irradiation. (After Martín-Aranda, R.M. et al., *J. Mol. Catal. A: Chem.*, 124, 115–121, 1997; Martín-Aranda, R.M. et al., *Catalysis Lett.*, 84, 201–204, 2002.)

Soriente et al. (1999) have reported that K10 and KSF montmorillonites can effectively catalyze the conjugate addition and alkoxyalkylation of 1,3-dicarbonyl compounds, such as ethyl benzoyl acetate, to an α,β-unsaturated carbonyl compound. K10 montmorillonite can also promote the solvent-free, one-pot synthesis of 3-(furan-2-yl)-4H-chromen-4-ones from 1-(2-hydroxyphenyl) butane-1,3-diones and 2,5-dimethoxy-2,5-dihydrofuran through an alkoxyalkylation reaction (Han et al. 2016). These commercially available clay catalysts were also used by Sharma et al. (2007) for the chemoselective carbon-sulfur bond formation via the Michael addition of thiols to α,β-unsaturated carbonyl compounds. Similarly, Cu^{2+}- and Sc^{3+}-montmorillonites can efficiently promote the Michael reaction of 1,3-dicarbonyls with enones as well as of various β-keto esters with 3-buten-2-one either in organic solvents or under solvent-free conditions (Kawabata et al. 2003, 2005). Interestingly, Shimizu et al. (2003, 2005) observed that Fe^{3+}-fluorotetrasilicic mica was more active than Fe^{3+}-montmorillonite in catalyzing the solvent-free Michael reaction of β-ketoesters with vinyl ketones.

Using an unspecified clay mineral as catalyst, Shaikh et al. (2001) were able to carry out a Michael type addition of aliphatic amines to α,β-ethylenic compounds but curiously enough, they failed to do the same with aromatic amines. Not long after, Chakrabarty and Sarkar (2002) reported that K10 montmorillonite could catalyze the dry tandem addition-elimination (Michael) reaction of indoles with 3-formylindole to yield tri-indolylmethanes. This catalyst was similarly efficient in promoting the solvent-free Michael addition of indoles (and pyrrole) to nitroolefins (Chen and Li 2009; An et al. 2010). Nikpassand et al. (2010) obtained 3,3-di(indolyl)indolin-2-one by refluxing isatin and indole in ethanol in the presence of KSF montmorillonite. Singh et al. (2006) also found that the Fe^{3+}-exchanged form of K10 was remarkably active in catalyzing the conjugate addition of indoles to α,β-unsaturated carbonyls. De Paolis et al. (2009) used K10 montmorillonite and microwave irradiation to promote the Michael type addition of aniline derivatives to cinnamaldehyde, followed by cyclization and oxidation to yield quinolines.

5.23 MUKAIYAMA ALDOL REACTION

The Mukaiyama aldol reaction refers to the Lewis acid-catalyzed condensation of silyl enol ethers and aldehydes.

Using Al^{3+}-montmorillonite as a catalyst, Kawai et al. (1986, 1988a) condensed silyl enol ethers, such as 1-phenyl silyl enol ether (**1**) and aldehydes, such as benzaldehyde (**2**) to obtain the corresponding adducts (**3**) where R = $Si(CH_3)_3$ through the Mukaiyama aldol reaction (Scheme 5.20). The same clay was used by Onaka et al. (1987) to promote the reaction of silyl ketene acetals with carbonyl compounds. Subsequently, Loh and Li (1999) reported that K10 montmorillonite was equally efficient in catalyzing a series of similar reactions, while Takehira et al. (2014) found Sn^{4+}-exchanged montmorillonite to be an excellent catalyst for the Mukaiyama aldol reaction of congested ketones with silicon enolates to give the corresponding silylated aldol products.

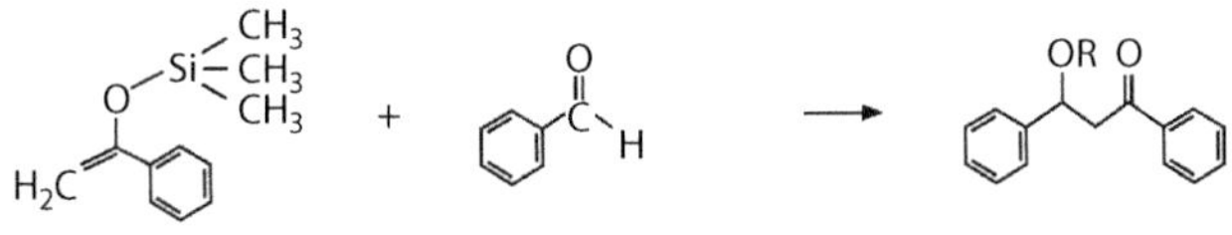

SCHEME 5.20 Mukaiyama aldol reaction of 1-phenyl silyl enol ether (**1**) and benzaldehyde (**2**) to yield the corresponding adducts (**3**), catalyzed by Al^{3+}-montmorillonite. (After Kawai, M. et al., *Chemistry Lett.*, 1581–1584, 1986; Kawai, M. et al., *Bull. Chem. Soc. Japan*, 61, 1237–1245, 1988a.)

Using the chiral bis(oxazoline)-copper complexes, immobilized in the interlayers of Laponite, Fraile and coworkers (Fabra et al. 2008; Fraile et al. 2009, 2013) made the interesting observation that the stereoselectivity and stereochemical course of the Mukaiyama aldol reaction were markedly different from those found under homogeneous conditions. Thus, in the reaction of 2-(trimethylsilyloxy)furan with α-ketoesters, a large excess of one enantiomer of the aldol product was obtained. Similarly, in the reaction of 2-(trimethylsilyloxy)furan with electron-deficient alkenes, such as diethyl benzylidenemalonate, the major diastereomer has a *syn* configuration whereas the *anti* isomer is formed under homogeneous (solution) conditions.

5.24 NICHOLAS REACTION

The Nicholas reaction refers to the capture of a dicobalt hexacarbonyl-stabilized propargylic cation by a nucleophile to yield the propargylated (alkylated alkyne) product following oxidative demetallation.

The work by Crisóstomo et al. (2005) on the synthesis of 2-ethynyl-tetrahydrofuran from 6-(tetrahydro-2*H*-pyran-2-ylox)hex-1-yn-3-ol in the presence of K10 montmorillonite provides a rare example of a clay-catalyzed Nicholas reaction. The reaction involved adding the clay catalyst to a solution of the $Co_2(CO)_6$-alkyne complex in dichloromethane, filtering, and decomplexing the alkyne with ceric ammonium nitrate. The synthesis of α,α′-disubstituted linear ethers by an intermolecular Nicholas reaction involving the K10-induced isomerization of $Co_2(CO)_6$ cycloalkyne ether complexes has been reported by Ortega et al. (2009).

5.25 OPPENAUER OXIDATION

The Oppenauer oxidation refers to the catalyzed oxidation of secondary alcohols by an alkoxide.

Here we might mention the work by Bigi et al. (2000) who observed that the reaction between phenols and formaldehyde over an interlayer KSF montmorillonite-triethylamine complex yielded 2-hydroxy benzaldehydes, which then underwent an Oppenhauer oxidation with a further portion of formaldehyde to give substituted salicylic aldehydes. On the other hand, K10 montmorillonite with a lower iron content than KSF only yielded 2-hydroxy benzaldehyde, suggesting that structural iron was involved in the oxidation step.

5.26 PAAL–KNORR SYNTHESIS

The Paal–Knorr (pyrrole) synthesis describes the acid-catalyzed formation of pyrroles from dicarbonyl compounds and primary amines.

For example, Texier-Boullet et al. (1986) used K10 montmorillonite to synthesize pyrroles (and pyrazoles) from primary amines (or hydrazine) and 1,4- or 1,3-diketones under solvent-free conditions. Banik et al. (2004) have described the synthesis of substituted pyrroles by a modified Paal–Knorr method in the presence of KSF montmorillonite. Similarly, Abid et al. (2006) used K10 montmorillonite and microwave irradiation to catalyze the solvent-free synthesis of substituted pyrroles, obtaining quantitative yields with 100% selectivity.

Song et al. (2005) reacted 2,5-hexanedione (**1**) with amines (**2**) in dichloromethane in the presence of various cation-exchanged montmorillonites (Mt) to obtain the corresponding pyrrole derivatives (**3**) in 92%–96% isolated yields, according to Scheme 5.21, where R denotes *n*-alkyl, benzyl, or phenyl. Yields fell in the order Fe^{3+}-Mt > Zn^{2+}-Mt > Co^{2+}-Mt > Cu^{2+}-Mt > K10-Mt. Since the ionic potential of the exchangeable cations decreases in the same order, the reaction appears to involve Brønsted acid sites on the catalyst (cf. Equation 2.20).

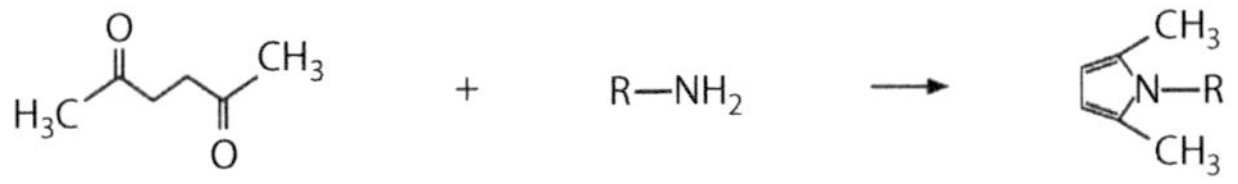

SCHEME 5.21 Paal–Knorr synthesis of pyrrole derivatives (**3**) by reacting 2,5-hexanedione (**1**) with amines (**2**) in dichloromethane, in the presence of polyvalent cation-exchanged montmorillonites. (After Song, G. et al., *Synthetic Commun.*, 35, 1051–1057, 2005.)

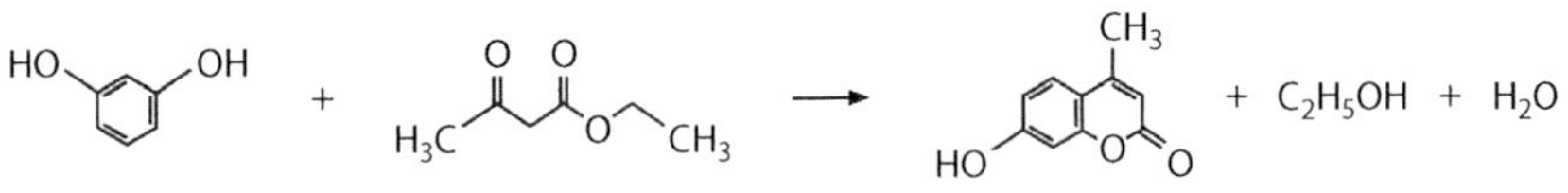

SCHEME 5.22 Pechmann condensation of resorcinol (**1**) with ethyl acetoacetate (**2**) to yield 7-hydroxy-4-methylcoumarin (**3**), catalyzed by K10 or KSF montmorillonite. (After Li, T.-S. et al., *J. Chem. Res.*, (*S*), 38–39, 1998b.)

5.27 PECHMANN CONDENSATION

The Pechmann condensation, or Pechmann coumarin synthesis, refers to the acid-catalyzed formation of coumarins from a phenol and a carboxylic acid or ester containing a β-carbonyl group.

Li et al. (1998b) used K10 or KSF montmorillonite to promote the synthesis of coumarins through the Pechmann condensation by refluxing various phenols and ethyl acetoacetate in toluene. An example is the condensation of resorcinol (**1**) with ethyl acetoacetate (**2**) to yield 7-hydroxy-4-methylcoumarin (**3**) according to Scheme 5.22.

Similarly, Kolancilar (2010) was able to obtain 7-hydroxy-4-methylcoumarin by heating the reactants in the presence of an interlayer K10-polyaniline complex, while Vijayakumar and Rao (2012) used K10-supported dodecatungstophosphoric acid under solvent-free conditions to effect the same synthesis. Earlier, Frère et al. (2001) reported the Pechmann synthesis of rare aminocoumarins by exposing the reactants, supported on a mixed graphite/K10 montmorillonite, to microwave radiation.

5.28 PRINS REACTION

The Prins (cyclization) reaction refers to the acid-catalyzed addition/condensation of alkenes/alcohols with carbonyl compounds, notably formaldehyde.

Tateiwa et al. (1996) reported that styrene could add to paraformaldehyde or 1,3,5-trioxane (in toluene) to yield 4-aryl-1,3-dioxanes selectively in the presence of various cation-exchanged montmorillonites acting as Brønsted acids. By the same token, Yadav et al. (2001b) found KSF montmorillonite to be an efficient catalyst for the Prins-type cyclization reaction of homoallylic alcohols with aldehydes to give 4-hydroxy-2,6-disubstituted tetrahydropyrans in high yields and diastereoselectivity. Subsequently, Yadav and Jasra (2006) used montmorillonite-supported $ZnCl_2$ to promote the Prins condensation of β-pinene with formaldehyde, formed in situ from paraformaldehyde, to yield nopol.

More recently, Baishya et al. (2013) prepared octahydro-2H-chromen-4-ols with high cis-selectivity through a Prins cyclization of (−)-isopulegol using K10 montmorillonite under solvent-free conditions. Timofeeva et al. (2015) obtained the same product by reacting (−)-isopulegol with vanillin (in toluene at 35°C) in the presence of acid-treated calcium-rich montmorillonite. They proposed that the cyclization reaction was controlled by the Brønsted acidity and microporosity of the catalyst. In this context, we might also mention the synthesis of tetrahydro-2-(4-nitrophenyl)-4-phenyl-2H-pyran by a tandem Prins–Friedel–Crafts multicomponent reaction of p-nitrobenzaldehyde with methanol, 3-buten-1-ol, and benzene in the presence of K10 montmorillonite (Dintzner et al. 2012).

5.29 RITTER REACTION

The reaction of alcohols (or alkenes) with nitriles in the presence of a strong Brønsted acid, to give amides, is known as the Ritter reaction. The work by Kumar et al. (1999) using KSF montmorillonite ($H_o < -8.2$) provides a rare example of a clay-catalyzed Ritter reaction. They were able to synthesize a variety of amides by reacting benzyl, allyl, and tertiary alcohols with various nitriles in the presence of KSF clay under microwave irradiation. One such reaction, shown in Scheme 5.23, is that between 1-phenylethanol (**1**) and 3-ethoxypropionitrile (**2**) to yield the corresponding amide (**3**).

5.30 SAKURAI ALLYLATION REACTION

The Sakurai or Hosomi–Sakurai allylation reaction is a Lewis acid catalyzed addition of allylsilanes to electron-deficient carbonyl compounds (aldehydes, ketones).

Kawabata et al. (2005) used Cu^{2+}-montmorillonite to catalyze the solvent-free Hosomi–Sakurai reaction between benzaldehyde (**1**) and allyltrimethylsilane (**2**), obtaining trimethyl[(1-phenyl-3-butenyl)oxy]silane (**3**) in 98% yield together with 1% of 4-phenyl-1-buten-4-ol (**4**), derived from the hydrolysis of the silane product (Scheme 5.24). Similarly, Dintzner et al. (2009) were able to prepare homoallylic silyl ethers by reacting various aromatic aldehydes with allyltrimethylsilane in dichloromethane, using K10 montmorillonite as catalyst. More recently, Elizarov et al. (2016) found montmorillonite-supported bismuth(III) salts to be highly efficient in catalyzing the Hosomi–Sakurai allylation of aromatic aldehydes.

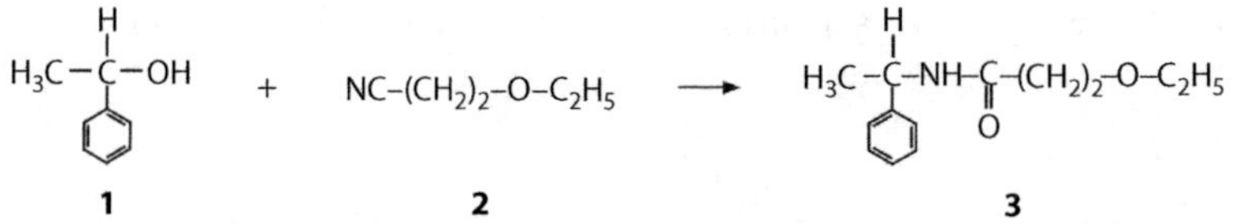

SCHEME 5.23 Ritter reaction between 1-phenylethanol (**1**) and 3-ethoxypropionitrile (**2**) to yield the corresponding amide (**3**) in the presence of KSF montmorillonite under microwave irradiation. (After Kumar, H.M.S. et al., *New J. Chem.*, 23, 955–956, 1999.)

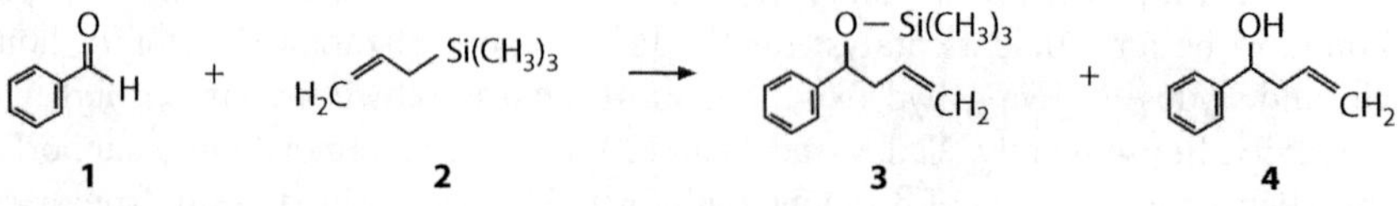

SCHEME 5.24 Solvent-free Hosomi–Sakurai reaction between benzaldehyde (**1**) and allyltrimethylsilane (**2**) to yield trimethyl[(1-phenyl-3-butenyl)oxy]silane (**3**), catalyzed by Cu^{2+}-montmorillonite. Hydrolysis of (**3**) gives rise to a minor amount of 4-phenyl-1-buten-4-ol (**4**). (After Kawabata, T. et al., *Chem.—A Europ. J.*, 11, 288–297, 2005.)

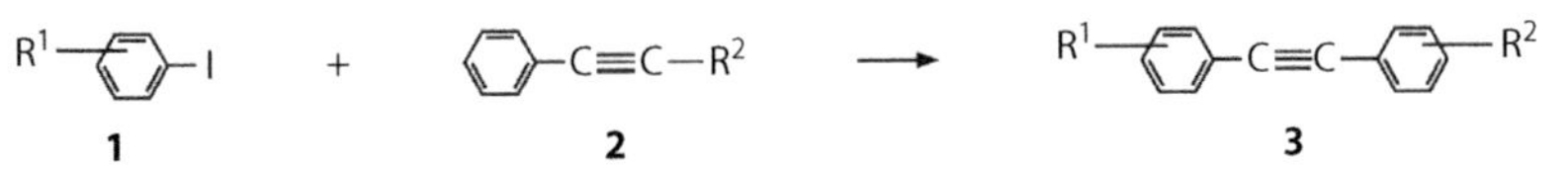

SCHEME 5.25 Sonogashira reaction in an ammonium-based organic liquid between haloarenes (**1**) and aryl acetylenes (**2**) to yield the corresponding products (**3**), catalyzed by an organically modified montmorillonite. (After Singh, V. et al., *J. Mol. Catal. A: Chem.*, 334, 13–19, 2011.)

5.31 SONOGASHIRA REACTION

The Sonogashira reaction refers to the palladium/copper catalyzed cross-coupling of organohalides with terminal alkynes.

Mas-Marzá et al. (2003) intercalated complexes of palladium with tridentate pincer bis-carbene ligands into K10 montmorillonite and bentonite, and used the resultant organopalladium-clays as catalysts for the Sonogashira reaction of aryl halides with terminal acetylenes. The catalysts could be used over several runs without appreciable loss of activity.

Similarly, Singh et al. (2011) reported that the catalyst formed by intercalating an ammonium-tagged oxime carbapalladacycle into montmorillonite could retain its activity over many reaction cycles. Scheme 5.25 shows the use of this catalyst for the Sonogashira reaction between haloarenes (**1**) and aryl acetylenes (**2**) where R^1 = alkoxy, keto, or alkyl, and R^2 = NH_2, alkyl, halogen, or alkoxy, in an ammonium-based ionic liquid, to yield the corresponding products (**3**) in 82%–95% yield. More recently, Xu et al. (2014) used montmorillonite-supported Pd/Cu nanoparticles to promote the Sonogashira cross-coupling reaction.

Similarly, Chavan et al. (2015) reported that palladium, immobilized on K10 montmorillonite that had been grafted with 3-aminopropyl triethoxysilane (APTES), could efficiently catalyze the Sonogashira reaction of aryl and hetero aryl iodides with terminal alkynes to yield the corresponding alkynyl ketones. Chavan et al. (2016) further noted that cyclization of 2-iodophenol with terminal alkynes was dependent on the solvent used; in 1,2-dimethylethane, the formation of the 5-*exo* products (aurones) was favored over the 6-endo products (flavones), while in dimethylformamide the reverse regioselectivity was observed. We might also add that copper and nickel ions, coordinated to the amine groups of 3-aminopropyl trimethoxysilane in K10 montmorillonite, could effectively catalyze the aryl-sulfur coupling reaction of thiophenol with aryl iodide (Varadwaj et al. 2013b).

5.32 STRECKER REACTION

The Strecker reaction, also known as the Strecker amino acid synthesis, refers to the condensation of an aldehyde and an amine to form an α-aminonitrile, which is then hydrolyzed to an amino acid.

Yadav et al. (2004a) synthesized α-aminonitriles in 85%–94% yields by reacting aryl imines, formed *in situ* from aldehydes and amines, with trimethylsilyl cyanide (TMSCN) (**3**) in the presence of KSF montmorillonite. Scheme 5.26 shows an example of such a synthesis, using benzaldehyde (**1**) and aniline (**2**) as reactants, and yielding 2-anilino-2-phenylacetronitrile (**4**) as product. Sometime

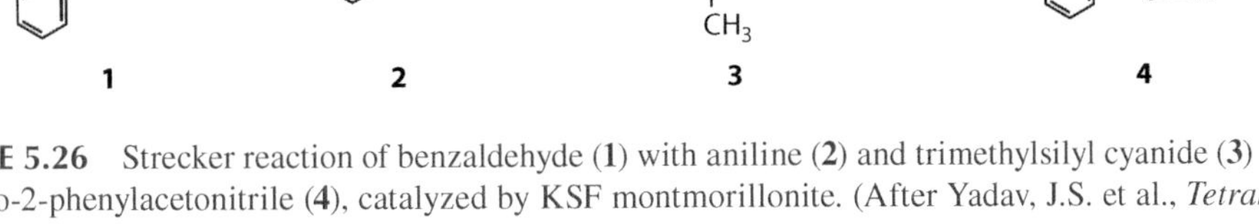

SCHEME 5.26 Strecker reaction of benzaldehyde (**1**) with aniline (**2**) and trimethylsilyl cyanide (**3**) to yield 2-anilino-2-phenylacetonitrile (**4**), catalyzed by KSF montmorillonite. (After Yadav, J.S. et al., *Tetrahedron*, 60, 1767–1771, 2004a.)

later, Wang et al. (2010, 2011a) reported that Sn-exchanged montmorillonite was similarly efficient in catalyzing the formation of α-aminonitriles by simply mixing a variety of aromatic and aliphatic carbonyl compounds (including sterically hindered ketones) with amines and TMSCN. The high activity of the catalyst was ascribed to its having strong Brønsted acid sites. These workers subsequently prepared nitriles by reacting various benzylic and allylic alcohols with trialkylsilyl cyanide in the presence of tin- and titanium-exchanged montmorillonites (Wang et al. 2011a).

5.33 SUZUKI REACTION

The Suzuki reaction, also known as the Suzuki-Miyaura or Suzuki coupling reaction, refers to the cross-coupling of boronic/organoboronic acid with organohalides, catalyzed by a palladium complex.

Shimizu et al. (2002, 2004) used $[Pd(NH_3)_4]^{2+}$-exchanged sepiolite as the catalyst for the Suzuki reaction of 4-bromophenol (**1**) with either phenylboronic acid (**2**) or sodium tetraphenylborate in water (or dimethylformamide), and K/Na carbonate as the base, to yield 4-hydroxybiphenyl (**3**), also known as 4-phenylphenol, according to Scheme 5.27. Strong electrostatic interactions between the $[Pd(NH_3)_4]^{2+}$ complex and sepiolite surface inhibited Pd metal precipitation, giving rise to a stable catalyst and high turnover numbers. Na^+-sepiolite impregnated with $PdCl_2$ was similarly efficient in promoting the cross-coupling reaction of halobenzenes with phenylboronic acid to yield the corresponding biphenyl compounds (Corma et al. 2004). Similarly, Scheuermann et al. (2009) have obtained a biphenyl by reacting 4-nitro bromobenzene with phenylboronic acid in the presence of an organophilic Pd^0-montmorillonite catalyst. In this context, we might mention the facile conversion of arylboronic acids to phenols by aqueous H_2O_2 in the presence of K10 montmorillonite (Sudhakar et al. 2013).

More recently, Firouzabadi et al. (2011, 2014) used palladium nanoparticles supported on an organoclay, while Gogoi et al. (2014) employed K10 montmorillonite-supported palladium triphenylphosphine, to catalyze similar reactions in water. Kaur and Singh (2015) described the use of an interlayer montmorillonite complex with an ammonium-tagged oxime carbapalladacycle to catalyze the Suzuki reaction in ionic liquid media. An interesting variant was the introduction of Pd^{2+} (by cation exchange) into a montmorillonite complex with L-cystine to yield an effective catalyst for the Suzuki reaction (Li et al. 2017).

5.34 WACKER OXIDATION

The Wacker oxidation refers to the palladium-catalyzed conversion of terminal alkenes to ketones in the presence of a copper salt (as a cocatalyst) under aerobic conditions. The initial process, named after a company (Wacker Chemie), described the oxidation of ethylene to acetaldehyde.

A rare example of a clay-catalyzed Wacker oxidation was provided by Mitsudome et al. (2006) who used Pd^{2+}-exchanged montmorillonite to promote the transformation of terminal olefins to the corresponding methyl ketone. Thus, using $CuCl_2$ as cocatalyst and N,N-dimethylacetamide as a solvent (at 80°C), 1-decene was converted into 2-decanone in 85% yield with 99% selectivity.

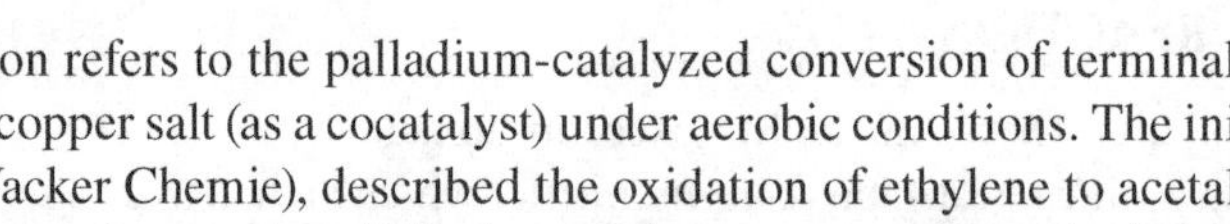

SCHEME 5.27 Suzuki (coupling) reaction of 4-bromophenol (**1**) with phenylboronic acid (**2**) to yield 4-hydroxybiphenyl (**3**), using $[Pd(NH_3)_4]^{2+}$-exchanged sepiolite as catalyst. (After Shimizu, K. et al., *Tetrahedron Lett.*, 43, 5653–5655, 2002; Shimizu, K. et al., *J. Catal.*, 227, 202–209.)

Earlier, Morikawa (1993) reported that Pd^{2+}-fluorotetrasilicic mica (TSM; cf. Chapter 1) was highly efficient in catalyzing the oxidation of propylene, while the same reaction over a mixed Cu^{2+}/Pd^{2+}-exchanged TSM yielded acetone selectively.

5.35 WITTIG REACTION

The Wittig reaction refers to the olefination of an aldehyde or ketone using a triphenyl phosphonium ylide to yield an alkene and triphenylphosphine oxide.

Upadhyay and Srivastava (2015) used montmorillonite-intercalated Ru^0 nanoparticles to promote the solvent-free hydrogenation of alkenes. At the same time, they tested the ability of the clay to promote the synthesis of dehydrobrittonin A, the intermediate of brittonin A, via a Wittig-type reaction.

REFERENCES

Abid, M., A. Spaeth, and B. Török. 2006. Solvent-free solid acid-catalyzed electrophilic annelations: A new green approach for the synthesis of substituted five-membered *N*-heterocycles. *Advanced Synthesis & Catalysis* 348: 2191–2196.

Adams, J.M. and T.V. Clapp. 1986. Reactions of the conjugated dienes butadiene and isoprene alone and with methanol over ion-exchanged montmorillonites. *Clays and Clay Minerals* 34: 287–294.

Adams, J.M., K. Martin, and R.W. McCabe. 1987a. Catalysis of Diels–Alder cycloaddition reactions by ion-exchanged montmorillonites. In *Proceedings of the International Clay Conference, Denver, 1985* (Eds.) L.G. Schultz, H. van Olphen, and F.A. Mumpton, pp. 324–328. Bloomington, IN: The Clay Minerals Society.

Adams, J.M., K. Martin, and R.W. McCabe. 1987b. Clays as selective catalysts in organic synthesis. *Journal of Inclusion Phenomena* 5: 663–674.

Adams, J.M., S.E. Davies, S.H. Graham, and J.M. Thomas. 1982. Catalyzed reactions of organic molecules at clay surfaces: Ester breakdown, dimerizations and lactonizations. *Journal of Catalysis* 78: 197–208.

Adams, J.M., S. Dyer, K. Martin, W.A. Matear, and R.W. McCabe. 1994. Diels–Alder reactions catalysed by cation-exchanged clay minerals. *Journal of the Chemical Society, Perkin Transactions 1* 1994(6): 761–765.

Ahmad, N., Z. Ali, N. Ali, N. Shahzad, F. Hussain, and S.M. Abbas. 2015. Cerium modified pillared montmorillonite supported cobalt catalysts for Fischer–Tropsch synthesis. *Journal of the Chemical Society of Pakistan* 37: 687–695.

Ahmad, N., S.T. Hussain, B. Muhammad, N. Ali, S.M. Abbas, and Z. Ali. 2013. Zr-pillared montmorillonite supported cobalt nanoparticles for Fischer–Tropsch synthesis. *Progress in Natural Science: Materials International* 23: 374–381.

Akiyama, T., K. Matsuda, and K. Fuchibe. 2002. Montmorillonite K10-catalyzed aza Diels–Alder reaction of Danishefsky's diene with aldimines, generated in situ from aliphatic aldehydes and amine, in aqueous media. *Synlett* 1898–1900.

An, L.-T., L.-L. Zhang, J.-P. Zou, and G.-L. Zhang. 2010. Montmorillonite K10: Catalyst for Friedel–Crafts alkylation of indoles and pyrrole with nitroalkenes under solventless conditions. *Synthetic Communications* 40: 1978–1984.

Arienti, A., F. Bigi, R. Maggi et al. 1997. Regioselective electrophilic alkylation of anilines with phenylacetylene in the presence of montmorillonite KSF. *Tetrahedron* 53: 3795–3804.

Avalos, M., R. Babiano, J.L. Bravo, P. Cintas, J.L. Jiménez, and J.C. Palacios. 1998a. Clay-catalyzed solventless addition reactions of furan with α,β-unsaturated carbonyl compounds. *Tetrahedron Letters* 39: 9301–9304.

Avalos, M., R. Babiano, J.L. Bravo et al. 1998b. Cycloadditions with clays and alumina without solvents. *Tetrahedron Letters* 39: 2013–2016.

Azimi, S.C. and E. Abbaspour-Gilandeh. 2014. Li^+ modified nanoporous Na^+-montmorillonite an efficient novel catalytic system for synthesis of quinolines. *Journal of Nanostructures* 4: 335–346.

Baishya, G., B. Sarmah, and N. Hazarika. 2013. An environmentally benign synthesis of octahydro-2*H*-chromen-4-ols via modified montmorillonite K10 catalyzed Prins cyclization reaction. *Synlett* 24: 1137–1141.

Ballantine, J.A., W. Jones, J.H. Purnell, D.T.B. Tennakoon, and J.M. Thomas. 1985. The influence of interlayer water on clay catalysis: Interlamellar conversion of 2-methylpropene. *Chemistry Letters* 14: 763–766.

Ballini, R., A. Palmieri, M. Petrini, and E. Torregiani. 2006. Solventless clay-promoted Friedel–Crafts reaction of indoles with α-amido sulfones: Unexpected synthesis of 3-(1-arylsulfonylalkyl) indoles. *Organic Letters* 8: 4093–4096.

Balogh, M. and P. Laszlo. 1993. *Organic Chemistry Using Clays*. Berlin, Germany: Springer-Verlag.

Banik, B.K., S. Samajdar, and I. Banik. 2004. Simple synthesis of substituted pyrroles. *Journal of Organic Chemistry* 69: 213–216.

Barlow, S.J., T.W. Bastock, J.H. Clark, and S.R. Cullen. 1993. Explanation of an unusual substituent effect in the benzylation of anisole and identification of the origin of the active sites in clayzic. *Tetrahedron Letters* 34: 3339–3342.

Basavaiah, D., A.J. Rao, and T. Satyanarayana. 2003. Recent advances in the Baylis–Hillman reaction and applications. *Chemical Reviews* 103: 811–891.

Békássy, S., J. Farkas, B. Ágai, and F. Figueras. 2000. Selectivity of *C*- versus *O*-acylation of diphenols by clay catalysts. I. Acylation of resorcinol with phenylacetyl chloride. *Topics in Catalysis* 13: 287–290.

Belaroui, L.S. and A. Bengueddach. 2012. Study of the catalytic activity of Al-Fe pillared clays in the Baeyer–Villiger oxidation. *Clay Minerals* 47: 275–284.

Belaroui, L.S., A.B. Sorokin, F. Figueras, A. Bengueddach, and J-M.M. Millet. 2010. Comparative Baeyer-Villiger oxidation of cyclohexanone on Fe-pillared clays and iron tetrasulfophthalocyanine covalently supported on silica. *Comptes Rendus Chimie* 13: 466–472.

Bigi, F., S. Carloni, B. Frullanti, R. Maggi, and G. Sartori. 1999b. A revision of the Biginelli reaction under solid acid catalysis: Solvent-free synthesis of dihydropyrimidines over montmorillonite KSF. *Tetrahedron Letters* 40: 3465–3468.

Bigi, F., L. Chesini, R. Maggi, and G. Sartori. 1999a. Montmorillonite KSF as an inorganic, water-stable, and reusable catalyst for the Knoevenagel synthesis of coumarin-3-carboxylic acids. *Journal of Organic Chemistry* 64: 1033–1035.

Bigi, F., M.L. Conforti, R. Maggi, and G. Sartori. 2000. Trialkylamine controlled phenol-formaldehyde reaction over clay catalysts: Selective and environmentally benign synthesis of salicylic aldehydes. *Tetrahedron* 56: 2709–2712.

Bolognini, M., F. Cavani, M. Cimini et al. 2004. An environmentally friendly synthesis of 2,4-dydroxybenzophenone by the single-step O-mono-benzoylation of 1,3-dihydroxybenzene (resorcinol) and Fries rearrangement of intermediate resorcinol monobenzoate: The activity of acid-treated montmorillonite clay catalysts. *Comptes Rendus Chimie* 7: 143–150.

Borah, B.J. and D.K. Dutta. 2013. In situ stabilization of Pd^0-nanoparticles into the nanopores of modified montmorillonite: Efficient heterogeneous catalysts for Heck and Sonogashira coupling reactions. *Journal of Molecular Catalysis A: Chemical* 366: 202–209.

Bosch, A.I., P. de la Cruz, E. Diez-Barra, A. Loupy, and F. Langa. 1995. Microwave assisted Beckmann rearrangement of ketoximes in dry media. *Synlett* 1995: 1259–1260.

Brochette, S., G. Descotes, A. Bouchu, Y. Queneau, N. Monnier, and C. Pétrier. 1997. Effect of ultrasound on KSF/0 mediated glycosylations. *Journal of Molecular Catalysis A: Chemical* 123: 123–130.

Cativiéla, C., J.M. Fraile, J.I. García, J.A. Mayoral, and F. Figueras. 1991. A study on the role of solvent in clay-catalysed Diels–Alder reactions. *Journal of Molecular Catalysis* 68: L31–L34.

Cativiéla, C., J.M. Fraile, J.I. García, J.A. Mayoral, and E. Pires. 1992a. Effect of clay calcination on clay-catalysed Diels–Alder reactions of cyclopentadiene with methyl and (–)-methyl acrylates. *Tetrahedron* 48: 6467–6476.

Cativiéla, C., J.M. Fraile, J.I. García et al. 1992b. Factors influencing the K10 montmorillonite-catalyzed Diels–Alder reaction between methyl acrylate and cyclopentadiene. *Journal of Catalysis* 137: 394–407.

Cativiéla, C., F. Figueras, J.M. Fraile et al. 1993. Comparison of the catalytic properties of protonic zeolites and exchanged clays for Diels–Alder synthesis. *Applied Catalysis A: General* 101: 253–267.

Cativiéla, C., J.I. García, M. García-Matres et al. 1995. Clay-catalyzed Friedel–Crafts alkylation of anisole with dienes. *Applied Catalysis A: General* 123: 273–287.

Centi, G. and S. Perathoner. 2008. Catalyis by layered materials: A review. *Microporous and Mesoporous Materials* 107: 3–15.

Chakrabarty, M., A. Mukherji, S. Arima, Y. Harigaya, and G. Pilet. 2009. Expeditious reaction of ninhydrin with active methylene compounds on montmorillonite K10. *Monatshefte für Chemie* 140: 189–197.

Chakrabarty, M. and S. Sarkar. 2002. Novel clay-mediated, tandem addition-elimination-(Michael) addition reactions of indoles with 3-formylindole: An eco-friendly route to symmetrical and unsymmetrical triidolylmethanes. *Tetrahedron Letters* 43: 1351–1353.

Chakraborty, S., G. Chattopadhyay, and C. Saha. 2011. Montmorillonite-KSF induced Fischer indole cyclization under microwave towards a facile entry to 1-keto-1,2,3,4-tetrahydrocarbazoles. *Indian Journal of Chemistry* 50B: 201–206.

Chavan, S.P., G.B.B. Varadwaj, and K.M. Parida. 2016. Solvent-switchable regioselective synthesis of aurones and flavones using palladium-supported amine-functionalized montmorillonite as a heterogeneous catalyst. *ChemCatChem* 8: 2649–2658.

Chavan, S.P., G.B.B. Varadwaj, K. Parida, and B.M. Bhanage. 2015. Palladium anchored on amine-functionalized K10 as an efficient, heterogeneous and reusable catalyst for carbonylative Sonogashira reaction. *Applied Catalysis A: General* 506: 237–245.

Chen, W.-Y. and X.-S. Li. 2009. Montmorillonite K10-catalyzed Michael addition of indoles to nitroolefins under solvent-free conditions. *Synthetic Communications* 39: 2014–2021.

Chiba, K., T. Hirano, Y. Kitano, and M. Tada. 1999. Montmorillonite-mediated hetero-Diels–Alder reaction of alkenes and *o*-quinomethanes generated *in situ* by dehydration of *o*-hydroxybenzyl alcohols. *Chemical Communications* 691–692.

Chiche, B., A. Finiels, C. Gauthier, P. Geneste, J. Graille, and D. Pioch. 1987. Acylation over cation-exchanged montmorillonite. *Journal of Molecular Catalysis* 42: 229–235.

Choudary, B.M., V. Bhaskar, M.L. Kantam, K.K. Rao, and K.V. Raghavan. 2001. Acylation of amines with carboxylic acids: The atom economic protocal catalysed by Fe(III)-montmorillonite. *Catalysis Letters* 74: 207–211.

Choudary, B.M., N.S. Chowdari, and M.L. Kantam. 2000. Friedel–Crafts sulfonylation of aromatics catalysed by solid acids: An eco-friendly route for sulfone synthesis. *Journal of the Chemical Society, Perkin Transactions 1* 2000(16): 2689–2693.

Choudary, B.M., N.S. Chowdari, M.L. Kantam, and R. Kannan. 1999. Fe(III) exchanged montmorillonite: A mild and eco-friendly catalyst for sulfonylation of aromatics. *Tetrahedron Letters* 40: 2859–2862.

Choudary, B.M., M.L. Kantam, M. Sateesh, K.K. Rao, and P.L. Santhi. 1997. Iron pillared clays—Efficient catalysts for Friedel–Crafts reactions. *Applied Catalysis A: General* 149: 257–264.

Choudary, B.M., B.P.C. Rao, N.S. Chowdari, and M.L. Kantam. 2002. Fe^{3+}-montmorillonite: A bifunctional catalyst for one pot Friedel–Crafts alkylation of arenes with alcohols. *Catalysis Communications* 3: 363–367.

Choudary, B.M., M. Sateesh, M.L. Kantam, and K.V.R. Prasad. 1998. Acylation of aromatic ethers with acid anhydrides in the presence of cation-exchanged clays. *Applied Catalysis A: General* 171: 155–160.

Choudhary, V.R., V.H. Tillu, V.S. Narkhede, H.B. Borate, and R.D. Wakharkar. 2003. Microwave assisted solvent-free synthesis of dihydropyrimidones by Biginelli reaction over Si-MCM-41 supported $FeCl_3$ catalyst. *Catalysis Communications* 4: 449–453.

Čičel, B., P. Komadel, and M. Nigrin. 1992. Catalytic activity of smectites on dimerization of oleic acid. *Collection of Czechoslovak Chemical Communications* 57: 1666–1671.

Clark, J.H. and D.J. Macquarrie. 1996. Environmentally friendly catalytic methods. *Chemical Society Reviews* 25: 303–310.

Clark, P.D., A. Kirk, and R.A. Kydd. 1994. Benzylation of thiophene using zinc and iron chloride modified montmorillonite clay. *Catalysis Letters* 25: 163–168.

Corma, A., H. García, A. Leyva, and A. Primo. 2004. Alkali-exchanged sepiolites containing palladium as bifunctional (basic ad noble metal) catalysts for the Heck and Suzuki reactions. *Applied Catalysis A: General* 257: 77–83.

Cornélis, A., A. Gerstmans, P. Laszlo, A. Mathy, and I. Zieba. 1990. Friedel–Crafts acylations with modified clays as catalysts. *Catalysis Letters* 6: 103–110.

Cornélis, A., P. Laszlo, and S.-F. Wang. 1993. On the transition state for "clayzic"-catalyzed Friedel–Crafts reactions upon anisole. *Tetrahedron Letters* 34: 3849–3852.

Crisóstomo, F.R.P., R. Carrillo, T. Martín, and V.S. Martín. 2005. Montmorillonite K-10 as a mild acid for the Nicholas reaction. *Tetrahedron Letters* 46: 2829–2832.

Cseri, T., S. Békássy, F. Figueras, E. Cseke, L.-C. de Menorval, and R. Dutartre. 1995b. Characterization of clay-based K catalysts and their application in Friedel–Crafts alkylation of aromatics. *Applied Catalysis A: General* 132: 141–155.

Cseri, T., S. Békássy, F. Figueras, and S. Rizner. 1995a. Benzylation of aromatics on ion-exchanged clays. *Journal of Molecular Catalysis A: Chemical* 98: 101–107.

Dar, B.A., P. Patidar, S. Kumar et al. 2013. Fe-Al/clay as an efficient heterogeneous catalyst for solvent-free synthesis of 3,4-dihydropyrimidones. *Journal of Chemical Sciences* 125: 545–553.

Darehkordi, A., S.M. Sadegh Hosseini, and M. Tahmooresi. 2012. Montmorillonite modified as an efficient and environment friendly catalyst for one-pot synthesis of 3,4-dihydropyrimidine-2(1H) ones. *Iranian Journal of Materials Science & Engineering* 9: 49–57.

de Faria, E.H., G.P. Ricci, L. Marçal et al. 2012. Green and selective oxidation reactions catalyzed by kaolinite covalently grafted with Fe(III) pyridine-carboxylate complexes. *Catalysis Today* 187: 135–149.

Delaude, L. and P. Laszlo. 1991. Regioselective hydrochlorination of olefins is favored by an acidic solid catalyst. *Tetrahedron Letters* 32: 3705–3708.

Delgado, F., J. Tamaríz, G. Zepeda, M. Landa, R. Miranda, and J. García. 1995. Knoevenagel condensation catalysed by a Mexican bentonite using infrared irradiation. *Synthetic Communications* 25: 753–759.

de Oliveira, R.N., J.R. de Freitas Filho, and R.M. Srivastava. 2002. Microwave-induced synthesis of 2,3-unsaturated *O*-glycosides under solvent-free conditions. *Tetrahedron Letters* 43: 2141–2143.

De Paolis, O., L. Teixeira, and B. Török. 2009. Synthesis of quinolones by a solid acid-catalyzed microwave-assisted domino cyclization-aromatization approach. *Tetrahedron Letters* 50: 2939–2942.

den Otter, M.J.A.M. 1970. Dimerization of oleic acid with a montmorillonite catalyst. I. Important process parameters; some main reactions. *Fette, Seifen, Anstrichmittel* 72: 667–673.

Deshmukh, A.R.A.S., D.G. Panse, and B.M. Bhawal. 1999. A clay catalyzed method for diethyl 2,2,2-trichloroethylidene propanedioate, an efficient intermediate for the synthesis of enamino esters. *Synthetic Communications* 29: 1801–1809.

Dhakshinamoorthy, A. and K. Pitchumani. 2005. Facile clay-induced Fischer indole synthesis: A new approach to synthesis of 1,2,3,4-tetrahydrocarbazole and indoles. *Applied Catalysis A: General* 292: 305–311.

Dintzner, M.R., A.J. Little, M. Pacilli et al. 2007. Montmorillonite clay-catalyzed hetero-Diels–Alder reaction of 2,3,-dimethyl-1,3-butadiene with benzaldehydes. *Tetrahedron Letters* 48: 1577–1579.

Dintzner, M.R., J.J. Maresh, C.R. Kinzie, A.F. Arena, and T. Speltz. 2012. A research-based undergraduate organic laboratory project: Investigation of a one-pot, multicomponent, environmentally friendly Prins–Friedel–Crafts-type reaction. *Journal of Chemical Education* 89: 265–267.

Dintzner, M.R., Y.A. Mondjinou, and B. Unger. 2009. Montmorillonite K10 clay-catalyzed synthesis of homoallylic silyl ethers: An efficient and environmentally friendly Hosomi-Sakurai reaction. *Tetrahedron Letters* 50: 6630–6641.

Dry, M.E. 2002. The Fischer–Tropsch process: 1950–2000. *Catalysis Today* 71: 227–241.

Dutta, D.K., B.J. Borah, and P.P. Sarmah. 2015. Recent advances in metal nanoparticles stabilization into nanopores of montmorillonite and their catalytic applications for fine chemicals synthesis. *Catalysis Reviews: Science and Engineering* 57: 257–305.

Ebitani, K., M. Kato, K. Motokura, T. Mizugaki, and K. Kaneda. 2006. Highly efficient heterogeneous acylations of aromatic compounds with acid anhydrides and carboxylic acids by montmorillonite-enwrapped titanium as a solid acid catalyst. *Research on Chemical Intermediates* 32: 305–315.

Elizarov, N., M. Pucheault, and S. Antoniotti. 2016. Highly efficient Hosomi-Sakurai reaction of aromatic aldehydes catalyzed by montmorillonite doped with simple bismuth(III) salts: Batch and continuous flow studies. *ChemistrySelect* 1: 3219–3222.

Fabra, M.J., J.M. Fraile, C.I. Herrerías, F.J. Lahoz, J.A. Mayoral, and I. Pérez. 2008. Surface-enhanced stereoselectivity in Mukaiyama aldol reactions catalyzed by clay-supported bis(oxazoline)-copper complexes. *Chemical Communications* 5402–5404.

Fernandes, C.I., C.D. Nunes, and P.D. Vaz. 2012. Clays in organic synthesis—Preparation and catalytic applications. *Current Organic Synthesis* 9: 670–694.

Figueras, F., C. Cativiéla, J.M. Fraile et al. 1994. Diels–Alder condensation of methyl and (–)-methyl acrylates with cyclopentadiene over zeolites and cation exchanged clays. *Studies in Surface Science and Catalysis* 83: 391–398.

Firouzabadi, H., N. Iranpoor, A. Ghaderi, M. Ghavami, and S.J. Hoseini. 2011. Palladium nanoparticles supported on aminopropyl-functionalized clay as efficient catalysts for phoshine-free C–C bond formation via Mizoroki–Heck and Suzuki–Miyaura reactions. *Bulletin of the Chemical Society of Japan* 84: 100–109.

Firouzabadi, H., N. Iranpoor, A. Ghaderi, M. Gholinejad, S. Rahimi, and S. Jokar. 2014. Design and synthesis of a new phosphinite-functionalized clay composite for the stabilization of palladium nanoparticles: Application of a recoverable catalyst for C–C bond formation reactions. *RSC Advances* 4: 27674–27682.

Fraile, J.M., N. García, and C.I. Herrerías. 2013. Support effect on stereoselectivities of vinylogous Mukaiyama–Michael reactions catalyzed by immobilized chiral copper complexes. *ACS Catalysis* 3: 2710–2718.

Fraile, J.M., J.I. García, C.I. Herrerías, J.A. Mayoral, E. Pires, and L. Salvatella. 2009. Beyond reuse in chiral immobilized catalysis: The bis(oxazoline) case. *Catalysis Today* 140: 44–50.

Fraile, J.M., J.I. García, D. Gracia, J.A. Mayoral, T. Tarnai, and F. Figueras. 1997. Contribution of different mechanisms and different active sites to the clay-catalyzed Diels–Alder reactions. *Journal of Molecular Catalysis A: Chemical* 121: 97–102.

Frère, S., V. Thiéry, and T. Besson. 2001. Microwave acceleration of the Pechmann reaction on graphite/montmorillonite K10: Application to the preparation of 4-substituted 7-aminocoumarins. *Tetrahedron Letters* 42: 2791–2794.

Giannelis, E.P., E.G. Rightor, and T.J. Pinnavaia. 1988. Reactions of metal-cluster carbonyls in pillared clay galleries: Surface coordination chemistry and Fischer–Tropsch catalysis. *Journal of the American Chemical Society* 110: 3880–3885.

Gogoi, A., S.J. Chutia, P.K. Gogoi, and U. Bora. 2014. A highly efficient heterogeneous montmorillonite K-10-supported palladium catalyst for Suzuki-Miyaura cross-coupling reaction in aqueous medium. *Applied Organometallic Chemistry* 28: 839–844.

Gómez, A.M., F. Lobo, C. Uriel, and J.C. López. 2013. Recent developments in the Ferrier rearrangement. *European Journal of Organic Chemistry* 2013: 7221–7262.

Gómez-Sanz, F., M.V. Morales-Vargas, B. González-Rodríguez, M.L. Rojas-Cervantes, and E. Pérez-Mayoral. 2017. Acid clay minerals as eco-friendly and cheap catalysts for the synthesis of β-amino ketones by Mannich reaction. *Applied Clay Science* 143: 250–257.

Gültekín, Z. 2004. Iron(III)-doped montmorillonite catalysis of alkenes bearing sulphoxide groups in Diels–Alder reactions. *Clay Minerals* 39: 345–348.

Han, J., T. Wang, S. Feng, C. Li, and Z. Zhang. 2016. One-pot synthesis of 3-(furan-2-yl)-4H-chromen-4-ones from 1-(2-hydroxyphenyl)butane-1,3-diones and 2,5-dimethoxy-2,5-dihydrofuran catalyzed *via* K10 montmorillonite under solvent-free conditions. *Green Chemistry* 18: 4092–4097.

Hao, Q.-Q., Z.-W. Liu, B. Zhang et al. 2012. Porous montmorillonite heterostructures directed by a single alkyl ammonium template for controlling the product distribution of Fischer–Tropsch synthesis over cobalt. *Chemistry of Materials* 24: 972–974.

Hao, Q.-Q., G.-W. Wang, Z.-T. Liu, J. Lu, and Z.-W. Liu. 2010. Co/pillared clay bifunctional catalyst for controlling the product distribution of Fischer–Tropsch synthesis. *Industrial & Engineering Chemistry Research* 49: 9004–9011.

Hao, Q.-Q., G.-W. Wang, Y.-H. Zhao, Z.-T. Liu, and Z.-W. Liu. 2013. Fischer–Tropsch synthesis over cobalt/montmorillonite promoted with different interlayer cations. *Fuel* 109: 33–42.

Hara, T., M. Hatekeyama, A. Kim, N. Ichikuni, and S. Shimazu. 2012. Preparation of clay-supported Sn catalysts and application to Baeyer–Villiger oxidation. *Green Chemistry* 14: 771–777.

Heravi, M.M., M. Haghighi, F. Derikvand, K. Bakhtiari, and S. Khaleghi. 2009. Rapid Knoevenagel condensation: Unique reactivity of kaolin in aqueous suspension. *Iranian Journal of Organic Chemistry* 3: 164–167.

Heravi, M.M., M. Tajbakhsh, B. Mohajerani, and M. Ghassemzadeh. 1999. Montmorillonite K10 catalyzed Knoevenagel condensation under microwave irradiation in solventless system. *Zeitschrift für Naturforschung* 54b: 541–543.

Huang. M.-Y., J.-C. Wu, F.-S. Shieu, and J.-J. Lin. 2010. Isomerization of *endo*-tetrahydrodicyclopentadiene over clay-supported chloroaluminate ionic liquid catalysts. *Journal of Molecular Catalysis A: Chemical* 315: 69–75.

Ikeda, K., Y. Ueno, S. Kitani, R. Nishino, and M. Sato. 2008. Ferrier glycosylation reaction catalyzed by Bi(OTf)3-montmorillonite K-10: Efficient synthesis of 3,4-unsaturated sialic acid derivatives. *Synlett* 1027–1030.

Iqbal, Z., A.H. Jackson, and K.R.N. Rao. 1988. Reactions on solid supports part IV: Reactions of αβ-unsaturated carbonyl compounds with indoles using clay as catalyst. *Tetrahedron Letters* 29: 2577–2580.

Izumi, Y. and M. Onaka. 1992. Organic syntheses using aluminosilicates. *Advances in Catalysis* 38: 245–282.

James, B., E. Suresh, and M.S. Nair. 2007. Friedel–Crafts alkylation of a cage enone: Synthesis of aralkyl substituted tetracyclo[5.3.1.0$^{[2,6]}$.0$^{[4,8]}$]undeca-9,11-diones and the formation of fascinating novel cage compounds with pyrrole and thiophene using montmorillonite K-10. *Tetrahedron Letters* 48: 6059–6063.

Jankovič, L. and P. Komadel. 2003. Microwave assisted synthesis of substituted indoles using montmorillonite as catalyst. *Solid State Phenomena* 90–91: 481–486.

Jørgensen, K.A. 2000. Catalytic asymmetric hetero-Diels–Alder reactions of carbonyl compounds and imines. *Angewandte Chemie International Edition* 39: 3558–3588.

Joseph, T., G.V. Shanbhag, and S.B. Halligudi. 2005. Copper(II) ion-exchanged montmorillonite as catalyst for the direct addition of N–H bond to CC triple bond. *Journal of Molecular Catalysis A: Chemical* 236: 139–144.

Joseph, T., G.V. Shanbhag, D.P. Sawant, and S.B. Halligudi. 2006. Chemoselective *anti*-Markovnikov hydroamination of α,β-ethylenic compounds with amines using montmorillonite clay. *Journal of Molecular Catalysis A: Chemical* 250: 210–217.

Kalbasi, R.J., A.R. Massah, and B. Daneshvarnejad. 2012. Preparation and characterization of bentonite/PS-SO$_3$H nanocomposites as an efficient acid catalyst for the Biginelli reaction. *Applied Clay Science* 55: 1–9.

Kamath, C.R. and S.D. Samant. 1999. Diels–Alder reaction of pyran-2(*H*)-ones: Part III–The Diels–Alder reactions of 4,6-disubstituted pyran-2(*H*)-ones with naphthoquinone and N-phenylmaleimide on montmorillonite K10, Engelhard filtrol-24 and pyrophyllite clays. *Indian Journal of Chemistry* 38B: 1214–1217.

Kamath, C.R., A.B. Shinde, and S.D. Samant. 2000. Diels–Alder reaction of pyran-2(*H*)-ones: Part V. Diels–Alder reactions of 4,6-disubstituted pyran-2(*H*)-ones with 1,4-naphthoquinone and *N*-phenylmaleimide under dry state adsorbed condition (DSAC) on montmorillonite K10, filtrol-24, bentonite, pyrophyllite; and Al^{3+}, Zn^{2+}, Fe^{3+} exchanged montmorillonite K10 and bentonite. *Indian Journal of Chemistry* 39B: 270–276.

Kamble, R.D., B.S. Dawane, O.S. Yemul, A.B. Kale, and S.D. Patil. 2013. Bleaching earth clay (pH 12.5): A green catalyst for rapid synthesis of pyranopyrazole derivatives via a tandem three-component reaction. *Research on Chemical Intermediates* 39: 3859–3866.

Kanagasabapathy, S., A. Sudalai, and B.C. Benicewicz. 2001. Montmorillonite K10-catalyzed regioselective addition of thiols and thiobenzoic acids onto olefins: An efficient synthesis of dithiocarboxylic esters. *Tetrahedron Letters* 42: 3791–3794.

Kannan, P. and K. Pitchumani. 1997. Clay-catalysed radical addition of aliphatic thiols to styrene. *Catalysis Letters* 45: 271–273.

Kannan, P., K. Pitchumani, S. Rajagopal, and C. Srinivasan. 1997. Sheet silicate catalysed demethylation and Fischer–Hepp rearrangement of *N*-methyl-*N*-nitrosoaniline. *Journal of Molecular Catalysis A: Chemical* 118: 189–193.

Kaur, N. and D. Kishore. 2012. Montmorillonite: An efficient, heterogeneous and green catalyst for organic synthesis. *Journal of Chemical and Pharmaceutical Research* 4: 991–1015.

Kaur, A. and V. Singh. 2015. Efficient Suzuki reaction catalyzed by a recyclable clay carbapalladacycle nanocomposite in ionic liquid media. *Tetrahedron Letters* 56: 1128–1131.

Kawabata, T., M. Kato, T. Mizugaki, K. Ebitani, and K. Kaneda. 2005. Monomeric metal aqua complexes in the interlayer space of montmorillonites as strong Lewis acid catalysts for heterogeneous carbon-carbon bond-forming reactions. *Chemistry—A European Journal* 11: 288–297.

Kawabata, T., T. Mizugaki, K. Ebitani, and K. Kaneda. 2003. A novel montmorillonite-enwrapped scandium as a heterogeneous catalyst for Michael reaction. *Journal of the American Chemical Society* 125: 10486–10487.

Kawai, M., M. Onaka, and Y. Izumi. 1986. Clay montmorillonite-catalyzed aldol reactions of silyl enol ethers with aldehydes and acetals. *Chemistry Letters* 1581–1584.

Kawai, M., M. Onaka, and Y. Izumi. 1987. Clay montmorillonite-catalysed Michael reactions of silyl ketene acetals and a silyl enol ether with α,β-unsaturated carbonyl compounds. *Journal of the Chemical Society, Chemical Communications* 1203–1204.

Kawai, M., M. Onaka, and Y. Izumi. 1988a. New application of solid acid to carbon-carbon bond formation reactions: Clay montmorillonite-catalyzed aldol reactions of silyl enol ethers with aldehydes and acetals. *Bulletin of the Chemical Society of Japan* 61: 1237–1245.

Kawai, M., M. Onaka, and Y. Izumi. 1988b. Clay montmorillonite: An efficient, heterogeneous catalyst for Michael reactions of silyl ketene acetals and silyl enol ethers with α,β-unsaturated carbonyl compounds. *Bulletin of the Chemical Society of Japan* 61: 2157–2164.

Kim, H.S., H.S. Lee, and J.N. Kim. 2009. Synthesis of rearranged *N*-tosyl *aza*-Baylis–Hillman adducts under acidic conditions catalyzed by CH_3SO_3H or montmorillonite K10. *Bulletin of the Korean Chemical Society* 30: 941–944.

Kolancilar, H. 2010. Synthesis of 7-hydroxy-4-methylcoumarin using polyaniline/montmorillonite K10 composite as catalyst. *Asian Journal of Chemistry* 22: 5694–5698.

Krishnakumar, V., F-R.N. Khan, B.K. Mandal, E-D. Jeong, and J.S. Jin. 2015. Montmorillonite-KSF-catalyzed synthesis of 4-heteroarylidene-*N*-arylhomophthalimides by Knoevenagel condensation. *Research on Chemical Intermediates* 41: 5509–5519.

Kumar, H.M.S., B.V.S. Reddy, S. Anjaneyulu, E.J. Reddy, and J.S. Yadav. 1999. Clay catalysed amidation of alcohols with nitriles in dry media. *New Journal of Chemistry* 23: 955–956.

Kumaran, E., M. Santhi, K.K. Balasubramanian, and S. Bhagavathy. 2011. Montmorillonite K-10 clay-catalyzed Ferrier rearrangement of 2-*C*-hydroxymethyl-D-glycals, 3,4,6-tri-*O*-alkyl-D-glycals, and 3,4(dihydro-2*H*-pyran-5-yl)methanol: A few unexpected domino transformations. *Carbohydrate Research* 346: 1654–1661.

Kurian, M. and S. Sugunan. 2003. Liquid phase benzylation of *o*-xylene over pillared clays. *Indian Journal of Chemistry* 42A: 2480–2486.

Kurian, M. and S. Sugunan. 2005. Selective benzylation of benzene over alumina pillared clays. *Indian Journal of Chemistry* 44A: 1772–1781.

Kurian, M. and S. Sugunan. 2006. *tert*-Butylation of phenol catalysed by metal exchanged iron pillared montmorillonites. *Catalysis Communications* 7: 417–421.

Lanke, S.R., Z.S. Qureshi, A.B. Patil, D.S. Patil, and B.M. Bhanage. 2012. Hydroarylation of arenes with styrenes using montmorillonite K-10 as an efficient, selective, and recyclable catalyst. *Green Chemistry Letters and Reviews* 5: 621–632.

Laszlo, P. 1986. Catalysis of organic reactions by inorganic solids. *Accounts of Chemical Research* 19: 121–127.

Laszlo, P. and J. Luchetti. 1984a. Catalysis of the Diels–Alder reaction in the presence of clays. *Tetrahedron Letters* 25: 1567–1570.

Laszlo, P. and J. Luchetti. 1984b. Acceleration of the Diels–Alder reaction of clays suspended in organic solvents. *Tetrahedron Letters* 25: 2147–2150.

Laszlo, P. and J. Luchetti. 1984c. Easy formation of Diels–Alder cycloadducts between furans and α,β-unsaturated aldehydes and ketones at normal pressure. *Tetrahedron Letters* 25: 4387–4388.

Laszlo, P. and A. Mathy. 1987. Catalysis of Friedel–Crafts alkylation by a montmorillonite doped with transition metal cations. *Helvetica Chimica Acta* 70: 577–586.

Laszlo, P. and H. Moison. 1989. Catalysis of Diels–Alder reactions with acrolein as dienophile by iron(III)-doped montmorillonite. *Chemistry Letters* 8: 1031–1034.

Lee, K.Y., S. Gowrisankar, and J.N. Kim. 2005. Baylis–Hillman reactions and chemical transformation of Baylis–Hillman adducts. *Bulletin of the Korean Chemical Society* 26: 1481–1490.

Lee, K.-Y. and K.-Y. Ko. 2004. Envirocat EPZ10: A recyclable solid acid catalyst for the synthesis of Biginelli-type 3,4-dihydropyrimidin-2(1*H*)-ones. *Bulletin of the Korean Chemical Society* 25: 1929–1931.

Lee, S.W., H.B. Lee, B.C. Kim, K. Sadaiah, K. Lee, and H. Shin. 2013. A large scale formal synthesis of CoQ_{10}: Highly stereoselective Friedel–Crafts allylation reaction of tetramethoxytoluene with (*E*)-4-chloro-2-methyl-1-phenylsulfonyl-2-butene in the presence of montmorillonite K-10. *Bulletin of the Korean Chemical Society* 34: 1257–1259.

Lei, Z., G. Ma, and C. Jia. 2007. Montmorillonite (MMT) supported tin(II) chloride: An efficient and recyclable heterogeneous catalyst for clean and selective Baeyer–Villiger oxidation with hydrogen peroxide. *Catalysis Communications* 8: 305–309.

Lei, Z., Q. Zhang, R. Wang, G. Ma, and C. Jia. 2006. Clean and selective Baeyer–Villiger oxidation of ketones with hydrogen peroxide catalyzed by Sn-palygorskite. *Journal of Organometallic Chemistry* 691: 5767–5773.

Letaïef, S., Y. Lu, and C. Detellier. 2011. Zirconium oxide nanoparticles coated on sepiolite by sol-gel process–Their application as a solvent-free catalyst for condensation reactions. *Canadian Journal of Chemistry* 89: 280–288.

Li, J.J. 2006. *Name Reactions: A Collection of Detailed Reaction Mechanisms*, 3rd ed. Berlin, Germany: Springer-Verlag.

Li, J.-T., C.-Y. Xing, and T.-S. Li. 2004. An efficient and environmentally friendly method for synthesis of arylmethylenemalononitrile catalyzed by montmorillonite $K10-ZnCl_2$ under ultrasound irradiation. *Journal of Chemical Technology and Biotechnology* 79: 1275–1278.

Li, L.-J., B. Lu, T.-S. Li, and J.T. Li. 1998a. Montmorillonite clay catalysis. Part 8. Synthesis of arylcholestenes by Friedel–Crafts reaction catalysed by montmorillonite K-10. *Synthetic Communications* 28: 1439–1449.

Li, T.-S. and A.-X. Li. 1998. Montmorillonite clay catalysis. Part 10. K-10 and KSF-catalysed acylation of alcohols, phenols, thiols and amines: Scope and limitation. *Journal of the Chemical Society, Perkin Transactions 1* 54(12): 1913–1918.

Li, T., Y. Liu, and F-S. Liu. 2017. Efficient preparation and application of palladium loaded montmorillonite as a reusable and effective heterogeneous catalyst for Suzuki cross-coupling reaction. *Applied Clay Science* 136: 18–25.

Li, T.-S., Z.-H. Zhang, F. Yang, and C.-G. Fu. 1998b. Montmorillonite clay catalysis. Part 7. An environmentally friendly procedure for the synthesis of coumarins *via* Pechmann condensation of phenols with ethyl acetoacetate. *Journal of Chemical Research (S)* 38–39.

Li, Y.X. and W.I. Bao. 2003. Microwave-assisted solventless Biginelli reaction catalyzed by montmorillonite clay-$SmCl_3 \cdot 6H_2O$ system. *Chinese Chemical Letters* 14: 993–995.

Lima-Junior, C.G. and M.L.A.A. Vasconcellos. 2012. Morita–Baylis–Hillman adducts: Biological activities and potentialities to the discovery of new cheaper drugs. *Bioorganic & Medicinal Chemistry* 20: 3954–3871.

Lin, H., J. Ding, X. Chen, and Z. Zhang. 2000. An efficient synthesis of 5-alkoxycarbonyl-4-aryl-3,4-dihydropyrimidin-2(1*H*)-ones catalyzed by KSF montmorillonite. *Molecules* 5: 1240–1243.

Lipińska, T., E. Guibé-jampel, A. Petit, and A. Loupy. 1999. 2-(2-pyridyl)indole derivatives preparation via Fischer reaction on montmorillonite K10/zinc chloride under microwave irradiation. *Synthetic Communications* 29: 1349–1354.

Liu, Y.-H., Q.-S. Liu, and Z.-H. Zhang. 2009. An efficient Friedel–Crafts alkylation of nitrogen heterocycles catalyzed by antimony trichloride/montmorillonite K-10. *Tetrahedron Letters* 50: 916–921.

Liu, Y., K. Murata, and M. Inaba. 2014. Synthesis of mixed alcohols from synthesis gas over alkali and Fischer–Tropsch metals modified MoS_2/Al_2O_3-montmorillonite catalysts. *Reaction Kinetics, Mechanisms and Catalysis* 113: 187–200.

Loh, T.-P. and X.-R. Li. 1999. Clay montmorillonite K10 catalyzed aldol-type reaction of aldehydes with silyl enol ethers in water. *Tetrahedron* 55: 10789–10802.

López, I., G. Silvero, M.J. Arévalo, R. Babiano, J.C. Palacios, and J.L. Bravo. 2007. Enhanced Diels–Alder reactions: On the role of mineral catalysts and microwave irradiation in ionic liquids as recyclable media. *Tetrahedron* 63: 2901–2906.

Lotz, W.W. and J. Gosselck. 1979. Friedel–Crafts-Alkylierung mit Cu^{2+}-Bentonit. *Zeitschrift für Naturforschung* 34b: 121–122.

Manju, K. and S. Sankaran. 2004. Selective benzylation of *o*-xylene over transition metal doped montmorillonites. *Reaction Kinetics and Catalysis Letters* 81: 57–64.

Maquestiau, A., A. Mayeence, and J-J. Vanden Eynde. 1991. Ultrasound-promoted aromatization of Hantzsch 1,4-dihydropyridines by clay-supported cupric nitrate. *Tetrahedron Letters* 32: 3839–3840.

Marco-Contelles, J., E. Pérez-Mayoral, A. Samadi, M. do C. Carreiras, and E. Soriano. 2009. Recent advances in the Friedländer Reaction. *Chemical Reviews* 109: 2652–2671.

Martín-Aranda, R.M., E. Ortega-Cantero, M.L. Rojas-Cervantes, M.A. Vicente-Rodríguez, and M.A. Bañares-Muñoz. 2002. Sonocatalysis and basic clays: Michael addition between imidazole and ethyl acrylate. *Catalysis Letters* 84: 201–204.

Martín-Aranda, R.M., E. Ortega-Cantero, M.L. Rojas-Cervantes, M.A. Vicente-Rodríguez, and M.A. Bañares-Muñoz. 2005. Ultrasound-activated Knoevenagel condensation of malenonitrile with carbonylic compounds catalysed by alkaline-doped saponites. *Journal of Chemical Technology and Biotechnology* 80: 234–238.

Martín-Aranda, R.M., M.A. Vicente-Rodríguez, J.M. López-Pestaña et al. 1997. Application of basic clays in microwave activated Michael additions: Preparation of *N*-substituted imidazoles. *Journal of Molecular Catalysis A: Chemical* 124: 115–121.

Martínez, A.V., F. Invernizzi, A. Leal, J.A. Mayoral, and J.I. García. 2015b. Microwave-promoted solventless Mizoroki–Heck reactions catalysed by Pd nanoparticles supported on laponite clay. *RSC Advances* 5: 10102–10109.

Martínez, A.V., A. Leal-Duaso, J.I. García, and J.A. Mayoral. 2015a. An extremely highly recoverable clay-supported Pd nanoparticle catalyst for solvent-free Heck–Mizoroki reactions. *RSC Advances* 5: 59983–59990.

Martínez, A.V., J.A. Mayoral, and J.I. García. 2014. Pd nanoparticles immobilized in [bmim][PF_6] supported on laponite clay as highly recyclable catalysts for the Mizoroki–Heck reaction. *Applied Catalysis A: General* 472: 21–28.

Mas-Marzá, E., A.M. Segarra, C. Claver, E. Peris, and E. Fernandez. 2003. Improved Sonogashira C–C coupling through clay supported palladium complexes with tridentate pincer bis-carbene ligands. *Tetrahedron Letters* 44: 6595–6599.

Massah, A.R., R.J. Kalbasi, and M. Toghyani. 2012. Sulfonated polystyrene/montmorillonite nanocomposite as a new and efficient catalyst for the solvent-free Mannich reaction. *Iranian Journal of Catalysis* 2: 41–49.

McCabe, R.W. and J.M. Adams. 2013. Clay minerals as catalysts. In *Handbook of Clay Science,* 2nd ed. *Developments in Clay Science, Vol. 5B* (Eds.) F. Bergaya and G. Lagaly, 491–538. Amsterdam, the Netherlands: Elsevier.

Meshram, H.M. 1990. Clay catalysed facile Beckmann rearrangement of ketoximes. *Synthetic Communications* 20: 3253–3258.

Miles, W.H., E.A. Dethoff, H.H. Tuson, and G. Ulas. 2005. Kishner's reduction of 2-furylhydrazone gives 2-methylene-2,3-dihydrofuran, a highly reactive ene in the ene reaction. *Journal of Organic Chemistry* 70: 2862–2865.

Mitsudome, T., T. Matsuno, S. Sueoka, T. Mizugaki, K. Jitsukawa, and K. Kaneda. 2012. Titanium cation-exchanged montmorillonite as an active heterogeneous catalyst for the Beckmann rearrangement under mild reaction conditions. *Tetrahedron Letters* 53: 5211–5214.

Mitsudome, T., K. Nose, T. Mizugaki, K. Jitsukawa, and K. Kaneda. 2008. Reusable montmorillonite-entrapped organocatalyst for asymmetric Diels–Alder reaction. *Tetrahedron Letters* 49: 5464–5466.

Mitsudome, T., T. Umetani, K. Mori, T. Mizugaki, K. Ebitane, and K. Kaneda. 2006. Highly efficient Wacker oxidation catalyzed by heterogeneous Pd montmorillonite under acid-free conditions. *Tetrahedron Letters* 47: 1425–1428.

Mogilaiah, K. and B. Sakram. 2006. Montmorillonite K10 clay catalyzed Friedländer synthesis of 1,8-naphthyridines in dry media under microwave irradiation. *Indian Journal of Chemistry* 45B: 2749–2750.

Morikawa, Y. 1993. Catalysis by metal ions intercalated in layer lattice silicates. In *Advances in Catalysis, Vol. 39* (Eds.) D.D. Eley, H. Pines, and P.B. Weisz, pp. 302–327. New York: Academic Press.

Moussaoui, Y. and R.B. Salem. 2007. Catalyzed Knoevenagel reactions on inorganic solid supports: Application to the synthesis of coumarine compounds. *Comptes Rendus Chimie* 10: 1162–1169.

Naicker, K.P., K. Pitchumani, and R.S. Varma. 1998. The catalytic influence of clays on Bamberger rearrangement: unexpected formation of *p*-nitrosodiphenyl amine from *N*-phenylhydroxylamine. *Catalysis Letters* 54: 165–167.

Narayanan, S. and K. Deshpande. 1996. A comparative aniline alkylation activity of montmorillonite and vanadia-montmorillonite with silica and vanadia-silica. *Applied Catalysis A: General* 135: 125–135.

Nikpassand, M., M. Mamaghani, K. Tabatabaeian, and M.K. Abiazi. 2009. KSF: An efficient catalyst for the regioselective synthesis of 1,5-diaryl pyrazoles using Baylis–Hillman adducts. *Molecular Diversity* 13: 389–393.

Nikpassand, M., M. Mamaghani, K. Tabatabaeian, and H.A. Samimi. 2010. An efficient and clean synthesis of symmetrical and unsymmetrical 3,3-di(indolyl)indolin-2-ones using KSF. *Synthetic Communications* 40: 3552–3560.

Obrador, E., M. Castro, J. Tamaríz, G. Zepeda, R. Miranda, and F. Delgado. 1998. Knoevenagel condensation in heterogeneous phase catalyzed by IR radiation and Tonsil Actisil FF. *Synthetic Communications* 28: 4649–4663.

Onaka, M., R. Ohno, M. Kawai, and Y. Izumi. 1987. Clay montmorillonite-catalyzed aldol reactions of silyl ketene acetals with carbonyl compounds. *Bulletin of the Chemical Society of Japan* 60: 2689–2691.

Ortega, N., T. Martín, and V.S. Martín. 2009. Synthesis of α,α'-disubstituted linear ethers by an intermolecular Nicholas reaction—Application to the synthesis of (+)-*cis*/(−)-*trans*-lauthisan and (+)-*cis*/(+)-*trans*-obtusan. *European Journal of Organic Chemistry* 554–563.

Osnaya, R., G.A. Arroyo, L. Parada et al. 2003. Biginelli vs Hantzsch esters study under infrared radiation and solventless conditions. *Arkivoc* 2003: 112–117.

Pai, S.G., A.R. Bajpai, A.B. Deshpande, and S.D. Samant. 1997a. Beckmann rearrangement of substituted diaryl ketoximes using FeCl$_3$ impregnated montmorillonite K10. *Synthetic Communications* 27: 379–384.

Pai, S.G., A.R. Bajpai, A.B. Deshpande, and S.D. Samant. 1997b. Friedel–Crafts benzylation of arenes using FeCl$_3$ impregnated montmorillonite K10. *Synthetic Communications* 27: 2267–2273.

Pai, S.G., A.R. Bajpai, A.B. Deshpande, and S.D. Samant. 2000. Benzylation of arenes in the presence of montmorillonite K10 modified using aqueous and acetonitrile solutions of FeCl$_3$. *Journal of Molecular Catalysis A: Chemical* 156: 233–243.

Pinnavaia, T.J., M. Rameswaran, E.D. Dimotakis, E.P. Giannelis, and E.G. Rightor. 1989. Carbon monoxide hydrogenation selectivity of catalysts derived from ruthenium clusters on acidic pillared clay and basic layered double-hydroxide supports. *Faraday Discussions of the Chemical Society* 87: 227–237.

Pitchumani, K. and A. Pandian. 1990. Cation-exchanged montmorillonite clays as Lewis acid catalysts in the Fries rearrangement of phenyl toluene-*p*-sulphonate. *Journal of the Chemical Society, Chemical Communications* 1613–1614.

Poupaert, J.H., J. Bukuru, and A. Gozzo. 1999. Clay (montmorillonite K10) catalysis of the *Michael* addition of α,β-unsaturated carbonyl compounds to indoles: The beneficial role of alcohols. *Monatshefte für Chemie* 130: 929–932.

Poyatos, M., F. Márquez, E. Peris, C. Claver, and E. Fernandez. 2003. Preparation of a new clay-immobilized highly stable palladium catalyst and its efficient recyclability in the Heck reaction. *New Journal of Chemistry* 27: 425–431.

Ramchandani, R.K., M.P. Vinod, R.D. Wakharkar, V.R. Choudhary, and A. Sudalai. 1997. Pd-Cu-exchanged montmorillonite K10 clay: An efficient and reusable heterogeneous catalyst for vinylation of aryl halides. *Chemical Communications* 2071–2072.

Ramesh, E. and R. Raghunathan. 2009. Microwave-assisted K-10 monmorillonite clay-mediated Knoevenagel hetero-Diels–Alder reactions: A novel protocol for the synthesis of polycyclic pyrano[2,3,4-*kl*]xanthene derivatives. *Synthetic Communications* 39: 613–625.

Rao, Y.V.S. and B.M. Choudary. 1991. Knoevenagel condensation catalysed by new montmorillonite silylpropylethylenediamine. *Synthetic Communications* 21: 1163–1166.

Rightor, E.G., M.-S. Tzou, and T.J. Pinnavaia. 1991. Iron oxide pillared clay with large gallery height: Synthesis and properties as a Fischer–Tropsch catalyst. *Journal of Catalysis* 130: 29–40.

Rocchi, D., J.F. González, and J.C. Menéndez. 2014. Montmorillonite clay-promoted, solvent-free cross-aldol condensation under focused microwave irradiation. *Molecules* 19: 7317–7326.

Roudier, J.F. and A. Foucaud. 1984. Clay catalysed ene-reactions. Synthesis of γ-lactones. *Tetrahedron Letters* 25: 4375–4378.

Roudier, J.F. and A. Foucaud. 1986. Clay catalysed ene-reactions. Synthesis of γ-lactones. In *Chemical Reactions in Organic and Inorganic Constrained Systems* (Ed.) R. Setton, pp. 229–235. Dordrecht, the Netherlands: D. Reidel.

Sabitha, G., R.S. Babu, B.V.S. Reddy, and J.S. Yadav. 1999. Microwave assisted Friedländer condensation catalysed by clay. *Synthetic Communications* 29: 4403–4408.

Sadjadi, S., M.M. Heravi, and M. Daraie. 2017. Heteropolyacid supported on amine-functionalized halloysite nano clay as an efficient catalyst for the synthesis of pyrazolopyranopyrimidines via four-component domino reaction. *Research on Chemical Intermediates*, 43(4): 2201–2214. doi:10.1007/s11164-016-2756-8.

Saikia, L., D. Dutta, and D.K. Dutta. 2012. Efficient clay supported Ni^0 nanoparticles as heterogeneous catalyst for solvent-free synthesis of Hantzsch polyhydroquinoline. *Catalysis Communications* 19: 1–4.

Saikia, P.K., P.P. Sarmah, B.J. Borah, L. Saikia, K. Saikiah, and D.K. Dutta. 2016. Stabilized Fe_3O_4 magnetic nanoparticles into nanopores of modified montmorillonite clay: A highly efficient catalyst for the Baeyer–Villiger oxidation under solvent free conditions. *Green Chemistry* 18: 2843–2850.

Salmón, M., R. Osnaya, L. Gómez, G. Arroyo, F. Delgado, and R. Miranda. 2001. Contribution to the Biginelli reaction, using a bentonitic clay as catalyst and a solvent-free procedure. *Revista de la Sociedad Química de México* 45: 206–207.

Samantaray, S., S.K. Sahoo, and B.G. Mishra. 2011. Phosphomolybdic acid dispersed in the micropores of sulfate-treated Zr-pillared clay as efficient heterogeneous catalyst for the synthesis of β-aminocarbonyl compounds in aqueous media. *Journal of Porous Materials* 18: 573–580.

Sartori, G., F. Bigi, R. Maggi, A. Mazzacani, and G. Oppici. 2001. Clay/water mixtures—A heterogeneous and ecologically efficient catalyst for the three-component stereoselective synthesis of tetrahydroquinolines. *European Journal of Organic Chemistry* 2001: 2513–2518.

Sayyahi, S., S. Jahanbakhshi, and Z. Dehghani. 2013. A green and efficient method for the preparation of 3,4-dihydropyrimidin-2(1H)-ones using quaternary ammonium-treated clay in water. *Journal of Chemistry Volume 2013*, Article ID 605324, 5 pages.

Sayyahi, S. 2013. Organoclay as a highly efficient catalyst for three-component one-pot Biginelli synthesis of octahydroquinazolinone derivatives. *Asian Journal of Chemistry* 25: 6471–6472.

Scheuermann, G.M., R. Thomann, and R. Mülhaupt. 2009. Catalysts based upon organoclay with tunable polarity and dispersion behavior: New catalysts for hydrogenation, C–C coupling reactions and fluorous biphase catalysis. *Catalysis Letters* 132: 355–362.

Shaikh, N.S., V.H. Despande, and A.V. Bedekar. 2001. Clay catalyzed chemoselective Michael type addition of aliphatic amines to α,β-ethylenic compounds. *Tetrahedron* 57: 9045–9048.

Shanbhag, G.V. and S.B. Halligudi. 2004. Intermolecular hydroamination of alkynes catalyzed by zinc-exchanged montmorillonite clay. *Journal of Molecular Catalysis A: Chemical* 222: 223–228.

Shanbhag, G.V., S.M. Kumbar, T. Joseph, and S.B. Halligudi. 2006. Heterogeneous intermolecular hydroamination of terminal alkynes with aromatic amines. *Tetrahedron Lettters* 47: 141–143.

Shanmugam, P. and V. Nair. 1996. Montmorillonite K10 clay catalysed glycosidation of 1-O-acetyl-2,3-dideoxy-dl-pent-2-enopyrano-4-ulose. *Synthetic Communications* 26: 3007–3013.

Shanmugam, P. and P. Rajasingh. 2001. Montmorillonite K10 clay-microwave assisted isomerisation of acetates of the Baylis–Hillman adducts: A facile method of stereoselective synthesis of (E)-trisubstituted alkenes. *Synlett* 2001: 1314–1316.

Shanmugam, P. and P. Rajasingh. 2002. Montmorillonite K10 clay catalyzed mild, clean, solvent free one-pot protection-isomerisation of the Baylis–Hillman adducts with alcohols. *Chemistry Letters* 31: 1212–1213.

Shanmugam, P. and P. Rajasingh. 2004. Studies on montmorillonite K10-microwave assisted isomerisation of Baylis–Hillman adduct: Synthesis of E-trisubstituted alkenes and synthetic application to lignan core structures by vinyl radical cyclization. *Tetrahedron* 60: 9283–9295.

Shanmugam, P. and P. Rajasingh. 2005. Montmorillonite K10 clay-catalyzed synthesis of substituted 1-aryl indenes from Baylis–Hillman adducts. *Chemistry Letters* 34: 1494–1495.

Shanmugam, P. and L. Varma. 2001. Montmorillonite K10 clay catalysed Baeyer condensation of heterocyclic aldehydes with N,N-dimethylaniline: Synthesis and photo irradiation studies of heteroaryldiarylmethanes. *Indian Journal of Chemistry* 40B: 1258–1263.

Shanmugasundaram, B., A.K. Bose, and K.K. Balasubramanian. 2002. Microwave-induced, montmorillonite K10-catalyzed Ferrier rearrangement of tri-O-acetyl-D-galactal: Mild, eco-friendly, rapid glycosidation with allylic rearrangement. *Tetrahedron Letters* 43: 6795–6798.

Shao, L.-X. and M. Shi. 2003. Montmorillonite KSF-catalyzed one-pot, three-component, aza-Diels–Alder reactions of methylenecyclopropanes with arenecarbaldehydes and arylamines. *Advanced Synthesis & Catalysis* 345: 963–966.

Sharma, G., R. Kumar, and A.K. Chakraborti. 2007. A novel environmentally friendly process for carbon-sulfur bond formation catalyzed by montmorillonite clays. *Journal of Molecular Catalysis A: Chemical* 263: 143–148.

Sharma, G.V.M., K.C.V. Ramanaiah, and K. Krishnudu. 1994. Clay montmorillonite in carbohydrates: Use of "claysil" as an efficient heterogeneous catalyst for the intramolecular Ferrier reactions leading to 1,6-anhydro rare saccharides. *Tetrahedron: Asymmetry* 5: 1905–1908.

Shimizu, K., T. Kan-no, T. Kodama, H. Hagiwara, and Y. Kitayama. 2002. Suzuki cross-coupling reaction catalyzed by palladium-supported sepiolite. *Tetrahedron Letters* 43: 5653–5655.

Shimizu, K., R. Maruyama, S. Komai, T. Kodama, and Y. Kitayama. 2004. Pd-sepiolite catalyst for Suzuki coupling reaction in water: Structural and catalytic investigations. *Journal of Catalysis* 227: 202–209.

Shimizu, K., M. Miyagi, T. Kan-no, T. Hatamachi, T. Kodama, and Y. Kitayama. 2005. Michael reaction of β-ketoesters with vinyl ketones by iron(III)-exchanged fluorotetrasilicic mica: Catalytic and spectroscopic studies. *Journal of Catalysis* 229: 470–479.

Shimizu, K., M. Miyagi, T. Kan-no, T. Kodama, and Y. Kitayama. 2003. Fe^{3+}-exchanged fluorotetrasilicic mica as an active and reusable catalyst for Michael reaction. *Tetrahedron Letters* 44: 7421–7424.

Shinde, A.B., N.B. Shrigadi, and S.D. Samant. 2004. *tert*-Butylation of phenols using *tert*-butyl alcohol in the presence of $FeCl_3$-modified montmorillonite K10. *Applied Catalysis A: General* 276: 5–8.

Sieskind, O. and P. Albrecht. 1993. Synthesis of alkylbenzenes by Friedel–Crafts reactions catalysed by K10 montmorillonite. *Tetrahedron Letters* 34: 1197–1200.

Singh, A.K., S.K. Shukla, I. Ahamad, and M.A. Quraishi. 2009. Solvent-free microwave-assisted synthesis of 1*H*-indole-2,3-dione derivatives. *Journal of Heterocyclic Chemistry* 46: 571–574.

Singh, D.U. and S.D. Samant. 2004. Comparative study of benzylation of benzene using benzyl chloride in the presence of pillared bentonite; ion-exchanged and pillaring solution impregnated montmorillonite K10. *Journal of Molecular Catalysis A: Chemical* 223: 111–116.

Singh, D.U., P.R. Singh, and S.D. Samant. 2006. Conjugate addition of indoles with electron-deficient olefins catalyzed by Fe-exchanged montmorillonite K10. *Synthetic Communications* 36: 1265–1271.

Singh, V., R. Ratti, and S. Kaur. 2011. Synthesis and characterization of recyclable and recoverable MMT-clay exchanged ammonium tagged carbapalladacycle catalyst for Mizoroki–Heck and Sonogashira reactions in ionic liquid media. *Journal of Molecular Catalysis A: Chemical* 334: 13–19.

Slimi, H., Y. Moussaoui, and R. Ben Salem. 2012. Synthesis of 3,4-dihydropyrimidinones catalyzed by ammonium chloride or montmorillonite KSF without solvent under ultrasonic irradiation. *Journal de la Société Chimique de Tunisie* 14: 1–5.

Song, G., B. Wang, G. Wang, Y. Kang, T. Yang, and L. Yang. 2005. Fe^{3+}-montmorillonite as effective, recyclable catalyst for Paal-Knorr pyrrole synthesis under mild conditions. *Synthetic Communications* 35: 1051–1057.

Soriente, A., R. Arienzo, M. De Rosa, L. Palombi, A. Spinella, and A. Scettri. 1999. K10 montmorillonite catalysis. C–C bond formation by catalyzed conjugate addition and alkoxyalkylation of 1,3-dicarbonyl compounds. *Green Chemistry* 1: 157–162.

Sowmiya, M., A. Sharma, S. Parsodkar, B.G. Mishra, and A. Dubey. 2007. Nanosized sulfated SnO_2 dispersed in the micropores of Al-pillared clay as an efficient catalyst for the synthesis of some biologically important molecules. *Applied Catalysis A: General* 333: 272–280.

Su, H., S. Zeng, H. Dong, Y. Du, Y. Zhang, and R. Hu. 2009. Pillared montmorillonite supported cobalt catalysts for the Fischer–Tropsch reaction. *Applied Clay Science* 46: 325–329.

Sudhakar, H., G.P. Kumari, and N. Mulakayala. 2013. Montmorillonite K10 as highly efficient catalyst for the synthesis of phenols from arylboronic acids. *Indian Journal of Advances in Chemical Science* 2: 57–61.

Suresh and J.S. Sandhu. 2012. Past, present and future of the Biginelli reaction: A critical perspective. *Arkivoc* 66–133.

Takehira, S., Y. Masui, and M. Onaka. 2014. The Mukaiyama aldol reaction for congested ketones catalyzed by solid acid of tin(IV) ion-exchanged montmorillonite. *Chemistry Letters* 43: 498–500.

Tao, R., S. Miao, Z. Liu et al. 2009. Pd nanoparticles immobilized on sepiolite by ionic liquids: Efficient catalysts for hydrogenation of alkenes and Heck reactions. *Green Chemistry* 11: 96–101.

Tateiwa, J.-I., K. Hashimoto, T. Yamauchi, and S. Uemura. 1996. Cation-exchanged montmorillonite (M^{n+}-mont)-catalyzed Prins reaction. *Bulletin of the Chemical Society of Japan* 69: 2361–2368.

Tateiwa, J.-I., H. Horiuchi, K. Hashimoto, T. Yamauchi, and S. Uemura. 1994. Cation-exchanged montmorillonite-catalyzed facile Friedel–Crafts alkylation of hydroxy and methoxy aromatics with 4-hydroxybutan-2-one to produce raspberry ketone and some pharmaceutically active compounds. *Journal of Organic Chemistry* 59: 5901–5904.

Teimouri, A. and A.N. Chermahini. 2016. A mild and highly efficient Friedländer synthesis of quinolines in the presence of heterogeneous solid acid nano-catalyst. *Arabian Journal of Chemistry* 9: S433–S439.

Texier-Boullet, F., B. Klein, and J. Hamelin. 1986. Pyrrole and pyrazole ring closure in heterogeneous media. *Synthesis* 409–411.

Theng, B.K.G. 2012. *Formation and Properties of Clay-Polymer Complexes*, 2nd ed. Amsterdam, the Netherlands: Elsevier.

Theng, B.K.G. and Scharpenseel, H.-W. 1976. The adsorption of ^{14}C-labelled humic acid by montmorillonite. In *Proceedings of the International Clay Conference 1975* (Ed.) S.W. Bailey, pp. 643–653. Wilmette, IL: Applied Publishing.

Thomas, B. and S. Sugunan. 2006. Rare-earth (Ce^{3+}, La^{3+}, Sm^{3+}, and RE^{3+}) exchanged Na-Y zeolites and K10 clay as solid acid catalysts for the synthesis of benzoxazole via Beckmann rearrangement of salicylaldoxime. *Microporous and Mesoporous Materials* 96: 55–64.

Timofeeva, M.N., K.P. Volcho, O.S. Mikhalchenko et al. 2015. Synthesis of octahydro-2*H*-chromen-4-ol from vanillin and isopulegol over acid modified montmorillonite clays: Effect of acidity on the Prins cyclization. *Journal of Molecular Catalysis A: Chemical* 398: 26–34.

Toma, Š., P. Elečko, J. Gažová, and E. Solčániová. 1987. Diels–Alder reaction of acryloylferrocene with 1-phenyl-1,3-butadiene catalysed by homoionic forms of montmorillonite. *Collection of Czechoslovak Chemical Communications* 52: 391–394.

Toshima, K., T. Ishizuka, G. Matsuo, and M. Nakata. 1995. Practical glycosidation method of glycals using montmorillonite K-10 as an environmentally acceptable and inexpensive industrial catalyst. *Synlett* 26: 306–308.

Toshima, K., N. Miyamoto, G. Matsuo, M. Nakata, and S. Matsumura. 1996. Environmentally compatible *C*-glycosidation of glycals using montmorillonite K-10. *Chemical Communications* 1379–1380.

Upadhyay, P. and V. Srivastava. 2015. Ruthenium nanoparticle-intercalated montmorillonite clay for solvent-free alkene hydrogenation reaction. *RSC Advances* 5: 740–745.

Varadwaj, G.B.B., S. Rana, and K.M. Parida. 2013a. Amine functionalized K10 montmorillonite: A solid acid-base catalyst for the Knoevenagel condensation reaction. *Dalton Transactions* 42: 5122–5129.

Varadwaj, G.B.B., S. Rana, and K.M. Parida. 2013b. A stable amine functionalized montmorillonite supported Cu, Ni catalyst showing synergistic and cooperative effectiveness towards C–S coupling reactions. *RSC Advances* 3: 7570–7578.

Varma, R.S., K.P. Naicker, and P.J. Liesen. 1999. Palladium chloride and tetraphenylphosphonium bromide intercalated clay as a new catalyst for the Heck reaction. *Tetrahedron Letters* 40: 2075–2078.

Venkatachalapathy, C. and K. Pitchumani. 1997. Fries rearrangement of esters in montmorillonite clays: Steric control and selectivity. *Tetrahedron* 53: 17171–17176.

Venkatesha, N.J., B.M. Chandrashekara, B.S. Jai Prakash, and Y.S. Bhat. 2014. Active and deactive modes of modified montmorillonite in *p*-cresol acylation. *Journal of Molecular Catalysis A: Chemical* 392: 181–187.

Venkatesha, N.J., B.S. Jai Prakash, and Y.S. Bhat. 2015. The active site accessibility aspect of montmorillonite for ketone yield in ester rearrangement. *Catalysis Science & Technology* 5: 1629–1637.

Vijayakumar, B. and G.R. Rao. 2012. PWA/montmorillonite K10 catalyst for synthesis of coumarins under solvent-free conditions. *Journal of Porous Materials* 19: 233–242.

Villemin, D., B. Labiad, and Y. Ouhilal. 1989. One-pot synthesis of indoles catalysed by montmorillonite under microwave irradiation. *Chemistry & Industry* 607–608.

Wang, G.-W., Q.Q. Hao, Z.-T. Liu, and Z.-W. Liu. 2011b. Fischer–Tropsch synthesis over Co/montmorillonite–Insights into the role of interlayer exchangeable cations. *Applied Catalysis A: General* 405: 45–54.

Wang, J., Y. Masui, and M. Onaka. 2010. Synthesis of α-amino nitriles from carbonyl compounds, amines, and trimethylsilyl cyanide: Comparison between catalyst-free conditions and the presence of tin ion-exchanged montmorillonite. *European Journal of Organic Chemistry* 2010: 1763–1771.

Wang, J., Y. Masui, and M. Onaka. 2011a. Direct synthesis of nitriles from alcohols with trialkylsilyl cyanide using Brønsted acid montmorillonite catalysts. *ACS Catalysis* 1: 446–454.

Waterlot, C., D. Couturier, and B. Rigo. 2000. Montmorillonite-palladium-copper catalysed cross-coupling of methyl acrylate with aryl amines. *Tetrahedron Letters* 41: 317–319.

Xu, W., H. Sun, B. Yu, G. Zhang, W. Zhang, and Z. Gao. 2014. Sonogashira couplings on the surface of montmorillonite-supported Pd/Cu nanoalloys. *ACS Applied Materials & Interfaces* 6: 20261–20268.

Yadav, G.D. and G. George. 2008. Single step synthesis of 4-hydroxybenzophenone via esterification and Fries rearrangement: Novelty of cesium substituted heteropoly acid supported on clay. *Journal of Molecular Catalysis A: Chemical* 292: 54–61.

Yadav, G.D. and S.R. More. 2011. Green alkylation of anisole with cyclohexene over 20% cesium modified heteropoly acid on K-10 acidic montmorillonite clay. *Applied Clay Science* 53: 254–262.

Yadav, G.D., T.S. Thorat, and P.S. Kumbhar. 1993. Inversion of the relative reactivities and selectivities of benzyl chloride and benzyl alcohol in Friedel–Crafts alkylation with toluene using different solid acid catalysts: An adsorption related phenomenon. *Tetrahedron Letters* 34: 529–532.

Yadav, J.S., B.V.S. Reddy, B. Eeshwaraiah, and M. Srinivas. 2004a. Montmorillonite KSF clay catalyzed one-pot synthesis of α-aminonitriles. *Tetrahedron* 60: 1767–1771.

Yadav, J.S., B.V.S. Reddy, G.M. Kumar, and Ch. V.S.R. Murthy. 2001b. Montmorillonite clay catalyzed in situ Prins-type cyclisation reaction. *Tetrahedron Letters* 42: 89–91.

Yadav, J.S., B.V.S. Reddy, and C. Madan. 2001a. Montmorillonite clay-catalyzed stereoselective syntheses of aryl-substituted (*E*)- and (*Z*)-allyl iodides and bromides. *New Journal of Chemistry* 25: 1114–1117.

Yadav, J.S., B.V.S. Reddy, K. Sadasiv, and P.S.R. Reddy. 2002. Montmorillonite clay-catalyzed [4+2] cycloaddition reactions: A facile synthesis of pyrano- and furanoquinolines. *Tetrahedron Letters* 43: 3853–3856.

Yadav, J.S., B.V.S. Reddy, V. Sunitha, K.S. Reddy, and K.V.S. Ramakrishna. 2004b. Montmorillonite KSF-catalyzed one-pot synthesis of hexahydro-1*H*-pyrrolo[3,2-*c*]quinoline derivatives. *Tetrahedron Letters* 45: 7947–7950.

Yadav, M.K. and R.V. Jasra. 2006. Synthesis of nopol from β-pinene using $ZnCl_2$ impregnated Indian montmorillonite. *Catalysis Communications* 7: 889–895.

Yang, Q.J., H.S. Chen, C.R. Liu et al. 2013. Research on dimerization of oleic acid catalyzed by organic montmorillonite. *Advanced Materials Research* 791–793: 120–123.

Zeng, M., Y. Wang, Q. Liu et al. 2016. Encaging palladium nanoparticles in chitosan modified montmorillonite for efficient, recyclable catalysts. *ACS Applied Materials & Interfaces* 8: 33157–33164.

Zhang, Z.-H., F. Yang, T.-S. Li, and C.-G. Fu. 1997. Montmorillonite clay catalysis VI: Synthesis of triarylmethanes via Baeyer condensation of aromatic aldehydes with *N,N*-dimethylaniline catalysed by montmorillonite K-10. *Synthetic Communications* 27: 3823–3828.

Zhao, Y.-H., Q.-Q. Hao, Y.-H. Song et al. 2013. Cobalt supported on alkaline-activated montmorillonite as an efficient catalyst for Fischer–Tropsch synthesis. *Energy Fuels* 27: 6362–6371.

Zhao, Y.-H., Y.-H. Song, Q.-Q. Hao et al. 2015b. Cobalt-supported carbon and alumina co-pillared montmorillonite for Fischer–Tropsch synthesis. *Fuel Processing Technology* 138: 116–124.

Zhao, Y.-H., Y.-J. Wang, Q.-Q. Hao, Z.-T. Liu, and Z.-W. Liu. 2105a. Effective activation of montmorillonite and its application for Fischer–Tropsch synthesis over ruthenium promoted cobalt. *Fuel Processing Technology* 136: 87–95.

Zhu, Z.-B., L.-X. Shao, and M. Shi. 2009. Brønsted acid or solid acid catalyzed aza-Diels–Alder reactions of methylenecyclopropanes with ethyl (arylimino)acetates. *European Journal of Organic Chemistry* 2009: 2576–2580.

Zonouz, A.M. and S.B. Hosseini. 2008. Montmorillonite K10 clay: An efficient catalyst for Hantzsch synthesis of 1,4-dihydropyridine derivatives. *Synthetic Communications* 38: 290–296.

Zonouz, A.M. and N. Sahranavard. 2010. Synthesis of 1,4-dihydropyridine derivatives under aqueous media. *E-Journal of Chemistry* 7: S372–S376.

6 Clay Mineral Catalysis of Isomerization, Dimerization, Oligomerization, and Polymerization Reactions

6.1 INTRODUCTION

The ability of clay minerals to catalyze or promote the isomerization, dimerization, oligomerization, and polymerization of organic compounds has been known for a long time (Theng and Walker 1970; Theng 1974, 1982; Solomon and Hawthorne 1983; Adams 1987; Rupert et al. 1987; Chitnis and Sharma 1997; Eastman and Porter 2000). Some examples of these reactions have been given in the preceding chapters, and listed in Table 3.3. Here we survey and summarize the scattered literature on these topics. Related reactions involved in the clay-catalyzed transformations of organic compounds in soil and sediment are referred to in Chapter 8.

The *isomerization* of an organic molecule/compound refers to its rearrangement into a new molecule/compound with the same number and type of atoms but differing in bonding arrangement. As the term suggests, *dimerization* is the process of combining two structurally identical compounds (*monomers*) through covalent bonding or other bonding modes. For our purposes, *oligomerization* denotes the process of combining a few (3–50) monomers into a molecular entity, while *polymerization* refers to the process of converting a monomer, or a mixture of monomers, into a large chain-like or network molecule. Since organic reactions over acid-activated clay minerals, such as K10 montmorillonite, take place in the proton-rich mesopores and interlayers, there is scope for shape-selective and dimensionally-confined transformations.

6.2 ISOMERIZATION

6.2.1 Hydrocarbons

The isomerization of normal hydrocarbons to their respective branched-chain structures is denoted as skeletal rearrangement, while the isomerization of alkenes involves double-bond migration or conjugation (Kellendonk et al. 1987; Rupert et al. 1987). A (de)hydrogenation step is required for the closely similar process of hydroisomerization. It is generally accepted that the clay-catalyzed isomerization of alkenes involves carbocation intermediates, formed by transfer of protons from the clay mineral surface to the reactants, which are subsequently rearranged and then deprotonated (by proton transfer to the catalyst) to yield the corresponding isomers.

Because of its importance to the flavor and fragrance industry, the transformation of terpenes over clay mineral (and other heterogeneous) catalysts has received much attention (Swift 2004; Volcho and Salakhutdinov 2008; Kumar and Agarwal 2014). One of the earliest examples reported in the *open* literature is the isomerization of terpenes over partially dehydrated palygorskite (Gurwitsch 1923). In this connection, the focus of research has been on the clay-catalyzed isomerization of α-pinene to camphene (Kellendonk et al. 1987).

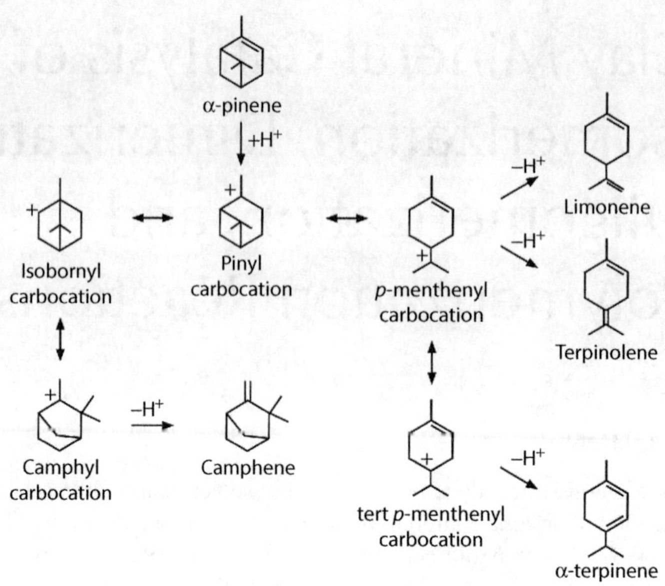

FIGURE 6.1 Diagram showing the Brønsted acid catalyzed isomerization of α-pinene through either the *p*-menthenyl carbocation pathway to give monocyclic ring compounds, such as limonene and terpinolene, or the isobornyl carbocation pathway to yield bicyclic compounds, such as camphene.

Figure 6.1 shows that α-pinene isomerization is a Brønsted acid catalyzed process, initiated by the formation of the pinyl carbocation that rearranges into two main isomers: (a), the *p*-menthenyl carbocation, which gives rise to monocyclic ring compounds, including limonene and terpinolene; and (b), the isobornyl carbocation, which forms bicyclic compounds, such as camphene (Yadav et al. 2004; Flores-Holguín et al. 2012).

As mentioned in Chapter 2, the Brønsted acid sites in clay minerals, such as smectites, arise from the dissociation of water molecules around the interlayer counterions (cf. Equation 2.20). The extent of dissociation is thus dependent on the polarizing power of the counterion: the higher its valency (z), and the smaller its radius (r), the greater is the polarizing effect. In other words, the Brønsted acidity of smectites is positively correlated with the ionic potential (z/r ratio) of the counterion.

For example, Breen and Moronta (2001) found that the activity of two montmorillonites (SWy-2, SAz-1) and a saponite (Sap-Ca), exchanged with different cations, in catalyzing the isomerization of α-pinene (80°C, 2 h) decreased in the order Al > Ni > Mg > Ca > Na as did the polarizing power or ionic potential of the cations. By the same token, Sap-Ca, whose layer charge arose from isomorphous substitution in the tetrahedral sheet, was more active than SWy-2. Intercalation of the polycation *magnafloc* into SWy-2, SAz-1, followed by treatment with 6 M HCl (95°C), greatly enhanced the activity of the samples in catalyzing the isomerization of α-pinene to camphene and limonene because of the increased hydrophobicity of the interlayer space (Breen and Watson 1998). Opening up the montmorillonite interlayers by pillaring with Al_{13} cation (cf. Figure 3.2) was also conducive to α-pinene isomerization (Yao et al. 2013). Earlier examples of hydrocarbon isomerization reactions, catalyzed by pillared smectites, are given in Table 3.5.

Likewise, acid treatment of tetramethylammonium (TMA^+)-intercalated SWy-2 gave rise to a catalytically active material for the isomerization of α-pinene (Breen and Moronta 1999). The activity of mixed Al^{3+}/TMA^+-montmorillonites was similar to that of the fully Al^{3+}-exchanged samples when less than 20% of the exchange sites were occupied by TMA^+ ions. Beyond this level of occupancy, however, the presence of TMA^+ had a depressing effect on α-pinene conversion (Breen and Moronta 2000). In this context, it is relevant to mention the work by Moronta et al. (2002) on the catalytic activity of acid-treated TMA^+-exchanged montmorillonites (for the isomerization of

1-butene to *cis*- and *trans*-2-butene). Montmorillonite that is first treated with acid (12 M HCl) and then intercalated with TMA$^+$ is apparently less active than the material obtained by reversing the order of treatment.

Volzone et al. (2001) reported that the activity of acid-treated smectites in promoting the isomerization of α-pinene was positively correlated with the surface area, pore volume, and surface acidity of the clay samples. They further noted that the reaction was more selective to camphene than limonene. Using sulfuric acid-activated montmorillonites to catalyze the liquid phase isomerization of α-pinene, Gündüz and Murzin (2002) found a nearly constant (initial) selectivity to camphene and limonene. In terms of camphene production, optimum activity was observed at 200°C (4 h) at an acid/clay ratio between 0.2 and 0.4. Under these conditions, the catalysts had a high Brønsted acidity and a homogeneous mesopore (2 nm) distribution. A heterogeneous mesoporosity, on the other hand, appeared to favor limonene formation (Beşün et al. 2002). Similarly, Agabekov et al. (2011) reported that the activity of a natural aluminosilicate for α-pinene isomerization and its selectivity for camphene production were dependent on the ratio of acid (10% HCl) volume to catalyst weight.

Using sulfuric acid-modified montmorillonites, exchanged with Ce^{3+}, Fe^{3+}, La^{3+} and Ag$^+$ cations, Yadav et al. (2004) also observed more than 93% conversion of α-pinene with 43%–49% selectivity to camphene. It is not clear, however, why the Ca^{2+}- and Mg^{2+}-exchanged clays are relatively inactive. Limonene was the principal product formed using clays that had been treated with 0.5–2 M H$_2$SO$_4$, while α-terpinene (Figure 6.1) was mainly formed in the presence of clays that had been pretreated with 2.5–4.5 M H$_2$SO$_4$.

The role of interlayer counterions in *directing* hydrocarbon reactions has also been described by Frenkel and Heller-Kallai (1983) with reference to the transformation of limonene over some cation-exchanged montmorillonites. They found that the clay-catalyzed disproportionation of limonene (**1**) to *p*-menthene (**2**) and *p*-cymene (**3**), depicted in Scheme 6.1, was influenced by the surface acidity of the clay catalysts, increasing in the order Na$^+$ < Mg^{2+} < Al^{3+} < H$^+$. In a follow-up study, Catrinescu et al. (2006) described the isomerization of limonene (to terpinolene, isoterpinolene, α-terpinene, γ-terpinene), and its disproportionation (to *p*-menthene and *p*-cymene) in the presence of Serra de Dentro (SD) bentonite from Portugal and SAz-1 montmorillonite from Arizona (USA), exchanged with Al^{3+}, Ni^{2+}, Cr^{3+}, and Na$^+$ ions. Because of its higher surface acidity, structural iron content, and specific surface area, the SD bentonite was more active than the SAz-1 material. Here again, the nature of the interlayer counterion has a significant effect on catalytic activity, which in the case of the SD clay decreases in the order Ni^{2+} > Al^{3+} > Cr^{3+} > Na$^+$. We might also add that montmorillonite can effectively promote the isomerization of 1,2-limonene oxide to carvenone under solvent-free conditions (Thi Nguyen et al. 2013).

The transformation of trimethylbenzene over pillared montmorillonite and other smectites has been the subject of many investigations. Using Al- and Zr-PILC as catalysts, Kikuchi and coworkers (Kikuchi et al. 1984, 1985; Matsuda et al. 1988) have noted that the more common conversion is one of disproportionation rather than isomerization. Thus, 1,2,4-trimethylbenzene selectively yields 1,2,4,5-tetramethylbenzene (the smallest of the tetramethylbenzene isomers) and *o*-xylene, allowing the former product to isomerize (Kojima et al. 1991). The isomerization reaction, however, was greatly suppressed when the catalyst was heated above 550°C due to a reduction in strong Brønsted acid sites.

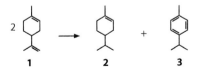

SCHEME 6.1 Conversion by disproportionation of limonene (**1**) to *p*-menthene (**2**) and *p*-cymene (**3**) catalyzed by various cation-exchanged montmorillonites. (After Frenkel, M. and Heller-Kallai, L., *Clays Clay Min.*, 31, 92–96, 1983.)

Ko and Cheng (1992) further showed that the selectivity for trimethylbenzene (TMB) isomerization over Al-PILC decreased in the order 1,2,3-TMB ≥ 1,2,4-TMB > 1,3,5-TMB.

The isomerization of butenes, in relation to the surface acidity of the clay catalyst, has been the topic of many investigations. The early work by Sohn and Ozaki (1980) on the isomerization *n*-butene (1-butene) over a synthetic nickel-substituted montmorillonite has been described in some detail in Chapter 2 (cf. Figure 2.14). Moronta et al. (2005) used two commercial acid-activated montmorillonites (F24 and F124) exchanged with Al^{3+}, Cr^{3+}, Fe^{3+}, and Ni^{2+} ions to catalyze the isomerization of *cis*-2- and *trans*-2-butenes (at 300°C). They found that catalytic activity was correlated with the ionic potential of the exchangeable cations and with the Brønsted acidity of the clay samples (as determined by cyclohexylamine desorption). Like F24 and F124, K10 montmorillonite can efficiently promote the isomerization of cholest-5-ene to a mixture of (20R and 20S) cholest-13(17)-enes (Sieskind and Albrecht (1985), and that of 5-vinyl-2-norbornene to 5-ethylidene-2-norbornene (Pillai et al. 1995) as well as that of 2-ethyl-1-hexene to a mixture of the *cis*- and *trans*-isomers of 3-methyl-2-heptene and 3-methyl-3-heptene (Harvey and Quintana 2010).

In a follow-up study, Moronta et al. (2008) used a montmorillonite from Texas, USA (STx-1), exchanged with Al^{3+} and Fe^{3+}, or pillared with polyoxo cations of aluminum and iron, to isomerize 1-butene at 300°C. The performance of the pillared montmorillonite is superior to that of the cation-exchanged samples, while the efficiency of the Al-pillared material (Al-PILC) is related to its high acidity and specific surface area. Singh et al. (2007) have also found that Al- and Zr-PILC are efficient in converting longifolene to isolongifolene. Subsequently, Radwan et al. (2009) used a sulfated zirconia-pillared bentonite to promote the transformation of *n*-hexane into mono- and di-branched isomers. González et al. (2009) prepared a disordered mesoporous Al-PILC (cf. Figure 3.3) by intercalating Al_{13} into a synthetic trioctahedral smectite in the presence of a non-ionic surfactant (Igepal CO-720) and then used the material to catalyze the isomerization of 1-butene into *cis*- and *trans*-butene. A montmorillonite-supported chloroaluminate ionic liquid, prepared from a quaternary ammonium chloride and $AlCl_3$, was used by Huang et al. (2010) to catalyze the transformation of *endo*-tetrahydrodicyclo-pentadiene into the corresponding *exo* isomer.

We might also mention that clay minerals can catalyze the isomerization of highly branched isoprenoid (HBI) alkenes, extracted from the diatom *Haslea ostrearia* (Belt et al. 2000; Belt and Cabedo-Sanz 2015). In the case of 25-carbon HBI dienes (**1**), the Brønsted acid-catalyzed reaction over K10 montmorillonite involved only double-bond migration and geometric isomerization to give the *E*- and *Z*-isomers. On the other hand, the corresponding trienes (**2**) underwent isomerization, followed by rapid cyclization, to yield the substituted cyclohexenes (**3**). In both instances, the clay catalyst acted as a Brønsted acid by protonating one of the double bonds to yield the corresponding carbocation intermediate (Scheme 6.2).

Similar reactions are involved in the transformation of hydrocarbons, such as steranes and triterpanes (Tannenbaum et al. 1986), and other organic compounds in sediments (Chapter 8). Interestingly, Lao et al. (1988, 1989) found kaolinite to be more efficient than bentonite in promoting the low-temperature double-bond isomerization of 1-pristene to 2-pristene, especially after drying the clay at 100°C, and even when the mineral has previously been coated with char (at 370°C). Above 100°C, pristane was formed by hydrogen transfer. Lao et al. further noted that the isomerization of 1-undecane in the presence of various clay minerals took place side by side with the formation of methyldecanes, undecane, and some cracking products.

Although smectites account for the majority of clay-catalyzed hydrocarbon transformations, other clay mineral species are known to be similarly active. Nazir et al. (1976), for example, used sulfuric acid-treated kaolinite and fire clay to catalyze the isomerization of α-pinene to camphene. Likewise, Perissinotto et al. (1997) have found that acid-treated metakaolinite is as active as, if not more so than, K10 montmorillonite in promoting the isomerization of 1-butene to isobutene. Using a natural kaolinite that has been heated and then treated with 3 M sulfuric acid as catalyst, Volzone et al. (2005) could achieve a 67%–94% transformation of α-pinene with a selectivity to camphene of 65% and to limonene of 23%. Imamura et al. (1993) have suggested that the strong (Brønsted)

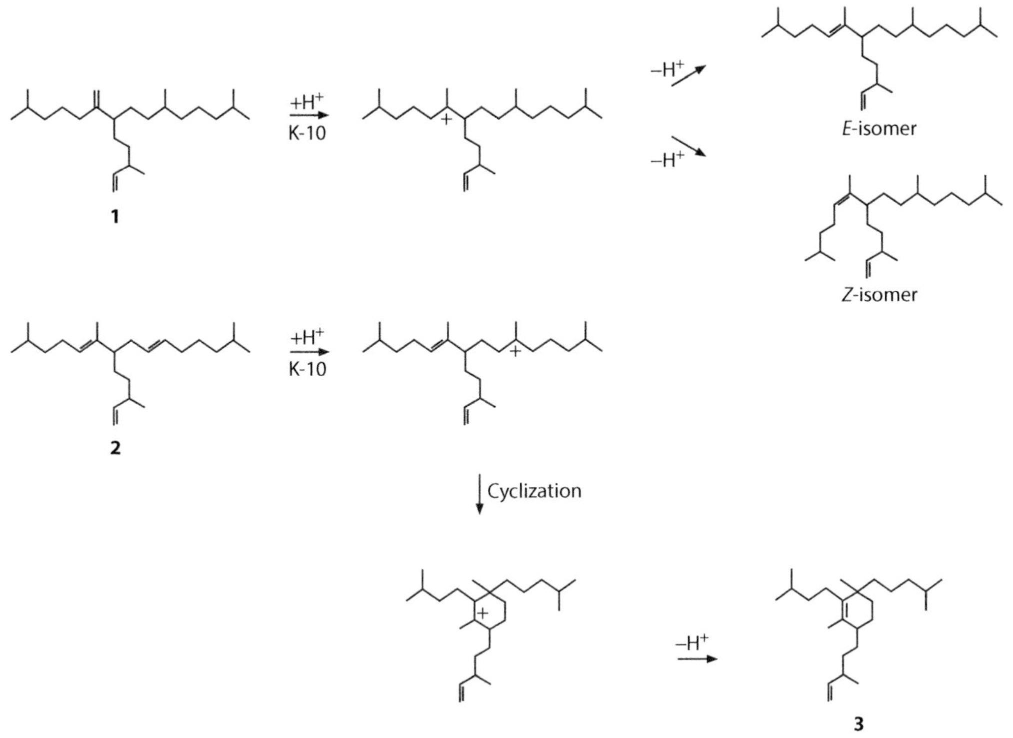

SCHEME 6.2 K10 montmorillonite-catalyzed isomerization of highly branched isoprenoid (25-carbon) dienes (**1**) to yield the *E*- and *Z*-isomers. Isomerization of the corresponding trienes (**2**), followed by rapid cyclization, yields substituted cyclohexenes (**3**). (After Belt, S.T. et al., *Geochimica et Cosmochimica Acta*, 64, 3337–3345, 2000; Belt, S.T. and Cabedo-Sanz, P., *Org. Geochem.*, 87, 55–67, 2015.)

acid sites, formed by heating at 400°C–500°C, lie behind the ability of imogolite to catalyze the isomerization of 1-butene to *cis*-2- and *trans*-2-butene. The efficiency of an acid-treated natural aluminosilicate (illite) in promoting the isomerization of α-pinene (Sidorenko et al. 2014) may be explained in similar terms.

Following Sohn and Ozaki (1980), van Santen et al. (1985) used a synthetic nickel-substituted mica montmorillonite (Ni SMM) (cf. Chapter 1) before and after impregnation with Pd (nanoparticles) as a catalyst for the hydroisomerization of *n*-alkanes. They noted that isomorphous substitution of Al^{3+} for Ni^{2+} in octahedral positions, and subsequent reduction of structural nickel ions, led to an increase in both Brønsted acidity and isomerization activity. The de(hydrogenation) reaction, on the other hand, was controlled by Pd nanoparticles. In other words, Pd-impregnated Ni SMM can act as a bifunctional catalyst in paraffin hydroisomerization. The efficiency of Al-pillared saponites and synthetic saponites in promoting the hydroisomerization of *n*-heptane may be explained in similar terms (Moreno et al. 1997; Vogels et al. 2005).

van Santen et al. (1985) further reported that pillaring of Ni SMM with Al_{13} cations led to an appreciable increase in both surface area (from 170 to 230 m²/g) and catalytic activity. Molina et al. (1994) have similarly suggested that alumina-pillared synthetic beidellites can act as bifunctional catalysts for the hydroisomerization (and hydrocracking) of decane. Kooli et al. (2008) have also noted that alumina-pillared montmorillonite (Al-PILC) and saponites, especially the latter, are effective in catalyzing the hydroisomerization of *n*-heptane (at 300°C). A Pt-Re impregnated alumina pillared bentonite was used by Parulekar and Hightower (1987) to catalyze the hydroisomerization of *n*-pentane, *n*-hexane, and *n*-heptane, yielding single-branched isomers as the major products. Al-PILC that has been impregnated with Pt, Pd, and Ni nanoparticles is similarly effective (Zakarina et al. 2008). This observation is in keeping with the general observation that metal impregnation increases the Brønsted acidity, catalytic activity, and isomer selectivity of pillared montmorillonites (Issaadi et al. 2001, 2006; Issaadi and Garin 2003). Even the incorporation of some metals (Zr, Si) into the pillars of Al-PILC has a positive effect on (n-heptane) hydroconversion (Molina et al. 2005). More recently, Kooli (2014) reported that porous clay heterostructures (cf. Figure 3.4) prepared from an Al_{13}-intercalated montmorillonite and Al-PILC were moderately effective in catalyzing the hydroisomerization of n-heptane, giving an isomer yield of 60% with a 50% selectivity at 350°C. Other examples of alkane hydroisomerization, catalyzed by various pillared clays, are listed in Table 3.5.

6.2.2 Non-Hydrocarbons

Among non-hydrocarbons, epoxides are attractive substrates for isomerization as both Brønsted and Lewis acids can apparently promote the opening of the epoxide ring (Nagendrappa 2011). An example is the room temperature isomerization of diastereomeric (R)-(+)-limonene diepoxide (**1**) over K10 montmorillonite (in dichloromethane) to yield compounds (**2**), (**3**), (**4**), and (**5**) in the ratio of 4:3:7:2, according to Scheme 6.3 (Salomatina et al. 2005).

In investigating the isomerization of allyl alcohols (of the pinane series) and their epoxides in the presence of a bentonite clay, Il'ina et al (2007) found that (+)-*trans*-pinocarveol was converted to the (−)-myrtenol isomer (and a dimeric ether) whereas under homogeneous (acid) conditions, the former compound underwent a rearrangement reaction through protonation of the exocyclic double bond. The respective epoxides of these alcohols were transformed into different isomeric compounds although the same intermediate might have been involved in both cases. Thus, pinocarveol epoxide gives the corresponding α,β-unsaturated ketone, while myrtenol epoxide yields 4-isopropyl benzylalcohol, as the main product. Biermann and Metzger (2008) reported that the clay-catalyzed isomerization of oleic acid yielded a complex mixture of dimeric fatty acids and the monomeric alkyl-branched fatty compounds (*isostearic acid*) whereas the zeolite-mediated reaction gave rise to a mixture isostearic acids with the methyl branch on the 8–14 positions of the alkyl chain (See Figure 6.3).

A striking example of a shape- or stereo-selective conversion was provided by Ortega et al. (2006) who reported that the *cis*-isomer of lauthisan (**1**) was enriched from a *cis*:*trans* ratio of 1:1.7 to one of 17:1 when a mixture of *cis*- and *trans*-lauthisan (**2**) (Scheme 6.4) was exposed

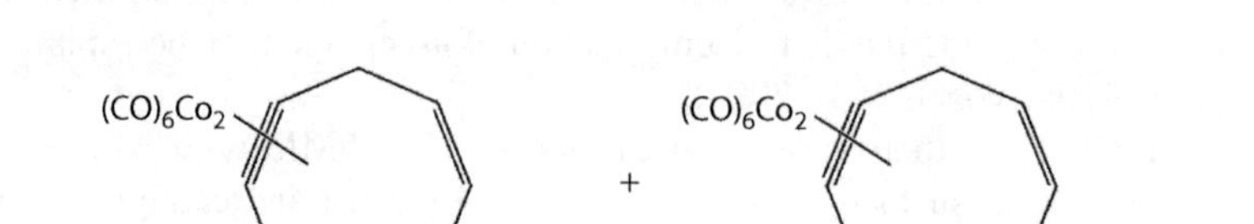

SCHEME 6.3 Isomerization of diastereomeric (R)-(+)-limonene diepoxide (**1**) over K10 montmorillonite to yield compounds (**2**), (**3**), (**4**), and (**5**) in certain proportions. (After Salomatina, O.V. et al., *Mendeleev Commun.*, 59–61, 2005.)

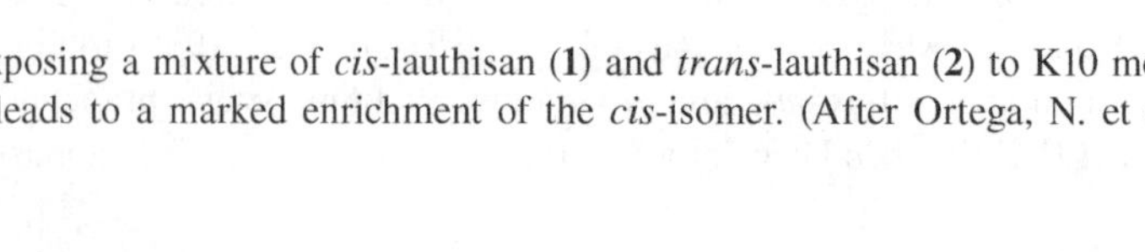

SCHEME 6.4 Exposing a mixture of *cis*-lauthisan (**1**) and *trans*-lauthisan (**2**) to K10 montmorillonite (in dichloromethane) leads to a marked enrichment of the *cis*-isomer. (After Ortega, N. et al., *Org. Lett.*, 8, 871–873, 2006.)

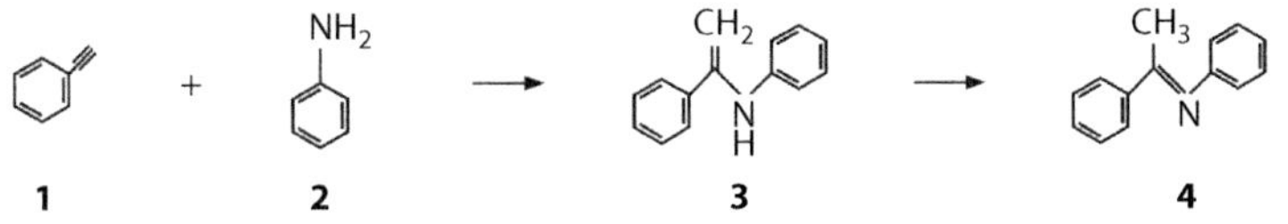

SCHEME 6.5 Hydroamination of phenylacetylene (**1**) with aniline (**2**), in the presence of Zn^{2+}-K10 montmorillonite, to give the corresponding enamine (**3**) which then isomerizes to phenyl-(1-phenylethylidene) amine (**4**). (After Shanbhag, G.V. and Halligudi, S.B., *J. Mol. Catal. A: Chem.*, 222, 223–228, 2004.)

to K10 montmorillonite in dichloromethane at room temperature. Such a reaction was not observed in the presence of acidic catalysts under homogeneous conditions. Earlier, Larcher et al. (1986) described the occurrence of alkyl hydrogen exchange and configurational isomerization in acyclic isoprenoid acids when these compounds were heated at 160°C in the presence of montmorillonite. The former reaction took place between clay-adsorbed water and the α-position of the organic acid, while the latter transformation was observed when this position was chiral.

An interesting reaction, described by Shanbhag and Halligudi (2004), is the conversion of the enamine (**3**) derived from the hydroamination of phenylacetylene (**1**) with aniline (**2**) into the phenyl-(1-phenylethylidene) amine (**4**) isomer in the presence of Zn^{2+}-K10 montmorillonite (Scheme 6.5). The high efficiency and regioselectivity of the intermolecular hydroamination reaction may be ascribed to an interplay of Brønsted and Lewis acidities. Shanbhag and Halligudi suggested that the reaction was facilitated through activation of the triple bond in (**1**) by π-complexation with exchangeable Zn^{2+} cations acting as Lewis acid sites.

As already referred to in Chapter 5, phenyl toluene-*p*-sulfonate can undergo a Fries rearrangement in the presence of various cation-exchanged K10 montmorillonites to give 2- and 4-hydroxyphenyl-*p*-tolyl sulfones with the former isomer being selectively formed (Pitchumani and Pandian 1990). The same catalysts were also used by Kannan et al. (1999) to promote the (Orton) rearrangement of *N*-chloroacetanilide (**1**) in carbon tetrachloride (60°C, 1 h) to yield *p*-chloroacetanilide (**2**) as the major product and a small amount of the corresponding *o*-isomer (**3**) (Scheme 6.6). In the absence of clay, on the other hand, the conversion was incomplete even after a few days' immersion in the solvent. Kumar et al. (2014) also made the interesting observation that the rearrangement of benzyl phenyl ether over H^+-, Al^{3+}-, and Fe^{3+}-K10 montmorillonites under solvent-free conditions yielded the *o*-benzylphenol isomer as the sole product within 3 h of contact. Under the same conditions, diazoaminobenzene was also completely and selectively converted to its *p*-isomer. Kumar et al. suggested that the conversion involved protonation of diazoaminobenzene, dissociation into diazonium cation and aniline, and attack by the protonated compound of aniline at the *para* position.

Assisted by microwave irradiation (2 min), Das et al. (2000) were able to epimerize naturally occurring furofuran lignans, such as (+)-sesamin, (+)-eudesmin, (+)-syringaresinol, and (+)-yangambin, in

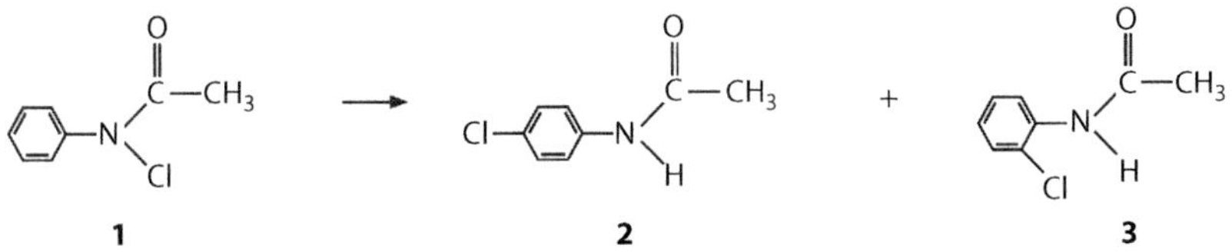

SCHEME 6.6 Rearrangement of *N*-chloroacetanilide (**1**) to give *p*-chloroacetanilide (**2**) together with a small amount of the corresponding *o*-isomer (**3**), catalyzed by various cation-exchanged K10 montmorillonites. (After Kannan, P. et al., *Indian J. Chem.*, 38B, 384–386, 1999.)

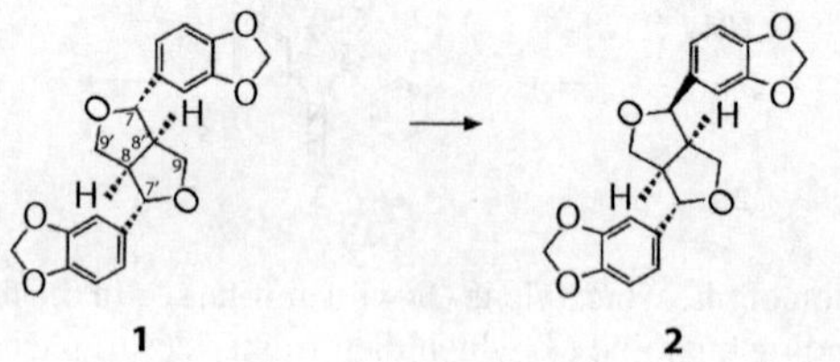

SCHEME 6.7 Conversion of (+)-sesamin (**1**) to the corresponding C-7 epimer (**2**) in the presence of KSF montmorillonite under microwave irradiation. (After Das, B. et al., *Synthetic Commun.*, 30, 4001–4006, 2000.)

the presence of KSF montmorillonite. Scheme 6.7 illustrates the conversion of (+)-sesamin (**1**) to its corresponding C-7 epimer (**2**). Earlier, Varma and Saini (1997) reported that K10 montmorillonite could catalyze the isomerization of 2′-aminochalcones to 2-aryl-1,2,3,4-tetrahydro-4-quinolones under mild and solvent-free conditions, especially when assisted by microwave irradiation.

To end this section, we wish to mention the clay-catalyzed *cis-trans* isomerization of $Ru(bpy)_2(H_2O)_2^{2+}$ where bpy denotes 2,2′-bipyridine (Cruz et al. 1982). The isomerization of this positively charged compound in aqueous solution can be induced by light irradiation, the *cis* isomer being the stable form. In the interlayer space of hectorite, however, the reaction is highly dependent on the hydration status of the clay mineral. Surprisingly, the *cis*-isomer is the preferred form when the clay-organic complex is dehydrated whereas the *trans*-isomer is most stable when the interlayers are hydrated, indicating that the surface conformation of the adsorbed complex is water-dependent.

Another interesting reaction is the *sensitized* photoisomerization of *cis*-stilbazolium cations (cf. Figure 6.2) by $Ru(bpy)_3^{2+}$ in the interlayer space of saponite in that the yield is 100 times higher than that observed in a homogeneous solution (Usami et al. 1998). More recently, Menéndez-Rodríguez et al. (2013) reported that the K10-supported $RuCl_2(\eta^6\text{-}C_6H_6)$(PTA-Me) complex where PTA-Me stood for the ionic phosphine ligand, 1-methyl-3,5-diaza-1-azonia-7-phosphaadamantane chloride, was an efficient and reusable catalyst for the (redox) isomerization of various allylic alcohols (e.g., 1-octene-3-ol) to the corresponding carbonyl compounds (e.g., octane-3-one).

6.3 DIMERIZATION

Like isomerization, the dimerization of organic species in the presence of clay minerals is more often than not a Brønsted acid catalyzed reaction, involving carbocation intermediates. Some conjugated dienes can also form dimers through a Diels–Alder cycloaddition reaction (Adams and Clapp 1986). Under certain conditions and over transition metal-exchanged smectites, dimerization may be mediated by electron transfer from the adsorbed monomer to the metal ion, acting as a Lewis acid site, to yield the corresponding radical-cation as indicated in Figure 2.8 for the conversion of triphenylamine to *N,N,N′,N′*-tetraphenylbenzidine (Tricker et al. 1975a), and in Figure 2.9 for the dimerization of 1,1-diphenylethylene to 1-methyl-1,3,3-triphenylindan (Adams et al. 1977; Thomas et al. 1977). The dimerization of dibenzo-*p*-dioxin, pentachlorophenol, 3- and 4-chloroanisoles, aniline, *N,N*-dimethylaniline, and pyrene has been referred to in Chapter 2.

6.3.1 Hydrocarbons

Tricker et al. (1975b) have suggested that the thermal dimerization of *trans*-stilbene over dehydrated Cu^{2+}-exchanged montmorillonite is facilitated by the interaction of the central olefinic double bond of the monomer with the interlayer Cu^{2+} ion. More recently, Joseph-Ezra et al. (2014) reported that pyrene could undergo both oxidation and dimerization on the surface of

Cu^{2+}-montmorillonite. On the other hand, the solvent-free dimerization (90°C–115°C, 10–15 min) of *trans*-stilbene and other olefins, such as 1,1-diphenylethylene and α-methylstyrene, in the presence of Fe^{3+}- and Al^{3+}-exchanged K10 montmorillonite, is a Brønsted acid catalyzed process. Derivatives of indane and tetralin are also formed (Madhavan et al. 2001). These observations are consistent with Hiscock and Porter's (1972) proposal that the formation of indane and tetralin from *trans*-stilbene in the presence of $SbCl_3$ involves a dimeric stilbene carbocation. The dimerization of butadiene to 4-vinylcyclohexene and of isoprene to limonene (*p*-methadiene) over Cr^{3+}-montmorillonite similarly involves carbocation intermediates. When the reaction temperature is raised, the yield of dimers increases and oligomers are formed (Adams and Clapp 1986). Brønsted acid sites are also involved in the selective dimerization of ethylene to *n*-butenes over Ni^{2+}-montmorillonite (Sohn and Park 1996). Similarly, Ballantine et al. (1985) have identified the dimers 2,4,4-trimethylpent-2-ene and 2,4,4-trimethylpent-1-ene by intercalating 2-methylpropene into dehydrated Al^{3+}-montmorillonite and Al^{3+}-Laponite. According to Shah and Sharma (1993) and Shah et al. (1994), the rate of dimerization of isoamylene (2-methyl-2-butene) over a commercial acid-treated (Engelhard F-24) clay at 60°C–100°C follows first-order (Rideal-Eley) kinetics as does the cross-dimerization of methylstyrene with isoamylene. Interestingly, Chaudhuri and Sharma (1989) have noted that different acid-treated clays can effectively promote the dimerization of α-methylstyrene with a high selectivity for the industrially important dimer, 2,4-diphenyl-4-methyl-1-pentene.

The dimerization of 1-dodecene in the presence of Mg^{2+}-Laponite was reported by Koster et al. (1998). Subsequently, Pillai and Ravindranathan (1994) used K10 montmorillonite to promote the liquid-phase dimerization (and oligomerization) of 1-decene. The same clay can also catalyze the dimerization (and trimerization) of α-methyl styrene in the presence of such solvents as pentane and dichloromethane (Pillai et al. 1995). Similarly, Harvey et al. (2010) found K10 montmorillonite to be an efficient catalyst for the dimerization of β-pinene, and the formation of *p*-cymene by opening the β-pinene ring, followed by dehydrogenation.

The synthesis of high-density fuels through the selective dimerization of α-pinene, camphene, limonene, and crude turpentine over various solid acid catalysts, including K10 montmorillonite, has been reported by Meylemans et al. (2012). The dimerization and trimerization of aromatic hydrocarbons (e.g., naphthalene), hydroxylated aromatics (e.g., 2-naphthol), and methyl-substituted aromatics (e.g., 2-methylnaphthalene) in the presence of Fe^{3+}-montmorillonite may be ascribed to oxidative coupling, which involves the reduction of exchangeable Fe^{3+} to Fe^{2+} (Watson and Sephton 2015). This process and related clay-catalyzed Fenton-like oxidation reactions will be described in Chapter 7. Wiederrecht et al. (2001) have also noted that pyrene can form face-to-face dimers in the interpillar space of alumina-pillared montmorillonite (cf. Figure 3.2), allowing 5%–6% w/w of the monomer to be loaded whereas other aromatic hydrocarbons such as benzene, naphthalene, and perylene show only limited incorporation.

6.3.2 Non-Hydrocarbons

As we have mentioned in Chapter 2, intercalated anisole in Cu^{2+}-hectorite can dimerize to form 4,4′-dimethoxybiphenyl (Fenn et al. 1973). Adams et al. (1982) have also noted that divalent and trivalent cation-exchanged montmorillonites can efficiently promote the dimerization of anethole (*p*-methoxyphenyl propene), when refluxed in xylene, to give a mixture of metanethole and isomethole. On the other hand, the conversion of isohomogenol (1,2-dimethoxy-4-propenylbenzene) into di-isohomogenol requires the presence of highly acidic Al^{3+}- or Fe^{3+}-montmorillonite. Cu^{2+}-montmorillonite is similarly effective in converting adsorbed phenol and 2,6-xylenol into the corresponding biphenyl-diol dimers (Soma et al. 1986).

Al^{3+}-montmorillonite is efficient in dimerizing 2-(trifluoromethyl) aniline, while the Cu^{2+}-exchanged counterpart is highly active in catalyzing the formation of the 12-membered cyclic trimer (Kowalska et al. 2001). Likewise, structural or exchangeable Fe^{3+} in smectites can induce the dimerization of 1-aminonaphthalene, presumably by the reaction between the radical-cation and the neutral form of the molecule (Ainsworth et al. 1991). Al^{3+}-exchanged bentonite is similarly efficient in promoting the dimerization of ethylene oxide to 1,4-dioxan and 2-methyl-1,3-dioxolane (Ballantine et al. 1983). Using the superacidic montmorillonite-supported trifluoromethanesulfonic acid ($H_o < -12.75$), Salmón et al. (1997) were able to dimerize propylene oxide and its reaction products with various alcohols.

The photocyclodimerization of stilbazolium cations in the interlayer space of saponite has been investigated by Usami and coworkers (Takagi et al. 1989; Usami et al. 1990, 1992a). When an aqueous suspension of the clay-organic complex is irradiated with ultraviolet light, the intercalated (*E*)-stilbazolium ions (**1**) are regioselectively converted to the *syn* head-to-tail dimers (**2**) together with a small amount of *syn* head-to-head dimers (**3**) and the corresponding (*Z*)-stilbazolium ions (**4**) as shown in Scheme 6.8. In homogeneous solutions, on the other hand, (*E-Z*) or *cis-trans* isomerization is the preferred transformation.

Nakahira et al. (1995) have reported similarly for the interlayer complex with hectorite. These observations would indicate that the interlayer stilbazolium ions (in saponite) adopt an antiparallel orientation (Figure 6.2), allowing intermolecular van der Waals interactions to be established in dependence on the negative layer charge separation. Similarly, Aldersley and Joshi (2013) have found that the surface layer charge density of montmorillonite plays an important role in directing the synthesis of RNA dimers from a nucleoside and an activated nucleotide. Interestingly, the cointercalation of long-chain alkylammonium ions suppresses the cyclodimerization of stilbazolium ions, making *cis-trans* isomerization to be the dominant reaction (Usami et al. 1990).

Similarly, Madhavan and Pitchumani (2002, 2003) have observed that the photo-induced dimerization of intercalated enones in bentonite is sensitive to the interlayer arrangement of the reactants,

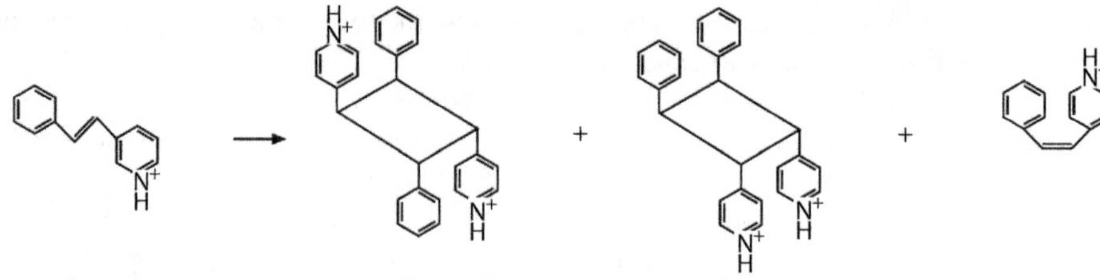

SCHEME 6.8 Regioselective conversion of saponite-intercalated (*E*)-stilbazolium ions (**1**) to the *syn* head-to-tail dimers (**2**) under ultraviolet irradiation. A small amount of *syn* head-to-head dimers (**3**) and (*Z*)-stilbazolium ions (**4**). (After Takagi, K. et al., *J. Chem. Soc. Chem. Commun.*, 1174–1175, 1989; Usami, H. et al., *J. Chem. Soc., Perkin Trans.*, 2, 1723–1728, 1990; Usami, H. et al., *J. Chem. Soc. Faraday Trans.*, 88, 77–81, 1992a.)

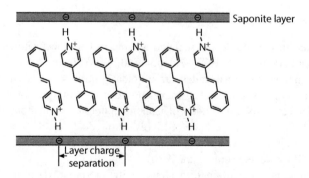

FIGURE 6.2 Diagram showing the arrangement of stilbazolium ions in the interlayer space of saponite. (Adapted from Usami, H. et al., *J. Chem. Soc., Perkin Trans.*, 2, 1723–1728, 1990.)

which in this case would be controlled by ion-dipole interactions between the carbonyl group of the compounds and the exchangeable cations at the clay surface (Mortland 1970; Theng 1974). Thus, the clay-catalyzed photodimerization of 2-cyclohexenone is highly regioselective, favoring the formation of head-to-head dimers whereas 2-cyclopentenone dimerizes in a head-to-tail fashion. Interestingly, the photodimerization of acenaphthylene in the interlayer space of monovalent cation-exchanged bentonites preferentially leads to the formation of the *cis*-dimer, while the formation of the *trans*-dimer is favored with polyvalent cation-exchanged bentonites. The regioselective photocyclodimerization of intercalated 2-cyclohexenones, yielding *anti*-head-to-head dimers, has also been reported by Usami et al. (1992b) using saponite as a catalyst.

The dimerization (and oligomerization) of unsaturated fatty acids has received much attention because the products have applications in the adhesive, paint, and plastic industries. The reaction is commonly carried out in an autoclave at a high temperature and under pressure (Adams 1987). Figure 6.3 shows a simplified diagram for the dimerization of oleic acid ($C_{18}H_{34}O_2$) together with the principal side products. As Koster et al. (1998) have pointed out, each product comprises a variety of compounds due to the occurrence of many side reactions, notably hydrogen transfer, double-bond shift, *cis-trans* isomerization, and chain branching. Since the oleic acid feedstock commonly contains other fatty acids, such as elaidic acid (the *trans* isomer of oleic acid), palmitic acid, and stearic acid, the results of various dimerization reactions are not always self-consistent.

In an early systematic investigation into the montmorillonite-catalyzed dimerization of oleic acid, den Otter (1970a, 1970b, 1970c) reported a yield limit of ca. 60% w/w of dimers and trimers, and the apparent formation of cyclic dimers (and bicyclic trimers). Dimeric cyclic and acyclic structures were also detected by McMahon and Crowell (1974), and more recently by Park et al. (2015) using proton NMR spectroscopy. When Nakano et al. (1985) heated oleic acid at 230°C for 3 h, in the presence of montmorillonite, they obtained a 35% yield of dimer and trimer acids, while the 27% yield of branched-chain isomers of oleic acid increased nearly twofold when the clay catalyst was previously impregnated with methanesulfonic or phosphoric acid.

These observations have been rationalized in terms of dehydrogenation (hydrogen transfer) combined with a Diels–Alder type reaction (den Otter 1970a, 1970b, 1970c). Mass spectrometric

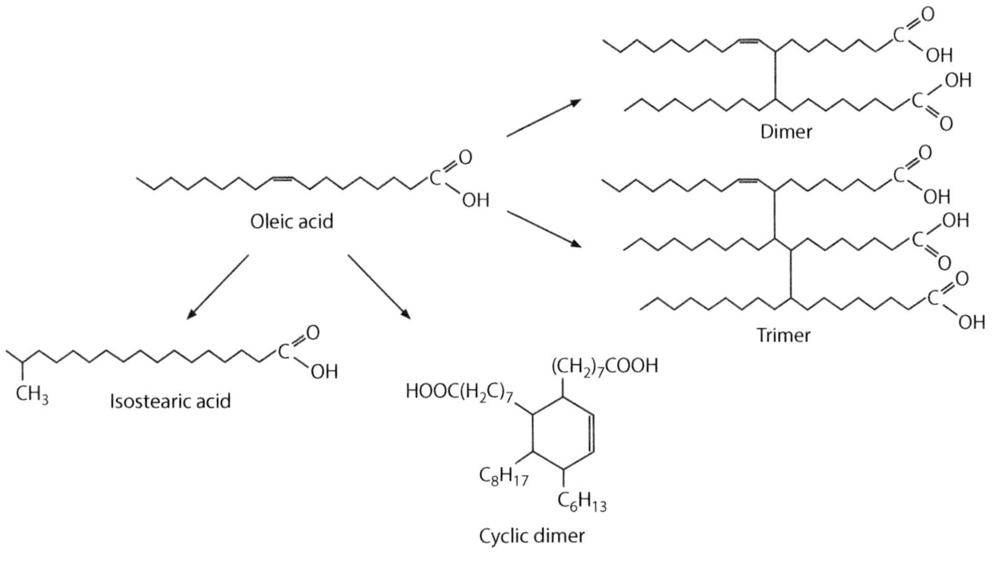

FIGURE 6.3 The clay-catalyzed conversion of oleic acid to the corresponding dimer and trimer, and to isostearic acid. Also shown is the formation of a cyclic dimer from the reaction of oleic and linoleic acids. (Adapted from Koster, R.M. et al., *J. Mol. Catal. A: Chem.*, 134, 159–169, 1998; Park, K.J. et al., *Spectrochim. Acta Part A* 149, 402–407, 2015.)

analysis of the clay-catalyzed dimer of linoleic acid by Wheeler et al. (1970) indicates the presence of monocyclic, bicyclic, and tricyclic structures. The monocyclic structure presumably arises through a Diels–Alder type addition reaction, while the bicyclic structure forms by free radical coupling followed by intramolecular ring closure. Using an apparently pure oleic acid and an unspecified organoclay, Yang et al. (2013) found that the reactant was dehydrogenated to the corresponding dienoic acid, which then dimerized through a Diels–Alder reaction. On the other hand, Čičel et al. (1992) have proposed that the dimerization of oleic acid in the presence of cation-exchanged and acid-treated bentonites is a Brønsted acid-catalyzed process.

Subsequent measurements by Koster et al. (1998) would indicate that Brønsted acid sites, associated with isomorphous substitution in the tetrahedral sheet of montmorillonite (commercial F160 grade), play a determining role in the dimerization reaction. The importance of tetrahedral surface charge density has already been remarked on in connection with the conversion of tertiary butylalcohol into isobutylene (Davidtz 1976; cf. Figure 2.11). Koster et al. (1998) have further suggested that oleic acid dimerization over montmorillonite is influenced by the nature of the counterions (in the clay catalyst), which controls interlayer swelling (cf. Table 1.2) and hence the accessibility of the interlayer space to the reactant. By increasing interlayer expansion, the addition of water to the reaction mixture would also enhance dimerization activity.

6.4 OLIGOMERIZATION

6.4.1 Hydrocarbons

The oligomerization of alkenes over clay minerals is by and large a Brønsted acid catalyzed reaction involving carbocation intermediates. Under certain conditions, however, Lewis acid sites may also play a part in the reaction. The oligomerization of toluene over Cu^{2+}-montmorillonite, for example, involved the formation of both benzyl cations and radical cations (Tipton and Gerdom 1992). The initially formed carbocation is subjected to nucleophilic attack by the alkene to yield a higher carbocation that may add another alkene or lose a proton (Purnell 1990), according to Scheme 6.9. As might be expected, catalytic activity is influenced by the size and valency of the exchangeable cations, hydration status, acid pretreatment of the clay mineral, reaction temperature, and solvent type (Adams and Clapp 1986; Adams 1987).

Thus, the conversion of dec-1-ene to its trimeric (and oligomeric) forms, in the presence of various cation-exchanged montmorillonites (Mt) at 140°C *in vacuo*, decreased in the order: H^+-Mt > Zr^{2+}-Mt > Al^{3+}-Mt > K10-Mt > Na^+-Mt (Pillai and Ravindranathan 1994). Using Al^{3+}-montmorillonite as catalyst, Ballantine et al. (1985) have reported that intercalated 2-methylpropene converts into a mixture of oligomers when the clay sample is previously dehydrated. Kaolinite is similarly efficient in catalyzing the conversion of α-pinene to oligo-α-pinene (Cheshchevoi et al. 1989).

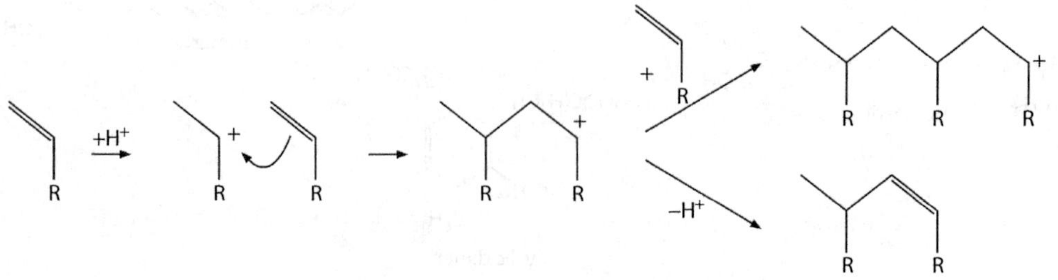

SCHEME 6.9 Brønsted acid-catalyzed oligomerization of alkenes, involving the formation of carbocation intermediates in the presence of cation-exchanged smectites. The carbocations can undergo a nucleophilic attack by the alkene to yield higher molecular weight species which can either add further alkene or deprotonate. (After Purnell, J.H., *Catal. Lett.* 5, 203–210, 1990.)

Zubkov et al. (1994) have suggested that the oligomerization of ethylene at 27°C over alumina-pillared montmorillonite (Al-PILC) is a Brønsted acid-catalyzed process. However, both Brønsted and Lewis acid sites appear to be involved in the oligomerization of 1-pentene over alumina-pillared montmorillonite and saponite (Casagrande et al. 2005). On the other hand, Occelli et al. (1985) have proposed that the oligomerization of propylene over Al-PILC and mixed Zr–Al-pillared bentonites, after degassing at 300°C, is catalyzed by Lewis acid sites associated with undercoordinated aluminum ions in the pillars (cf. Chapter 3). Because of its inherent acidity and mesoporous nature, Ni-exchanged K10 is superior to Ni-PILC in catalyzing ethylene oligomerization, showing a high selectivity to linear C4 and C6 olefins (Aid et al. 2017).

6.4.2 Non-Hydrocarbons

The cyclic (and linear) oligomerization of benzylic alcohols (in dichloromethane or carbon disulfide) in the presence of a bentonite catalyst has been investigated by Salmón and coworkers (Salmón et al. 1994, 1995; Cruz-Almanza et al. 1997). In the case of 3,4,5-trimethoxybenzyl alcohol (**1**), the reaction yields the trimeric cyclotriveratrylene (**2**) plus other cyclic and linear oligomers of benzylic ether (**3**) serving as key intermediates (Scheme 6.10). Compound **3** arises by protonation of **1**, yielding a carbocation that is then subjected to a nucleophilic attack by the alcohol or by intermolecular condensation of **1**. Either way, this compound becomes a precursor of the corresponding oligomers formed through an intramolecular electrophilic aromatic substitution reaction between **3** and the benzyl cation from **1**.

Cabrera et al. (1995) also used a bentonite to promote the cyclo-oligomerization of olefin oxides (in benzene at 150°C) to yield crown ethers, while Dintzner et al. (2010) chose K10 montmorillonite to catalyze the cyclo-trimerization of aliphatic aldehydes. Similarly, Miranda et al. (1999) were able to obtain trimethylendioxyorthocyclophane through the cyclo-oligomerization of piperonyl alcohol in the presence of a commercial bentonite (Tonsil Actisil FF) or HCl. The same clay can also promote the oligomerization of toluenes and benzyltoluenes when thermal energy and ultrasound are applied (Miranda et al. 2003) as well as the conversion of *p*-methoxybenzyl acetate (in carbon disulfide at room temperature) into the corresponding linear isomeric dimers, trimers, tetramers, and a pentamer (Salmón et al. 2011). More recently, Miranda et al. (2013) reported that Tonsil Actisil FF was similarly effective in promoting the solvent-free synthesis of cycloveratrylene macrocycles and benzyl oligomers from the corresponding benzyl alcohols under microwave and infrared irradiation.

In investigating the decomposition of 2-propanol over synthetic smectites, containing structural nickel ions, Nishiyama et al. (1993) obtained acetone as the main product together with the dimers and trimers of propene. Similarly, Yong et al. (1997) observed the formation of trimers and tetramers by oxidative coupling of 2,6-dimethylphenol and *o*-methylphenol adsorbed to cation-exchanged montmorillonites. Desjardins et al. (1999) have suggested that the activity

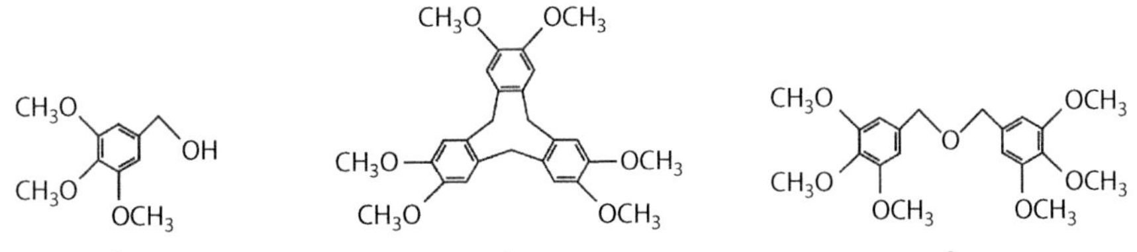

SCHEME 6.10 Oligomerization of 3,4,5-trimethoxybenzyl alcohol (**1**) in the presence of a commercial acid-activated bentonite clay to yield the trimeric cyclotriveratrylene (**2**) plus other cyclic and linear oligomers of benzylic ether (**3**) serving as intermediates. (After Salmón, M. et al., *Tetrahedron Lett.*, 35, 5797–5800, 1994; Cruz-Almanza, R. et al., *J. Mol. Catal. A: Chem.*, 126, 161–168, 1997.)

of (anhydrous) Fe^{3+}-montmorillonite in catalyzing the oxidative oligomerization of chloro- and methoxy-phenol is due to the ability of exchangeable ferric ions to act as electron acceptors (Lewis acid sites). The presence of water, acting as a Lewis base, markedly reduces catalytic activity with respect to chlorophenol oxidation. Methoxyphenol, on the other hand, can successfully compete with water for Lewis acid sites to form oligomers. Fe^{3+}-montmorillonite can also promote the oxidative transformation and oligomerization of phenol (Soma et al. 1986), phenolic acids (Polubesova et al. 2010), triclosan (Liyanapatirana et al. 2010), and 17β-estradiol (Qin et al. 2015).

Using K10 montmorillonite, Erhan and Isbell (1997) were able to prepare oligomeric fatty acid esters (estolides) from oleic acid and meadowfoam oil fatty acids. Prior treatment of the clay with a ferric salt or a cationic surfactant enhanced catalytic activity, increasing estolide yields by nearly 30%. Earlier, Weiss (1981) made the interesting observation that the oligomerization of oleic acid (to di-, tri-, and oligo-carboxylic acids) in the interlayer space of tetramethylammonium-montmorillonites was influenced by the charge density of the clay mineral support.

To round off this section, we would mention the activity of montmorillonite in catalyzing the formation of RNA oligomers from adenosine-5′-phosphorimidazolide (ImpA) (cf. Figure 8.2) or cytidine-5′-phosphorimidazolide (ImpC) (Ferris 2005; Joshi et al. 2009; Ertem and Gan 2014). The clay that has been treated with 0.5 M HCl (4°C, 30 min), washed, and then titrated with NaOH to pH 6–7 (catalyst A) is a much better catalyst (yielding oligomers with up to 50 monomer units) than the Na^+-exchanged clay mineral (B). The high concentration of sodium ions in B would prevent the activated monomers from entering the clay interlayers, and hence only short-chain oligomers are formed. However, since only a proportion of the exchange sites in A is occupied by protons, some activated monomers can intercalate to form relatively long oligomeric chains. We suggest, however, that the negatively charged ImpA/ImpC would largely be repelled by the montmorillonite surface so that little, if any, adsorption or intercalation can occur. On the other hand, the particle edges of the acid-treated clay sample, even after titration with NaOH, would retain some positive charges to which activated RNA monomers can bind. The clay-catalyzed formation of ribonucleotide and amino acid oligomers is further described in Chapter 8.

6.5 POLYMERIZATION

The polymerization of organic compounds at clay mineral surfaces may be initiated and catalyzed by either Brønsted or Lewis acid sites. With transition metal cation-exchanged clay minerals, the reaction commonly involves radical-cation intermediates, formed by electron transfer from the monomer to the (exchangeable) counterion (cf. Equation 2.22). Adsorbed and intercalated monomers that do not polymerize spontaneously can be induced to do so by chemical initiators or ionizing radiation in which case the clay mineral acts more like a reaction-controlling/directing agent than a catalyst. In some instances, the clay mineral surface needs to be modified or rendered organophilic prior to adding or introducing the organic monomer.

Some monomers, such as hydroxyethyl methacrylate, can polymerize spontaneously on intercalation into montmorillonite (Solomon and Loft 1968). The reaction, however, can only occur when the mineral contains octahedrally coordinated transition metal ions in the lower valency state, notably Fe^{2+}. When the clay mineral is previously heated (in air), or treated with an oxidizing agent, its ability to initiate polymerization is lost although catalytic activity may be restored by treatment with aqueous hydrazine. In this instance, the underlying mechanism would appear to be one of electron donation by the clay to the monomer to yield the corresponding radical-anion, which then converts into a propagating free radical by protonation (Figure 6.4a). Interestingly, Oral et al. (2008) have obtained polyhydroxyethyl methacrylate by atom transfer radical polymerization (ATRP), initiated by a synthetic cationic organic compound intercalated

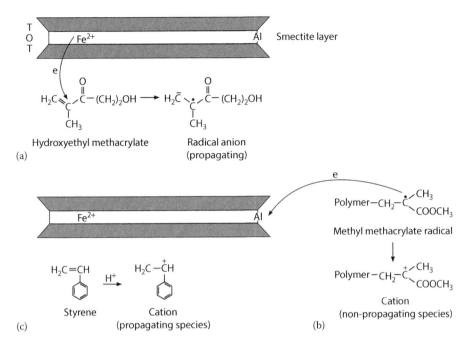

FIGURE 6.4 Diagram illustrating the influence of clay minerals (e.g., smectite) on the polymerization of some organic monomers. (a) interlayer polymerization of hydroxyethyl methacrylate, involving a radical anion intermediate, formed by electron transfer from octahedrally coordinated ferrous ion to the monomer; (b) inhibition of methyl methacrylate polymerization by electron transfer from the propagating radical to edge aluminum; and (c) external surface polymerization of styrene by transfer of protons (from exchange sites or through dissociation of water molecules coordinated to exchangeable aluminum) to the monomer. T = tetrahedral sheet; O = octahedral sheet. (Adapted from Theng, B.K.G. and Walker, G.F., *Isr. J. Chem.*, 8, 417–424, 1970.)

into K10 montmorillonite. Other examples of clay-catalyzed ATRP reactions are described in the following, and some are listed in Table 6.1.

We might also add that clay minerals may inhibit the polymerization of adsorbed monomers. This point may be illustrated by the polymerization of methyl methacrylate, which normally is initiated by free radicals formed either by heat treatment or decomposition of chemical initiators. Methyl methacrylate, however, fails to polymerize in the presence of montmorillonite even after heating at 100°C, while in the presence of kaolinite the yield of poly(methyl methacrylate) is reduced, and its molecular weight increases (Solomon and Swift 1967; Theng and Walker 1970). These observations are indicative of a reduction in the concentration of free radical intermediates available for chain initiation or propagation, presumably because the reactive intermediate is converted into the non-propagating methyl methacrylate cation through electron transfer to aluminum ions exposed at the clay particle edge surface (Figure 6.4b). During the *in situ* intercalative polymerization of ethylene (catalyzed by a montmorillonite-supported metallocene), Wang et al. (2004) have also observed that the rates of propagation, chain transfer, and termination decrease, yielding a polymer of high molecular weight.

Theng and Walker (1970) have also pointed out that the integrity of chemical initiators may be affected by clay minerals. Thus, benzoyl peroxide decomposes to an unreactive species when it binds to clays from a nonpolar solvent in the absence of monomers. Other peroxides and hydroperoxides become inactive whether or not monomer is present although azo compounds are apparently unaffected (Solomon et al. 1971). Earlier, Dekking (1965) reported that 2,2′azobisisobutyramidinium, adsorbed to kaolinite and bentonite, decomposed twice as fast as the azo compound itself. Fan et al. (2003) further noted that the valency and interlayer conformation of cationic chemical initiators

TABLE 6.1
Synthesis of Various Polymer-Clay Nanocomposites through *In Situ* Polymerization of the Corresponding Intercalated Monomers

Polymer (in Alphabetical Order)	Clay Mineral Component[a]	Polymerization Method[b]	Catalyst/Initiator/Cocatalyst	Reference (in Chronological Order)
Polyaniline	Na⁺-MMT	Emulsion	Dodecylbenzene sulfonic acid or DBSA (dopant/emulsifier)	Kim et al. (1999, 2002)
	Na⁺-MMT	Oxidative	Ammonium persulfate (oxidant)	Lee et al. (2000); Wu et al. (2000)
	Organically modified MMT	Oxidative	Ammonium persulfate (oxidant)	Yeh et al. (2001)
	Organically modified MMT	Oxidative	DBSA and ammonium persulfate (oxidant)	Jia et al. (2002)
	Na⁺-bentonite	Oxidative	3-pentadecylphenol 4-sulfonic acid (dopant)	Sudha et al. (2007)
	Organically modified attapulgite[c]	Oxidative	DBSA and ammonium persulfate (oxidant)	Lei et al. (2008)
	Acid-treated montmorillonite	Conventional	Supported organo–metal complexes	Scott et al. (2008)
	Pristine (Na⁺-MMT)	Emulsion	Sodium lauryl sulfate and pentanol (dopant/emulsifier); sonication	Song et al. (2008)
	Synthetic mica	Oxidative	Ammonium persulfate (oxidant)	Wu et al. (2008)
	Commercial organoclay	Oxidative	Amphiphilic dopants; DBSA and ammonium persulfate (oxidant)	Reena et al. (2009)
	Organically modified MMT	Emulsion	Sodium dodecyl sulfate (emulsifier; ammonium persulfate (oxidant/initiator)	Sedláková et al. (2009)
	Silylated attapulgite[c]	Graft/oxidative	Ammonium persulfate (oxidant)	Shao et al. (2010)
	Na⁺-MMT	Oxidative	Ammonium persulfate (oxidant)	Shakoor et al. (2011)
	Palygorskite[c]	Oxidative	Ammonium persulfate (oxidant)	Chae et al. (2015)
Poly(butylene terephthalate)	Organically modified MMT	Copolymerization of dimethyl terephthalate with 1,4-butanediol	Isopropyl titanate; heating	Chang et al. (2003)
Polycaprolactone	Fluorohectorite	Ring opening; conventional	Heat treatment	Messersmith and Giannelis (1993)
	Na⁺-MMT and organically modified MMT	Ring opening; conventional	Dibutyltin dimethoxide (initiator/catalyst)	Lepoittevin et al. (2002); Gorrasi et al. (2003)
Poly(ethyl acrylate)	Bentonite	Emulsion	Sodium dodecyl sulfate (emulsifier); potassium persulfate (oxidant)	Tong et al. (2002)
	Commercial organoclay with bromopropionyl bromide initiator	Atom transfer radical polymerization (ATRP)	CuBr/2,2′bipyridyl	Datta et al. (2008a)

(*Continued*)

TABLE 6.1 (Continued)
Synthesis of Various Polymer-Clay Nanocomposites through in situ Polymerization of the Corresponding Intercalated Monomers

Polymer (in Alphabetical Order)	Clay Mineral Component[a]	Polymerization Method[b]	Catalyst/Initiator/Cocatalyst	Reference (in Chronological Order)
Polyethylene	Palygorskite[c]	Ziegler–Natta	Supported $TiCl_4$ with trialkylaluminum as cocatalyst	Rong et al. (2001a, 2001b)
	Organically modified MMT	Ziegler–Natta	Supported $TiCl_4$ or $MgCl_2$–$TiCl_4$ with triethylaluminum as cocatalyst	Ma et al. (2001); Jin et al. (2002); Yang et al. (2003)
	MMT with bifunctional organic modifiers	Conventional	Supported organo-metal catalyst with methylaluminoxane (MAO)	Shin et al. (2003)
	Organically modified MMT	Conventional	Supported zirconocene with MAO as cocatalyst, or in reverse order	Liu et al. (2003b); Leone et al. (2008)
	Organically modified MMT (Cloisite 20A)	Conventional	Supported bis(imino)pyridine iron(II) with MAO as cocatalyst	Ray et al. (2005)
	Organically modified MMT	Conventional	Supported nickel-α-diimine with triethylaluminum as cocatalyst	He et al. (2007)
	Bentonite	Ziegler–Natta	Supported $Mg(OEt)_2$/$TiCl_4$ with triisobutylaluminum as cocatalyst	Nikkhah et al. (2009)
	KSF-MMT; commercial organoclay (Cloisite 15A)	Conventional	Supported nickel-α-diimine with MAO or trimethylaluminum as cocatalyst	Mignoni et al. (2011)
	Clay (unspecified)	Ziegler–Natta	Supported butyloctyl magnesium with triethylaluminum as cocatalyst	Abedi et al. (2013)
	Sepiolite[d]	Conventional	MAO grafted to sepiolite surface by an ether bond	Núñez et al. (2014)
Poly(ethylene) terephthalate	Montmorillonite-rich clay	Conventional	Autoclaving	Ke et al. (1999)
	Organically modified MMT	Conventional	Thermal	Chang et al. (2004); Zeng and Bai (2005)
Polyfuran	Pristine MMT	Oxidative	Without extraneous oxidant	Ballav and Biswas (2004b)
Poly(2-hydroxy ethylmethacrylate)	Organoclay with an initiator	Atom transfer radical polymerization (ATRP)	CuBr/2,2′ bipyridyl	Oral et al. (2008)
Polylactide	Organically modified MMT	Ring opening	Thermal and triethylaluminum as cocatalyst	Paul et al. (2003)

(Continued)

TABLE 6.1 (Continued)
Synthesis of Various Polymer-Clay Nanocomposites through *in situ* Polymerization of the Corresponding Intercalated Monomers

Polymer (in Alphabetical Order)	Clay Mineral Component[a]	Polymerization Method[b]	Catalyst/Initiator/Cocatalyst	Reference (in Chronological Order)
Poly(o-methoxy aniline)	Organically modified MMT	Oxidative	Ammonium persulfate (oxidant)	Yeh and Chin (2003)
Polymethyl methacrylate	Na$^+$-MMT	Conventional (free radical)	γ-Irradiation; benzoyl peroxide	Blumstein and Watterson (1968)
	Na$^+$-MMT	Emulsion	Sodium lauryl sulfate (dopant/emulsifier); potassium persulfate	Lee and Jang (1996)
	Organically modified MMT	Conventional (free radical)	*t*-Butyl peroxy-2-ethylhexanate; thermal	Okamoto et al. (2000)
	Organically modified MMT	Conventional (free radical)	Thermal	Yeh et al. (2002); Nikolaidis et al. (2011)
	Silylated montmorillonite	Conventional (free radical)	2,2′-azobisisobutyronitrile (AIBN)	Xie et al. (2003b)
	Organically modified MMT	Emulsion (free radical)	Benzoyl peroxide; thermal	Yeh et al. (2004)
	MMT modified with a redox cationic surfactant	Conventional (free radical)	Benzoyl peroxide; room temperature	Stadtmueller et al. (2005)
	Triethylaluminum-grafted (K10) MMT	Conventional	Methylaluminoxane (MAO) cocatalyst	Cui et al. (2008)
	Organically modified MMT	Conventional (free radical)	Photo-initiated	Oral et al. (2009)
	Commercial organoclay (Cloisite 30B) with bromopropionyl bromide (initiator)	Atom transfer radical polymerization (ATRP)	CuBr/2,2′bipyridine and hexamethyltriethylene tetraamine as catalyst	Amin et al. (2013)
Polypropylene	Organo-fluorohectorite	Conventional	Supported cationic palladium complex	Bergman et al. (1999)
	Organically modified MMT	Ziegler–Natta	Supported TiCl$_4$ or MgCl$_2$/TiCl$_4$ with triethylaluminum as cocatalyst	Ma et al. (2001); He et al. (2006); Du et al. (2007, 2013); Dias et al. (2011); Wang and He (2011)
	Organically modified MMT	Conventional	Supported metallocene without MAO cocatalyst	Sun and Garcés (2002)

(*Continued*)

TABLE 6.1 (Continued)
Synthesis of Various Polymer-Clay Nanocomposites through *in situ* Polymerization of the Corresponding Intercalated Monomers

Polymer (in Alphabetical Order)	Clay Mineral Component[a]	Polymerization Method[b]	Catalyst/Initiator/Cocatalyst	Reference (in Chronological Order)
Polypropylene (continued)	Organically modified MMT	Ziegler–Natta	Supported metallocene plus MAO	Yang et al. (2007)
	Na⁺-MMT; triethylaluminum-treated MMT	Ziegler–Natta	Supported MgCl$_2$/TiCl$_4$	Marques et al. (2011)
	Na⁺-MMT	Ziegler–Natta	Supported Mg(OEt)$_2$/TiCl$_4$ with triethylaluminum or triisobutylaluminum as cocatalyst, and an organosilane as external electron donor	Ramazani et al. (2010)
	MMT (pristine?)	Conventional	Supported metallocene with MAO or mixed MAO/triisobutylaluminum as cocatalyst	Marques and Moreira (2017)
Polypyrrole	Na⁺-MMT	Emulsion	DBSA (dopant/emulsifier)	Hong et al. (2001); Kim et al. (2003)
	Na⁺-MMT	Oxidative	Potassium persulfate; FeCl$_3$ (oxidant)	Boukerma et al. (2006); Kassim et al. (2007)
	Na⁺-MMT, organoclay; K10-MMT	Emulsion/oxidative	DBSA (emulsifier) and FeCl$_3$ (oxidant)	Mravčáková et al. (2006, 2007); Peighambardoust and Pourabbas (2007)
	Acid-treated MMT	Oxidative	Ammonium persulfate (oxidant)	Karim and Yeum (2008)
	Na⁺-MMT	Oxidative	FeCl$_3$ (oxidant)	Han (2009)
	Al-pillared MMT	Oxidative	FeCl$_3$ (oxidant)	Shakoor et al. (2012)
	Organically modified K10-MMT	Emulsion/oxidative	DBSA (emulsifier), ammonium persulfate (oxidant)	Pontes et al. (2013)
	Na⁺-MMT	Oxidative	Anionic and cationic surfactants; FeCl$_3$ (oxidant)	Ramôa et al. (2015)
	Fe³⁺-exchanged illite-MMT	Oxidative	FeCl$_3$ (oxidant)	Zidi et al. (2016)

(Continued)

TABLE 6.1 (Continued)
Synthesis of Various Polymer-Clay Nanocomposites through *in situ* Polymerization of the Corresponding Intercalated Monomers

Polymer (in Alphabetical Order)	Clay Mineral Component[a]	Polymerization Method[b]	Catalyst/Initiator/Cocatalyst	Reference (in Chronological Order)
Polystyrene	Organically modified MMT	Conventional (free radical)	*t*-Butyl peroxy-2-ethylhexanate; thermal	Okamoto et al. (2000)
	Kaolinite-DMSO intercalate	Conventional (free radical)	Benzoyl peroxide	Elbokl and Detellier (2006)
	MMT modified with vinylbenzyltrimethyl ammonium (initiator)	Atom transfer radical polymerization (ATRP)	CuBr/pentamethyldiethylene triamine as catalyst	Roghani-Mamaqani et al. (2011)
	MMT modified with cationic styrene oligomer	Conventional (radical) in dispersion	2,2′-azobisisobutyronitrile (AIBN)	Greesh et al. (2012)
	Commercial organoclay (Cloisite 30B) with bromopropionyl bromide (initiator)	Atom transfer radical polymerization (ATRP)	CuBr/2,2′bipyridine and hexamethyltriethylene tetraamine as catalyst	Amin et al. (2013)
	Silylated palygorskite	Conventional (free radical)	2,2′-azobisisobutyronitrile (AIBN)	Liu et al. (2014)
Poly(urethane acrylate)	MMT intercalated with a cationic photo-initiator	Photo-induced polymerization (free radical)	2-benzyl-2-dimethylamino-1-(4-morpholinephenyl) butanone HCl (initiator)	Qin et al. (2009)
Poly(4-vinylpyridine)	Organically modified and silylated MMT	Conventional	2,2′-azobisisobutyronitrile (AIBN)	Şen et al. (2006)

Note: The extent of clay particle exfoliation is controlled by experimental conditions.

[a] MMT stands for montmorillonite.
[b] The *conventional* method of preparing polymer-clay nanocomposites is to intercalate the monomer into a clay mineral, or its organically modified form, followed by its polymerization induced by a suitable initiator, heat treatment, or irradiation.
[c] The term *attapulgite* has been discredited, and should be replaced by *palygorskite* (cf. Chapter 1).
[d] The synthesized polymers are presumably associated with external particle and intra-channel surfaces.

affected the properties of the polymers formed as well as the extent of clay particle exfoliation in the polymer matrix. Zeng and Lee (2001) have suggested that particle dispersion is profoundly affected by the compatibility of the initiator and monomer with the clay surface. The ratio of organoclay to initiator can also influence the rate of polymerization (Liu et al. 2003a).

Thus, besides being dependent on catalyst pretreatment and experimental conditions, the clay-catalyzed polymerization of organic molecules may be affected by various factors, and involve different mechanisms.

6.5.1 Hydrocarbons

Gurwitsch (1923) made the early investigation into the polymerization of pentenes and hexenes by palygorskite. Heat-activated palygorskite was also used by Gayer (1933) to polymerize propylene through a carbocation mechanism, while Salt (1948) turned to activated (presumably acid-treated) clays for the polymerization of α-methyl styrene. Similarly, Harrane et al. (2002) have proposed that carbocation intermediates are involved in the polymerization of isobutylene by an acid-activated (proton-exchanged) montmorillonite.

The (oxidative) polymerization of arenes in the interlayer space of smectites, exchanged with transition metal cations, has attracted a great deal of attention during the 1970s and 1980s (Mortland and Halloran 1976; Pinnavaia 1977; Stoessel et al. 1977; Walter et al. 1990; Eastman et al. 1996). Using resonance Raman spectroscopy, Soma et al. (1983, 1984, 1985, 1986) showed that benzene, biphenyl, and p-terphenyl, intercalated into Cu^{2+}-, Fe^{3+}-, and Ru^{3+}-montmorillonites, were converted into the poly(p-phenylene) cation which, in the presence of water vapor, was reduced to the corresponding uncharged polymer. Similarly, Sohn and Park (1993) have ascribed the activity of Cr^{3+}-montmorillonite in polymerizing adsorbed ethylene to the ability of chromium ions to serve as Lewis acid sites.

The use of clay-supported metallocene catalysts for ethylene and propylene polymerization has been the focus of attention since the late 1990s. Weiss et al. (2002), for example, reported that K10 montmorillonite, impregnated with titanocene dichloride (Cp_2TiCl_2) or zirconocene dichloride (Cp_2ZrCl_2), was effective in catalyzing ethylene and propylene polymerization in the presence of trimethylaluminum (TMA) or triisobutylaluminum as cocatalysts. The kaolinite-supported counterpart, however, was less active in polymerizing ethylene than the comparable homogeneous system and failed to polymerize propylene. The superior performance of the K10-supported catalyst is related to the mesoporous structure and high acidity of this acid-treated montmorillonite. Jeong et al. (2003) have suggested similarly for Cp_2ZrCl_2 supported on K10-montmorillonite that has been modified with methylaluminoxane formed by partial hydrolysis of TMA, while Li et al. (2010) used palygorskite-supported Cp_2TiCl_2 to initiate ethylene polymerization. K10 and K30 montmorillonites, modified with trialkyl aluminum compounds and a $ZrCp^*Cp'Cl_2$ complex where $Cp^* = C_5Me_5$ and $Cp' = C_5H_4(SiMe_2CH_2CH=CH_2)$, are similarly active in promoting ethylene polymerization (Tabernero et al. 2010). In an attempt to rationalize the high activity of the hectorite supported catalyst, Camejo-Abreu et al. (2014) have proposed that TMA provides Lewis acid sites (in the form of 5-coordinated Al), enhancing the acidity of structural hydroxyls (Brønsted acid sites) through an inductive effect (cf. Scheme 2.1).

Liu et al. (2002, 2003b) intercalated zirconocene into an organo-montmorillonite obtained by replacing the sodium counterions, initially present, with methylglycinate (MeGlyH⁺) cations. In the presence of methylaluminoxane (MAO) as cocatalyst, the organoclay-supported Cp_2ZrCl_2 could effectively promote ethylene polymerization as well as the copolymerization of ethylene with 1-octene. Similarly, Lee et al. (2005) were able to prepare a polyethylene-montmorillonite nanocomposite by *in situ* polymerization of the monomer, using Cp_2ZrCl_2 and MAO, supported on a Na⁺-exchanged clay and a commercial organoclay (Cloisite 25A). In a variant approach, Liu et al. (2004) introduced the zirconocene into montmorillonite that had been silylated with 3-aminopropyl) triethoxysilane to yield a material capable of catalyzing the formation of polymers with a bimodal molecular weight distribution. Bimodal high-density polyethylenes were also obtained by Yamamoto et al. (2010) using Cr^{3+}-exchanged montmorillonite as support.

We might add that zirconocenes, supported on cation-exchanged fluorotetrasilicic mica or TSM (cf. Chapter 1), can catalyze ethylene polymerization in the presence of triisobutylaluminum (Kurokawa et al. 2009). Earlier, Tudor et al. (1996) intercalated the Ziegler–Natta catalyst, [Zr(η-C_5H_5Me(thf))]$^+$, into TSM and Laponite by cation exchange and used the clay-supported organometallic complex to catalyze the polymerization of propylene in the presence of MAO. A similar approach was adopted by Hwu and Jiang (2005) to obtain polypropylene-clay nanocomposites using *rac*-Et(Ind)$_2$ZrCl$_2$ supported on montmorillonite.

Tayano et al. (2016) have attempted to identify the active sites in metallocene and their location in acid-treated (Kunipia F) montmorillonite, using various instrumental techniques. X-ray diffraction and transmission electron microscopy indicate that the metallocene complex is located and activated (by Lewis acid sites) at clay particle edges. This finding is consistent with the concept that acid treatment leads to the migration of Al^{3+} to interlayer and particle edge positions, and the formation of a protonated *amorphous* silica phase (cf. Chapter 3). Since neither the metallocene nor the monomer can penetrate the interlayer space, the polypropylene formed is largely associated with the edge surface of the montmorillonite support. Interestingly, Sun and Garcés (2002) were able to obtain polypropylene-clay nanocomposites by *in situ* polymerization using metallocene-clay catalysts in the absence of a cocatalyst such as methylaluminoxane. The preparation of clay-supported Ziegler–Natta catalysts for olefin polymerization, with particular reference to the synthesis of polyolefin-clay nanocomposites, has been reviewed by Qian and Guo (2010) and Abedi and Abdouss (2014). Other examples of the *in situ* polymerization of ethylene and propylene, in the presence of clay-supported catalysts to yield the corresponding polymer-clay nanocomposites, are listed in Table 6.1.

The interplay of Brønsted and Lewis acidity, and the difficulty of assigning catalytic activity to either type of acid sites, are also evident in the clay-catalyzed polymerization of styrene (cf. Figure 2.10). Both Bittles et al. (1964) and Pezerat and Mantin (1967) have proposed that the polymerization reaction in the presence of acid-treated montmorillonite is initiated by a carbocation arising from the transfer of a proton from exchange sites, or from the dissociation of water molecules coordinated to Al^{3+} counterions, to the styrene monomer (Figure 6.4c). Kusnitsyna and Ostrovskaya (1967) also reported that montmorillonites, containing various proportions of exchangeable hydrogen and aluminum ions, could initiate styrene polymerization. Similarly, Matsumoto et al. (1969) have favored a Brønsted acid catalyzed (cationic) mechanism for the initiation process since replacement of the exchangeable protons by sodium or ammonium ions almost completely inhibits polystyrene formation, while catalytic activity is restored by rewashing the clay with acid. Curiously, however, treatment of the clay with trityl chloride, which selectively adsorbs to Lewis acid sites, leads to an appreciable decrease in polymer formation.

On the other hand, Solomon and Rosser (1965) have rationalized the results on styrene polymerization, in the presence of various dry cation-exchanged kaolinite and palygorskite, in terms of electron transfer from the monomer to octahedrally coordinated aluminum at the edge surface, acting as an electron acceptor (Lewis acid) as illustrated in Figure 6.5. Thus, when this surface is masked by pretreating

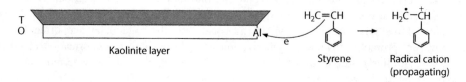

FIGURE 6.5 Diagram illustrating the influence of clay minerals (e.g., kaolinite) on the polymerization of styrene, involving a radical cation intermediate, formed by electron transfer from the monomer to (undercoordinated) aluminum at the clay particle edge. T = tetrahedral sheet; O = octahedral sheet. (Adapted from Theng, B.K.G. and Walker, G.F., *Isr. J. Chem.*, 8, 417–424, 1970.)

the clays with polyphosphate, there is a marked reduction of catalytic activity. We might also mention that the polymerization of hydroquinone in the presence of nontronite and kaolinite is suppressed when the catalysts have previously been treated with sodium metaphosphate (Wang and Huang 1989).

Solomon and Rosser (1965) have further suggested that the reaction is initiated by the formation of radical-cations, which can rapidly dimerize (cf. Figure 2.10). Both the radical-cation and its dimer are involved in the initiation step, but propagation is presumably cationic in nature. The addition of water and amines inhibits polymerization because such Lewis bases would be preferentially adsorbed over styrene. ^{13}C NMR spectroscopic measurements by Njopwouo et al. (1987, 1988) also indicate that the thermal polymerization of styrene, in the presence of acid-treated kaolinite, halloysite, and a mixed kaolinite-palygorskite-illite-montmorillonite mineral, involves both Lewis and Brønsted acid sites.

Styrene is not capable of intercalating into inorganic cation-exchanged smectites or does so only with difficulty (Solomon and Rosser 1965). By leaving the dry montmorillonite in contact with excess styrene for two weeks, however, Friedlander and Frink (1964) claimed to have obtained partial intercalation. A similar claim was made by Pezerat and Mantin (1967) with respect to acid-treated montmorillonite. Little doubt, however, attaches to the ability of styrene in penetrating the interlayer space of *organoclays*—that is, montmorillonites that have been modified by intercalation of quaternary ammonium cations and related cationic surfactants (Churchman et al. 2006; Theng et al. 2008; He et al. 2014). The influence of surfactant design and architecture on the properties of the resultant polystyrene-clay complexes has been investigated by Simons et al. (2010). For example, Kato et al. (1981) intercalated styrene into montmorillonite that had been exchanged with various quaternary ammonium cations, and then polymerized the interlayer monomer using benzoyl peroxide (BPO) as initiator. The basal spacing of ~2 nm for the interlayer complex of stearyltrimethylammonium-montmorillonite with styrene increased to 3.2 nm following polymerization of the intercalated monomer. On the other hand, Hwu et al. (2004) observed extensive exfoliation of the montmorillonite (host) particles following the suspension free-radical polymerization of the intercalated styrene.

Similarly, Moet and Akelah (1993) and Akelah and Moet (1996) were able to intercalate styrene into a vinylbenzyltrimethylammonium-montmorillonite complex. When the interlayer monomer was polymerized (by a free radical process), the resultant polystyrene was partly intercalated and partly attached to external particle surfaces. In a variant of this approach, Weimer et al. (1999) replaced the interlayer Na^+ ions in montmorillonite with nitroxyl-substituted ammonium cations and used the resultant organoclay to initiate the *living* free radical polymerization of styrene. The nitroxide-mediated polymerization of styrene on the surface of Laponite has also been reported by Konn et al. (2007), using an intercalated synthetic quaternary ammonium alkoxyamine as initiator. Earlier, Fan et al. (2002) used an interlayer complex of montmorillonite with cationic 1,1-diphenylethylene to initiate the living anionic polymerization of styrene. Similarly, Akat et al. (2008) treated montmorillonite with diethyloctylammonium ethylmercaptan bromide and used the product as a chain-transfer agent for the free radical polymerization of styrene (and methyl methacrylate). The living polymerization of organic monomers in the presence of clay minerals has been summarized by Tasdelen et al. (2010).

Following Böttcher et al. (2002), many investigators have used the atom transfer radical polymerization (ATRP) method (Wang and Matyjaszewski 1995) to synthesize various polymers in the presence of clay minerals and organoclays acting as hosts and supports of the initiator species. Zhao et al. (2004), for example, intercalated the combination of a quaternary ammonium salt and 2-bromo-2methyl propionate (as ATRP initiator) into montmorillonite by cation exchange. The product and Cu(I)Br/Cl-organo complexes were used to catalyze the synthesis of polystyrene and poly(methyl methacrylate) with a narrow range of molecular weights. The initiator used by Abbasian and Mahi (2012) for the ATRP synthesis of polystyrene also consists of a quaternary ammonium salt moiety and an α-phenyl chloroacetyl chloride moiety, while the catalyst is Cu(I)Cl/2,2′-bipyridine. With respect to the ATRP synthesis of ethyl acrylate in the presence of an organically modified clay, Datta et al. (2008b) have suggested that the organoclay interacts with the carbonyl group of the monomer. As a result, the electron density in the conjugated C=C bond is reduced, and monomer reactivity increases. Other examples of clay-polymer nanocomposites obtained by *in situ* ATRP are listed in Table 6.1.

More recently, Greesh et al. (2012) prepared montmorillonite, modified with oligomeric styrene bearing a quaternary ammonium end groups as a stabilizer, for the dispersion polymerization of styrene, using 2,2'-azobisisobutyronitrile (AIBN) as initiator, while Kim et al. (2012) intercalated styrene and divinyl benzene into montmorillonite, exchanged with 1-[4-ethylphenyl)methyl-3-butyl-imidazolium ions, and polymerized the mixture using γ-irradiation. In comparing the performance of intercalated mono- and bi-cationic free radical initiators with respect to styrene polymerization, Fan et al. (2003) noted that the mono-cationic species give a larger basal spacing, facilitating interlayer entry of monomer and formation of a relatively high molecular weight polymer.

Liu et al. (2005) have investigated the bulk polymerization of styrene initiated by BPO in the presence of Cloisite 6A, a commercial organoclay prepared by replacing the sodium ions in montmorillonite with dimethyl dihydrogenated tallow ammonium, $[(CH_3)_2N^+(HT)_2]$, where HT denotes hydrogenated tallow, consisting largely of octadecyl chains. Similarly, Siddiqui et al. (2013) were able to synthesize a copolymer of styrene and butyl methacrylate by intercalating the monomers into Cloisites 15A and 10A, and using BPO as initiator. The intercalation of styrene into montmorillonite exchanged with various quaternary ammonium ions (surfactants) has also been described by Doh and Cho (1998), Xie et al. (2003a), Essawy et al. (2004), and Qi et al. (2005).

The BPO- and heat-initiated polymerization of interlayer styrene (monomer) commonly leads to a marked increase in basal spacing as well as appreciable delamination of montmorillonite particles, the extent of which is influenced by BPO concentration and surfactant type. Some layer exfoliation was also reported by Mrah et al. (2015) during the polymerization of styrene, intercalated into a cetyltrimethylammonium-exchanged montmorillonite. We refer to the review by Carastan and Demarquette (2007) for more details.

A variant of the method is to intercalate styrene into montmorillonite that has been exchanged with a cationic surfactant containing a polymerizable group (Zeng and Lee 2001), such as vinylbenzyldimethyldodecylammonium ions (Fu and Qutubuddin 2000, 2001), vinylbenzyldimethylethanolammonium ions (Tseng et al. 2002), a styryl monomer on the ammonium ion (Wang et al. 2002a), or N,N-dimethyl-n-octadecyl-4-vinylbenzylammonium and tri-n-butyl-4-vinylbenzylphosphonium (Akelah et al. 2006). Polymerization of the interlayer styrene, using AIBN or BPO as initiator, yields the corresponding exfoliated clay-polystyrene nanocomposites. Alternatively, the Na$^+$ ions in montmorillonite may be replaced with R-$CH_2N^+(CH_3)_3$ ions where R denotes styrene, and abbreviated to styrene-N$^+$ as described by Khalil et al. (2005a, 2005b). They then used the styrene-N$^+$-montmorillonite complex as a comonomer in the bulk polymerization of styrene-acrylonitrile, using AIBN as initiator. An appreciable amount of the styrene-acrylonitrile copolymer is apparently attached to the styrene-N$^+$-montmorillonite complex, forming an envelope around the clay. Transmission electron microscopy also shows highly exfoliated clay layers and 3–5 nm nanospheres of copolymer containing 2% clay-monomer complex. An exfoliated polyethylene-clay nanocomposite was also obtained by Ren et al. (2010) through the *in situ* polymerization of the monomer, using montmorillonite exchanged with a multifunctional ammonium modifier containing carbonyl and vinyl groups.

6.5.2 Non-Hydrocarbons

Early on, Friedlander and Frink (1964) reported that 4-vinyl pyridine and acrylamide could polymerize spontaneously after intercalation into (dry) sodium-montmorillonite. Likewise, Brønsted acid sites are apparently involved in the spontaneous polymerization of diazomethane in the interlayers of a bentonite (Bart et al. 1979). Liu et al. (2001) have proposed similarly for intercalated 2-ethynylpyridine in montmorillonite although solid-state NMR spectroscopy would indicate that this polymer is associated with either Brønsted or Lewis acid sites on the surface of Ca^{2+}-montmorillonite (Sahoo et al. 2003).

On the other hand, the polymerization of adsorbed phenol in Fe^{3+}- or Cu^{2+}-montmorillonite involves a radical-cation intermediate, formed by electron transfer from the monomer to the exchangeable counterion (Mortland and Halloran 1976). A similar mechanism is likely to be operative during the polymerization of 2,6,-dimethylphenol (Sawhney et al. 1984) and in the formation of polyethylene glycol from ethylene oxide (Pusino et al. 1990). Na^+- and H^+-montmorillonite can also promote the reaction of phenol with formaldehyde in the presence or absence of oxalic acid to yield novolac resins (Wang et al. 2002b; Huskić and Žagar 2017). Birkel et al. (2002) have reported that montmorillonite can catalyze the oxidation of adsorbed phenols (catechol, pyrogallol, 2,6-dimethylphenol), allowing the oxidation products to polymerize into a humic-like substance. We might also add that different clay minerals can effectively catalyze the formation of humic-like polymers (melanoidins) through the *Maillard* reaction between amino acids (e.g., arginine, glycine, L-tyrosine) and glucose (Arfaioli et al. 1999; Bosetto et al. 2002; Gonzalez and Laird 2004).

As we have mentioned in Chapter 2, thiophene and methylthiophenes can yield the corresponding polymers in the presence of transition metal-exchanged montmorillonites (Soma et al. 1987). Under certain conditions, even Na^+-montmorillonite can apparently mediate the polymerization of intercalated thiophene (Ballav and Biswas 2004a) and furan (Ballav and Biswas 2004b). As might be expected, Fe^{3+}- or Cu^{2+}-montmorillonite as well as K10 montmorillonite are effective in promoting the polymerization of intercalated pyrrole and toluene (Saehr et al. 1991; Faguy et al. 1994; Ballav and Biswas 2004c; Letaïef et al. 2005).

Likewise, polyaniline is formed when aniline is introduced from the vapor phase or an aqueous solution into Fe^{3+}-exchanged montmorillonite (Cloos et al. 1979), synthetic Cu^{2+}-fluorohectorite (Mehrota and Giannelis 1991), Cu^{2+}-hectorite (Eastman et al. 1996; Porter et al. 1997; Eastman and Porter 2000), Cu^{2+}-montmorillonite (Moreale et al. 1985; Ilic et al. 2000; Zeng et al. 2002), and Cu^{2+}-halloysite (Luca and Thomson 2000). Montmorillonite-intercalated aniline could also be polymerized by electrochemical means (Inoue and Yoneyama 1987). More often than not, polyaniline is obtained by direct intercalation of anilinium cations into Na^+-montmorillonite (through a cation exchange process), or by adsorption in the case of palygorskite and halloysite nanotubes, followed by oxidation with ammonium persulfate, according to Figure 6.6 (Chang et al. 1992; Wu et al. 2000; Lee et al. 2003; do Nascimento et al. 2004, 2006; Liu et al. 2007; Zhang et al. 2008;

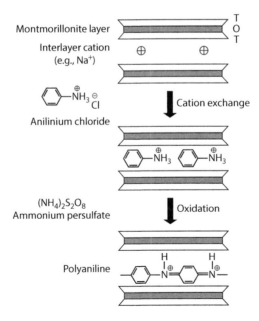

FIGURE 6.6 Diagram showing the formation of polyaniline-montmorillonite nanocomposite through *in situ* polymerization of anilinium ions, involving cation exchange and oxidation. T = tetrahedral sheet; O = octahedral sheet. (Adapted from Ruiz-Hitzky, E. and Van Meerbeek, A., Clay mineral- and organoclay-polymer nanocomposite, In *Handbook of Clay Science*, F. Bergaya, B.K.G. Theng, and G. Lagaly (Eds.), Elsevier, Amsterdam, the Netherlands, pp. 583–621, 2006.)

Bekri-Abbes and Srasra 2010a). A variant of the method is to grind anilinium chloride and Cu^{2+}-montmorillonite in an agate mortar and pestle at room temperature (Bekri-Abbes and Srasra 2010b). The polymerization of aniline intercalated into acid-treated montmorillonite (Soundararajah et al. 2009; Narayanan et al. 2010) or an organo-montmorillonite (Zeng et al. 2002), and adsorbed to palygorskite (Chae et al. 2015) or an organically modified palygorskite (Lei et al. 2008) can also be initiated by ammonium persulfate, or by co-intercalated Fe^{3+} ions acting as oxidizing agents (Gupta et al. 2014). Other examples are described in the following.

The polymerization of interlayer methyl methacrylate in montmorillonite, like that of styrene, has attracted much attention. The reaction is commonly initiated by free radicals formed by heat treatment, γ-ray irradiation, or decomposition of intercalated chemical and photochemical initiators (Blumstein 1965b; Dekking 1967; Malhotra et al. 1972; Lee and Jang 1996; Zhu et al. 2002; Liu et al. 2003a; Qu et al. 2005; Wang et al. 2005; Nese et al. 2006). Interestingly, Eastman et al. (1999) failed to detect the formation of interlayer poly(methyl methacrylate) with Cu^{2+}-hectorite or the formation of free radicals. They suggested that polymerization was effected through an unusual cationic mechanism involving the formation of a positively charged species, stabilized by interaction with the clay surface. On the other hand, methyl methacrylate can penetrate the interlayer space of VO^{2+}- and Fe^{3+}-exchanged hectorite to yield the corresponding quasi-two dimensional polymer (Porter et al. 2002).

Blumstein (1961) did some of the early investigations into the *in situ* polymerization of intercalated monomers, focusing more on the properties of the isolated polymers than on the nature of the resultant polymer-clay composite materials. In the case of methyl methacrylate, the interlayer polymer resisted extraction by common solvents and could only be isolated by dissolving the clay mineral component with hydrofluoric acid (Blumstein 1965a). The heat stability and dilute solution properties of the isolated poly(methyl methacrylate) were quite different from those of the bulk (pristine) polymer, suggesting that the former material had a sheet-like structure (Blumstein 1965b; Blumstein and Billmeyer 1966). Similarly, the spontaneous polymerization of ethynylpyridine in the interlayer space of various cation-exchanged montmorillonites yields an extended-chain polymer of high thermal stability (Kim et al. 2001). More direct evidence for the synthesis of a two-dimensional polymer was provided by Theng (1970) who polymerized 1,3-butyleneglycol dimethacrylate in the interlayer space of a synthetic calcium aluminate hydrate, using γ-ray irradiation. Scanning electron microscope examination of the interlayer complex, following dissolution of the host mineral with dilute HCl, showed the formation of ~20 nm thick platy nanoparticles of the corresponding polymer.

Using a similar approach, Glavati et al. (1963) and Glavati and Polak (1964) claimed to have obtained syndiotactic poly(acrylonitrile) and poly(acrylic acid) although their evidence was not unambiguous. More recently, Han and coworkers (He et al. 2006; Du et al. 2007, 2013) as did Yang et al. (2007) and Dias et al. (2011) reported the formation of highly isotactic polypropylene through *in situ* polymerization of the monomer using montmorillonite-supported metallocene and $MgCl_2$/$TiCl_4$ catalysts. The formation of stereoregular macromolecules by *in situ* intercalative polymerization of the corresponding monomers, however, is by no means unlikely because organic molecules are generally oriented and packed in some particular arrangement when they enter the interlayer space of montmorillonite (Mortland 1970; Theng, 1974).

The formation of a structurally ordered poly(acrylonitrile) by *in situ* polymerization of the interlayer monomer was reported by Blumstein et al. (1974). Similarly, Fan et al. (2005) obtained narrowly polydispersed poly(methyl methacrylates) by mixing the monomer with clay minerals containing monocationic 2,2′-azobisisobutyronitrile. Furthermore, the molecular weight of the resultant polymers was higher than that of the materials formed in the presence of clay-intercalated bicationic initiators (Fan et al. 2003). Likewise, Shen et al. (2005) could prepare syndiotactic polystyrene, using a monotitanocene complex, supported on an organically modified montmorillonite as a catalyst by activation with methylaluminoxane (MAO) and triisobutylaluminum.

The radical polymerization of acrylonitrile in the interlayer space of montmorillonite and its silylated derivative can be initiated by chemical initiators, such as BPO and AIBN, or by mild heating (Kato et al. 1979a; Bergaya and Kooli 1991; Seçkin et al. 2002). Fernández-Saavedra et al. (2004), for example, used AIBN to initiate acrylonitrile polymerization in the channels of sepiolite. More recently, Liu et al. (2014) synthesized palygorskite-polystyrene nanocomposites through the radical polymerization of styrene in the presence of an organically modified palygorskite.

Sonobe et al. (1988) obtained thin films of carbon by heating an interlayer montmorillonite-polyacrylonitrile complex at 700°C and liberating the carbonized product by acid treatment. Likewise, isoprene, pyrrole, styrene, and thiophene can polymerize within the channel structure of sepiolite as does ethylene under special conditions (Inagaki et al. 1995; Sandi et al. 1999). Pyrolysis of the resultant polymers, followed by dissolution of the sepiolite host with HCl or HF, yields carbon nanofibers of 20–30 nm in diameter and about 1 μm in length. Even kaolinite is capable of forming an interlayer complex with polyacrylonitrile by first intercalating the monomer into the clay that has been expanded with ammonium acetate, and then heating the intercalate at 50°C for 24 h (Sugahara et al. 1988). For example, Sugahara et al. (1990) introduced acrylamide into an interlayer complex of kaolinite with N-methylformamide and heated the mixture to obtain polyacrylamide. Subsequently, Komori et al. (1999) were able to intercalate poly(vinylpyrrolidone), dissolved in methanol, into an interlayer kaolinite-methanol complex.

Earlier, Dekking (1965, 1967) intercalated the cationic form of 2,2′-azobisisobutyramidine (AIBA) into montmorillonite and used the resultant complex to polymerize methyl methacrylate, acrylamide, vinyl acetate, 4-vinylpyridine, and styrene. The rate of polymer formation was greater than that in the presence of AIBA alone, presumably because the rate of decomposition of adsorbed AIBA was higher than that of the free azo compound. A similar approach was adopted by Uthirakumar et al. (2005) who used the cationic radical initiator 2,2′-azobis[2-methyl-*N*-(2-acetoxy-2-*N,N,N*-tributylammonium bromide) ethyl propionamide, intercalated in montmorillonite, for the *in situ* polymerization of styrene.

6.5.3 *In Situ* Polymerization of Monomers and Polymer-Clay Nanocomposite Formation

Much of the research into the polymerization of various monomers in the presence of clay minerals has focused on the formation and properties of the corresponding polymer-clay nanocomposites, a few examples of which have already been given. Since polymer-clay nanocomposites generally show marked improvements in thermo-mechanical, gas barrier, and fire retardant properties (over the corresponding polymers on their own), these hybrid materials have a huge potential for industrial and technological applications. The large volume of accumulated literature on this topic has been periodically reviewed (Giannelis et al. 1999; Gilman 1999; LeBaron et al. 1999; Alexandre and Dubois 2000; Biswas and Ray 2001; Fischer 2003; Ray and Okamoto 2003; Chen 2004; Zeng et al. 2005; Nguyen and Baird 2006; Okada and Usuki 2006; Ruiz-Hitzky and Van Meerbeek 2006; Tjong 2006; Liu 2007; Paul and Robeson 2008; Pavlidou and Papaspyrides 2008; Calabi-Floody et al. 2009; Theng 2012; Alateyah et al. 2013; Abedi and Abdouss 2014; Ray 2014).

Many examples relating to the *in situ* polymerization of clay-intercalated monomers have been given in the preceding sections. What follows is an attempt at summarizing the information on the intercalative polymerization of various monomers in smectites to yield the corresponding polymer-clay nanocomposites, a large number of which are listed in Table 6.1.

Dekking (1964) did some of the pioneering work on the preparation of polymer-clay *compounds* by neutralizing H^+-exchanged montmorillonite (and kaolinite) with amine-terminated polystyrene in an organic solvent. Sometime later, Kato et al. (1979b) obtained montmorillonite-nylon complexes by heating the clay-aminocaproic acid intercalates at 240°C–250°C (1 h) in a flow of N_2.

However, it is the nylon 6-clay nanocomposite, developed by workers at Toyota Central R & D Laboratories, Japan (Fukushima and Inagaki, 1987; Fukushima et al., 1988), that has often been credited as being the first *hybrid* material to have found commercial and industrial applications. The original nanocomposite was prepared by the *in situ* polymerization of ε-caprolactam, involving three successive steps: (1), replacing the interlayer Na^+ ions in montmorillonite with protonated 12-aminododecanoic (12-aminolauric) acid; (2), intercalating ε-caprolactam into the organically modified montmorillonite; and (3), heating the organoclay-monomer complex to induce the ring-opening polymerization of ε-caprolactam. Since the first report by Usuki et al. (1993), the method has been modified, improved, and refined to the extent that nylon 6-clay nanocomposites can now be obtained through a single-step (*one-pot*) synthesis (Kojima et al., 1993; Yasue et al., 2000). Scientists at Toyota Central R & D Laboratories (Kato et al. 2004) have also developed a novel method for producing polypropylene-clay nanocomposites from pristine Na^+-montmorillonite, thereby eliminating the requirement for a preliminary synthesis of the corresponding organoclay. Similarly, Greenwell et al. (2006) reported the synthesis of poly(alkylene oxide) by *in situ* polymerization in the interlayer space of Li^+- and K^+-montmorillonite. By contrast, Wang et al. (2013) used a *doubly* organo-modified vermiculite, obtained by intercalating cetyltrimethylammonium cations and subsequent grafting with 3-(trimethoxysilyl)propyl methacrylate, to prepare a thermostable polystyrene-vermiculite nanocomposite.

The *in situ* intercalative polymerization technique is also applicable to the preparation of thermoset and elastomeric nanocomposites (Alexandre and Dubois, 2000; Ray and Okamoto, 2003; Azeez et al. 2013). In investigating the polymerization of diglycidyl ether of bisphenol A (DGEBA) in the interlayer space of an organoclay, Messersmith and Giannelis (1994) found that the structures of the resultant epoxy-clay nanocomposites were greatly influenced by the nature of the (amine) curing agent as well as the reaction temperature, while the degree of clay particle exfoliation was dependent on the accessibility of the clay interlayers to the epoxy and diamine monomers (Lan et al. 1995). The detailed studies by Pinnavaia and coworkers regarding the synthesis of thermoset epoxy nanocomposites using octadecylammonium-exchanged montmorillonite have been summarized by Wang et al. (2000). The key to obtaining an exfoliated structure is to intercalate both the epoxide and the amine curing agent into the organoclay, allowing the acidic alkylammonium ion to catalyze the interlayer ring-opening polymerization of DGEBA at a rate that is competitive with the rate of polymerization on external particle surfaces.

A rather elegant means of preparing thermoset epoxy-clay nanocomposites was described by Triantafillidis et al. (2002) who used smectites exchanged with the diprotonated forms of polyoxypropylene diamines capable of functioning as a clay surface modifier, polymerization catalyst, and curing agent. Choi et al. (2009) have also reported that the mechanical properties of epoxy nanocomposites, formed using silylated montmorillonite, are superior to those shown by the samples produced from a commercial organo-montmorillonite (Cloisite 15A). The structures of thermoset polyurethane nanocomposites, obtained using organically modified montmorillonites, are influenced by the molecular mass of the modifying agent and the degree of surface modification (Salahuddin et al. 2010).

The synthesis of electro-active polymers and the corresponding composites through the *in situ* intercalative polymerization of suitable monomers has also been the focus of many investigations. A prime example is the formation of polyaniline by polymerizing the monomer in the interlayer space of Cu^{2+}-exchanged smectites (e.g., synthetic fluorohectorite). The resultant polymer becomes electrically conductive when the emeraldine base form is converted into the corresponding salt form through exposure to HCl vapor (Figure 6.7). Being constrained by the two-dimensional interlayer space geometry, the (cationic) polyaniline shows a highly anisotropic conductivity. Thus, the electrical conductivity parallel to the basal silicate layer (~10^{-3} S/cm) is many orders of magnitude larger than that measured in a perpendicular direction (Mehrota and Giannelis 1991; Ruiz-Hitzky and Van Meerbeek 2006). Interestingly, an electrical conductivity of 10^{-3} S/cm was also measured by Chae et al. (2015) for a surface complex of palygorskite with polyaniline. Besides preventing segmental

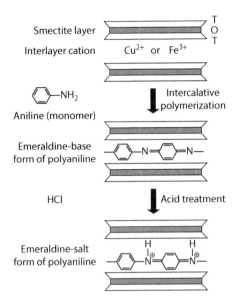

FIGURE 6.7 Diagram showing the various steps involved in the synthesis of conducting polyaniline (in the form of its emeraldine salt) in the interlayer space of a transition metal ion-exchanged smectite. T = tetrahedral sheet; O = octahedral sheet. (Adapted from Ruiz-Hitzky, E. and Van Meerbeek, A., Clay mineral- and organoclay-polymer nanocomposite, In *Handbook of Clay Science*, F. Bergaya, B.K.G. Theng, and G. Lagaly (Eds.), Elsevier, Amsterdam, the Netherlands, pp. 583–621, 2006.)

chain motion, confinement of polypyrrole in the interlayer space of montmorillonite brings about a large increase in the glass transition temperature of intercalated polypyrrole over the value measured for the pristine polymer (Yeh et al. 2003). Other examples of polyaniline-clay nanocomposite synthesis are listed in Table 6.1.

Similarly, the spontaneous polymerization of pyrrole intercalated into Cu^{2+}-fluorohectorite (Mehrota and Giannelis 1992) or Fe^{3+}- and Cu^{2+}-smectites (Faguy et al. 1994; Letaïef et al. 2005) yields conducting polypyrrole-clay complexes with a basal spacing of 1.41–1.51 nm, indicating that the heterocyclic rings of the polymer are oriented parallel to the interlayer surface. Doping of the complex (nanocomposite) with iodine vapor increases the in-plane electrical conductivity from 2×10^{-5} (for the undoped material) to 1.2×10^{-2} S/cm (Mehrota and Giannelis 1992). Rajapakse et al. (2008) reported electrical conductivities of about 3×10^{-4} S/cm and ionic conductivities of about 9×10^{-3} S/cm for intercalated Cu(I)/polypyrrole-montmorillonite nanocomposites, while Yang and Liu (2009) measured a conductivity of 50 S/cm for a surface-adsorbed (non-intercalated) composite of polypyrrole with palygorskite (*attapulgite*). We refer to Table 6.1 for other examples of polypyrrole-clay nanocomposite synthesis.

Interestingly, the conductivity of a polypyrrole-clay nanocomposite, obtained by *in situ* polymerization of the monomer in the interlayers of Na^+-montmorillonite, is much lower than that of the material prepared using the corresponding organically modified clay (Mravčáková et al. 2006). This observation may be explained in terms of the extent to which the clay mineral layers are dispersed in the nanocomposite, as Okamoto et al. (2001) have suggested for composites of polystyrene and poly(methyl methacrylate)-*co*-acrylamide. Earlier, Biswas and Ray (1998) measured an electrical conductivity of 10^{-6} S/cm for a poly(*N*-vinylcarbazole)-montmorillonite composite whereas the value given by the pristine polymer was only 10^{-16} S/cm. Ballav and Biswas (2004a) further observed that the conductivity of a polythiophene-montmorillonite nanocomposite was 10^3 times higher than that of the clay mineral component. More recently, García-González et al. (2016) described the facile polymerization of ethylenedioxythiophene (dissolved in chloroform) in the presence of Fe^{3+}-exchanged montmorillonite to give the electrically conducting poly(3,4-ethylenedioxythiophene).

REFERENCES

Abbasian, M. and R. Mahi. 2012. Synthesize of polymer-silicate nanocomposites by in situ metal-catalyzed living radical polymerization. *Polymer Composites* 33: 933–939. doi:10.1002/pc.22216.

Abedi, S. and M. Abdouss. 2014. A review of clay-supported Ziegler–Natta catalysts for production of polyolefin/clay nanocomposites through in situ polymerization. *Applied Catalysis A: General* 475: 386–409.

Abedi, S., M. Abdouss, M. Nekoomanesh-Haghighi, and N. Sharifi-Sanjani. 2013. New clay-supported Ziegler–Natta catalyst for preparation of PE/clay nanocomposites via *in situ* polymerization. *Journal of Applied Polymer Science* 128: 1879–1884.

Adams, J.M. 1987. Synthetic organic chemistry using pillared, cation-exchanged and acid-treated montmorillonite catalysts—A review. *Applied Clay Science* 2: 309–342.

Adams, J.M. and T.V. Clapp. 1986. Reactions of the conjugated dienes butadiene and isoprene alone and with methanol over ion-exchanged montmorillonites. *Clays and Clay Minerals* 34: 287–294.

Adams, J.M., E. Davies, S.H. Graham, and J.M. Thomas. 1982. Catalyzed reactions of organic molecules at clay surfaces: Ester breakdown, dimerizations, and lactonizations. *Journal of Catalysis* 78: 197–208.

Adams, J.M., S.H. Graham, P.I. Reid, and J.M. Thomas. 1977. Chemical conversions using sheet silicates: Ready dimerization of diphenylethylene. *Journal of the Chemical Society, Chemical Communications* 67–68.

Agabekov, V.E., G.M. Sen'kov, A.Y. Sidorenko, N.D. Tuyen, and V.A. Tuan. 2011. New α-pinene isomerization catalysts. *Catalysis in Industry* 3: 319–330.

Aid, A., R.D. Andrei, S. Amokrane, C. Cammarano, D. Nibou, and V. Hulea. 2017. Ni-exchanged cationic clays as novel heterogeneous catalysts for ethylene oligomerization. *Applied Clay Science* 146: 432–438.

Ainsworth, C.C., B.D. McVeety, S.C. Smith, and J.M. Zachara. 1991. Transformation of 1-aminonaphthalene at the surface of smectite clays. *Clays and Clay Minerals* 39: 416–427.

Akat, H., M.A. Tasdelen, F.D. Prez, and Y. Yagci. 2008. Synthesis and characterization of polymer/clay nanocomposites by intercalated chain transfer agent. *European Polymer Journal* 44: 1949–1954.

Akelah, A. and A. Moet. 1996. Polymer-clay nanocomposites: Free-radical grafting of polystyrene on to organophilic montmorillonite interlayers. *Journal of Materials Science* 31: 3589–3596.

Akelah, A., A. Rehab, T. Agag, and M. Betiha. 2006. Polystyrene nanocomposite materials by *in situ* polymerization into montmorillonite-vinyl monomer interlayers. *Journal of Applied Polymer Science* 103: 3739–3750.

Alateyah, A.I., H.N. Dhakal, and Z.Y. Zhang. 2013. Processing, properties, and applications of polymer nanocomposites based on layer silicates: A review. *Advances in Polymer Technology* 32: 21368 (36 pages).

Aldersley, M.F. and P.C. Joshi. 2013. RNA dimer synthesis using montmorillonite as a catalyst: The role of surface layer charge. *Applied Clay Science* 83–84: 77–82.

Alexandre, M. and P. Dubois. 2000. Polymer-layered silicate nanocomposites: Preparation, properties and uses of a new class of materials. *Materials Science and Engineering* 28: 1–63.

Amin, A., R. Sarkar, C.N. Moorefield, and G.R. Newkome. 2013. Synthesis of polymer-clay nanocomposites of some vinyl monomers by surface-initiated atom transfer radical polymerization. *Designed Monomers and Polymers* 16: 528–536.

Arfaioli, P., O.L. Pantani, M. Bosetto, and G.G. Ristori. 1999. Influence of clay minerals and exchangeable cations on the formation of humic-like substances (melanoidins) from D-glucose and L-tyrosine. *Clay Minerals* 34: 487–497.

Azeez, A.A., K.Y. Rhee, S.J. Park, and D. Hui. 2013. Epoxy clay nanocomposites—Processing, properties and applications: A review. *Composites: Part B* 45: 308–320.

Ballantine, J.A., W. Jones, J.H. Purnell, D.T.B. Tennakoon, and J.M. Thomas. 1985. The influence of interlayer water on clay catalysis: Interlamellar conversion of 2-methylpropene. *Chemistry Letters* 14: 763–766.

Ballantine, J.A., J.H. Purnell, and J.M. Thomas. 1983. Organic reactions in a clay micro-environment. *Clay Minerals* 18: 347–356.

Ballav, N. and M. Biswas. 2004a. A conducting nanocomposite via intercalative polymerisation of thiophene in montmorillonite clay. *Synthetic Metals* 142: 309–315.

Ballav, N. and M. Biswas. 2004b. Preparation and evaluation of nanocomposites of polyfuran with Al_2O_3 and montmorillonite clay. *Polymer International* 53: 1467–1472.

Ballav, N. and M. Biswas. 2004c. High yield polymerisation of aniline and pyrrole in presence of montmorillonite clay and formation of nanocomposites thereof. *Polymer Journal* 36: 162–166.

Bart, J.C., F. Cariati, L. Erre, C. Gessa, G. Micera, and P. Piu. 1979. Formation of polymeric species in the interlayer of bentonite. *Clays and Clay Minerals* 27: 429–432.

Bekri-Abbes, I. and E. Srasra. 2010a. Characterization and AC conductivity of polyaniline-montmorillonite nanocomposites synthesized by mechanical/chemical reaction. *Reactive & Functional Polymers* 70: 11–18.

Bekri-Abbes, I. and E. Srasra. 2010b. Solid-state synthesis and electrical properties of polyaniline/Cu-montmorillonite nanocomposite. *Materials Research Bulletin* 45: 1941–1947.

Belt, S.T., W.G. Allard, J. Rintatalo, L.A. Johns, A.C.T. van Duin, and S.J. Rowland. 2000. Clay and acid catalysed isomerisation and cyclisation reactions of highly branched isoprenoid (HBI) alkenes: Implications for sedimentary reactions and distributions. *Geochimica et Cosmochimica Acta* 64: 3337–3345.

Belt, S.T. and P. Cabedo-Sanz. 2015. Characterisation and isomerisation of mono- and di-unsaturated highly branched isoprenoid (HBI) alkenes: Considerations for palaeoenvironment studies. *Organic Geochemistry* 87: 55–67.

Bergaya, F. and F. Kooli. 1991. Acrylonitrile-smectite complexes. *Clay Minerals* 26: 33–41.

Bergman, J.S., H. Chen, E.P. Giannelis, M.G. Thomas, and G.W. Coates. 1999. Synthesis and characterization of polyolefin-silicate nanocomposites: A catalyst intercalation and *in situ* polymerization approach. *Chemical Communications* 2179–2180.

Beşün, N., F. Özkan, and G. Gündüz. 2002. Alpha-pinene isomerization on acid-treated clays. *Applied Catalysis A: General* 224: 285–297.

Biermann, U. and J.O. Metzger. 2008. Synthesis of alkyl-branched fatty acids. *European Journal of Lipid Science and Technology* 110: 805–811.

Birkel, U., G. Gerold, and J. Niemeyer. 2002. Abiotic reactions of organics on clay mineral surfaces. In *Developments in Soil Science, Vol. 28A* (Eds.) A. Violante, P.M. Huang, J.-M. Bollag, and L. Gianfreda, pp. 437–447. Amsterdam, the Netherlands: Elsevier.

Biswas, M. and S.S. Ray. 1998. Preparation and evaluation of composites from montmorillonite and some heterocyclic polymers. 1: Poly(*N*-vinylcarbazole)-montmorillonite nanocomposite system. *Polymer* 39: 6423–6428.

Biswas, M. and S.S. Ray. 2001. Recent progress in synthesis and evaluation of polymer-montmorillonite nanocomposites. *Advances in Polymer Science* 155: 170–221.

Bittles, J.A., A.K. Chaudhuri, and S.W. Benson. 1964. Clay-catalyzed reactions of olefins. I. Polymerization of styrene. *Journal of Polymer Science: Part A: General Papers* 2: 1221–1231.

Blumstein, A., 1961. Étude des polymérisations en couche adsorbée I. *Bulletin de la Société Chimique de France* 899–906.

Blumstein, A. 1965a. Polymerization of adsorbed monolayers. I. Preparation of the clay-polymer complex. *Journal of Polymer Science Part A: Polymer Chemistry* 3: 2653–2664.

Blumstein, A., 1965b. Polymerization of adsorbed monolayers. II. Thermal degradation of the inserted polymer. *Journal of Polymer Science Part A: Polymer Chemistry* 3: 2665–2672.

Blumstein, A. and F.W. Billmeyer, Jr. 1966. Polymerization of adsorbed monolayers. III. Preliminary structure studies in dilute solution of the insertion polymers. *Journal of Polymer Science Part B: Polymer Physics* 4: 465–474.

Blumstein, A. and A.C. Watterson. 1968. Tacticity of poly(methyl methacrylate) prepared by radical polymerization within a monolayer of methyl methacrylate adsorbed on montmorillonite. *Journal of Polymer Science Part C: Polymer Letters* 6: 69–74.

Blumstein, R., A. Blumstein, and K. Parikh. 1974. Polymerization of monomolecular layers adsorbed on montmorillonite: Cyclization in polyacrylonitrile and polymethacrylonitrile. *Applied Polymer Symposia* 25: 81–88.

Bosetto, M., P. Arfaioli, and O.L. Pantani. 2002. Study of the Maillard reaction products formed by glycine and D-glucose on different mineral substrates. *Clay Minerals* 37: 195–204.

Böttcher, H., M.L. Hallensleben, S. Nuβ, H. Wurm, J. Bauer, and P. Behrens. 2002. Organic/inorganic hybrids by "living"/controlled ATRP grafting from layered silicates. *Journal of Materials Chemistry* 12: 1351–1354.

Boukerma, K., J.-Y. Piquemal, M.M. Chehimi, M. Mravčáková, M. Omastová, and P. Beaunier. 2006. Synthesis and interfacial properties of montmorillonite/polypyrrole nanocomposites. *Polymer* 47: 569–576.

Breen, C. and A. Moronta. 1999. Influence of layer charge on the catalytic activity of mildly acid-activated tetramethylammonium-exchanged bentonites. *Journal of Physical Chemistry B* 103: 5675–5680.

Breen, C. and A. Moronta. 2000. Characterization and catalytic activity of aluminum- and aluminum/tetramethylammonium-exchanged bentonites. *Journal of Physical Chemistry B* 104: 2702–2708.

Breen, C. and A. Moronta. 2001. Influence of exchange cation and layer charge on the isomerization of α-pinene over SWy-2, SAz-1 and Sap-Ca. *Clay Minerals* 36: 467–472.

Breen, C. and R. Watson. 1998. Acid-activated organoclays: Preparation, characterisation and catalytic activity of polycation-treated bentonites. *Applied Clay Science* 12: 479–494.

Cabrera, A., J. Peón, L. Velasco, R. Miranda, A. Salmón, and M. Salmón. 1995. Clay-mediated cyclooligomerization of olefin oxides: A one-pot route to crown ethers. *Journal of Molecular Catalysis A: Chemical* 104: L5–L7.

Calabi-Floody, M., B.K.G. Theng, P. Reyes, and M.L. Mora. 2009. Natural nanoclays: Application and future trends—A Chilean perspective. *Clay Minerals* 44: 161–176.

Camejo-Abreu, C., V. Tabernero, M.D. Alba, T. Cuenca, and P. Terreros. 2014. Enhanced activity of clays and its crucial role for the activity in ethylene polymerization. *Journal of Molecular Catalysis A: Chemical* 393: 96–104.

Carastan, D.J. and N.R. Demarquette. 2007. Polystyrene/clay nanocomposites. *International Materials Reviews* 52: 345–380.

Casagrande, M., L. Storaro, M. Lenarda, and S. Rossini. 2005. Solid acid catalysts from clays: Oligomerization of 1-pentene on Al-pillared smectites. *Catalysis Communications* 6: 568–572.

Catrinescu, C., C. Fernandes, P. Castilho, and C. Breen. 2006. Influence of exchange cations on the catalytic conversion of limonene over Serra de Dentro (SD) and SAz-1 clays: Correlations between acidity and catalytic activity/selectivity. *Applied Catalysis A: General* 311: 172–184.

Chae, H.S., W.L. Zhang, S.H. Piao, and H.J. Choi. 2015. Synthesized palygorskite/polyaniline nanocomposite particles by oxidative polymerization and their electrorheology. *Applied Clay Science* 107: 165–172.

Chang, J.H., Y.U. An, S.J. Kim, and S. Im. 2003. Poly(butylene terephthalate)/organoclay nanocomposites prepared by in situ interlayer polymerization and its fiber (II). *Polymer* 44: 5655–5661.

Chang, J.-H., S.J. Kim, Y.L. Joo, and S. Im. 2004. Poly(ethylene terephthalate) nanocomposites by *in situ* interlayer polymerization: The thermo-mechanical properties and morphology of the hybrid fibers. *Polymer* 45: 919–926.

Chang, T.-C., S.-Y. Ho, and K.-J. Chao. 1992. Intercalation of polyaniline in montmorillonite and zeolite. *Journal of the Chinese Chemical Society* 39: 209–212.

Chaudhuri, B. and M.M. Sharma. 1989. Some novel aspects of dimerization of alpha-methylstyrene with acidic ion-exchange resins, clays and other acidic materials as catalysts. *Industrial & Engineering Chemistry Research* 28: 1757–1763.

Chen, B. 2004. Polymer-clay nanocomposites: An overview with emphasis on interaction mechanisms. *British Ceramic Transactions* 103: 241–249.

Cheshchevoi, V.N., D.N. Palamarev, V.A. Diner, and V.A. Polushkin. 1989. Spectral study of the oligomerization products of α-pinene on kaolinite. *Polymer Science U.S.S.R.* 31: 2274–2279.

Chitnis, S.R. and M.M Sharma. 1997. Industrial applications of acid-treated clays as catalysts. *Reactive & Functional Polymers* 32: 93–115.

Choi, Y.Y., S.H. Lee, and S.H. Ryu. 2009. Effect of silane functionalization of montmorillonite on epoxy/montmorillonite nanocomposite. *Polymer Bulletin* 63: 47–55.

Churchman, G.J., W.P. Gates, B.K.G. Theng, and G. Yuan. 2006. Clays and clay minerals for pollution control. In *Handbook of Clay Science, Developments in Clay Science, Vol. 1* (Eds.) F. Bergaya, B.K.G. Theng, and G. Lagaly, 625–675. Amsterdam, the Netherlands: Elsevier.

Čičel, B., P. Komadel, and M. Nigrin. 1992. Catalytic activity of smectiteson dimerization of oleic acid. *Collection of Czechoslovak Chemical Communications* 57: 1666–1671.

Cloos, P., A. Moreale, C. Broers, and C. Badot. 1979. Adsorption and oxidation of aniline and p-chloroaniline by montmorillonite. *Clay Minerals* 14: 307–321.

Cruz, M.I., H. Nijs, J.J. Fripiat, and H. Van Damme. 1982. Photochemical and photocatalytic properties of adsorbed coordination compounds. 3. Cis ↔ trans isomerization of $Ru(bpy)_2(H_2O)_2^{2+}$ on layer-lattice silicates. *Journal de Chimie Physique* 79: 753–757.

Cruz-Almanza, R., I. Shiba-Matzumoto, A. Fuentes et al. 1997. Oligomerization of benzylic alcohols and its mechanism. *Journal of Molecular Catalysis A: Chemical* 126: 161–168.

Cui, L., N.H. Tarte, and S.I. Woo. 2008. Effects of modified clay on the morphology and properties of PMMA/clay nanocomposites synthesized by *in situ* polymerization. *Macromolecules* 41: 4268–4274.

Das, B., P. Madhusudhan, and B. Venkataiah. 2000. Clay catalysed convenient isomerization of natural furofuran lignans under microwave irradiation. *Synthetic Communications* 30: 4001–4006.

Datta, H., A.K. Bhowmick, and N.K. Singha. 2008a. Tailor-made hybrid nanostructure of poly(ethyl acrylate)/clay by surface-initiated atom transfer radical polymerization. *Journal of Polymer Science: Part A: Polymer Chemistry* 46: 5014–5027.

Datta, H., N.K. Singha, and A.K. Bhowmick. 2008b. Beneficial effect of nanoclay in atom transfer radical polymerization of ethyl acrylate: A one pot preparation of tailor-made polymer nanocomposite. *Macromolecules* 41: 50–57.

Davidtz, J.C. 1976. The acid activity of 2:1 layer silicates. *Journal of Catalysis* 43: 260–263.

Dekking, H.G.G. 1964. Preparation and properties of some polymer-clay compounds. *Clays and Clay Minerals* 13: 603–616.

Dekking, H.G.G. 1965. Propagation of vinyl polymers on clay surfaces. I. Preparation, structure, and decomposition of clay initiators. *Journal of Applied Polymer Science* 9: 1641–1651.

Dekking, H.G.G. 1967. Propagation of vinyl polymers on clay surfaces. II. Polymerization of monomers initiated by free radicals attached to clay. *Journal of Applied Polymer Science* 11: 23–36.

den Otter, M.J.A.M. 1970a. The dimerization of oleic acid with a montmorillonite catalyst. I: Important process parameters; some main reactions. *Fette Seifen Anstrichmittel* 72: 667–673.

den Otter, M.J.A.M. 1970b. The dimerization of oleic acid with a montmorillonite catalyst. II: Glc analysis of the monomer; the structure of the dimer; a reaction model. *Fette Seifen Anstrichmittel* 72: 875–883.

den Otter, M.J.A.M. 1970a. The dimerization of oleic acid with a montmorillonite catalyst. III: Test of the reaction model. *Fette Seifen Anstrichmittel* 72: 1056–1066.

Desjardins, S., J.A. Landry, and J.P. Farant. 1999. Effects of water and pH on the oxidative oligomerization of chloro and methoxyphenol by a montmorillonite clay. *Journal of Soil Contamination* 8: 175–195.

Dias, M.L., R.M. Fernandes, R.H. Cunha, S. Jaconis, and A.C. Silvino. 2011. Highly filled clay polypropylene nanocomposites prepared by in situ polymerization with clay-supported magnesium/titanium catalysts. *Applied Catalysis A: General* 403: 48–57.

Dintzner, M.R., Y.A. Mondjinou, and D.J. Pileggi. 2010. Montmorillonite clay catalyzed cyclotrimerization and oxidation of aliphatic aldehydes, *Tetrahedron Letters* 51: 826–827.

Doh, J.G. and I. Cho. 1998. Synthesis and properties of polystyrene-organoammonium montmorillonite hybrid. *Polymer Bulletin* 41: 511–518.

do Nascimento, G.M., V.R.L. Constantino, R. Landers, and M.L.A. Temperini. 2004. Aniline polymerization into montmorillonite clay: A spectroscopic investigation of the intercalated conducting polymer. *Macromolecules* 37: 9373–9385.

do Nascimento, G.M., V.R.L. Constantino, R. Landers, and M.L.A. Temperini. 2006. Spectroscopic characterization of polyaniline formed in the presence of montmorillonite clay. *Polymer* 47: 6131–6139.

Du, K., A.H. He, F.-Y. Bi, and C.C. Han. 2013. Synthesis of exfoliated isotactic polypropylene/functional alkyl-triphenylphosphonium-modified clay nanocomposites by *in situ* polymerization. *Chinese Journal of Polymer Science* 31: 1501–1508.

Du, K., A.H. He, X. Liu, and C.C. Han. 2007. High-performance exfoliated poly(propylene)/clay nanocomposites by in situ polymerization with a novel Z–N/clay compound catalyst. *Macromolecular Rapid Communications* 28: 2294–2299.

Eastman, M.P., E. Bain, T.L. Porter et al. 1999. The formation of poly(methyl-methacrylate) on transition metal-exchanged hectorite. *Applied Clay Science* 15: 173–185.

Eastman, M.P., M.E. Hagerman, J.L. Attuso, E.D. Bain, and T.L. Porter. 1996. Polymerization of benzene and aniline on Cu(II)-exchanged hectorite clay films: A scanning force microscope study. *Clays and Clay Minerals* 44: 769–773.

Eastman, M.P. and T.L. Porter. 2000. Polymerization of organic monomers and biomolecules on hectorite. In *Polymer-Clay Nanocomposites* (Eds.) T.J. Pinnavaia and G.W. Beall, 65–93. Chichester, UK: John Wiley & Sons.

Elbokl, T.A. and C. Detellier. 2006. Aluminosilicate nanohybrid materials: Intercalation of polystyrene in kaolinite. *Journal of Physics and Chemistry of Solids* 67: 950–955.

Erhan, S.M. and T.A. Isbell. 1997. Estolide production with modified clay catalysts and process conditions. *Journal of the American Oil Chemists' Society* 74: 249–254.

Ertem, G. and Z. Gan. 2014. Role of preparation method on the extent of montmorillonite catalyst for oligomer formation. *Applied Clay Science* 101: 90–93.

Essawy, H.A., A.S. Badran, A.M. Youssef, A. El-Fettouh, and A.A. El-Hakim. 2004. Polystyrene/montmorillonite nanocomposites prepared by in situ intercalative polymerimerzation: Influence of the surfactant type. *Macromolecular Chemistry and Physics* 205: 2366–2370.

Faguy, P.W., W. Ma, J.A. Lowe, W.-P. Pan, and T. Brown. 1994. Conducting polymer-clay composites for electrochemical applications. *Journal of Materials Chemistry* 4: 771–772.

Fan, X., C. Xia, and R.C. Advincula. 2003. Grafting of polymers from clay nanoparticles via in situ free radical surface-initiated polymerization: Monocationic versus bicationic initiators. *Langmuir* 19: 4381–4389.

Fan, X., C. Xia, and R.C. Advincula. 2005. On the formation of narrowly polydispersed PMMA by surface initiated polymerization (SIP) from AIBN-coated/intercalated clay nanoparticle platelets. *Langmuir* 21: 2537–2544.

Fan, X., Q. Zhou, C. Xia, W. Cristofoli, J. Mays, and R. Advincula. 2002. Living anionic surface-initiated polymerization (LASIP) of styrene from clay nanoparticles using surface bound 1,2,diphenylethylene (DPE) initiators. *Langmuir* 18: 4511–4518.

Fenn, D.B., M.M. Mortland, and T.J. Pinnavaia. 1973. The chemisorption of anisole on Cu(II) hectorite. *Clays and Clay Minerals* 21: 315–322.

Fernández-Saavedra, R., P. Aranda, and E. Ruiz-Hitzky. 2004. Templated synthesis of carbon nanofibers using sepiolite. *Advanced Functional Materials* 14: 77–82.

Ferris, J.P. 2005. Mineral catalysis and prebiotic synthesis: Montmorillonite-catalyzed formation of RNA. *Elements* 1: 145–149.

Fischer, H. 2003. Polymer nanocomposites: From fundamental research to specific applications. *Materials Science and Engineering C* 23: 763–772.

Flores-Holguín, N., A. Aquilar-Elguézabal, L.-M. Rodríguez-Valdes, and D. Glossman-Mitnik. 2012. A theoretical study of the carbocation formation energy involved in the isomerization of α-pinene. *Chemical Physics Letters* 546: 168–170.

Frenkel, M. and L. Heller-Kallai. 1983. Interlayer cations as reaction directors in the transformation of limonene on montmorillonite. *Clays and Clay Minerals* 31: 92–96.

Friedlander, H.Z. and C.R. Frink. 1964. Organized polymerization. III. Monomers intercalated in montmorillonite. *Journal of Polymer Science. Part C: Polymer Letters* 2: 475–479.

Fu, X. and S. Qutubuddin. 2000. Synthesis of polystyrene-clay nanocomposites. *Materials Letters* 42: 12–15.

Fu, X. and S. Qutubuddin. 2001. Polymer-clay nanocomposites: Exfoliation of organophilic montmorillonite nanolayers in polystyrene. *Polymer* 42: 807–813.

Fukushima, Y. and S. Inagaki. 1987. Synthesis of an intercalated compound of montmorillonite and 6-polyamide. *Journal of Inclusion Phenomena* 5: 473–482.

Fukushima, Y., A. Okada, M. Kawasumi, T. Kurauchi, and O. Kamigaito. 1988. Swelling behavior of montmorillonite by poly-6-amide. *Clay Minerals* 23: 27–34.

García-González, N., B.A. Frontana-Uribe, E. Ordoñez-Regil, J. Cárdenas, and J.A. Morales-Serna. 2016. Evaluation of Fe^{3+} fixation into montmorillonite clay and its application in the polymerization of ethylenedioxythiophene. *RSC Advances* 6: 95879–95887.

Gayer, F.H. 1933. The catalytic polymerization of propylene. *Industrial & Engineering Chemistry* 25: 1122–1127.

Giannelis, E.P., R. Krishnamoorti, and E. Manias. 1999. Polymer-silicate nanocomposites: Model systems for confined polymers and polymer brushes. *Advances in Polymer Science* 138: 107–147.

Gilman, J.W. 1999. Flammability and thermal stability studies of polymer layered-silicate (clay) nanocomposites. *Applied Clay Science* 15: 31–49.

Glavati, O.L. and L.S. Polak. 1964. Kinetics and mechanisms of radiation polymerization in montmorillonite inclusion compounds. *Neftekhimiya* 4: 77–81.

Glavati, O.L., L.S. Polak, and V.V. Shchekin. 1963. The radiational stereospecific polymerization of acrylonitrile and acrylic acid in montmorillonite compounds. *Neftekhimiya* 3: 905–910.

González, E., D. Rodriguez, L. Huerta, and A. Moronta. 2009. Isomerization of 1-butene catalyzed by surfactant-modified Al_2O_3-pillared clays. *Clays and Clay Minerals* 57: 383–391.

Gonzalez, J.M. and D.A. Laird. 2004. Role of smectites and Al-substituted goethites in the catalytic condensation of arginine and glucose. *Clays and Clay Minerals* 52: 443–450.

Gorrasi, G., M. Tortora, and V. Vittoria et al. 2003. Vapor barrier properties of polycaprolactone montmorillonite nanocomposites: Effect of clay dispersion. *Polymer* 44: 2271–2279.

Greenwell, H.C., A.A. Bowden, B. Chen et al. 2006. Intercalation and *in situ* polymerization of poly(alkylene oxide) derivatives within M^+-montmorillonite (M = Li, Na, K). *Journal of Materials Chemistry* 16: 1082–1094.

Greesh, N., R. Sanderson, and P. Hartmann. 2012. Preparation of polystyrene-clay nanocomposites via dispersion polymerization using oligomeric styrene-montmorillonite as stabilizer. *Polymer International* 61: 834–843.

Gündüz, G. and D.Y. Murzin. 2002. Influence of catalytic pretreatment on α-pinene isomerization over natural clays. *Reaction Kinetics and Catalysis Letters* 75: 231–237.

Gupta, B., A. Rakesh, A.A. Melvin, A.C. Pandey, and R. Prakash. 2014. In-situ synthesis of polyaniline coated montmorillonite (Mt) clay using Fe^{+3} intercalated Mt as oxidizing agent. *Applied Clay Science* 95: 50–54.

Gurwitsch, L. 1923. Zur Kenntnis der heterogenen Katalyse. *Zeitschrift für physikalische Chemie Frankfurt* 107: 235–248.

Han, Y. 2009. Synthesis and characterization of montmorillonite/polypyrrole nanocomposites. *Polymer Composites* 30: 66–69.

Harrane, A., R. Meghabar, and M. Belbachir. 2002. A protons exchanged montmorillonite clay as an efficient catalyst for the reaction of isobutylene polymerization. *International Journal of Molecular Sciences* 3: 790–800.

Harvey, B.G. and R.L. Quintana. 2010. Synthesis of renewable jet and diesel fuels from 2-ethyl-1-hexene. *Energy & Environmental Science* 3: 352–357.

Harvey, B.G., M.E. Wright, and R.L. Quintana. 2010. High-density renewable fuels based on the selective dimerization of pinenes. *Energy & Fuels* 24: 267–273.

He, A., L. Wang, J. Li, J. Dong, and C.C. Han. 2006. Preparation of exfoliated isotactic polypropylene/alkyl-triphenylphosphonium-modified montmorillonite nanocomposites via in situ intercalative polymerization. *Polymer* 47: 1767–1771.

He, F.-A., L.-M. Zhang, H.-L. Jiang, L.-S. Chen, Q. Wu, and H.-H. Wang. 2007. A new strategy to prepare polyethylene nanocomposites by using a late-transition-metal catalyst supported on AlEt$_3$-activated organoclay. *Composites Science and Technology* 67: 1727–1733.

He, H., L. Ma, J. Zhu, R.L. Frost, B.K.G. Theng, and F. Bergaya. 2014. Synthesis of organoclays: A critical review and some unresolved issues. *Applied Clay Science*, 100: 22–28.

Hiscock, M. and G.B. Porter. 1972. Antimony halides as solvents. Part IX. The dimerization of *trans*-stilbene in molten antimony trichloride. *Journal of the Chemical Society, Perkin Transactions* 2: 79–83.

Hong, S.H., B.H. Kim, J. Joo, J.W. Kim, and H.J. Choi. 2001. Polypyrrole-montmorillonite nanocomposites synthesized by emulsion polymerization. *Current Applied Physics* 1: 447–450.

Huang, M.-Y., J.-C. Wu, F.-S. Shieu, and J.-J. Lin. 2010. Isomerization of *endo*-tetrahydrodicyclopentadiene over clay-supported chloroaluminate ionic liquid catalysts. *Journal of Molecular Catalysis A: Chemical* 315: 69–75.

Huskić, M. and E. Žagar. 2017. Catalytic activity of mineral montmorillonite on the reaction of phenol with formaldehyde. *Applied Clay Science* 136: 158–163.

Hwu, J.-M. and G.-J. Jiang. 2005. Preparation and characterization of polypropylene-montmorillonite nanocomposites generated by *in situ* metallocene catalyst polymerization. *Journal of Applied Polymer Science* 95: 1228–1236.

Hwu, J.-M., T.H. Ko, W.-T. Yang et al. 2004. Synthesis and properties of polystyrene-montmorillonite nanocomposites by suspension polymerization. *Journal of Applied Polymer Science* 91: 101–109.

Ilic, M., E. Koglin, A. Pohlmeier, H.D. Narres, and M.J. Schwuger. 2000. Adsorption and polymerization of aniline on Cu(II)-montmorillonite: Vibrational spectroscopy and ab initio calculation. *Langmuir* 16: 8946–8951.

Il'ina, I.V., K.P. Volcho, D.V. Korchagina, V.A. Barkhash, and N.F. Salakhutdinov. 2007. Reactions of allyl alcohols of the pinane series and of their epoxides in the presence of montmorillonite clay. *Helvetica Chimica Acta* 90: 353–367.

Imamura, S., Y. Hayashi, K. Kajiwara, H. Hoshino, and C. Kaito. 1993. Imogolite: A possible new type of shape-selective catalyst. *Industrial & Engineering Chemistry Research* 32: 600–603.

Inagaki, S., Y. Fukushima, and M. Miyata. 1995. Inclusion polymerization of isoprene in the channels of sepiolite. *Research on Chemical Intermediates* 21: 167–180.

Inoue, H. and H. Yoneyama. 1987. Electropolymerization of aniline intercalated in montmorillonite. *Journal of Electroanalytical Chemistry* 233: 291–294.

Issaadi, R. and F. Garin. 2003. Catalytic behaviour of acid catalysts supported palladium: Use of Al and Zr-pillared montmorillonite as supports. Part II: Kinetic study. *Applied Catalysis A: General* 243: 367–377.

Issaadi, R., F. Garin, and C.E. Chitour. 2006. Study of the acid character of some palladium-modified pillared clay catalysts: Use of isopropanol decomposition as test reaction. *Catalysis Today* 113: 166–173.

Issaadi, R., F. Garin, C.E. Chitour, and G. Maire. 2001. Catalytic behaviour of combined palladium-acid catalysts: Use of Al and Zr-pillared montmorillonite as supports. Part I. Reactivity of linear, branched and cyclic hexane hydrocarbons. *Applied Catalysis A: General* 207: 323–332.

Jeong, D.W., D.S. Hong, H.Y. Cho, and S.I. Woo. 2003. The effect of water and acidity of the clay for ethylene polymerization over Cp$_2$ZrCl$_2$ supported on TMA-modified clay minerals. *Journal of Molecular Catalysis A: Chemical* 206: 205–211.

Jia, W., E. Segal, D. Kornemandel, Y. Lamhot, M. Narkis, and A. Siegmann. 2002. Polyaniline–DBSA/organophilic clay nanocomposites: Synthesis and characterization *Synthetic Metals* 128: 115–120.

Jin, Y.-H., H.-J. Park, S.-S. Im, S.-Y. Kwak, and S. Kwak. 2002. Polyethylene/clay nanocomposites by in-situ exfoliation of montmorillonite during Ziegler–Natta polymerization of ethylene. *Macromolecular Rapid Communications* 23: 135–140

Joseph-Ezra, H., A. Nasser, and U. Mingelgrin. 2014. Surface interactions of pyrene and phenanthrene on Cu-montmorillonite. *Applied Clay Science* 95: 348–356.

Joshi, P.C., M.F. Aldersley, J.W. Delano, and J.P. Ferris. 2009. Mechanism of montmorillonite catalysis in the formation of RNA oligomers. *Journal of the American Chemical Society* 131: 13369–13374.

Kannan, P., C. Venkatachalapathy, and K. Pitchumani. 1999. Orton rearrangement of N-chloroacetanilide in clay microenvironment. *Indian Journal of Chemistry* 38B: 384–386.
Karim, M.R. and J.H. Yeum. 2008. In situ intercalative polymerization of conducting polypyrrole/montmorillonite nanocomposites. *Journal of Polymer Science: Part B: Polymer Physics* 46: 2279–2285.
Kassim, A., H.N.M. Ekramul Mahmud, and F. Adzmi. 2007. Polypyrrole-montmorillonite clay composites: An organic semiconductor. *Materials Science in Semiconductor Processing* 10: 246–251.
Kato, C., K. Kuroda, and K. Hasegawa. 1979a. Electrical conductivity of a montmorillonite-organic complex. *Clay Minerals* 14: 13–20.
Kato, C., K. Kuroda, and M. Misawa. 1979b. Preparation of montmorillonite-nylon complexes and their thermal properties. *Clays and Clay Minerals* 27: 129–136.
Kato, C., K. Kuroda, and H. Takahara. 1981. Preparation and electrical properties of quaternary ammonium montmorillonite-polystyrene complexes. *Clays and Clay Minerals* 29: 294–298.
Kato, M., M. Matsushita, and K. Fukumori. 2004. Development of a new production method for a polypropylene-clay nanocomposite. *Polymer Engineering and Science* 44: 1205–1211.
Ke, Y., C. Long, and Z. Qi. 1999. Crystallization, properties, and crystal and nanoscale morphology of PET-clay nanocomposites. *Journal of Applied Polymer Science* 71: 1139–1146.
Kellendonk, F.J.A., J.J.L. Heinerman, and R.A. van Santen. 1987. Clay-activated isomerization reactions. In *Preparative Chemistry Using Supported Reagents* (Ed.) P. Laszlo, pp. 455–468. New York: Academic Press.
Khalil, H., D. Mahajan, and M. Rafailovich. 2005a. Polymer-montmorillonite clay nanocomposites. Part 1: Complexation of montmorillonite clay with a vinyl monomer. *Polymer International* 54: 423–427.
Khalil, H., D. Mahajan, and M. Rafailovich. 2005b. Complex formation of montmorillonite clay with polymers. Part 2: The use of montmorillonite clay-vinyl monomer complex as a comonomer in the copolymerization reaction of styrene-acrylonitrile monomers. *Polymer International* 54: 428–436.
Kikuchi, E., T. Matsuda, H. Fujiki, and Y. Morita. 1984. Disproportionation of 1,2,4,-trimethylbenzene over montmorillonite pillared by aluminium oxide. *Applied Catalysis* 11: 331–340.
Kikuchi, E., T. Matsuda, H. Fujiki, and Y. Morita. 1985. Conversion of trimethylbenzenes over montmorillonites pillared by aluminium and zirconium oxides. *Applied Catalysis* 16: 401–410.
Kim, B.-H., J.-H. Jung, S.-H. Hong et al. 2002. Nanocomposite of polyaniline and Na^+-montmorillonite clay. *Macromolecules* 35: 1419–1423.
Kim, D.W., A. Blumstein, H. Liu, M.J. Downey, J. Kumar, and S.K. Tripathy. 2001. Organic/inorganic nanocomposites prepared by spontaneous polymerization of ethynylpyridine within montmorillonite. *Journal of Macromolecular Science–Pure and Applied Chemistry* A38: 1405–1415.
Kim, J.W., S.G. Kim, H.J. Choi, and M.S. Jhon. 1999. Synthesis and electrorheological properties of polyaniline-Na^+-montmorillonite suspensions. *Macromolecular Rapid Communications* 20: 450–452.
Kim, J.W., F. Liu, H.J. Choi, S.H. Hong, and J. Joo. 2003. Intercalated polypyrrole/Na^+-montmorillonite nanocomposite via an inverted emulsion pathway method. *Polymer* 44: 289–293.
Kim, S.-K., H.-D. Kwen, and S.-H. Choi. 2012. Radiation-induced synthesis of vinyl copolymer based nanocomposites filed with reactive organic montmorillonite clay. *Radiation Physics and Chemistry* 81: 519–523.
Ko, A.-N. and H.-C. Cheng. 1992. Reaction of trimethylbenzene over alumina-pillared montmorillonite. *Journal of the Chinese Chemical Society* 39: 81–86.
Kojima, M., R. Hartford, and C.T. O'Connor. 1991. The effect of pillaring montmorillonite and beidellite on the conversion of trimethyl benzenes. *Journal of Catalysis* 128: 487–498.
Kojima, Y., A. Usuki, M. Kawasumi, A. Okada, T. Kurauchi, and O. Kamigaito. 1993. One-pot synthesis of nylon 6-clay hybrid. *Journal of Polymer Science: Part A: Polymer Chemistry* 31: 1755–1758.
Komori, Y., Y. Sugahara, and K. Kuroda. 1999. Direct intercalation of poly(vinylpyrrolidone) into kaolinite by a refined guest displacement method. *Chemistry of Materials* 11: 3–6.
Konn, C., F. Morel, E. Beyou, P. Chaumont, and E. Bourgeat-Lami. 2007. Nitroxide-mediated polymerization of styrene initiated from the surface of Laponite clay platelets. *Macromolecules* 40: 7464–7472.
Kooli, F. 2014. Porous clay heterostrustures (PCHs) from Al_{13}-intercalated and Al_{13}-pillared montmorillonites: Properties and heptane hydro-isomerization catalytic activity. *Microporous and Mesoporous Materials* 184: 184–192.
Kooli, F., Y. Liu, S.F. Alshahateet, P. Siril, and R. Brown. 2008. Effect of pillared clays on the hydroisomerization of n-heptane. *Catalysis Today* 131: 244–249.
Koster, R.M., M. Bogert, B. de Leeuw, E.K. Poels, and A. Bliek. 1998. Active sites in the clay catalysed dimerization of oleic acid. *Journal of Molecular Catalysis A: Chemical* 134: 159–169.
Kowalska, M.W., J.D. Ortego, and A. Jezierski. 2001. Transformation of 2-(trifluoromethyl)aniline over ion-exchanged montmorillonites: Formation of a dimer and cyclic trimer. *Applied Clay Science* 18: 233–243.

Kumar, B.S., A. Dhakshinamoorthy, and K. Pitchumani. 2014. K10 montmorillonite clays as environmentally benign catalysts for organic reactions. *Catalysis Science & Technology* 4: 2378–2396.

Kumar, V. and A.K. Agarwal. 2014. A review on catalytic terpene transformation over heterogeneous catalyst. *International Journal of Current Research in Chemistry and Pharmaceutical Sciences* 1: 78–88.

Kurokawa, H., S. Morita, M. Matsuda, H. Suzuki, M-a. Ohshima, and H. Miura. 2009. Polymerization of ethylene using zirconocenes supported on swellable cation-exchanged fluorotetrasilicic mica. *Applied Catalysis A: General* 360: 192–198.

Kusnitsyna, T.A. and I.K. Ostrovskaya. 1967. Catalytic activity of acid aluminosilicates in the styrene polymerization reaction. *Vysokomolekulyarnye Soedineniya* A9: 2510–2514.

Lan, T., P.D. Kaviratna, and T.J. Pinnavaia. 1995. Mechanism of clay tactoid exfoliation in epoxy-clay nanocomposites. *Chemistry of Materials* 7: 2144–2150.

Lao, Y., J. Korth, J. Ellis, and P.T. Crisp. 1988. Mineral-catalysed transformations of terminal alkenes during pyrolysis. *Journal of Analytical and Applied Pyrolysis* 14: 191–201.

Lao, Y., J. Korth, J. Ellis, and P.T. Crisp. 1989. Heterogeneous reactions of 1-pristene catalysed by clays under simulated geological conditions. *Organic Geochemistry* 14: 375–379.

Larcher, A.V., R. Alexander, S.J. Rowland, and R.I. Kagi. 1986. Acid catalysis of alkyl hydrogen exchange and configurational isomerisation reactions: Acyclic isoprenoid acids. *Organic Geochemistry* 10: 1015–1021.

LeBaron, P.C., Z. Wang, and T.J. Pinnavaia. 1999. Polymer-layered silicate nanocomposites: An overview. *Applied Clay Science* 15: 11–29.

Lee, D., K. Char, S.W. Lee, and Y.W. Park. 2003. Structural changes of polyaniline/montmorillonite nanocomposites and their effects on physical properties. *Journal of Materials Chemistry* 13: 2942–2947.

Lee, D., S.-H. Lee, K. Char, and J. Kim. 2000. Expansion distribution of basal spacing of the silicate layers in polyaniline/Na$^+$-montmorillonite nanocomposites monitored with X-ray diffraction. *Macromolecular Rapid Communications* 21: 1136–1139.

Lee, D.C. and L.W. Jang. 1996. Preparation and characterization of PMMA-clay hybrid composite by emulsion polymerization. *Journal of Applied Polymer Science* 61: 1117–1122.

Lee, D.-H., H.-S. Kim, K.-B. Yoon, K.E. Min, K.H. Seo, and S.K. Noh. 2005. Polyethylene/MMT nanocomposites prepared by in situ polymerization using supported catalyst systems. *Science and Technology of Advanced Materials* 6: 457–462.

Lei, X., Y. Liu, and Z. Su. 2008. Synthesis and characterization of organo-attapulgite/polyaniline-dodecylbenzenesulfonic acid based on emulsion polymerization method. *Polymer Composites* 29: 240–244.

Leone, G., F. Bertini, M. Canetti, L. Boggioni, P. Stagnaro, and I. Tritto. 2008. *In situ* polymerization of ethylene using metallocene catalysts: Effect of clay pretreatment on the properties of highly filled polyethylene nanocomposites. *Journal of Polymer Science: Part A: Polymer Chemistry* 46: 5390–5403.

Lepoittevin, B., N. Pantoustier, M. Devalckenaere et al. 2002. Poly(ε-caprolactone)/clay nanocomposites by in-situ intercalative polymerization catalyzed by dibutyltin dimethoxide. *Macromolecules* 35: 8385–8390.

Letaïef, S., P. Aranda, and E. Ruiz-Hitzky. 2005. Influence of iron in the formation of conductive polypyrrole-clay nanocomposites. *Applied Clay Science* 28: 183–198.

Li, W., A. Adams, J. Wang, B. Blümich, and Y. Yang. 2010. Polyethylene/palygorskite nanocomposites: Preparation by *in situ* polymerization and their characterization. *Polymer* 51: 4686–4697.

Liu, C., T. Tang, and B. Huang. 2004. Zirconocene catalyst well-spaced inside modified montmorillonite for ethylene polymerization: Role of pretreatment and modification of montmorillonite in tailoring polymer properties. *Journal of Catalysis* 221: 162–169.

Liu, C., T. Tang, D. Wang, and B. Huang. 2003b. In situ ethylene homopolymerization and copolymerization catalyzed by zirconocene catalysts entrapped inside functionalized montmorillonite. *Journal of Polymer Science: Part A: Polymer Chemistry* 41: 2187–2196.

Liu, C., T. Tang, Z. Zhao, and B. Huang. 2002. Preparation of functionalized montmorillonites and their application in supported zirconocene catalysts for ethylene polymerization. *Journal of Polymer Science: Part A: Polymer Chemistry* 40: 1892–1898.

Liu, G.-D., L.-C. Zhang, X.-W. Qu, B.-T. Wang, and Y. Zhang. 2003a. Tentative study on kinetics of bulk polymerization of methyl methacrylate in presence of montmorillonite. *Journal of Applied Polymer Science* 90: 3690–3695.

Liu, G., L. Zhang, D. Zhao, and X. Qu. 2005. Bulk polymerization of styrene in the presence of organomodified montmorillonite. *Journal of Applied Polymer Science* 96: 1146–1152.

Liu, H., D.W. Kim, A. Blumstein, J. Kumar, and S.K. Tripathy. 2001. Nanocomposite derived from intercalative spontaneous polymerization of 2-ethynylpyridine within layered aluminosilicate: Montmorillonite. *Chemistry of Materials* 13: 2756–2758.

Liu, P. 2007. Polymer modified clay minerals: A review. *Applied Clay Science* 38: 64–76.

Liu, Y., P. Liu, and Z. Su. 2007. Core-shell attapulgite@polyaniline composite particles via in situ oxidative polymerization. *Synthetic Metals* 157: 585–591.

Liu, P., L. Zhu, J. Guo, A. Wang, Y. Zhao, and Z. Wang. 2014. Palygorskite/polystyrene nanocomposites via facile in-situ bulk polymerization: Gelation and thermal properties. *Applied Clay Science* 100: 95–101.

Liyanapatirana, C., S.R. Gwaltney, and K. Xia. 2010. Transformation of triclosan by Fe(III)-saturated montmorillonite. *Environmental Science & Technology* 44: 668–674.

Luca, V. and S. Thomson. 2000. Intercalation and polymerisation of aniline within a tubular aluminosilicate. *Journal of Materials Chemistry* 10: 2121–2126.

Ma, J., Z. Qi, and Y. Hu. 2001. Synthesis and characterization of polypropylene/clay nanocomposites. *Journal of Applied Polymer Science* 82: 3611–3617.

Madhavan, D., M. Murugalakshmi, A. Lalitha, and K. Pitchumani. 2001. Dimerisation of olefins catalysed by K10-montmorillonite clays. *Catalysis Letters* 73: 1–4.

Madhavan, D. and K. Pitchumani. 2002. Photodimerisation of enones in a clay microenvironment. *Photochemical & Photobiological Sciences* 1: 991–995.

Madhavan, D. and K. Pitchumani. 2003. Photodimerisation of acenaphthylene in a clay microenvironment. *Photochemical & Photobiological Sciences* 2: 95–97.

Malhotra, S.I., K.K. Parikh, and A. Blumstein. 1972. Polymerization of monolayers. VII. Influence of the exchangeable cation on the polymerization rate of methyl methacrylate monolayers adsorbed on montmorillonite. *Journal of Colloid and Interface Science* 41: 318–327.

Marques, M. de F.V. and R. Moreira. 2017. Influence of polypropylene reaction time on the clay exfoliation process by *in situ* polymerization. *Journal of Nanoscience and Nanotechnology* 17: 5095–5103.

Marques, M., J. Rosa, and K. Cruz. 2011. Influence of chemical treatment on sodium clay for obtaining polypropylene nanocomposites. *Chemistry & Chemical Technology* 5: 433–438.

Matsuda, T., M. Asanuma, and E. Kikuchi. 1988. Effect of high-temperature treatment on the activity of montmorillonite pillared by alumina in the conversion of 1,2,4-trimethylbenzene. *Applied Catalysis* 38: 289–299.

Matsumoto, T., I. Sakai, and M. Arihara. 1969. Polymerization of styrene by acid clay. *The Chemistry of High Polymers Japan* 26: 378–384.

McMahon, D.H. and E.P. Crowell. 1974. Characterization of products from clay catalyzed polymerization of tall oil fatty acids. *Journal of the American Oil Chemists' Society* 51: 522–527.

Mehrota, V. and E.P. Giannelis. 1991. Metal-insulator molecular multilayers of electroactive polymers: Intercalation of polyaniline in mica-type layered silicates. *Solid State Communications* 77: 155–158.

Mehrota, V. and E.P. Giannelis. 1992. Nanometer scale multilayers of electroactive polymers: Intercalation of polypyrrole in mica-type silicates. *Solid State Ionics* 51: 115–122.

Menéndez-Rodríguez, L., P. Crochet, and V. Cadierno. 2013. Catalytic isomerization of allylic alcohols promoted by complexes [RuCl$_2$(η^6-C$_6$H$_6$)(PTA-Me)] under homogeneous conditions and supported on montmorillonite K-10. *Journal of Molecular Catalysis A: Chemical* 366: 390–399.

Messersmith, P.B. and E.P. Giannelis. 1993. Polymer-layered silicate nanocomposites: In situ intercalative polymerization of ε-caprolactone in layered silicates. *Chemistry of Materials* 5: 1064–1066.

Messersmith, P.B. and E.P. Giannelis. 1994. Synthesis and characterization of layered silicate-epoxy nanocomposites. *Chemistry of Materials* 6: 1719–1725.

Meylemans, H.A., R.L. Quintana, and B.G. Harvey. 2012. Efficient conversion of pure and mixed terpene feedstocks to high density fuels. *Fuel* 97: 560–568.

Mignoni, M.L., J.V.M. Silva, M.O. de Souza, R.dos S. Mauler, R.F. de Souza, and K. Bernardo-Gusmão. 2011. Polyethylene-montmorillonite nanocomposites obtained by *in situ* polymerization of ethylene with nickel-diimine catalysts. *Journal of Applied Polymer Science* 122: 2159–2165.

Miranda, R., J. Escobar, F. Delgado, M. Salmón, and A. Cabrera. 1999. Catalytic promotion of piperonyl alcohol to trimethylendioxyorthocyclophane by bentonitic earth, or by hydrochloric acid. *Journal of Molecular Catalysis A: Chemical* 150: 299–305.

Miranda, R., H. Ríos, F. Delgado, M. Castro, A. Cogordán, and M. Salmón. 2003. Characterization of a bentonite clay and its application as catalyst in the preparation of benzyltoluenes and oligotoluenes. *Applied Catalysis A: General* 244: 217–233.

Miranda, R., O. Valencia-Vázquez, C.A. Maya-Vega et al. 2013. Synthesis of cycloveratrylene macrocycles and benzyl oligomers catalysed by bentonite under microwave/infrared and solvent-free conditions. *Molecules* 18: 12820–12844.

Moet, A.S. and A. Akelah. 1993. Polymer-clay nanocomposites: Polystyrene grafted onto montmorillonite interlayers. *Materials Letters* 18: 97–102.

Molina, M.F., R. Molina, and S. Moreno. 2005. Hydroconversion of heptane over a Colombian montmorillonite modified with mixed pillars of Al-Zr and Al-Si. *Catalysis Today* 107–108: 426–430.

Molina, R., S. Moreno, A. Vieira-Coelho, J.A. Martens, P.A. Jacobs, and G. Poncelet. 1994. Hydroisomerization–hydrocracking of decane over Al- and Ga-pillared clays. *Journal of Catalysis* 148: 304–314.

Moreale, A., P. Cloos, and C. Badot. 1985. Differential behaviour of Fe(III)- and Cu(II)-montmorillonite with aniline: I. Suspensions with constant solid:liquid ratio. *Clay Minerals* 20: 29–37.

Moreno, S., E. Gutierrez, A. Alvarez, N.G. Papayannakos, and G. Poncelet. 1997. Al-pillared clays: From lab syntheses to pilot scale production: Characterization and catalytic properties. *Applied Catalysis A: General* 165: 103–114.

Moronta, A., V. Ferrer, J. Quero, G. Arteaga, and E. Choren. 2002. Influence of preparation method on the catalytic properties of acid-activated tetramethylammonium-exchanged clays. *Applied Catalysis A: General* 230: 127–135.

Moronta, A., J. Luengo, Y. Ramírez, J. Quiñónez, E. González, and J. Sánchez. 2005. Isomerization of *cis*-2-butene and *trans*-2-butene catalyzed by acid- and ion-exchanged smectite-type clays. *Applied Clay Science* 29: 117–123.

Moronta, A., T. Oberto, G. Carruyo et al. 2008. Isomerization of 1-butene catalyzed by ion-exchanged, pillared and ion-exchanged/pillared clays. *Applied Catalysis A: General* 334: 173–178.

Mortland, M.M. 1970. Clay-organic complexes and interactions. *Advances in Agronomy* 22: 75–117.

Mortland, M.M. and L.J. Halloran. 1976. Polymerization of aromatic molecules on smectite. *Soil Science Society of America Journal* 40: 367–370.

Mrah, L., R. Meghabar, and M. Belbachir. 2015. In-situ polymerization of styrene to produce polystyrene/montmorillonite nanocomposites. *Bulletin of Chemical Reaction Engineering & Catalysis* 10: 249–255.

Mravčáková, M., K. Boukerma, M. Omastová, and M.M. Chehimi. 2006. Montmorillonite/polypyrrole nanocomposites: The effect of organic modification of clay on the chemical and electrical properties. *Materials Science and Engineering C* 26: 306–313.

Mravčáková, M., M. Omastová, K. Olejníková, B. Pukánszky, and M.M. Chehimi. 2007. The preparation and properties of sodium and organomodified-montmorillonite/polypyrrole composites: A comparative study. *Synthetic Metals* 157: 347–357.

Nagendrappa, G. 2011. Organic synthesis using clay and clay-supported catalysts. *Applied Clay Science* 53: 106–138.

Nakahira, T., M. Hama, O. Fukuchi et al. 1995. Control of photochemistry of stilbazolium ion by adsorption to poly(potassium vinylsulfate) and to hectorite clay. *Macromolecular Rapid Communications* 16: 717–723.

Nakano, Y., T.A. Foglia, H. Kohashi, T. Perlstein, and S. Serota. 1985. Thermal alteration of oleic acid in the presence of clay catalysts with co-catalysts. *Journal of the American Oil Chemists' Society* 62: 888–891.

Narayanan, B.N., R. Koodathil, T. Gangadharan, Z. Yaakob, F.K. Saidu, and S. Chandralayam. 2010. Preparation and characterization of exfoliated polyaniline/montmorillonite nanocomposites. *Materials Science and Engineering B* 168: 242–244.

Nazir, M., M. Ahmed, and F.M. Chaudhary. 1976. Isomerization of α-pinene to camphene using indigeneous clays as catalysts. *Pakistan Journal of Scientific and Industrial Research* 19: 175–178.

Nese, A., S. Sen, M.A. Tasdelen, N. Nugay, and Y. Yagci. 2006. Clay-PMMA nanocomposites by photoinitiated radical polymerization using intercalated phenacyl pyridinium salt initiators. *Macromolecular Chemistry and Physics* 207: 820–826.

Nguyen, Q.T. and D.G. Baird. 2006. Preparation of polymer-clay nanocomposites and their properties. *Advances in Polymer Technology* 25: 270–285.

Nikkhah, S.J., S. A. Ramazani, H. Baniasadi, and F. Tavakolzadeh. 2009. Investigation of properties of polyethylene/clay nanocomposites prepared by new in situ Ziegler–Natta catalyst. *Materials and Design* 30: 2309–2315.

Nikolaidis, A.K., D.S. Achilias, and G.P. Karayannides. 2011. Synthesis and characterization of PMMA/organomodified montmorillonite nanocomposites prepared by in situ bulk polymerization. *Industrial & Engineering Chemistry Research* 50: 571–579.

Nishiyama, Y., M. Arai, S.-L. Guo, N. Sonehara, T. Naito, and K. Torii. 1993. Catalytic properties of hectorite-like smectites containing nickel. *Applied Catalysis A: General* 95: 171–181.

Njopwouo, D., G. Roques, and R. Wandji. 1987. A contribution to the study of the catalytic action of clays on the polymerization of styrene: I. Characterization of polystyrenes. *Clay Minerals* 22: 145–156.

Njopwouo, D., G. Roques, and R. Wandji. 1988. A contribution to the study of the catalytic action of clays on the polymerization of styrene: II. Reaction mechanism. *Clay Minerals* 23: 35–43.

Núñez, K., R. Gallego, J.M. Pastor, and J.C. Merino. 2014. The structure of sepiolite as support of metallocene co-catalyst during in situ polymerization of polyolefin (nano)composites. *Applied Clay Science* 101: 73–81.

Occelli, M.L., J.T. Hsu, and L.G. Galya. 1985. Propylene oligomerization with pillared clays. *Journal of Molecular Catalysis* 33: 371–389.

Okada, A. and A. Usuki. 2006. Twenty years of polymer-clay nanocomposites. *Macromolecular Materials and Engineering* 291: 1449–1476.

Okamoto, M., S. Morita, and T. Kotaka. 2001. Dispersed structure and ionic conductivity of smectic clay/polymer nanocomposites. *Polymer* 42: 2685–2688.

Okamoto, M., S. Morita, H. Taguchi, Y.H. Kim, T. Kotaka, and H. Tateyama. 2000. Synthesis and structure of smectic clay/poly(methyl methacrylate) and clay/polystyrene nanocomposites via *in situ* intercalative polymerization. *Polymer* 41: 3887–3890.

Oral, A., T. Shahwan, and C. Güler. 2008. Synthesis of poly-2-hydroxyethyl methacrylate-montmorillonite nanocomposite via in situ atom transfer radical polymerization. *Journal of Materials Research* 23: 3316–3322.

Oral, A., M.A. Tasdelen, A.L. Demirel, and Y. Yagci. 2009. Poly(methyl methacrylate)/caly nanocomposite by photoinitiated free radical polymerization using intercalated monomer. *Polymer* 50: 3905–3910.

Ortega, N., T. Martin, and V.S. Martin. 2006. Stereoselective synthesis of eight-membered cyclic ethers by tandem Nicholas reaction/ring closing metathesis: A short synthesis of (+)-*cis*-lauthesan. *Organic Letters* 8: 871–873.

Park, K.J., M. Kim, S. Seok, Y.-W. Kim, and D.H. Kim. 2015. Quantitative analysis of cyclic dimer fatty acid content in the dimerization product by proton NMR spectroscopy. *Spectrochimica Acta Part A: Molecular and Biomolecular Spectroscopy* 149: 402–407.

Parulekar, V.N. and J.W. Hightower. 1987. Hydroisomerization of n-paraffins on a platinum-rhenium/pillared clay mineral catalyst. *Applied Catalysis* 35: 249–262.

Paul, D.R. and L.M. Robeson. 2008. Polymer nanotechnology: Nanocomposites. *Polymer* 49: 3187–3204.

Paul, M.-A., M. Alexandre, P. Degée, C. Calberg, R. Jérôme, and P. Dubois. 2003. Exfoliated polylactide/clay nanocomposites by in-situ coordination-insertion polymerization. *Macromolecular Rapid Communications* 24: 561–566.

Pavlidou, S. and C.D. Papaspyrides. 2008. A review on polymer-layered silicate nanocomposites. *Progress in Polymer Science* 33: 1119–1198.

Peighambardoust, S.J. and B. Pouabbas. 2007. Synthesis and characterization of conductive polypyrrole/montmorillonite nanocomposites via one-pot emulsion polymerization. *Macromolecular Symposium* 247: 99–109.

Perissinotto, M., M. Lenarda, L. Storaro, and R. Ganzerla. 1997. Solid acid catalysts from clays: Acid leached metakaolin as isopropanol dehydration and 1-butene isomerization catalyst. *Journal of Molecular Catalysis A: Chemical* 121: 103–109.

Pezerat, H. and I. Mantin. 1967. Polymerisation cationic du styrène entre les feuillets d'une montmorillonite acide. *Comptes rendus de l'Académie des Sciences* 265: 941–944.

Pillai, M., A. Wali, and S. Satish. 1995. Oligomerization of vinylic compounds on montmorillonite clay catalysts. *Reaction Kinetics and Catalysis Letters* 55: 251–257.

Pillai, S.M. and M. Ravindranathan. 1994. Oligomerization of dec-1-ene over montmorillonite clay catalysts. *Journal of the Chemical Society, Chemical Communications* 1813–1814.

Pinnavaia, T.J. 1977. Metal-catalyzed reactions in the intracrystal space of layer lattice silicates. In *Catalysis in Organic Syntheses 1977* (Ed.) G.V. Smith, pp. 131–137. New York: Academic Press.

Pitchumani, K. and A. Pandian. 1990. Cation-exchanged montmorillonite clays as Lewis acid catalysts in the Fries rearrangement of phenyl toluene-*p*-sulphonate. *Journal of the Chemical Society, Chemical Communications* 1613–1614.

Polubesova, T., S. Eldad, and B. Chefetz. 2010. Adsorption and oxidative transformation of phenolic acids by Fe(III)-montmorillonite. *Environmental Science & Technology* 44: 4203–4209.

Pontes, L.F.B.L., J.E.G. de Souza, A. Galembeck, and C.P. de Melo. 2013. Gas sensor based on montmorillonite/polypyrrole composites prepared by *in situ* polymerization in aqueous medium. *Sensors and Actuators B: Chemical* 177: 1115–1121.

Porter, T.L., M.P. Eastman, D.Y. Zhang, and M.E. Hagerman. 1997. Surface polymerization of organic monomers on Cu(II)-exchanged hectorite. *Journal of Physical Chemistry B* 101: 11106–11111.

Porter, T.L., D. Pace, R. Whitehorse, M.P. Eastman, and E. Bain. 2002. Formation of poly(methylmethacrylate) on the surface and intergallery regions of Cu^{2+}-, Fe^{3+}, and VO^{2+}-exchanged hectorite thin films. *Materials Chemistry and Physics* 76: 92–98.

Purnell, J.H. 1990. Catalysis by ion-exchanged montmorillonites. *Catalysis Letters* 5: 203–210.

Pusino, A., M. Gennari, A. Premoli, and C. Gessa. 1990. Formation of polyethylene glycol on montmorillonite by sterilization with ethylene oxide. *Clays and Clay Minerals* 38: 213–215.

Qi, R., X. Jin, J. Nie, W. Yu, and C. Zhou. 2005. Synthesis and properties of polystyrene-clay nanocomposites via *in situ* intercalative polymerization. *Journal of Applied Polymer Science* 97: 201–207.

Qian, J. and C.-Y. Guo. 2010. Polyolefin-clay nanocomposites from olefin polymerization between clay layers. *The Open Macromolecules Journal* 4: 1–14.

Qin, C., D. Troya, C. Shang, S. Hildreth, R. Helm, and K. Xia. 2015. Surface catalyzed oxidative oligomerization of 17β-estradiol by Fe^{3+}-saturated montmorillonite. *Environmental Science & Technology* 49: 956–964.

Qin, X., Y. Wu, K. Wang, H. Tan, and J. Nie. 2009. In-situ synthesis of exfoliated nanocomposites by photopolymerization using a novel montmorillonite-anchored initiator. *Applied Clay Science* 45: 133–138.

Qu, X., T. Guan, G. Liu, Q. She, and L. Zhang. 2005. Preparation, structural characterization, and properties of poly(methyl methacrylate)/montmorillonite nanocomposites by bulk polymerization. *Journal of Applied Polymer Science* 97: 348–357.

Radwan, D., L. Saad, S. Mikhail, and S.A. Selim. 2009. Catalytic evaluation of sulfated zirconia pillared clay in n-hexane transformation. *Journal of Applied Sciences Research* 5: 2332–2342.

Rajapakse, R.M.G., R.M.M.Y. Rajapakse, H.M.N. Bandara, and B.S.B. Karunarathne. 2008. Electrically conducting polypyrrole-fuller's earth nanocomposites: Their preparation and characterisation. *Electrochimica Acta* 53: 2946–2952.

Ramazani, S.A.A., F. Tavakolzadeh, and H. Baniasadi. 2010. Synthesis of polypropylene/clay nanocomposites using bisupported Ziegler–Natta catalyst. *Journal of Applied Polymer Science* 115: 308–314.

Ramôa, S.D.A.S., G.M.O. Barra, C. Merlini, W.H. Schreiner, S. Livi, and B.G. Soares. 2015. Production of montmorillonite/polypyrrole nanocomposites through *in situ* oxidative polymerization of pyrrole: Effect of anionic and cationic surfactants on structure and properties. *Applied Clay Science* 104: 160–167.

Ray, S.S. 2014. Recent trends and future outlooks in the field of clay-containing polymer nanocomposites. *Macromolecular Chemistry and Physics* 215: 1162–1179.

Ray, S.S. and M. Okamoto. 2003. Polymer/layered silicate nanocomposites: A review form preparation to processing. *Progress in Polymer Science* 28: 1539–1641.

Ray, S., G. Galgali, A. Lele, and S. Sivaram. 2005. *In situ* polymerization of ethylene with bis(imino)pyridine iron (II) catalysts supported on clay: The synthesis and characterization of polyethylene-clay nanocomposites. *Journal of Polymer Science: Part A: Polymer Chemistry* 43: 304–318.

Reena, V., J.D. Sudha, and C. Pavithran. 2009. Role of amphiphilic dopants on the shape and properties of electrically conducting polyaniline-clay nanocomposite. *Journal of Applied Polymer Science* 113: 4066–4076.

Ren, C., X. Du, L. Ma, Y. Wang, J. Zheng, and T. Tang. 2010. Preparation of multifunctional supported metallocene catalyst using organic multifunctional modifier for synthesizing polyethylene/clay nanocomposites via *in situ* intercalative polymerization. *Polymer* 51: 3416–3424.

Roghani-Mamaqani, H., V. Haddadi-Asl, M. Najafi, and M. Salami-Kalajahi. 2011. Preparation of tailor-made polystyrene nanocomposite with mixed clay-anchored and free chains via atom transfer radical polymerization. *American Institute of Chemical Engineers Journal* 57: 1873–1881.

Rong, J., Z. Jing, X.H. Li, and M. Sheng. 2001a. A polyethylene nanocomposite prepared via *in-situ* polymerization. *Macromolecular Rapid Communications* 22: 329–334.

Rong, J., H. Li, Z. Jing, X. Hong, and M. Sheng. 2001b. Novel organic/inorganic nanocomposite of polyethylene. I. Preparation via in situ polymerization approach. *Journal of Applied Polymer Science* 82: 1829–1837.

Ruiz-Hitzky, E. and A. Van Meerbeek. 2006. Clay mineral- and organoclay-polymer nanocomposite. In *Handbook of Clay Science* (Eds.) F. Bergaya, B.K.G. Theng, and G. Lagaly, pp. 583–621. Amsterdam, the Netherlands: Elsevier.

Rupert, J.P., W.T. Granquist, and T.J. Pinnavaia. 1987. Catalytic properties of clay minerals. In *Chemistry of Clays and Clay Minerals, Monograph No. 6* (Ed.) A.C.D. Newman, pp. 289–318. London, UK: Mineralogical Society.

Saehr, D., D. Walter, and R. Wey. 1991. Fixation de toluène dans une montmorillonite-Cu(II). *Clay Minerals* 26: 43–48.

Sahoo, S.K., D.W. Kim, J. Kumar, A. Blumstein, and A.L. Cholli. 2003. Nanocomposites from in-situ polymerization of substituted polyacetylene within lamellar surface of the montmorillonite: A solid-state NMR study. *Macromolecules* 36: 2777–2784.

Salahuddin, N., S.A. Abo-El-Enein, A. Selim, and O. Salah-El-Dien. 2010. Synthesis and characterization of polyurethane/organo-montmorillonite nanocomposites. *Applied Clay Science* 47: 242–248.

Salmón, M., A. Cabrera, N. Zavala et al. 1995. X-ray crystal structure and ^{1}H-NMR chiral shift reagent study of a crown trimer. *Journal of Chemical Crystallography* 25: 759–763.

Salmón, M., R. Miranda, I. Nicolás-Vázquez et al. 2011. Effects of bentonite on *p*-methoxybenzyl acetate: A theoretical model for oligomerization via an electrophilic substitution mechanism. *Molecules* 16: 1761–1775.

Salmón, M., M. Pérez-Luna, C. Lopéz-Franco et al. 1997. Catalytic conversion of proypylene oxide on a super acid sulfonic clay (SASC) system. *Journal of Molecular Catalysis A: Chemical* 122: 169–174.

Salmón, M., N. Zavala, M. Martínez et al. 1994. Cyclic and linear oligomerization reaction of 3,4,5-trimethoxybenzyl alcohol with a bentonite-clay. *Tetrahedron Letters* 35: 5797–5800.

Salomatina, O.V., O.I. Yarovaya, D.V. Korchagina, M.P. Polovinka, and V.A. Barkhash. 2005. Solid acid-catalysed isomerization of $R(+)$-limonene diepoxides. *Mendeleev Communications* 59–61.

Salt, F.E. 1948. The use of activated clays as catalysts in polymerisation processes, with particular reference to polymers of alpha methyl styrene. *Clay Minerals Bulletin* 55–57.

Sandi, G., K.A. Carrado, R.E. Winans, C.S. Johnson, and R. Csencsits. 1999. Carbons for lithium battery applications prepared using sepiolite as inorganic template. *Journal of the Electrochemical Society* 146: 3644–3648.

Sawhney, B.L., R.K. Kozloski, P.J. Isaacson, and M.P.N. Gent. 1984. Polymerization of 2,6-dimethylphenol on smectite surfaces. *Clays and Clay Minerals* 32: 108–114.

Scott, S.L., B.C. Peoples, C. Yung et al. 2008. Highly dispersed clay-polyolefin nanocomposites free of compatibilizers, *via* the *in situ* polymerization of α-olefins by clay-supported catalysts. *Chemical Communications* 4186–4188.

Seçkin, T., A. Gültek, M. G. Içduygu, and Y. Önal. 2002. Polymerization and characterization of acrylonitrile with γ-methacryloxypropyltrimethoxy-silane grafted bentonite clay. *Journal of Applied Polymer Science* 84: 164–171.

Sedláková, Z., J. Pleštil, J. Baldrian, M. Šlouf, and P. Holub. 2009. Polymer-clay nanocomposites prepared via in situ emulsion polymerization. *Polymer Bulletin* 63: 365–384.

Şen, S., M. Memeşa, N. Nugay, and T. Nugay. 2006. Synthesis of effective poly(4-vinylpyridine) nanocomposites: In situ polymerization from edges/surfaces and interlayer galleries of clay. *Polymer International* 55: 216–221.

Shah, N.F., M.S. Bhagwat, and M.M. Sharma. 1994. Cross-dimerization of α-methylstyrene with isoamylene and aldol condensation of cyclohexanone using a cation-exchange resin and acid-treated clay catalysts. *Reactive Polymers* 22: 19–34.

Shah, N.F. and M.M. Sharma. 1993. Dimerization of isoamylene: Ion exchange resin and acid-treated clay as catalysts. *Reactive Polymers* 19: 181–190.

Shakoor, A., T.Z. Rizvi, and M. Hina. 2012. Charge transport mechanism in intercalated polypyrrole aluminum-pillared montmorillonite clay nanocomposites. *Journal of Applied Polymer Science* 124: 3434–3439.

Shakoor, A., T.Z. Rizvi, and A. Nawaz. 2011. Raman spectroscopy and AC conductivity of polyaniline montmorillonite (PANI-MMT) nanocomposites. *Journal of Materials Science: Materials in Electronics* 22: 1076–1080.

Shanbhag, G.V. and S.B. Halligudi. 2004. Intermolecular hydroamination of alkynes catalyzed by zinc-exchanged montmorillonite clay. *Journal of Molecular Catalysis A: Chemical* 222: 223–228.

Shao, L., J. Qiu, M. Liu et al. 2010. Preparation and characterization of attapulgite/polyaniline nanofibers via self-assembling and graft polymerization. *Chemical Engineering Journal* 161: 301–307.

Shen, Z., F. Zhu, D. Liu, X. Zeng, and S. Lin. 2005. Preparation of syndiotactic polystyrene/montmorillonite nanocomposites via *in situ* intercalative polymerization of styrene with monotitanocene catalyst. *Journal of Applied Polymer Science* 95: 1412–1417.

Shin, S.-Y.A., L.C. Simon, J.B.P. Soares, and G. Scholz. 2003. Polyethylene-clay hybrid nanocomposites: In situ polymerization using bifunctional organic modifiers. *Polymer* 44: 5317–5321.

Siddiqui, M.H., H.H. Redhwi, K. Gkinis, and D.S. Achilias. 2013. Synthesis and characterization of novel nanocomposite materials based on poly(styrene-co-butyl methacrylate) copolymers and organomodified clay. *European Polymer Journal* 49: 353–365.

Sidorenko, A.Y., G.M. Sen'kov, and V.E. Agabekov. 2014. Effect of acid treatment on the composition and structure of a natural aluminosilicate and on its catalytic properties in α-pinene isomerization. *Catalysis in Industry* 6: 94–104.

Sieskind, O. and P. Albrecht. 1985. Efficient synthesis of rearranged choles-13(17)-enes catalysed by montmorillonite-clay. *Tetrahedron Letters* 26: 2135–2136.

Simons, R., G.G. Qiao, C.E. Powell, and S.A. Bateman. 2010. Effect of surfactant architecture on the properties of polystyrene-montmorillonite nanocomposites. *Langmuir* 26: 9023–9031.

Singh, B., J. Patial, P. Sharma, S.G. Agarwal, G.N. Qazi, and S. Maity. 2007. Influence of acidity of montmorillonite and modified montmorillonite clay minerals for the conversion of longifolene to isolongifolene. *Journal of Molecular Catalysis A: Chemical* 266: 215–220.

Sohn, J.R. and A. Ozaki. 1980. Acidity of nickel silicate and its bearing on the catalytic activity for ethylene dimerization and butene isomerization. *Journal of Catalysis* 61: 29–38.

Sohn, J.R. and M.Y. Park. 1993. Ethylene polymerization over transition metal-exchanged montmorillonite catalysts. *Applied Catalysis A: General* 101: 129–142.

Sohn, J.R. and M.Y. Park. 1996. Ethylene dimerization and polymerization over transition metal-exchanged montmorillonite catalysts. *Reaction Kinetics and Catalysis Letters* 57: 361–373.

Solomon, D.H. and D.G. Hawthorne. 1983. *Chemistry of Pigments and Fillers*. New York: John Wiley & Sons.

Solomon, D.H. and B.C. Loft. 1968. Reactions catalyzed by minerals. III. The mechanisms of spontaneous interlamellar polymerizations in aluminosilicates. *Journal of Applied Polymer Science* 12: 1253–1262.

Solomon, D.H. and M.J. Rosser. 1965. Reactions catalyzed by minerals. I. Polymerization of styrene. *Journal of Applied Polymer Science* 9: 1261–1271.

Solomon, D.H. and J.D. Swift. 1967. Reactions catalyzed by minerals. II. Chain termination in free radical polymerizations. *Journal of Applied Polymer Science* 11: 2567–2575.

Solomon, D.H., J.D. Swift, G. O'Leary, and I.G. Treeby. 1971. The mechanism of the decomposition of peroxides and hydroperoxides by mineral fillers. *Journal of Macromolecular Science: Part A. Chemistry* 5: 995–1005.

Soma, Y., M. Soma, Y. Furukawa, and I. Harada. 1987. Reactions of thiophene and methylthiophenes in the interlayer of transition-metal ion-exchanged montmorillonite studied by resonance Raman spectroscopy. *Clays and Clay Minerals* 35: 53–59.

Soma, Y., M. Soma, and I. Harada. 1983. Resonance Raman spectra of benzene adsorbed on Cu^{2+}-montmorillonite: Formation of poly(*p*-phenylene) cations in the interlayer of the clay mineral. *Chemical Physics Letters* 99: 153–156.

Soma, Y., M. Soma, and I. Harada. 1984. The reaction of aromatic molecules in the interlayer of transition-metal ion-exchanged montmorillonite studied by resonance Raman spectroscopy. 1. Benzene and *p*-phenylenes. *Journal of Physical Chemistry* 88: 3034–3038.

Soma, Y., M. Soma, and I. Harada. 1985. Reactions of aromatic molecules in the interlayer of transition-metal ion-exchanged montmorillonite studied by resonance Raman spectroscopy. 2. Monosubstituted benzenes and 4,4′-disubstituted biphenyls. *Journal of Physical Chemistry* 89: 738–742.

Soma, Y., M. Soma, and I. Harada. 1986. The oxidative polymerization of aromatic molecules in the interlayer of montmorillonites studied by resonance Raman spectroscopy. *Journal of Contaminant Hydrology* 1: 95–106.

Song, D.H., H.M. Lee, K.-H. Lee, and H.J. Choi. 2008. Intercalated conducting polyaniline-clay nanocomposites and their electrical characteristics. *Journal of Physics and Chemistry of Solids* 69: 1383–1385.

Sonobe, N., T. Kyotani, and A. Tomita. 1988. Carbonization of polyacrylonitrile in a two-dimensional space between montmorillonite lamellae. *Carbon* 26: 573–578.

Soundararajah, Q.Y., B.S.B. Karunaratne, and R.M.G. Rajapakse. 2009. Montmorillonite polyaniline nanocomposites: Preparation, characterization and investigation of mechanical properties. *Materials Chemistry and Physics* 113: 850–855.

Stadtmueller, L.M., K.R. Ratinac, and S.P. Ringer. 2005. The effects of intragallery polymerization on the structure of PMMA-clay nanocomposites. *Polymer* 46: 9574–9584.

Stoessel, F., J.L. Guth, and R. Wey. 1977. Polymerisation de benzene en polyparaphenylene dans une montmorillonite cuivrique. *Clay Minerals* 12: 255–259.

Sudha, J.D., V.L. Reena, and C. Pavithran. 2007. Facile green strategy for micro/nano structured conducting polyaniline-clay nanocomposite via template polymerization using amphiphilic dopant, 3-pentadecyl phenol-4-sulphonic acid. *Journal of Polymer Science: Part B: Polymer Physics* 45: 2664–2673.

Sugahara, Y., S. Satokawa, K. Kuroda, and C. Kato. 1988. Evidence for the formation of interlayer polyacrylonitrile in kaolinite. *Clays and Clay Minerals* 36: 343–348.

Sugahara, Y., S. Satokawa, K. Kuroda, and C. Kato. 1990. Preparation of a kaolinite-polyacrylamide intercalation compound. *Clays and Clay Minerals* 38: 137–143.

Sun, T. and J.M. Garcés. 2002. High-performance polypropylene-clay nanocomposites by in-situ polymerization with metallocene/clay catalysts. *Advanced Materials* 14: 128–130.

Swift, K.A.D. 2004. Catalytic transformations of the major terpene feedstocks. *Topics in Catalysis* 27: 143–155.

Tabernero, V., C. Camejo, P. Terreros, M.D. Alba, and T. Cuenca. 2010. Silicoaluminates as "support activator" systems in olefin polymerization processes. *Materials* 3: 1015–1030.

Takagi, K., H. Usami, H. Fukaya, and Y. Sawaki.1989. Spatially controlled photocycloaddition of a clay-intercalated stilbazolium cation. *Journal of the Chemical Society, Chemical Communications* 1174–1175.

Tannenbaum, E., E. Ruth, and I.R. Kaplan. 1986. Steranes and triterpanes generated from kerogen pyrolysis in the absence and presence of minerals. *Geochimica et Cosmochimica Acta* 50: 805–812.

Tasdelen, M.A., J. Kreutzer, and Y. Yagci. 2010. In situ synthesis of polymer/clay nanocomposites by living and controlled/living polymerization. *Macromolecular Chemistry and Physics* 211: 279–285.

Tayano, T., H. Uchino, T. Sagae et al. 2016. Locating the active sites of metallocene catalysts supported on acid-treated montmorillonite. *Journal of Molecular Catalysis A: Chemical* 420: 228–236.

Theng, B.K.G. 1970. Formation of two-dimensional organic polymers on a mineral surface. *Nature* 228: 853–854.

Theng, B.K.G. 1972. Formation, properties, and practical applications of clay-organic complexes. *Journal of the Royal Society of New Zealand* 2: 437–457.

Theng, B.K.G. 1974. *The Chemistry of Clay-Organic Reactions.* London, UK: Adam Hilger.

Theng, B.K.G. 1982. Clay-activated organic reactions. In *International Clay Conference 1981. Developments in Sedimentology 35* (Eds.) H. van Olphen and F. Veniale, pp. 197–238. Amsterdam, the Netherlands: Elsevier.

Theng, B.K.G. 2012. *Formation and Properties of Clay-Polymer Complexes,* 2nd ed. Amsterdam, the Netherlands: Elsevier.

Theng, B.K.G., G.J. Churchman, W.P. Gates, and G. Yuan. 2008. Organically modified clays for pollutant uptake and environmental protection. In *Soil Mineral-Microbe-Organic Interactions* (Eds.) Q. Huang, P.M. Huang, and A. Violante, pp. 145–174. Berlin, Germany: Springer-Verlag.

Theng, B.K.G. and G.F. Walker. 1970. Interactions of clay minerals with organic monomers. *Israel Journal of Chemistry* 8: 417–424.

Thi Nguyen, T.-T., D.-K. Nguyen Chau, F. Duus, and T.N. Le. 2013. Green synthesis of carvenone by montmorillonite-catalyzed isomerization of 1,2-limonene oxide. *International Journal of Organic Chemistry* 3: 206–209.

Thomas, J.M., J.M. Adams, S.H. Graham, and D.T.B. Tennakoon. 1977. Chemical conversions using sheet-silicate intercalates. In *Solid State Chemistry of Energy Conversion and Storage. Advances in Chemistry Series 163* (Eds.) J.B. Goodenough and M.S. Whittingham, pp. 298–315. Washington, DC: American Chemical Society.

Tipton, T. and L.E. Gerdom. 1992. Polymerization and transalkylation reactions of toluene on Cu(II)-montmorillonite. *Clays and Clay Minerals* 40: 429–435.

Tjong, S.C. 2006. Structural and mechanical properties of polymer nanocomposites. *Materials Science and Engineering R* 53: 73–197.

Tong, X., H. Zhao, T. Tang, Z. Feng, and B. Huang. 2002. Preparation and characterization of poly(ethyl acrylate)/bentonite nanocomposites by *in situ* emulsion polymerization. *Journal of Polymer Science: Part A: Polymer Chemistry* 40: 1706–1711.

Triantafillidis, C.S., P.C. LeBaron, and T.J. Pinnavaia. 2002. Thermoset epoxy–clay nanocomposites: The dual role of α,ω-diamines as clay surface modifiers and polymer curing agents. *Journal of Solid State Chemistry* 167: 354–362.

Tricker, M.J., D.T.B. Tennakoon, J.M. Thomas, and S.H. Graham. 1975b. Novel reactions of hydrocarbon complexes of metal-substituted sheet silicates; thermal dimerisation of *trans*-stilbene. *Nature* 253: 110–111.

Tricker, M.J., D.T.B. Tennakoon, J.M. Thomas, and J. Heald. 1975a. Organic reactions in clay-mineral matrices: Mass-spectrometric study of the conversion of triphenylamine to N,N,N',N'-tetraphenylbenzidine. *Clays and Clay Minerals* 23: 77–82.

Tseng, C.-R., J.-Y. Wu, H.-Y. Lee, and F.-C. Chang. 2002. Preparation and characterization of polystyrene-clay nanocomposites by free-radical polymerization. *Journal of Applied Polymer Science* 85: 1370–1377.

Tudor, J., L. Willington, D. O'Hare, and B. Royan. 1996. Intercalation of catalytically active metal complexes in phyllosilicates and their application as propene polymerisation catalysts. *Chemical Communications* 2031–2032.

Usami, H., T. Nakamura, T. Makino, H. Fujimatsu, and S. Ogasawara. 1998. Sensitized photoisomerization of *cis*-stilbazolium ions intercalated in saponite clay layers. *Journal of the Chemical Society, Faraday Transactions* 94: 83–87.

Usami, H., K. Takagi, and Y. Sawaki. 1990. Controlled photocycloaddition of stilbazolium ions intercalated in saponite clay layers. *Journal of the Chemical Society, Perkin Transactions* 2: 1723–1728.

Usami, H., K. Takagi, and Y. Sawaki. 1992a. Clay-inclusion photocyclodimerization: Intercalation and migration of stilbazolium ions. *Journal of the Chemical Society, Faraday Transactions* 88: 77–81.

Usami, H., K. Takagi, and Y. Sawaki. 1992b. Regioselective photocyclodimerization of cyclohexenones intercalated on clay layers. *Chemistry Letters* 21: 1405–1408.

Usuki, A., Y. Kojima, M. Kawasumi et al. 1993. Synthesis of nylon 6-clay hybrid. *Journal of Materials Research* 8: 1179–1184.

Uthirakumar, P., M.-K. Song, C. Nah, and Y.-S. Lee. 2005. Preparation and characterization of exfoliated polystyrene/clay nanocomposites using a cationic radical initiator-MMT hybrid. *European Polymer Journal* 41: 211–217.

van Santen, R.A., K.-H.W. Röbschläger, and C.A. Emeis. 1985. The hydroisomerization activity of nickel-substituted mica montmorillonite clay. In *Solid State Chemistry in Catalysis. ACS Symposium Series 279* (Eds.) R.K. Grasselli and J.F. Brazdil, pp. 275–291. Washington, DC: American Chemical Society.

Varma, R.S. and R.K. Saini. 1997. Microwave-assisted isomerization of 2'-aminochalcones on clay: An easy route to 2-aryl-1,2,3,4-tetrahydro-4-quinolones. *Synlett* 857–858.

Vogels, R.J.M.J., J.T. Kloprogge, and J.W. Geus. 2005. Catalytic activity of synthetic saponite clays: Effects of tetrahedral and octahedral composition. *Journal of Catalysis* 231: 443–452.

Volcho, K.P. and N.F. Salakhutdinov. 2008. Transformations of terpenoids on acidic clays. *Mini-Reviews in Organic Chemistry* 5: 345–354.

Volzone, C., O. Masini, N.A. Comelli, L.M. Grzona, E.N. Ponzi, and M.I. Ponzi. 2001. Production of camphene and limonene from pinene over acid di- and trioctahedral smectite clays. *Applied Catalysis A: General* 214: 213–218.

Volzone, C., O. Masini, N.A. Comelli, L.M. Grzona, E.N. Ponzi, and M.I. Ponzi. 2005. α-Pinene conversion by modified-kaolinitic clay. *Materials Chemistry and Physics* 93: 296–300.

Walter, D., D. Saehr, and R. Wey. 1990. Les complexes montmorillonite-Cu(II)-benzene: Une contribution. *Clay Minerals* 25: 343–354.

Wang, D., J. Zhu, Q. Yao, and C.A. Wilkie. 2002a. A comparison of various methods for the preparation of polystyrene and poly(methyl methacrylate) clay nanocomposites. *Chemistry of Materials* 14: 3837–3843.

Wang, H., T. Zhao, L. Zhi, Y. Yan, and Y. Yu. 2002b. Synthesis of novolac/layered silicate nanocomposites by reaction exfoliation using acid-modified montmorillonite. *Macromolecular Rapid Communications* 23: 44–48.

Wang, H.-W., C.-F. Shieh, K.-C. Chang, and H.-C. Chu. 2005. Synthesis and dielectric properties of poly(methyl methacrylate)-clay nanocomposite materials. *Journal of Applied Polymer Science* 97: 2175–2181.

Wang, J.-S. and K. Matyjaszewski. 1995. Controlled/"living" radical polymerization: Atom transfer radical polymerization in the presence of transition-metal complexes. *Journal of the American Chemical Society* 117: 5614–5615.

Wang, L., X. Wang, Z. Chen, and P. Ma. 2013. Effect of doubly organo-modified vermiculite on the properties of vermiculite/polystyrene nanocomposites. *Applied Clay Science* 75–76: 74–81.

Wang, L.-M. and A.-H. He. 2011. Preparation of polypropylene/clay nanocomposites by *in situ* polymerization with $TiCl_4/MgCl_2$/clay compound catalyst. *Chinese Journal of Polymer Science* 29: 597–601.

Wang, M.C. and P.M. Huang. 1989. Catalytic power of nontronite, kaolinite and quartz and their reaction sites in the formation of hydroquinone-derived polymers. *Applied Clay Science* 4: 43–57.

Wang, Q., Z. Zhou, L. Song, H. Xu, and L. Wang. 2004. Nanoscopic confinement effects on ethylene polymerization by intercalated silicate with metallocene catalyst. *Journal of Polymer Science: Part A: Polymer Chemistry* 42: 38–43.

Wang, Z., J. Massam, and T.J. Pinnavaia. 2000. Epoxy-clay nanocomposites. In *Polymer-Clay Nanocomposites* (Eds.) T.J. Pinnavaia and G.W. Beall, pp. 127–149. Chichester, UK: John Wiley & Sons.

Watson, J.S. and M.A. Sephton. 2015. Heat, aromatic units, and iron-rich phyllosilicates: A mechanism for making macromolecules in the early solar system. *Astrobiology* 15: 787–792.

Weimer, M.W., H. Chen, E.P. Giannelis, and D.Y. Sogah. 1999. Direct synthesis of dispersed nanocomposites by in situ living free radical polymerization using a silicate-anchored initiator. *Journal of the American Chemical Society* 121: 1615–1616.

Weiss, A. 1981. Replication and evolution in inorganic systems. *Angewandte Chemie International Edition in English* 20: 850–860.

Weiss, K., C. Wirth-Pfeifer, M. Hofmann et al. 2002. Polymerisation of ethylene or propylene with heterogeneous metallocene catalysts on clay minerals. *Journal of Molecular Catalysis A: Chemical* 182–183: 143–149.

Wheeler, D.H., A. Milun, and F. Linn. 1970. Dimer acid structures: Cyclic structures of clay-catalyzed dimers of normal linoleic acid, 9-*cis*, 12-*cis*-octadecadienoic acid. *Journal of the American Oil Chemists Society* 47: 242–244.

Wiederrecht, G.P., G. Sandi, K.A. Carrado, and S. Seifert. 2001. Intermolecular dimerization within pillared, layered clay templates. *Chemistry of Materials* 13: 4233–4238.

Wu, C.-S., Y.-J. Huang, T.-H. Hsieh et al. 2008. Studies on the conducting nanocomposite prepared by in situ polymerization of aniline monomers in a neat (aqueous) synthetic mica clay. *Journal of Polymer Science: Part A: Polymer Chemistry* 46: 1800–1809.

Wu, Q., Z. Xue, Z. Qi, and F. Wang. 2000. Synthesis and characterization of Pan/clay nanocomposite with extended chain conformation of polyaniline. *Polymer* 41: 2029–2032.

Xie, T., G. Yang, X. Fang, and Y. Ou. 2003b. Synthesis and characterization of poly(methyl methacrylate)/montmorillonite nanocomposites by *in situ* bulk polymerization. *Journal of Applied Polymer Science* 89: 2256–2260.

Xie, W., J.M. Hwu, G.J. Jiang, T.M. Buthelezi, and W.-P. Pan. 2003a. A study of the effect of surfactants on the properties of polystyrene-montmorillonite nanocomposites. *Polymer Engineering and Science* 43: 214–222.

Yadav, M.K., C.D. Chudasama, and R.V. Jasra. 2004. Isomerisation of α-pinene using modified montmorillonite clays. *Journal of Molecular Catalysis A: Chemical* 216: 51–59.

Yamamoto, K., Y. Ishihama, and K. Sakata. 2010. Preparation of bimodal HDPEs with metallocene on Cr-montmorillonite support. *Journal of Polymer Science: Part A: Polymer Chemistry* 48: 3722–3728.

Yang, C. and P. Liu. 2009. Core-shell attapulgite@polypyrrole composite with well-defined corn cob-like morphology via self-assembling and in situ oxidative polymerization. *Synthetic Metals* 159: 2056–2062.

Yang, F., X. Zhang, H. Zhao, B. Chen, B. Huang, and Z. Feng. 2003. Preparation and properties of polyethylene/montmorillonite nanocomposites by *in situ* polymerization. *Journal of Applied Polymer Science* 89: 3680–3684.

Yang, K., Y. Huang, and J.-Y. Dong. 2007. Efficient preparation of isotactic polypropylene/montmorillonite nanocomposites by in situ polymerization technique via a combined use of functional surfactant and metallocene catalysis. *Polymer* 48: 6254–6261.

Yang, Q., S. Chen, C. Liu et al. 2013. Research on dimerization of oleic acid catalyzed by organic montmorillonite. *Advanced Materials Research* 791–794: 120–123.

Yao, P.Q., L.H. Zhu, J. Yang, and T. Si. 2013. Isomerization of α-pinene over alumina-pillared montmorillonite. *Advanced Materials Research* 746: 49–52.

Yasue, K., Katahira, S., Yoshikawa, M., Fujimoto, K. 2000. *In situ* polymerization route to nylon 6-clay nanocomposites. In *Polymer-Clay Nanocomposites* (Eds.) T.J. Pinnavaia and G.E. Beall, pp. 111–126. Chichester, UK: John Wiley & Sons.

Yeh, J.-M. and C.-P. Chin. 2003. Structure and properties of poly(o-methoxyaniline)-clay nanocomposite materials. *Journal of Applied Polymer Science* 88: 1072–1080.

Yeh, J.-M., C.-P. Chin, and S. Chang. 2003. Enhanced corrosion protection coatings prepared from soluble electronically conductive polypyrrole-clay nanocomposite materials. *Journal of Applied Polymer Science* 88: 3264–3272.

Yeh, J.-M., S.-J. Liou, M.-C. Lai et al. 2004. Comparative studies of the properties of poly(methyl methacrylate)-clay nanocomposite materials prepared by *in situ* emulsion polymerization and solution dispersion. *Journal of Applied Polymer Science* 94: 1936–1946.

Yeh, J.-M., S.-J. Liou, C.-Y. Lin, C.-Y. Cheng, Y.-W. Chang, and K.-R. Lee. 2002. Anticorrosive enhanced PMMA-clay nanocomposite materials with quaternary alkylphosphonium salt as an intercalating agent. *Chemistry of Materials* 14: 154–161.

Yeh, J.-M., S.-J. Liu, C.-Y. Chiung, P.C. Wu, and T.-Y. Tsai. 2001. Enhancement of corrosion protection effect in polyaniline via the formation of polyaniline-clay nanocomposite materials. *Chemistry of Materials* 13: 1131–1136.

Yong, R.N., S. Desjardins, J.P. Farant, and P. Simon. 1997. Influence of pH and exchangeable cation on oxidation of methylphenols by a montmorillonite clay. *Applied Clay Science* 12: 93–110.

Zakarina, N.A., L.D. Volkova, A.K. Akurpekova, and L.V. Komashko. 2008. Isomerization of *n*-hexane on platinum, palladium, and nickel catalysts deposited on columnar montmorillonite. *Petroleum Chemistry* 48: 186–192.

Zeng, C. and L.J. Lee. 2001. Poly(methyl methacrylate) and polystyrene/clay nanocomposites prepared by in-situ polymerization. *Macromolecules* 34: 4098–4103.

Zeng, K. and Y. Bai. 2005. Improve the gas barrier property of PET film with montmorillonite by in situ interlayer polymerization. *Materials Letters* 59: 3348–3351.

Zeng, Q.H., D.Z. Wang, A.B. Yu, and G.Q. Lu. 2002. Synthesis of polymer-montmorillonite nanocomposites by *in situ* intercalative polymerization. *Nanotechnology* 13: 549–553.

Zeng, Q.H., A.B. Yu, G.Q. (Max) Lu, and D.R. Paul. 2005. Clay-based polymer nanocomposites: Research and commercial development. *Journal of Nanoscience and Nanotechnology* 5: 1574–1592.

Zhang, L., T. Wang, and P. Liu. 2008. Polyaniline-coated halloysite nanotubes via in-situ chemical polymerization. *Applied Surface Science* 255: 2091–2097.

Zhao, H., S.D. Argoti, B.P. Farrell, and D.A. Shipp. 2004. Polymer-silicate nanocomposites produced by *in situ* atom transfer radical polymerization. *Journal of Polymer Science: Part A: Polymer Chemistry* 42: 916–924.

Zhu, J., P. Start, K.A. Mauritz, and C.A. Wilkie. 2002. Thermal stability and flame retardancy of poly(methyl methacrylate)-clay nanocomposites. *Polymer Degradation and Stability* 77: 253–258.

Zidi, R., I. Bekri-Abbes, N. Sdiri, A. Vimalanandan, M. Rohwerder, and E. Srasra. 2016. Electrical and dielectric investigation of intercalated polypyrrole montmorillonite nanocomposite prepared by spontaneous polymerization of pyrrole into Fe(III)-montmorillonite. *Materials Science and Engineering B* 212: 14–23.

Zubkov, S.A., L.M. Kustov, V.B. Kazansky, G. Fetter, D. Tichit, and F. Figueras. 1994. Study of the nature of acid sites of montmorillonites pillared with aluminium and oligosilsesquioxane complex cations 1. Brönsted acidity. *Clays and Clay Minerals* 42: 421–427.

7 Clay Mineral Catalysis of Redox, Asymmetric, and Enantioselective Reactions

7.1 INTRODUCTION

The ability of clay minerals and their surface-modified forms to catalyze the oxidation and reduction of organic compounds has already been mentioned in Chapter 3, and some examples of such reactions are listed in Table 3.3. For more details of these and related conversions, we refer to the reviews by Nikalje et al. (2000), Nagendrappa (2011), Fernandes et al. (2012), and McCabe and Adams (2013). Here, we focus on oxidation reactions that take place in the presence of hydrogen peroxide, *tert*-butyl hydroperoxide, ozone, molecular oxygen, or a suitable compound acting as an oxidizing agent or cocatalyst. Likewise, the clay-catalyzed reduction and hydrogenation of organic compounds generally require dihydrogen as a coreactant (Kotkar and Thakkar 1995, 1997; Aldea and Alper 2000).

More often than not, clay-catalyzed redox reactions involve the participation of transition metal ions, such as iron, occupying structural or interlayer exchange sites, or associated with metal oxide pillars (Cheng 1999; Gorski et al. 2013; Kurian and Kavitha 2016). In the case of structural iron, the concomitant reduction of Fe^{3+} (to Fe^{2+}) can affect clay surface hydration and acidity as well as enhance the negative layer charge and basicity of structural oxygens (Stucki 2006; Latta et al. 2017). Furthermore, clay minerals containing ferrous iron are generally more efficient than their ferric iron-rich counterparts in degrading pesticides, extracting chlorine from chlorinated hydrocarbons, and reducing nitroamines (Schoonheydt 2016).

As indicated in Tables 7.1 through 7.3, surface-modified clay minerals, and more so pillared interlayered clays (PILC), have been widely used as heterogeneous catalysts in Fenton-like and photo-Fenton oxidation reactions (Herney-Ramirez et al. 2010; Navalon et al. 2010; Garrido-Ramirez et al. 2010; Sanabria et al. 2012; Shahidi et al. 2015a). Similarly, clay-supported transition metal salts and metal chelates together with Fe- and Ti-PILC can serve as efficient photo-assisted catalysts for the oxidation and degradation of organic compounds (Van Damme et al. 1984; Shichi and Takagi 2000; Cieśla et al. 2004; Catrinescu et al. 2012; Ben Achma et al. 2014; Liu and Zhang 2014; Li et al. 2015).

Being a negatively charged porous solid with a layer structure, clay minerals are well suited to serving as a support and/or intercalating agent of chiral metal-organic complexes capable of catalyzing asymmetric organic conversions and syntheses (Li 2004; Fraile et al. 2009a). An example to which we have already referred (Chapter 5) is the activity of Laponite-immobilized chiral bis(oxazoline)-copper complexes in promoting the enantioselective Mukaiyama aldol reaction (Fabra et al. 2008; Fraile et al. 2008a, 2009b). The review by Thomas and Raja (2008) provides a fascinating account of asymmetric catalysis by silica-immobilized chiral catalysts. Here we summarize the information on asymmetric organic reactions and enantioselective syntheses catalyzed by clay minerals and clay-supported reagents.

TABLE 7.1
Fenton and Photo-Fenton Oxidation and Degradation of Phenol and Substituted Phenols Catalyzed by Pillared Interlayered Clays (PILC) and Surface-Modified Clay Minerals

Organic Compound	Clay Catalyst/Support	Oxidant/ Assisting Agent	Process/ Reaction	Reference (in Chronological Order)
Phenol	Al–Cu, Al–Fe, and Al–Ce–Fe-PILC	H_2O_2	Wet oxidation	Frini et al. (1997); Barrault et al. (1998, 2000a, 2000b); Carriazo et al. (2003, 2005a, 2005b, 2007); Kiss et al. (2003); Tatibouët et al. (2005); Sanabria et al. (2009)
	Ti-PILC; Ti-pillared beidellite, hectorite, Laponite and saponite	Oxygen/UV light	Wet oxidation	Ding et al. (1999); Zhu et al. (2005); Yang et al. (2010)
	Fe-exchanged Al-pillared beidellite	H_2O_2	Wet oxidation	Catrinescu et al. (2003)
	Al- or Al–Fe-PILC	H_2O_2	Wet oxidation	Guélou et al. (2003); Guo and Al-Dahhan (2003a, 2003b, 2005)
	K10 montmorillonite	Sodium perborate	Wet oxidation	Hashemi et al. (2005)
	Al-PILC, Al-PILC with adsorbed Fe, Al–Fe- and Fe–Cu–Al-PILC	H_2O_2	Wet oxidation	Timofeeva et al. (2005, 2009)
	Al–Fe-PILC	H_2O_2	Wet oxidation	Kurian and Sugunan (2006); Sanabria et al. (2008); Banković et al. (2009)
	Al–Fe- and Zr–Fe-PILC	H_2O_2	Wet oxidation	Molina et al. (2006)
	Al–Fe-PILC	H_2O_2/ultrasound	Wet oxidation	Nikolopoulos et al. (2006)
	Fe-pillared Laponite	H_2O_2/UV light	Photo-Fenton	Iurascu et al. (2009)
	Al–Fe-PILC	H_2O_2	Fenton-like	Luo et al. (2009)
	Al–Fe- and Al–Fe–Ce-PILC	H_2O_2	Wet oxidation	Olaya et al. (2009); Sanabria et al. (2009)
	Fe- and Al–Fe-PILC	H_2O_2	Wet oxidation	Timofeeva and Khankhasaeva (2009)
	Rectorite containing dispersed hematite	Air	Wet oxidation	Xu et al. (2009)
	Anatase nanoparticles supported on acid-leached Laponite	Air/UV light	Wet oxidation	Yang et al. (2009)
	Ti- and Fe–Ti-PILC	H_2O_2/UV light	Photo-Fenton	Carriazo et al. (2010)
	Fe–Al-PILC	H_2O_2	Wet oxidation	Guélou et al. (2010)
	Al–Fe-PILC	H_2O_2/solar light	Photo-Fenton	Cam et al. (2011)
	Fe-pillared bentonite, KSF and K10 montmorillonite	H_2O_2	Fenton-like	Platon et al. (2011, 2013)
	Trinuclear iron cluster-montmorillonite interlayer complex	H_2O_2/UV light	Photo-Fenton	Zhang et al. (2011d)
	Palygorskite-supported SnO_2–TiO_2	Air/UV light	Wet oxidation	Zhang et al. (2011e)
	Rare earth exchanged Al-PILC and Al–Fe-PILC	H_2O_2	Wet oxidation	Kurian et al. (2012); Kurian and Babu (2013)

(Continued)

TABLE 7.1 (Continued)
Fenton and Photo-Fenton Oxidation and Degradation of Phenol and Substituted Phenols Catalyzed by Pillared Interlayered Clays (PILC) and Surface-Modified Clay Minerals

Organic Compound	Clay Catalyst/Support	Oxidant/Assisting Agent	Process/Reaction	Reference (in Chronological Order)
Phenol (continued)	Iron/copper oxide-supported on synthetic allophane	H_2O_2	Wet oxidation	Garrido-Ramirez et al. (2012)
	Iron-rich red clay	H_2O_2	Fenton-like	Djeffal et al. (2014b)
	Iron oxide-impregnated clay	H_2O_2/UV light	Photo-Fenton	Hadjltaief et al. (2014)
	Cu-doped Fe-PILC	H_2O_2/UV light	Photo-Fenton	Hadjltaief et al. (2015)
	Al- and Ce-impregnated Zr-PILC	H_2O_2	Wet oxidation	Mnasri-Ghnimi and Frini-Srasra (2016)
	Fe–Cr-PILC	H_2O_2/ultrasound	Wet oxidation	Tomul (2016)
	Al–Fe–Cu-PILC	H_2O_2	Wet oxidation	Zhou et al. (2016)
	Maghemite-montmorillonite composite	H_2O_2	Fenton-like	Jin et al. (2017)
	Al–Fe-PILC	O_2	Wet oxidation	Sassi et al. (2018)
Catechol	Fe-modified montmorillonite	H_2O_2	Fenton-like	Baizig et al. (2013)
2-Chlorophenol	Copper-impregnated kaolinite and montmorillonite	H_2O_2	Wet oxidation	Khanikar and Bhattacharyya (2013)
4-Chlorophenol	Ti-PILC	Air/UV light	Wet oxidation	Pichat et al. (2005)
	Fe^{3+}-montmorillonite Al–Fe-PILC, Fe- and Fe–Al-PILC	H_2O_2	Wet oxidation	Catrinescu and Teodosiu (2007); Khankhasaeva et al. (2008); Arsene et al. (2010); Catrinescu et al. (2011, 2012)
	Copper-impregnated kaolinite and montmorillonite	H_2O_2	Wet oxidation	Khanikar and Bhattacharyya (2013)
	Al–Fe-, Al–Cu-, Al–Fe–Cu-PILC	H_2O_2	Wet oxidation	Zhou et al. (2014)
Dichlorophenol	Copper-impregnated kaolinite and montmorillonite	H_2O_2	Wet oxidation	Khanikar and Bhattacharyya (2013)
	Fe-pillared K10 montmorillonite	Peroxymono sulfate, H_2O_2, peracetic acid	Wet oxidation	Virkutyte and Varma (2014)
Hydroquinone	Fe-modified montmorillonite	H_2O_2	Fenton-like	Baizig et al. (2013)
2-Nitrophenol	Cu-doped Al-PILC	H_2O_2/visible light	Photo-Fenton	Najjar et al. (2001)
4-Nitrophenol	Fe^{3+}-montmorillonite; Al–Fe-PILC	H_2O_2	Wet oxidation	Chirchi and Ghorbel (2002)
	Fe-pillared rectorite	H_2O_2/visible light	Photo-Fenton	Zhang et al. (2010)
	TiO_2-kaolinite composite	Air/UV light	Wet oxidation	Zhang et al. (2011a)
	TiO_2-sepiolite composite	Air/UV light	Wet oxidation	Zhang et al. (2011b)
	Cu-PILC supported ferrioxalate complex	H_2O_2/visible light	Photo-Fenton	Ayodele and Hameed (2013)

(Continued)

TABLE 7.1 (Continued)
Fenton and Photo-Fenton Oxidation and Degradation of Phenol and Substituted Phenols Catalyzed by Pillared Interlayered Clays (PILC) and Surface-Modified Clay Minerals

Organic Compound	Clay Catalyst/Support	Oxidant/Assisting Agent	Process/Reaction	Reference (in Chronological Order)
4-Nitrophenol (continued)	Poly-hydroxy iron/sepiolite	H_2O_2/visible light	Photo-Fenton	Gao et al. (2013)
	Ti-PILC	Air/UV light	Wet oxidation	Baizig et al. (2015)
Resorcinol	Iron-impregnated bentonite plate	H_2O_2/UV light	Photo-Fenton	González-Bahamón et al. (2011)
	Fe-modified montmorillonite	H_2O_2	Fenton-like	Baizig et al. (2013)
2,4,6 Trichlorophenol	Organoclay modified with palladium phthalocyanine-sulfonate	Singlet molecular oxygen/visible light	Photocatalysis	Xiong et al. (2005a)
Tyrosol	Cu-impregnated Al-PILC	H_2O_2	Wet oxidation	Ben Achma et al. (2008)
	Iron-rich red clay	H_2O_2	Fenton-like	Djeffal et al. (2014a)

Note: Unless otherwise specified, the clay mineral species in PILC is montmorillonite or montmorillonite-rich bentonite.

TABLE 7.2
Wet Catalytic and Photocatalytic Oxidation and Degradation of Organic Contaminants/Pollutants (Other than Phenol and Substituted Phenols) in the Presence of Pillared Interlayered Clays (PILC) and Surface-Modified Clay Minerals

Organic Contaminant or Pollutant (in Alphabetical Order)	Clay Catalyst/Support	Oxidant/Assisting Agent	Process/Reaction	Reference
Acetaldehyde	Allophane-supported titania	Air/UV irradiation	Photocatalytic degradation	Nishikiori et al. (2017)
Acetic acid	Montmorillonite-supported iron oxide	Air/light	Wet oxidation	Miyoshi and Yoneyama (1989); Miyoshi et al. (1991)
Amoxicillin	Ferric oxalate-PILC; kaolinite-supported ferric oxalate	H_2O_2; H_2O_2/UV light	Photo-Fenton	Ayodele et al. (2012); Ayodele (2013)
Aniline	Ti-PILC	H_2O_2	Wet oxidation	Jagtap and Ramaswamy (2006)
Benzene	Pd supported on Zr-, Ce-, and Al-pillared Laponite	Air	Deep oxidation	Li et al. (2005)
Benzylic alcohols	$CoCl_2$ supported on K10 montmorillonite	H_2O_2	Wet oxidation	Ezabadi et al. (2008)
Bisphenol A	KSF montmorillonite; montmorillonite-supported nano-MnO_2	Air/UV light	Photo-Fenton	Liu et al. (2008); Fang et al. (2016)

(Continued)

TABLE 7.2 (Continued)
Wet Catalytic and Photocatalytic Oxidation and Degradation of Organic Contaminants/Pollutants (Other than Phenol and Substituted Phenols) in the Presence of Pillared Interlayered Clays (PILC) and Surface-Modified Clay Minerals

Organic Contaminant or Pollutant (in Alphabetical Order)	Clay Catalyst/Support	Oxidant/Assisting Agent	Process/Reaction	Reference
Bisphenol A (continued)	Cu-phthalocyanine intercalated into a TiO_2-organoclay complex	Air/visible light irradiation	Wet oxidation	Sasai et al. (2014)
	Ti-PILC; Fe- and Cu-impregnated Ti-PILC	H_2O_2	Wet oxidation	Tomul et al. (2016)
n-Butylamine	Cr/Ce supported on Al- and Zr-PILC	Air	Deep oxidation	Huang et al. (2010)
Carbamazepine	Montmorillonite-supported nano-magnetite	H_2O_2	Fenton-like	Sun et al. (2013)
Chlorobenzene	Cr or Pd-impregnated Al-PILC; Al- and Fe-PILC	Air; O_2/Ar mixture	Deep oxidation (high temperature)	Oliveira et al. (2008); Li et al. (2013b)
	Pd- and Pt-supported on Al-PILC	Oxygen-hydrocarbon	Deep oxidation (high temperature)	Aznárez et al. (2015)
	Cr- and Ce-oxides supported on Ti-PILC	Air	Deep oxidation (high temperature)	Zuo et al. (2015)
Cinnamic acid	Fe-PILC	H_2O_2	Wet oxidation	Tabet et al. (2006)
Ciprofloxacin	Fe-pillared Laponite	H_2O_2/UV light	Photo-Fenton	Bobu et al. (2008)
p-Coumaric acid	Cu-PILC	H_2O_2	Wet oxidation	Caudo et al. (2006, 2007, 2008)
Decabromodiphenyl ether (BDE 209)	TiO_2 immobilized on an organoclay	Air/UV light	Wet oxidation	An et al. (2008)
Dichloroacetic acid	TiO_2-montmorillonite composite	Air/UV light	Wet oxidation	Kun et al. (2006b)
Dimethachlor	Ti-pillared montmorillonite and Laponite	Air/UV light	Wet oxidation	Belessi et al. (2007)
1,4-Dioxane	TiO_2-montmorillonite composite	Air/UV light	Wet oxidation	Kameshima et al. (2009)
Esfenvalerate	Kaolinite and montmorillonite suspensions	Air/visible light (Xenon lamp)	Fenton-like	Katagi (1993)
Ethylenediamine	Cr/Ce supported on Al- and Zr-PILC	Air	Deep oxidation	Huang et al. (2010)
Herbicides	TiO_2-Laponite composite	Air/UV light	Wet oxidation	Paul et al. (2012)
γ-Hexachlorocyclohexane	TiO_2-montmorillinite composite	Air/UV light	Wet oxidation	Zhao et al. (2007)
p-Hydroxybenzoic acid	Cu- and Fe-PILC	H_2O_2	Wet oxidation	Caudo et al. (2007)
Humic acid	Montmorillonite-supported TiO_2	Air/UV light	Wet oxidation	Kavurmaci and Bekbolet (2013)
Ibuprofen	Montmorillonite-supported nano-magnetite	H_2O_2	Fenton-like	Sun et al. (2013)
D-Limonene	Hectorite- and kaolinite-TiO_2 complexes	Air/UV light	Wet oxidation	Kibanova et al. (2009)

(Continued)

TABLE 7.2 (*Continued*)
Wet Catalytic and Photocatalytic Oxidation and Degradation of Organic Contaminants/ Pollutants (Other than Phenol and Substituted Phenols) in the Presence of Pillared Interlayered Clays (PILC) and Surface-Modified Clay Minerals

Organic Contaminant or Pollutant (in Alphabetical Order)	Clay Catalyst/Support	Oxidant/ Assisting Agent	Process/Reaction	Reference
β-*Naphthol*	TiO_2-sepiolite composite	Air/UV light	Wet oxidation	Karamanis et al. (2011)
Phenanthrene	Fe^{3+}-exchanged smectites, vermiculite, kaolinite	Air/visible light	Fenton-like	Jia et al. (2012)
Phenolic acids	Fe^{3+}-montmorillonite	Air	Fenton-like	Polubesova et al. (2010)
Phthalate esters	Ti pillared saponite, hectorite, montmorillonite, fluoromica	Air/UV light	Wet oxidation	Ooka et al. (2004)
Polyphenols (*olive oil mill wastewater*)	Cu-PILC	H_2O_2	Wet oxidation	Giordano et al. (2007)
2-Propanol	Ti-PILC	Air	Wet oxidation	Yoneyama et al. (1989)
Salicylic acid	Interlayer TiO_2/ZnO-smectite composites	Air	Wet oxidation	Mogyorósi et al. (2001)
Sulfanilamide	Al–Fe-PILC and Fe–Cu–Al-PILC	H_2O_2	Fenton-like and wet oxidation	Khankhasaeva et al. (2015, 2017)
Toluene	Montmorillonite-supported iron oxide; Al- and Fe-PILC;	Air; O_2/Ar mixture	Deep oxidation (high temperature)	Nogueira et al. (2011); Li et al. (2013b)
	Hectorite- and kaolinite-TiO_2 complexes	Air/UV light	Wet oxidation	Kibanova et al. (2009)
	Al–Cu PILC	H_2O_2	Wet oxidation	Mojović et al. (2009)
	TiO_2-palygorskite composite	Air/UV-and solar-light irradiation	Wet oxidation	Papoulis et al. (2013b)
	Fe-impregnated Ti-PILC	Air	Deep oxidation (high temperature)	Liang et al. (2016)
Trichloroethylene	Allophane-supported titania	Air/UV irradiation	Photocatalytic degradation	Nishikiori et al. (2017)
Xylene	Cr or Pd-impregnated Al-PILC	Air	Deep oxidation (high temperature)	Oliveira et al. (2008)

Note: Unless otherwise specified, the clay mineral species in PILC is montmorillonite or (montmorillonite-rich) bentonite.

TABLE 7.3
Wet Catalytic and Photocatalytic Degradation and Decoloration of Organic Dyes, in the Presence of Pillared Interlayered Clays (PILC) and Surface-Modified Clay Minerals

Organic Dye (in Alphabetical Order)	Clay Catalyst/Support	Oxidant/Assisting Agent	Process/ Reaction	Reference
Acid Black 1	Fe^{3+}-Laponite; Cu-impregnated acid-activated bentonite;	H_2O_2/UV irradiation	Photo-Fenton	Sum et al. (2004, 2005); Yip et al. (2005)
Acid Chrome Dark Blue	Fe–Cu–Al-PILC	H_2O_2	Wet oxidation	Khankhasaeva et al. (2013)
Acid Fuchsin	Iron-impregnated kaolinite	Air	Fenton-like	Xu et al. (2009)
Acid Green 25	Fe-impregnated clay (calcined)	H_2O_2/UV irradiation	Photo-Fenton	Azmi et al. (2014)
Acid Light Yellow G	Fe-impregnated bentonite	H_2O_2/UV irradiation	Photo-Fenton	Chen and Zhu (2007)
Acid Red G	Ti-PILC	Air/UV irradiation	Wet oxidation	Zhang et al. (2008b)
	TiO_2-kaolinite composite	Air/UV irradiation	Wet oxidation	Zhang et al. (2011a)
Acid Yellow 17	Fe-immobilized K10 montmorillonite	H_2O_2/UV irradiation	Photo-Fenton	Muthuvel et al. (2012)
Azo dye X-3B	Fe- and Al–Fe-PILC	H_2O_2/UV irradiation	Photo-Fenton	Li et al. (2006)
Basic Blue 41	TiO_2-palygorskite composite	Air/UV irradiation	Wet oxidation	Stathatos et al. (2012)
Brilliant Orange X-GN	Hydroxy Fe–Al-montmorillonite complex	H_2O_2/visible light	Photo-Fenton	Li et al. (2011)
	Iron-pillared vermiculite	H_2O_2/UV irradiation	Photo-Fenton	Chen et al. (2010)
Cationic Red GTL	Ti-PILC	Air/UV irradiation	Wet oxidation	Sun et al. (2002)
Congo Red	TiO_2-impregnated kaolinite	Air/UV irradiation	Wet oxidation	Chong et al. (2009b); Vimonses et al. (2010)
	Ti-PILC	Air/UV irradiation	Wet oxidation	Dvininov et al. (2009)
	Fe-PILC	H_2O_2/UV irradiation	Photo-Fenton	Hadjltaief et al. (2013)
	Bi_2O_3-impregnated organoclay	Air/visible light	Wet oxidation	Patil et al. (2015)
Crystal Violet	Fe^{3+}-montmorillonite	H_2O_2/UV irradiation	Photo-Fenton	Guz et al. (2014)
	Montmorillonite-TiO_2 composite	Air/UV irradiation	Wet oxidation	Djellabi et al. (2014)
Direct Pure Blue	Fe- and Fe–Al-PILC	H_2O_2	Wet oxidation	Dashinamzhilova and Khankhasaeva (2011)
Malachite Green	Nontronite and Fe^{3+}-exchanged montmorillonite and Laponite	H_2O_2/visible light	Photo-Fenton	Cheng et al. (2008)
	Fe-PILC	H_2O_2/UV irradiation	Photo-Fenton	Hadjltaief et al. (2013)
Methylene Blue	TiO_2 supported on Ag- and Cu-modified palygorskite	Air/UV irradiation	Wet oxidation	Zhao et al. (2006, 2007)
	Ag-doped TI-PILC	Air/UV irradiation	Wet oxidation	Liu et al. (2007)
	Fe-PILC	H_2O_2/UV irradiation	Photo-Fenton	De Léon et al. (2008); Tirelli et al. (2015)
	Hectorite- and montmorillonite-TiO_2 composite;	Air/UV irradiation	Wet oxidation	Ma et al. (2009); Djellabi et al. (2014)
	Ti-PILC	Air/UV irradiation	Wet oxidation	Liu et al. (2009); Fatimah (2012); Yang et al. (2013b); Chen et al. (2012, 2014); Sahel et al. (2014)
	Kaolinite-TiO_2 composite	Air/UV irradiation	Wet oxidation	Barbosa et al. (2015); Shao et al. (2015)

(Continued)

TABLE 7.3 (Continued)
Wet Catalytic and Photocatalytic Degradation and Decoloration of Organic Dyes, in the Presence of Pillared Interlayered Clays (PILC) and Surface-Modified Clay Minerals

Organic Dye (in Alphabetical Order)	Clay Catalyst/Support	Oxidant/Assisting Agent	Process/Reaction	Reference
Methylene Blue (continued)	Clay-TiO_2 composite; mixture of kaolinite and P25 TiO_2	Air/UV irradiation	Wet oxidation	Hajjaji et al. (2013, 2016)
	Montmorillonite-supported Cu nanoparticles	H_2O_2	Wet oxidation	Mekawi et al. (2016)
Methyl Orange	Si–Ti-PILC	Air/UV-visible irradiation	Wet oxidation	Liu et al. (2006)
	Ti-PILC	Air/UV or visible light irradiation	Wet oxidation	Sun et al. (2006); Yuan et al. (2011)
	Pt-doped Ti-PILC	Air/UV irradiation	Wet oxidation	Ding et al. (2008)
	SiO_2/TiO_2-organoclay composite	Air/UV irradiation	Wet oxidation	Meng et al. (2008)
	Attapulgite-SnO_2-TiO_2 nanocomposite	Air/UV irradiation	Wet oxidation	Zhang et al. (2009)
	Al–Fe, Al–Cu- and Al–Fe–Cu-PILC	H_2O_2	Wet oxidation	Galeano et al. (2010)
	Mn-impregnated Al-PILC, Al–Mn-PILC, intercalated MnS clusters	H_2O_2	Wet oxidation	Galeano et al. (2011b)
	Kaolinite-TiO_2 composite	Air/UV irradiation	Wet oxidation	Wang et al. (2011a)
	Pd-doped palygorskite-CuO composite	Air/UV irradiation	Wet oxidation	Huo and Yang (2013)
	Montmorillonite-TiO_2 composite	Air/UV irradiation	Wet oxidation	Djellabi et al. (2014)
	Halloysite-CeO_2-AgBr composite	Air/solar light irradiation	Wet oxidation	Li et al. (2015b)
Orange G	Sepiolite-TiO_2 composite	Air/UV irradiation	Wet oxidation	Zhou et al. (2017)
Orange II	Bentonite- and Laponite-Fe oxide/silicate composites; Fe-PILC	H_2O_2/UV irradiation	Photo-Fenton	Feng et al. (2003a, 2003b, 2004a, 2004b, 2005, 2006); Chen and Zhu (2006)
	Iron-impregnated Al-pillared saponite; Fe-pillared saponite	H_2O_2	Fenton-like	Herney-Ramirez et al. (2007, 2008, 2011)
	Kaolinite-TiO_2 composite	Air/UV irradiation	Wet oxidation	Kutlákova et al. (2011)
	Fe–Ni–Al-PILC	H_2O_2	Wet oxidation	Gao et al. (2014)
	Al-pillared Fe^{3+}-smectite	H_2O_2/UV irradiation	Photo-Fenton	Li et al. (2015a)
	Fe–Zn–Al-PILC	H_2O_2	Wet oxidation	Ye et al. (2016)
Reactive Black 5	Kaolinite-TiO_2 composite	Air	Wet oxidation	Henych and Štengl (2013)
Reactive Blue 4	Fe-impregnated ball clay	H_2O_2	Fenton-like	Hassan and Hameed (2011)
Reactive Red HE-3B	Fe-impregnated Laponite	H_2O_2/UV irradiation	Photo-Fenton	Feng et al. (2003c)
Remazol Black	Ti-PILC, Ag-doped Ti-PILC	Air/UV irradiation	Wet oxidation	Sahel et al. (2014)

(Continued)

TABLE 7.3 (Continued)
Wet Catalytic and Photocatalytic Degradation and Decoloration of Organic Dyes, in the Presence of Pillared Interlayered Clays (PILC) and Surface-Modified Clay Minerals

Organic Dye (in Alphabetical Order)	Clay Catalyst/Support	Oxidant/Assisting Agent	Process/ Reaction	Reference
Rhodamine B	Fe-pillared rectorite	H_2O_2/visible light	Photo-Fenton	Zhang et al. (2010)
	AgBr-palygorskite composite	Air/visible light	Wet oxidation	Zhang et al. (2012)
	Halloysite-supported TiO_2	Air/visible light	Wet oxidation	Li et al. (2013a)
	Montmorillonite-TiO_2 composite	Air/UV irradiation	Wet oxidation	Djellabi et al. (2014)
	Polyaniline-TiO_2-halloysite composite	Air/visible light	Wet oxidation	Li et al. (2015c)
	γ-Fe_2O_3-PILC	H_2O_2/UV irradiation	Photo-Fenton	Chen et al. (2017)
Rosso Zetanyl B-NG	Clay-supported Fe^0 nanoparticles	H_2O_2	Wet oxidation	Kerkez et al. (2014)
Solophenyl Red 3BL	Ti-PILC	Air/visible light	Wet oxidation	Damardji et al. (2009)
Sulforhodamine B	V-doped Ti-PILC	Air/visible light	Wet oxidation	Chen et al. (2011)
Tartrazine	Al–Fe-PILC	H_2O_2	Wet oxidation	Banković et al. (2012)

Note: Unless otherwise specified, the clay mineral species in PILC is montmorillonite or (montmorillonite-rich) bentonite.

7.2 OXIDATION REACTIONS

As already remarked on, clay-supported organometallic complexes and zero-valent metal nanoparticles can effectively catalyze the oxidation and epoxidation of a large variety of organic compounds (Chapter 4). We have also referred to some clay-catalyzed *name* reactions, such as the Baeyer–Villiger, Oppenauer, and Wacker oxidations (Chapter 5). Here we focus attention on the ability of clays and modified clays in promoting advanced oxidation processes (AOP) involving the *in situ* generation of radicals (primarily hydroxyl) in the presence of suitable oxidants. Such processes have attracted much attention because of their potential in degrading and mineralizing organic contaminants and pollutants in water, wastewater, and soil as well as in industrial and municipal effluents (Pignatello et al. 2006; Bautista et al. 2008; Primo et al. 2008; Mohajerani et al. 2009; Chong et al. 2010; Herney-Ramírez and Madeira 2010; Perathoner and Centi 2005, 2010; Hartmann et al. 2010; Chelba et al. 2014; Galeano et al. 2011, 2012, 2014; Usman et al. 2016; Rodriguez-Narvaez et al. 2017). Typical AOP include ozonation, wet air oxidation, and wet peroxide oxidation commonly in conjunction with Fenton and photo-Fenton reactions, which in turn may be assisted (*intensified*) by UV/visible light irradiation or sonication (Pliego et al. 2015). Although wet peroxide oxidation has been the process of choice, catalytic ozonation (in the presence of clay minerals) is a promising method for degrading and mineralizing organic pollutants in water and wastewater (Azzouz et al. 2010; Shahidi et al. 2015a, 2015b).

Following Garrido-Ramirez et al. (2010), the Fenton process is initiated by the reaction of Fe^{2+} species with H_2O_2 under acid conditions (pH ~ 3.0) to yield hydroxyl radicals (HO•), as shown in Equation 7.1, and propagated by the reduction of Fe^{3+} to Fe^{2+} (Equation 7.2) with the formation of more radicals, including the perhydroxyl ($HO_2^•$) species, as shown in Equations 7.3 through 7.5.

$$Fe^{2+} + H_2O_2 \rightarrow Fe^{3+} + OH^- + HO^• \tag{7.1}$$

$$Fe^{3+} + H_2O_2 \rightarrow Fe^{2+} + HO_2^{\bullet} + H^+ \tag{7.2}$$

$$Fe^{2+} + HO_2^{\bullet} \rightarrow Fe^{3+} + HO_2^- \tag{7.3}$$

$$Fe^{3+} + HO_2^{\bullet} \rightarrow Fe^{2+} + O_2 + H^+ \tag{7.4}$$

$$H_2O_2 + HO^{\bullet} \rightarrow HO_2^{\bullet} + H_2O \tag{7.5}$$

It seems hardly surprising, therefore, that nanocrystallline Fe(hydr)oxides have usefully served as catalysts in the Fenton-like reactions and degradation of organic contaminants (Pereira et al. 2012). Similarly, clay supported iron oxides and iron oxide pillared interlayered clays (Fe-PILC) can actively promote a variety of organic conversions and reactions. Pillai and Sahle-Demessie (2003), for example, used iron-impregnated K10 montmorillonite to catalyze the Fenton-like oxidation of secondary alcohols to the corresponding ketones. As already indicated (cf. Table 3.5), Fe-PILC, rare earth exchanged Fe-PILC, and iron-impregnated Si-PILC are effective in promoting the hydroxylation of phenol to dihydroxybenzene in the presence of H_2O_2 (Letaïef et al. 2003; Kurian et al. 2012; Yang et al. 2013a), while an iron-containing clay can catalyze the conversion of tetrahydrofuran into butyrolactone (Ausavasukhi and Sooknoi 2015). Earlier, Ebitani et al. (2002) used a chain-like Fe^{3+}-montmorillonite intercalate to catalyze the oxidation of cyclohexane to cyclohexyl hydroperoxide in the presence of H_2O_2. Similarly, Nogueira et al. (2011) have noted that a montmorillonite-rich clay, impregnated with iron oxides, can promote the oxidation of toluene in air.

Other transition metals (cerium, cobalt, copper, manganese, vanadium) can also generate hydroxyl radicals from hydrogen peroxide (Barrault et al. 2000a; Carriazo et al. 2005a, 2005b; Garrido-Ramirez et al. 2012). Thus, copper-doped Al-PILC can promote the wet peroxide oxidation of toluene, xylenes (Bahranowski et al. 1999), and 2-nitrophenol (Najjar et al. 2001), while a cerium-impregnated Al- and mixed Fe–Al-PILC are effective in catalyzing the wet peroxide oxidation of phenol (Ellias and Sugunan 2014). Similarly, cobalt- and copper-impregnated Al-PILC can mediate the oxidative degradation of ethanol in air (Pérez et al. 2014). In the presence of H_2O_2, vanadium-doped Fe-, Al- and Fe–Al-PILC are effective in promoting the oxidation of cyclohexanol (Manju and Sugunan 2005), while clay-supported vanadium oxide can catalyze the selective hydroxylation of benzene to phenol (Gao and Xu 2006). More recently, Huo and Yang (2013) have reported that copper oxide nanoparticles deposited on palygorskite fibers are more active than the pristine (unsupported) CuO in degrading methyl orange (under UV light). Also interesting is the finding by Kshirsagar et al. (2009) that Co^{2+} ions incorporated in the tetrahedral and octahedral sheet of saponite can efficiently mediate the air oxidation of p-cresol. Similarly, Garade et al. (2011) noted that a cobalt-rich synthetic saponite could selectively convert p-vanillyl alcohol to p-vanillin. A synthetic saponite intercalated with an organometallic complex and TiO_2 particles can promote the oxidation of benzene to phenol under visible light irradiation (Goto and Ogawa 2016). Although impregnation of Pd into Al-pillared bentonite may lead to the destruction of the alumina pillars, the activity of the clay in catalyzing the oxidation (in oxygen) of chlorobenzene or xylene is not necessarily impaired (Oliveira et al. 2008).

The photo-Fenton process refers to the irradiation of the classical Fenton reaction system with UV or visible light to yield additional hydroxyl radicals according to Equation 7.6, while Fe^{3+} is photochemically reduced to Fe^{2+} according to Equation 7.7:

$$H_2O_2 + h\nu \rightarrow HO^{\bullet} + HO^{\bullet} \tag{7.6}$$

$$Fe^{3+} + H_2O + h\nu \rightarrow Fe^{2+} + HO^{\bullet} + H^+ \tag{7.7}$$

Both the homogeneous Fenton and the photo-Fenton reaction can be carried out under ambient conditions of temperature and pressure, and in the absence of special equipment (e.g., Zazo et al. 2006). The requirement for an acid medium, however, is a definite drawback as is the production of an iron-rich sludge although sludge formation is negligible in the photo-Fenton process (Catrinescu et al. 2003;

Navalon et al. 2010). These disadvantages may be overcome by using heterogeneous catalysts such as clay minerals, pillared interlayered clays, and clay-supported metal oxides since these materials can operate over a wide range of pH, are easily recoverable, and remain active during successive operations. Furthermore, the parent (raw) clay materials are inexpensive and environmentally benign (Barrault et al. 1998, 2000a, 2000b; Carriazo et al. 2003; Tatibouët et al. 2005; Garrido-Ramirez et al. 2012). We might also add that photo-irradiation of aqueous suspensions of iron-rich montmorillonite can generate hydroxyl radicals capable of converting benzene to phenol (Wu et al. 2008). An interesting variant of the photo-Fenton process is the introduction of an iron-rich compound into the interlayers of an organoclay acting as a catalyst support (Chen et al. 2017), or the intercalation of an organometallic photosensitizer capable of generating singlet oxygen *in situ* (Xiong et al. 2005a, 2005b).

The wet oxidation (and hydroxylation) of various organic compounds at near-ambient temperature and pressure, catalyzed by clay minerals and surface-modified clays in the presence of oxidizing agents (air, hydrogen peroxide, ozone, singlet oxygen), has been the subject of many investigations. For a general overview of heterogeneous photocatalysis, the reader is referred to the article by Ibhadon and Fitzpatrick (2013). The clay-catalyzed oxidative degradation of phenol(s) and substituted phenols has received much attention (Table 7.1) as these highly toxic priority pollutants are present in the effluents from various industrial processes (Liotta et al. 2009; Lee and Tiwari 2012; Arena et al. 2015). The ability and use of organically modified clays in removing phenols and many other non-ionic organic pollutants from aqueous solutions have been described by Churchman et al. (2006), Lee and Tiwari (2012), and Zhu et al. (2016).

As indicated in Figure 7.1, the intermediate oxidation products of phenol are catechol, hydroquinone (and pyrogallol) together with a number of carboxylic acids such as maleic, oxalic, and formic acids (Carriazo et al. 2005b; Ellias and Sugunan 2014; Tomul 2016; Wei et al. 2017). Being resistant to wet oxidation, acetic acid has also been detected (Guo and Al-Dahhan 2005). Catechol and hydroquinone can convert Fe^{3+} to Fe^{2+} and hence enhance the Fenton reaction rate (Garrido-Ramirez et al. 2012; Jin et al. 2017; Wei et al. 2017). As Zhou et al. (2016) have pointed out, the presence of chloride ions would increase the rate of phenol oxidation by accelerating the decomposition of H_2O_2 according to Equation 7.2. In common with most organic contaminants, phenol is eventually oxidized (*degraded*) into CO_2 and water (Carriazo et al. 2005b; Guo et al. 2006; Wei et al. 2017; Hadjltaief et al. 2018) as indicated in Figure 7.1.

Earlier, Del Castillo et al. (1996) have noted that the hydroxylation of phenol over Ti-PILC is dependent on the type of solvent used. With methanol as a solvent, hydroquinone is the principal product formed, while catechol is preferentially formed in the presence of acetone. Zhou et al. (2014) have identified 4-chlorocatechol and 5-chloro-1,2,4-benzenetriol as intermediates in the catalytic wet peroxide oxidation of 4-chlorophenol (over and Al–Fe-, Al–Cu-, and Al–Fe–Cu-PILC), yielding 2,4-dioxopentanedioic acid and a ferric-oxalate complex as end products in addition to CO_2, H_2O,

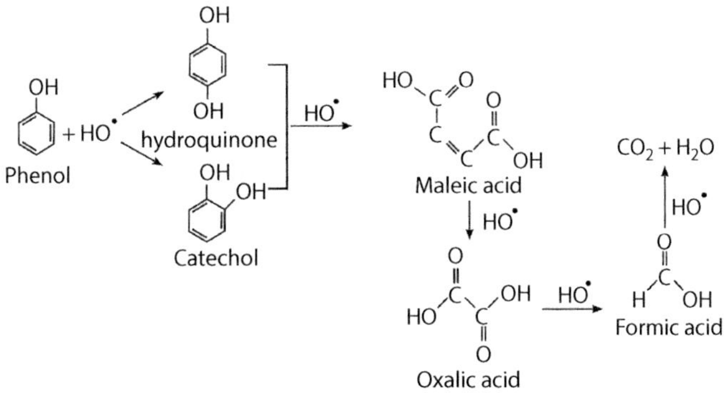

FIGURE 7.1 Diagram showing the pathway of phenol degradation through the Fenton and photo-Fenton oxidation over pillared interlayered clays (PILC) and clay-supported catalysts.

and Cl⁻. We might add that the reaction of phenol with hydroxyl radicals can lead to polymer formation (Voudrias and Reinhard 1987; Birkel et al. 2002), while Fe^{3+}-exchanged montmorillonite can effectively promote the surface oligomerization of phenolic compounds and estrogens (Polubesova et al. 2010; Qin et al. 2015).

As indicated in Table 7.1, the most commonly used oxidant is H_2O_2 although *tert*-butylhydroperoxide (TBHP) is also effective in the clay-catalyzed oxidation of alkyl arenes (Nikalje and Sudalai 1999), secondary benzylic alcohols (Palombi et al. 1999), and α-pinene (Romanenko et al. 2006) as well as in the (oxidative) coupling of naphthols (Bhor et al. 2006). Earlier, Nishimura et al. (1998) observed that incorporation of $[RuCl_2(CO)_2]_n$ into a bipyridylsilylated montmorillonite (bpy-mont) yielded an interlayer Ru(II)-bpy-mont (where Ru stands for $RuCl_2(CO)_2$), capable of catalyzing the oxidation of aromatic alkenes in the presence of TBHP and triethylamine. Khedher et al. (2006) have reported that vanadium-impregnated K10 montmorillonite (VK10) can promote the oxidation of sulfides (to sulfoxides) in the presence of TBHP. Although the activity of VK10 for the asymmetric oxidation of methyl phenyl sulfide declines after being modified with chiral ligands, the treated catalyst still shows a respectable 9%–11% enantiomeric excess.

Being capable of generating hydroxyl radicals in the presence of an oxidant (e.g., H_2O_2), Fe^{3+}-exchanged and iron oxide-impregnated clays as well as iron-rich clays (e.g., nontronite) and Fe-PILC are efficient in catalyzing the Fenton-like and photo-Fenton degradation of organic contaminants (antibiotics, hormones, pesticides, pharmaceuticals) and organic dyes. The degradation of cationic Rhodamine B dye by H_2O_2 (under visible light irradiation) in the presence of a synthetic nontronite (Liu et al. 2014) is diagrammatically illustrated in Figure 7.2. In this instance, the *active* iron is located in the octahedral sheet of the layer structure. Iron species occupying interlayer exchange sites, or associated with metal oxide pillars, can clearly perform a similar function. Indeed, all things being equal, Fe^{3+}-exchanged montmorillonite is superior to nontronite in catalyzing the photodegradation of malachite green (Cheng et al. 2008). This finding is not altogether surprising, however, since in the case of montmorillonite both the iron species and the substrate (dye) are located in the interlayer space.

Among the various metal oxide semiconductors, titanium dioxide (commonly in the form of anatase rather than rutile or brookite) has received special attention because of its high photocatalytic activity. Briefly, irradiation of TiO_2 with light of an energy greater than the band gap energy of titania induces the transfer of an electron from the valence band to the conduction band, leaving a valence band hole (h^+), which in turn can react with OH^- or H_2O to yield hydroxyl radicals ($HO^•$). At the same time, the photoelectron can reduce oxygen (e.g., in water) to give superoxide radical anions ($O_2^{•-}$), which then react with protons to yield perhydroxyl radicals ($HO_2^•$). The different radicals and valence band holes are capable of oxidizing and degrading a large range and variety of

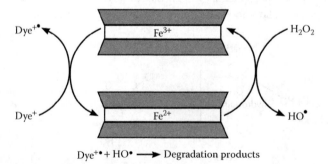

FIGURE 7.2 Diagram showing the photo-Fenton degradation of a cationic dye, such as Rhodamine B, by hydrogen peroxide under visible light irradiation, catalyzed by a synthetic nontronite with iron occupying octahedral positions in the layer structure. The process involves electron transfer from the excited dye molecule to structural ferric ion, and the formation of a hydroxyl radical ($HO^•$) from H_2O_2. For the sake of clarity, tetrahedrally coordinated ferric and ferrous ions are separately placed (in two adjacent silicate layers). (from Liu, R. et al., *RSC Advances*, 4, 12958–12963, 2014.)

organic compounds. For more details of the process, and the accompanying literature, we refer to the reviews by Akpan and Hameed (2009) and Szczepanik (2017). We might also mention that clay-supported TiO_2 can show antibacterial activity (Fatimah 2012; Dědková et al. 2014).

An added advantage is that photoactive clay-TiO_2 composites may be obtained without recourse to sophisticated equipment. Ménesi et al. (2008), for example, reported that grinding a 10% suspension of commercial P25 TiO_2 (Degussa) with Ca^{2+}-montmorillonite (in an agate mill) gave a composite material capable of catalyzing the photodegradation of phenol. Similarly, Wang et al. (2011a) obtained an efficient photocatalyst by mixing a kaolinite powder with an acidified solution of $TiCl_4$, while Ma et al. (2009) and Djellabi et al. (2014) added $TiCl_4$ (as a solid and in solution) to a stirred suspension of hectorite and montmorillonite to achieve the same objective. It is worth recalling that clay-TiO_2 composites can also mediate the formation—rather than the decomposition—of organic compounds. For example, Selvam and Swaminathan (2007) synthesized 2-alkylbenzimidazoles by cyclization of 1,2-phenylenediamine with propylene glycol (or primary alcohols) in the presence of a silver-doped montmorillonite (K10)-TiO_2 composite and air.

Within the smectite group of phyllosilicates (cf. Table 1.1), *montmorillonite* has been the single, most widely used species that can support or intercalate titania photocatalysts. The photocatalytic activity of TiO_2 nanoparticles, associated with external particle surfaces of *kaolinite* (Chong et al. 2009b; Vimonses et al. 2010; Kutláková et al. 2011; Zhang et al. 2011a; Henych and Štengl 2013; Barbosa et al. 2015; Sia et al. 2015; Hajjaji et al. 2016; Mishra et al. 2017), *halloysite* (Papoulis et al. 2010, 2013a; Wang et al. 2011b; Li et al. 2013a; Zheng et al. 2015), *palygorskite* (Zhang et al. 2011d; Stathatos et al. 2012; Papoulis et al. 2010, 2013b, 2013c; Shi et al. 2016), *sepiolite* (Suárez et al. 2008; Karamanis et al. 2011; Zhang et al. 2011b; Zhou et al. 2017), and *allophane* (Nishikiori et al. 2017), or interlayer surfaces of *rectorite* (Zhang et al. 2010, 2011c) and *mica* (Shimizu et al. 2002; Ooka et al. 2003, 2004; Yang et al. 2015) has also been investigated.

Irrespective of the clay mineral species, supported TiO_2 is generally much more efficient in promoting the photo-assisted degradation of organic contaminants and dyes than *free* P25 TiO_2 or anatase (Mogyorósi et al. 2001; Zhao et al. 2006; Liu et al. 2007; Zhang et al. 2008a; Kameshima et al. 2009; Ouidri and Khalaf 2009; Yuan et al. 2011; Li et al. 2013a; Papoulis et al. 2013c; Djellabi et al. 2014; Mishra et al. 2017) as Figure 7.3 would indicate. This observation may be explained in terms of the stabilization and reduced aggregation of surface-adsorbed TiO_2 nanoparticles, increasing their accessibility to the substrate (and photons). The same is true, if not more so, for TiO_2 intercalated into organoclays or immobilized as pillars in the interlayer space of smectites (Ilisz et al. 2004; An et al. 2008; Yang et al. 2008).

Table 7.2 lists the scattered information on the wet catalytic and photocatalytic oxidation and degradation of organic contaminants/pollutants in the presence of pillared interlayered clays (PILC) and surface-modified clay minerals, while Table 7.3 summarizes the large volume of literature on the clay-catalyzed degradation and decolorization of organic dyes that has accumulated over the past two decades.

The rate of photodegradation for organic contaminants and dyes, in the presence of clay-supported TiO_2 and pillared interlayered clays, generally obeys the Langmuir–Hinshelwood equation (Xiong et al. 2005b; Kumar et al. 2008; Szczepanik 2017),

$$r = -\frac{dC}{dt} = \frac{k_r KC}{1+KC} \tag{7.8}$$

where:
  r denotes the reaction rate
  C is the substrate concentration
  t is the reaction (illumination) time
  k_r is the reaction rate constant
  K is the (Langmuir) adsorption constant

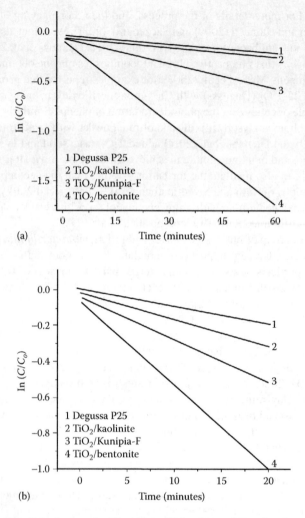

FIGURE 7.3 Application of the Langmuir–Hinshelwood kinetics model to the photocatalyzed degradation of methylene blue (a) and chlorobenzene (b) in the presence of P25 titania and various TiO_2/clay composites. NB: Kunipia-F is a Na^+-exchanged montmorillonite. (Adapted from Mishra, A. et al., *J. Alloys Compd.*, 694, 574–580, 2017.)

As Chong et al. (2010) have pointed out, the applicability of Equation 7.8 rests on the assumption that the reaction is surface mediated and that competition for catalytically active surface sites by the substrate intermediates is not limiting.

Equation 7.8 can be linearized to give Equation 7.9,

$$\frac{1}{r_o} = \frac{1}{k_r} + \frac{1}{k_r K C_o} \tag{7.9}$$

where:
 r_o is the initial rate of reaction
 C_o denotes the initial substrate concentration

 ...s, a plot of $1/r_o$ against $1/C_o$ would be linear (Xiong et al. 2005b). Further, if in Equation 7.8 ... then $r = k_r KC$, which after integration with respect to the following limits: $C = C_o$ at $t_o = 0$ ... at time t, the relationship simplifies to the following first-order kinetics equation,

$$-\ln\left(\frac{C}{C_o}\right) = k_l t \qquad (7.10)$$

where $k_l = k_r K$. According to Equation 7.10, a plot of $\ln(C/C_o)$ against t would give a straight line the slope of which equals $-k_l$ where k_l denotes the apparent rate constant (Konstantinou and Albanis 2004). Figure 7.3 shows such a plot for the photocatalytic degradation of methylene blue and chlorobenzene (Mishra et al. 2017).

Likewise, the kinetics of the wet (air) oxidation of phenol by hydrogen peroxide over Al–Fe-PILC may be described by the Langmuir–Hinshelwood equation, if one assumes that the rate-controlling step is the reaction between surface-adsorbed reactant species (Guo and Al-Dahhan 2003a, 2003b; Sobczyński et al. 2004). The same applies to the photocatalytic degradation kinetics of dimethachlor (Belessi et al. 2007) and solophenyl red 3BL (Damardji et al. 2009) in the presence of Ti-PILC, of bisphenol AF in KSF montmorillonite dispersions (Liu et al. 2010), of Congo red over kaolinite-supported TiO_2 (Vimonses et al. 2010), of methylene blue over Zn-PILC (Fatimah et al. 2011), of acid red G over TiO_2/sepiolite composite (Zhang et al. 2011b), of Reactive Black 5 dye in the presence of kaolinite-supported TiO_2 (Henych and Štengl 2013), of β-naphthol over an iron-doped TiO_2-montmorillonite complex (Ökte et al. 2014), of methyl green over Zn–Ti-PILC (Hadjltaief et al. 2016), of tetracycline over palygorskite-supported TiO_2 and Cu_2O–TiO_2, and of Rhodamine B in the presence of Fe-PILC (Chen et al. 2017).

Like Ti-PILC (Sterte 1986; Yamanaka et al. 1987; Ding et al. 1999), TiO_2-clay complexes are commonly obtained through the *sol-gel* or *acid hydrolysis* technique by mixing titanium(IV) isopropoxide/butoxide (often in alcohol) with excess water and adding an acid (HCl, HNO_3) to produce a titania sol (Liu and Zhang 2014). This sol is then added to an aqueous suspension of the clay mineral (with vigorous stirring), and the resultant hetero-coagulate is centrifuged and washed to yield the desired photocatalytically active TiO_2-clay *nanocomposite* material (Mogyorósi et al. 2003; Pichat et al. 2005; Zhu et al. 2005; Kun et al. 2006a, 2006b; Chong et al. 2009a; Kameshima et al. 2009; Kibanova et al. 2009; Vimonses et al. 2010; Barbosa et al. 2015). Besides titanium alkoxide, tetrabutyl titanate (Na et al. 2009), titanium tetrachloride (Ma et al. 2009; Wang et al. 2011a; Yuan et al. 2006; 2011), titanium oxychloride (Shao et al. 2015), and titanium oxysulfate (Yang et al. 2009; Paul et al. 2012) have served as the titania source. Like those of TiO_2, nanoparticles of ZnO and SnO_2 can be immobilized on clay mineral surfaces to act as photocatalysts (Mogyorósi et al. 2001; Körösi et al. 2004).

The intercalation into smectites of semiconducting metal oxides (before conversion to the corresponding pillars) may be facilitated by first expanding the mineral interlayers with cationic and non-ionic surfactants (Zhu et al. 2005; An et al. 2008; Ding et al. 2008; Meng et al. 2008; Chen et al. 2014). An added advantage of this approach is the creation of a hydrophobic interlayer space (e.g., by intercalation of hexadecylpyridinium chloride) that is conducive to the uptake and subsequent degradation of the contaminant molecules (Mogyorósi et al. 2002). For example, Yuan et al. (2011) intercalated TiO_2 nanoparticles into montmorillonite from their mixture with cetyltrimethylammonium bromide, and Yang et al. (2013b) obtained a series of Ti-PILC with uniform pores but different particle sizes (2.1–3.4 nm diameter) by adding a mixture of tetrabutyl titanate and surfactants with varying alkyl chain length to a montmorillonite suspension. Mesoporous Ti-PILC can also be prepared by reacting titanium oxysulfate with Laponite, hectorite, beidellite, and saponite under hydrothermal conditions (Yang et al. 2010).

7.2.1 EPOXIDATION AND OXYGENATION

As already referred to (cf. Chapter 4), clay-supported metal-Schiff base complexes, notably Mn(III) salen compounds, are effective in catalyzing the epoxidation of various alkenes (olefins), even with kaolinite as support (Dixit and Srinivasan 1988). A number of such reactions and transformations are listed in Table 4.3. The epoxidation of alkenes has attracted much attention since such epoxides are useful intermediates in the synthesis of organic species for the pharmaceutical and chemical industries (Vicente et al. 2010). Here we outline the oxidation and epoxidation of organic species by hydrogen peroxide and other oxidizing agents in the presence of surface-modified clay minerals and pillared interlayered clays (PILC). The asymmetric epoxidation and oxidation of organic compounds by clay-supported catalysts are also described. Early on, Bouhlel et al. (1993) noted that nickel acetate supported on K10 montmorillonite could catalyze the epoxidation of olefins by molecular oxygen (compressed air) under ambient conditions. Similarly, Mitsudome et al. (2005) reported that molecular oxygen could epoxidize cyclooctene (and related alkenes) as well as oxygenate adamantane in the presence of V^{3+}-montmorillonite. Arfaoui et al. (2006) used vanadia-doped Ti-PILC to catalyze the epoxidation of allylic alcohol, (E)-2-hexen-1-ol, by *tert*-butylhydroperoxide. The catalytic activity of unsulfated and sulfated Ti-PILC for the reaction was influenced by the sulfate-vanadia interaction (Arfaoui et al. 2010). The selective epoxidation of cyclohexene to cyclohexene epoxide by molecular oxygen in the presence of Ti-PILC under UV irradiation has also been reported by Ouidri et al. (2010). Although P25 TiO_2 can promote the same conversion, it is less effective in terms of product yield and selectivity. Boudjema et al. (2015) used 11-molybdo-vanado-phosphoric acid supported on an acid-activated bentonite to catalyze the epoxidation of cyclohexene, while Faria et al. (2011) immobilized a molybdenum acetylacetonate complex on K10 montmorillonite to obtain an efficient catalyst for the epoxidation of castor and soybean oils.

Trujillano et al. (2011) have also observed that synthetic saponites, containing Mg^{2+}, Ni^{2+}, or Fe^{2+} in the octahedral sheet, and having tetrahedrally coordinated Al^{3+} or Fe^{3+}, can effectively catalyze the selective epoxidation of (Z)-cyclooctene. Indeed, the same oxidation by either iodosylbenzene or hydrogen peroxide in the presence of nickel-impregnated Al-PILC is 100% selective for the epoxide (Mata et al. 2009). Kaolinite, grafted with Fe(III) picolinate and Fe(III) dipicolinate complexes, is similarly efficient in promoting the epoxidation of *cis*-cyclooctene (to *cis*-cyclooctenoxide) as well as the oxidation of cyclohexane (to cyclohexanol and cyclohexanone) by H_2O_2 under ambient conditions (de Faria et al. 2012). A vanadia-pillared montmorillonite was used by Choudary et al. (1990a) to promote the regioselective epoxidation of allylic alcohols. More recently, Bhuyan et al. (2014) used an acid-activated montmorillonite to catalyze the synthesis of β-amino alcohols and β-alkoxy alcohols through the regioselective ring opening of epoxides, such as styrene oxide. We might also mention that an iron-pillared Venezuelan smectite (Huerta et al. 2003) and a cobalt-impregnated Al-PILC (Gonzalez and Moronta 2004) can catalyze the dehydrogenation of ethylbenzene to styrene.

Mandelli et al. (2009) used K10 montmorillonite (containing structural iron) to catalyze the hydrogen peroxide oxidation of saturated and unsaturated hydrocarbons. Alkanes gave alkyl hydroperoxides as the main products, presumably through a Fenton-like reaction, while styrene yielded benzaldehyde and styrene epoxide. Furthermore, the oxidation of *cis*-1,2-dimethylcyclohexane (DMCH) preferentially yields the alcohol with a *cis*-orientation of the methyl groups whereas that of the *trans* isomer of DMCH is selective for the *trans*-alcohol. More recently, Belaidi et al. (2015) have proposed that impregnation of chromium and vanadium into an acid-activated bentonite enhances the Brønsted (and Lewis) acidity of the sample. Refluxing cyclohexene with *tert*-butylhydroperoxide and heptane, in the presence of the surface-modified clay, yields the corresponding epoxide as the major product (71%) together with some cyclohexanone. The oxidation of styrene by *tert*-butylhydroperoxide, catalyzed by palygorskite-supported Pd nanoparticles, is also highly selective for the epoxide because of strong interaction between metal particles and clay mineral support (Wang et al. 2015).

7.3 REDUCTION, HYDROGENATION, AND DEOXYGENATION

An early example of a clay-catalyzed organic reduction is the hydrazine-assisted conversion of various aromatic nitroarenes to the corresponding anilines in the presence of montmorillonite (Byung and Dong 1990). Subsequently, Onaka et al. (1993) used strongly acidic clays to catalyze the reduction of aldehydes and ketones with triethylsilane to yield symmetrical ethers or hydrocarbons. Clay-supported metal-organic catalysts were used by Crocker et al. (1993) to promote the shape-selective hydrogenation of cyclic and mono-substituted alkenes. They first intercalated montmorillonite with positively charged [Pd(PPh$_3$)(NCMe)$_3$]$^{2+}$ and [Pd(NCMe)$_4$]$^{2+}$ complexes (PPh$_3$ = triphenylphosphine; Me = methyl), which on reduction gave rise to catalytically active Pd nanoparticles anchored to the mineral surface. More recently, Soni and Sharma (2016) used montmorillonite-supported Pd nanoparticles to catalyze the complete solvent-free hydrogenation of squalene to squalane. A variant approach is to incorporate [Rh(phen)$_3$]Cl$_3$.3H$_2$O (phen = 1,10 phenanthroline) into Al-PILC and reduce the metal-organic complex with hydrogen gas to yield Rh^0 (Kotkar and Thakkar 1995, 1997). Similarly, Rh^0 nanoparticles on K10 montmorillonite obtained by *in situ* reduction of impregnated RhCl$_3$.3H$_2$O with NaBH$_4$ are effective in catalyzing the selective hydrogenation of monoterpenes such as limonene, geraniol, linalool, and citronellal (Agrawal and Ganguli 2013). Likewise, Szöllösi et al. (2005) observed that Pt^0 nanoparticles immobilized on an interlayer complex of bentonite with a chiral cinchonidine could promote the selective hydrogenation of 2-cyclohexene-1-one to cyclohexanone, while a portion of the product was further converted to cyclohexanol.

Intercalation of a Schiff base Pt(II) complex into montmorillonite by cation exchange also yields an efficient catalyst for the hydrogenation of aromatic nitro compounds under ambient conditions (Parida et al. 2011). For example, Leitmannová and Červený (2010) introduced a Ru-organic compound into a bentonite, K10 montmorillonite, and a hectorite by cation exchange, and used the resultant interlayer complexes to catalyze the selective hydrogenation of sorbic acid to *cis*-hex-3-enoic acid. Earlier, Szöllösi et al. (1998) prepared a 5% Pt/K10 montmorillonite catalyst for the selective hydrogenation of cinnamyl aldehyde to cinnamyl alcohol. The approach adopted by Miao and coworkers (Miao et al. 2006, 2007; Tao et al. 2009) was to introduce Ru^{3+} into an intercalate of montmorillonite with an ionic liquid (e.g., 1,1,3,3-tetramethylguanidium trifluoroacetate), or Pd^{2+} and Rh^{3+} into palygorskite and sepiolite via an ionic liquid, and then to reduce the metal ions with hydrogen. The resultant immobilized zero-valent metal nanoparticles can effectively promote the hydrogenation of various alkenes, including benzene, cyclohexene, hexene, and styrene.

Upadhyay and Srivastava (2015) reported that a Ru^0-montmorillonite intercalate could catalyze the solvent-free hydrogenation of 16 different alkenes. Ru^0 particles, impregnated into the nanopores of acid-activated montmorillonite, can similarly catalyze the reduction of substituted nitrobenzenes to the corresponding anilines with 56%–97% conversion and 91%–100% selectivity (Sarmah and Dutta 2012). Even higher levels of conversion and selectivity have been recorded for the reduction of aromatic carbonyl compounds to the corresponding alcohols in the presence of isopropanol, acting as both solvent and reductant. We should point out that hydrogen (gas), the commonly used reducing agent in organic hydrogenation reactions, can also induce the hydrogenolysis of organic substrates as illustrated by the conversion of glycerol to 1,2-propanediol in the presence of montmorillonite-supported Ru-Cu catalysts (Jiang et al. 2009). Other examples of organic hydrogenation reactions, catalyzed by metal nanoparticles, immobilized or intercalated in montmorillonite and organoclays, are listed in Table 7.4. The asymmetric hydrogenation of organic compounds by clay-supported (chiral) catalysts is described in Section 7.4.

Louloudi and Papayannakos (2000) impregnated Ni into Al-PILC and used the product to convert benzene into cyclohexane. Subsequently, Dhakshinamoorthy and Pitchumani (2008) reported that montmorillonite-supported nickel nanoparticles obtained by treating the Ni^{2+}-exchanged (K10) clay with hydrazine could efficiently catalyze the hydrogenation of alkenes and alkynes. Ni nanoparticles supported on synthetic saponites are similarly active in promoting the hydrogenation of styrene oxide to 2-phenylethanol (Vicente et al. 2011). We refer to Table 3.5 for other examples of organic hydrogenation reactions catalyzed by metal-impregnated pillared smectites.

TABLE 7.4
Hydrogenation (and Semihydrogenation) of Organic Compounds Catalyzed by Platinum Group Metal Nanoparticles Immobilized/Intercalated in Clay Minerals and Surface-Modified Clays at Controlled Temperatures and Hydrogen Pressures

Organic Compound/Substrate (in Alphabetical Order)	Product	Metal/Clay Mineral	Selectivity	Reference
Benzaldehyde	Benzyl alcohol	Pd/vermiculite	High	Divakar et al. (2007
		Pd/bentonite	High	Divakar et al. (2008)
Benzene	Cyclohexane	Rh/MMT	Complete (100%)	Sidhpuria et al. (2009)
	Cyclohexane	Ru/MMT	High	Miao et al. (2006)
	Cyclohexene	Ru/bentonite	Moderate	Wang et al. (2012)
Carbon dioxide	Formic acid	Ru^{3+}-exchanged MMT	High	Srivastava (2014)
Cinnamyl aldehyde (cinnamaldehyde)	Cinnamyl alcohol	Pt and Ru on MMT or hectorite	Variable; higher with MMT than hectorite	Dhanagopal et al. (2010)
		Pd/vermiculite	High	Divakar et al. (2007)
		Organophilic Pt/MMT and Pt/hectorite	High	Manikandan et al. (2007a, 2007b)
Citral	Geraniol and nerol	Organophilic Pt/MMT and Ru/MMT	Moderate	Manikandan et al. (2008); Dhanagopal et al. (2010)
Croton aldehyde	Crotyl alcohol	Pt and Ru on MMT or hectorite;	Variable; higher with MMT than with hectorite;	Dhanagopal et al. (2010)
		Pd/vermiculite	High	Divakar et al. (2007)
Cyclohexadiene	Cyclohexene	Pd/sepiolite	High	Tao et al. (2009)
Cyclohexene	Cyclohexane	Pd/MMT	High to nearly complete	Király et al. (2001); Scheuermann et al. (2009)
		Pd/sepiolite	High to nearly complete	Tao et al. (2009)
		Ru/K10 MMT	High	Agarwal and Ganguli (2014)
1-Hexene	Hexane	Pd/MMT	High to nearly complete	Király et al. (2001)
2-Methylbenzaldehyde	2-Methylbenzyl alcohol	Pd/Fluoro-tetrasilicic mica	High	Sivakumar et al. (2001)
Naphthalene	1,2,3,4-Tetrahydro-naphthelene	Rh/MMT	Complete (100%)	Sidhpuria et al. (2009)
Nitrobenzene	Aniline	Pt/MMT	High	Pan et al. (2009)
1-Octene	1-Octane	Organophilic Pd/MMT	High	Király et al. (1996); Mastalir et al. (1997)
Phenylacetylene	Styrene (main), ethylbenzene	Organophilic Pd/MMT	High	Mastalir et al. (2004)

(*Continued*)

TABLE 7.4 (Continued)
Hydrogenation (and Semihydrogenation) of Organic Compounds Catalyzed by Platinum Group Metal Nanoparticles Immobilized/Intercalated in Clay Minerals and Surface-Modified Clays at Controlled Temperatures and Hydrogen Pressures

Organic Compound/ Substrate (in Alphabetical Order)	Product	Metal/Clay Mineral	Selectivity	Reference
1-Phenyl-1-butyne	Corresponding *cis-* and *trans-*alkenes	Pd/MMT; Pd-impregnated Al-PILC	High selectivity toward the *cis* stereoisomer	Mastalir et al. (2000); Marín-Astorga et al. (2005)
1-Phenyl-1-pentyne	1-Phenyl-cis-1-pentene	Pd/MMT; organophilic Pd/MMT	High selectivity toward the *cis*-alkene stereoisomer	Király et al. (2001); Mastalir et al. (2001)
1-Phenyl-1-propyne	Corresponding *cis-* and *trans-*alkenes	Pd-impregnated Al-PILC	High selectivity toward the *cis* stereoisomer	Marín-Astorga et al. (2005)
Styrene	Ethylbenzene	Pd/MMT	High to complete	Király et al. (2001); Scheuermann et al. (2009)
		Organophilic Pd/MMT	High	Király et al. (1996); Mastalir et al. (1997)
Styrene oxide	Phenyl ethanol (phenethyl alcohol)	Ni/synthetic saponites	High	Vicente et al. (2011)
Toluene	Methylcyclohexane	Rh/MMT	Complete (100%)	Sidhpuria et al. (2009)
o-, m-, and p-Xylene	1,2-, 1,3-, 1,4-dimethylhexane	Rh/MMT	Complete (100%)	Sidhpuria et al. (2009)

Note: Substrate conversion and product selectivity are temperature-dependent. The clay mineral species is commonly montmorillonite (MMT).

Xu et al. (2014) reported that Ru nanoparticles, immobilized on montmorillonite (with the aid of an acidic ionic liquid), were highly efficient in catalyzing the hydrodeoxygenation of phenolic compounds to cycloalkanes. Similarly, Soni et al. (2017) noted that a cobalt-impregnated natural clay could promote the hydrodeoxygenation of methyl oleate (to diesel-grade hydrocarbons). Tandiary et al. (2014) used Sn^{4+}-exchanged montmorillonite to catalyze the deoxygenation of tertiary and secondary benzylic alcohols to alkanes with the assistance of triethylsilane. On the other hand, Pd-impregnated (Brazilian) montmorillonite and its corresponding pillared interlayered variety are apparently inactive in deoxygenating soybean fatty acids (Detoni et al. 2014). By contrast, the 1% Pd/K10 montmorillonite complex shows nearly complete selectivity to aliphatic hydrocarbons, presumably because the clay mineral support is highly acidic and mesoporous.

In closing this section, we would mention that transition metal-doped pillared montmorillonites (PILC) can catalyze the selective reduction of NO with ammonia (to give dinitrogen and water) although this reaction is not strictly *organic* in nature. Thus, Bahranowski et al. (1997) used V-doped Ti-PILC to carry out the earlier reduction, while Chmielarz et al. (2007) chose Cu- and Fe-modified porous clay heterostructures (cf. Chapter 3) as catalysts. Using Ti-PILC doped or exchanged with iron-, chromium-, or vanadium-oxide, Yang and coworkers (Cheng et al. 1996; Long and Yang 1999,

2000a, 2000b) have proposed that the process involves the surface adsorption and oxidation (by O_2) of NO molecules, and it is controlled more by the Brønsted than Lewis acidity of the (pillared) clay supports. Impregnated V_2O_5 can also promote the reaction of NO with NH_3 occupying acid sites on sulfated Ti-PILC (Boudali et al. 2005). Besides ammonia, ethylene (Yang et al. 1998) and propylene (Valverde et al. 2003) have been used as reducing agents.

7.4 ASYMMETRIC/ENANTIOSELECTIVE REACTIONS AND SYNTHESES

Much of the background work on the formation and properties of complexes between clay minerals and chiral organic species, in terms of asymmetric synthesis, is attributable to Yamagishi and coworkers (Yamagishi 1987; Yamagishi and Sato 2012). These workers observed that interlayer complexes of montmorillonite with chiral transition metal ion (M) chelates, such as $[M(phen)_3]^{2+}$ or $[M(bpy)_3]^{2+}$ where M = Fe, Ni, Ru; phen = 1,10-phenanthroline; and bpy = 2,2'-bipyridine, could show chirality recognition and selection. Asymmetric recognition has also been observed for hectorite intercalated with chiral arylethylammonium ions (Iwai et al. 1993).

This facility arises from the manner in which the metal chelate guests intercalate into the clay mineral host. When adsorbed from a solution of a single enantiomer, the metal chelate can occupy two cation exchange sites. Because of steric interference between the same enantiomers, however, only one such site is occupied when the chiral guest compound is introduced from a racemic solution (Yamagishi and Soma 1981; Yamagishi 1982). In line with these observations, X-ray diffraction analysis of oriented films of montmorillonite-$[Ru(bpy)_3]^{2+}$ complexes indicated that enantiomeric cations intercalated as a bilayer whereas the racemic cation formed only a single layer (Villemure 1991).

Thus, intercalation of a racemic pair gives rise to a vacancy into which the antipodal enantiomer can be accommodated, opening a pathway to asymmetric and enantioselective synthesis. For example, when a phenyl alkyl sulfide is introduced into an interlayer montmorillonite complex with $[Ru(phen)_3]^{2+}$ or $[Ru(bpy)_3]^{2+}$, the corresponding chiral sulfoxide is formed with an enantiomeric excess of ~43% under visible light irradiation in the presence of dibenzoyl-D(+)-tartaric acid (Fujita et al. 2006). Earlier, Yamagishi (1986) reported that (prochiral) cyclohexyl phenyl sufide could convert into the optically active sulfoxide by oxidation with sodium metaperiodate in the presence of a montmorillonite-$[\Delta\text{-}Ni(phen)_3]^{2+}$ complex. Similarly, chiral titania pillared montmorillonite can promote the asymmetric oxidation of sulfides to sulfoxides (Choudary et al. 1993). We might also add that chromatographic columns, packed with $\Lambda\text{-}[Ru(phen)_3]^{2+}$- and $\Lambda\text{-}[Ru(bpy)_3]^{2+}$-exchanged montmorillonite and Laponite, are capable of separating diastereoisomers of organic species (Yamagishi 1983; Yamagishi et al. 1996).

Simulations of the adsorption characteristics and stereochemistry of clay-$[Ru(bpy)_3]^{2+}$ complexes by Monte Carlo techniques and computer modeling (Sato et al. 1992; Breu and Catlow 1995) indicate that the racemic mixture is more compact and stable than the enantiomeric pair in that the mean binding energy for the racemic pair is 1.5 kJ/mol lower than that for the enantiomeric pair. Subsequently, Török et al. (1999a, 2000) synthesized Laponite complexes with chiral n-alkyl phenylethylammonium ions. Interestingly, the short-chain n-alkyl derivatives intercalate as a single layer whereas their long-chain (C_{12}–C_{18}) counterparts adopt a bilayer configuration. The resultant organoclays can be expanded to different extents by imbibing ethanol, toluene, and their binary mixtures. Since the alkyl chains can take up a variety of orientations, voids are created for potential shape- and enantio-selective organic synthesis. A more striking example, reported by Hara et al. (2010), is the asymmetric aldol reaction of isatins with acetone or acetaldehyde. Catalyzed by an interlayer complex of montmorillonite with chiral N-(2-thiophenesulfonyl)prolinamide, the reaction yields products with an enantiomeric excess of >90%.

In a series of papers, Mayoral and coworkers (Cornejo et al. 2003; Fraile et al. 1998a, 2002, 2008a, 2008b; García et al. 2008) described the immobilization of cationic chiral complexes of copper(II) with bis(oxazoline) (**5**) or pyridine-oxazoline(**6**) ligands on Laponite, bentonite, and

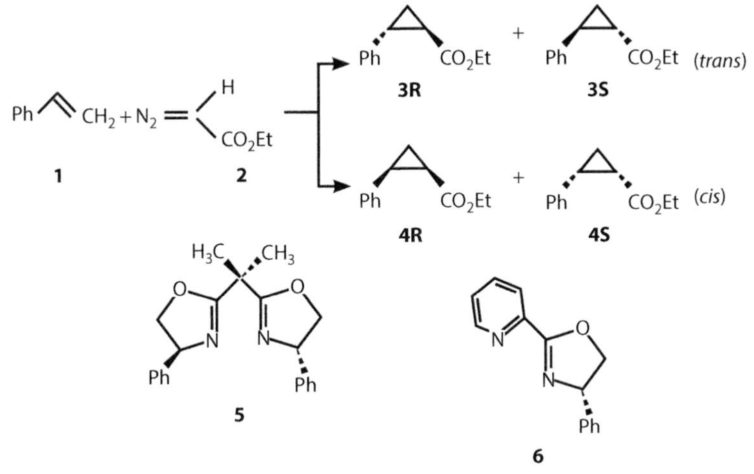

SCHEME 7.1 Enantioselective cyclopropanation of styrene (**1**) with ethyl diazoacetate (**2**), catalyzed by cationic chiral complexes of Cu(II) with bis(oxazoline) (**5**) or pyridine-oxazoline (**6**) immobilized on Laponite. The yield of products **3R**, **3S**, **4R**, **4S** and their respective *trans/cis* ratios are influenced by the type of solvent used. (After Cornejo, A. et al., *J. Mol. Catal. A: Chem.*, 196, 101–108, 2003; Fraile, J.M. et al., *Tetrahedron: Asym.*, 9, 3997–4008, 1998a.)

K10 montmorillonite by electrostatic interactions (cation exchange), and their use to mediate the enantioselective cyclopropanation of styrene (**1**) with ethyl diazoacetate (**2**) according to Scheme 7.1.

For both Cu-**5** and Cu-**6** complexes (with Laponite), the yield of products (**3R**, **3S**, **4R**, **4S**) and their respective *trans/cis* ratios are influenced by the type of solvent used. Since the Cu-bis(oxazoline) complex is apparently immobilized as an almost square-planar structure, a reduction in solvent polarity would make it easier for the complex to approach the clay surface and establish close surface-complex interactions (Alonso et al. 2000; Fernandez et al. 2001). Interestingly, Bigi et al. (2001) also noted that the regioselective conversion of 2,4-di-*tert*-butylphenol and phenol to 4-*tert*-butylphenol over KSF montmorillonite was solvent-dependent. Furthermore, the *trans/cis* diastereoselectivity of the reaction between styrene and ethyl diazoacetate (Scheme 7.1), catalyzed by the Laponite-supported Cu complex with pyridine-oxazoline (**6**), is the reverse of that observed with the bis(oxazoline) complex (**5**) as ligand (Table 7.5). Using Cu^{2+}-exchanged K10 montmorillonite as a catalyst, Fraile et al. (1996)

TABLE 7.5
Cyclopropanation Reaction of Styrene with Ethyl Diazoacetate Catalyzed by Chiral Complexes of Copper(II) with Bis(oxazoline) and Pyridineoxazoline Ligands Immobilized on Laponite

Ligand	Solvent	Yield (%)	trans/cis	Enantiomeric excess (%)[a]	
				trans	cis
Bis(oxazoline)	Dichloromethane	28	61/39	49 (**3R**)	24 (**4R**)
	Styrene	40	31/69	7 (**3R**)	34 (**4S**)
Pyridine-oxazoline	Dichloromethane	59	45/55	27 (**3R**)	27 (**4R**)
	Styrene	68	31/69	65 (**3R**)	24 (**4R**)

Source: Cornejo, A. et al., *J. Mol. Catal. A: Chem.*, 196, 101–108, 2003.

[a] **3R**, **4R**, and **4S** refer to cyclopropane products as shown in Scheme 7.1.

also found that the normal *trans/cis* selectivity in solution was reversed, giving the *cis*-cyclopropane as the major product. The immobilization of enantioselective (chiral) catalysts on solid supports, including clay minerals, has been reviewed by Fraile et al. (2009b).

Chiral Cu-bis(oxazoline) complexes, supported on Laponite, are equally active in promoting the enantioselective aziridination of styrene with *p*-toluensulfonylimino phenyliodinane although their performance is comparable with, if not lower than, that observed under homogeneous conditions (Fraile et al. 2004). Bahulayan et al. (2003) have also noted that (acid-activated) K10 montmorillonite can efficiently catalyze the stereoselective synthesis of β-acetamido ketones by reacting acetophenone with substituted benzaldehydes. Subsequently, Nowrouzi et al. (2009) used the potassium salts of organotrifluoroborate to promote the diastereoselective allylation and crotylation of ketones in the presence of K10 montmorillonite. The same clay catalyst was used by Zurita et al. (2015) to prepare chiral dioxa-caged compounds from levoglucosenone through a cascade 3-step cationic cyclization. Interestingly, Silva et al. (2004) were able to obtain chiral β-enamino esters from β-keto esters, carbohydrate derivatives, and amines in the presence of natural smectite, palygorskite (*attapulgite*), and vermiculite.

Yadav et al. (2004, 2006) have reported that KSF montmorillonite is effective in promoting the synthesis of enantiometrically pure 5-substituted pyrazoles from 2,3-dihydro-4*H*-pyran-4-ones and phenyl hydrazines, and of chiral tetrahydroquinolines by cyclization of arylamines with 2-deoxy-D-ribose. Also noteworthy is the ability of (natural) kaolinite to catalyze the reaction of aromatic and aliphatic nitriles, and 2,2-disubstituted malononitriles, with 1,2-amino alcohol to yield 2-oxazolines (Jnaneshwara et al. 1998, 1999). The same clay can also promote the conversion of isatoic anhydride into 2-(*o*-aminophenyl) oxazolines (Gajare et al. 2000).

Earlier, Cativiéla et al. (1993) used Fe^{3+}-, Ti^{4+}-, and Zn^{2+}-exchanged K10 montmorillonites to catalyze the asymmetric Diels–Alder reaction between cyclopentadiene and chiral acrylates to yield the corresponding cycloadducts with high endo/exo selectivity. Mitsudome et al. (2008) have also observed that the Diels–Alder reaction of cyclohexene with acrolein (in acetonitrile/water), in the presence of an interlayer complex of montmorillonite with chiral (5*S*)-2,2,3-trimethyl-5-phenylmethyl-4-imidazolidinone (cf. Figure 5.1), affords the corresponding cycloadduct with an endo enantiomeric excess of 86% and an endo/exo ratio of 93/7. Although the result is comparable to that obtained using the chiral organic compound on its own, the organoclay catalyst can be reused up to four times without loss of activity.

That chiral Mn(salen) complexes, supported on montmorillonite and Laponite, show lower enantioselectivity in olefin epoxidation than their counterparts in solution, has also been reported by Fraile et al. (1998b). Earlier, Choudary et al. (1990b) used titania-pillared montmorillonite in the presence of (+)-diisopropyl or (+)-diethyl tartrate to catalyze the asymmetric epoxidation of allylic alcohols, obtaining 72%–91% isolated yields and an enantiomeric excess of >90%. Kureshy et al. (2003, 2004) intercalated dicationic chiral Mn(III) salen complexes into montmorillonite for the enantioselective oxidation of styrene, indene, and 6-nitro-2,2-dimethyl chromene, attaining nearly complete conversion and selectivity except for styrene. The reaction with styrene, however, gave a significantly higher enantiomeric excess relative to the conversion under homogeneous conditions. Chiral Mn(III) salen complexes, grafted to porous clay heterostructures (cf. Chapter 3) are similarly efficient in catalyzing the asymmetric epoxidation of olefins in the presence of an organic oxidant (Kuźniarska-Biernacka et al. 2010, 2011).

Complexes of clay minerals with chiral metal chelates have also featured in the asymmetric hydrogenation of organic species. Mazzei et al. (1980), for example, used chiral ruthenium-aminophosphine catalysts supported on smectite clays for the asymmetric hydrogenation (reduction) of substituted acrylic acids, such as α-acetamido acrylic and α-acetamido cinnamic acids. Optical yield (enantiomeric excess) is dependent on the mineral support, decreasing in the

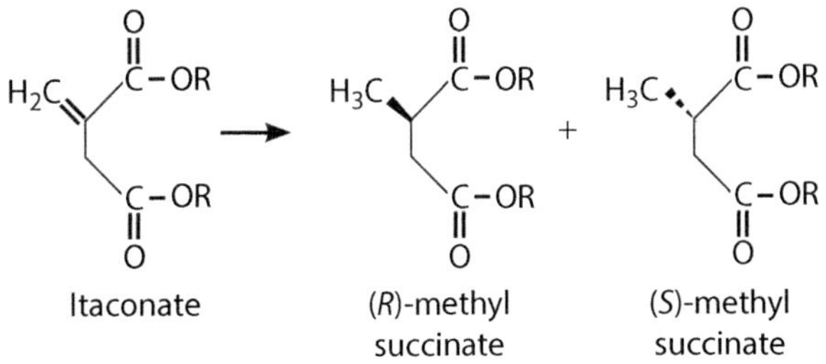

FIGURE 7.4 Asymmetric hydrogenation of itaconate to yield (R)- and (S)-methyl succinates over a chiral rhodium-phosphine complex, intercalated into a commercial synthetic hectorite; R denotes methyl, ethyl, 1-propyl, 1-butyl, 1-pentyl, or 1-hexyl. (Modified from Sento, T. et al., *J. Mol. Catal. A: Chem.*, 137, 263–267.)

sequence hectorite > bentonite > nontronite, which reflects the order of catalyst-substrate interactions. In a follow-up investigation, Shimazu and coworkers (Shimazu et al. 1996; Sento et al. 1999) intercalated chiral rhodium(I)-phosphine complexes into a (synthetic) hectorite by cation exchange. The resultant organically modified clay was then used to catalyze the asymmetric hydrogenation of α,β-unsaturated carboxylic acid esters (itaconates), giving rise to the corresponding (R)- and (S)-methyl succinates (Figure 7.4). The dependence of enantioselectivity on the solvent and ester group size may be explained in terms of the interaction between substrate and metal-chelate in the interlayer space.

Balázsik et al. (1999) and Török et al. (1999b) introduced chiral cinchonidine into K10 montmorillonite (by cation exchange) together with Pt, Pd, and Rh metal nanoparticles, and then used the resultant (clay-organic-metal) complex to promote the enantioselective hydrogenation (by H_2) of ethyl pyruvate and 2-methyl-2-pentenoic acid. The clay-supported Pt-cinchonidine catalyst was most effective, yielding the (R)- and (S)-ethyl lactates (Figure 7.5) with up to 75% enantiomeric excess (ee) for the R isomer as estimated from ee (%) = ([R] − [S])/([R] + [S]) × 100. Subsequently, Mastalir et al. (2002) and Szöllősi et al. (2005) extended the investigation using Na^+-montmorillonite, synthetic Na^+-hectorite, and bentonite as supports. Differences in performance among the clay mineral hosts are related to variation in the size and distribution of the supported Pt nanoparticles.

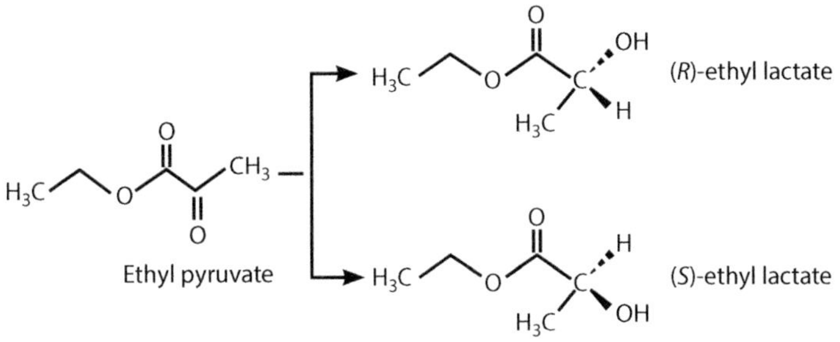

FIGURE 7.5 Asymmetric hydrogenation of ethyl pyruvate to (R)- and (S)-ethyl lactate catalyzed by platinum nanoparticles, immobilized on an interlayer bentonite-cinchonidine complex. (Modified from Szöllősi, G. et al., *J. Mater. Chem.*, 15, 2464–2469, 2005.)

REFERENCES

Agarwal, S. and J.N. Ganguli. 2013. Selective hydrogenation of monoterpenes on Rhodium(0) nanoparticles stabilized in montmorillonite K-10 clay. *Journal of Molecular Catalysis A: Chemical* 372: 44–50.

Agarwal, S. and J.N. Ganguli. 2014. Hydrogenation by nanoscale ruthenium embedded into the nanopores of K10 clay. *RSC Advances* 4: 11893–11898.

Akpan, U.G. and B.H. Hameed. 2009. Parameters affecting the photocatalytic degradation of dyes using TiO_2-based photocatalysts: A review. *Journal of Hazardous Materials* 170: 520–529.

Aldea, R. and H. Alper. 2000. Ruthenium clay catalyzed reduction of α-iminoesters and α-iminoketones, and the reductive amination of α-ketoesters. *Journal of Organometallic Chemistry* 593–594: 454–457.

Alonso, P.J., J.M. Fraile, J. García et al. 2000. Spectroscopic study of the structure of bis(oxazoline)copper complexes in solution and immobilized on laponite clay: Influence of the structure on the catalytic performance. *Langmuir* 16: 5607–5612.

An, T., J. Chen, G. Li et al. 2008. Characterization and photocatalytic activity of TiO_2 immobilized hydrophobic montmorillonite photocatalysts: Degradation of decabromodiphenyl ether (BDE 209). *Catalysis Today* 139: 69–76.

Arena, F., R. di Chio, B. Gumina, L. Spadaro, and G. Trunfio. 2015. Recent advances on wet air oxidation catalysts for treatment of industrial wastewaters. *Inorganica Chimica Acta* 431: 101–109.

Arfaoui, J., L.K. Boudali, and A. Ghorbel. 2006. Vanadia-doped titanium-pillared clay: Preparation, characterization and reactivity in the epoxidation of allylic alcohol (E)-2-hexen-1-ol. *Catalysis Communications* 7: 86–90.

Arfaoui, J., L.K. Boudali, and A. Ghorbel. 2010. Catalytic epoxidation of allylic alcohol (E)-hexen-1-ol over vanadium supported on unsulfated and sulfated titanium pillared montmorillonite catalysts: Effect of sulfate groups and vanadium loading. *Applied Clay Science* 48: 171–178.

Arsene, D., C. Catrinuscu, B. Drăgoi, and C. Teodosiu. 2010. Catalytic wet hydrogen peroxide oxidation of 4-chlorophenol over iron-exchanged clays. *Environmental Engineering and Management Journal* 9: 7–16.

Ausavasukhi, A. and T. Sooknoi. 2015. Oxidation of tetrahydrofuran to butyrolactone catalyzed by iron-containing clay. *Green Chemistry* 17: 435–441.

Ayodele, O.B. 2013. Effect of phosphoric acid treatment on kaolinite supported ferrioxalate catalyst for the degradation of amoxicillin in batch photo-Fenton process. *Applied Clay Science* 72: 74–83.

Ayodele, O.B. and B.H. Hameed. 2013. Synthesis of copper pillared bentonite ferrioxalate catalyst for degradation of 4-nitrophenol in visible light assisted Fenton process. *Journal of Industrial and Engineering Chemistry* 19: 966–974.

Ayodele, O.B., J.K. Lim, and B.H. Hameed. 2012. Pillared montmorillonite supported ferric oxalate as heterogeneous photo-Fenton catalyst for degradation of amoxicillin. *Applied Catalysis A: General* 413–414: 301–309.

Azmi, N.H.M., O.B. Ayodele, V.M. Vadivelu, M. Asif, and B.H. Hameed. 2014. Fe-modified local clay as effective and reusable heterogeneous photo-Fenton catalyst for the decolorization of Acid Green 25. *Journal of the Taiwan Institute of Chemical Engineers* 45: 1459–1467.

Aznárez, A., R. Delaigle, P. Eloy, E.M. Gaigneaux, S.A. Korili, and A. Gil. 2015. Catalysts based on pillared clays for the oxidation of chlorobenzene. *Catalysis Today* 246: 15–27.

Azzouz, A., A. Kotbi, P. Niquette et al. 2010. Ozonation of oxalic acid catalyzed by ion-exchanged montmorillonite in moderately acidic media. *Reaction Kinetics, Mechanisms and Catalysis* 99: 289–302.

Bahranowski, K., M. Gasior, A. Kielski et al. 1999. Copper-doped alumina-pillared montmorillonites as catalysts for oxidation of toluene and xylenes with hydrogen peroxide. *Clay Minerals* 34: 79–87.

Bahranowski, K., J. Janas, T. Machej, E.M. Serwicka, and L.A. Vartikian. 1997. Vanadium-doped titania-pillared montmorillonite clay as a catalyst for selective catalytic reduction of NO by ammonia. *Clay Minerals* 32: 665–672.

Bahulayan, D., S.K. Das, and J. Iqbal. 2003. Montmorillonite K10 clay: An efficient catalyst for the one-pot stereoselective synthesis of β-acetamido ketones. *Journal of Organic Chemistry* 68: 5735–5738.

Baizig, M., B. Jamoussi, and N. Batis. 2013. Optimization by RSM of the degradation of three phenolic compounds—hydroquinone, resorcinol and catechol—on Fe-modified clays. *Water Quality Research Journal* 48: 171–179.

Baizig, M., S. Khalfallah, B. Jamoussi, N. Batis, and R. Trujillano. 2015. Preparation and characterization of different TiO_2-modified montmorillonite meso-microporous materials, with enhanced photocatalytic activity. *International Journal of Engineering Research & Technology* 4: 182–192.

Balázsik, K., B. Török, G. Szakonyi, and M. Bartók. 1999. Homogeneous and heterogeneous asymmetric reactions. Part X: Enantioselective hydrogenations over K-10 montmorillonite supported noble metal catalysts with immobilized modifier. *Applied Catalysis A: General* 182: 53–63.

Banković, P., A. Milutinović, N. Jović-Jovićić et al. 2009. Synthesis, characterization and application of Al, Fe-pillared clays. *Acta Physica Polonica A* 115: 811–815.

Banković, P., A. Milutinović-Nikolić, Z. Mojović et al. 2012. Al, Fe-pillared clays in catalytic decolorization of aqueous tartrazine solutions. *Applied Clay Science* 58: 73–78.

Barbosa, L.V., L. Marçal, E.J. Nassar et al. 2015. Kaolinite-titanium oxide nanocomposites prepared via sol-gel as heterogeneous photocatalysts for dyes degradation. *Catalysis Today* 246: 133–142.

Barrault, J., M. Abdellaoui, C. Bouchoule et al. 2000b. Catalytic wet peroxide oxidation over mixed (Al-Fe) pillared clays. *Applied Catalysis B: Environmental* 27: 225–230.

Barrault, J., C. Bouchoule, K. Echachoui, N. Frini-Srasra, M. Trabelsi, and F. Bergaya. 1998. Catalytic wet peroxide oxidation (CWPO) of phenol over mixed (Al-Cu)-pillared clays. *Applied Catalysis B: Environmental* 15: 269–274.

Barrault, J., J.M. Tatibouët, and N. Papayannakos. 2000a. Catalytic wet peroxide oxidation of phenol over pillared clays containing iron or copper species. *Comptes Rendus de l'Académie des Sciences Séries II C, Chimie* 3: 777–783.

Bautista, P., A.F. Mohedano, J.A. Casas, J.A. Zazo, and J.J. Rodroguez. 2008. An overview of the application of Fenton oxidation to industrial wastewater treatment. *Journal of Chemical Technology and Biotechnology* 83: 1323–1338.

Belaidi, N., S. Bedrane, A. Choukchou-Braham, and R. Bachir. 2015. Novel vanadium-chromium-bentonite green catalyst for cyclohexene epoxidation. *Applied Clay Science* 107: 14–20.

Belessi, V., D. Lambropoulou, I. Konstantinou et al. 2007. Structure and photocatalytic performance of TiO_2/clay nanocomposites for the degradation of dimethachlor. *Applied Catalysis B: Environmental* 73: 292–299.

Ben Achma, R., A. Ghorbel, A. Dafinov, and F. Medina. 2008. Copper-supported pillared clay catalyst for the wet hydrogen peroxide catalytic oxidation of model pollutant tyrosol. *Applied Catalysis A: General* 349: 20–28.

Ben Achma, R., A. Ghorbel, A. Dafinov, and F. Medina. 2014. Anaerobic digestion of olive oil mill wastewater pre-treated with catalytic wet peroxide photo-oxidation using copper supported pillared clay catalysts. *Journal of Materials Science and Chemical Engineering* 2: 9–17.

Bhor, M.D., N.S. Nandurkar, M.J. Bhanushali, and B.M. Bhanage. 2006. An efficient oxidative coupling of naphthols catalyzed by Fe impregnated pillared montmorillonite K10. *Catalysis Letters* 112: 45–50.

Bhuyan, D., L. Saikia, and D.K. Dutta. 2014. Modified montmorillonite clay catalysed regioselective ring opening of epoxide with amines and alcohols under solvent-free conditions. *Applied Catalysis A: General* 487: 195–201.

Bigi, F., M.L. Conforti, R. Maggi, A. Mazzacani, and G. Sartori. 2001. Montmorillonite KSF-catalysed regioselective *trans-tert*-butylation of *tert*-butylphenols. *Tetrahedron Letters* 42: 6543–6545.

Birkel, U., G. Gerold, and J. Niemeyer. 2002. Abiotic reactions of organics on clay mineral surfaces. In *Soil Mineral–Organic Matter–Microorganism Interactions and Ecosystem Health, Developments in Soil Science, Vol. 28A* (Eds.) A. Violante, P.M. Huang, J.-M. Bollag, and L. Gianfreda, pp. 437–447. Amsterdam, the Netherlands: Elsevier.

Bobu, M., A. Yediler, I. Siminiceanu, and S. Schulte-Hostede. 2008. Degradation studies of ciprofloxacin on a pillared iron catalyst. *Applied Catalysis B: Environmental* 83: 15–23.

Boudali, L.K., A. Ghorbel, P. Grange, and F. Figueras. 2005. Selective catalytic reduction of NO with ammonia over V_2O_5 supported sulfated titanium-pillared clay catalysts: Influence of V_2O_5 content. *Applied Catalysis B: Environmental* 59: 105–111.

Boudjema, S., E. Vispe, A. Choukchou-Braham, J.A. Mayoral, R. Bachir, and J.M. Fraile. 2015. Preparation and characterization of activated montmorillonite clay supported 11-molybdo-vanado-phosphoric acid for cyclohexene oxidation. *RSC Advances* 5: 6853–6863.

Bouhlel, E., P. Laszlo, M. Levant, M-T. Montaufier, and G.P. Singh. 1993. Epoxidation of olefins by molecular oxygen with clay-impregnated nickel catalysts. *Tetrahedron Letters* 34: 1123–1126.

Breu, J. and C.R.A. Catlow. 1995. Chiral recognition among tris(diimine)-metal complexes. 4. Atomistic modeling of a monolayer of $[Ru(bpy)_3]^{2+}$ intercalated into a smectite clay. *Inorganic Chemistry* 34: 4504–4510.

Byung, H.H. and G.J. Dong. 1990. Montmorillonite catalyzed reduction of nitroarenes with hydrazine. *Tetrahedron Letters* 31: 1181–1182.

Cam, N.T.D., D.T. Phuong, H.V. Tai, N.D. Bang, and N.V. Noi. 2011. Degradation of phenol using the mixed (Al–Fe) pillared bentonite as a heterogeneous photo-Fenton catalyst. *e–Journal of Surface Science and Nanotechnology* 9: 490–493.

Carriazo, J.G., M.A. Centeno, J.A. Odriozola, S. Moreno, and R. Molina. 2007. Effect of Fe and Ce on Al-pillared bentonite and their performance in catalytic oxidation reactions. *Applied Catalysis A: General* 317: 120–128.

Carriazo, J.G., E. Guélou, J. Barrault, J.M. Tatibouët, R. Molina, and S. Moreno. 2005a. Synthesis of pillared clays containing Al, Al–Fe or Al–Ce–Fe from a bentonite: Characterization and catalytic activity. *Catalysis Today* 107–108: 126–132.

Carriazo, J.G., E. Guélou, J. Barrault, J.M. Tatibouët, R. Molina, and S. Moreno. 2005b. Catalytic wet peroxide oxidation of phenol by pillared clays containing Al–Ce–Fe. *Water Research* 39: 3891–3899.

Carriazo, J.G., E. Guélou, J. Barrault, J.M. Tatibouët, and S. Moreno. 2003. Catalytic wet peroxide oxidation of phenol over Al–Cu or Al–Fe modified clays. *Applied Clay Science* 22: 303–308.

Carriazo, J.G., M. Moreno-Forero, R.A. Molina, and S. Moreno. 2010. Incorporation of titanium and titanium-iron species inside a smectite-type mineral for photocatalysis. *Applied Clay Science* 50: 401–408.

Cativiéla, C., F. Figueras, J.M. Fraile, J.I. García, and J.A. Mayoral. 1993. Clay-catalysed asymmetric Diels–Alder reaction of cyclopentadiene with chiral acrylates. *Tetrahedron Asymmetry* 4: 223–228.

Catrinescu, C., D. Arsene, P. Apopei, and C. Teodosiu. 2012. Degradation of 4-chlorophenol from wastewater through heterogeneous Fenton and photo-Fenton process, catalyzed by Al–Fe PILC. *Applied Clay Science* 58: 96–101.

Catrinescu, C., D. Arsene, and C. Teodosiu. 2011. Catalytic wet hydrogen peroxide oxidation of *para*-chlorophenol over Al/Fe pillared clays (AlFePILC) prepared from different host clays. *Applied Catalysis B: Environmental* 101: 451–460.

Catrinescu, C. and C. Teodosiu. 2007. Wet hydrogen peroxide catalytic oxidation of para-chlorophenol over clay based catalysts. *Environmental Engineering and Management Journal* 6: 405–412.

Catrinescu, C., C. Teodosiu, M. Macoveanu, J. Miehe-Brendlé, and R. Le Dred. 2003. Catalytic wet peroxide oxidation of phenol over Fe-exchanged pillared beidellite. *Water Research* 37: 1154–1160.

Caudo, S., G. Centi, C. Genovese, and S. Perathoner. 2006. Homogeneous versus heterogeneous catalytic reactions to eliminate organics from waste water using H_2O_2. *Topics in Catalysis* 40: 207–219.

Caudo, S., G. Centi, C. Genovese, and S. Perathoner. 2007. Copper- and iron-pillared clay catalysts for the WHPCO of model and real wastewater streams from olive oil milling production. *Applied Catalysis B: Environmental* 70: 437–446.

Caudo, S., C. Genovese, S. Perathoner, and G. Centi. 2008. Copper-pillared clays (Cu-PILC) for agro-food wastewater purification with H_2O_2. *Microporous and Mesoporous Materials* 107: 46–57.

Chelba, A., D. Arsene, I. Morosanu, L. Tofan, and C. Teodosiu. 2014. Secondary municipal effluent treatment by catalytic wet hydrogen peroxide oxidation. *Environmental Engineering and Management Journal* 13: 2401–2410.

Chen, D., H. Zhu, and X. Wang. 2014. A facile method to synthesize the photocatalytic TiO_2/montmorillonite nanocomposites with enhanced photoactivity. *Applied Surface Science* 319: 158–166.

Chen, D., Q. Zhu, F. Zhou, X. Deng, and F. Li. 2012. Synthesis and photocatalytic performance of the TiO_2 pillared montmorillonite. *Journal of Hazardous Materials* 235–236: 186–193.

Chen, J. and L. Zhu. 2006. Catalytic degradation of Orange II by UV-Fenton with hydroxyl-Fe-pillared bentonite in water. *Chemosphere* 65: 1249–1255.

Chen, J. and L. Zhu. 2007. Heterogeneous UV-Fenton catalytic degradation of dyestuff in water with hydroxyl-Fe pillared bentonite. *Catalysis Today* 126: 463–470.

Chen, K., J. Li, W. Wang, Y. Zhang, X. Wang, and H. Su. 2011. The preparation of vanadium-doped TiO_2-montmorillonite nanocomposites and the photodegradation of sulforhodamine B under visible light irradiation. *Applied Surface Science* 257: 7276–7285.

Chen, Q., P. Wu, Z. Dang et al. 2010. Iron pillared vermiculite as a heterogeneous photo-Fenton catalyst for photocatalytic degradation of azo dye reactive brilliant orange X-GN. *Separation and Purification Technology* 71: 315–323.

Chen, S., Y. Wu, G. Li et al. 2017. A novel strategy for preparation of an effective and stable heterogeneous photo-Fenton catalyst for the degradation of dye. *Applied Clay Science* 136: 103–111.

Cheng, L.S., R.T. Yang, and N. Chen. 1996. Iron oxide and chromia supported on titania-pillared clay for selective catalytic reduction of nitric oxide with ammonia. *Journal of Catalysis* 164: 70–81.

Cheng, M., W. Song, W. Ma et al. 2008. Catalytic activity of iron species in layered clays for photodegradation of organic dyes under visible irradiation. *Applied Catalysis B: Environmental* 77: 355–363.

Cheng, S. 1999. From layer compounds to catalytic materials. *Catalysis Today* 49: 303–312.

Chirchi, L. and A. Ghorbel. 2002. Use of various Fe-modified montmorillonite samples for 4-nitrophenol degradation by H_2O_2. *Applied Clay Science* 21: 271–276.

Chmielarz, L., P. Kuśtrowski, R. Dziembaj, P. Cool, and E.F. Vansant. 2007. Selective catalytic reduction of NO with ammonia over porous clay heterostructures modified with copper and iron species. *Catalysis Today* 119: 181–186.

Chong, M.N., B. Jin, C.W.K. Chow, and C. Saint. 2010. Recent developments in photocatalytic water treatment technology: A review. *Water Research* 44: 2997–3027.

Chong, M.N., S. Lei, B. Jin, C. Saint, and C.W.K. Chow. 2009a. Optimisation of an annular photoreactor process for degradation of Congo Red using a newly synthesized titania impregnated kaolinite nano-photocatalyst. *Separation and Purification Technology* 67: 355–363.

Chong, M.N., V. Vimonses, S. Lei, B. Jin, C. Chow, and C. Saint. 2009b. Synthesis and characterisation of novel titania impregnated kaolinite nano-photocatalyst. *Microporous and Mesoporous Materials* 117: 233–242.

Choudary, B.M., S.S. Rani, and N. Narender. 1993. Asymmetric oxidation of sulfides to sulfoxides by chiral titanium pillared montmorillonite. *Catalysis Letters* 19: 299–307.

Choudary, B.M., V.L.K. Valli, and A.D. Prasad. 1990a. A new vanadium-pillared montmorillonite catalyst for the regioselective epoxidation of allylic alcohols. *Journal of the Chemical Society, Chemical Communications* 721–722.

Choudary, B.M., V.L.K. Valli, and A.D. Prasad. 1990b. An improved asymmetric epoxidation of allyl alcohols using titanium-pillared montmorillonite as a heterogeneous catalyst. *Journal of the Chemical Society, Chemical Communications* 1186–1187.

Churchman, G.J., W.P. Gates, B.K.G. Theng, and G. Yuan. 2006. Clays and clay minerals for pollution control. In *Handbook of Clay Science, Developments in Clay Science, Vol. 1* (Eds.) F. Bergaya, B.K.G. Theng, and G. Lagaly, pp. 625–675. Amsterdam, the Netherlands: Elsevier.

Cieśla, P., P. Kocot, P. Mytych, and Z. Stasicka. 2004. Homogeneous photocatalysis by transition metal complexes in the environment. *Journal of Molecular Catalysis A: Chemical* 224: 17–33.

Cornejo, A., J.M. Fraile, J.I. García et al. 2003. Surface-mediated improvement of enantioselectivity with clay-immobilized copper catalysts. *Journal of Molecular Catalysis A: Chemical* 196: 101–108.

Crocker, M., R.H.M. Herold, J.G. Buglass, and P. Companje. 1993. Preparation and characterization of montmorillonite-supported palladium hydrogenation catalysts possessing molecular sieving properties. *Journal of Catalysis* 141: 700–712.

Damardji, B., H. Khalaf, L. Duclaux, and B. David. 2009. Preparation of TiO_2-pillared montmorillonite as photocatalyst. Part II. Photocatalytic degradation of a textile azo dye. *Applied Clay Science* 45: 98–104.

Dashinamzhilova, E.Ts. and S.Ts. Khankhasaeva. 2011. Use of intercalated clays in oxidation of organic dyes. *Russian Journal of Applied Chemistry* 84: 1207–1212.

de Faria, E.H., G.P. Ricci, L. Marçal et al. 2012. Green and selective oxidation reactions catalyzed by kaolinite covalently grafted with Fe(III) pyridine-carboxylate complexes. *Catalysis Today* 187: 135–149.

Dědková, K., K. Matějová, J. Lang et al. 2014. Antibacterial activity of kaolinite/nanoTiO_2 composites in relation to irradiation time. *Journal of Photochemistry and Photobiology B: Biology* 135: 17–22.

De León, M.A., J. Castiglioni, J. Bussi, and M. Sergio. 2008. Catalytic activity of an iron-pillared montmorillonite clay mineral in heterogeneous photo-Fenton process. *Catalysis Today* 133–135: 600–605.

del Castillo, H.L., A. Gil, and P. Grange. 1996. Hydroxylation of phenol on titanium pillared montmorillonite. *Clays and Clay Minerals* 44: 706–709.

Detoni, C., F. Bertella, M.M.V.M. Souza, S.B.C. Pergher, and D.A.G. Aranda. 2014. Palladium supported on clys to catalytic deoxygenation of soybean fatty acids. *Applied Clay Science* 95: 388–395.

Dhakshinamoorthy, A. and K. Pitchumani. 2008. Clay entrapped nickel nanoparticles as efficient and recyclable catalysts for hydrogenation of olefins. *Tetrahedron Letters* 49: 1818–1823.

Dhanagopal, M., D. Divakar, R.A. Valentine, M.R. Viswanathan, and S. Thiripuranthagan. 2010. Nanosize noble metals intercalated in clay as catalysts for selective hydrogenation. *Chinese Journal of Catalysis* 31: 1200–1208.

Ding, X., T. An, G. Li et al. 2008. Preparation and characterization of hydrophobic TiO_2 pillared clay: The effect of acid hydrolysis catalyst and doped Pt amount on photocatlytic activity. *Journal of Colloid and Interface Science* 320: 501–507.

Ding, Z., H.Y. Zhu, G.Q. Lu, and P.F. Greenfield. 1999. Photocatalytic properties of titania pillared clays by different drying methods. *Journal of Colloid and Interface Science* 209: 193–199.

Divakar, D., D. Manikandan, G. Kalidoss, and T. Sivakumar. 2008. Hydrogenation of benzaldehyde over palladium intercalated bentonite catalysts: Kinetic studies. *Catalysis Letters* 125: 277–282.

Divakar, D., D. Manikandan, V. Rupa, E.L. Preethi, R. Chandrasekar, and T. Sivakumar. 2007. Palladium-nanoparticles intercalated vermiculite for selective hydrogenation of α,β-unsaturated aldehydes. *Journal of Chemical Technology and Biotechnology* 82: 253–258.

Dixit, P.S. and K. Srinivasan. 1988. The effect of clay-support on the catalytic epoxidation activity of a manganese(III)-Schiff base complex. *Inorganic Chemistry* 27: 4507–4509.

Djeffal, L., S. Abderrahmane, M. Benzina, M. Fourmentin, S. Siffert, and S. Fourmentin. 2014b. Efficient degradation of phenol using natural clay as heterogeneous Fenton-like catalyst. *Environmental Science and Pollution Research* 21: 3331–3338.

Djeffal, L., S. Abderrahmane, M. Benzina, S. Siffert, and S. Fourmentin. 2014a. Efficiency of natural clay as heterogeneous Fenton and photo-Fenton catalyst for phenol and tyrosol degradation. *Desalination and Water Treatment* 52: 2225–2230.

Djellabi, R., M.F. Ghorab, G. Cerrato et al. 2014. Photoactive TiO_2-montmorillonite composite for degradation of organic dyes in water. *Journal of Photochemistry and Photobiology A: Chemistry* 295: 57–63.

Dvininov, E., E. Popovici, R. Pode, L. Cocheci, P. Barvinschi, and V. Nica. 2009. Synthesis and characterization of TiO_2-pillared Romanian clay and their application for azoic dyes photodegradation. *Journal of Hazardous Materials* 167: 1050–1056.

Ebitani, K., M. Ide, T. Misudome, T. Mizugaki, and K. Kaneda. 2002. Creation of a chain-like cationic iron species in montmorillonite as a highly active heterogeneous catalyst for alkane oxygenations using hydrogen peroxide. *Chemical Communications* 690–691.

Ellias, N. and S. Sugunan. 2014. Wet peroxide oxidation of phenol over cerium impregnated aluminium and iron-aluminium pillared clays. *IOSR Journal of Applied Chemistry* 7: 80–85.

Ezabadi, A., G.R. Najafi, and M.M. Hashemi. 2008. A green and efficient oxidation of benzylic alcohols using H_2O_2 catalyzed by montmorillonite K-10 supported $CoCl_2$. *Chinese Chemical Letters* 19: 1277–1280.

Fabra, M.J., J.M. Fraile, C.I. Herrerías, F.J. Lahoz, J.A. Mayoral, and I. Pérez. 2008. Surface-enhanced stereoselectivity in Mukaiyama aldol reactions catalyzed by clay-supported bis(oxazoline)-copper complexes. *Chemical Communications* 5402–5404.

Fang, L., R. Hong, J. Gao, and C. Gu. 2016. Degradation of bisphenol A by nano-sized manganese dioxide synthesized using montmorillonite as template. *Applied Clay Science* 132–133: 155–160.

Faria, M., M. Matinelli, and G.K. Rolim. 2011. Immobilized molybdenum acetylacetonate complex on montmorillonite K-10 as catalyst for epoxidation of vegetable oils. *Applied Catalysis A: General* 403: 119–127.

Fatimah, I. 2012. Composite of TiO_2-montmorillonite from Indonesia and its photocatalytic properties in methylene blue and *E. coli* reduction. *Journal of Materials and Environmental Science* 3: 983–992.

Fatimah, I., S. Wang, and D. Wulandari. 2011. ZnO/montmorillonite for photocatalytic and photochemical degradation of methylene blue. *Applied Clay Science* 53: 553–560.

Feng, J., X. Hu, and P.L. Yue. 2004a. Novel bentonite clay-based Fe–nanocomposite as a heterogeneous catalyst for photo-Fenton discoloration and mineralization of Orange II. *Environmental Science & Technology* 38: 269–275.

Feng, J., X. Hu, and P.L. Yue. 2005. Discoloration and mineralization of Orange II by using a bentonite clay-based Fe nanocomposite film as a heterogeneous photo-Fenton catalyst. *Water Research* 39: 89–96.

Feng, J., X. Hu, and P.L. Yue. 2006. Effect of initial solution pH on the degradation of Orange II using clay-based Fe nanocomposites as heterogeneous photo-Fenton catalyst. *Water Research* 40: 641–646.

Feng, J., X. Hu, P.L. Yue, H.Y. Zhu, and G.Q. Lu. 2003b. Degradation of azo dye Orange II by a photoassisted Fenton reaction using a novel composite of iron oxide and silicate nanoparticles as a catalyst. *Industrial & Engineering Chemistry Research* 42: 2058–2066.

Feng, J., X. Hu, P.L. Yue, H.Y. Zhu, and G.Q. Lu. 2003c. Discoloration and mineralization of Reactive Red HE-3B by heterogeneous photo-Fenton reaction. *Water Research* 37: 3776–3784.

Feng, J., X. Hu, P.L. Yue, H.Y. Zhu, and G.Q. Lu. 2003a. A novel laponite clay-based Fe nanocomposite and its photo-catalytic activity in photo-assisted degradation of Orange II. *Chemical Engineering Science* 58: 679–685.

Feng, J., R.S.K. Wong, X. Hu, and P.L. Yue. 2004b. Discoloration and mineralization of Orange II by using Fe^{3+}-doped TiO_2 and bentonite clay-based Fe nanocatalysts. *Catalysis Today* 98: 441–446.

Fernández, A.I., J.M. Fraile, J.I. García, C.I. Herrerías, J.A. Mayoral, and L. Salvatella. 2001. Reversal of enantioselectivity by change of solvent with clay-immobilized bis(oxazoline)-copper catalysts. *Catalysis Communications* 2: 165–170.

Fernandes, C.I., C.D. Nunes, and P.D. Vaz. 2012. Clays in organic synthesis–Preparation and catalytic applications. *Current Organic Synthesis* 9: 670–694.

Fraile, J.M., J.I. García, M.A. Harmer et al. 2002. Immobilisation of bis(oxazoline)-copper complexes on clays and nanocomposites: Influence of different parameters on activity and selectivity. *Journal of Materials Chemistry* 12: 3290–3295.

Fraile, J.M., J.I. García, C.I. Herrerías, J.A. Mayoral, and E. Pires. 2009a. Enantioselective catalysis with chiral complexes immobilized on nanostructured supports. *Chemical Society Reviews* 38: 695–706.

Fraile, J.M., J.I. García, G. Lafuente, J.A. Mayoral, and L. Salvatella. 2004. Bis(oxazoline)-copper complexes, immobilized by electrostatic interactions, as catalysts for enantioselective aziridination. *Arkivoc* 67–73.

Fraile, J.M., J.I. García, J. Massam, and J.A. Mayoral. 1998b. Clay-supported non-chiral and chiral Mn(salen) complexes as catalysts for olefin epoxidation. *Journal of Molecular Catalysis A: Chemical* 136: 47–57.

Fraile, J.M., J.I. García, and J.A. Mayoral. 1996. Cyclopropanation reactions catalysed by copper(II)-exchanged clays and zeolites: Influence of the catalyst on the selectivity. *Chemical Communications* 1319–1320.

Fraile, J.M., J.I. García, and J.A. Mayoral. 2008a. Recent advances in the immobilization of chiral catalysts containing bis(oxazolines) and related ligands. *Coordination Chemistry Reviews* 252: 624–646.

Fraile, J.M., J.I. García, and J.A. Mayoral. 2009b. Noncovalent immobilization of enantioselective catalysts. *Chemical Reviews* 109: 360–417.

Fraile, J.M., J.I. García, J.A. Mayoral, and T. Tarnai. 1998a. Clay-supported bis(oxazoline)-copper complexes as heterogeneous catalysts of enantioselective cyclopropanation reactions. *Tetrahedron: Asymmetry* 9: 3997–4008.

Frini, N., M. Crespin, M. Trabelsi, D. Messad, H. Van Damme, and F. Bergaya. 1997. Preliminary results on the properties of pillared clays by mixed Al–Cu solutions. *Applied Clay Science* 12: 281–292.

Fujita, S., H. Sato, N. Kakegawa, and A. Yamagishi. 2006. Enantioselective photooxidation of a sulfide by a chiral ruthenium(II) complex immobilized on a montmorillonite clay surface: The role of weak interactions in asymmetric induction. *Journal of Physical Chemistry B* 110: 2533–2540.

Gajare, A.S., N.S. Shaik, G.K. Jnaneshwara, V.H. Deshpande, T. Ravindranathan, and A.V. Bedekar. 2000. Clay catalyzed conversion of isatoic anhydride to 2-(o-amino-phenyl)oxazolines. *Journal of the Chemical Society, Perkin Transactions 1* 999–1001.

Galeano, L.A., P.F. Bravo, C.D. Luna, M.A. Vicente, and A. Gil. 2012. Removal of natural organic matter for drinking water production by Al/Fe-PILC-catalyzed wet peroxide oxidation: Effect of the catalyst preparation from concentrated precursors. *Applied Catalysis B: Environmental* 111–112: 527–535.

Galeano, L.A., A. Gil, and M.A. Vicente. 2010. Effect of the atomic metal ratio in Al/Fe-, Al/Cu and Al/(Fe–Cu)-intercalation solutions on the physicochemical properties and catalytic activity of pillared clays in the CWPO of methyl orange. *Applied Catalysis B: Environmental* 100: 271–281.

Galeano, L.A., A. Gil, and M.A. Vicente. 2011a. Treatment of municipal leachate of landfill by Fenton-like heterogeneous catalytic wet peroxide oxidation using an Al/Fe-pillared montmorillonite as active catalyst. *Chemical Engineering Journal* 178: 146–153.

Galeano, L.A., A. Gil, and M.A. Vicente. 2011b. Strategies for immobilization of manganese on expanded natural clays: Catalytic activity in the CWPO of methyl orange. *Applied Catalysis B: Environmental* 104: 252–260.

Galeano, L.A., M.A. Vicente, and A. Gil. 2014. Catalytic degradation of organic pollutants in aqueous streams by mixed Al/M-pillared clays (M = Fe, Cu, Mn). *Catalysis Reviews* 56: 239–287.

Gao, H., B.-X. Zhao, J.-C. Luo et al. 2014. Fe–Ni–Al pillared montmorillonite as a heterogeneous catalyst for the catalytic wet peroxide oxidation degradation of Orange Acid II: Preparation condition and properties study. *Microporous and Mesoporous Materials* 196: 208–215.

Gao, X. and J. Xu. 2006. A new application of clay-supported vanadium oxide catalyst to selective hydroxylation of benzene to phenol. *Applied Clay Science* 33: 1–6.

Gao, Y., H. Gan, G. Zhang, and Y. Guo. 2013. Visible light assisted Fenton-like degradation of rhodamine B and 4-nitrophenol solutions with a stable poly-hydroxyl-iron/sepiolite catalyst. *Chemical Engineering Journal* 217: 221–230.

Garade, A.C., N.S. Biradar, S.M. Joshi, V.S. Kshirsagar, R.K. Jha, and C.V. Rode. 2011. Liquid phase oxidation of p-vanillyl alcohol over synthetic Co-saponite catalyst. *Applied Clay Science* 53: 157–163.

García, J.I., B. López-Sánchez, J.A. Mayoral, E. Pires, and I. Villalba. 2008. Surface confinement effects in enantioselective catalysis: Design of new heterogeneous chiral catalysts based on C_1-symmetric bisoxazolines and their application in cyclopropanation reactions. *Journal of Catalysis* 258: 378–385.

Garrido-Ramírez, E.G., M.V. Sivaiah, J. Barrault et al. 2012. Catalytic wet peroxide oxidation of phenol over iron or copper oxide-supported allophane clay materials: Influence of catalyst SiO_2/Al_2O_3 ratio. *Microporous and Mesoporous Materials* 162: 189–198.

Garrido-Ramírez, E.G., B.K.G. Theng, and M.L. Mora. 2010. Clays and oxide minerals as catalysts in Fenton-like reactions. *Applied Clay Science* 47: 182–192.

Giordano, G., S. Perathoner, G. Centi et al. 2007. Wet hydrogen peroxide catalytic oxidation of olive oil mill wastewater using Cu-zeolite and Cu-pillared clay catalysts. *Catalysis Today* 124: 240–246.

Gonzalez, E. and A. Moronta. 2004. The dehydrogenation of ethylbenzene to styrene catalyzed by a natural and Al-pillared clays impregnated with cobalt compounds: A comparative study. *Applied Catalysis A: General* 258: 99–105.

González-Bahamón, L.F., D.F. Hoyos, N. Benitez, and C. Pulgarín. 2011. New Fe-immobilized natural bentonite plate used as photo-Fenton catalyst for organic pollutant degradation. *Chemosphere* 82: 1185–1189.

Gorski, C.A., L.E. Klüpfel, A. Voegelin, M. Sander, and T.B. Hofstetter. 2013. Redox properties of structural Fe in clay minerals: 3. Relationships between smectite redox and structural properties. *Environmental Science & Technology* 47: 13477–13485.

Goto, T. and M. Ogawa. 2016. Efficient photocatalytic oxidation of benzene to phenol my metal complex-clay/TiO_2 hybrid photocatalyst. *RSC Advances* 6: 23794–23797.

Guélou, E., J. Barrault, J. Fournier, and J.-M. Tatibouët. 2003. Active iron species in the catalytic wet peroxide oxidation of phenol over pillared clays containing iron. *Applied Catalysis B: Environmental* 44: 1–8.

Guélou, E., J.-M. Tatibouët, and J. Barrault. 2010. Fe–Al pillared clays: Catalysts for wet peroxide oxidation of phenol. In *Pillared Clays and Related Catalysts* (Eds.) A. Gil, S.A. Korili, R. Trujillano, and M.A. Vicente, pp. 201–224. New York: Springer.

Guo, J. and M. Al-Dahhan. 2003a. Catalytic wet oxidation of phenol by hydrogen peroxide over pillared clay catalyst. *Industrial & Engineering Chemistry Research* 42: 2450–2460.

Guo, J. and M. Al-Dahhan. 2003b. Kinetics of wet air oxidation of phenol over a novel catalyst. *Industrial & Engineering Chemistry Research* 42: 5473–5481.

Guo, J. and M. Al-Dahhan. 2005. Catalytic wet air oxidation of phenol in concurrent downflow and upflow packed-bed reactors over pillared clay catalyst. *Chemical Engineering Science* 60: 735–746.

Guo, Z., R. Ma, and G. Li. 2006. Degradation of phenol by nanomaterial TiO_2 in wastewater. *Chemical Engineering Journal* 119: 55–59.

Guz, L., G. Curutchet, R.M. Torres Sánchez, and R. Candal. 2014. Adsorption of crystal violet on montmorillonite (or iron modified montmorillonite) followed by degradation through Fenton or photo-Fenton type reactions. *Journal of Environmental Chemical Engineering* 2: 2344–2351.

Hadjltaief, H.B., P. Da Costa, P. Beaunier, M.E. Gálvez, and M.B. Zina. 2014. Fe-clay-plate as a heterogeneous catalyst in photo-Fenton oxidation of phenol as probe molecule for water treatment. *Applied Clay Science* 91–92: 46–54.

Hadjltaief, H.B., P. Da Costa, M.E. Galvez, and M.B. Zina. 2013. Influence of operational parameters in the heterogeneous photo-Fenton discoloration of wastewaters in the presence of an iron-pillared clay. *Industrial & Engineering Chemistry Research* 52: 16656–16665.

Hadjltaief, H.B., A. Sdiri, W. Ltaief, P. Da Costa, M.E. Gálvez, and M.B. Zina. 2018. Efficient removal of cadmium and 2-chlorophenol in aqueous systems by natural clay: Adsorption and photo-Fenton degradation processes. *Comptes Rendus Chimie* 21: 253–262.

Hadjltaief, H.B., M.B. Zina, M.E. Galvez, and P. Da Costa. 2015. Photo-Fenton oxidation of phenol over a Cu-doped Fe-pillared clay. *Comptes Rendus Chimie* 18: 1161–1169.

Hadjltaief, H.B., M.B. Zina, M.E. Galvez, and P. Da Costa. 2016. Photocatalytic degradation of methyl green dye in aqueous solution over natural clay supported ZnO–TiO_2 catalysts. *Journal of Photochemistry and Photobiology A: Chemistry* 315: 25–33.

Hajjaji, W., S. Andrejkovičová, R.C. Pullar et al. 2016. Effective removal of anionic and cationic dyes by kaolinite and TiO_2/kaolinite composites. *Clay Minerals* 51: 19–27.

Hajjaji, W., S.O. Ganiyu, D.M. Tobaldi et al. 2013. Natural Portuguese clayey materials and derived TiO_2-containing composites used for decolouring methylene blue (MB) and orange II (OII) solutions. *Applied Clay Science* 83–84: 91–98.

Hara, N., S. Nakamura, N. Shibata, and T. Toru. 2010. Enantioselective aldol reaction using recyclable montmorillonite-entrapped N-(2-thiophenesulfonyl) prolinamide. *Advanced Synthesis & Catalysis* 352: 1621–1624.

Hartmann, M., S. Kullmann, and H. Keller. 2010. Wastewater treatment with heterogeneous Fenton-type catalysts based on porous materials. *Journal of Materials Chemistry* 20: 9002–9017.

Hashemi, M.M., B. Eftekhari-Sis, B. Khalili, and Z. Karimi-Jaberi. 2005. Solid-state oxidation of phenols to quinones with sodium perborate on wet montmorillonite K10. *Journal of the Brazilian Chemical Society* 16: 1082–1084.

Hassan, H. and B.H. Hameed. 2011. Fe-clay as effective heterogeneous Fenton catalyst for the decolorization of Reactive Blue 4. *Chemical Engineering Journal* 171: 912–918.

Henych, J. and V. Štengl. 2013. Feasible synthesis of TiO_2 deposited on kaolin for photocatalytic applications. *Clays and Clay Minerals* 61: 165–176.

Herney-Ramírez, J., C.A. Costa, L.M. Madeira et al. 2007. Fenton-like oxidation of Orange II solutions using heterogeneous catalysts based on saponite clay. *Applied Catalysis B: Environmental* 71: 44–56.

Herney-Ramírez, J., M. Lampinen, M.A. Vicente, C.A. Costa, and L.M. Madeira. 2008. Experimental design to optimize the oxidation of Orange II dye solution using a clay-based Fenton-like catalyst. *Industrial & Engineering Chemistry Research* 47: 284–294.

Herney-Ramírez, J. and L.M. Madeira. 2010. Use of pillared clay-based catalysts for wastewater treatment through Fenton-like processes. In *Pillared Clays and Related Catalysts* (Eds.) A. Gil, S.A. Korili, R. Trujillano, and M.A. Vicente, pp. 129–165. New York: Springer.

Herney-Ramírez, J., A.M.T. Silva, M.A. Vicente, C.A. Costa, and L.M. Madeira. 2011. Degradation of Acid Orange 7 using a saponite-based catalyst in wet hydrogen peroxide oxidation: Kinetic study with the Fermi's equation. *Applied Catalysis B: Environmental* 101: 197–205.

Herney-Ramírez, J., M.A. Vicente, and L.M. Madeira. 2010. Heterogeneous photo-Fenton oxidation with pillared clay-based catalysts for wastewater treatment: A review. *Applied Catalysis B: Environmental* 98: 10–26.

Huang, Q., S. Zuo, and R. Zhou. 2010. Catalytic performance of pillared interlayered clays (PILCs) supported CrCe catalysts for deep oxidation of nitrogen-containing VOCs. *Applied Catalysis B: Environmental* 95: 327–334.

Huerta, L., A. Meyer, and E. Choren. 2003. Synthesis, characterization, and catalytic application for ethylbenzene dehydrogenation of an iron-pillared clay. *Microporous and Mesoporous Materials* 57: 219–227.

Huo, C. and H. Yang. 2013. Preparation and enhanced photocatalytic activity of Pd-CuO/palygorskite nanocomposites. *Applied Clay Science* 74: 87–94.

Ibhadon, A.O. and P. Fitzpatrick. 2013. Heterogeneous photocatalysis: Recent advances and applications. *Catalysis* 3: 189–218.

Ilisz, I., A. Dombi, K. Mogyorósi, and I. Dékány. 2004. Photocatalytic water treatment with different TiO_2 nanoparticles and hydrophilic/hydrophobic layer silicate adsorbents. *Colloids and Surfaces A: Physicochemical and Engineering Aspects* 230: 89–97.

Iurascu, B., I. Siminiceanu, D. Vione, M.A. Vicente, and A. Gil. 2009. Phenol degradation in water through a heterogeneous photo-Fenton process catalyzed by Fe-treated laponite. *Water Research* 43: 1313–1322.

Iwai, M., H. Shoji, S. Shimazu, and T. Uematsu. 1993. Asymmetric recognition of hectorite modified with chiral arylethylammonium. *Chemistry Letters* 989–992.

Jagtap, N. and V. Ramaswamy. 2006. Oxidation of aniline over titania pillared montmorillonite clays. *Applied Clay Science* 33: 89–98.

Jia, H., J. Zhao, X. Fan, K. Dilimulati, and C. Wang. 2012. Photodegradation of phenanthrene on cation-modified clays under visible light. *Applied Catalysis B: Environmental* 123–124: 43–51.

Jiang, T., Y. Zhou, S. Liang, H. Liu, and B. Han. 2009. Hydrogenolysis of glycerol catalyzed by Ru-Cu bimetallic catalysts supported on clay with the aid of ionic liquids. *Green Chemistry* 11: 1000–1006.

Jin, M., M. Long, H. Su et al. 2017. Magnetically separable maghemite/montmorillonite composite as an efficient heterogeneous Fenton-like catalyst for phenol degradation. *Environmental Science and Pollution Research* 24: 1926–1937.

Jnaneshwara, G.K., V.H. Deshpande, and A.V. Bedekar. 1999. Clay-catalyzed conversion of 2,2-disubstituted malononitriles to 2-oxazolines: Towards unnatural amino acids. *Journal of Chemical Research (S)* 252–253.

Jnaneshwara, G.K., V.H. Deshpande, M. Lalithambika, T. Ravindranathan, and A.V. Bedekar. 1998. Natural kaolinitic clay catalyzed conversion of nitriles to 2-oxazolines. *Tetrahedron Letters* 39: 459–462.

Kameshima, Y., Y. Tamura, A. Nakajima, and K. Okada. 2009. Preparation and properties of TiO_2/montmorillonite composites. *Applied Clay Science* 45: 20–23.

Karamanis, D., A.N. Ökte, E. Vardoulakis, and T. Vaimakis. 2011. Water vapor adsorption and photocatalytic pollutant degradation with TiO_2-sepiolite nanocomposites. *Applied Clay Science* 53: 181–187.

Katagi, T. 1993. Photodegradation of esfenvalerate in clay suspensions. *Journal of Agricultural and Food Chemistry* 41: 2178–2183.

Kavurmaci, S.S. and M. Bekbolet. 2013. Photocatalytic degradation of humic acid in the presence of montmorillonite. *Applied Clay Science* 75–76: 60–66.

Kerkez, D.V., D.D. Tomašević, G. Kozma et al. 2014. Three different clay-supported nanoscale iron materials for industrial azo dye degradation: A comparative study. *Journal of the Taiwan Institute of Chemical Engineers* 45: 2451–2461.

Khanikar, N. and K.G. Bhattacharyya. 2013. Cu(II)-kaolinite and Cu(II)montmorillonite as catalysts for wet oxidative degradation of 2-chlorophenol, 4-chlorophenol and 2,4-dichlorophenol. *Chemical Engineering Journal* 233: 88–97.

Khankhasaeva, S.Ts., S.V. Badmaeva, and E.Ts. Dashinamzhilova. 2008. Preparation, characterization and catalytic application of Fe- and Fe/Al-pillared clays in the catalytic wet peroxide oxidation of 4-chlorophenol. *Studies in Surface Science and Catalysis* 174: 1311–1314.

Khankhasaeva, S.Ts., D.V. Dambueva, E.Ts. Dashinamzhilova, A. Gil, M.A. Vicente, and M.N. Timofeeva. 2015. Fenton degradation of sulfanilamide in the presence of Al, Fe-pillared clay: Catalytic behavior and identification of the intermediates. *Journal of Hazardous Materials* 293: 21–29.

Khankhasaeva, S.Ts., E.Ts. Dashinamzhilova, and D.V. Dambueva. 2017. Oxidative degradation of sulfanilamide catalyzed by Fe/Cu/Al-pillared clays. *Applied Clay Science* 146: 92–99.

Khankhasaeva, S.Ts., E.Ts. Dashinamzhilova, D.V. Dambueva, and M.N. Timofeeva. 2013. Catalytic properties of Fe–Cu–Al-montmorillonites in the oxidation of acid chrome dark blue azo dye. *Kinetics and Catalysis* 54: 307–313.

Khedher, I., A. Ghorbel, J.M. Fraile, and J.A. Mayoral. 2006. Vanadium sites in V-K10: Characterization and catalytic properties in liquid-phase sulfide oxidation. *Journal of Molecular Catalysis A: Chemical* 255. 92–96.

Kibanova, D., M. Trejo, H. Destaillats, and J. Cervini-Silva. 2009. Synthesis of hectorite-TiO_2 and kaolinite-TiO_2 nanocomposites with photocatalytic activity for the degradation of model air pollutants. *Applied Clay Science* 42: 563–568.

Király, Z., I. Dékány, A. Mastalir, and M. Bartók. 1996. In situ generation of palladium nanoparticles in smectite clays. *Journal of Catalysis* 161: 401–408.

Király, Z., B. Veisz, A. Mastalir, and Gy. Köfaragó. 2001. Preparation of ultrafine palladium particles on cationic and anionic clays, mediated by oppositely charged surfactants: Catalytic probes in hydrogenations. *Langmuir* 17: 5381–5387.

Kiss, E.F., J.G. Ranogajec, R.P. Marinković-Nedučin, and T.J. Vulić. 2003. Catalytic wet peroxide oxidation of phenol over AlFe-pillared montmorillonite. *Reaction Kinetics and Catalysis Letters* 80: 255–260.

Konstantinou, I.K. and T.A. Albanis. 2004. TiO_2-assisted photocatalytic degradation of azo dyes in aqueous solution: Kinetic and mechanistic investigations. A review. *Applied Catalysis B: Environmental* 49: 1–14.

Körösi, L., K. Mogyorósi, R. Kun, J. Németh, and I. Dékány. 2004. Preparation and photooxidation properties of metal oxide semiconductors incorporated in layer silicates. *Progress in Colloid and Polymer Science* 125: 27–33.

Kotkar, D. and N.V. Thakkar. 1995. Hydrogenation of olefins on Ru/pillared montmorillonite. *Proceedings of the Indian Academy of Sciences (Chemical Sciences)* 107: 39–42.

Kotkar, D. and N.V. Thakkar. 1997. Hydrogenation of cycloalkenes on Rh/montmorillonite. *Proceedings of the Indian Academy of Sciences (Chemical Sciences)* 109: 99–104.

Kshirsagar, V.S., A.C. Garade, K.R. Patil, R.K. Jha, and C.V. Rode. 2009. Heterogeneous cobalt-saponite catalyst for liquid phase air oxidation of *p*-cresol. *Industrial & Engineering Chemistry Research* 48: 9423–9427.

Kumar, K.V., K. Porkodi, and F. Rocha. 2008. Langmuir-Hinshelwood kinetics—A theoretical study. *Catalysis Communications* 9: 82–84.

Kun, R., K. Mogyorósi, and I. Dékány. 2006a. Synthesis and structural and photocatalytic properties of TiO_2/montmorillonite nanocomposites. *Applied Clay Science* 32: 99–110.

Kun, R., M. Szekeres, and I. Dékány. 2006b. Photooxidation of dichloroacetic acid controlled by pH-stat technique using TiO_2/layer silicate nanocomposites. *Applied Catalysis B: Environmental* 68: 49–58.

Kureshy, R.I., N.H. Khan, S.H.R. Abdi, I. Ahmad, S. Singh, and R.V. Jasra. 2003. Immobilization of dicationic Mn(III) salen in the interlayers of montmorillonite clay for enantioselective epoxidation of nonfunctionalized alkenes. *Catalysis Letters* 91: 207–210.

Kureshy, R.I., N.H. Khan, S.H.R. Abdi, I. Ahmad, S. Singh, and R.V. Jasra. 2004. Dicationic chiral Mn(III) salen complex exchanged in the interlayers of montmorillonite clay: A heterogeneous enantioselective catalyst for epoxidation of nonfunctionalized alkenes. *Journal of Catalysis* 221: 234–240.

Kurian, M. and R. Babu. 2013. Iron aluminium mixed pillared montmorillonite and the rare earth exchanged analogues as efficient catalyst for phenol oxidation. *Journal of Environmental Chemical Engineering* 1: 86–91.

Kurian, M., A. Eldhose, and R.M. Thasleenabi. 2012. Mild temperature oxidation of phenol over rare earth exchanged aluminum pillared montmorillonites. *International Journal of Environmental Research* 6: 669–676.

Kurian, M., M. Joy, and D. Raj. 2012. Hydroxylation of phenol over rare earth exchanged iron pillared montmorillonites. *Journal of Porous Materials* 19: 633–640.

Kurian, M. and S. Kavitha. 2016. A review on the importance of pillared interlayered clays in green chemical catalysis. *IOSR Journal of Applied Chemistry* 1: 47–54.

Kurian, M. and S. Sugunan. 2006. Wet peroxide oxidation of phenol over mixed pillared montmorillonites. *Chemical Engineering Journal* 115: 139–146.

Kutláková, K.M., J. Tokarský, P. Kovář et al. 2011. Preparation and characterization of photoactive composite kaolinite/TiO_2. *Journal of Hazardous Materials* 188: 212–220.

Kuźniarska-Biernacka, I., C. Pereira, A.P. Carvalho, J. Pires, and C. Freire. 2011. Epoxidation of olefins catalyzed by manganese(III) salen complexes grafted to porous heterostructured clays. *Applied Clay Science* 53: 195–203.

Kuźniarska-Biernacka, I., A.R. Silva, A.P. Carvalho, J. Pires, and C. Freire. 2010. Anchoring of chiral manganese(III) salen complex onto organoclay and porous clay heterostructure and catalytic activity in alkene epoxidation. *Catalysis Letters* 134: 63–71.

Latta, D.E., A. Neumann, W.A.P.J. Premaratne, and M.M. Scherer. 2017. Fe(II)–Fe(III) electron transfer in a clay mineral with low Fe content. *ACS Earth and Space Chemistry* 1: 197–208. doi:10.102/acsearthspacechem.7b00013.

Lee, S.M. and D. Tiwari. 2012. Organo and inorgano-organo-modified clays in the remediation of aqueous solutions: An overview. *Applied Clay Science* 59–60: 84–102.

Leitmannová, E. and L. Červený. 2010. Ruthenium complex immobilization on layered silicate materials. *Reaction Kinetics, Mechanisms and Catalysis* 99: 79–84.

Letaïef. S., B. Casal, P. Aranda, M.A. Martín-Luengo, and E. Ruiz-Hitzky. 2003. Fe-containing pillared clays as catalysts for phenol hydroxylation. *Applied Clay Science* 22: 263–277.

Li, C. 2004. Chiral synthesis on catalysts immobilized in microporous and mesoporous materials. *Catalysis Reviews* 46: 419–492.

Li, C., J. Wang, S. Feng, Z. Yang, and S. Ding. 2013a. Low-temperature synthesis of heterogeneous crystalline TiO_2-halloysite nanotubes and their visible light photocatalytic activity. *Journal of Materials Chemistry A* 1: 8045–8054.

Li, C., J. Wang, H. Guo, and S. Ding. 2015c. Low temperature synthesis of polyaniline-crystalline TiO_2-halloysite composite nanotubes with enhanced visible light photocatalytic activity. *Journal of Colloid and Interface Science* 458: 1–13.

Li, D., C. Li, and K. Suzuki. 2013b. Catalytic oxidation of VOCs over Al- and Fe-pillared montmorillonite. *Applied Clay Science* 77–78: 56–60.

Li, H., Y. Li, L. Xiang et al. 2015a. Heterogeneous photo-Fenton decolorization of Orange II over Al-pillared Fe-smectite: Response surface approach, degradation pathway, and toxicity evaluation. *Journal of Hazardous Materials* 287: 32–41.

Li, H., P. Wu, Z. Dang, N. Zhu, P. Li, and J. Wu. 2011. Synthesis, characterization, and visible-light photo-Fenton catalytic activity of hydroxyl Fe/Al-intercalated montmorillonite. *Clays and Clay Minerals* 59: 466–477.

Li, L., Z. Jiang, Z. Hao, X. Xu, and Y. Zhuang. 2005. Pillared laponite clays-supported palladium catalysts for the complete oxidation of benzene. *Journal of Molecular Catalysis A: Chemical* 225: 173–179.

Li, X., C. Yao, X. Lu, Z. Hu, Y. Yin, and C. Ni. 2015b. Halloysite–CeO_2–AgBr nanocomposite for solar light photodegradation of methyl orange. *Applied Clay Science* 104: 74–80.

Li, Y., Y. Lu, and X. Zhu. 2006. Photo-Fenton discoloration of the azo dye X-3B over pillared bentonites containing iron. *Journal of Hazardous Materials* B132: 196–201.

Liang, X., F. Qi, P. Liu et al. 2016. Performance of Ti-pillared montmorillonite supported Fe catalysts for toluene oxidation: The effect of Fe on catalytic activity. *Applied Clay Science* 132–133: 96–104.

Liotta, L.F., M. Gruttadauria, G. di Carlo, G. Perrini, and V. Librando. 2009. Heterogeneous catalytic degradation of phenolic substrates: Catalysts activity. *Journal of Hazardous Materials* 162: 588–606.

Liu, J.L. and G. Zhang. 2014. Recent advances in synthesis and applications of clay-based photocatalysts: A review. *Physical Chemistry Chemical Physics* 16: 8178–8192.

Liu, J., M. Dong, S. Zuo, and Y. Yu. 2009. Solvothermal preparation of TiO_2/montmorillonite and photocatalytic activity. *Applied Clay Science* 43: 156–159.

Liu, J., X. Li, S. Zuo, and Y. Yu. 2007. Preparation and photocatalytic activity of silver and TiO_2 nanoparticles/montmorillonite composites. *Applied Clay Science* 37: 275–280.

Liu, R., D. Xiao, Y. Guo, Z. Wang, and J. Liu. 2014. A novel photosensitized Fenton reaction catalysed by sandwiched iron in synthetic nontronite. *RSC Advances* 4: 12958–12963.

Liu, S., J.-H. Yang, and J.-H. Choy. 2006. Microporous SiO_2-TiO_2 nanosols pillared montmorillonite for photocatalytic decomposition of methyl orange. *Journal of Photochemistry and Photobiology A: Chemistry* 179: 75–80.

Liu, Y.X., X. Zhang, L. Guo, F. Wu, and N.S. Deng. 2008. Photodegradation of bisphenol A in the montmorillonite KSF suspended solutions. *Kinetics, Catalysis, and Reaction Engineering* 47: 7141–7146.

Liu, Y., X. Zhang, and F. Wu. 2010. Photodegradation of bisphenol AF in montmorillonite dispersions: Kinetics and mechanism study. *Applied Clay Science* 49: 182–186.

Long, R.Q. and R.T. Yang. 1999. Selective catalytic reduction of nitrogen oxides by ammonia over Fe^{3+}-exchanged TiO_2-pillared clay catalysts. *Journal of Catalysis* 186: 254–268.

Long, R.Q. and R.T. Yang. 2000a. FTIR and kinetic studies of the mechanism of Fe^{3+}-exchanged TiO_2-pillared clay catalyst for selective catalytic reduction of NO with ammonia. *Journal of Catalysis* 190: 22–31.

Long, R.Q. and R.T. Yang. 2000b. Catalytic performance and characterization of VO^{2+}-exchanged pillared clay catalyst for selective catalytic reduction of NO with ammonia. *Journal of Catalysis* 196: 73–85.

Louloudi, A. and N. Papayannakos. 2000. Hydrogenation of benzene on Ni/Al-pillared montmorillonite catalysts. *Applied Catalysis A: General* 204: 167–176.

Luo, M., D. Bowden, and P. Brimblecombe. 2009. Catalytic property of Fe–Al pillared clay for Fenton oxidation of phenol by H_2O_2. *Applied Catalysis B: Environmental* 85: 201–206.

Ma, J., Y. Jia, Y. Jing, J. Sun, and Y. Yao. 2009. Synthesis and photocatalytic activity of TiO_2-hectorite composites. *Applied Clay Science* 46: 114–116.

Mandelli, D., A.C.N. do Amaral, Y.N. Kozlov et al. 2009. Hydrogen peroxide oxygenation of saturated and unsaturated hydrocarbons catalyzed by montmorillonite and aluminum oxide. *Catalysis Letters* 132: 235–243.

Manikandan, D., D. Divakar, A.V. Rupa, S. Revathi, M.E.L. Preethi, and T. Sivakumar. 2007b. Synthesis of platinum nanoparticles in montmorillonite and their catalytic behaviour. *Applied Clay Science* 37: 193–200.

Manikandan, D., D. Divakar, and T. Sivakumar. 2007a. Utilization of clay minerals for developing Pt nanoparticles and their catalytic activity in the selective hydrogenation of cinnamaldehyde. *Catalysis Communications* 8: 1781–1786.

Manikandan, D., D. Divakar, and T. Sivakumar. 2008. Selective hydrogenation of citral over noble metals intercalated montmorillonite catalysts. *Catalysis Letters* 123: 197–114.

Manju, K. and S. Sugunana. 2005. Pillared clays as efficient catalysts for cyclohexanol oxidation. *Reaction Kinetics and Catalysis Letters* 85: 37–44.

Marín-Astorga, N., G. Alvez-Manoli, and P. Reyes. 2005. Stereoselective hydrogenation of phenyl alkyl acetylenes on pillared clays supported palladium catalysts. *Journal of Molecular Catalysis A: Chemical* 226: 81–88.

Mastalir, A., Gy. Szöllõsi, Z. Király, and Zs. Rázga. 2002. Preparation and characterization of platinum nanoparticles immobilized in dihydrocinchonidine-modified montmorillonite and hectorite. *Applied Clay Science* 22: 9–16.

Mastalir, A., Z. Király, and F. Berger. 2004. Comparative study of size-quantized Pd-montmorillonite catalysts in liquid-phase semihydrogenation of alkynes. *Applied Catalysis A: General* 269: 161–168.

Mastalir, A., Z. Király, Gy. Szöllösi, and M. Bartók. 2000. Preparation of organophilic Pd-montmorillonite, an efficient catalyst in alkyne semihydrogenation. *Journal of Catalysis* 194: 146–152.

Mastalir, A., Z. Király, Gy. Szöllõsi, and M. Bartók. 2001. Stereoselective hydrogenation of 1-phenyl1-1-pentyne over low-loaded Pd-montmorillonite catalysts. *Applied Catalysis A: General* 213: 133–140.

Mastalir, A., F. Notheisz, Z. Király, M. Bartók, and I. Dékány. 1997. Novel clay intercalated metal catalysts: A study of the hydrogenation of styrene and 1-octene on clay intercalated Pd catalysts. *Studies in Surface Science and Catalysis* 108: 477–484.

Mata, G., R. Trujillano, M.A. Vicente et al. 2009. (Z)-cyclooctene epoxidation and cyclohexane oxidation on Ni/alumina-pillared clay catalysts. *Microporous and Mesoporous Materials* 124: 218–226.

Mazzei, M., W. Marconi, and M. Riocci. 1980. Asymmetric hydrogenation of substituted acrylic acids by Rh'-aminophosphine chiral complex supported on mineral clays. *Journal of Molecular Catalysis* 9: 381–387.

McCabe, R.W. and J.M. Adams. 2013. Clay minerals as catalysts. In *Handbook of Clay Science, 2nd edition. Developments in Clay Science, Vol. 5B* (Eds.) F. Bergaya and G. Lagaly, pp. 491–538. Amsterdam, the Netherlands: Elsevier.

Mekawi, M.A., A.S. Darwish, M.S. Amin, Gh. Eshaq, and H.A. Bourazan. 2016. Copper nanoparticles supported onto montmorillonite clays as efficient catalyst for methylene blue dye degradation. *Egyptian Journal of Petroleum* 25: 269–279.

Ménesi, J., L. Kőrösi, E. Bazsó, V. Zöllmer, A. Richardt, and I. Dékány. 2008. Photocatalytic oxidation of organic pollutants on titania-clay composites. *Chemosphere* 70: 538–542.

Meng, X., Z. Qian, H. Wang, X. Gao, S. Zhang, and M. Yang. 2008. Sol-gel immobilization of SiO_2/TiO_2 on hydrophobic clay and its removal of methyl orange from water. *Journal of Sol-Gel Science and Technology* 46: 195–200.

Miao, S., Z. Liu, B. Han et al. 2006. Ru nanoparticles immobilized on montmorillonite by ionic liquids: A highly efficient heterogeneous catalyst for the hydrogenation of benzene. *Angewandte Chemie International Edition* 45: 266–269.

Miao, S., Z. Liu, Z. Zhang et al. 2007. Ionic liquid-assisted immobilization of Rh on attapulgite and its application in cyclohexene hydrogenation. *Journal of Physical Chemistry C* 111: 2185–2190.

Mishra, A., A. Mehta, M. Sharma, and S. Basu. 2017. Enhanced heterogeneous photodegradation of VOC and dye using microwave synthesized TiO_2/clay nanocomposites: A comparison study of different type of clays. *Journal of Alloys and Compounds* 694: 574–580.

Mitsudome, T., N. Nosaka, K. Mori, T. Mizugaki, K. Ebitani, and K. Kaneda. 2005. Liquid-phase epoxidation of alkenes using molecular oxygen catalyzed by vanadium cation-exchanged montmorillonite. *Chemistry Letters* 34: 1626–1627.

Mitsudome, T., K. Nose, T. Mizugaki, K. Jitsukawa, and K. Kaneda. 2008. Reusable montmorillonite-entrapped organocatalyst for asymmetric Diels–Alder reaction. *Tetrahedron Letters* 49: 5464–5466.

Miyoshi, H., H. Mori, and H. Yoneyama. 1991. Light-induced decomposition of saturated carboxylic acids on iron oxide incorporated clay suspended in aqueous solutions. *Langmuir* 7: 503–507.

Miyoshi, H. and H. Yoneyama. 1989. Photochemical properties of iron oxide incorporated in clay interlayers. *Journal of the Chemical Society, Faraday Transactions 1* 85: 1873–1880.

Mnasri-Ghnimi, S. and N. Frini-Srasra. 2016. Effect of Al and Ce on Zr-pillared bentonite and their performance in catalytic oxidation of phenol. *Russian Journal of Physical Chemistry A* 90: 1766–1773.

Mogyorósi, K., I. Dékány, and J.H. Fendler. 2003. Preparation and characterization of clay mineral intercalated titanium dioxide nanoparticles. *Langmuir* 19: 2938–2946.

Mogyorósi, K., A. Farkas, I. Dékány, I. Ilisz, and A. Dombi. 2002. TiO_2-based photocatalytic degradation of 2-chlorophenol adsorbed on hydrophobic clay. *Environmental Science & Technology* 36: 3618–3624.

Mogyorósi, K., J. Németh, I. Dékány, and J.H. Fendler. 2001. Preparation, characterization, and photocatalytic properties of layered-silicate-supported TiO_2 and ZnO nanoparticles. *Progress in Colloid and Polymer Science* 117: 88–93.

Mohajerani, M., M. Mehrvar, and F. Ein-Mozaffari. 2009. An overview of the integration of advanced oxidation technologies and other processes for water and wastewater treatment. *International Journal of Engineering* 3: 120–146.

Mojović, Z., P. Banković, A. Milutinović-Nikolić, J. Dostanić, N. Jović-Jovičić, and D. Jovanović. 2009. Al, Cu-pillared clays as catalysts in environmental protection. *Chemical Engineering Journal* 154: 149–155.

Molina, C.B., J.A. Casas, J.A. Zazo, and J.J. Rodríguez. 2006. A comparison of Al–Fe and Zr–Fe pillared clays for catalytic wet peroxide oxidation. *Chemical Engineering Journal* 118: 29–35.

Muthuvel, I., B. Krishnakumar, and M. Swaminathan. 2012. Novel Fe encapsulated montmorillonite K10 clay for photo-Fenton mineralization of Acid Yellow 17. *Indian Journal of Chemistry* 51A: 800–806.

Na, P., B. Zhao, L. Gu, J. Liu, and J. Na. 2009. Deep desulfurization of model gasoline over photoirradiated titanium-pillared montmorillonite. *Journal of Physics and Chemistry of Solids* 70: 1465–1470.

Nagendrappa, G. 2011. Organic synthesis using clay and clay-supported catalysts. *Applied Clay Science* 53: 106–138.

Najjar, W., L. Chirchi, E. Santos, and A. Ghorbel. 2001. Kinetic study of 2-nitrophenol photodegradation on Al-pillared montmorillonite doped with copper. *Journal of Environmental Monitoring* 3: 697–701.

Navalon, S., M. Alvaro, and H. Garcia. 2010. Heterogeneous Fenton catalysts based on clays, silicas and zeolites. *Applied Catalysis B: Environmental* 99: 1–26.

Nikalje, M.D., P. Phukan, and A. Sudalai. 2000. Recent advances in clay-catalyzed organic transformations. *Organic Preparations and Procedures International* 32: 1–40.

Nikalje, M.D. and A. Sudalai. 1999. Catalytic selective oxidation of alkyl arenes to aryl *tert*. butylperoxides with TBHP over Ru-exchanged montmorillonite K10. *Tetrahedron* 55: 5903–5908.

Nikolopoulos, A.N., O. Igglessi-Markopoulou, and N. Papayannakos. 2006. Ultrasound assisted catalytic wet peroxide oxidation of phenol: Kinetics and intraparticle diffusion effect. *Ultrasonics Sonochemistry* 13: 92–97.

Nishikiori, H., S. Matsunaga, N. Furuichi et al. 2017. Influence of allophane distribution on photocatalytic activity of allophane-titania composite films. *Applied Clay Science* 146: 43–49.

Nishimura, T., T. Onoue, K. Ohe, J-I. Tateiwa, and S. Uemura. 1998. Ruthenium(II)-bipyridine anchored montmorillonite-catalyzed oxidation of aromatic alkenes with *tert*-butyl hydroperoxide. *Tetrahedron Letters* 39: 4359–4362.

Nogueira, F.G.E., J.H. Lopes, A.C. Silva, R.M. Lago, J.D. Fabris, and L.C.A. Oliveira. 2011. Catalysts based on clay and iron oxide for oxidation of toluene. *Applied Clay Science* 51: 385–389.

Nowrouzi, F., A.N. Thadani, and R.A. Batey. 2009. Allylation and crotylation of ketones and aldehydes using potassium organotriflouroborate salts under Lewis acid and montmorillonite K10 catalyzed conditions. *Organic Letters* 11: 2631–2634.

Ökte, A.N., D. Tuncel, A.H. Pekcan, and T. Özden. 2014. Characteristics of iron-loaded TiO_2-supported montmorillonite catalysts: β-Naphthol degradation under UV-A irradiation. *Journal of Chemical Technology and Biotechnology* 89: 1155–1167.

Olaya, A., G. Blanco, S. Bernal, S. Moreno, and R. Molina. 2009. Synthesis of pillared clays with Al–Fe and Al–Fe–Ce starting from concentrated suspensions of clay using microwaves or ultrasound, and their catalytic activity in the phenol oxidation reaction. *Applied Catalysis B: Environmental* 93: 56–65.

Oliveira, L.C.A., R.M. Lago, J.D. Fabris, and K. Sapag. 2008. Catalytic oxidation of aromatic VOCs with Cr or Pd-impregnated Al-pillared bentonite: Byproduct formation and deactivation studies. *Applied Clay Science* 39: 218–222.

Onaka, M., K. Higuchi, H. Nanami, and Y. Izumi. 1993. Reduction of carbonyl compounds with hydrosilanes on solid acid and solid base. *Bulletin of the Chemical Society of Japan* 66: 2638–2645.

Ooka, C., H. Yoshida, K. Suzuki, and T. Hattori. 2003. Adsorption and photocatalytic degradation of toluene vapor in air on highly hydrophobic TiO_2 pillared clay. *Chemistry Letters* 32: 896–897.

Ooka, C., H. Yoshida, K. Suzuki, and T. Hattori. 2004. Highly hydrophobic TiO_2 pillared clay for photocatalytic degradation of organic compounds in water. *Microporous and Mesoporous Materials* 67: 143–150.

Ouidri, S., C. Guillard, V. Caps, and H. Khalaf. 2010. Epoxidation of olefins on photoirradiated TiO_2-pillared clays. *Applied Clay Science* 48: 431–437.

Ouidri, S. and H. Khalaf. 2009. Synthesis of benzaldehyde from toluene by a photocatalytic oxidation using TiO_2-pillared clays. *Journal of Photochemistry and Photobiology A: Chemistry* 207: 268–273.

Palombi, L., F. Bonadies, and A. Scettri. 1999. Aluminosilicate-catalyzed oxidation of alcohols by *t*-butyl hydroperoxide. *Journal of Molecular Catalysis A: Chemical* 140: 47–53.

Pan, J., J. Liu, S. Guo, and Z. Yang. 2009. Preparation of platinum/montmorillonite nanocomposites in supercritical methanol and their application in the hydrogenation of nitrobenzene. *Catalysis Letters* 131: 179–183.

Papoulis, D., S. Komarneni, A. Nikolopoulos et al. 2010. Palygorskite- and halloysite-TiO_2 nanocomposites: Synthesis and photocatalytic activity. *Applied Clay Science* 50: 118–124.

Papoulis, D., S. Komarneni, D. Panagiotaras et al. 2013a. Halloysite-TiO_2-nanocomposites: Synthesis, characterization and photocatalytic activity. *Applied Catalysis B: Environmental* 132–133: 416–422.

Papoulis, D., S. Komarneni, D. Panagiotaras et al. 2013b. Palygorskite-TiO_2 nanocomposites: Part 1. Synthesis and characterization. *Applied Clay Science* 83–84: 191–197.

Papoulis, D., S. Komarneni, D. Panagiotaras et al. 2013c. Palygorskite-TiO_2 nanocomposites: Part 2. Photocatalytic activities in decomposing air and organic pollutants. *Applied Clay Science* 83–84: 198–202.

Parida, K., G.B.B. Varadwaj, S. Sahu, and P.C. Sahoo. 2011. Schiff base Pt(II) complex intercalated montmorillonite: A robust catalyst for hydrogenation of aromatic nitro compounds at room temperature. *Industrial & Engineering Chemistry Research* 50: 7849–7856.

Patil, S.P., V.S. Shrivastava, G.H. Sonawane, and S.H. Sonawane. 2015. Synthesis of novel Bi_2O_3-montmorillonite nanocomposite with enhanced photocatalytic performance in dye degradation. *Journal of Environmental Chemical Engineering* 3: 2597–2603.

Paul, B., W.N. Martens, and R.L. Frost. 2012. Immobilized anatase on clay mineral particles as a photocatalyst for herbicides degradation. *Applied Clay Science* 57: 49–54.

Perathoner, S. and G. Centi. 2005. Wet hydrogen peroxide catalytic oxidation (WHPCO) of organic waste in agro-food and industrial streams. *Topics in Catalysis* 33: 207–224.

Perathoner, S. and G. Centi. 2010. Catalytic wastewater treatment using pillared clays. In: *Pillared Clays and Related Catalysts* (Eds.) A. Gil, S.A. Korili, R. Trujillano, and M.A. Vicente, pp. 167–200. New York: Springer.

Pereira, M.C., L.C. Oliveira, and E. Murad. 2012. Iron oxide catalysts: Fenton and Fenton-like reactions—A review. *Clay Minerals* 47: 285–302.

Pérez, A., M. Montes, R. Molina, and S. Moreno. 2014. Modified clays as catalysts for the catalytic oxidation of ethanol. *Applied Clay Science* 95: 18–24.

Pichat, P., H. Khalaf, D. Tabet, M. Houari, and M. Saidi. 2005. Ti-montmorillonite as photocatalyst to remove 4-chlorophenol in water and methanol in air. *Environmental Chemistry Letters* 2: 191–194.

Pignatello, J.J., E. Oliveros, and A. MacKay. 2006. Advanced oxidation processes for organic contaminant destruction based on the Fenton reaction and related chemistry. *Critical Reviews in Environmental Science and Technology* 36: 1–84.

Pillai, U.R. and E. Sahle-Demessie. 2003. Oxidation of alcohols over Fe^{3+}/montmorillonite-K10 using hydrogen peroxide. *Applied Catalysis A: General* 245: 103–109.

Platon, N., I. Siminiceanu, I.D. Nistor, N.D. Miron, G. Muntianu, and A.M. Mares. 2011. Fe-pillared clay as an efficient Fenton-like heterogeneous catalyst for phenol degradation. *Revista de Chimie* 62: 676–679.

Platon, N., I. Siminiceanu, I.D. Nistor et al. 2013. Catalytic wet oxidation of phenol with hydrogen peroxide over modified clay minerals. *Revista de Chimie* 64: 1459–1464.

Pliego, G., J.A. Zazo, P. Garcia-Muñoz, M. Munoz, J.A. Casas, and J.J. Rodriguez. 2015. Trends in the intensification of the Fenton process for wastewater treatment: An overview. *Critical Reviews in Environmental Science and Technology* 45: 2611–2692.

Polubesova, T., S. Eldad, and B. Chefetz. 2010. Adsorption and oxidative transformation of phenolic acids by Fe(III)-montmorillonite. *Environmental Science & Technology* 44: 4203–4209.

Primo, O., M.J. Rivero, and I. Ortiz. 2008. Photo-Fenton process as an efficient alternative to the treatment of landfill leachates. *Journal of Hazardous Materials* 153: 834–842.

Qin, C., D. Troya, C. Shang, S. Hildreth, R. Helm, and K. Xia. 2015. Surface catalyzed oxidative oligomerization of 17β-estradiol by Fe^{3+}-saturated montmorillonite. *Environmental Science & Technology* 49: 956–964.

Rodriguez-Narvaez, O.M., J.M. Peralta-Hernandez, A. Goonetilleke, and E.R. Bandala. 2017. Treatment technologies for emerging contaminants in water: A review. *Chemical Engineering Journal* 323: 361–380.

Romanenko, E.P., E.A. Taraban, and A.V. Tkachev. 2006. Catalytic oxidation of α-pinene with *tert*-butyl hydroperoxide in the presence of Fe-pillared montmorillonite. *Russian Chemical Bulletin* 55: 993–998.

Sahel, K., M. Bouhent, F. Belkhadem et al. 2014. Photocatalytic degradation of anionic and cationic dyes over TiO_2, P25, and Ti-pillared clays and Ag-doped Ti-pillared clays. *Applied Clay Science* 95: 205–210.

Sanabria, N., A. Álvarez, R. Molina, and S. Moreno. 2008. Synthesis of pillared bentonite starting from the Al–Fe polymeric precursor in solid state, and its catalytic evaluation in the phenol oxidation reaction. *Catalysis Today* 133–135: 530–533.

Sanabria, N., M.A. Centeno, R. Molina, and S. Moreno. 2009. Pillared clays with Al–Fe and Al–Ce–Fe in concentrated medium: Synthesis and catalytic activity. *Applied Catalysis A: General* 356: 243–249.

Sanabria, N.R., R. Molina, and S. Moreno. 2012. Development of pillared clays for wet hydrogen peroxide oxidation of phenol and its application in the posttreatment of coffee wastewater. *International Journal of Photoenergy* 1–17.

Sarmah, P.P. and D.K. Dutta. 2012. Chemoselective reduction of a nitro group through transfer hydrogenation catalysed by Ru^0-nanoparticles stabilized on modified montmorillonite clay. *Green Chemistry* 14: 1086–1093.

Sarmah, P.P. and D.K. Dutta. 2014. Stabilized Rh^0-nanoparticles-montmorillonite clay composite: Synthesis and catalytic transfer hydrogenation reaction. *Applied Catalysis A: General* 470: 355–360.

Sasai, R., R. Watanabe, and T. Yamada. 2014. Preparation and characterization of titania- and organo-pillared clay hybrid photocatalysts capable of oxidizing aqueous bisphenol A under visible light. *Applied Clay Science* 93–94: 72–77.

Sassi, H., G. Lafaye, H.B. Amor, A. Gannouni, M.R. Jeday, and J. Barbier, Jr. 2018. Wastewater treatment by catalytic wet air oxidation process over Al–Fe pillared clays synthesized using microwave irradiation. *Frontiers of Environmental Science & Engineering* 12: 2. doi:10.1007/s11783-017-0971-1.

Sato, H., A. Yamagishi, and S. Kato. 1992. Monte Carlo simulations of the interactions of metal complexes with the silicate sheets of clay: Comparison of binding states between tris(1,10-phenanthroline)metal(II) and tris(2,2′-bipyridyl)metal(II) chelates. *Journal of the American Chemical Society* 114: 10933–10940.

Scheuermann, G.M., R. Thomann, and R. Mülhaupt. 2009. Catalysts based upon organoclay with tunable polarity and dispersion behavior: New catalysts for hydrogenation, C–C coupling reactions and fluorous biphase catalysis. *Catalysis Letters* 132: 355–362.

Schoonheydt, R.A. 2016. Reflections on the material science of clay minerals. *Applied Clay Science* 131: 107–112.

Selvam, K. and M. Swaminathan. 2007. A green chemical synthesis of 2-alkylbenimidazoles from 1,2-phenylenediamine and propylene glycol, or alcohols mediated by Ag-TiO_2/clay composite photocatalyst. *Chemistry Letters* 36: 1060–1061.

Sento, T., S. Shimazu, N. Ichikumi, and T. Uematsu. 1999. Asymmetric hydrogenation of itaconates by hectorite-intercalated Rh-DIOP complex. *Journal of Molecular Catalysis A: Chemical* 137: 263–267.

Shahidi, D., A. Moheb, R. Abbas, S. Larouk, R. Roy, and A. Azzouz. 2015b. Total mineralization of sulfamethoxazole and aromatic pollutants through Fe^{2+}-montmorillonite catalyzed ozonation. *Journal of Hazardous Materials* 298: 338–350.

Shahidi, D., R. Roy, and A. Azzouz. 2015a. Advances in catalytic oxidation of organic pollutants—Prospects for thorough mineralization by natural clay catalysts. *Applied Catalysis B: Environmental* 174–175: 277–292.

Shao, G.N., M. Engole, S.M. Imran, S.J. Jeon, and H.T. Kim. 2015. Sol-gel synthesis of photoactive kaolinite-titania: Effect of the preparation method and their photocatalytic properties. *Applied Surface Science* 331: 98–107.

Shi, Y., Z. Yang, B. Wang, H. An, Z. Chen, and H. Cui. 2016. Adsorption and photocatalytic degradation of tetracycline hydrochloride using a palygorskite-supported Cu_2O–TiO_2 composite. *Applied Clay Science* 119: 311–320.

Shichi, T. and K. Takagi. 2000. Clay minerals as photochemical reaction fields. *Journal of Photochemistry and Photobiology C: Photochemistry Reviews* 1: 113–130.

Shimizu, K.-I., T. Kaneko, T. Fujishima, T. Kodama, H. Yoshida, and Y. Kitayama. 2002. Selective oxidation of liquid hydrocarbons over photoirradiated TiO_2 pillared clays. *Applied Catalysis A: General* 225: 185–191.

Shimazu, S., K. Ro, T. Sento, N. Ichikuni, and T. Uematsu. 1996. Asymmetric hydrogenation of α,β-unsaturated carboxylic acid esters by rhodium(I)-phosphine complexes supported on smectites. *Journal of Molecular Catalysis A: Chemical* 107: 297–303.

Sia, T.H., S. Dai, B. Jin, M. Biggs, and M.N. Chong. 2015. Hybridising nitrogen doped titania with kaolin: A feasible catalyst for a semi-continuous photo-degradation reactor system. *Chemical Engineering Journal* 279: 939–947.

Sidhpuria, K.B., H.A. Patel, P.A. Parikh, P. Bahadur, H.C. Bajaj, and R.V. Jasra. 2009. Rhodium nanoparticles intercalated into montmorillonite for hydrogenation of aromatic compounds in the presence of thiophene. *Applied Clay Science* 42: 386–390.

Silva, F.C., M.C.B.V. de Souza, V.F. Ferreira, S.J. Sabino, and O.A.C. Antunes. 2004. Natural clays as efficient catalysts for obtaining chiral β-enamino esters. *Catalysis Communications* 5: 151–155.

Sivakumar, T., T. Mori, J. Kubo, and Y. Morikawa. 2001. Selective hydrogenation of 2-methylbenzaldehyde using palladium particles generated in situ in surfactant exchanged fluorotetrasilicic mica. *Chemistry Letters* 30: 860–861.

Sobczyński, A., L. Duczmal, and W. Zmudziński. 2004. Phenol destruction by photocatalysis on TiO_2: An attempt to solve the reaction mechanism. *Journal of Molecular Catalysis A: Chemical* 213: 225–230.

Soni, V.K. and R.K. Sharma. 2016. Palladium-nanoparticles-intercalated montmorillonite clay: A green catalyst for the solvent-free chemoselective hydrogenation of squalene. *ChemCatChem* 8: 1763–1768.

Soni, V.K., P.R. Sharma, G. Choudhary, S. Pandey, and R.K. Sharma. 2017. Ni/Co-natural clay as green catalysts for microalgae oil to diesel-grade hydrocarbons conversion. *ACS Sustainable Chemistry & Engineering* 5: 5351–5359.

Srivastava, V. 2014. Ru-exchanged MMT clay with functionalized ionic liquid for selective hydrogenation of CO_2 to formic acid. *Catalysis Letters* 144: 2221–2226.

Stathatos, E., D. Papoulis, C.A. Aggelopoulos, D. Panagiotaras, and A. Nikolopoulos. 2012. TiO_2/palygorskite composite nanocrystalline films prepared by surfactant templating route: Synergistic effect to the photocatalytic degradation of an azo-dye in water. *Journal of Hazardous Materials* 211–212: 68–76.

Sterte, J. 1986. Synthesis and properties of titanium oxide cross-linked montmorillonite. *Clays and Clay Minerals* 34: 658–664.

Stucki, J.W. 2006. Properties and behaviour of iron in clay minerals. In *Handbook of Clay Science, Developments in Clay Science, Vol. 1* (Eds.) F. Bergaya, B.K.G. Theng, and G. Lagaly, pp. 423–475. Amsterdam, the Netherlands: Elsevier.

Suárez, S., J.M. Coronado, R. Portela et al. 2008. On the preparation of TiO_2-sepiolite hybrid materials for the photocatalytic degradation of TCE: Influence of TiO_2 distribution in the mineralization. *Environmental Science & Technology* 42: 5892–5896.

Sum, O.S.N., J. Feng, X. Hu, and P.L. Yue. 2004. Pillared laponite clay-based Fe nanocomposites as heterogeneous catalysts for photo-Fenton degradation of acid black 1. *Chemical Engineering Science* 59: 5269–5275.

Sum, O.S.N., J. Feng, X. Hu, and P.L. Yue. 2005. Photo-assisted fenton mineralization of an azo-dye acid black 1 using a modified laponite clay-based Fe nanocomposite as a heterogeneous catalyst. *Topics in Catalysis* 33: 233–242.

Sun, S., Y. Jiang, L. Yu et al. 2006. Enhanced photocatalytic activity of microwave treated TiO_2 pillared montmorillonite. *Materials Chemistry and Physics* 98: 377–381.

Sun, S.-P., X. Zeng, and A.T. Lemley. 2013. Nano-magnetite catalyzed heterogeneous Fenton-like degradation of emerging contaminants carbamazepine and ibuprofen in aqueous suspensions and montmorillonite clay slurries at neutral pH. *Journal of Molecular Catalysis A: Chemical* 371: 94–103.

Sun, Z., Y. Chen, Q. Ke, Y. Yang, and J. Yuan. 2002. Photocatalytic degradation of cationic azo dye by TiO_2/bentonite nanocomposite. *Journal of Photochemistry and Photobiology A: Chemistry* 149: 169–174.

Szczepanik, B. 2017. Photocatalytic degradation of organic contaminants over clay-TiO_2 nanocomposites: A review. *Applied Clay Science* 141: 227–239.

Szöllõsi, G., Á. Mastalir, Z. Király, and I. Dékány. 2005. Preparation of Pt nanoparticles in the presence of a chiral modifier and catalytic applications in chemoselective and asymmetric hydrogenations. *Journal of Materials Chemistry* 15: 2464–2469.

Szöllõsi, G., B. Török, L. Baranyi, and M. Bartók. 1998. Chemoselective hydrogenation of cinnamaldehyde to cinnamyl alcohol over Pt/K-10 catalyst. *Journal of Catalysis* 179: 619–623.

Tabet, D., M. Saidi, M. Houari, P. Pichat, and H. Khalaf. 2006. Fe-pillared clay as a Fenton-type heterogeneous catalyst for cinnamic acid degradation. *Journal of Environmental Management* 80: 342–346.

Tandiary, M.A., Y. Masui, and M. Onaka. 2014. Deoxygenation of tertiary and secondary alcohols into alkanes with triethylsilane catalyzed by solid acid tin(IV) ion-exchanged montmorillonite. *Tetrahedron Letters* 55: 4160–4162.

Tao, R., S. Miao, Z. Liu et al. 2009. Pd nanoparticles immobilized on sepiolite by ionic liquids: Efficient catalysts for hydrogenation of alkenes and Heck reactions. *Green Chemistry* 11: 96–101.

Tatibouët, J.M., E. Guélou, and J. Fournier. 2005. Catalytic oxidation of phenol by hydrogen peroxide over a pillared clay containing iron: Active species and pH effect. *Topics in Catalysis* 33: 225–232.

Thomas, J.M. and R. Raja. 2008. Exploiting nanospace for asymmetric catalysis: Confinement of immobilized, single-site chiral catalysts enhances enantioselectivity. *Accounts of Chemical Research* 41: 708–720.

Timofeeva, M.N. and S.Ts. Khankhasaeva. 2009. Regulating the physicochemical and catalytic properties of layered aluminosilicates. *Kinetics and Catalysis* 50: 57–64.

Timofeeva, M.N., S.Ts. Khankhasaeva, S.V. Badmaeva et al. 2005. Synthesis, characterization and catalytic application for wet oxidation of phenol of iron-containing clays. *Applied Catalysis B: Environmental* 59: 243–248.

Timofeeva, M.N., S.Ts. Khankhasaeva, E.P. Talsi et al. 2009. The effect of Fe/Cu ratio in the synthesis of mixed Fe, Cu, Al-clays used as catalysts in phenol peroxide oxidation. *Applied Catalysis B: Environmental* 90: 618–627.

Tirelli, A.A., I. do R. Guimarães, J.C. de Souza Terra, R.R. da Silva, and M.C. Guerreiro. 2015. Fenton-like processes and adsorption using iron-pillared clay with magnetic properties for organic compound mitigation. *Environmental Science and Pollution Research* 22: 870–881.

Tomul, F. 2016. The effect of ultrasound treatment on iron–chromium pillared bentonite synthesis and catalytic wet peroxide oxidation of phenol. *Applied Clay Science* 120: 121–134.

Tomul, F., F.T. Basoglu, and H. Canbay. 2016. Determination of adsorptive and catalytic properties of copper, silver and iron contain titanium-pillared bentonite for the removal bisphenol A from aqueous solution. *Applied Surface Science* 360: 579–593.

Török, B., K. Balázsik, I. Dékány, and M. Bartók. 2000. Preparation and characterization of chirally modified Laponites. *Molecular Crystals and Liquid Crystals* 341: 339–344.

Török, B., K. Balázsik, I. Kun, G. Szöllõsi, G. Szakonyi, and M. Bartók. 1999b. Homogeneous and heterogeneous asymmetric reactions. Part 13. Clay-supported noble metal catalysts in enantioselective hydrogenations. *Studies in Surface Science and Catalysis* 125: 515–522.

Török, B., M. Bartók, and I. Dékány. 1999a. The structure of chiral phenyl-ethylammonium montmorillonites in ethanol-toluene mixtures. *Colloid and Polymer Science* 277: 340–346.

Trujillano, R., E. Rico, M.A. Vicente et al. 2011. Rapid microwave-assisted synthesis of saponites and their use as oxidation catalysts. *Applied Clay Science* 53: 326–330.

Upadhyay, P. and V. Srivastava. 2015. Ruthenium nanoparticle-intercalated montmorillonite clay for solvent-free alkene hydrogenation reaction. *RSC Advances* 5: 740–745.

Usman, M., K. Hanna, and S. Haderlein. 2016. Fenton oxidation to remediate PAHs in contaminated soils: A critical review of major limitations and counter-strategies. *Science of the Total Environment* 569–570: 179–190.

Valverde, J.L., A. de Lucas, P. Sánchez, F. Dorado, and A. Romero. 2003. Cation exchanged and impregnated Ti-pillared clays for selective catalytic reduction of NO_x by propylene. *Applied Catalysis B: Environmental* 43: 43–56.

Van Damme, H., H. Nijs, and J.J. Fripiat. 1984. Photocatalytic reactions on clay surfaces. *Journal of Molecular Catalysis* 27: 123–142.

Vicente, I., P. Salagre, and Y. Cesteros. 2011. Ni nanoparticles supported on microwave-synthesised saponite for the hydrogenation of styrene oxide. *Applied Clay Science* 53: 212–219.

Vicente, M.A., R. Trujillano, K.J. Ciuffi, E.J. Nassar, S.A. Korili, and A. Gil. 2010. Pillared clay catalysts in green oxidation reactions. In: *Pillared Clays and Related Catalysts* (Eds.) A. Gil, S.A. Korili, R. Trujillano, and M.A. Vicente, pp. 301–318. New York: Springer.

Villemure, G. 1991. X-ray diffraction patterns of montmorillonite oriented films exchanged with enantiomeric and racemic tris(2,2'-bipyridyl)ruthenium(II). *Clays and Clay Minerals* 39: 580–585.

Vimonses, V., M.N. Chong, and B. Jin. 2010. Evaluation of the physical properties and photodegradation ability of titania nanocrystallline impregnated onto modified kaolin. *Microporous and Mesoporous Materials* 132: 201–209.

Virkutyte, J. and R.S. Varma. 2014. Eco-friendly magnetic iron oxide-pillared montmorillonite for advanced catalytic degradation of dichlorophenol. *ACS Sustainable Chemistry & Engineering* 2: 1545–1550.

Voudrias, E.A. and M. Reinhard. 1987. Abiotic organic reactions at mineral surfaces. In *Geochemical Processes at Mineral Surfaces, Chapter 22, ACS Symposium Series* 323: 462–486.

Wang, C., H. Shi, P. Zhang, and Y. Li. 2011a. Synthesis and characterization of kaolinite/TiO_2 nano-photocatalysts. *Applied Clay Science* 53: 646–649.

Wang, F., J. Zhang, C. Liu, and J. Liu. 2015. Pd-palygorskite catalysts: Preparation, characterization and catalytic performance for the oxidation of styrene. *Applied Clay Science* 105–106: 150–155.

Wang, R., G. Jiang, Y. Ding et al. 2011b. Photocatalytic activity of heterostructures based on TiO_2 and halloysite nanotubes. ACS *Applied Materials & Interfaces* 3: 4154–4158.

Wang, W., H. Liu, T. Wu et al. 2012. Ru catalyst supported on bentonite for partial hydrogenation of benzene to cyclohexene. *Journal of Molecular Catalysis A: Chemical* 355: 174–179.

Wei, X., H. Wu, G. He, and Y. Guan. 2017. Efficient degradation of phenol using iron-montmorillonite as a Fenton catalyst: Importance of visible light irradiation and intermediates. *Journal of Hazardous Materials* 321: 408–416.

Wu, F., J. Li, Z. Peng, and N. Deng. 2008. Photochemical formation of hydroxyl radicals catalyzed by montmorillonite. *Chemosphere* 72: 407–413.

Xiong, Z., Y. Xu, L. Zhu, and J. Zhao. 2005a. Enhanced photodegradation of 2,4,6-trichlorophenol over palladium phthalocyaninesulfonate modified organobentonite. *Langmuir* 21: 10602–10607.

Xiong, Z., Y. Xu, L. Zhu, and J. Zhao. 2005b. Photosensitized oxidation of substituted phenols on aluminum phthalocyanine-intercalated organoclay. *Environmental Science & Technology* 39: 651–657.

Xu, A., M. Yang, H. Yao, H. Du, and C. Sun. 2009. Rectorite as catalyst for wet air oxidation of phenol. *Applied Clay Science* 43: 435–438.

Xu, H.-Y., X.-L. He, Z. Wu, L.-W. Shan, and W.-D. Zhang. 2009. Iron-loaded natural clay as heterogeneous catalyst for Fenton-like discoloration of dyeing wastewater. *Bulletin of the Korean Chemical Society* 30: 2249–2252.

Xu, H., K. Wang, H. Zhang, L. Hao, J. Xu, and Z. Liu. 2014. Ionic liquid modified montmorillonite-supported Ru nanoparticles: Highly efficient heterogeneous catalysts for the hydrodeoxygenation of phenolic compounds to cycloalkanes. *Catalysis Science & Technology* 4: 2658–2663.

Yadav, J.S., B.V.S. Reddy, S. Meraj, V. Vishnumurthy, K. Narsimulu, and A.C. Kunwar. 2006. Montmorillonite clay catalyzed synthesis of enantiomerically pure 1,2,3,4-tetrahydroquinolines. *Synthesis* 2923–2926.

Yadav, J.S., B.V.S. Reddy, M. Srinivas, A. Prabhakar, and B. Jagadeesh. 2004. Montmorillonite KSF clay-promoted synthesis of enantiomerically pure 5-substituted pyrrazoles from 2,3-dihydro-4H-pyran-4-ones. *Tetrahedron Letters* 45: 6033–6036.

Yamagishi, A. 1982. Racemic adsorption of iron(II) tris(1,10-phenanthroline) chelate on a colloidal clay. *Journal of Physical Chemistry* 86: 2472–2479.

Yamagishi, A. 1983. Clay as a medium for optical resolution: Chromatographic resolution of 2,3-dihydro-2-methyl-5,6-diphenylpyrazine on a Λ-[Ru(phen)$_3$]$^{2+}$ (phen = 1,10-phenanthroline) montmorillonite column. *Journal of the Chemical Society, Chemical Communications* 9–10.

Yamagishi, A. 1986. Template effects of a clay-chelate adduct on the asymmetric oxidation of alkyl phenyl sulphide by sodium metaperiodate. *Journal of the Chemical Society, Chemical Communications* 290–291.

Yamagishi, A. 1987. Optical resolution and asymmetric syntheses by use of adsorption on clay minerals. *Journal of Coordination Chemistry* 16: 131–211.

Yamagishi, A. and H. Sato. 2012. Stereochemistry and molecular recognition on the surface of a smectite clay mineral. *Clays and Clay Minerals* 60: 411–419.

Yamagishi, A. and M. Soma. 1981. Optical resolution of metal chelates by use of adsorption on a colloidal clay. *Journal of the American Chemical Society* 103: 4640–4642.

Yamagishi, A., M. Taniguchi, Y. Imamura, and H. Sato. 1996. Clay column chromatography for optical resolution: Selectivities of Λ-[Ru(phen)$_3$]$^{2+}$- and Λ-[Ru(bpy)$_3$]$^{2+}$ laponite columns toward 1,1'-binaphthol. *Applied Clay Science* 11: 1–10.

Yamanaka, S., T. Nishihara, and M. Hattori. 1987. Preparation and properties of titania pillared clay. *Materials Chemistry and Physics* 17: 87–101.

Yang, J.-H., H. Piao, A. Vinu et al. 2015. TiO$_2$-pillared clays with well-ordered porous structure and excellent photocatalytic activity. *RSC Advances* 5: 8210–8215.

Yang, R.T., N. Tharappiwattananon, and R.Q. Long. 1998. Ion-exchanged pillared clays for selective catalytic reduction of NO by ethylene in the presence of oxygen. *Applied Catalysis B: Environmental* 19: 289–304.

Yang, S., G. Liang, A. Gu, and H. Mao. 2013a. Synthesis of mesoporous iron-incorporated silica-pillared clay and catalytic performance for phenol hydroxylation. *Applied Surface Science* 285P: 721–726.

Yang, S., G. Liang, A. Gu, and H. Mao. 2013b. Synthesis of TiO$_2$ pillared montmorillonite with ordered interlayer mesopores structure and high photocatalytic activity by an intra-gallery templating method. *Materials Research Bulletin* 48: 3948–3954.

Yang, X., X. Ke, D. Yang et al. 2010. Effect of ethanol washing of titania clay mineral composites on photocatalysis for phenol decomposition. *Applied Clay Science* 49: 44–50.

Yang, X., D. Yang, H. Zhu et al. 2009. Mesoporous structure with size controllable anatase attached on silicate layers for efficient photocatalysis. *Journal of Physical Chemistry C* 113: 8243–8248.

Yang, X., H. Zhu, J. Liu et al. 2008. A mesoporous structure for efficient photocatalysts: Anatase nanocrystals attached to leached clay layers. *Microporous and Mesoporous Materials* 112: 32–44.

Ye, W., B. Zhao, H. Gao, J. Huang, and X. Zhang. 2016. Preparation of highly efficient and stable Fe, Zn, Al-pillared montmorillonite as heterogeneous catalyst for catalytic wet peroxide oxidation of Orange II. *Journal of Porous Materials* 23: 301–310.

Yip, A.C.-K., F.L.-Y. Lam, and X. Hu. 2005. A novel heterogeneous acid-activated clay supported copper catalyst for the photobleaching and degradation of textile organic pollutant using photo-Fenton-like reaction. *Chemical Communication* 3218–3220.

Yoneyama, H., S. Haga, and S. Yamanaka. 1989. Photocatalytic activities of microcrystalline titania incorporated in sheet silicates of clay. *Journal of Physical Chemistry* 93: 4833–4837.

Yuan, L., D. Huang, W. Guo, Q. Yang, and J. Yu. 2011. TiO$_2$/montmorillonite nanocomposites for removal of organic pollutant. *Applied Clay Science* 53: 272–278.

Yuan, P., X. Yin, H. He, D. Yang, L. Wang, and J. Zhu. 2006. Investigation on the delaminated-pillared structure of TiO$_2$-PILC synthesized by TiCl$_4$ hydrolysis method. *Microporous and Mesoporous Materials* 93: 240–247.

Zazo, J.A., J.A. Casas, A.F. Mohedano, and J.J. Rodríguez. 2006. Catalytic wet peroxide oxidation of phenol with a Fe/active carbon catalyst. *Applied Catalysis B: Environmental* 65: 261–268.

Zhang, G., Y. Gao, Y. Zhang, and Y. Guo. 2010. Fe$_2$O$_3$-pillared rectorite as an efficient and stable Fenton-like heterogeneous catalyst for photodegradation of organic contaminants. *Environmental Science & Technology* 44: 6384–6389.

Zhang, G., X. Ding, F.S. He et al. 2008b. Low-temperature synthesis and photocatalytic activity of TiO$_2$ pillared montmorillonite. *Langmuir* 24: 1026–1030.

Zhang, G., X. Ding, Y. Hu et al. 2008a. Photocatalytic degradation of 4BS dye by N, S-codoped TiO$_2$ pillared montmorillonite photocatalysts under visible-light irradiation. *Journal of Physical Chemistry C* 112: 17994–17997.

Zhang, L., J. Liu, C. Tang et al. 2011e. Palygorskite and SnO$_2$–TiO$_2$ for the photodegradation of phenol. *Applied Clay Science* 51: 68–73.

Zhang, L., F. Lv, W. Zhang et al. 2009. Photo degradation of methyl orange by attapulgite–SnO$_2$–TiO$_2$ nanocomposites. *Journal of Hazardous Materials* 171: 294–300.

Zhang, S., S. Liang, X. Wang, J. Long, Z. Li, and L. Wu. 2011d. Trinuclear iron cluster intercalated montmorillonite catalyst: Microstructure and photo-Fenton performance. *Catalysis Today* 175: 362–369.

Zhang, X., J. Li, X. Lu, C. Tang, and G. Lu. 2012. Visible light induced CO$_2$ reduction and Rh B decolorization over electrostatic-assembled AgBr/palygorskite. *Journal of Colloid and Interface Science* 377: 277–283.

Zhang, Y., H. Gan, and G. Zhang. 2011a. A novel mixed-phase TiO$_2$/kaolinite composites and their photocatalytic activity for degradation of organic contaminants. *Chemical Engineering Journal* 172: 936–943.

Zhang, Y., Y. Guo, G. Zhang, and Y. Gao. 2011c. Stable TiO$_2$/rectorite: Preparation, characterization and photocatalytic activity. *Applied Clay Science* 51: 335–340.

Zhang, Y., D. Wang, and G. Zhang. 2011b. Photocatalytic degradation of organic contaminants by TiO_2/sepiolite composites prepared at low temperature. *Chemical Engineering Journal* 173: 1–10.

Zhao, D., J. Zhou, and N. Liu. 2006. Characterization of the structure and catalytic activity of copper modified palygorskite/TiO_2 (Cu^{2+}-PG/TiO_2) catalysts. *Materials Science and Engineering A* 431: 256–262.

Zhao, D., J. Zhou, and N. Liu. 2007. Surface characteristics and photoactivity of silver-modified palygorskite clays coated with nanosized titanium dioxide particles. *Materials Characterization* 58: 249–255.

Zhao, X., X. Quan, S. Chen, H.-M. Zhao, and Y. Liu. 2007. Photocatalytic remediation of γ-hexachlorocyclohexane contaminated soils using TiO_2 and montmorillonite composite photocatalyst. *Journal of Environmental Sciences* 19: 358–361.

Zheng, P., Y. Du, P.R. Zhang, and X. Ma. 2015. Amylose-halloysite-TiO_2 composites: Preparation, characterization and photodegradation. *Applied Surface Sciene* 329: 256–261.

Zhou, F., C. Yan, H. Wang, S. Zhou, and S. Komarneni. 2017. Sepiolite-TiO_2 nanocomposites for photocatalysis: Synthesis by microwave hydrothermal treatment versus calcination. *Applied Clay Science* 146: 246–253.

Zhou, S., C. Zhang, X. Hu et al. 2014. Catalytic wet peroxide oxidation of 4-chlorophenol over Al–Fe-, Al–Cu-, and Al–Fe–Cu-pillared clays: Sensitivity, kinetics and mechanism. *Applied Clay Science* 95: 275–283.

Zhou, S., C. Zhang, R. Xu, C. Gu, Z. Song, and M. Xu. 2016. Chloride ions promoted the catalytic wet peroxide oxidation of phenol over clay-base catalysts. *Water Science & Technology* 73: 1025–1032.

Zhu, H.Y., J.-Y. Li, J.C. Zhao, and G.J. Churchman. 2005. Photocatalysts prepared from layered clays and titanium hydrate for degradation of organic pollutants in water. *Applied Clay Science* 28: 79–88.

Zhu, R., Q. Chen, Q. Zhou, Y. Xi., J. Zhu, and H. He. 2016. Adsorbents based on montmorillonite for contaminant removal from water: A review. *Applied Clay Science* 123: 239–258.

Zuo, S., M. Ding, J. Tong, L. Feng, and C. Qi. 2015. Study on the preparation and characterization of a titanium-pillared clay-supported CrCe catalyst and its application to the degradation of a low concentration of chlorobenzene. *Applied Clay Science* 105–106: 118–123.

Zurita, M.A., A. Avila, R.A. Spanevello, A.G. Suárez, and A. Sarotti. 2015. Montmorillonite K-10 promoted synthesis of chiral dioxa-caged compounds derived from levoglucosenone. *Carbohydrate Research* 402: 67–70.

8 Clay Mineral Catalysis of Natural Processes and Prebiotic Organic Reactions

8.1 INTRODUCTION

Clay minerals are known to play an important role in *geocatalysis*, which in a broad sense is concerned with heterogeneous catalysis at mineral surfaces (Schoonen et al. 1998). Here we use the term in a more restricted sense to denote the role of clay minerals in catalyzing natural processes, with particular reference to the (abiotic) transformations of organic compounds in sediments and soils (Fripiat and Cruz-Cumplido 1974; Theng 1974; Mayer 1994; Pal et al. 1994; Birkel et al. 2002; Booth et al. 2004). In the case of sediments, geocatalysis is commonly discussed in relation to hydrocarbon (petroleum) formation, conversion, and migration (Galwey 1969a, 1972; Johns and Shimoyama 1972; Goldstein 1982, 1983; Wu et al. 2012). Although some mineral-catalyzed organic conversions proceed at a very low rate under ambient environmental temperatures, these processes are important from a geochemical perspective (Voudrias and Reinhard 1987). This chapter also describes the possible role of clay minerals in mediating and promoting prebiotic organic reactions (Katchalsky 1973; Theng 1974; Paecht-Horowitz 1976, 1977; Negrón-Mendoza et al. 1996; Ferris et al. 2004; Ferris 2005; Brack 2006; Maurel and Leclerc 2016).

8.2 CLAYS AND CLAY MINERALS AS GEOCATALYSTS

8.2.1 Hydrocarbon Cracking

The cracking of petroleum in the presence of clays and clay minerals is a well-known example of a geocatalytic (degradation) reaction. Normal paraffins may be cracked through either a thermal or a catalytic mechanism (Brooks 1948, 1950; Eisma and Jurg 1969). The former process involves passing preheated oil under pressure through a stationary bed of granular clay (commonly palygorskite) at ~500°C (Robertson 1948). The reaction is initiated by the loss of a hydrogen atom from the hydrocarbon molecule. Cracking occurs at the C–C bond in the β position to the carbon atom lacking the hydrogen atom, yielding a primary radical and an α-olefin. Thus, the major products of thermal cracking consist of normal hydrocarbons, α-olefins, and ethylene. These findings are consistent with the model described by Xiao et al. (1997) using an *ab initio* approach.

The thermal process has largely been replaced by catalytic cracking using solid acids, such as (acid-activated) clays, pillared interlayered clays, and synthetic zeolites, giving rise to improved efficiency (Haresnape 1948; Thomas et al. 1950; Hansford 1952; Milliken et al. 1955; Voge 1958; Occelli 1983). For example, Goldstein (1983) reported that the activation energy, derived from the Arrhenius plot (Equation 8.2) for the acid-catalyzed cracking of n-hexadecane was 1.7 times lower than the value obtained for thermal cracking.

The Houdry catalytic cracking process (originally using lignite as feedstock) was carried out in vessels containing the solid acid catalyst and the preheated petroleum, in the presence of a kaolinite-supported nickel. This catalyst was later (in 1927) replaced by acid-activated fuller's earth (Hettinger 1991). Synthetic silica-alumina, followed by halloysite- and kaolinite-based catalysts and zeolites were developed during and after World War II (Heller-Kallai 2002).

Modern petroleum cracking is still the largest user of kaolin as catalyst and catalyst support/matrix (Rong and Xiao 2002; Emam 2013). Irrespective of the type of solid acid, however, the catalyst needs to be regularly regenerated by burning off the accumulated coke in the air at a high temperature (Mosely, 1984).

Since the early 1980s, zeolite ZSM-5 has been widely used as an additive because of its efficiency in converting petroleum to light (C_3–C_4) olefins and suppressing hydrogen transfer reactions (Buchanan 2000; den Hollander et al. 2002). For example, a ZSM-5/rectorite composite, formed by depositing 2–3 µm particles of the zeolite on the surface of a calcined rectorite, could appreciably enhance the yield of, and selectivity to, propylene (Wei et al. 2010; Liu et al. 2012b). Earlier, Min (1994) prepared a pillared rectorite and a palladium-impregnated Al-PILC as hydrocracking catalysts. Volkova et al. (2014) reported that a zeolite-impregnated alumina pillared montmorillonite (Al-PILC) was highly efficient in cracking heavy vacuum gas oils, whereas an acid-activated kaolinite mainly yielded a light cycle oil. The chromia and tin-oxide pillared montmorillonites and laponites, prepared by Gyftopoulou et al. (2005), were similarly efficient in mediating the hydrocracking of heavy liquid fuels derived from coal. Despite appreciable coke deposition on their surface, the catalysts remained active after 4 h of use. An acid-treated sepiolite that was highly resistant to heavy metal poisoning was used by Zheng et al. (2009) as a fluid cracking catalyst. Much of the early literature on the application of clays to the hydrotreatment of heavy crude oils has been summarized by Rosa-Brussin (1995).

The Houdry process has been superseded by that of fluid catalytic cracking (FCC) in which the preheated heavy gas oil feedstock is brought into contact with the hot fluidized catalyst (in the feed *riser* or reactor), causing the hydrocarbons to vaporize and then crack into low molecular weight alkenes, branched alkenes, and aromatics (Occelli 1983; Occelli et al. 1984; Corma 1989; Biswas and Maxwell 1990). As the largest extant industrial chemical process, FCC is apparently initiated by abstraction of a hydride ion from the *n*-paraffin molecule. The corresponding carbocation produced then undergoes a number of rearrangements and transformations such as β-scission, hydrogen and methyl group shift/transfer, and hydride abstraction (Hattori 2010; Vogt and Weckhuysen 2015).

Likewise, the cracking of vegetable oils (to yield biofuels) by pillared interlayered clays appears to involve β-scission and hydrogen transfer reactions (Kloprogge et al. 2005). Using a ruthenium-impregnated Al-PILC as catalyst, Liu et al. (2012a) were able to convert waste cooking oil to biodiesel through the cracking of C_{15}–C_{18} *n*-paraffins to iso-paraffins and light alkanes. More recently, Soni et al. (2017) used a nickel- or cobalt-impregnated natural clay to convert microalgae oil into diesel-grade hydrocarbons. Using methyl oleate as a model compound, they proposed that the nickel-rich catalyst promoted decarboxylation (and decarbonylation) whereas the cobalt-rich sample was highly selective to hydrodeoxygenation. For more details regarding biodiesel production over heterogeneous catalysts, including clays and modified clay minerals, we refer to the reviews by Semwal et al. (2011), Borges and Diaz (2012), and Sani et al. (2014).

In an attempt at simulating catalytic cracking, Henderson et al. (1968) heated long-chain *n*-alkanes (e.g., *n*-octacosane) in the presence of montmorillonites and analyzed the decomposition products by gas-liquid chromatography. Heating at 375°C for varying periods of time yielded a series of *n*-alkanes and *n*-alkenes, centering around nC_{21} (alkanes) and at C_{22} and C_{18} (alkenes). The yield of *n*-alkanes increased with heating time, while the proportion of branched and cyclic alkanes as well as the yield of alkenes and aromatics tended to fall. Heating at 200°C for ~42 days under vacuum, in the presence of bentonite, led to a tenfold increase in the conversion of *n*-octacosane to alkanes, alkenes, and aromatics with a high concentration of branched and cyclic alkanes.

Earlier, Greensfelder et al. (1949) reported that alkanes and alkenes could undergo thermal cracking at the same rate. In the presence of bentonite, on the other hand, alkenes readily form carbocations, causing an increase in the rate of cracking. They suggested that thermal cracking generated a high concentration of *n*-alkanes whereas catalytic cracking largely produced branched and cyclic alkanes. Skoularikis et al. (1988) used cation-exchanged and alumina-pillared bentonites to promote the hydrocracking of decane at 400°C. Incorporation of Fe^{3+} and Cr^{3+} ions into the pillared clay increased catalytic activity, producing lighter hydrocarbons and coke. Manos et al. (2001) have compared the performance of two natural clays and their pillared derivatives in the catalytic cracking of polyethylene with that of ultrastable Y zeolite. Both the yield and the molecular weight of the liquid products are higher with clays than with the zeolite-like materials. Since the surface acidity of the clay catalysts is relatively mild, the occurrence of secondary hydrogen transfer reactions is limited.

8.2.2 Carboxylic Acid Transformation

The hypothesis that fatty acids are the precursors of *n*-paraffins in petroleum (Cooper 1962; Cooper and Bray 1963; Kvenvolden 1966) has stimulated a great deal of research into the role of clay minerals in catalyzing the degradation of adsorbed fatty acids and related compounds (Brooks 1950; Galwey 1969b, 1970; Wilson and Galwey 1976; Lao et al. 1989; Bu et al. 2017a). In reviewing the early literature, Hunt (1979) referred to the work by Engler (1911–1912) who obtained hydrocarbons by heating oleic acid over clays. Other early references on the transformation of hydrocarbons in *nature*, especially with respect to publications from the former USSR, are to be found in the book by Andreev et al. (1968). The catalytic activity of clay minerals in the formation, accumulation, and migration of oil has been reviewed by Johns (1979) and more recently by Wu et al (2012).

Jurg and Eisma (1964) made an early attempt at testing the hypothesis by heating behenic acid ($C_{21}H_{43}COOH$) with bentonite at 200°C for 89 and 760 h in the presence and absence of water. The reaction yielded hydrocarbons containing three, four, or five carbon units, while no hydrocarbons were produced in the absence of bentonite. Further, the concentration of saturated hydrocarbons increased with time of heating, while that of the olefinic type declined. In a follow-up study, Eisma and Jurg (1969) heated behenic acid in the presence of kaolinite at different temperatures and for varying periods of time, and they determined the hydrocarbons produced. The C_{21} *n*-alkane was always the major compound formed, and the yield of low molecular weight hydrocarbons decreased with increasing heating time. Eisma and Jurg suggested that the process was initiated by decarboxylation of the fatty acid to yield the corresponding alkyl radical. The latter then reacted with the original acid to yield an *n*-alkane and a secondary radical of the acid, which split into four components through scission. In a subsequent study by Sieskind and Ourisson (1972), stearic acid was heated at 250°C in the presence of montmorillonite, yielding *n*-heptadecane together with some branched and cyclic alkanes.

It was left to Johns and coworkers (Shimoyama and Johns 1971; Johns and Shimoyama 1972; Almon and Johns 1976; Johns 1982) to make a systematic investigation into the clay-catalyzed decarboxylation of behenic acid and the subsequent cracking of the corresponding *n*-alkane product. In agreement with Eisma and Jurg (1969), $C_{21}H_{44}$ *n*-alkane is the major product of decarboxylation. The formation of minor amounts of C_{16}–C_{20} hydrocarbons (in the presence of Ca^{2+}-montmorillonite at ~260°C) indicates that the C_{16} and C_{17} *n*-alkanes arise from the decomposition of *n*-C_{21}, while the C_{18}, C_{19}, and C_{20} paraffins apparently derive from both the original behenic acid and the subsequent cracking of $C_{21}H_{44}$ *n*-alkane.

The kinetics of decarboxylation and the dependence of reaction rate on temperature may be described by the Arrhenius equation (Almon and Johns 1976; Johns 1982),

$$k = Ae^{-E_a/RT} \tag{8.1}$$

where:
- k is the rate constant
- A is the Arrhenius constant (*frequency factor*)
- E_a is the activation energy for the reaction
- R is the universal gas constant
- T is the absolute temperature (in Kelvin)

TABLE 8.1
Activation Energy Values and Reaction Times for the Thermal and Clay-Catalyzed Decarboxylation of Behenic Acid Derived from the Corresponding Arrhenius Plots

Process	Activation Energy (kJ/mol)	Reaction Time[a] (years)
Thermal	244.3	2.9×10^{20}
Clay-catalyzed		
KGa-1 kaolinite	181.2	N.a.
SWy-1 montmorillonite	130.1	413
SWa-1 montmorillonite	111.3	10
NG-1 nontronite	103.3	0.03

Sources: Almon, W.R. and Johns, W.D., Petroleum forming reactions: Clay catalyzed fatty acid decarboxylation, In *Proceedings of the International Clay Conference 1975*, S.W. Bailey (Ed.), Applied Publishing, Wilmette, IL, pp. 399–409, 1976 and Johns, W.D., The role of the clay mineral matrix in petroleum generation during burial diagenesis, In *International Clay Conferences 1981*, H. van Olphen and F. Veniale (Eds.), Elsevier, Amsterdam, the Netherlands, pp. 655–664, 1982.

[a] For 90% conversion of behenic acid at 60°C; N.a. = Not available.

Taking the natural logarithm, Equation 8.1 transforms to

$$\ln(k) = \ln(A) - \frac{E_a}{RT} \quad (8.2)$$

According to Equation 8.2, a plot of $\ln(k)$ against $1/T$ would yield a straight line whose slope is equal to E_a/R, while the activation energy of the reaction is given by (−R) times the slope. Table 8.1 lists the E_a values and reaction times for the decarboxylation of behenic acid in the presence of different clay minerals, including KGa-1 kaolinite (Fe_2O_3 = 0.13%), SWy-1 montmorillonite (Fe_2O_3 = 3.35%), SWa-1 montmorillonite (Fe_2O_3 = 25.25%), and NG-1 nontronite (Fe_2O_3 = 32.2%). Besides showing the superiority of heterogeneous (clay) catalysis over the thermal process, the data indicate that catalytic activity increases with the concentration of structural ferric ions. In assessing the effect of clay addition on crude oil combustion, using thermogravimetry and differential thermal analysis, Kok and Gundogar (2010) and Kok (2006, 2012) have also found that the activation energy of the reaction decreases as the amount of added clay increases.

Almon and Johns (1976) further noted that the rate of (behenic acid) decarboxylation at 250°C was appreciably enhanced by addition of hydrogen peroxide but was markedly reduced when the clay catalysts had been pretreated with polyphosphate to *mask* the clay particle edge surface (Solomon 1968; Theng and Walker 1970). Similarly, Negrón-Mendoza et al. (1997) found that the decarboxylation of α-ketoglutaric acid over montmorillonite was greatly inhibited when the edge surface of the clay catalyst had been blocked by phosphate.

It seems relevant to mention here that prior treatment of a Ca^{2+}-nontronite with sodium metaphosphate leads to a marked decrease in the ability of the mineral to catalyze the polymerization of hydroquinone (Wang and Huang 1987). Subsequently, Wang and Huang (1989) reported that the activity of Ca^{2+}-exchanged nontronite, bentonite, and kaolinite in promoting the ring cleavage of pyrogallol (to yield a humic-like substance) decreased in the order nontronite > kaolinite > bentonite. This finding is consistent with the decreased occupancy of octahedral and particle edge sites by Fe^{3+} ions and the increased involvement of edge-surface Al^{3+} ions, acting as Lewis acid sites in the reaction. Interestingly, Wallis et al. (2006) did not find Fe(III)-treated K10 montmorillonite to be very effective in promoting the oxidative coupling of phenol substrates.

As mentioned in Chapter 2, the decarboxylation reaction involves electron transfer from the fatty acid (RCOOH) to structural Fe^{3+} ions and/or incompletely coordinated Al^{3+} ions at the clay particle edge, acting as Lewis acid sites. The resultant acyl radical (RCO•) then decomposes to an alkyl radical, which in turn forms a normal alkane by hydrogenation. In the presence of water, the ratio of branched to normal alkanes was about 0.1, which increased to about 4.5 under anhydrous conditions. The preference for straight-chain products when free water is present would indicate that the cracking process is largely catalyzed by Lewis acid sites with the formation of free radicals (Seewald 2003). On the other hand, the preferential formation of branched to normal alkanes in the absence of free water is indicative of an ionic or carbocation mechanism involving Brønsted acid sites, arising from dissociation of cation-coordinated interlayer water molecules (cf. Equation 2.20). A similar mechanism was proposed by Galwey (1969a, 1970) to account for the predominant formation of branched alkanes during the montmorillonite-catalyzed reaction of normal 1-alcohols, and by Tannenbaum and Kaplan (1985) during the pyrolysis of kerogen in the presence of montmorillonite.

Tong et al. (2014) have reported that the Brønsted acidity of acid-treated montmorillonite is well correlated with its activity in catalyzing the cracking of rosin, which is rich in abietic acid ($C_{19}H_{29}COOH$). Similarly, oleic acid [$CH_3(CH_2)_7CH=CH(CH_2)_7COOH$] yields a C_9 monocarboxylic acid (which converts into a C_6 γ-lactone by thermal oxidation) on heating at 200°C under oxygen, in the presence of Na^+-montmorillonite (Shimoyama et al. 1993). The Brønsted acid-catalyzed lactonization of cyclo-octene-5 carboxylic acid in the presence of some cation-exchanged montmorillonites has been previously reported by Adams et al. (1982).

Johns (1982) has pointed out that the measured decarboxylation reaction times (Table 8.1) greatly exceed the corresponding values observed for natural shale kerogen systems. Li et al. (1998) have also suggested that the degradation of immature kerogen in the presence of montmorillonite involves both electron transfer and the formation of carbocation intermediates. The role of clay minerals in converting kerogen into petroleum hydrocarbons is described in the following section.

The mechanisms underlying the decarboxylation of propionic acid (to ethane and CO_2) have been modeled by Geatches et al. (2010, 2011), using a periodic cell, planewave, and *ab initio* density functional theory (DFT) computation. The results suggest that the decarboxylation reaction can proceed in the presence of Al-substituted pyrophyllite (cf. Table 1.1) with sodium as a counterion, and it involves the formation of an alcohol intermediate. Johns and McKallip (1989) have also noted that the catalytic activity of a cation site, formed by substitution of Al^{3+} for Si^{4+} in the tetrahedral sheet of illite-smectite clays, is 40 times greater than that of a corresponding octahedral site. Similarly, Wilson et al. (1986) have reported that tetrahedral aluminum in montmorillonites and short-range order aluminosilicates is very efficient in altering the ratio of 1-alkenes to internal alkenes in the pyrolyzate of an aliphatic oil shale. Moreover, the activity of Ti- and Al-pillared beidellites in catalyzing cumene cracking is superior to that shown by their montmorillonite counterparts (Swarnakar et al. 1996). Likewise, the beidellite-based samples in which the layer charge arises from substitution of Al^{3+} for Si^{4+} in the tetrahedral sheet are highly selective to yielding benzene.

Some time ago, Heller-Kallai et al. (1984) investigated the thermal decomposition of stearic acid under *bulk flow* conditions in the presence of various clay mineral species. Besides failing to find a correlation between decarboxylation and hydrocarbon formation, the former reaction occurred more readily when the octahedral sheet of the catalysts was occupied by Mg^{2+} and Li^+ rather than by Al^{3+} and Fe^{3+} ions. Heller-Kallai et al. suggested that catalysis by Lewis acid sites, exposed on clay particle edges, might not be the dominant mechanism controlling the reaction, and decarboxylation did not necessarily precede hydrocarbon transformation, at least under the experimental conditions used. Similarly, Miloslavski et al. (1991) observed that heating of long-chain *n*-alkanes in the presence of clay volatiles or condensates gave rise to cracked products resembling those formed on direct contact with the clay minerals. Subsequently, Heller-Kallai et al. (1996) reported that the condensate obtained by condensing the vapor evolved on heating montmorillonite was more active in catalyzing the cracking of *n*-octane than the parent clay.

Yuan et al. (2013) have investigated the thermal decomposition of 12-aminolauric (ALA) or 12-aminododecanoic acid [($H_2N(CH_2)_{10}CH_2COOH$)], either mixed with or intercalated into, montmorillonite (MMT). The amount of C_1–C_5 hydrocarbons released (by pyrolysis) from the interlayer MMT-ALA complex is nearly nine times greater than that measured for the physical mixture of MMT and ALA. The degradation products of MMT-intercalated ALA include ammonia and n-alkanes (Liu et al. 2013). In combination with Fourier transform infrared spectroscopy, these findings indicate that the reaction is catalyzed by both Lewis and Brønsted acid sites.

The results of high-pressure pyrolysis experiments by Bu et al. (2017b) using montmorillonite complexes with octadecanoic acid, octadecyltrimethylammonium bromide, and octadecylamine may also be explained in terms of the combined involvement of Lewis and Brønsted acid sites. Following Almon and Johns (1976), these workers have suggested that Lewis acid sites are primarily responsible for the decarboxylation of the clay-adsorbed organics and the resultant release of CO_2. On the other hand, the cracking of the hydrocarbon chains into C_1–C_5 fragments and the isomerization of normal hydrocarbons are probably promoted by Brønsted acid sites, involving the formation of carbocations.

Naidja and Siffert (1990) have investigated the oxygen-induced decarboxylation of isocitric acid in the presence of Na^+-, Cu^{2+}-, and Mn^{2+}-montmorillonites. Following intercalation into the sodium-exchanged clay, the organic acid converts into α-ketoglutaric acid (and CO_2), probably by a Brønsted acid-catalyzed process involving a carbocation intermediate. With divalent cation-exchanged montmorillonites, on the other hand, the isocitrate anion apparently forms a complex with the counterion, inhibiting its decarboxylation. For the decarboxylation of oxaloacetic acid (to pyruvic acid and CO_2) over Na^+-, Cu^{2+}-, and Mn^{2+}-montmorillonites, Siffert and Naidja (1987) measured activation energy (E_a) values of 82.8, 73.6, and 68.6 kJ/mol, respectively (cf. Table 8.1). Earlier, Wilson and Galwey (1976) derived E_a values of 82–177 kJ/mol for the formation of 2-methylhexane from stearic acid in the presence of illite, kaolinite, and montmorillonite.

It is worth recalling that long-chain fatty acids can intercalate into montmorillonite when the silicate layers have previously been expanded by treatment with n-hexanol or n-octanol (Brindley and Moll 1965; Theng 1974). Yariv and Shoval (1982) have also reported that montmorillonite is capable of intercalating lauric and stearic acids from a CCl_4 solution. On the other hand, pentadecanoic, docosanoic, 5β-cholanic and ursolic acids fail to penetrate the interlayer space of synthetic saponites from their corresponding solutions in seawater (Drouin et al. 2010).

Negrón-Mendoza et al. (1995, 2015) used γ-ray irradiation to induce the decarboxylation of low molecular weight fatty acids (acetic, malonic, succinic) in aqueous solution in the presence of montmorillonite. In common with Siffert and Naidja (1987) and Naidja and Siffert (1990), the clay-catalyzed reaction yields the corresponding acids with one less carbon than the parent substrates whereas in the absence of clay, the main reaction is one of fatty acid dimerization. Faure et al. (2003) also reported on the air oxidation of n-alkanes at low temperature (100°C) in the presence of Na^+-montmorillonite. The reaction produced alcohols and ketones (with an aliphatic chain length similar to that of the parent hydrocarbon) as well as carboxylic acids and 3-substituted γ-lactones through chain (C–C bond) cleavage.

8.2.3 Kerogen Transformation and Pyrolysis

The formation, preservation, and stability of organic matter (OM)/kerogen in sediment and soil, and especially its interaction with clay minerals in relation to hydrocarbon (petroleum) generation, have attracted a great deal of attention. The extensive literature on this topic has been reviewed by Durand (1980), Tissot and Welte (1984), Seewald (2003), Vandenbroucke and Largeau (2007), and Arndt et al. (2013). Because of the intimate association between clay minerals and OM/kerogen (Keil et al. 1994; Mayer 1994), it is not always easy to differentiate the *protective* from the *catalytic* effect of clay on OM/kerogen transformation (Kennedy et al. 2014).

Collins et al. (1995) have suggested that adsorption and condensation of organic matter at clay surfaces, operating in concert, are involved in the formation and preservation of kerogen, while Salmon et al. (2000) have argued for physical protection by minerals, using Cenomanian black shale as an example. Even in contemporary soils, OM can be preserved (for millennia) by intercalation into clay minerals (Theng et al. 1992). More recently, Galwey (2015) proposed that initial adsorption to (acidic) clay minerals would protect OM from rapid decay, allowing slow (thermal) desorption of petroleum hydrocarbons to occur after sediment burial. It would appear that the clay mineral matrix can continue to function as sorbent and catalyst during the transformation of OM by diagenesis (Wu et al. 2012).

The hypothesis that the inhibitory versus promoting influence of clay minerals on hydrocarbon generation in shales is determined by the clay-kerogen interaction has been tested by Rahman et al. (2017), using samples from two shale formations. In the Monterey formation (California, USA), the nanoparticles of organic matter (OM) are closely associated with clay minerals in the form of clay-OM nanocomposites, giving a positive relationship ($R^2 = 0.91$) between total organic carbon and mineral surface area. On the other hand, the pore-filling clays in the Stuart Range Formation (South Australia) are only loosely associated with discrete OM particles ($R^2 = 0.54$). Thus, the shale samples from the Monterey formation generate maximum hydrocarbons, and at a higher pyrolysis temperature, as compared with samples from the Stuart Range Formation. The *free* mineral surfaces in the latter samples, however, can retain more hydrocarbons. Thus, hydrocarbon generation is influenced not only by the absence or presence of certain clay mineral species but also by the physical association of clay surfaces with organic matter.

Perhaps the single most versatile technique for assessing the clay-OM/kerogen interaction and the role of clay minerals in the transformation/degradation of the associated OM/kerogen is that of pyrolysis (Py) under hydrous or anhydrous (dry) conditions, in combination with gas chromatography (GC), and often mass spectroscopy (MS). Supplementary instrumental methods of analysis such as Fourier-transform infrared (FTIR) and nuclear magnetic resonance (NMR) spectroscopies, and thermal analysis can provide additional information.

Horsfield and Douglas (1980) made an early attempt at investigating the effect of various minerals, including bentonite and kaolinite, on the pyrolysis of *kerogen*—that is, the high molecular weight carbon-rich organic material in sedimentary rocks that is insoluble in water and common organic solvents. They noted that the yield of low molecular weight pyrolysis products was higher in the presence of minerals than that measured for kerogen by itself. In a follow-up study, Davis and Stanley (1982) have compared the pyrolyzate of smectite-kerogen mixtures with that of the isolated kerogen. The degree of catalytic cracking is influenced by the ratio of clay to organic matter (OM). The results further indicate that smectite can efficiently catalyze C–C cleavage of high molecular weight hydrocarbons to yield short-chain species. Al-Harahsheh et al. (2011) also found that the activation energy (cf. Equation 8.2) associated with the pyrolysis of a Jordanian oil shale was appreciably lower than the value measured for the isolated kerogen. This finding is consistent with the capacity of the mineral component in the shale oil in mediating catalytic cracking.

Regtop et al. (1986) reported that pyrolysis of an oil shale (from Condor, Australia), its demineralized kerogen, and a kerogen-kaolinite mixture yielded 1-pristene as the main product from which 2-pristene was derived by clay-catalyzed isomerization. In a follow-up study, Lao et al. (1989) chemisorbed 1-pristene (at different temperatures and periods) to various minerals and Condor oil shale, and extracted the products with dichloromethane. With kaolinite at 25°C and 50°C, 2-pristene is formed by double-bond isomerization, and 2-pristanol through hydration, while at 150°C pristane forms by hydrogen transfer. The catalytic activity of kaolinite increases after heating the mineral at 100°C. Saxby et al. (1992) later noted that the presence of bentonite during heating a bituminous coal (200°C–300°C for 4 days) led to the catalytic cracking of volatile matter and the formation of polycyclic aromatic hydrocarbons in the liberated oil. Volatile, low molecular weight organic acids have also been identified by Kawamura and Kaplan (1986) and Kawamura et al. (1986) during pyrolysis of kerogens and humic acids in the presence and absence of water and clay minerals.

Kaplan and coworkers (Tannenbaum and Kaplan 1985; Tannenbaum et al. 1986a, 1986b; Huizinga et al. 1987a, 1987b) have published a series of papers on the role of clay minerals in the formation, degradation, and transformation of organic compounds during the pyrolysis of two immature kerogens under dry and hydrous conditions. For both the Monterey formation kerogen (type II) and Green River formation kerogen (type I), the presence of montmorillonite yields up to five times more C_1–C_6 hydrocarbons than is observed in its absence, with a dominance of branched hydrocarbons in the C_4–C_6 range. The results are indicative of a Brønsted acid-catalyzed cracking process involving carbocation intermediates (Tannenbaum and Kaplan 1985). A similar mechanism has also been proposed by Alexander et al. (1982, 1984) and Larcher et al. (1985) for the clay-catalyzed alkyl and aromatic hydrogen exchange reactions between acidic clay surfaces and various organic compounds.

Unlike montmorillonite, illite shows little, if any, catalytic activity although both minerals can retain up to 80% of the generated *bitumen*—that is, the soluble component of kerogen. Consisting of polar compounds and asphaltenes, bitumen can be cracked to yield low molecular weight compounds and insoluble pyrobitumen (Tannenbaum et al. 1986a). Pyrolysis of the Monterey formation kerogen (at 300°C for 2 h; with or without minerals) also produces steranes and triterpanes. The extent of isomerization of these *biomarkers* at C-20 in 14α(H),17α(H) steranes, at C-22 in 17α(H), 21β(H)-hopanes, and of 17β(H),21β(H)-hopanes corresponds to an early diagenetic stage and increases with heating time. The catalytic activity of montmorillonite for the reaction accords with its high Brønsted acidity since isomerization is inhibited in the presence of calcite (Tannenbaum et al. 1986b).

The distribution and maturation of hopanoids and steranes released from (Estonian Kukersite) kerogen during hydrous or anhydrous pyrolysis in the presence or absence of kaolinite, montmorillonite, calcite, and dolomite have been investigated by Pan et al. (2010a). Maturation rates of these compounds increase with mineral acidity and decline in the presence of water. The stability of hopenes relative to 17β(H),21β(H)-hopanes varies with pyrolysis conditions as hopenes are quite sensitive to the amount of added water. The relative abundance of *n*-alkanes decreases, while that of isoalkanes, cycloalkanes, and light alkylbenzenes increases with an increase in mineral acidity and water/organic carbon ratio (Pan et al. 2009). Analysis of the gases released during pyrolysis indicates that the ratios of ethene/ethane, propene/propane, isobutane/*n*-butane, and isopentane/*n*-pentane increase with a rise in water/organic carbon ratio (Pan et al. 2008).

Pan et al. (2010b) further observed that the ratios isobutane/*n*-butane and isopentane/*n*-pentane in the pyrolyzate of a crude oil from the Tarim Basin, China, were appreciably higher in the presence of (acidic) montmorillonite than when calcite was added. Indeed, Hu et al. (2014) found calcite to have an inhibitory effect on oil formation during pyrolysis of a kerogen (from the Huadian oil shale, China). The activity of montmorillonite and kaolinite in promoting the formation of C_7–C_{12} hydrocarbons, and increasing isoalkanes/*n*-alkanes and alkanes/olefin ratios, is consistent with a Lewis acid catalyzed process.

Wei et al. (2006a, 2006b) have looked into the formation of diamondoids during (anhydrous and hydrous) pyrolysis at 340°C of organic-rich modern sediments and six kerogens (types I, II, II-S, III) in the presence of K10 montmorillonite and an acidic aluminosilicate (MS-25). By comparison, kaolinite is less active and illite is inactive, while calcite appears to inhibit the process. The catalytic activity of the K10 and MS-25 materials may be explained in terms of a carbocation mechanism associated with Lewis acid sites. The overall experimental results, however, would indicate that the reactions proceed through cooperation (synergy) between Brønsted and Lewis acid sites (cf. Chapter 2) similar to what Sieskind et al. (1979) have earlier suggested with respect to the clay-catalyzed transformation of sterols.

Changes in hopane composition have also been recorded when a sample of immature hopanes is heated at 75°C in the presence of Al^{3+}-montmorillonite (Larcher et al. 1988). This finding is ascribed to the selective removal of 17β(H),21β(H)-hopanes rather than their conversion to the corresponding diastereomers. Montmorillonite (and illite) can also take up the polar constituents of bitumen as a result of which the concentration of hydrocarbons in the pyrolyzate increases (Tannenbaum et al. 1986b; Huizinga et al. 1987a). Kaolinite, halloysite, and muscovite, but more so montmorillonite, are similarly efficient in adsorbing organic nitrogen compounds from a kerogen pyrolyzate (Charlesworth 1986).

Huizinga et al. (1987b) further reported that the influence of clay minerals on the formation of n-alkanes, acyclic isoprenoids, and alkenes was strongly dependent on the water content of the system. Subsequently, Tang and Stauffer (1995) used the Arrhenius equation (Equation 8.1) to model the formation of the isoprenoids, pristane, and phytane in Monterey kerogen under both open and closed pyrolysis conditions. In the case of low maturity oils, the pristane/phytane ratio does not change with maturity. We might mention here that K10 montmorillonite can effectively catalyze the isomerization and cyclization of highly branched isoprenoid alkenes. With dienes, double-bond migration and geometric isomerization prevail whereas trienes are isomerized and undergo rapid cyclization to yield substituted cyclohexenes through a Brønsted acid catalyzed process involving carbocation intermediates (Belt et al. 2000).

Huizinga et al. (1987b) have also noted that the presence of montmorillonite during dry pyrolysis (when only pyrolyzate water is available) suppresses the formation of C_{12+} n-alkanes and acyclic isoprenoids although illite is quite ineffective in this regard. Interestingly, alkene formation is significantly reduced in the presence of either clay mineral species. Under hydrous pyrolysis conditions, on the other hand, addition of montmorillonite and illite does not significantly influence the earlier reactions probably because of partial adsorption of the substrate to the added clay. Burkow et al. (1990), for example, reported that during hydrous pyrolysis of naphthalene, methylnaphthalene, dimethylnaphthalene, and dibenzothiophene in the presence of montmorillonite, illite, and chlorite, more than 60% of these aromatic compounds became strongly bound to the mineral surface. Hetényi (1995) also found that montmorillonite could retain both soluble and volatilized bitumen during pyrolysis of type III/a and III/b kerogens, causing a reduction in yields.

Lu and Kaplan (1989) and Lu et al. (1989) pyrolyzed the bitumens and kerogens, extracted from immature Cretaceous Black Shale (CBS) and Rocky Mountain Coal (RMC). The formation of biomarkers during pyrolysis together with their isomerization and degradation were significantly influenced by the presence of montmorillonite. Irrespective of kerogen type, montmorillonite could effectively catalyze the release of biomarkers during pyrolysis at 200°C and control the rate of isomerization and degradation at 300°C. More hopanes than steranes were found in the original bitumen and pyrolyzates of type III RMC kerogen as compared with type II CBS kerogen. Anhydrous pyrolysis at 200°C and 300°C (for 2–100 h) yields triaromatic hydrocarbons by partial cracking of asphaltene and kerogen. Although the presence of montmorillonite does not enhance aromatization in either kerogen type, it promotes the cracking and degradation of aromatic steranes, giving rise to either phenanthrene or hydrocarbons of low molecular weight.

Rose et al. (1992) have reported that an appreciable proportion of the aliphatic carbon in the Rundle shale (Queensland, Australia) and the isolated kerogen can convert into aromatic carbon during pyrolysis at 450°C. The addition of various minerals, including montmorillonite, does not enhance kerogen aromatization although oil yield and the ratio of external to internal alkene protons in the oil show an increase. Faure et al. (2006) have cautioned that the aromatization of raw organic matter during flash Py/GC-MS in the presence of clays (illite, montmorillonite) may give rise to artifacts. In order to inhibit or minimize this effect, they suggest applying thermally assisted hydrolysis methylation-GC-MS and using samples that have not been demineralized.

In following up this suggestion, Schulten et al. (1996) chose pyrolysis methylation-MS to characterize the organic matter in an interlayer clay-organic complex from soil. Analysis of the pyrolyzate indicates that the intercalated organic material is greatly depleted in lignin-derived aromatics. That intensive aromatization can occur when humic acid is pyrolyzed in the presence of clays has also been reported by Faure et al. (2006). A similar transformation was observed by Theng and Hashizume (1996) during the hydrothermal (300°C, 100 MPa) treatment of humic acid in the presence of montmorillonite. Likewise, Hayatsu et al. (1984) reported that lignin transformed into an insoluble material, resembling low rank coal, when heated at 150°C for 2–8 months in the presence of acid-activated montmorillonite.

In comparing the pyrolyzates of some shales with those of the separated kerogens, Espitalié et al. (1980, 1984) noted that the shales yielded more hydrocarbons than the associated kerogens because the former materials could selectively retain the high molecular weight hydrocarbon fractions. The results of pyrolysis experiments using mixtures of kerogens and different clay minerals are consistent with the proposed mechanism. As would be expected, hydrocarbon adsorption rises as the specific surface area of the mineral component increases. Besides providing an adsorbing surface, montmorillonite can catalyze the formation of light hydrocarbons and aromatic compounds. Although illite and palygorskite (*attapulgite*) can similarly retain the hydrocarbons formed during pyrolysis, these minerals are relatively inactive as catalysts.

8.3 CLAY MINERAL CATALYSIS OF PREBIOTIC ORGANIC REACTIONS

Clay minerals, older than 3.7 Ga when *life* on Earth began to emerge (Brooks 1981; Moorbath 1994; Mojzsis et al. 1996; Joyce 2002), have been detected in early Archaean rocks from Isua, West Greenland (Appel 1980), and the Pilbara, Western Australia (Cullers et al. 1993; Schopf 1993). Besides the likelihood of being present in the primitive Earth (Cleaves et al. 2012; Zaia 2012), clay minerals have a propensity for adsorbing and intercalating simple and polymeric organic compounds (Mortland 1970; Theng 1974, 2012; Yariv and Cross 2002; Lagaly et al. 2006).

It was Bernal (1951), however, who first proposed that clay minerals played an important role in chemical evolution and life's origin because of their propensity for taking up and concentrating extraneous molecules, protecting them against ultraviolet/cosmic radiation, and allowing the adsorbed species to be transformed or polymerized. In addition, clay minerals may promote the synthesis of biomonomers from gaseous constituents as well as provide an environment in which such monomers can adopt a specific orientation, condense, and serve as templates for organic replication (Ponnamperuma et al. 1982). Directly or indirectly, Bernal's hypothesis has stimulated a great deal of research into the ability of clay minerals to mediate chemical evolution and prebiotic organic synthesis (Degens and Mathéja 1970; Rao et al. 1980; Cairns-Smith and Hartman 1986; Ferris 1993, 2006; Eastman and Porter 2000; Negrón-Mendoza and Ramos-Bernal 2004; Lambert 2008; Cleaves et al. 2012; Hashizume et al. 2013; Yang et al. 2013).

Pavlovskaya et al. (1960) made an early report on the formation of amino acids by exposing solutions of formaldehyde and ammonium salts to ultra-violet radiation in the presence of clays. Similarly, Yoshino et al. (1971) observed the formation of various amino acids, together with purines and pyrimidines, by reacting CO, H_2, and NH_3 at 200°C to 700°C, in the presence of montmorillonite. Degens and Mathéja (1968, 1970) and Harvey et al. (1972a) used kaolinite to catalyze the formation of polypeptides (plus detectable amounts of cytosine, uracil) from CO and NH_3. Schwartz and Chittenden (1977) subsequently obtained 5,6-dihydrouracil (DHU) by evaporating and mildly heating solutions of β-alanine and urea in the presence of montmorillonite. Ultraviolet irradiation then transformed the DHU product to uracil, which in turn was converted to thymine by the addition of acetate.

Gabel and Ponnamperuma (1967) used kaolinite to promote the synthesis of monosaccharides from formaldehyde, and Harvey et al. (1972b) obtained polysaccharides by incubating kaolinite with paraformaldehyde in water at 80°C for 160 days. Similarly, Saladino et al. (2010) were able to synthesize amino sugar derivatives from a mixture of formamide and formaldehyde in the presence of KSF montmorillonite.

Earlier, Hanafusa and Akabori (1959) reported the formation of di- and tri-glycine by heating aminoacetonitrile sulfate at 120°C–140°C for 5 h in the presence of kaolinite or a Japanese acid clay. An acidic clay was also used by Losse and Anders (1961) to catalyze the synthesis of an alanine polymer from α-aminopropionitrile. Likewise, Ventilla and Egami (1976) were able to obtain amino acids and peptides by shaking a mixture of formaldehyde, hydroxylamine, various metal salts, and clay minerals (acid clay, kaolinite, montmorillonite) at 80°C for 120 h. Not long after, Shimoyama et al. (1978) prepared a range of amino acids through an electric discharge on a mixture of CH_4 and N_2, noting that product yields were markedly higher in the presence of Na^+-montmorillonite than in its absence.

Saladino et al. (2004) and Costanzo et al. (2007) have described the formation of nucleic acid bases from formamide in the presence of various minerals, including the K-series of montmorillonites and an alumina pillared interlayered clay (Al-PILC). More recently, Hashizume et al. (2018) were able to obtain adenosine by subjecting adenine and ribose (cf. Figure 8.2) to repeated wetting and drying (35°C–50°C in air or N_2 gas) with or without the addition of kaolinite or montmorillonite.

8.3.1 POLYPEPTIDE SYNTHESIS

The use of clays as supports and catalysts for polypeptide synthesis has received a great deal of attention. In this context, we should recall that the condensation of unsubstituted (*non-activated*) amino acids in water is energetically unfavorable. Furthermore, the peptide bond is thermodynamically unstable and prone to hydrolysis (Brack 2006; Marshall-Bowman et al. 2010). The investigation by Flegmann and Scholefield (1978) on the lysine/dilysine condensation (in water up to 90°C) also indicates that the thermodynamic barrier to the reaction in the presence of clay minerals is comparable to that in a homogeneous solution. Nevertheless, selective condensation of amino acids in water can be achieved via the intermediary formation of N-carboxy anhydrides (Brack 2006, 2007) or under high-pressure conditions, even in the absence of a catalyst (Otake et al. 2011).

When Fripiat et al. (1966) heated a montmorillonite complex with glycine and β-alanine (in the zwitterionic form), the adsorbed amino acids appeared to have formed a peptide bond. Mifsud et al. (1970) have observed similarly for L-ornithine adsorbed in the interlayer space of vermiculite. More interestingly, Degens and Mathéja (1970) and Degens et al. (1970) have obtained high molecular weight polypeptides by condensing L-amino acids on the surface of some minerals, notably kaolinite, at temperatures below the boiling point of water. The formation of peptides when mixtures of amino acids are heated at 65°C–85°C for up to 81 days has also been reported by Rohlfing (1976). Likewise, Yanagawa et al. (1990) and Ito et al. (1990) were able to synthesize high molecular weight polypeptides by subjecting various amino acid amides to hydration-dehydration cycles and microwave heating, noting that the presence of kaolinite enhances polypeptide yields.

Pant et al. (2009) have revisited the heat-induced formation of peptides from aqueous glycine/aspartic acid and glycine/valine systems with and without clay. The oligomerization of glycine is influenced by the duration of heating (at 85°C), the nature of the reactants, and the presence of (divalent cation-exchanged) montmorillonite clay. When Fuchida et al. (2014) heated a montmorillonite-glycine complex at 150°C for 336 h under dry conditions, close to 14% of the adsorbed monomer condensed into peptides among which diketopiperazine was the major product. However, glycine failed to oligomerize in the absence of montmorillonite, and no peptides were formed under wet conditions in the presence of clay.

The clay-catalyzed formation of oligopeptides from pristine (nonactivated) amino acids under conditions of repeated wetting, drying, and/or heating has been the topic of many investigations. Simulating the cyclic wetting and evaporation of lagoons and rock pools, such a *fluctuating environment* is considered to be conducive to oligomerization as the adsorbed monomers and oligomers can be released and redistributed on the mineral surface during the following hydration-dehydration cycle (Lahav and Chang 1976). Using this approach, Lahav and coworkers (Lahav 1978; Lahav et al. 1978; Lahav and White 1980) have successfully synthesized short-chain oligoglycines in the presence of kaolinite and montmorillonite. Likewise, Lawless and Levi (1979) could obtain alanine and glycine oligomers in the presence of various cation-exchanged bentonites, among which the Cu^{2+}-exchanged sample was especially effective.

Similar observations have been made by Porter et al. (1998a, 1998b, 1999a; 2000, 2001) using Cu^{2+}-hectorite as an adsorbent and catalyst. Density functional calculations by Rimola et al. (2007)

further suggest that synergy between the Cu^{2+}-glycine interaction and water, acting as a proton-transfer agent, can significantly decrease the activation barrier for peptide bond formation. More recently, Gururani et al. (2012) obtained glycine tetramers by heating mixtures of glycine-glutamic acid and glycine-leucine at 90°C under wetting-drying cycles in the presence of Ca^{2+}-, Cu^{2+}-, and Mg^{2+}-montmorillonites. The enhancing effect of divalent counterions on oligomerization is apparently related to their propensity for forming stable ion-dipole complexes with the zwitterions of the amino acid.

The role of smectites (and non-clay minerals) in the formation and chain elongation of peptides, under several cycles of wetting and drying at 80°C, has been described by Bujdák and coworkers in a series of papers. Besides promoting glycine oligomerization, Ca^{2+}- and Cu^{2+}-montmorillonites can catalyze peptide chain elongation (Bujdák et al. 1994). Thus, diglycine can yield hexaglycine, while tri- and tetra-glycine can form higher oligomers even after one cycle of wetting and drying (Bujdák et al. 1995). Following White et al. (1984), Bujdák et al. (1996b) suggest that the carboxyl groups of amino acids can condense with the hydroxyl groups of under-coordinated Al (and Si), exposed at clay particle edges, to yield the corresponding *activated anhydrides*. In conjunction with the formation of cyclic anhydride intermediates, notably diketopiperazine, this process plays a key role in the formation and elongation of peptides. High temperatures and a neutral pH appear to favor cyclic anhydride formation over glycine dimerization (Bujdák et al. 1996a) as well as diketopiperazine conversion into diglycine (Bujdák and Rode 1999). More recently, Jaber et al. (2014) have obtained cyclic dimers and trimers by condensing arginine with glutamic acid in the presence of montmorillonite.

Bujdák and Rode (1996) have also noted that trioctahedral smectites (e.g., hectorite) are more active than their dioctahedral counterparts (e.g., montmorillonite) in catalyzing the oligomerization of glycine and diglycine. The presence of Fe^{2+} in the silicate structure is especially conducive to amino acid dimerization. In addition, these workers have investigated peptide bond formation involving alanine, alanine + glycine, alanine + diglycine, and alanine + glycine cyclic anhydride in the presence of various minerals. Interestingly, silica and alumina are more effective than smectites in promoting amino acid dimerization, although smectites are superior in terms of peptide chain elongation (Bujdák and Rode 1997). Besides enhancing the salt-induced formation of peptides and peptide chain elongation, montmorillonite can stabilize the peptide oligomers against hydrolysis (Son et al. 1998; Rode et al. 1999).

Little doubt, however, attaches to the ability of clay minerals in catalyzing the oligomerization of *activated* amino acids. For example, Ferris et al. (1996) and Hill et al. (1998) have obtained short-chain peptides by incubating illite with glutamic acid and carbonyl diimidazole, separating the solid (by centrifugation), and adding fresh monomer and activating agent. When the *feeding* process is repeated up to 50 times, oligomers with 40 or more monomers are formed whereas only up to 10-mer long polypeptides are obtained in the absence of clay. Similarly, amino acid thioesters can oligomerize in the presence of montmorillonite according to Equation 8.3, while the formation of the cyclic dipeptide is inhibited (Bertrand et al. 2001; Brack 2006).

$$n NH_2-CHR-CO-S-C_2H_5 \rightarrow H-(NH-CHR-CO)_n-S-C_2H_5 + n HS-C_2H_5 \quad (8.3)$$

A well-known example of a clay-catalyzed amino acid oligomerization reaction is the formation of polypeptides from amino acid adenylates (AAA) in the presence of Na^+-montmorillonite (Paecht-Horowitz et al. 1970; Paecht-Horowitz 1976, 1977; Eirich 1981). An oligomerization degree of 30–50 was deduced, accompanied by the liberation of adenylic acid, whereas only up to 12 monomer units combined in the absence of clay. Furthermore, little monomer hydrolysis occurred in the presence of montmorillonite consistent with the protective effect offered by surface and interlayer adsorption. Interestingly, kaolinite showed little, if any, catalytic activity presumably because the surface concentration of activated monomers must reach a certain level before appreciable polycondensation could occur. Adsorption and X-ray diffraction studies suggest that the phosphate group

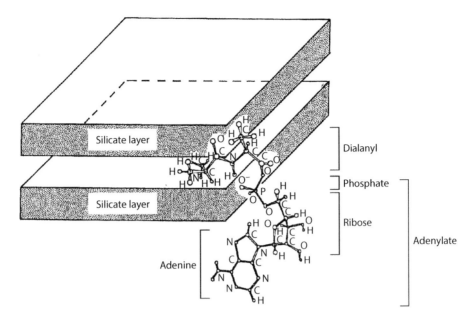

FIGURE 8.1 Diagram showing the interaction between an amino acid adenylate (monomer) and a montmorillonite particle, represented by two contiguous silicate layers. The phosphate group of the monomer is attached to the particle edge surface (presumably by ligand exchange), while the dialanyl group penetrates the interlayer space. (Modified from Paecht-Horowitz, M. and Lahav, N., *J. Mol. Evol.*, 10, 73–76, 1977.)

of AAA adsorbs on the edge surface of montmorillonite, presumably through ligand exchange with exposed $(OH)Al(OH_2)$ groups (Hashizume et al. 2010), while the terminal NH_3^+ group (of AAA) is attached to the (planar) interlayer surface, as illustrated in Figure 8.1. The peptide chain that forms in the interlayer space would therefore be protected against hydrolysis. A spatially similar process has been proposed by Jaber et al. (2014) for the condensation of intercalated arginine with edge-adsorbed glutamic acid on the surface of a synthetic montmorillonite.

Paecht-Horowitz and Lahav (1977) further observed that the oligomerization (of alanine adenylate) in the presence of non-expanding Al^{3+}-montmorillonite was much more restricted than was the case with the swelling Na^+-exchanged clay. The oligomerization of amino acid adenylates was also facilitated when the montmorillonite support had been *primed* by adsorbed peptides (Paecht-Horowitz and Eirich 1988). The enhancing effect of previously adsorbed species, such as histidyl-histidine, polyribonucleotides, and adenosine triphosphate on glycine oligomerization, under repeated wetting and drying conditions, has similarly been reported by White and Erickson (1980, 1981) and Rishpon et al. (1982).

8.3.2 SELECTIVE ADSORPTION AND POLYMERIZATION OF AMINO ACID ENANTIOMERS

Besides serving as a support and catalyst of polypeptide formation, clay minerals can conceivably discriminate between amino acid enantiomers because of the preferential adsorption of one optical isomer over the other. Although the published information remains controversial, and the underlying mechanism is not fully understood, the possible selective adsorption of amino acids and peptides by clay minerals may provide a clue to the origin of L-homochirality of amino acids in proteins of living organisms (Julg 1989).

Early on, Jackson (1971a, 1971b) reacted aspartic acid, serine, and phenylalanine with kaolinite at 60°C–90°C for varying periods of time. The L-optical isomers appeared to polymerize at a significantly higher rate, and to a larger extent, than did their corresponding D-enantiomers, while the racemic mixture polymerized at an intermediate rate. In the case of aspartic acid, for example, 25%

of the L-isomer was polymerized as against 3% of the D-isomer, and 14% of the racemic mixture. Although polypeptides also formed in the absence of kaolinite, no consistent relationship was observed between chirality and the extent of polymer formation. It seems possible that the L-enantiomers preferentially adsorb to the edge surface of kaolinite particles, which may be enantiomorphous due to interlayer displacements and the presence of octahedral site vacancies (Bailey 1963; Kameda et al. 2005).

Flores and Bonner (1974) and McCullough and Lemmon (1974), however, could not substantiate Jackson's (1971a, 1971b) finding with respect to the preferential adsorption and polymerization of aspartic acid by kaolinite. Similarly, Bonner and Flores (1973) failed to confirm that kaolinite could take up D- and L-phenylalanine to different extents (from aqueous solutions at acid pH) although differential adsorption of 1.0%–1.8% was measured for D- and L-alanine on d- and l-quartz (Bonner et al. 1975). The subsequent finding by Bondy and Harrington (1979) that montmorillonite could preferentially bind L-leucine and L-aspartate (as well as D-glucose) has also been called into question by Youatt and Brown (1981).

On the other hand, Friebele et al. (1981a, 1981b) could detect a small (0.2%–2.0%) but statistically significant preferential adsorption of the L-enantiomer (over the D-optical isomer) by incubating a racemic mixture of α-alanine, α-aminobutyric acid, valine, and norvaline with Na^+-montmorillonite at different pH values. Fraser et al. (2011) have also reported that the adsorption of alanine, lysine, and histidine to a high-swelling interlayer complex of vermiculite with n-propylammonium leads to a significant chiral enrichment of either the L- or D-enantiomer, depending on the amino acid. Earlier, Hashizume et al. (2002) made the interesting observation that an allophane clay from a volcanic ash bed near Te Kuiti, New Zealand, showed a clear preference for L-alanyl-L-alanine over the D-enantiomer. By way of explanation, they suggested that the size, intramolecular charge separation, and surface orientation of L-alanyl-L-alanine zwitterions combined to confer *structural chirality* to the allophane-dipeptide complex. In keeping with this suggestion, the *ab initio* study by Yu et al. (2001) on the adsorption of D- and L-enantiomers of N-formylalanine amide by nontronite indicates that the interlayer complex with the L-enantiomeric form of the dipeptide is more stable (by 6 kcal/mol) than its D-counterpart. Bujdák et al. (2006) subsequently noted that montmorillonite could preferentially take up the diastereomers of dialanine, containing either L- or D-amino acids, over the stereoisomers, containing a racemic mixture of the dipeptide. Montmorillonite has also been reported to promote the epimerization of isoleucine (Frenkel and Heller-Kallai 1977; Akiyama 1978). The ability or otherwise of certain clay minerals to discriminate between optical isomers of organic compounds clearly merits further investigation.

8.3.3 Dimerization and Oligomerization of Nucleotides

Ribonucleic acid (RNA) may well be the most important biopolymer in the early life on Earth since in the *RNA world* (Gilbert 1986; Orgel 1988; Joyce 2002; Srivatsan 2004) this molecule could act as both an enzyme (capable of catalyzing protein synthesis) and a store house of genetic information (Zaug and Cech 1986; Ferris 2005; Brack 2006). This hypothesis has stimulated much research into the clay-catalyzed synthesis of RNA oligomers from the corresponding nucleotides.

Early on, Ibanez et al. (1971) reported that the condensation of thymidine monophosphate with cyanamide at neutral pH, in the presence of montmorillonite, gave rise to oligodeoxyribonucleotides with longer chains than those formed in homogeneous aqueous solutions. By adsorbing nucleoside phosphoramidates to montmorillonite, and then heating the dried clay-organic complex, Burton et al. (1974) were also able to obtain the corresponding dinucleotides. Subsequently, Strigunkova et al. (1986) described the formation of oligonucleotides by exposing adenosine monophosphate, adsorbed on a kaolinite surface, to UV radiation, while Odom et al. (1979) obtained oligonucleotide and dinucleotide pyrophosphate by subjecting kaolinite, water, a nucleotide, and cyanamide to repeated drying and wetting cycles.

The clay-catalyzed dimerization and oligomerization of nucleotides have been systematically investigated by Ferris and coworkers. An interesting example is the *in situ* oxidation of diaminomaleonitrile to diiminosuccinonitrile (DISN) involving structural Fe^{3+} in montmorillonite acting as a Lewis acid. The condensing agent, DISN, then reacts with 3'-adenosine monophosphate to yield the 2',3'-cyclic derivative (Ferris et al. 1988). Treatment of 5'-adenosine monophosphate (pA) with [1-ethyl-3-(3-dimethylaminopropyl) carbodiimide (EDAC), another condensing agent, in the presence of Na^+-montmorillonite, gave rise to 2',5'-$(pA)_2$, 3',5'-$(pA)_2$, and AppA dimers (Ferris et al. 1989). Molecular dynamics simulation studies by Mathew and Luthey-Schulten (2010) using nucleotides activated with 1-methyladenine indicate that dimerization occurs more efficiently in the interlayer space than on external basal surfaces. Furthermore, 3',5' linkages are formed in preference to those involving the 2',5' positions as marked in Figure 8.2.

In the presence of Na^+-montmorillonite (at pH 8 and room temperature), the self-condensation of an RNA nucleotide activated on the phosphate group with imidazole (ImpA) (Figure 8.2) can yield oligomers of up to 10 monomer units (Ferris and Ertem 1993). Purine nucleotides predominantly yield 3',5'-linked oligomers whereas the pyrimidine counterparts largely give rise to oligomers with 2',5' linkages (Ertem 2004).

The yield of RNA oligomers is enhanced in a solution of high ionic strength but decreases when the temperature is raised to 50°C (Miyakawa et al. 2006). More interestingly, the decanucleotide adsorbed to montmorillonite can be elongated by adding fresh ImpA, allowing the mixture to stand for one day, centrifuging, and repeating the process. Thus, Ferris et al. (1996) could obtain polyadenylates containing 40–50 units after 14 *feedings* with ImpA. Joshi et al. (2000) further reported that montmorillonite can promote the formation of RNA dimers that are predominantly D-D or L-L from a mixture of D- and L-nucleotides (containing D- and L-ribose).

Following Ferris et al. (1996), Porter et al. (1999b) synthesized adenylic acid oligomers with eight or more monomers in length by repeatedly exposing a Cu^{2+}-hectorite film to a (15 mM) solution of ImpA at moderate ionic strength and pH 8. Electron spin resonance spectroscopy and X-ray diffraction suggest that the reaction involves the formation of a complex between interlayer Cu^{2+}

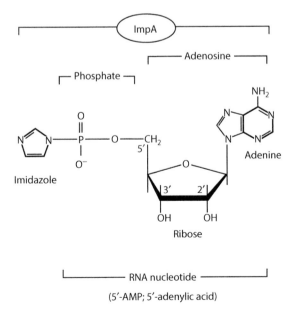

FIGURE 8.2 Chemical structures of the various units making up an RNA nucleotide, represented here by 5'-adenosine monophosphate (5'-AMP) or 5'adenylic acid. The 2' and 3' positions on the ribose ring are also indicated. The nucleotide is *activated* by attachment of imidazole to the phosphate group, giving rise to the phosphoroimidazolide of adenosine (ImpA).

ions and ImpA (Figure 8.2). Thus, the formation of RNA oligomers through self-condensation of the 5′phosphorimidazolide of nucleosides, is strongly inhibited when the montmorillonite interlayers have previously been blocked by intercalation of alkylammonium or aluminum polyoxo cations (Ertem and Ferris 1998).

Large-scale simulations of 25-mer sequences of RNA in water and aqueous suspensions of montmorillonite using molecular dynamics techniques suggest that the oligomer can bind to the silicate surface through a nucleotide base. As a result, the 3′-end of the molecule is exposed for regioselective adsorption and chain elongation (Swadling et al. 2010). We should add that montmorillonites vary widely in their ability to catalyze RNA dimerization and oligomerization. This observation is apparently related to the surface charge density of the minerals inasmuch as the catalytic activity of low-charge montmorillonites is superior to that of their high-charge counterparts (Joshi et al. 2009; Ertem et al. 2010; Aldersley and Joshi 2013).

REFERENCES

Adams, J.M., S.E. Davies, S.H. Graham, and J.M. Thomas. 1982. Catalyzed reactions of organic molecules at clay surfaces: Ester breakdown, dimerization, and lactonizations. *Journal of Catalysis* 78: 197–208.

Akiyama, M. 1978. Epimerization of L-isoleucine on Na-montmorillonite and its implication to Precambrian chemical fossils. In *Origin of Life* (Ed.) H. Noda, pp. 541–545. Tokyo, Japan: Center for Academic Publications Japan/Japan Scientific Societies Press.

Aldersley, M.F. and P.C. Joshi. 2013. RNA dimer synthesis using montmorillonite as a catalyst: The role of surface layer charge. *Applied Clay Science* 84–84: 77–82.

Alexander, R., R.I. Kagi, and A.V. Larcher. 1982. Clay catalysis of aromatic hydrogen-exchange reactions. *Geochimica et Cosmochimica Acta* 46: 219–222.

Alexander, R., R.I. Kagi, and A.V. Larcher. 1984. Clay catalysis of alkyl hydrogen exchange reactions–Reaction mechanisms. *Organic Geochemistry* 6: 755–760.

Al-Harahsheh, M., O. Al-Ayed, J. Robinson et al. 2011. Effect of demineralization and heating rate on the pyrolysis kinetics of Jordanian oil shales. *Fuel Processing Technology* 92: 1805–1811.

Almon, W.R. and W.D. Johns. 1976. Petroleum forming reactions: Clay catalyzed fatty acid decarboxylation. In *Proceedings of the International Clay Conference 1975* (Ed.) S.W. Bailey, pp. 399–409. Wilmette, IL: Applied Publishing.

Andreev, P.F., A.I. Bogomolov, A.F. Dobryanskii, and A.A. Kartsev. 1968. *Transformation of Petroleum in Nature*. Oxford, UK: Pergamon Press.

Appel, P.W.U. 1980. On the early Archaean Isua iron-formation, West Greenland. *Precambrian Research* 11: 73–87.

Arndt, S., B.B. Jørgensen, D.E. LaRow, J.J. Middelburg, R.D. Pancost, and P. Regnier. 2013. Quantifying the degradation of organic matter in marine sediments: A review and synthesis. *Earth Science Reviews* 123: 53–86.

Bailey, S.W. 1963. Polymorphism of the kaolin minerals. *American Mineralogist* 4: 1196–1209.

Belt, S.T., W.G. Allard, J. Rintatalo, L.A. Johns, A.C.T. van Duin, and S.J. Rowland. 2000. Clay and acid catalysed isomerisation and cyclisation reactions of highly branched isoprenoid (HBI) alkenes: Implications for sedimentary reactions and distributions. *Geochimica et Cosmochimica Acta* 64: 3337–3345.

Bernal, J.D. 1951. *The Physical Basis of Life*. London, UK: Routledge and Kegan Paul.

Bertrand, M., C. Bure, F. Fleury, and A. Brack. 2001. Prebiotic polymerisation of amino acid thioesters on mineral surfaces. In *Geochemistry and the Origin of Life* (Eds.) S. Nakashima, S. Maruyama, A. Brack, and B.F. Windley, pp. 51–60. Tokyo, Japan: Universal Academy Press.

Birkel, U., G. Gerold, and J. Niemeyer. 2002. Abiotic reactions of organics on clay mineral surfaces. *Developments in Soil Science* 28, Part A: 437–447.

Biswas, J. and I.E. Maxwell. 1990. Recent process- and catalyst-related developments in fluid catalytic cracking. *Applied Catalysis* 63: 197–258.

Bondy, S.C. and M.E. Harrington. 1979. L-amino acids and D-glucose bind stereospecifically to a colloidal clay. *Science* 203: 1243–1244.

Bonner, W.A. and J. Flores. 1973. On the asymmetric adsorption of phenylalanine enantiomers by kaolin. *Biosystems* 5: 103–113.

Bonner, W.A., P.R. Kavasmaneck, and F.S. Martin. 1975. Asymmetric adsorption by quartz: A model for the prebiotic origin of optical activity. *Origins of Life* 6: 367–376.

Booth, K.J., A.F. Patti, J.L. Scott, and P.J. Wallis. 2004. Organic matter transformation catalysed by clays: Model reactions for carbon sequestration in soils. *SuperSoil 2004, 3rd Australian New Zealand Soils Conference*, 1–8.

Borges, M.E. and L. Díaz. 2012. Recent developments on heterogeneous catalysts for biodiesel production by oil esterification and transesterification reactions: A review. *Renewable and Sustainable Energy Reviews* 16: 2839–2849.

Brack, A. 2006. Clay minerals and the origin of life. In *Handbook of Clay Science, Developments in Clay Science, Vol. 1* (Eds.) F. Bergaya, B.K.G. Theng, and G. Lagaly, pp. 379–391. Amsterdam, the Netherlands: Elsevier.

Brack, A. 2007. From interstellar amino acids to prebiotic catalytic peptides: A review. *Chemistry & Biodiversity* 4: 665–679.

Brindley, G.W. and W.F. Moll, Jr. 1965. Complexes of natural and synthetic Ca-montmorillonite with fatty acids (Clay-organic studies IX). *American Mineralogist* 50: 1355–1370.

Brooks, B.T. 1948. Active-surface catalysis in formation of petroleum. *American Association of Petroleum Geologists Bulletin* 32: 2269–2286.

Brooks, B.T. 1950. Catalysis and carbonium ions in petroleum formation. *Science* 111: 648–650.

Brooks, J. 1981. Organic matter in meteorites and Precambrian rocks: Clues about the origin and development of living systems. *Philosophical Transactions of the Royal Society of London A* 303: 595–609.

Bu, H., P. Yuan, H. Liu, D. Liu, and X. Zhou. 2017a. Thermal decomposition of long-chain fatty acids and its derivative in the presence of montmorillonite. *Journal of Thermal Analysis and Calorimetry* 128: 1661–1669.

Bu, H., P. Yuan, H. Liu et al. 2017b. Effects of complexation between organic matter (OM) and clay mineral on OM pyrolysis. *Geochimica et Cosmochimica Acta* 212: 1–15.

Buchanan, J.S. 2000. The chemistry of olefin production by ZSM-5 addition to catalytic cracking units. *Catalysis Today* 50: 207–212.

Bujdák, J., A. Eder, Y. Yongyai, K. Faybíková, and B.M. Rode. 1996b. Investigation on the mechanism of peptide chain prolongation on montmorillonite. *Journal of Inorganic Biochemistry* 61: 69–78.

Bujdák, J., K. Faybíková, A. Eder, Y. Yongyai, and B.M. Rode. 1995. Peptide chain elongation: A possible role of montmorillonite in prebiotic synthesis of protein precursors. *Origins of Life and Evolution of the Biosphere* 25: 431–441.

Bujdák, J., M. Remko, and B.M. Rode. 2006. Selective adsorption and reactivity of dipeptide stereoisomers in clay mineral suspension. *Journal of Colloid and Interface Science* 294: 304–308.

Bujdák, J. and B.M. Rode. 1996. The effect of smectite composition on the catalysis of peptide bond formation. *Journal of Molecular Evolution* 43: 326–333.

Bujdák, J. and B.M. Rode. 1997. Silica, alumina, and clay-catalyzed alanine peptide bond formation. *Journal of Molecular Evolution* 45: 457–466.

Bujdák, J. and B.M. Rode. 1999. The effect of clay structure on peptide bond formation catalysis. *Journal of Molecular Catalysis A: General* 144: 129–136.

Bujdák, J., H. Slosiarikova, N. Texler, M. Schwndinger, and B.M. Rode. 1994. On the possible role of montmorillonites in prebiotic peptide formation. *Monatshefte für Chemie* 125: 1033–1039.

Bujdák, J., H.L. Son, Y. Yongyai, and B.M. Rode. 1996a. The effect of reaction conditions on montmorillonite-catalysed peptide formation. *Catalysis Letters* 37: 267–272.

Burkow, I.C., E. Jørgensen, T. Meyer, Ø. Rekdal, and L.K. Sydnes. 1990. Experimental simulation of chemical transformations of aromatic compounds in sediments. *Organic Geochemistry* 15: 101–108.

Burton, F.G., R. Lohrmann, and L.E. Orgel. 1974. On the possible role of crystals in the origins of life VII. The adsorption and polymerization of phosphoramidates by montmorillonite clay. *Journal of Molecular Evolution* 3: 141–150.

Cairns-Smith, A.G. and H. Hartman. 1986. *Clay Minerals and the Origin of Life*. Cambridge, UK: Cambridge University Press.

Charlesworth, J.M. 1986. Interaction of clay minerals with organic nitrogen compounds released by kerogen pyrolysis. *Geochimica et Cosmochimica Acta* 50: 1431–1435.

Cleaves, H.J. II, A.M. Scott, F.C. Hill, J. Leszczcynski, N. Sahai, and R. Hazen. 2012. Mineral-organic interfacial processes: Potential roles in the origins of life. *Chemical Society Reviews* 41: 5502–5525.

Collins, M.J., A.N. Bishop, and P. Farrimond. 1995. Sorption by mineral surfaces: Rebirth of the classical condensation pathway for kerogen formation? *Geochimica et Cosmochimica Acta* 59: 2387–2391.

Cooper, J.E. 1962. Fatty acids in recent and ancient sediments and petroleum reservoir waters. *Nature* 193: 744–746.

Cooper, J.E. and E.E. Bray. 1963. A postulated role of fatty acids in petroleum formation. *Geochimica et Cosmochimica Acta* 27: 1113–1127.

Corma, A. 1989. Application of zeolites in fluid catalytic cracking and related processes. *Studies in Surface Science and Catalysis* 49: 49–67.

Costanzo, G., R. Saladino, C. Crestini, F. Ciciriello, and E. Di Mauro. 2007. Formamide as the main building block in the origin of nucleic acids. *BMC Evolutionary Biology* 7 (Suppl. 2): S1.

Cullers, R.L., M.J. DiMarco, D.R. Lowe, and J. Stone. 1993. Geochemistry of a silicified, felsic volcaniclastic suite from the early Archaean panorama formation, Pilbara block, Western Australia: An evaluation of depositional and post-depositional processes with special emphasis on the rear-earth elements. *Precambrian Research* 60: 99–116.

Davis, J.B. and J.P. Stanley. 1982. Catalytic effect of smectite clays in hydrocarbon generation revealed by pyrolysis–gas chromatography. *Journal of Analytical and Applied Pyrolysis* 4: 227–240.

Degens, E.T. and J. Mathéja. 1968. Origin, development, and diagenesis of biogeochemical compounds. *Journal of the British Interplanetary Society* 21: 52–82.

Degens, E.T. and J. Mathéja. 1970. Formation of organic polymers on inorganic templates. In *Prebiotic and Biochemical Evolution* (Eds.) A.P. Kimball and J. Oró, pp. 39–69. Amsterdam, the Netherlands: North Holland.

Degens, E.T., J. Mathéja, and T.A. Jackson. 1970. Template catalysis: Asymmetric polymerization of amino acids on clay minerals. *Nature* 227: 492–493.

den Hollander, M.A., M. Wissink, M. Makkee, and J.A. Moulijn. 2002. Synergy effects of ZSM-5 addition in fluid catalytic cracking of hydrotreated flashed distillate. *Applied Catalysis A: General* 223: 103–119.

Drouin, S., M. Boussafir, J.-L. Robert, P. Alberic, and A. Durand. 2010. Carboxylic acid sorption on synthetic clays in sea water: In vitro experiments and implications for organic-clay behaviour under marine conditions. *Organic Geochemistry* 41: 192–199.

Durand, B. (Ed.). 1980. *Kerogen, Insoluble Organic Matter from Sedimentary Rocks*. Paris, France: Editions Technip.

Eastman, M.P. and T.L. Porter. 2000. Polymerization of organic monomers and biomolecules on hectorite. In *Polymer–Clay Nanocomposites* (Eds.) T.J. Pinnavaia and G.W. Beall, pp. 65–93. Chichester, UK: John Wiley & Sons.

Eirich, F.R. 1981. On the polycondensation of amino acid adenylates on montmorillonites. In *Origin of Life* (Ed.) Y. Wolman, pp. 285–293. Dordrecht, the Netherlands: D. Reidel.

Eisma, E. and J.W. Jurg. 1969. Fundamental aspects of the generation of petroleum. In *Organic Geochemistry* (Eds.) G. Eglinton and M.J.T. Murphy, pp. 676–698. Berlin, Germany: Springer-Verlag.

Emam, E.A. 2013. Clays as catalysts in petroleum refining industry. *ARPN Journal of Science and Technology* 3: 356–375.

Engler, C. 1911–1912. Die Bildung der Hauptbestandteile des Erdöls. *Petrol, Zhurnal* 7: 399.

Ertem, G. 2004. Montmorillonite, oligonucleotides, RNA and origin of life. *Origins of Life and Evolution of the Biosphere* 34: 549–570.

Ertem, G. and J.P. Ferris. 1998. Formation of RNA oligomers on montmorillonite: Site of catalysis. *Origins of Life and Evolution of the Biosphere* 28: 485–499.

Ertem, G., A. Steudel, K. Emmerich, G. Lagaly, and R. Schuhmann. 2010. Correlation between the extent of catalytic activity and charge density of montmorillonites. *Astrobiology* 10: 743–749.

Espitalié, J., M. Madec, and B. Tissot. 1980. Role of mineral matrix in kerogen pyrolysis: Influence on petroleum generation and migration. *American Association of Petroleum Geologists Bulletin* 64: 59–66.

Espitalié, J., K.S. Makadi, and J. Trichet. 1984. Role of mineral matrix during kerogen pyrolysis. *Organic Geochemistry* 6: 365–382.

Faure, P., L. Schlepp, V. Burkle-Vitzthum, and M. Elie. 2003. Low temperature air oxidation of *n*-alkanes in the presence of Na-smectite. *Fuel* 82: 1751–1762.

Faure, P., L. Schlepp, L. Mansuy-Huault, M. Elie, E. Jardé, and M. Pelletier. 2006. Aromatization of organic matter induced by the presence of clays during flash pyrolysis-gas chromatography-mass spectrometry (PyGC-MS). A major analytical artifact. *Journal of Analytical and Applied Pyrolysis* 75: 1–10.

Ferris, J.P. 1993. Prebiotic synthesis on minerals: RNA oligomer formation. In *The Chemistry of Life's Origins* (Eds.) J.M. Greenberg, C.X. Mondoza-Gómez, and V. Pirronello, pp. 301–322. Dordrecht, the Netherlands: Kluwer Academic Publishers.

Ferris, J.P. 2005. Mineral catalysis and prebiotic synthesis: Montmorillonite-catalyzed formation of RNA. *Elements* 1: 145–149.

Ferris, J.P. 2006. Montmorillonite-catalysed formation of RNA oligomers: The possible role of catalysis in the origins of life. *Philosophical Transactions of the Royal Society B* 361: 1777–1786.

Ferris, J.P. and G. Ertem. 1993. Montmorillonite catalysis of RNA oligomer formation in aqueous solution, a model for the prebiotic formation of RNA. *Journal of the American Chemical Society* 115: 12270–12275.

Ferris, J.P., G. Ertem, and V.K. Agarwal. 1989. Mineral catalysis of the formation of dimers of 5'-AMP in aqueous solution: The possible role of montmorillonite clays in the prebiotic synthesis of RNA. *Origins of Life and Evolution of the Biosphere* 19: 165–178.

Ferris, J.P., A.R. Hill, Jr., R. Liu, and L.E. Orgel. 1996. Synthesis of long prebiotic oligomers on mineral surfaces. *Nature* 381: 59–61.

Ferris, J.P., C.-H. Huang, and W.J. Hagan, Jr. 1988. Montmorillonite: A multifunctional mineral catalyst for the prebiological formation of phosphate esters. *Origins of Life and Evolution of the Biosphere* 18: 121–133.

Ferris, J.P., P.C. Joshi, K.-J. Wang, S. Miyakawa, and W. Huang. 2004. Catalysis in prebiotic chemistry: Application to the synthesis of RNA oligomers. *Advances in Space Research* 33: 100–105.

Flegmann, A.W. and D. Scholefield. 1978. Thermodynamics of peptide bond formation at clay mineral surfaces. *Journal of Molecular Evolution* 12: 101–112.

Flores, J.J. and W.A. Bonner. 1974. On the asymmetric polymerization of aspartic acid enantiomers by kaolin. *Journal of Molecular Evolution* 3: 49–56.

Fraser, D.G., D. Fitz, T. Jakschitz, and B.M. Rode. 2011. Selective adsorption and chiral amplification of amino acids in vermiculite clay—Implications for the origin of biochirality. *Physical Chemistry Chemical Physics* 13: 831–838.

Frenkel, M. and L. Heller-Kallai. 1977. Racemisation of organic compounds by montmorillonite—Implications for age determinations and stereocatalysis. *Chemical Geology* 19: 161–166.

Friebele, E., A. Shimoyama, P.E. Hare, and C. Ponnamperuma. 1981b. Adsorption of amino acid enantiomers by Na-montmorillonite. *Origins of Life and Evolution of the Biosphere* 11: 173–184.

Friebele, E., A. Shimoyama, and C. Ponnamperuma. 1981a. Possible selective adsorption of enantiomers by Na-montmorillonite. In *Origin of Life* (Ed.) Y. Wolman, pp. 337–346. Dordrecht, the Netherlands: Springer.

Fripiat, J.J., P. Cloos, B. Calicis, and K. Makay. 1966. Adsorption of amino acids and peptides by montmorillonite. II. Identification of adsorbed species and decay products by infra-red spectroscopy. *Proceedings of the International Clay Conference, Jerusalem* 1: 233–245.

Fripiat, J.J. and M.I. Cruz-Cumplido. 1974. Clays as catalysts for natural processes. *Annual Review of Earth and Planetary Sciences* 2: 239–256.

Fuchida, S., H. Masuda, and K. Shinoda. 2014. Peptide formation mechanism on montmorillonite under thermal conditions. *Origins of Life and Evolution of the Biosphere* 44: 13–28.

Gabel, N. and C. Ponnamperuma. 1967. Model for origin of monosaccharides. *Nature* 216: 453–455.

Galwey, A.K. 1969a. Heterogeneous reactions in petroleum genesis and maturation. *Nature* 223: 1257–1260.

Galwey, A.K. 1969b. Reactions of alcohols adsorbed on montmorillonite and the role of minerals in petroleum genesis. *Journal of the Chemical Society D: Chemical Communications* 577–578.

Galwey, A.K. 1970. Reactions of alcohols and of hydrocarbons on montmorillonite surfaces. *Journal of Catalysis* 19: 330–342.

Galwey, A.K. 1972. The rate of hydrocarbon desorption from mineral surfaces and the contribution of heterogeneous catalytic-type processes to petroleum genesis. *Geochimica et Cosmochimica Acta* 36: 1115–1130.

Galwey, A. 2015. Some chemical and mechanistic aspects of petroleum formation. *Transactions of the Royal Society of South Africa* 70: 9–24.

Geatches, D.L., S.J. Clark, and H.C. Greenwell. 2010. Role of clay minerals in oil-forming reactions. *Journal of Physical Chemistry A* 114: 3569–3575.

Geatches, D.L., H.C. Greenwell, and S.J. Clark. 2011. Ab initio transition state searching in complex systems: Fatty acid decarboxylation in minerals. *Journal of Physical Chemistry A* 115: 2658–2667.

Gilbert, W. 1986. The RNA world. *Nature* 319: 618.

Goldstein, T.P. 1982. Clays as acid geocatalysts in generation and maturation of petroleum: Abstract. *American Association of Petroleum Geologists Bulletin* 66: 1444.

Goldstein, T.P. 1983. Geocatalytic reactions in formation and maturation of petroleum. *American Association of Petroleum Geologists Bulletin* 67: 152–159.

Greensfelder, B.S., H.H. Voge, and G.M. Good. 1949. Catalytic and thermal cracking of pure hydrocarbons. *Industrial & Engineering Chemistry* 41: 2573–2584.

Gururani, K., C.K. Pant, N. Pandey, and P. Pandey. 2012. Heat induced formation of peptides from reaction mixture of glycine-glutamic acid and glycine-leucine in presence and absence of montmorillonite clay with or without metal ions under wetting drying cycles of primitive Earth. *International Journal of Scientific & Technology Research* 1: 159–164.

Gyftopoulou, M.E., M. Millan, A.V. Bridgwater, D. Dugwell, R. Kandiyoti, and J.A. Hriljac. 2005. Pillared clays as catalysts for hydrocracking of heavy liquid fuels. *Applied Catalysis A: General* 282: 205–214.

Hanafusa, H. and S. Akabori. 1959. Polymerization of aminoacetonitrile. *Bulletin of the Chemical Society of Japan* 32: 626–630.

Hansford, R.C. 1952. Chemical concepts of catalytic cracking. In *Advances in Catalysis and Related Subjects, Vol. IV* (Eds.) W.G. Frankenburg, E.K. Rideal, and V.I. Komarewsky, pp. 1–30. New York: Academic Press.

Haresnape, J.N. 1948. The use of montmorillonite catalysts in the cracking of petroleum fractions. *Clay Minerals Bulletin* No. 2: 59–64.

Harvey, G.R., E.T. Degens, and K. Mopper. 1972a. Synthesis of nitrogen heterocycles on kaolinite. *Naturwissenschaften* 58: 624.

Harvey, G.R., K. Mopper, and E.T. Degens. 1972b. Synthesis of carbohydrates and lipids on kaolinite. *Chemical Geology* 9: 79–87.

Hashizume, H., B.K.G. Theng, S. van der Gaast, and K. Fujii. 2018. Formation of adenosine from adenine and ribose under conditions of repeated wetting and drying. (to be submitted for publication).

Hashizume, H., B.K.G. Theng, and A. Yamagishi. 2002. Adsorption and discrimination of alanine and alanyl-alanine enantiomers by allophane. *Clay Minerals* 37: 551–557.

Hashizume, H., S. van der Gaast, and B.K.G. Theng. 2010. Adsorption of adenine, cytosine, uracil, and phosphate by Mg-exchanged montmorillonite. *Clay Minerals* 45: 469–475.

Hashizume, H., S. van der Gaast, and B.K.G. Theng. 2013. Interactions of clay minerals with RNA components. In *Evolutionary Biology: Exobiology and Evolutionary Mechanisms* (Ed.) P. Pontarotti, pp. 61–79. Berlin, Germany: Springer-Verlag.

Hattori, H. 2010. Solid acid catalysts: Roles in chemical industries and new concepts. *Topics in Catalysis* 53: 432–438.

Hayatsu, R., R.L. McBeth, R.G. Scott, R.E. Botto, and R.E. Winans. 1984. Artificial coalification study: Preparation and characterization of synthetic macerals. *Organic Geochemistry* 6: 463–471.

Heller-Kallai, L. 2002. Clay catalysis in reactions of organic matter. In: *Organo-Clay Complexes and Interactions* (Eds.) S. Yariv and H. Cross, pp. 567–613. New York: Marcel Dekker.

Heller-Kallai, L., Z. Aizenshtat, and I. Miloslavski. 1984. The effect of various clay minerals on the thermal decomposition of stearic acid under "bulk flow" conditions. *Clay Minerals* 19: 779–788.

Heller-Kallai, L., T.P. Goldstein, and A. Navrotsky. 1996. Active components in clay condensates and extracts as potential geocatalysts. *Clays and Clay Minerals* 44: 393–397.

Henderson, W., G. Eglinton, P. Simmonds, and J.E. Lovelock. 1968. Thermal alteration as a contributory process to the genesis of petroleum. *Nature* 219: 1012–1016.

Hetényi, M. 1995. Simulated thermal maturation of type I and III kerogens in the presence, and absence, of calcite and montmorillonite. *Organic Geochemistry* 23: 121–127.

Hettinger, W.P., Jr. 1991. Contribution to catalytic cracking in the petroleum industry. *Applied Clay Science* 5: 445–468.

Hill, A.R. Jr., C. Böhler, and L.E. Orgel. 1998. Polymerization on the rocks: Negatively charged α-amino acids. *Origins of Life and Evolution of the Biosphere* 28: 235–243.

Horsfield, B. and A.G. Douglas. 1980. The influence of minerals on the pyrolysis of kerogens. *Geochimica et Cosmochimica Acta* 44: 1119–1131.

Hu, M., Z. Cheng, M. Zhang et al. 2014. Effect of calcite, kaolinite, gypsum, and montmorillonite on Huadian oil shale kerogen pyrolysis. *Energy Fuels* 28: 1860–1867.

Huizinga, B.J., E. Tannenbaum, and I.R. Kaplan. 1987a. The role of minerals in the thermal alteration of organic matter—III: Generation of bitumen in laboratory experiments. *Organic Geochemistry* 11: 591–604.

Huizinga, B.J., E. Tannenbaum, and I.R. Kaplan. 1987b. The role of minerals in the thermal alteration of organic matter—IV: Generation of *n*-alkanes, acyclic isoprenoids, and alkenes in laboratory experiments. *Geochimica et Cosmochimica Acta* 51: 1083–1097.

Hunt, J.M. 1979. *Petroleum Geochemistry and Geology*. San Francisco, CA: W.H. Freeman.

Ibanez, J.D., A.P. Kimball, and J. Oró. 1971. Possible prebiotic condensation of mononucleotides by cyanamide. *Science* 173: 444116.

Ito, M., N. Handa, and H. Yanagawa. 1990. Synthesis of polypeptides by microwave heating. II. Function of polypeptides synthesized during repeated hydration-dehydration cycles. *Journal of Molecular Evolution* 31: 187–194.

Jaber, M., T. Georgelin, H. Bazzi et al. 2014. Selectivities in adsorption and peptidic condensation in the (arginine and glutamic acid)/monmorillonite clay system. *Journal of Physical Chemistry C* 118: 25447–25455.

Jackson, T.A. 1971a. Preferential polymerization and adsorption of L-optical isomers of amino acids relative to D-optical isomers on kaolinite templates. *Chemical Geology* 7: 295–306.

Jackson, T.A. 1971b. Evidence for selective adsorption and polymerization of the L-optical isomers of amino acids relative to the D-optical isomers on the edge faces of kaolinite. *Experientia* 27: 242–243.

Johns, W.D. 1979. Clay mineral catalysis and petroleum generation. *Annual Review of Earth and Planetary Sciences* 7: 1983–1988.

Johns, W.D. 1982. The role of the clay mineral matrix in petroleum generation during burial diagenesis. In *International Clay Conferences 1981* (Eds.) H. van Olphen and F. Veniale, pp. 655–664. Amsterdam, the Netherlands: Elsevier.

Johns, W.D. and T.E. McKallip. 1989. Burial diagenesis and specific catalytic activity of illite-smectite clays from Vienna basin, Austria. *American Association of Petroleum Geologists Bulletin* 73: 472–482.

Johns, W.D. and A. Shimoyama. 1972. Clay minerals and petroleum-forming reactions during burial and diagenesis. *American Association of Petroleum Geologists Bulletin* 56: 2160–2167.

Joshi, P.C., M.F. Aldersley, J.W. Delano, and J.P. Ferris. 2009. Mechanism of montmorillonite catalysis in the formation of RNA oligomers. *Journal of the American Chemical Society* 131: 13369–13374.

Joshi, P.C., S. Pitsch, and J.P. Ferris. 2000. Homochiral selection in the montmorillonite-catalyzed and uncatalyzed prebiotic synthesis of RNA. *Chemical Communications* 2497–2498.

Joyce, G.F. 2002. The antiquity of RNA-based evolution. *Nature* 418: 214–221.

Julg, A. 1989. Origin of the L-homochirality of amino acids in the proteins of living organisms. In *Molecules in Physics, Chemistry, and Biology. Topics in Molecular Organization and Engineering, Vol. 4* (Ed.) J. Maruani, pp. 33–52. Dordrecht, the Netherlands: Springer.

Jurg, J.W. and E. Eisma. 1964. Petroleum hydrocarbons: Generation from a fatty acid. *Science* 144: 1451–1452.

Kameda, J., A. Yamagishi, and T. Kogure. 2005. Morphological characteristics of ordered kaolinite: Investigations using electron back-scattered diffraction. *American Mineralogist* 90: 1462–1465.

Katchalsky, A. 1973. Prebiotic synthesis of biopolymers on inorganic templates. *Die Naturwissenschaften* 60: 215–220.

Kawamura, K., and I.R. Kaplan. 1986. Dicarboxylic acids generated by thermal alteration of kerogen and humic acids. *Geochimica et Cosmochimica Acta* 51: 3201–3207.

Kawamura, K., E. Tannenbaum, B.J. Huizinga, and I.R. Kaplan. 1986. Volatile organic acids generated from kerogen during laboratory heating. *Geochemical Journal* 20: 51–59.

Keil, R.G., D.B. Montluçon, F.G. Prahl, and J.I. Hedges. 1994. Sorptive preservation of labile organic matter in marine sediments. *Nature* 370: 549–552.

Kennedy, M.J., S.C. Löhr, S.A. Fraser, and E.T. Baruch. 2014. Direct evidence for organic carbon preservation as clay-organic nanocomposites in a Devonian black shale; from deposition to diagenesis. *Earth and Planetary Science Letters* 388: 59–70.

Kloprogge, J.T., L.V. Duong, and R.L. Frost. 2005. A review of the synthesis and characterisation of pillared clays and related porous materials for cracking of vegetable oils to produce biofuels. *Environmental Geology* 47: 967–981.

Kok, M.V. 2006. Effect of clay on crude oil combustion by thermal analysis techniques. *Journal of Thermal Analysis and Calorimetry* 84: 361–366.

Kok, M.V. 2012. Clay concentration and heating rate effect on crude oil combustion by thermogravimetry. *Fuel Processing Technology* 96: 134–139.

Kok, M.V. and A.S. Gundogar. 2010. Effect of different clay concentrations on crude oil combustion kinetics by thermogravimetry. *Journal of Thermal Analysis and Calorimetry* 99: 779–783.

Kvenvolden, K.A. 1966. Molecular distribution of normal fatty acids and paraffins in some lower Cretaceous sediments. *Nature* 209: 573–575.

Lagaly, G., M. Ogawa, and I. Dékány. 2006. Clay mineral organic interactions. In *Handbook of Clay Science, Developments in Clay Science, Vol. 1* (Eds.) F. Bergaya, B.K.G. Theng, and G. Lagaly, pp. 309–377. Amsterdam, the Netherlands: Elsevier.

Lahav, N. 1978. Peptide formation in the prebiotic era: Thermal condensation of glycine in fluctuating clay environments. *Science* 201: 67–69.

Lahav, N. and S. Chang. 1976. The possible role of solid surface area in condensation reactions during chemical evolution: Reevaluation. *Journal of Molecular Evolution* 8: 357–380.

Lahav, N. and D. White. 1980. A possible role of fluctuating clay-water systems in the production of ordered prebiotic oligomers. *Journal of Molecular Evolution* 16: 11–21.

Lahav, N., D. White, and S. Chang. 1978. Peptide formation in the prebiotic era: Thermal condensation of glycine in fluctuating clay environments. *Science* 201: 67–69.

Lambert, J.-F. 2008. Adsorption and polymerization of amino acids on mineral surfaces: A review. *Origins of Life and Evolution of the Biosphere* 38: 211–242.

Lao, Y., J. Korth, J. Ellis, and P.T. Crisp. 1989. Heterogeneous reactions of 1-pristene catalysed by clays under simulated geological conditions. *Organic Geochemistry* 14: 375–379.

Larcher, A.V., R. Alexander, and R.I. Kagi. 1988. Differences in reactivities of sedimentary hopane diastereomers when heated in the presence of clays. *Organic Geochemistry* 13: 665–669.

Larcher, A.V., R. Alexander, S.J. Rowland, and R.I. Kagi. 1985. Acid catalysis of alkyl hydrogen exchange and configurational isomerisation reactions: Acyclic isoprenoid acids. *Advances in Organic Geochemistry* 10: 1015–1021.

Lawless, J.G. and N. Levi. 1979. The role of metal ions in chemical evolution: Polymerization of alanine and glycine in a cation-exchanged clay environment. *Journal of Molecular Evolution* 13: 281–286.

Li, S., S. Guo, and X. Tan. 1998. Characteristics and kinetics of catalytic degradation of immature kerogen in the presence of mineral and salt. *Organic Geochemistry* 29: 1431–1439.

Liu, H., L. Cao, B. Wei, Y. Fan, G. Shi, and X. Bao. 2012. *In-situ* synthesis and catalytic properties of ZSM-5/rectorite composites as propylene boosting additive in fluid catalytic cracking process. *Chinese Journal of Chemical Engineering* 20: 158–166.

Liu, H., P. Yuan, Z. Qin et al. 2013. Thermal degradation of organic matter in the interlayer clay-organic complex: A TG-FTIR study on a montmorillonite/12-aminolauric acid system. *Applied Clay Science* 80–81: 398–406.

Liu, Y., R. Sotelo-Boyás, K. Murata, T. Minowa, and K. Sakanishi. 2012. Production of bio-hydrogenated diesel by hydrotreatment of high-acid-value waste cooking oil over ruthenium catalyst supported on Al-polyoxocation-pillared montmorillonite. *Catalysts* 2: 171–190.

Losse, G. and K. Anders. 1961. Die Polymerisation von α-Amino-propionitril an mineralischen Trägern als Modell für die primäre Bildung von Eiweißstoffen auf der Erde. *Hoppe-Seyler Zeitschrift für physiologische Chemie* 323: 111–115.

Lu, S.-T. and I.R. Kaplan. 1989. Pyrolysis of kerogens in the absence and presence of montmorillonite—II. Aromatic hydrocarbons generated at 200 and 300°C. *Organic Geochemistry* 14: 501–510.

Lu, S.-T., E. Ruth, and I.R. Kaplan. 1989. Pyrolysis of kerogens in the absence and presence of montmorillonite–I. The generation, degradation and isomerization of steranes and triterpanes at 200 and 300°C. *Organic Geochemistry* 14: 491–499.

Manos, G., I.Y. Yusof, N. Papayannakos, and N.H. Gangas. 2001. Catalytic cracking of polyethylene over clay catalysts: Comparison with an ultrastable Y zeolite. *Industrial & Engineering Chemistry Research* 40: 2220–2225.

Marshall-Bowman, K., S. Ohara, D.A. Sverjensky, R.M. Hazen, and H.J. Cleaves. 2010. Catalytic peptide hydrolysis by mineral surface: Implications for prebiotic chemistry. *Geochimica et Cosmochimica Acta* 74: 5852–5861.

Mathew, D.C. and Z. Luthey-Schulten. 2010. Influence of montmorillonite on nucleotide oligomerization reactions: A molecular dynamic study. *Origins of Life and Evolution of the Biosphere* 40: 303–317.

Maurel, M.-C. and F. Leclerc. 2016. From foundation stones to life: Concepts and results. *Elements* 12: 407–412.

Mayer, L.M. 1994. Relationships between mineral surfaces and organic carbon concentrations in soils and sediments. *Chemical Geology* 114: 347–363.

McCullough, J.J. and R.M. Lemmon. 1974. The question of the possible asymmetric polymerization of aspartic acid on kaolinite. *Journal of Molecular Evolution* 3: 57–61.

Mifsud, A., V. Fornes, and J.A. Rausell-Colom. 1970. Cationic complexes of vermiculite with L-ornithine. *Reunion Hispano-Belga de Minerales de la Arcilla*, Madrid, pp. 121–127.

Milliken, T.H., A.G. Oblad, and G.A. Mills. 1955. Use of clays as petroleum cracking catalysts. *Clays and Clay Minerals* 1: 314–326.

Miloslavski, I., L. Heller-Kallai, and Z. Aizenshtat. 1991. Reactions of clay condensates with *n*-alkanes: Comparison between clay volatiles and clay condensates. *Chemical Geology* 91: 287–296.

Min, E. 1994. Development of pillared clays for industrial catalysis. *Studies in Surface Science and Catalysis* 83: 443–452.

Miyakawa, S., P.C. Joshi, M.J. Gaffey et al. 2006. Studies in the mineral and salt-catalyzed formation of RNA oligomers. *Origins of Life and Evolution of the Biosphere* 36: 343–361.

Mojzsis, S.J., G. Arrhenius, K.D. McKeegan, T.M. Harrison, A.P. Nutman, and C.R.L. Friend. 1996. Evidence for life on Earth before 3800 million years ago. *Nature* 384: 55–59.

Moorbath, S. 1994. Age of the oldest rocks with biogenic components. *Journal of Biological Physics* 20: 85–94.

Mortland, M.M. 1970. Clay-organic complexes and interactions. *Advances in Agronomy* 22: 75–117.

Moseley, C.G. 1984. Eugene Houdry, catalytic cracking, and World War II aviation gasoline. *Journal of Chemical Education* 61: 65–66.

Naidja, A. and B. Siffert. 1990. Oxidative decarboxylation of isocitric acid in the presence of montmorillonite. *Clay Minerals* 25: 27–37.

Negrón-Mendoza, A., G. Albarran, and S. Ramos-Bernal. 1996. Clays as natural catalyst in prebiotic processes. In *Chemical Evolution: Physics of the Origin and Evolution of Life* (Eds.) J. Chela-Flores and F. Raulin, pp. 97–106. Dordrecht, the Netherlands: Springer.

Negrón-Mendoza, A., S. Ramos, and G. Albarrán. 1995. Enhanced decarboxylation reaction of carboxylic acids in clay minerals. *Radiation Physics and Chemistry* 46: 565–568.

Negrón-Mendoza, A. and S. Ramos-Bernal. 2004. The role of clays in the origin of life. In *Cellular Origins, Life in Extreme Habitats and Astrobiology, Vol. 6* (Ed.) J. Seckbach, pp. 181–194. Dordrecht, the Netherlands: Springer.

Negrón-Mendoza, A., S. Ramos-Bernal, G. Albarrán, and J. Reyes-Gasga. 1997. Selective blocking sites in the structure clay-nanometric montmorillonite. *NanoStructured Materials* 9: 209–212.

Negrón-Mendoza, A., S. Ramos-Bernal, M. Colín-Garcia, and F.G. Mosqueira. 2015. Radiation-induced catalysis of fatty acids adsorbed onto clay minerals. *Journal of Radioanalytical and Nuclear Chemistry* 303: 1507–1511.

Occelli, M.L. 1983. Catalytic cracking with an interlayered clay: A two-dimensional molecular sieve. *Industrial & Engineering Chemistry Product Research and Development* 22: 553–559.

Occelli, M.L., S.D. Landau, and T.J. Pinnavaia. 1984. Cracking selectivity of a delaminated clay catalyst. *Journal of Catalysis* 90: 256–260.

Odom, D.G., N. Lahav, and S. Chang. 1979. Prebiotic nucleotide oligomerization in a fluctuating environment: Effects of kaolinite and cyanamide. *Journal of Molecular Evolution* 12: 259–264.

Orgel, L.E. 1998. The origin of life—A review of facts and speculations. *Trends in Biochemical Sciences* 23: 491–495.

Otake, T., T. Taniguchi, Y. Furukawa, F. Kawamura, H. Nakazawa, and T. Kakegawa. 2011. Stability of amino acids and their oligomerization under high-pressure conditions: Implications for prebiotic chemistry. *Astrobiology* 11: 799–813.

Paecht-Horowitz, M. 1976. Clays as possible catalysts for peptide formation in the prebiotic era. *Origins of Life* 7: 369–381.

Paecht-Horowitz, M. 1977. The mechanism of clay catalyzed polymerization of amino acid adenylates. *Biosystems* 9: 93–98.

Paecht-Horowitz, M., J. Berger, and A. Katchalsky. 1970. Prebiotic synthesis of polypeptides by heterogeneous polycondensation of amino acid adenylates. *Nature* 228: 636–639.

Paecht-Horowitz, M. and F.R. Eirich. 1988. The polymerization of amino acid adenylates on sodium montmorillonite with preadsorbed polypeptides. *Origins of Life and Evolution of the Biosphere* 18: 359–387.

Paecht-Horowitz, M. and N. Lahav. 1977. Polymerization of alanine in the presence of a non-swelling montmorillonite. *Journal of Molecular Evolution* 10: 73–76.

Pal, S., J.M. Bollag, and P.M. Huang. 1994. Role of abiotic and biotic catalysts in the transformation of phenolic compounds through oxidative coupling reactions. *Soil Biology and Biochemistry* 26: 813–820.

Pan, C., A. Geng, N. Zhong, and J. Liu. 2010a. Kerogen pyrolysis in the presence and absence of water and minerals: Steranes and triterpenoids. *Fuel* 89: 336–345.

Pan, C., A. Geng, N. Zhong, J. Liu, and L. Yu. 2008. Kerogen pyrolysis in the presence and absence of water and minerals. 1. Gas components. *Energy Fuels* 22: 416–427.

Pan, C., A. Geng, N. Zhong, J. Liu, and L. Yu. 2009. Kerogen pyrolysis in the presence and absence of water and minerals: Amounts and compositions of bitumen and liquid hydrocarbons. *Fuel* 88: 909–919.

Pan, C., L. Jiang, J. Liu, S. Zhang, and G. Zhu. 2010b. The effects of calcite and montmorillonite on oil cracking in confined pyrolysis experiments. *Organic Geochemistry* 41: 611–626.

Pant, C.K., H. Lata, H.D. Pathak, and M.S. Mehata. 2009. Heat-initiated prebiotic formation of peptides from glycine/aspartic acid and glycine/valine in aqueous environment and clay suspension. *International Journal of Astrobiology* 8: 107–115.

Pavlovskaya, T.E., A.G. Pasinskyi, and A.I. Grebenikova. 1960. The production of amino acids under the influence of ultraviolet rays on solutions of formaldehyde and ammonium salts in the presence of adsorbents. *Doklady Akademii Nauk SSSR* 135: 743–746.

Ponnamperuma, C., A. Shimoyama, and E. Friebele. 1982. Clay and the origin of life. *Origins of Life* 12: 9–40.

Porter, T.L., M.P. Eastman, E. Bain, and S. Begay. 2001. Analysis of peptides synthesized in the presence of SAz-1 montmorillonite and Cu^{2+} exchanged hectorite. *Biophysical Chemistry* 91: 115–124.

Porter, T.L., M.P. Eastman, M.E. Hagerman, L.B. Price, and R.F. Shand. 1998a. Site-specific prebiotic oligomerization reactions of glycine on the surface of hectorite. *Journal of Molecular Evolution* 47: 373–377.

Porter, T.L., M.P. Eastman, L.B. Price, and R.F. Shand. 1999. Synthesis of prebiotic peptides and oligonucleotides on clay mineral surfaces: A scanning force microscopy study. In *Atomic Force Microscopy/Scanning Tunneling Microscopy* (Eds.) S.H. Cohen and M.L. Lightbody, pp. 189–196. Dordrecht, the Netherlands: Kluwer Academic/Plenum Publishers.

Porter, T.L., M.P. Eastman, R. Whitehorse, E. Bain, and K. Manygoats. 2000. The interaction of biological molecules with clay minerals: A scanning force microscopy study. *Scanning* 22: 1–5.

Porter, T.L., K. Manygoats, M. Bradley et al. 1998b. Surface and intergallery catalytic properties of Cu(II)-exchanged hectorite: A scanning force microscope study. *Journal of Vacuum Science & Technology A* 16: 926–931.

Porter, T.L., R. Whitehorse, M.P. Eastman, and E.D. Bain. 1999. Studies on the reaction of the 5′phosphorimidazolide of adenosine with Cu(II)-exchanged hectorite. *Applied Physics Letters* 75: 2674–2676.

Rahman, H.M., M. Kennedy, S. Löhr, and D.N. Dewhurst. 2017. Clay-organic association as a control on hydrocarbon generation in shale. *Organic Geochemistry* 105: 42–55.

Rao, M., D.G. Odom, and J. Oró. 1980. Clays in prebiological chemistry. *Journal of Molecular Evolution* 15: 317–331.

Regtop, R.A., P.T. Crisp, J. Ellis, and C.J.R. Fookes. 1986. 1-Pristene as a precursor for 2-pristene in pyrolysates of oil shale from Condor, Australia. *Organic Geochemistry* 9: 233–236.

Rimola, A., L. Rodriguez-Santiago, P. Ugliengo, and M. Sodupe. 2007. Is the peptide bond formation activated by Cu^{2+} interactions? Insights from density functional calculations. *Journal of Physical Chemistry B* 111: 5740–5747.

Rishpon, J., P.J. O'Hara, N. Lahav, and J.G. Lawless. 1982. Interaction between ATP, metal ions, glycine, and several minerals. *Journal of Molecular Evolution* 18: 179–184.

Robertson, R.H.S. 1948. Clay minerals as catalysts. *Clay Minerals Bulletin* 2 (1): 47–54.

Rode, B.M., H.L. Son, Y. Suwannachot, and J. Bujdák. 1999. The combination of salt induced formation reaction and clay catalysis: A way to higher peptides under primitive Earth conditions. *Origins of Life and Evolution of the Biosphere* 29: 273–286.

Rohlfing, D.L. 1976. Thermal polyamino acids: Synthesis at less than 100 degrees C. *Science* 193: 68–70.

Rong, T.-J. and J.-K. Xiao. 2002. The catalytic cracking activity of the kaolin-group minerals. *Materials Letters* 57: 297–301.

Rosa-Brussin, M.F. 1995. The use of clays for the hydrotreatment of heavy crude oils. *Catalysis Reviews–Science and Engineering* 37: 1–100.

Rose, H.R., D.R. Smith, J.V. Hanna, A.J. Palmisano, and M.A. Wilson. 1992. Comparison of the effect of minerals on aromatization reactions during kerogen pyrolysis. *Fuel* 71: 355–360.

Saladino, R., C. Crestini, U. Ciambecchini, F. Ciciriello, G. Costanzo, and E. Di Mauro. 2004. Synthesis and degradation of nucleobases and nucleic acids by formamide in the presence of montmorillonites. *ChemBioChem* 5: 1558–1566.

Saladino, R., V. Neri, and C. Crestini. 2010. Role of clays in the prebiotic synthesis of sugar derivatives from formamide. *Philosophical Magazine* 90: 2329–2337.

Salmon, V., S. Derenne, E. Lallier-Vergès, C. Largeau, and B. Beaudoin. 2000. Protection of organic matter by mineral matrix in a Cenomanian black shale. *Organic Geochemistry* 31: 463–474.

Sani, Y.M., W.M.A.W. Daud, and A.R.A. Aziz. 2014. Activity of solid acid catalysts for biodiesel production: A critical review. *Applied Catalysis A: General* 470: 140–161.

Saxby, J.D., P. Chatfield, G.H. Taylor, J.D. Fitzgerald, I.R. Kaplan, and S.-T. Lu. 1992. Effect of clay minerals on products from coal maturation. *Organic Geochemistry* 18: 373–383.

Schoonen, M.A.A., Y. Xu, and D.R. Strongin. 1998. An introduction to geocatalysis. *Journal of Geochemical Exploration* 62: 201–215.

Schopf, J.W. 1993. Microfossils of the early Archean apex chert: New evidence of the antiquity of life. *Science* 260: 640–646.

Schulten, H.-R., P. Leinweber, and B.K.G. Theng. 1996. Characterization of organic matter in an interlayer clay-organic complex from soil by pyrolysis methylation-mass spectrometry. *Geoderma* 69: 105–118.

Schwartz, A.W. and G.J.F. Chittenden. 1977. Synthesis of uracil and thymine under simulated prebiotic conditions. *Biosystems* 9: 87–92.

Seewald, J.S. 2003. Organic-inorganic interactions in petroleum-producing sedimentary basins. *Nature* 426: 327–333.

Semwal, S., A.K. Arora, R.P. Badoni, and D.K. Tuli. 2011. Biodiesel production using heterogeneous catalysts. *Biosource Technology* 102: 2151–2161.

Shimoyama, A., N. Blair, and C. Ponnamperuma. 1978. Synthesis of amino acids under primitive earth conditions in the presence of clays. In *Origin of Life* (Ed.) H. Noda, pp. 95–99. Tokyo, Japan: Center for Academic Publications Japan/Japan Scientific Societies Press.

Shimoyama, A., K. Hayakawa, and K. Harada. 1993. Conversion of oleic acid to monocarboxylic acids and γ-lactones by laboratory heating experiments in relation to organic diagenesis. *Geochemical Journal* 27: 59–70.

Shimoyama, A. and W.D. Johns. 1971. Catalytic conversion of fatty acids to petroleum-like paraffins and their maturation. *Nature* 232: 140–144.

Sieskind, O., G. Joly, and P. Albrecht. 1979. Simulation of the geochemical transformations of sterols: Superacid effect of clay minerals. *Geochimica et Cosmochimica Acta* 43: 1675–1679.

Sieskind, O. and G. Ourisson. 1972. Hydrocarbures formés par la craquage thermocatalytique de l'acide stearic en presence de montmorillonite. *Comptes Rendus de l'Académie des Sciences Série C* 274: 2186–2189.

Siffert, B. and A. Naidja. 1987. Decarboxylation catalytique de l'acide oxaloacetique en presence de montmorillonite. *Clay Minerals* 22: 435–446.

Skoularikis, N.D., R.W. Coughlin, A. Kostapapas, K. Carrado, and S.L. Suib. 1988. Catalytic performance of iron (III) and chromium (III) exchanged pillared clays. *Applied Catalysis* 39: 61–76.

Solomon, D.H. 1968. Clay minerals as electron acceptors and/or donors in organic reactions. *Clays and Clay Minerals* 16: 13–39.

Son, H.L., Y. Suwannachot, J. Bujdák, and B.M. Rode. 1998. Salt-induced peptide formation from amino acids in the presence of clays and related catalysts. *Inorganica Chimica Acta* 272: 89–94.

Soni, V.K., P.R. Sharma, G. Choudhary, S. Pandey, and R.K. Sharma. 2017. Ni/Co-natural clay as green catalyst for microalgae oil to diesel-grade hydrocarbons conversion. *ACS Sustainable Chemistry & Engineering* 5: 5351–5359.

Srivatsan, S.G. 2004. Modeling prebiotic catalysis with nucleic acid-like polymers and its implications for the proposed RNA world. *Pure and Applied Chemistry* 76: 2085–2099.

Strigunkova, T.F., G.A. Lavrentiev, and V.A. Otroshchenko. 1986. Abiogenic synthesis of oligonucleotides on kaolinite under the action of ultraviolet radiation. *Journal of Molecular Evolution* 23: 269–278.

Swadling, J.B., P.V. Coveney, and H.C. Greenwell. 2010. Clay minerals mediate folding and regioselective interactions of RNA: A large-scale atomistic simulation study. *Journal of the American Chemical Society* 132: 13750–13764.

Swarnakar, R., K.B. Brandt, and R.A. Kydd. 1996. Catalytic activity of Ti- and Al-pillared montmorillonite and beidellite for cumene cracking and hydrocracking. *Applied Catalysis A: General* 142: 61–71.

Tang, Y.C. and M. Stauffer. 1995. Formation of pristene, pristane and phytane: Kinetic study by laboratory pyrolysis of Monterey source rock. *Organic Geochemistry* 23: 451–460.

Tannenbaum, E., B.J. Huizinga, and I.R. Kaplan. 1986a. Role of minerals in thermal alteration of organic matter—II: A material balance. *American Association of Petroleum Geologists Bulletin* 70: 1156–1165.

Tannenbaum, E. and I.R. Kaplan. 1985. Role of minerals in the thermal alteration of organic matter—I: Generation of gases and condensates under dry condition. *Geochimica et Cosmochimica Acta* 49: 2589–2604.

Tannenbaum, E., E. Ruth, and I.R. Kaplan. 1986b. Steranes and triterpanes generated from kerogen pyrolysis in the absence and presence of minerals. *Geochimica et Cosmochimica Acta* 50: 805–812.

Theng, B.K.G. 1974. *The Chemistry of Clay-Organic Reactions*. London, UK: Adam Hilger.

Theng, B.K.G. 2012. *Formation and Properties of Clay-Polymer Complexes*, 2nd ed. Amsterdam, the Netherlands: Elsevier.

Theng, B.K.G. and H. Hashizume. 1996. Clay-activated transformation of humic acid at high pressure and temperature. *Applied Clay Science* 10: 431–437.

Theng, B.K.G., K.R. Tate, and P. Becker-Heidmann. 1992. Towards establishing the age, location, and identity of the inert soil organic matter of a Spodosol. *Zeitschrift für Pflanzenernährung und Bodenkunde* 155: 181–184.

Theng, B.K.G. and G.F. Walker. 1970. Interactions of clay minerals with organic monomers. *Israel Journal of Chemistry* 8: 417–424.

Thomas, C.L., J. Hickey, and G. Stecker. 1950. Chemistry of clay cracking catalysts. *Industrial & Engineering Chemistry* 42: 866–871.

Tissot, B.P. and D.H. Welte. 1984. *Petroleum Formation and Occurrence.* Berlin, Germany: Springer-Verlag.
Tong, D.S., Y.M. Zheng, W.H. Yu, L.M. Wu, and C.H. Zhou. 2014. Catalytic cracking of rosin over acid-activated montmorillonite catalysts. *Applied Clay Science* 100: 123–128.
Vandenbroucke, M. and C. Largeau. 2007. Kerogen origin, evolution and structure. *Organic Geochemistry* 38: 719–833.
Ventilla, M. and F. Egami. 1976. Formation of amino acids and polypeptides from formaldehyde and hydroxylamine in modified sea medium. *Proceedings of the Japan Academy* 52: 21–24.
Voge, H.H. 1958. Catalytic cracking. In *Catalysis Vol. 6* (Ed.) P.H. Emmett, pp. 407–499. New York: Reinhold.
Vogt, E.T.C. and B.M. Weckhuysen. 2015. Fluid catalytic cracking: Recent developments on the grand old lady of zeolite catalysis. *Chemical Society Reviews* 44: 7342–7370.
Volkova, L.D., N.A. Zakarina, and A.K. Akurpekova. 2014. Cracking of heavy vacuum gas oils over (HCeY) zeolite-containing, aluminum-pillared montmorillonite and activated kaolinite clays. *Petroleum Chemistry* 54: 37–41.
Voudrias, E.A. and M. Reinhard. 1987. Abiotic organic reactions at mineral surfaces. In *Geochemical Processes at Mineral Surfaces, Chapter 22, ACS Symposium Series* 323: 462–486.
Wallis, P.J., K.J. Booth, A.F. Pratt, and J.L. Scott. 2006. Oxidative coupling revisited: Solvent-free, heterogeneous and in water. *Green Chemistry* 8: 333–337.
Wang, M.C. and P.M. Huang. 1987. Catalytic polymerization of hydroquinone by nontronite. *Canadian Journal of Soil Science* 67: 867–875.
Wang, M.C. and P.M. Huang. 1989. Pyrogallol transformation as catalyzed by nontronite, bentonite, and kaolinite. *Clays and Clay Minerals* 37: 525–531.
Wei, B., H. Liu, T. Li, L. Cao, Y. Fan, and X. Bao. 2010. Natural rectorite mineral: A promising substitute of kaolin for in-situ synthesis of fluid catalytic cracking catalysts. *AIChE Journal* 56: 2913–2922.
Wei, Z., J.M. Moldowan, J. Dahl, T.P. Goldstein, and D.M. Jarvie. 2006b. The catalytic effects of minerals on the formation of diamondoids from kerogen macromolecules. *Organic Geochemistry* 37: 1421–1436.
Wei, Z., J.M. Moldowan, and A. Paytan. 2006a. Diamondoids and molecular biomarkers generated from modern sediments in the absence and presence of minerals during hydrous pyrolysis. *Organic Geochemistry* 37: 891–911.
White, D.H. and J.C. Erickson. 1980. Catalysis of peptide bond formation by histidyl-histidine in a fluctuating clay environment. *Journal of Molecular Evolution* 16: 279–290.
White, D.H. and J.C. Erickson. 1981. Enhancement of peptide formation by polyribonucleotides on clay surfaces in fluctuating environments. *Journal of Molecular Evolution* 17: 19–26.
White, D.H., R.M. Kennedy, and J. Macklin. 1984. Acyl silicates and acyl aluminates as activated intermediates in peptide formation on clays. *Origins of Life and Evolution of the Biosphere* 14: 273–278.
Wilson, M.A., S.A. McCarthy, P.J. Collin, and D.E. Lambert. 1986. Alkene transformations catalysed by mineral matter during oil shale pyrolysis. *Organic Geochemistry* 5: 245–253.
Wilson, M.C. and A.K. Galwey. 1976. Reactions of stearic acid, of *n*-dodecanol and of cyclohexanol on the clay minerals illite, kaolinite and montmorillonite. *Journal de Chimie Physique* 73: 441–446.
Wu, L.M., C.H. Zhou, J. Keeling, D.S. Tong, and W.H. Yu. 2012. Towards an understanding of the role of clay minerals in crude oil formation, migration and accumulation. *Earth-Science Reviews* 115: 373–386.
Xiao, Y., J.M. Longo, G.B. Hieshima, and R.J. Hill. 1997. Understanding the kinetics and mechanisms of hydrocarbon thermal cracking: An *ab initio* approach. *Industrial & Engineering Chemistry Research* 36: 4033–4040.
Yanagawa, H., K. Kojima, M. Ito, and N. Handa. 1990. Synthesis of polypeptides by microwave heating I. Formation of polypeptides during repeated hydration-dehydration cycles and their characterization. *Journal of Molecular Evolution* 31: 180–186.
Yang, D., S. Peng, M.R. Hartman et al. 2013. Enhanced transcription and translation in clay hydrogel and implications for early life evolution. *Scientific Reports* 3: 3165. doi:10.1038/srep03165.
Yariv, S. and H. Cross (Eds.). 2002. *Organo-Clay Complexes and Interactions.* New York: Marcel Dekker.
Yariv, S. and S. Shoval. 1982. The effects of thermal treatment on associations between fatty acids and montmorillonite. *Israel Journal of Chemistry* 22: 259–265.
Yoshino, D., R. Hayatsu, and E. Anders. 1971. Origin of organic matter in early solar systems—III. Amino acids: Catalytic synthesis. *Geochimica et Cosmochimica Acta* 35: 927–938.
Youatt, J.B. and R.D. Brown. 1981. Origins of chirality in nature: A reassessment of the postulated role of bentonite. *Science* 212: 1145–1146.

Yu, C.-H., S.Q. Newton, D.M. Miller, B.J. Teppen, and L. Schäfer. 2001. *Ab initio* study of the nonequivalence of adsorption of D- and L-peptides on clay mineral surfaces. *Structural Chemistry* 12: 393–398.

Yuan, P., H. Liu, D. Liu, D. Tan, W. Yan, and H. He. 2013. Role of the interlayer space of montmorillonite in hydrocarbon generation: An experimental study based on high temperature-pressure pyrolysis. *Applied Clay Science* 75–76: 82–91.

Zaia, D.A.M. 2012. Adsorption of amino acids and nucleic acid bases onto minerals: A few suggestions for prebiotic chemistry experiments. *International Journal of Astrobiology* 11: 229–234.

Zaug, A.J. and T.R. Cech. 1986. The intervening sequence RNA of tetrahymena is an enzyme. *Science* 231: 470–475.

Zheng, S.-Q., Y. Han, X-H. Huang et al. 2009. Possibility of using sepiolite as the matrix of FCC catalysts. *Kemija u Industriji* 58: 507–514.

Index

Note: Page numbers followed by f and t refer to figures and tables respectively.

1,4,-diazabicyclo[2,2,2]octane (Dabco) compound, 176–177, 263
1-ethyl-3-(3-dimethylaminopropyl) carbodiimide (EDAC), 403
2,2′-azobisisobutyramidine (AIBA), 325
3-aminopropyl triethoxysilane (APTES), 181–183, 182f, 238, 283

A

AAA. *See* Amino acid adenylates (AAA)
Acetic acid, 138t, 143t, 146t, 149t, 166t, 171t, 178t, 243t, 350t, 357
    alcohols acetylation with, 25t
    esterification, 25t, 177, 178t
    methanol degradation and, 19t
Acid activation, 133
    clay minerals, 135
    metakaolinite, 155
    MMT, 134, 158
    organic reactions catalyzed by, 137t–154t
    sepiolite, 157
    smectites, 131, 136
Acid and base, 85–86
Acidic aluminosilicate (MS-25), 396
Acid strength, 86–94, 88f
    distribution curves for MMT, 96f
    Hammett indicators for visual estimation, 88t, 89t
    representative clay minerals, 92t
Acylations, FC reaction, 222, 271
Advanced oxidation processes (AOP), 355
Aging process, 136
AIBA (2,2′-azobisisobutyramidine), 325
Al^{3+}-MMT, 156, 158, 307
    Brønsted acidity, 109
    as catalyst, 310
    MA, 278–279
    Mukaiyama aldol reaction, 279–280
    relative inactivity, 267
Aldehyde, 143t, 152t, 162t, 233t
    hydrogenation, 23t
    olefination, 285
Aldehydes/ketones, reduction, 137t, 141t, 143t, 153t, 159t, 161t, 363
Alder–ene reaction, 261–262, 262f
Alkenes, 20t, 24t, 94t, 139t, 140t, 142t, 146t, 233t
    epoxidation/oxidation, 362
    oligomerization, 310
Alkylation, 271
    anisole, 272
    with benzyl alcohol, 223
    Brønsted acidity, 272
    catalytic activity for arenes, 272
    FC. *See* Friedel–Crafts (FC) reaction

Allophane, 4, 48–53, 113
    clay minerals, 157
    composition and structure, 49
    halloysite- and imogolite-like, 50
    HRTEM, 49f
    nanoball, 50, 50f
    organic reactions catalyzed by, 26t
    PZC, 50
    stream-deposit, 50
    surface acidity, 50
α-pinene, isomerization, 17t, 19t, 21t–22t, 25t, 136, 151t, 180, 299–302, 300f
Al-PILC. *See* Alumina-pillared montmorillonite (Al-PILC)
Alumina octahedral sheet, 7, 8f
Alumina-pillared montmorillonite (Al-PILC), 164–165, 271, 311
    impregnation, 356, 362–363
    isomerization, 303–304
    ruthenium impregnation, 390
Aluminas, 7f, 8f, 52f, 96
Aluminum ions, 7, 36, 50, 134, 135
Amine titration, 109–114, 112f
Amino acid
    adenylates oligomerization, 401
    condensation of, 399
    enantiomers, 401–402
    thioesters, 400
Amino acid adenylates (AAA), 400
    MMT particle, 401f
    oligomerization, 401
Ammonia/pyridine, infrared bands for, 116, 116t
Ammonium-exchanged form, 44
Ammonium nitrate (clayan), 227–229
AOP (advanced oxidation processes), 355
APTES (3-aminopropyl triethoxysilane), 181–183, 182f, 238, 283
Aromatic aldehydes
    Baylis–Hillman reaction, 177
    Biginelli reaction, 265
    BV condensation, 262, 262f
    KC, 276
Aromatic hydrocarbon, 26t, 153t, 167t, 222, 307
Arrhenius equation, temperature rate, 391
Arylmethanol indicators, 90
    Brønsted acidity of solid catalysts, 90t
    with *n*-butylamine titration, 91t
Asymmetric/enantioselective synthesis, 366, 367f
Atom transfer radical polymerization (ATRP), 312–313, 314t, 318t, 321
Attapulgite, 45, 398. *See also* Palygorskite
Autotransformation process, 135–136, 136f
Aza-DA reaction, 266, 266f
    cycloaddition of imine, 269, 269f
    Danishefsky's diene, 269

417

Aza-DA reaction (*Continued*)
  endo:exo ratio, 266–268, 267f
  normal-demand/inverse-demand, 266, 266f

B

Baeyer–Villiger (BV) condensation/oxidation, 230, 262, 262f
Ball clay, 1, 354t
Bamberger rearrangement, 263, 263f
Baylis–Hillman (BH) reaction, 145t, 162t, 177, 263–264, 263f
Beckmann rearrangement (BR) reaction, 264, 264f
Beidellite, phyllosilicates, 5t, 16, 34, 35f, 165
Bentonite, phyllosilicates, 1, 100t, 137t–153t, 159t, 166t–168t, 174t, 233t–235t, 242t–243t, 314t–315t, 348t, 350t, 353t–354t, 364t, 391
Benzene reaction, 103, 222, 222f, 225
Benzidine compound, 101–102, 106
Benzoyl peroxide (BPO), 313, 316t, 318t, 321–322
Bernal's hypothesis, 398
BH (Baylis–Hillman) reaction, 145t, 162t, 177, 263–264, 263f
Biginelli reaction/pyrimidone synthesis, 265, 265f
Bitumen, 396–397
Bleaching earth, 1, 276
BPO (benzoyl peroxide), 313, 316t, 318t, 321–322
Brittle mica, 5t, 12, 13t, 16
Brønsted
  acid sites concentration, 133, 155–157, 165
  -Lewis acid combination/synergy, 106–109
Brønsted acid catalyzed process
  α-pinene isomerization, 300f
  Al-PILC, 311
  cis-2- and *trans*-2-butenes, 302
  cyclohexylamine desorption, 302
  dimerization, 307, 310
  isomorphous substitution, 303
  oligomerization, 310, 310f
  polymerization, 312
Brønsted acidity, 95–100, 99f
  arylmethanol indicators for, 90t
  and Lewis acid, 106–109
Brønsted acid-mediated process, 264
  BR, 264
  Fischer glycosidation/glycosylation, 270
  Prins (cyclization) reaction, 281–282
  Ritter reaction, 282
Brønsted, J.N., 85
BR (Beckmann rearrangement) reaction, 264, 264f
BV (Baeyer–Villiger) condensation/oxidation, 230, 262, 262f

C

C_{21} *n*-alkane, 391
Calcination process, 155–157, 163
Carbocation, 310–311, 319, 396
  colored, 89
  from ethanol, 272
  intermediate, 99–100, 113–114, 267, 299, 302, 310f
  protonation and, 105
  triphenylmethyl carbonium, 97f
Carboxylic acid, 137t–139t, 148t–149t, 153t, 159t
  esterification of, 19t
  regeneration of, 18t
  transformation, 391–394

Catalysis, 29, 221
  clay mineral. *See* Clay mineral catalysis
  geocatalysis, 389–398
  organic. *See* Organic catalysis, clay-supported reagents for
  shape-selective, 3, 228
Catalytic activity, surface acidity and, 85–118
Catalytic cracking, geocatalysts, 389
Catechol, 349t
  oxidation, 357
  polycondensation of, 26t
Cation exchange capacity (CEC), 12, 13t, 27, 163
  clay minerals, 12
  illite, 43
  MMT by, 134, 134t
Cationic surfactant, 20t, 175, 232
CBS (Cretaceous Black Shale), 397
CEC. *See* Cation exchange capacity (CEC)
Chain-layer silicates, 45
Chamosite chlorite, 41
Charge-balancing cations, 12
Chlorites, 41–42
  chamosite, 41
  organic reactions catalyzed by, 16, 21t–23t
  structure, 41f
Chrysotile, 32–34
  fibril structure, 33f
  Ni-substituted, 34
  organic reactions catalyzed by, 17t–20t
Clay(s), 1
  associated minerals, 4
  catalysis, 1, 4, 14, 43
  fraction, 1, 48
  as geocatalysts, 389–398
  layer-stacking arrangement, 10f
  non-clay constituents, 1–2
  organically modified, 133, 176–183, 232, 369
  supported catalysts, 133, 222, 320
Clayan (ammonium nitrate), 227–229
Clay-catalyzed, 392t
  decomposition, 100t
  examples of, 261
  isomerization of organic compounds, 154
  organic conversions, 109
  organic reactions, 95, 131, 159t–162t
  polymerization of styrene, 105f, 106
  redox reactions, 347
  Wacker oxidation, 284–285
Claycop (cupric nitrate), 227–228
Clayfen (ferric nitrate), 226
  nitrate reagents, 226–228, 226f
  nitration of phenolic compounds, 227t
Clay-kerogen interaction, 395
Clay mineral(s), 2f, 3, 132. *See also* Clays and clay minerals
  allophane, 157
  autotransformation, 136f
  building blocks, 7f
  CEC/layer charge density, 12, 13t
  charge per formula unit, 5t–6t, 10, 13
  classification scheme, 5t–6t
  definition, 3
  as geocatalysts, 389–398
  H^+-exchanged, 135–136, 136f
  imogolite, 157

interstratified structures, 10
MMT. *See* MMT, clay minerals
organically modified, 176–183
organic derivatives, 181
proton-donating capacity, 95
role, 131
silylation, 181–182
structural aspects, 7–16
structural-chemical properties, 5t–6t
structural/surface properties, 16–53
surface-modified, 348t–355t, 359, 362
surfaces, 133
swelling in water, 13t, 40
temperature ranges in, 155
Clay mineral catalysis
asymmetric reactions, 366–369
deoxygenation reactions, 362–366
dimerization reactions, 306–310
enantioselective reactions, 366–369
hydrogenation reactions, 362–366
isomerization reactions, 299–306
name reactions, 261–285
natural processes, 389–398
oligomerization reactions, 310–312
polymerization reactions, 312–313, 319–327
polypeptide synthesis, 399–401
prebiotic organic reactions, 398–404
redox reactions, 347
Clay-organic complexes/interactions, 15, 100, 236
aqueous suspension, 308
catalyze DA reaction, 268f
in sediment and soil, 299
Clays and clay minerals
activation energy, 392t
adsorption and polymerization, 401–402
dimerization and oligomerization, 402–404
geocatalysis, 389–398
polypeptide synthesis, 399–401
RNA nucleotide, 403f
Clay-supported
CdS, 230
HPA, 241, 242t–243t
iron(III)/metal nitrates, 226–229
metal nanoparticles, 230–236, 230f, 233t–235t, 236f
metal oxides/sulfides, 229–230
metal salt, 221–222
metal sulfides, 230
organic/metal-organic reagents, 236–238, 238f, 241
zinc(II)/metal chlorides, 222–226
Clayzic ($ZnCl_2$), 222–223
-catalyzed FC reactions, 273
Envirocat EPZ10, 225
Lewis acid catalyst, 224
structure, 224, 224f
uses, 107
CLS (cross-linked smectites), 163
Conjugate base of HA, 85
Counterions, 12. *See also* Exchangeable cations
hydration enthalpy, 98, 98t
ionic potential, 95, 98, 98t, 280, 302
CPD (cyclopentadiene) reaction, 152t, 266–267, 267f
Cracking, 100t, 168t
of biomass tars and benzene, 25t
of cumene, 17t, 23t–25t

of heavy oil, 18t–19t
hydrocarbon, 389–391
Cretaceous Black Shale (CBS), 397
Cross-coupling of boronic/organoboronic acid, 284, 284f
Cross-linked smectites (CLS), 163
Crystalline swelling, 13t, 14
Cupric nitrate (claycop), 227–228, 274
Cyclododecanone oxime, 264, 264f
Cyclopentadiene (CPD) reaction, 152t, 266–267, 267f
Cyclopropanation reaction, 367t

D

Dabco (1,4,-diazabicyclo[2,2,2]octane) compound, 176–177, 178t, 263
DA reaction. *See* Diels–Alder (DA) reaction
Decarboxylation reaction, 393
behenic acid, 106, 391, 392t
fatty acid, 391
propionic acid, 393
Dehydration, 17t–18t, 24t, 39, 45, 47, 146t–147t, 155–156, 169t, 243t
Dehydroxylation, 101, 107
kaolinite, 107
sepiolite, 157
structural, 47, 114
Density functional theory (DFT), 393
Dickite, particle, 5t, 27
Diels–Alder (DA) reaction, 182, 266–269
aza. *See* Aza-DA reaction
clay-organic complex, 268f
inverse-demand reaction, 266, 266f
KSF and K10 MMT, 268
normal-demand, 266, 266f
Dienophile, 266
Diiminosuccinonitrile (DISN), 403
Diketopiperazine peptides, 399–400
Dimerization, 299, 306
Al^{3+}-MMT, 307–308
butadiene, 307
cyclodimerization of stilbazolium, 308
DA cycloaddition reaction, 306, 309
1-dodecene, 307
Fenton-like oxidation, 307
hydrocarbons, 306–307
non-hydrocarbons, 307–310
oleic acid, 309, 309f
stilbazolium ions, 308f
trans-stilbene, 306–307
Dimethylsulfoxide (DMSO), 238, 318
Dioctahedral sheet, 7, 34, 43
smectite, 36f, 107–108, 108f
vermiculites, 38
Direct titration method, 109
DISN (diiminosuccinonitrile), 403
DMSO (dimethylsulfoxide), 238, 318
Dodecatungstophosphoric acid (DTP), 241, 242t–243t

E

EDAC (1-ethyl-3-(3-dimethylaminopropyl) carbodiimide), 403
Electron-donating group (EDG), 266
Electron-withdrawing group (EWG), 263, 266
Enantioselective reactions, 366–369

Ene reaction. *See* Alder–ene reaction
Envirocat EPZ10/EPZG, 225–226
Ethyl pyruvate, asymmetric hydrogenation, 233t, 369, 369f
EWG (electron-withdrawing group), 263, 266
Exchangeable cations, 12, 15, 92t, 94t, 95, 102, 112, 134t, 155

F

FCC (fluid catalytic cracking), 18t, 390
FC reaction. *See* Friedel–Crafts (FC) reaction
Fe^{3+}-MMT, 91, 102–103, 156, 307, 312
Fenton-like reaction, 12, 175, 229, 362
Fenton/photo-Fenton oxidative degradation, 347
    organic dyes, 358
    phenols/substituted phenols, 348t–350t
Fenton process, 355–356
Ferric nitrate (clayfen), 226, 226f
Ferrier (glycal allylic) rearrangement (FR), 269–270, 269f
Fire clay, 1, 302
First principle molecular dynamics (FPMD) simulations, 28, 37
Fischer glycosidation/glycosylation, 270, 270f
Fischer–Hepp rearrangement, 20t, 270
Fischer indole synthesis, 270
Fischer–Tropsch (FT) synthesis, 20t, 271
Fluid catalytic cracking (FCC), 18t, 390
Fluorotetrasilicic mica (TSM), 16, 44–45, 320
    interlayer complex, 237
    organic reactions catalyzed by, 21t–23t
Fourier-transform infrared spectroscopy (FTIR), 9, 164
    adsorbed ammonia and pyridine, 116
    Brønsted/Lewis acid sites, 115f
    pyridine analysis, 165
FPMD (first principle molecular dynamics) simulations, 28, 37
FR. *See* Ferrier (glycal allylic) rearrangement (FR)
Friedel–Crafts (FC) reaction, 271–272, 272f
    acylation, 115, 157, 226, 271
    alkylation, 155–156, 222–223, 241, 272f
    arylmethylation, 225
    benzene, 107, 155
    sulfonylation of arenes, 103
    toluene, 156
Friedländer (quinoline) synthesis, 273–274, 273f
Fries rearrangement, 138t, 241, 274, 305
FTIR. *See* Fourier-transform infrared spectroscopy (FTIR)
FT (Fischer–Tropsch) synthesis, 20t, 271
Fuller's earth and fulling, 1

G

Gas/liquid chromatography, 106
Geocatalysis, 389
    clays and clay minerals, 389
    hydrocarbon cracking, 389–391
    prebiotic organic reactions, 398–399
Glycine
    dimerization, 400
    oligomerization, 399–401

H

Halloysite, 30
    composition and structure, 31f
    hydrated, 30, 31f, 155
    intercalation, 30
    interlayer water, 30–31
    -like allophane, 49
    organic reactions catalyzed by, 17t–20t
    particle shape, 31–32
Haloarenes, 283, 283f
Hammett acidity function, 87, 92t
Hammett indicators, 87, 91, 110–111, 114
    clay minerals, acid strength, 92t
    with n-butylamine titration, 91t
    ultraviolet spectroscopy, 89t
    visual estimation, 88t
Hantzsch dihydropyridine synthesis, 274, 275f
HBD (hydrogen-bond donor), 86
HBI (highly branched isoprenoid) alkenes, 302, 303f
HDTMA (hexadecyltrimethylammonium), 178, 179f
Heating process, 155–157. *See also* Calcination process
Heck reaction, 24t, 237, 275–276, 275f
Hectorite, 15, 35
    hydrate of, 11f
    intercalation, 178–179
    interlayer space, 306
Heterogeneous mesoporosity, 301
Heteropoly acids (HPA), 221, 241, 242t–243t
Hexadecyltrimethylammonium (HDTMA), 178, 179f
Highly branched isoprenoid (HBI) alkenes, 302, 303f
High-resolution transmission electron microscopy (HRTEM), 49
    allophane, 49f
    imogolite, 51, 51f
HIV (hydroxy interlayered vermiculites), 40
Hollow chrysotile fibril, 33f
Hosomi–Sakurai reaction. *See* Sakurai allylation reaction
Houdry catalytic cracking process, 389
Houdry process, 131, 390
HPA (heteropoly acids), 221, 241, 242t–243t
HRTEM. *See* High-resolution transmission electron microscopy (HRTEM)
Hückel model, 10
HX (hydrogen halide), 277
Hydrated halloysite/halloysite, 30, 31f, 155
Hydrogenation/semihydrogenation, 364t–365t
Hydrogen-bond donor (HBD), 86
Hydrogen halide (HX), 277
Hydroisomerization process, 21t–23t, 171t–172t, 235t, 299, 303–304
Hydroquinone, 17t, 21t–22t, 242t–243t, 321, 349t, 357, 392
Hydrothermal treatment effect, 40
Hydroxy interlayered vermiculites (HIV), 40

I

Illite, clay mineral, 43–44, 396
    organic reactions catalyzed by, 21t–23t
Imogolite, 48–53
    catalytic activity, 53
    clay minerals, 157, 164, 181, 183
    composition and structure, 51, 52f
    HRTEM, 51, 51f
    nanotube, 51
    organic reactions catalyzed by, 26t
Incipient wetness technique, 241
Infrared (IR) spectroscopy, 15, 102, 114, 132t

Index

Intercalation process, 30, 163–164, 182
  halloysite, 30
  hectorite, 178–179
  organoclays, 133
  racemic pair, 366
  type II organoclays, 178
Inverse-demand DA reaction, 266, 266f
Iron(III) nitrate/metal nitrate, clay-supported, 226–229
IR (infrared) spectroscopy, 15, 102, 114, 132t
Isomerization, 299
  Al^{3+}/TMA^+, 300
  Al- and Zr-PILC, 302
  benzyl phenyl ether rearrangement, 305
  bifunctional catalyst, 303
  Brønsted acid catalyzed, 300f
  butenes, 302
  cis-2- and *trans*-2-butenes, 302–303
  cis-isomer, 304, 304f
  double-bond migration/conjugation, 299, 302
  epoxide ring, 304
  hydrocarbons, 299–304
  K10 MMT-catalyzed, 303f
  limonene conversion, 301, 301f
  MMT, 301
  non-hydrocarbons, 304–306
  skeletal rearrangement, 299
Itaconate, asymmetric hydrogenation, 369, 369f

K

K10 MMT catalyst, 136, 221, 228. *See also* KSF MMT
  $AlCl_3$ and $FeCl_3$, 222
  Alder–ene reaction, 261–262
  alkoxyalkylation reaction, MA, 279
  ammonium nitrate (clayan), 228
  Bamberger rearrangement, 263
  BH, 263–264
  Biginelli reaction, 265
  bis(trimethylsilyl) chromate on, 236
  BR, 264
  BV, 262–263
  CDP and MVK, 267
  DA, 266
  endo:exo ratios, 267–268
  FC alkylation, 272
  Fe_3O_4 nanoparticles, 230
  ferric nitrate (clayfen), 226, 226f
  FR, 269–270
  Friedländer condensation, 273
  Fries rearrangement, 305
  Heck reaction, 275–276
  hetero-DA reaction, 268
  $InCl_3$, 225
  Markovnikov adducts, 278
  metal chloride reagents, 225
  metal fluorides, 225
  $MnCl_2$, 223
  molybdenum acetylacetonate, 238
  Nicholas reaction, 280
  oligomeric, 312
  one-pot, solvent-free, 264
  Oppenauer oxidation, 280
  organic reactions catalyzed by, 159t–162t

$PdCl_2$ complexes, 237
  -supported ammonium nitrate. *See* Clayan (ammonium nitrate)
  -supported copper nitrate. *See* Claycop (cupric nitrate)
  -supported iron nitrate. *See* Clayfen (ferric nitrate)
  -supported zinc chloride. *See* Clayzic ($ZnCl_2$)
  Suzuki reaction, 284, 284f
  $ZnBr_2$, 224
  $ZnCl_2$, 222
K- and K-series catalysts, 131. *See also* K10 MMT catalyst
  organic reactions catalyzed by, 137t–154t
  surface properties, 132t
Kaolinite, 26–29, 95
  acid sites/concentration/relative strength, 111f, 112
  dehydroxylation, 155
  layer structure and stacking, 26
  organic reactions catalyzed by, 17t–20t
  positive/negative charges development, 28f
  scanning electron micrograph, 30f
  structure, 12f
  sulfuric acid/moisture content, 93, 93f
  surface properties, 27t
  transmission electron micrograph, 30f
  under-coordinated aluminium/aluminol, 28
KC (Knoevenagel condensation), 144t, 162t, 183, 276–277, 277f
Keggin HPA structure, 241
Kerogen, 44, 393
  diamondoids formation, 19t
  hydrocarbons formation during pyrolysis, 22t
  transformation and pyrolysis, 394–398
Ketone olefination, 285
Knoevenagel condensation (KC), 144t, 162t, 183, 276–277, 277f
KSF MMT, 136. *See also* K10 MMT catalyst
  alkoxyalkylation reaction, MA, 279
  Biginelli reaction, 265
  $Bi(NO_3)_3$, 228
  BR, 264, 264f
  hetero-DA reaction, 268
  KC, 276–277
  Oppenauer oxidation, 280
  pyrrole synthesis, 280, 281f
  TMSCN, 283

L

Langmuir–Hinshelwood kinetics, 359, 360f, 361
Laponite, 35, 164, 172t, 231, 238, 239t, 275, 348t, 353t, 367t, 368
  Al-pillared, 350t
  Fe-impregnated, 354t
  Fe-pillared, 351t
  immobilized on, 367f
  Mn(III)-salen on, 240t
  Ni^0/Zr-pillared, 232
  phyllosilicates, 16
  synthesized, 366
  Ti-pillared montmorillonite and, 351t
Layer-ribbon
  silicates, 45
  structures, 6t, 16

Layer silicates. *See also* Phyllosilicates
  basal spacing, 10, 37, 39
  interlayer distance/separation, 10
  interlayer space, 9–10
  isomorphous substitution, 10, 14
L/D-enantiomers and isomers, 401–402
Lewis acid-catalyzed, 261
  Alder–ene, 261–262
  clayzic, 273
  DA reaction, 266–269
  FC reaction, 271–272
  Ferrier (glycal allylic) rearrangement, 269–270
  Fries rearrangement, 274
  Hosomi–Sakurai allylation reaction, 282, 282f
  Mukaiyama aldol reaction, 279–280, 279f
Lewis concept, 85
  acidity, 101–107
  acids/acid sites, 12, 33, 155–157, 165, 223
  base, 14, 29, 312, 321
Ligand, 239t–240t, 367t
  axis, 15
  chiral, 238, 358
  exchange, 29
  non-chiral salen, 238f
  Schiff base, 238
Limonene, 147t, 307, 363
  camphene and, 151t
  disproportionation of, 301f
  formation, 300–301
  isomerization-disproportionation of, 106
Lithium ions, 134–135
Lysine/dilysine condensation, 399

M

MA (Michael addition), 139t, 278–279, 278f
Maillard reaction, 323
Mannich reaction, 277
MAO (methylaluminoxane), 315t–317t, 319, 325–326
Markovnikov addition rule, 277–278, 278f
Metakaolinite, 155
  acid-activated, 18t–19t, 158
  acid-treated, 223, 302
Metal chloride(s)
  clay-supported ZinC(II) chloride and, 222–226
  kaolinite-supported, 18t
  reagents, 225
Metal-doped pillared MMT, 365
Methylaluminoxane (MAO), 315t–317t, 319, 325–326
Methylvinyl ketone (MVK) reaction, 20t, 266–267, 267f
Mica, 42–43
  composition and structure, 42f
  organic reactions catalyzed by, 21t–23t
Michael addition (MA), 139t, 278–279, 278f
Microcrystalline forms, 38
Microwave-assisted reaction, 157–158, 159t–162t, 275
Microwave irradiation, 157–158, 163, 226, 229, 232, 269f, 270, 276, 282, 306, 306f
  assisted organic conversions and syntheses, 159t–162t
Mizoroki–Heck (MH) reaction. *See* Heck reaction
MM (Moosburg montmorillonite), 176, 177f
MMT, clay minerals, 155, 158, 176
  by CEC, 134, 134t
  dehydration in, 155
  Fe^{3+} counterions in, 180
  interlayer space, 155–156
  K10 and KSF, 136
  proton-exchanged, 156
  structural integrity, 158
Montmorillonite (MMT), 34, 110–111, 394. *See also* K10 MMT catalyst; KSF MMT
  acid-activated/acid-treated, 106, 134–135, 158, 362
  in acid sites, concentration, 117t
  acid strength distribution curves, 96f
  -catalyzed dimerization, 103, 104f
  cation-exchanged, 118f
  in hectorite and, 15
  layer structure, 35f
  MMT, 156
  in octahedral sheet, 14
  organic reactions catalyzed by, 137t–154t
  reduced charge, 34
  surfaces, organic species on, 94t
  synthetic nickel-substituted, 113f
  tactoids, average thickness, 38t
Moosburg montmorillonite (MM), 176, 177f
Morita–Baylis–Hillman reaction, 263–264
Mössbauer spectroscopy, 101
Mukaiyama aldol reaction, 279–280, 279f, 347
Mullite formation, 157
MVK (methylvinyl ketone) reaction, 20t, 266–267, 267f

N

Nanotechnological applications, 32
n-butylamine, 91, 91t, 110, 113–114, 117t, 351t
N-chloroacetanilide rearrangement, 305, 305f
Nicholas reaction, 280
Non-interstratified structures, 8
Non-ionic surfactants, 20t, 164, 302, 361
Nontronite layer structure, 5t, 12, 34, 35f, 353t, 392t, 402
Normal-demand DA reaction, 266, 266f
n-paraffins, 390–391
N-phenylhydroxylamine, 263, 263f
Nucleotides, dimerization/oligomerization, 402–404

O

Octahedral ("O") sheet, 7
  substitution, 13–14
  vacancy, 31, 107
Oleic acid, 233t, 393
  clay-catalyzed conversion, 309, 309f
  clay-catalyzed isomerization, 304
  dimerization, 266, 310
  feedstock, 309
  with methanol, 19t, 150t
  oligomerization, 312
Oligomerization, 299
  Cu^{2+}-MMT, 310
  dec-1-ene conversion, 310
  Fe^{3+}-MMT, 310
  hydrocarbons, 310–311
  non-hydrocarbons, 311–312
  3,4,5-trimethoxybenzyl alcohol, 311, 311f
Oppenauer oxidation, 280

Index

Organically modified clay minerals, 133, 176. *See also* Organoclays
  commercially available (*Cloisites*), 322
  metal precursors, intercalation, 232
  organoclays and materials, 176–181, 177f
  phase-transfer catalysis, 180
  surface grafting and silylation, 181–183
Organic catalysis, clay-supported reagents for, 221
  HPA, 241, 242t–243t
  metal nanoparticles, 230–236, 230f, 233t–235t, 236f
  metal oxides/sulfides, 229–230
  metal salts, 221–229
  organic/metal-organic reagents, 236–238, 238f, 241
  overview, 221
Organic matter (OM)/kerogen, 394
Organic reactions catalyst
  by 1:1 type layer silicates, 17t–20t
  by allophane and imogolite, 26t
  by sepiolite and palygorskite, 24t–25t
  by vermiculite/chlorite/mica/illite/SMM/TSM, 21t–23t
Organoclays, 133
  Dabco^{2+}-exchanged MMT, 176–177
  and materials, 176–181, 177f
  type I, 176, 180
  type II, 178, 180
Outer-sphere type, 14–15
Oxidation reactions, 355

P

PAAC (pillared acid-activated clays), 169t, 175
Paal–Knorr (pyrrole) synthesis, 142t–143t, 145t, 280–281, 281f
Palladium/copper catalyzed cross-coupling, 283
Palygorskite, 45–48, 48f, 389
  composition and structure, 46, 46f
  heat-activated, 319
  organic reactions catalyzed by, 24t–25t
  scanning electron micrograph, 45f
Paraffin hydroisomerization, 303
Particle edge, 29, 36, 104, 112, 312
PCH. *See* Porous clay heterostructures (PCH)
Pechmann condensation/coumarin synthesis, 281, 281f
Phenol degradation pathway, 357f
Photo-Fenton. *See also* Fenton/photo-Fenton oxidative degradation
  degradation, 358f
  oxidation reactions, 347
  process, 356
Phyllosilicates, 3–4, 359. *See also* Layer silicates
  beidellite, 16, 34, 35f, 165
  bentonite, 1, 391
  Laponite, 16, 35, 164, 368
PILC. *See* Pillared interlayered clays (PILC)
Pillared acid-activated clays (PAAC), 169t, 175
Pillared interlayered clays (PILC), 131–132, 262, 347
  application, 165, 175
  calcining, 163
  degradation/decoloration, 348t–350t, 350t–352t, 353t–355t
  FT synthesis, 271
  intercalating pillaring agent, 163
  organic reactions catalyzed by, 166t–174t
  PCH, 163–165, 165f, 175–176
  structure, 164f
  synthesis, 163–164
  by washing-centrifugation/dialysis, 163
Pillared layered solids (PLS), 163
Pillaring, 132, 232
  agent, 40, 163
Pinocarveol epoxide isomerization, 304
PLS (pillared layered solids), 163
Point of zero charge (PZC), 28, 52
  allophane, 50
  imogolite, 52
  kaolinite, 28–29
Polyaniline, 323
  -MMT nanocomposite, 323f
  synthesis, 327f
Polymer-clay nanocomposites, 181–182, 314t–318t, 325
Polymerization, 299, 312–319
  AIBA, 325
  atom transfer radical, 312, 321
  benzoyl peroxide, 313
  clay minerals on, 313f
  clay nanocomposites, 314t–318t
  hydrocarbons, 319–322
  hydroxyethyl methacrylate, 312
  intercalative, 313, 324–326
  kaolinite, 325
  non-hydrocarbons, 322–325
  Raman spectroscopy, 319
  in situ, 325–327
  styrene, 320f
  TMA, 319
  zirconocenes, 320
Polypyrrole-clay nanocomposite, 317t, 327
Porous clay heterostructures (PCH), 175
  formation, 175
  PILC. *See* Pillared interlayered clays (PILC)
  surface acidity, 176
  synthesis, 175f
Prins (cyclization) reaction, 281–282
Probe molecules/adsorption/desorption, 114–118
Proton, 108, 110–111, 165
  acceptor, 85
  adsorption, 50
  donor, 85
  exchange, 133
  interlayer space, 179–180
  NMR spectroscopy, 309
  smectite, tetrahedral sheet, 107f
  source of, 97
Protonation, 28
  adsorbed organic species, 95
  organic species on MMT surfaces, 94t
  triphenylcarbinol, 96, 97f
Pseudo-hexagonal platelets, 29
Purine nucleotides, 403
Pyridine, 94t, 116, 116t, 159t
  cyano, 152t
  FTIR analysis, 165
Pyrolysis, kerogen transformation and, 22t, 394–398
Pyrophyllite, clay mineral, 5t, 7, 13t, 14, 92t, 267
PZC. *See* Point of zero charge (PZC)

R

Racemic mixture, 366, 401–402
Radical-cation(s), 101, 103–104, 312, 321
Reaction types catalyzed by
    acid activation, 137t–154t
    allophane, 26t
    chlorites, 16, 21t–23t
    chrysotile, 17t–20t
    clay-supported HPA, 242t–243t
    clay-supported metal nanoparticles, 233t–235t
    clay-supported metal-Schiff base complexes, 239t–240t
    halloysite, 17t–20t
    illite, clay mineral, 21t–23t
    imogolite, 26t
    K10 MMT catalyst, 159t–162t
    K- and K-series catalysts, 137t–154t
    kaolinite, 17t–20t
    K-series catalysts, 137t–154t
    mica, 21t–23t
    MMT, 137t–154t
    palygorskite, 24t–25t
    PILC, 166t–174t
    sepiolite, 24t–25t
    SMM, 21t–23t
    TSM, 21t–23t
    vermiculites, 21t–23t
Rectorite, 10, 359
    dispersed hematite, 348t
    Fe-pillared, 349t, 355t
Reduced-charge MMT, 34
Reduced dimensionality, 3, 268
Resorcinol, 242t, 350t
    acylation, 138t
    monobenzoate synthesis, 274f
    Pechmann condensation, 281, 281f
Rheological method, PZC, 28
Ribonucleic acid (RNA), 402–403
Ritter reaction, 282, 282f
Rocky Mountain Coal (RMC), 397

S

Sakurai allylation reaction, 282, 282f
Saponites, 35, 175
    activity, 277
    Al-pillared, 303
    interlayer space, 306, 308, 308f
    isomerization, 303–304
    synthetic, 356, 362–363
SAz-1 MMT, 301
Schiff base complexes/ligands, 238, 239t–240t
SD (Serra de Dentro) bentonite, 301
Sensitized photoisomerization, 306
Sepiolite, 45–48, 48f
    composition and structure, 46, 46f
    dehydration and dehydroxylation, 47, 157
    organic reactions catalyzed by, 24t–25t
Serra de Dentro (SD) bentonite, 301
Sesamin conversion, 306
Shape-selective catalysis, 3, 228
Short-range order minerals, 4, 6t
Silica-aluminas, 96, 99, 112, 389

Silicate layer, 7. *See also* Layer silicates
    1:1 (T-O) and 2:1 (T-O-T) type, 7
    structure, 8f
    thickness, 10
Siloxane surface, 13–15
Smectite(s), 34–38, 37f, 400
    acid-activated, 115, 131, 175–176
    CLS, 3, 163
    deammoniated, 107
    dioctahedral and trioctahedral, 35–36
    layer charge distribution, 34–35
    layer structure, 176
    reduced charge, 34
    tetrahedral sheet, 107f
SMM. *See* Synthetic mica-montmorillonite (SMM)
Sn(IV)-tetrapyridyl-porphyrin complex, 93f, 94
Sol-gel/acid hydrolysis technique, 361
Solid acid catalysts, 87, 91, 307, 389
    ammonia and pyridine to, 116t
    reactions, 100
    ultraviolet spectroscopy, 89t
    visual estimation, 88t
Solid-state Al NMR spectroscopy, 135, 155
Sonogashira reaction, 283, 283f
Specific surface area (SSA), 50, 52, 132t, 135, 158
Strecker reaction, 283–284, 283f
Structural hydroxyl groups, acidity, 107, 107f
Structural/surface properties, clay minerals, 16–53
Stuart Range Formation, clay, 395
Styrene reaction, 368
    clay-catalyzed polymerization, 105f
    oxidation, 362
    polymerization, 322
    quantitative conversion, 228, 228f
Sulfonated polystyrene/MMT nanocomposite, 277
Sulfonylation, FC reaction, 174t, 271
Sulfuric acid-modified MMT, 301
Surface
    acidity/catalytic activity, 85–118
    acid sites concentration, 109–118
    activation/modification, 131–183
    charge density, 14, 108, 310, 404
Surface acidity/catalytic activity, 85
    acids and bases, 85–94
    Brønsted acidity, 95–100, 99f
    concentration and distribution, 109–118
    Lewis acidity, 101–107
Surface acid sites, concentration/distribution, 109
    amine titration, 109–114, 112f
    probe molecules, adsorption/desorption, 114–118
Surface activation and modification
    acid activation, 133–136, 137t–154t
    organically modified clay minerals, 176–183
    overview, 131–133, 132t
    PILC and PCH, 163–165, 175–176
    thermal activation and treatments, 155–158
Surface grafting/silylation process, 181–183
    by APTES, 181–182
    Ca^{2+}-MMT, 182f
    locking effect, 182
Suzuki-Miyaura/Suzuki coupling reaction, 284, 284f
Swelling
    crystalline, 14–15
    osmotic (macroscopic), 37

Index

Symmetrical arenes, 103
Synthesis gas (*syngas*), 271
Synthetic mica-montmorillonite (SMM), 16, 44
    nickel substituted, 107, 113f, 303
    organic reactions catalyzed by, 21t–23t

T

Talc, clay mineral, 5t, 7, 13t, 14, 92t
Tamele, M.W., 109
Temperature-programmed desorption (TPD), 114
tert-butylhydroperoxide (TBHP), 238, 358, 362
Tetraethylorthosilicate (TEOS), 175, 175f
Tetrahedral ("T") sheet, 7–8, 9f
    rotation, 8–9, 31–32, 38, 42
    substitution, 15, 27, 393
    tilting, 8
Tetramethoxytoluene, FC alkylation, 142t, 272, 272f
Tetramethylammonium (TMA), 176, 179–180
Thallium(III) nitrate (TTN), 221
Thermal activation and treatments
    heating/calcination, 155–157
    microwave/ultrasound irradiation, 157–158
Thiophenol, oxidative coupling, 153t, 226, 226f
Titration, amine, 109–114
TMSCN (trimethylsilyl cyanide), 139t, 152t, 283, 283f
TPD (temperature-programmed desorption), 114
Transmission electron microscopy, 231, 320, 322
Tricker, M.J., 101
Triethylene diamine, 176
Triisobutylaluminum catalysts, 315t, 317t, 319–320
Trimethylaluminum (TMA), 315t, 319
Trimethylsilyl cyanide (TMSCN), 139t, 152t, 283, 283f
Trioctahedral structure, 7
    chlorites, 41
    chrysotile, 32
    micas, 43
    smectite, 36f, 158, 400
Triphenylcarbinol protonation, 96, 97f
Triphenylmethylcarbonium (Ph_3C^+) ions, 97
TSM. *See* Fluorotetrasilicic mica (TSM)
TTN (thallium(III) nitrate), 221
Tubular halloysite, 32, 32f
Type II organoclays, 178, 180
Type I organoclays, 176, 180

U

Ultrasound irradiation, 158
    application, 163
    organic conversions/syntheses catalyzed by, 159t–162t

V

Vanadia-pillared MMT, 362
Vanadium-impregnated K10 MMT (VK10), 358
van der Waals
    cross-sectional area, 112
    diameter, 114
    interactions, 53, 176, 308
Vermiculites, 38–40
    HIV, 40
    hydrate arrangement, 11f, 15
    hydration states, 40
    intercalated with $Dabco^{2+}$, 176
    Mg, 39f
    organic reactions catalyzed by, 21t–23t

W

Wacker oxidation, 284–285
Walling, C., 109
Water-swollen hectorite arrangement, 11f
WB (Wyoming bentonite), 176, 177f
Wet catalytic oxidative degradation
    organic contaminants/pollutants, 350t–352t
    organic dyes, 353t–355t
Wittig reaction, 285
Wyoming bentonite (WB), 176, 177f

X

X-ray diffraction, 231, 403
    acid-activated clay minerals, 135
    adsorption and, 400–401
    allophane pattern, 4
    analysis of $H_3PW_{12}O_4$, 241
    macrocrystalline forms, 3
    titrated clays, 110
    transmission electron microscopy, 231, 320
X-ray diffractometry (XRD), 4, 10, 27, 31

Z

Zeolites, mineral, 44, 99, 113, 304, 389–390
Zinc(II) chloride/metal chloride, clay-supported, 222–226
    addition, 223
    deposition, 223
    impregnation of K10, 222–223
    preparation of clayzic, 223, 224f
$ZnCl_2$ (clayzic), 222
    Envirocat EPZ10, 225
    Lewis acid catalyst, 224
    preparation, 224f
    uses, 107